International Symposium on Runaway Reactions, Pressure Relief Design, and Effluent Handling

International Symposium on Runaway Reactions, Pressure Relief Design, and Effluent Handling

March 11–13, 1998 • New Orleans, Louisiana

Edited by **G. A. Melhem**
Arthur D. Little, Inc.
Cambridge, Massachusetts

H. G. Fisher
Union Carbide Corporation
South Charleston, West Virginia

AMERICAN INSTITUTE OF CHEMICAL ENGINEERS
345 East 47th Street, New York, New York 10017

Organized by the Design Institute for Emergency Relief Systems (DIERS) Users Group

Sponsored by
Arthur D. Little, Inc.
Center for Chemical Process Safety (CCPS)
AIChE Heat Transfer and Energy Conversion Division

American Institute of Chemical Engineers
345 East 47th Street
New York, New York 10017

ISBN 0-8169-0761-7

This book is dedicated to

W. B. Howard (Monsanto, retired)
Who identified emergency relief system design deficiencies and convened the meeting which ultimately lead to formation of DIERS.

H. S. Kemp (DuPont, retired)
Who chaired the administrative committee and raised the funds necessary to support the DIERS research project.

J. E. Huff (Dow Chemical, retired)
Who presented the initial state-of-the-art overview of the basis for sizing pressure relief systems and served as chair of key technical advisory committees during the DIERS research project.

H. K. Fauske (Fauske & Associates, Inc.)
Who served as the principal investigator and leader of the DIERS research project team.

Contents

Preface

The chemical, petrochemical, and pharmaceutical industries store, handle, process, and transport a wide range of reactive chemicals. Several incidents caused by runaway reactions have increased awareness of the severity and the potential losses that can result from such hazardous events. This awareness continues to catalyze research aimed at understanding reactivity hazards and providing tools for the effective design of mitigation systems.

The DIERS Users Group, established in 1986, works to reduce the frequency, severity, and consequences of pressure-producing accidents and to promote the development of new techniques to improve the design of emergency relief systems.

The purpose of this symposium is to provide a summary of the research and technologies that are applicable to the control of runaway reaction hazards and the design of pressure relief and effluent handling systems. This symposium compliments the 1995 symposium held in Boston, Massachusetts, and reports on recent research studies funded by the DIERS Users Group.

This book contains papers dealing with many aspects of chemical reactivity, pressure relief design, and effluent handling. These papers provide practical guidance on how to evaluate reactivity hazards and how to use the data for scaleup and design of emergency relief systems. The text generally addresses five topics:

- Experimental techniques for thermal hazards assessment
- Analytical/simplified tools for thermal hazards assessment and ERS design
- Elements of ERS design using DIERS technology
- Advanced ERS design
- Mitigation design

The editors hope that this book will prove useful to those who design or review emergency relief for both reactive and nonreactive systems.

G. A. Melhem
Arthur D. Little, Inc.
Cambridge, Massachusetts

H. G. Fisher
Union Carbide Corporation
South Charleston, West Virginia

Design Institute for Emergency Relief Systems (DIERS): The Dominant Force Over the Past Twenty Years for a Changed Emergency Relief System Design Paradigm

Harold G. Fisher
Union Carbide Corporation
P. O. Box 8361
South Charleston, West Virginia 25303

INTRODUCTION

The Design Institute for Emergency Relief Systems (DIERS) and DIERS Users Groups of the AIChE have over the past 20 years changed a chemical engineering paradigm—the approach to and techniques used for emergency relief system design.

Generalized techniques to size emergency relief devices for runaway reaction and/or two-phase flow emergency relief were almost absent from the chemical engineering literature before DIERS. Today both industrial practice and regulators expect consideration of these topics.

Manufacturing companies seeking engineering design and construction services are contractually requiring that "DIERS methods" be considered for emergency relief system designs. Contractors and consultants offer "DIERS" and "DIERS-related" services. Industrial advertisements seeking safety professionals mention "DIERS experience" as a prerequisite.

Four of the domestic safety relief valve manufacturers belong to and participate in DIERS Users Group activities. Several are conducting product-related research into two-phase flow through safety relief valves and are considering revisions of recommended engineering sizing practices.

Since emergency relief designs which consider runaway reactions and/or two-phase flow, as appropriate, are invariably larger than those not considering these phenomena, DIERS-related designs are more likely to stand the test of time. While such designs are perceived as time consuming and costly by some, failure to properly consider relevant design information places facilities at risk and will eventually require costly retrofits. First cost is usually less expensive than "do over" cost.

OSHA in its 1992 Process Safety Management (PSM) regulation (29 CFR 1910.119) has recognized technical reports published by the AIChE on topics such as two-phase flow for venting devices as constituting "generally recognized and accepted good engineering practice." An OSHA incident citation directing remedial action suggested "The employer only calculated the size of the emergency relief device for an external fire and did not take into consideration the potential for an internal reaction. Reevaluate the size and capacity of the emergency relief system using the methodology of AIChE's Design Institute for Emergency Relief Systems (DIERS) with the worst credible scenario."

In 1995 the AIChE Council presented the Gary Leach Award to the DIERS Users Group. Created in 1989, this award recognizes publicly those groups who actively work toward the institute's mission of advancing chemical engineering practices and serving society. The DIERS Users Group met this standard by advancing the technology of emergency relief systems (ERS) and providing a forum for over 100 companies worldwide to work together to make chemical plants safer. Since 1986, the group has implemented the findings of the DIERS-sponsored research to reduce the frequency, severity and consequences of pressure-producing accidents and to improve the design of ERS. The ERS technology that the group fosters is cited by the United States government as "good engineering practice."

DESIGN INSTITUTE FOR EMERGENCY RELIEF SYSTEMS (DIERS)

2.1. Introduction to DIERS

The Design Institute for Emergency Relief Systems (DIERS), a consortium of 29 companies (Table 1), was formed in 1976 under the auspices of the AIChE to develop methods for the design of emergency relief systems to handle runaway reactions. Of particular interest was the prediction of when two-phase flow venting would occur and the applicability of various sizing methods for two-phase vapor–liquid flashing flow. DIERS spent approximately $1.6MM to investigate two-phase vapor–liquid onset/disengagement dynamics and the hydrodynamics of emergency relief systems.

2.2. DIERS Research Program

The DIERS program evolved over time. The initial research focus involved an investigation of two-phase vapor–liquid

- ***onset and disengagement of two-phase flow from vessels***: This is concerned with methods for predicting when or if two-phase flow will

TABLE 1
Sponsors of the DIERS Program

Air Products and Chemicals, Inc.	Health and Safety Executive, UK
Allied Chemical Corporation	Hoffmann-LaRoche, Inc.
American Cyanamid Company	Occidental Chemical Corporation
Ashland Chemical Company	Imperial Chemical Industries PLC, UK
British Gas Corporation, UK	Industrial Risk Insurers
Ciba-Geigy Corporation	The Insurance Technical Bureau, UK
Dow Chemical Company	Mobil Research and Development Corporation
E. I. Du Pont De Nemours & Company	Monsanto Company
Dutch State Mines, Netherlands	Olin Corporation
Eastman Kodak Company	Phillips Petroleum Company
Factory Mutual Research Corporation	Rohm & Haas Company
FMC Corporation	Shell Oil Company
General Electric Company	Sandoz AG, Switzerland
The Goodyear Tire & Rubber Company	Union Carbide Corporation
Gulf Research and Development Company	

occur with runaway reactions where vapors or gases are generated within the reacting fluid in the vessel. Liquids with high viscosity and with a tendency to foam are more likely to cause two-phase flow. Analysis methods for characterizing flow and separation behavior within a reactor were validated in large-scale tests.

- ***relief system hydrodynamics:*** This consisted of a review of existing models and development of new techniques for analyzing various two-phase flow conditions to include flashing, non-flashing and highly viscous liquids in nozzles and piping. Experimental studies were used to verify the analysis methods.
- ***experimental verification tests of the proposed models.***

The second research phase consisted of

- both small- (32 liter) and large- (2200 liter) scale integral blowdown and runaway reaction experimental tests,
- computer simulation of the experimental results and
- preparation of technology reports.

The final research phase provided

- ***a design computer program (SAFIRE):*** The SAFIRE computer program was developed to analyze the two-phase vapor-liquid flow from the large-scale runaway reaction and blowdown experiments. This program is suitable for design and evaluation of pressure relief systems for runaway reactions and other situations involving two-phase flow.

- ***a bench-scale experimental apparatus:*** A test apparatus was developed which can provide information on chemical reaction dynamics (pressure versus time data), vapor-liquid disengagement within the vessel (indicate flow regime within the reactor) and required sizes of vents for runaway reactions.
- ***an independent review of the basic methodology***.

The DIERS research program came to a successful conclusion in 1985 with publication of the results by the AIChE.

DIERS research results available through AIChE include:

- ***SAFIRE Computer Program and Documentation:*** Version 10.0 of the SAFIRE (Systems Analysis for Integrated Relief Evaluation) computer program. An important tool used to estimate the pressure-temperature-time history and size emergency relief devices for a vessel containing a runaway reaction.
- ***Emergency Relief System Design Using DIERS Technology:*** The DIERS experimental program and technology is discussed in a 538-page book published by DIERS/AIChE. It includes chapters on vapor-liquid disengagement dynamics; pressure relief system flow in piping and nozzles; results of two-phase flow and large-scale venting tests; high viscosity flashing flow; containment, disposal and mechanical design; DIERS bench-scale apparatus for runaway reaction testing; and a description of the SAFIRE computer program.
- ***Technical information (19 manuals/1950 pages):*** Includes the DIERS theoretical/analytical methods, large-scale runaway reaction and blowdown test results, bench-scale apparatus design plans and safety valve two-phase flow test data.

For detailed information, contact the AIChExpress, 345 East 47th Street, New York, NY 10017-2395, 1-800-AIChemE, Fax: (212) 705-8400.

DIERS USERS GROUP

One hundred twenty-six (96 domestic/30 international) companies (Table 2) have formed the DIERS Users Group, a Design Institute of the AIChE, to cooperatively assimilate, implement, maintain and upgrade the DIERS methodology. The purpose of the group is to

- Reduce the frequency, severity and consequences of pressure-producing accidents.
- Develop new techniques which will improve the design of emergency relief systems.

TABLE 2
DIERS Users Group Member Companies

Year	Domestic	International	Total	Change
1987	32	8	40	—
1988	46	16	62	+22
1989	47	18	65	+ 3
1990	47	18	65	0
1991	53	19	72	+ 7
1992	56	19	75	+ 3
1993	73	23	96	+21
1994	83	24	107	+11
1995	89	27	116	+ 9
1996	93	30	123	+ 8
1997	96	30	126	+ 3

3.1 DIERS Users Group Membership

Membership in the DIERS Users Group offers access to refinements of the DIERS technology, participation in development of additional technology and an opportunity to share learning experiences. Membership is open to industrial or engineering organizations interested in the design, use or manufacture of emergency relief devices and systems.

Member companies (Table 3) vary from large to small manufacturing firms, engineering contractors and constructors, equipment manufacturers and suppliers and various consulting organizations. There are no fees for participation. Companies can join by contacting the AIChE or the author.

A European DIERS Users Group has also formed and meets regularly.

3.2. DIERS Users Group Technical Committees

Presentations at DIERS Users Group meetings are arranged by the various technical committees (Table 4), which work to advance emergency relief system design in the following areas:

- Venting Technology
- Emergency Relief System Design for Fire Exposure
- Emergency Relief System Design/Testing Methodology

TABLE 3

Design Institute for Emergency Relief Systems (DIERS) Users Group Member Companies

- ABB Lummus Global, Inc.
- Abbott Laboratories
- Air Products and Chemicals, Inc.
- Akzo Nobel Central Research BV
- Albright & Wilson
- ALLIANT—Allegheny Ballistics Lab
- AlliedSignal, Inc.
- American Cyanamid Company
- American Home Products
- Amoco Corporation
- Anderson Greenwood & Company
- ARCO Chemical Company
- ARINC, Inc.
- Aristech Chemical Corporation
- Arthur D. Little, Inc.
- Ashland Chemical Company
- Aspen Technology, Inc.
- Associated Chemical Engineering Services
- elf ATOCHEM
- BASF Corporation
- Bayer Corporation
- Bechtel Corporation
- BE&K-Delaware
- Berwanger, Inc.
- Borden Chemical, Inc.
- Celanese Chemical Division
- Centaurus Technology, Inc.
- Central Leather Research Institute
- Cheminform Ltd. St. Petersburg (CISP)
- Ciba Corporation
- CITGO Petroleum Corporation
- Clark, Richardson & Biskup, Inc.
- Compliance Engineering, Inc.
- Continental Disc Corporation
- Coors Brewing Company
- Crosby Valve, Inc.
- CYTEC Industries, Inc.
- Daelim America, Inc.
- The Dexter Corporation
- Dixie Chemical Company, Inc.
- Dow Chemical U.S.A.
- Dresser Valve and Controls
- DSM Research B.V.
- DuPont Company
- Eastman Kodak Company
- Eichleay Engineers, Inc.
- Eli Lilly and Company
- Enterprise Products Company
- Exxon Research & Engineering Company
- Factory Mutual Research Corporation
- Fauske & Associates, Inc.
- Fluor Daniel, Inc.
- FMC Corporation
- GE Plastics
- Georgia-Pacific Resins, Inc.
- The Goodyear Tire & Rubber Company
- Halliburton-NUS Corporation
- Hazard Evaluation Laboratories, Ltd.
- Health and Safety Executive
- Henkel Corporation
- Hercules Incorporated
- Hoechst AG
- Hoffman-LaRoche, Inc.
- Huff Consulting Services
- HÜLS America, Inc.
- Huntsman Chemical Corporation, Inc.
- Imes Engineering, Inc.
- Imperial Chemical Industries PLC
- Inburex GmbH
- Industrial Risk Insurers
- Industrial Technology Research Institute
- INGEMO IMPIANTI SaS di Pintabona Renato & Company
- Institute for Systems, Informatics and Safety, European Commission
- Institute of Occupational Safety and Health, Council of Labor Affairs, Executive Yuan
- Instituto Mexicano del Petroleo
- International Flavors & Fragrances Inc.
- ISP (International Specialty Products, Inc.)
- Jacobs Engineering Group, Inc.
- Javan & Walter, Inc.
- JBF Associates, Inc.
- Johnson & Higgins
- The M. W. Kellogg Company
- Kinetica, Inc.
- Kvaerner John Brown, Inc.
- Leung, Inc.
- Lockwood-Greene Engineers, Inc.
- Merck & Company, Inc.
- Middough Associates, Inc.
- 3M Company
- Morenco AGRA, Inc.
- Monsanto Company
- Montell Polyolefins
- Nalco Chemical Company
- NORAMCO, Inc.
- Norsk Hydro A/S
- NOVA Chemicals Ltd.
- Oxychem
- Olin Corporation
- Phillips Petroleum Company
- PPG Industries, Inc.
- Pred-Engineering, Inc.
- The Pritchard Corporation
- Process Development Corporation
- Process Safety & Design, Inc.
- The Procter & Gamble Company
- PROSAF, Inc.
- Raytheon Engineers & Constructors, Inc.
- Rhone Poulenc Industrialisation
- RMT/Jones & Neuse, Inc.
- Rohm and Haas Company
- Russian Scientific Center Applied Chemistry (RSCAP)
- S and B Engineers & Contractors, Ltd.
- SE Technologies, Inc.
- Shell Oil Company
- Simulation Sciences, Inc.
- Solutia, Inc.
- Stepan Company
- G. R. Stucker & Associates, Inc.
- Sumitomo Chemical Company, Ltd.
- DNV-Technica, Inc.
- Teledyne Fluid Systems
- Texaco, Inc.
- Toyo Engineering Corporation
- Union Carbide Corporation
- Westvaco Corporation
- Witco Corporation
- Zeneca, Inc.

TABLE 4
DIERS Users Group

Chair	H. G. Fisher	Union Carbide Corporation
Secretary	J. J. Sharkey	Merck & Company, Inc.
Council Liaison	R. Lee	Aces
COMMITTEE	CHAIR	AFFILIATION
1. Venting Technology	M. A. Grolmes H. S. Forrest	Centaurus Technology, Inc. Abb Lummus Global, Inc.
2. ERS Design for Fire Exposure	J. J. Hauser	Prosaf, Inc.
3. ERS Design/Testing Methodologies	J. C. Leung	Leung, Inc.
4. Case Histories/Incident Investigations	W. A. Greenfield K. B. Pearl	International Specialty Products Amoco Corporation
5. High Viscosity Hydrodynamics	S. G. Coats	Dupont, Inc.
6. Mathematical Modeling of ERS	D. A. Shaw M. E. Levin	Monsanto Company Shell Oil Company
7. New Technology Awareness	S. S. Grossel	Process Safety & Design, Inc.
8. Fire, Explosion and Runaway Risk Criteria	J. A. Noronha	Eastman Kodak Company
9. Process Safety Management	R. Lee	Aces
10. Effluent Handling	G. A. Melham	Arthur D. Little, Inc.
11. Information Retrieval	S. Kostos	Bayer Corporation
12. European DIERS Group Liaison	F. P. Nichols	Imperial Chemical Industries Plc

- Case Histories/Incident Investigations
- High Viscosity Hydrodynamics
- Mathematical Modeling of Emergency Relief Systems
- New Technology Awareness
- Fire, Explosion and Runaway Risk Criteria
- Process Safety Management
- Effluent Handling
- Information Retrieval
- European DIERS Users Group Liaison

3.3. DIERS Users Group Meetings

Semi-annual three-day technical meetings (Table 5) are held at various locations and feature presentations by member company representatives and

TABLE 5
DIERS Users Group Meetings

Year	Location	Attendance	Remarks (Plant Visits/Symposia)
1998	New Orleans, LA	—	2nd International Symposium on Runaway Reactions, Pressure Relief Design and Effluent Handling
	Charleston, SC	—	
1997	Scottsdale, AZ	69	
	Pittsburgh, PA	92	NIOSH Research Center (Bureau of Mines)
1996	Houston, TX	78	Anderson Greenwood Company
	Calgary, CN	66	Novacor Ethylene and Polyethylene Plants Westech Industrial Ltd.
1995	Orlando, FL	63	
	Boston, MA	165	International Symposium on Runaway Reactions and Pressure Relief Design
1994	San Francisco, CA	54	
	Boston, MA	61	Crosby Valve, Inc.
1993	Albuquerque, NM	56	Energetic Materials Research and Testing Center—New Mexico Tech
	Chicago, IL	65	FAI Laboratory
1992	Columbus, OH	70	
	Princeton, NJ	78	FMC Hazards Laboratory
1991	Pittsburgh, PA	—	RSST/VSP Users Symposium
	Toronto, CN	—	
1990	Austin, TX	50	Columbia Scientific, Inc.
	Denver, CO	50	Coors Brewing Company
1989	Seattle, WA	50	
	Cincinnati, OH	50	VSP Operation/Improvements Symposium
1988	Bethesda, MD	56	
	San Antonio, TX	50	
1987	San Diego, CA	—	
	Boston, MA	—	

invited speakers and discussion of the results of computational benchmark problems and experimental round-robins.

Approximately 28 technical presentations are given at each meeting. There have been a total of 616 technical presentations at the 22 semiannual meetings since 1986. Recent attendance averages 68. There are no fees for participation; however, an expense charge of approximately $40/day is collected to pay for lunch, coffee breaks and meeting expenses.

DIERS Users Group meetings, each a technical symposium, provide an annual week of job-related, continuing education for participants at minimum cost. Many of the significant emergency relief system design related papers have first been presented at DIERS Users Group meetings. Networking with true experts at meetings is invaluable. One soon learns who can both "talk the talk" and "walk the walk." Early access to and subsequent implementation of new design information and techniques as well as discussions with others in the field has more than paid the travel cost of meeting attendees.

Since each meeting is the equivalent of a three-day short course, those who have the background and/or regularly attend derive the maximum benefit. Comprehensive minutes that include every overhead shown at the meeting are sent to the principal representative of each member company with appropriate internal distribution recommended.

3.4. DIERS Users Group Experimental Round-Robin Testing

The ERS Design/Testing Methodologies Committee of the DIERS Users Group has organized and conducted eight experimental round-robins involving the following chemical systems:

Styrene Monomer Polymerization w/Benzoyl Peroxide Initiator
Di-t-butyl Peroxide Decomposition in Benzene Solvent
Acetic Anhydride and Methanol Reaction
Hydrogen Peroxide Decomposition in Water
Phlegmatized Benzoyl Peroxide Decomposition
Sodium Tetraethylene Glycolate/Tetraethylene Glycol Decomposition
Dicumyl Peroxide Decomposition in Ethylbenzene
Styrene/Butadiene Emulsion Polymerization

Testing protocols for various apparatus and the required chemicals (in most cases) were sent to registered participants. Comparisons by and across instruments were used to determine consensus and outlier results, which have been published in the DIERS Users Group meeting minutes and the open literature.

3.5. DIERS Users Group Emergency Relief Design Benchmarks

The Venting Technology Committee of the DIERS Users Group has organized and conducted five emergency relief design benchmark exercises involving the following chemical systems:

Hydrogen Peroxide Emergency Relief System Design
Viscous "Polywater" Pipe Flow (3 Cases)

Safety Relief Valve Capacity (15 Cases)
Vessel Two-Phase Flow Blowdown (7 Cases)
Dicumyl Peroxide Vent Sizing

Design information and instructions were sent to the membership. A modified Delphi methodology was used in which submitted results were compared to consensus results to discover the reasons for and to correct blunders.

The ability to produce the consensus results of the DIERS benchmarks is a useful training tool for both experienced and inexperienced relief system design engineers. Purchasers of emergency relief system design computer programs are encouraged to ensure that comparisons to the complete range of DIERS emergency relief design benchmarks are provided as part of the program documentation.

3.6. DIERS Users Group Research

"DIERS"-type research projects, conducted by both industrial consortia and universities, involving chemicals other than water have been completed or are underway in many countries. Research programs of others have supplemented and in some cases may supersede results from the original DIERS research program. The author, however, is not aware of definitive findings that presently invalidate any DIERS results.

The cooperative efforts of member companies of the DIERS Users Group over the past 12 years have expanded the body of knowledge produced by the original DIERS contractors. A recent deficiency analysis and ballot of the member companies of the DIERS Users Group, however, suggested additional areas for further research (Table 6). The accumulation of sufficient funds from the sale of research reports and the SAFIRE computer program by the AIChE and two contributions—$5000 by a group of duPont employees and $20000 by Arthur D. Little, Inc.—have allowed the presently funded research programs of the DIERS Users Group to be undertaken.

The initiation of these research programs has already prompted additional papers by several experts, which will be presented in part at this symposium. Preliminary, but unverified, results from the funded research will also be presented in order to stimulate contributions by others toward the final results. As always, the final reports of the research program will be made available to the CPI, HPI and general public by the AIChE after a short period of verification by the member companies of the DIERS Users Group.

3.7. DIERS Users Group Education Activities

In addition to participation in meetings, symposia, experimental round-robins and benchmark exercises organized by the DIERS Users Group,

TABLE 6
DIERS Users Group Funded Research Projects

Viscous Two-Phase Flow Through Safety Relief Valves
Texas Engineering Experimental Station (TEES)
Texas A&M University—Prof. R. Darby
Prairie View A&M University—Prof. J. G. Gabitto
Gulf Coast Relief Valve Consortium
Viscous Two-Phase Flow Through Pipe
Centaurus Technology, Inc.—Dr. M. A. Grolmes
Improve Computational Procedures for Emergency Relief System Design
Arthur D. Little, Inc.—Dr. G. A. Melhem
Reports and Data on Venting Tests Performed on the JRC DRACULA and COLUMBUS Facilities
Commission of the European Communities
Joint Research Center, ISPRA—Dr. S. D. Morris

member company representatives and many others have enrolled in and taken one or both of the AIChE Continuing Education courses:

Emergency Relief System (ERS) Design Using DIERS Technology (3 days)
Methods for Sizing Pressure Relief Vents (2 days).

These courses offer cost effective training by

- acquainting participants with reactive emergency relief system design principles, techniques and tools,
- providing a background on and understanding of the DIERS technology and
- illustrating emergency relief system design methodology by means of selected sample problems.

Over 430 students (Table 7) have taken advantage of the three-day training course.

3.8. DIERS Users Group Symposia (AIChE Topical Conferences)

The DIERS Users Group in conjunction with its co-sponsors have organized two symposia:

- International Symposium on Runaway Reactions and Pressure Relief Design; Boston, August 2–4, 1995
- 2nd International Symposium on Runaway Reactions, Pressure Relief Design and Effluent Handling; New Orleans, March 11–13, 1998.

TABLE 7
Emergency Relief System Design Using DIERS Technology (AIChE/DIERS Continuing Education Course)

Year	Courses	Participants	Average Attendance
1987	1	22	22.0
1988	4	57	14.3
1989	3	50	16.7
1990	4	41	10.3
1991	5	48	9.6
1992	3	35	11.7
1993	5	63	12.6
1994	5	55	11.0
1995	2	22	11.0
1996	3	25	8.3
1997	3	21	7.0
Total	**38**	**439**	**11.6**

Proceedings of both symposia are available from AIChE. Written by experts, the papers of the subject symposia provide both the novice and the expert with the information necessary to prevent runaway reactions and/or design appropriate emergency relief and effluent handling systems. The papers extend previously published DIERS and DIERS-related design methods. The hard cover volumes are organized into five separate, but complimentary, areas:

- Experimental Techniques for Thermal Hazards Assessment
- Analytical/Simplified Tools for Thermal Hazards Assessment and ERS Design
- Elements of ERS Design Using DIERS Technology
- Advanced ERS Design
- Mitigation Design

Sufficient review and tutorial material is provided to allow the reader to understand and use the design methods. Those charged with operating, designing or managing today's and tomorrow's CPI/HPI facilities will find that use of the information can lead to improved process safety.

SAFIRE COMPUTER PROGRAM

The SAFIRE computer program was originally developed by the DIERS principal contractor, Fauske and Associates, Inc., to analyze and model the DIERS-funded large-scale runaway reaction and blowdown experiments. Later, DIERS funded a project to enable the program to be more easily used for emergency relief system design.

The Mathematical Modeling of ERS Committee of the DIERS Users Group has over the years further enhanced the capabilities of the program and provided improved user-friendly interfaces and data-based physical property input capabilities. The program has also been migrated from a mainframe to a PC-based tool. The original DOS-based version has been converted to a WINDOWS 95 environment.

The twenty-nine DIERS sponsors have copies of the program. An additional 50 copies have been purchased since 1986. Extensive industrial use and benchmarking participation has helped to improve the functionality of the program. Use of and improvements to the program are discussed at meetings of the DIERS Users Group. Upgrades of the program are issued on a gratuitous basis as they become available. The SAFIRE computer program has proven to be a useful design tool for computer literate users accustomed to working in a simulation environment.

DIERS AND DIERS-RELATED PUBLICATIONS

The AIChE made the DIERS research reports available to the CPI, HPI, and to the general public in 1986. Many useful open literature articles and several books provide emergency relief system design guidance to the practicing engineer. *Emergency Relief System Design Using DIERS Technology—The DIERS Project Manual* was published by AIChE in 1992. The Center for Chemical Process Safety (CCPS) has published (1998) the *Guidelines for Pressure Relief and Effluent Handling Systems*. The British Health and Safety Executive has published (1998) a *Workbook for Chemical Reactor Relief System Sizing.* Since the conclusion (1985) of the DIERS research project, the AIChE and/or DIERS have organized three symposia (1989, 1995, 1998) related to runaway reactions, emergency relief system designs and effluent handling. The AIChE Loss Prevention and other symposia—both domestic and international—continue to feature DIERS and DIERS-related sessions. The International Standards Organization (ISO 185/SC1 and 2) has a joint working group of experts engaged in development of a standard for "Flashing Flow through Safety Devices." The American Petroleum Institute, based on the critiques and recommendations of others, will abandon its simple, but nonconservative, recommended design practice for two-phase flow from safety relief valves in the 1998 edition of API RP 520, Part I.

This paper contains a suggested, open-literature reading list (references) compiled by the author of approximately 100 articles related to emergency relief system design involving runaway reaction and two-phase flow. The list is not all inclusive, but does attempt to include significant information resulting from and/or related to the DIERS program. At least as many significant articles exist that are not on the list, which may simply reflect the author's personal preference for information or bias.

ACKNOWLEDGMENTS

J. J. Sharkey (88–present) and T. P. Nelson (86–87) served as the Secretary of the DIERS Users Group. Their administrative and technical skills have made significant contributions to the conduct of meetings and production/distribution of the meeting minutes. The committee chairs (Named in Table 4) organize the technical programs and provide the leadership for the continuing efforts of the member companies of the DIERS Users Group to develop and exchange technical information related to the design/use and manufacture of emergency relief devices and systems. The member companies of the DIERS Users Group have willingly shared nonproprietary, safety-related information and supported the participation of their representatives over the past 12 years.

SELECTED REFERENCES

DIERS/DIERS Users Group Programs

1. Kemp, H. S., "Formation and Management of AIChE Design Institutes." *CEP,* 79(6), 9–14 (June 1983).
2. Swift, I., "Developments in Emergency Relief Design and the Characterization of Accidental Discharges from Pressurized Vessels." *I. Chem. E. Symposium Series 85 (1984).*
3. Fisher, H. G., "DIERS Research Program on Emergency Relief Systems." CEP, 81(8), 33–36 (August 1985).
4. Anon., "AIChE's Design Institutes Help Reduce Industrial Risks." *CEP*, 84(9) 22–23 (September 1988).
5. Fisher, H. G., "The DIERS Users Group (A Forum for Development/Dissemination of Emergency Relief System Design Technology)." *Plant/Operations Progress,* 8(2) 70–72 (April 1989).
6. Fisher, H. G., "The DIERS Users Group: Status and Current Activities." *Process Safety Progress,* 14(2), 87–92 (April 1995).

Overview of Emergency Relief System Design

7. Fisher, H. G., "An Overview of Emergency Relief System Design Practice." *Plant/Operations Progress,* 10(1), 1–12 (January 1991).

8. Fisher, H. G., et al., "Emergency Relief System Design Using DIERS Technology—The DIERS Project Manual." AIChE/DIERS, New York, (1992).
9. Melhem, G. A., Fisher, H. G. and Shaw, D. A., "An Advanced Method for the Estimation of Reaction Kinetics, Scaleup, and Pressure Relief Design." *Process Safety Progress,* 14(1), 1–21 (January 1995).
10. Anon., "Guidelines for Pressure Relief and Effluent Handling Systems." AIChE/CCPS, New York (1998).
11. Etchells, J. C. and Wilday, A. J., "Workbook for Chemical Reactor Relief System Sizing." Health and Safety Executive, HSE Books (1998).

Two-Phase Flow Onset/Disengagement Dynamics

12. Grolmes, M. A., "A Simple Approach to Transient Two-Phase Level Swell." Multi-Phase Flow and Heat Transfer III. Part A: Fundamentals. Proceedings of the Third Multi-Phase Flow and Heat Transfer Symposium—Workshop, Miami Beach, FL (April 18–20, 1983).
13. Grolmes, M. A. and Fauske, H. K., "An Evaluation of Incomplete Vapor-Phase Separation in Freon-12 Top Vented Depressurization Experiments." Multi-Phase Flow and Heat Transfer III. Part A: Fundamentals. Proceedings of the Third Multi-Phase Flow and Heat Transfer Symposium——Workshop, Miami Beach, FL (April 18–20, 1983).
14. Grolmes, M. A. and Epstein, M., "Vapor–Liquid Disengagement in Atmospheric Liquid Storage Vessels Subjected to External Heat Source." *Plant/Operations Progress,* 4(4), 200–206 (October 1985).
15. Fauske, H. K., et al., "Emergency Relief Vent Sizing For Fire Emergencies Involving Liquid-Filled Atmospheric Storage Vessels." *Plant/Operations Progress,* 5(4), 205–208 (October 1986).
16. Epstein, M., Fauske, H. K. and Hauser, G.M., "The Onset of Two-Phase Venting Via Entrainment in Liquid-Filled Storage Vessels Exposed to Fire." *J. Loss Prev. Proc. Ind.,* 2(1), 45–49 (January 1989).
17. Grolmes, M. A. and Fisher, H. G., "Vapor–Liquid Onset/Disengagement Modeling for Emergency Relief Discharge Evaluation." Paper Presented at the AIChE 1994 Denver National Meeting, Denver, CO (August 16, 1994).
18. Sheppard, C. M., "Theoretical Basis for Two-Phase Disengagement and Vapor Fraction Evaluations." In Proceedings of the "International Symposium on Runaway Reactions and Pressure Relief Design." Melhem, G. A. and Fisher, H. G. (Ed), AIChE/DIERS, New York (1995).

Two-Phase Hydrodynamics

19. Fauske, H. K., "Flashing Flows or: Some Practical Guidelines for Emergency Releases." *Plant/Operations Progress,* 4(3), 132–134 (July 1985).
20. Huff, J. E., "Multi-Phase Flashing Flow in Pressure Relief Systems." *Plant/Operations Progress,* 4(4), 191–199 (October 1985).
21. Sallet, D. W. and Somers, G. W., "Flow Capacity and Response of Safety Relief Valves to Saturated Water Flow." *Plant/Operations Progress,* 4(4), 207–216 (October 1985).
22. Leung, J. C., "A Generalized Correlation for One-Component Homogeneous Equilibrium Flashing Choked Flow." *AIChEJ,* 32(10), 1743–1746 (October 1986).
23. Fauske, H. K., "Turbulent and Laminar Flashing Flow Considerations." Paper Presented at the 4th International Symposium on Multi-Phase Flow, Miami Beach, FL (December 15–17, 1986).
24. Leung, J. C. and Grolmes, M. A., "The Discharge of Two-Phase Flashing Flow In a Horizontal Duct." *AIChEJ,* 33(3), 524–527 (March 1987).

25. Fauske, H. K. and Epstein, M., "Source Term Considerations in Connection with Chemical Accidents and Vapor Cloud Modeling." CCPS International Conference on Vapor Cloud Modeling, Cambridge, MA (1987).

26. Leung, J. C. and Grolmes, M. A., "A Generalized Correlation for Flashing Choked Flow of Initially Subcooled Liquid." *AIChEJ,* 34(4), 688–691 (April 1988).

27. First, K. E. and Huff, J. E., "Design Charts for Two-Phase Flow in Emergency Pressure Relief Systems." *Plant/Operations Progress,* 8(1), 40–54 (January 1989).

28. Fauske and Associates, Inc., "Flashing Viscous Flow of Polystyrene Mixtures: Test Data and Scale-Up Methods." FAI/89–26 (April 1989).

29. Leung, J. C., "Two-Phase Flow Discharge in Nozzles and Pipes—A Unified Approach." *J. Loss Prev. Proc. Ind.,* 3(1), 27–32 (January 1990).

30. Huff, J. E., "Flow Through Emergency Relief Devices and Reaction Forces." *J. Loss Prev. Proc. Ind.,* (3), 43–49 (January 1990).

31. Grolmes, M. A., Leung, J. C. and Fauske, H. K., "A Review of Simplified Multiphase Flow Methods for Emergency Relief Calculations." Paper Presented at the AIChE 1990 Spring National Meeting, Orlando, FL (March 18–22, 1990).

32. Leung, J. C. and Nazario, F. N., "Two-Phase Flashing Flow Methods and Comparisons." *J. Loss Prev. Proc. Ind.,* (3), 253–260 (April 1990).

33. Leung, J. C. and Epstein, M., "A Generalized Correlation for Two-Phase NonFlashing Homogeneous Choked Flow." *Trans. ASME—Journal of Heat Transfer* 112(2), 528–530 (May 1990).

34. Leung, J. C. and Epstein, M., "The Discharge of Two-Phase Flashing Flow from an Inclined Duct." *Trans. ASME—Journal of Heat Transfer,* 112(2), 224–228 (May 1990).

35. Leung, J. C., "Similarity Between Flashing and NonFlashing Two-Phase Flows." *AIChEJ,* 36(5), 797–800 (May 1990).

36. Leung, J. C. and Epstein, M., "Flashing Two-Phase Flow Including the Effects of Noncondensible Gases." *Trans. ASME—Journal of Heat Transfer,* 113(2), 269–272 (February 1991).

37. Simpson, L. L., "Estimate Two-Phase Flow in Safety Devices."*Plant/Operations Progress,*, 98–102 (August 1991).

38. Leung, J. C., "Size Safety Relief Valves for Flashing Liquids." *CEP,* 70–75 (February 1992).

39. Nazario, F. N. and Leung, J. C., "Sizing Pressure Relief Valves in Flashing and Two-Phase Service: An Alternative Procedure." *J. Loss Prev. Proc. Ind.,* 5(5), 264–269 (December 1992).

40. Leung, J. E. and Ciolek, W. H., "Flashing Flow Discharge of Initially Subcooled Liquid in Pipes." *ASME Trans.—J. of Fluids Engineering,* 116(3), 643–645 (September 1994).

41. Boicourt, G. W., "Emergency Relief System (ERS) Design: An Integrated Approach Using DIERS Methodology." *Process Safety Progress,* 14(2), 93–106 (April 1995).

42. Leung, J. C., "The Omega Method for Discharge Rate Evaluation." In Proceedings of the "International Symposium on Runaway Reactions and Pressure Relief Design." Melhem, G. A. and Fisher, H. G. (Ed), AIChE/DIERS, New York (1995).

43. Simpson, L. L., "Navigating the Two-Phase Maze." In Proceedings of "International Symposium on Runaway Reactions and Pressure Relief Design." Melhem, G. A. and Fisher, H. G. (Ed), AIChE/DIERS, New York (1995).

44. Fisher, H. G., "Emergency Relief System Design: The DIERS Users Group Safety Relief Valve Example Problem." In Proceedings of the "International Symposium on Runaway Reactions and Pressure Relief Design." Melhem, G. A. and Fisher, H. G (Ed), AIChE/DIERS, New York (1995).

45. Leung, J. D., "Easily Size Relief Devices and Piping for Two-Phase Flow." *CEP,* 92(12), 28–50 (December 1996).

46. Leung, J. C. and Fisher, H. G., "A Method for Sizing Safety Valve Inlet Lines." Accepted for Publication in *Process Safety Progress* (1998).

Analysis of DIERS Large-Scale Experiments

47. Grolmes, M. A., Leung, J. C. and Fauske, H. K., "Large-Scale Experiments of Emergency Relief Systems." *CEP*, 81(8), 57–62 (August 1985).
48. Klein, H. H., "Analyses of DIERS Venting Tests: Validation of a Tool for Sizing Emergency Relief Systems for Runaway Chemical Reactions." *Plant/Operations Progress,* 5(1), 1–10 (January 1986).

Experimental/Analytical Techniques for Emergency Relief System Design

49. Townsend, D. I. and Tou, J. C., "Thermal Hazard Evaluation By An Accelerating Rate Calorimeter." Thermochimica Acta, 37, 1–30 (1980).
50. Huff, J. E., "Emergency Venting Requirements." *Plant/Operations Progress,* 1(4), 211–229 (October 1982).
51. Fauske, H. K., Grolmes, M. A. and Henry, R. E., "Emergency Relief Systems—Sizing and Scale-up." *Plant/Operations Progress,* 2(1), 21–26 (January 1983).
52. Fauske, H. K., "Scale-up for Safety Relief of Runaway Reactions." *Plant/Operations Progress,* 3(1), 7–11 (January 1984).
53. Harmon, G. W. and Stuper, W. W., "Sizing Emergency Relief Systems on Vessels Containing Monomers." *CEP*, 80(3), 53–59 (March 1984).
54. Fauske, H. K., "Generalized Vent Sizing Nomograph for Runaway Chemical Reactions." *Plant/Operations Progress,* 3(4), 213–214 (October 1984).
55. Fauske, H. K. and Leung, J. C., "New Experimental Technique for Characterizing Runaway Chemical Reactions." *CEP*, 81(8), 39–46 (August 1985).
56. Fauske, H. K., "Emergency Relief System (ERS) Design." *CEP*, 81(8), 53–56 (August 1985).
57. Leung, J. C., Fauske, H. K. and Fisher, H. G., "Thermal Runaway Reactions In A Low Thermal Inertia Apparatus." *Thermochimica Acta,* 104, 13–29 (1986).
58. Leung, J. C., "Simplified Vent Sizing Equations for Emergency Relief Requirements in Reactors and Storage Vessels." *AIChEJ,* 32(10), 1622–1634 (October 1986).
59. Fauske, H. K., "Venting of Runaway Chemical Reactions and Non-Equilibrium Effects." Paper Presented at the 4th International Symposium on Multi-Phase Flow, Miami Beach, FL (December 15–17, 1986).
60. Fauske, H. K., "Relief System Design for Runaway Chemical Reactions." CCPS International Symposium on Preventing Major Chemical Accidents, Washington, DC (February 3–5, 1987).
61. Leung, J. C. and Fauske, H. K., "Runaway System Characterization and Vent Sizing Based On DIERS Methodology." *Plant/Operations Progress,* 6(2), 77–83 (April 1987).
62. Leung, J. C., "Overpressure During Emergency Relief Venting In Bubbly and Churn-Turbulent Regimes." *AIChEJ,* 33(6), 952–958 (June 1987).
63. Fauske and Associates, Inc., "Vent Sizing Package System Manual—Methodology and General Operations." FAI/88–41 (February 1988).
64. Fauske, H. K., Grolmes, M. A. and Clare, G. H., "Process Safety Evaluation—Applying DIERS Methodology to Existing Plant Operations." Paper presented at the 1988 Spring National Meeting of the AIChE, New Orleans, LA (March 6–10, 1988).
65. Leung, J. C., et al., "Emergency Relief Requirements for Reactive Chemical Storage Tanks." *I. Chem. E. Symposium Series* 110 (1988).
66. Leung, J. C., Creed, M. J. and Fisher, H. G., "Round-Robin 'Vent Sizing Package' Results." Proceedings of the International Symposium on Runaway Reactions, CCPS, Cambridge, MA (March 1989).
66. Grolmes, M. A., Leung, J. C. and Fauske, H. K., "Reactive Systems Vent Sizing Requirements." Proceedings of the International Symposium on Runaway Reactions, CCPS, Cambridge, MA (March 1989).

68. Fauske, H. K., Clare, G. H. and Creed, M. J., "Laboratory Tool for Characterizing Chemical Systems." Proceedings of the International Symposium on Runaway Reactions, CCPS, Cambridge, MA (March 1989).

69. Grolmes, M. A., Leung, J. C. and Fauske, H. K., "Reactive Systems Vent Sizing Evaluations." Proceedings of the International Symposium on Runaway Reactions, CCPS, Cambridge, MA (March 1989).

70. Fauske, H. K., "Emergency Relief Systems Design For Runaway Chemical Reactions: Extension of The DIERS Methodology." *Chem. Eng. Res. Des.,* (67), 199–201 (March 1989).

71. Leung, J. C. and Fisher, H. G., "Two-Phase Flow Venting of Reactor Vessels." *J. Loss Prev. Proc. Ind.,* 2(2) 78–86 (April 1989).

72. Grolmes, M. A. and King, M. J., "Emergency Relief Requirements for Certain Organic Peroxides Based on DIERS Methodology." Paper Presented at the 23rd Annual Loss Prevention Symposium, AIChE Spring National Meeting, Houston, TX (April 1989).

73. Creed, M. J. and Fauske, H. K., "An Easy, Inexpensive Approach to the DIERS Procedure." *CEP*, 86(3), 45–49 (March 1990).

74. Leung, J. C., "Vent Sizing for Tempered Vapor Systems." In "Safety of Chemical Batch Reactors and Storage Tanks." Benuzzi, A. and Zaldivar, J. M. (Ed), 285–298, Kluwer Academic Publishers, Dordrecht (April 1991).

75. Leung, J. C., "Vent Sizing for Gassy and Hybrid Systems." In "Safety of Chemical Batch Reactors and Storage Tanks." Benuzzi, A. and Zaldivar, J. M. (Ed), 299–310, Kluwer Academic Publishers, Dordrecht (April 1991).

76. Leung, J. C., "Venting of Runaway Reactions with Gas Generation." *AIChEJ,* 38(5), 723–736 (May 1992).

77. Grolmes, M. A. and Yue, M. H., "Relief Vent Sizing and Location for Tubular Reactors." *J. Hazardous Mat.,* (33), 261–273 (1993).

78. Leung, J. C., "Simplified Vent Sizing Methods Incorporating Two-Phase Flow." In Proceedings of the "International Symposium on Runaway Reactions and Pressure Relief Design." Melhem, G. A. and Fisher, H. G. (Ed), AIChE/DIERS, New York (1995).

Computer Techniques for Emergency Relief System Design

79. Huff, J. E., "Computer Programming For Runaway Reaction Venting." *I. Chem. E. Symposium Series* 85, 109–129 (1984).

80. Grolmes, M. A. and Leung, J. C., "Code Method For Evaluating Integrated Relief Phenomena." *CEP*, 81(8), 47–52 (August 1985).

81. Shaw, D. A., "SAFIRE Program for Design of Emergency Pressure Relief Systems." *CEP*, 86(7), 14–17 (July 1990).

82. Shaw, D. A., and Levin, M. E., "Application of SAFIRE in the Design of an Emergency Relief Vent for a Storage Vessel." In Proceedings of the "International Symposium on Runaway Reactions and Pressure Relief Design." Melhem, G. A. and Fisher, H. G. (Ed), AIChE/DIERS, New York (1995).

83. Melhem, G. A., "Advanced ERS Design Using Computer Simulation." In Proceedings of the "International Symposium on Runaway Reactions and Pressure Relief Design." Melhem, G. A. and Fisher, H. G. (Ed), AIChE/DIERS, New York (1995).

84. Melhem, G. A. and Fisher, H. G., "An Overview of SuperChems for DIERS for Emergency Relief System and Effluent Handling Designs." *Process Safety Progress,* 16(3), 185–201 (Fall 1997).

Containment/Disposal of Two-Phase Effluent

85. Grossel, S. S., "An Overview of Equipment for Containment and Disposal of Emergency Relief System Effluents." *J. Loss Prev. Proc. Ind.,* (3), 112–124 (January 1990).
86. Huckins, H. E., "Selection and Design of Systems to Handle the Effluent from Emergency Relief Systems." In Proceedings of the "International Symposium on Runaway Reactions and Pressure Relief Systems." Melhem, G. A. and Fisher, H. G. (Ed), AIChE/DIERS, New York (1995).

Nonideal Gas Flow from Relief Devices

87. Leung, J. C. and Epstein, M., "A Generalized Critical Flow Model for NonIdeal Gases." *AIChEJ,* 34(9), 1568–1572 (September 1988).

Emergency Relief System Reaction Forces

88. Huff, J. E., "Relief System Reaction Forces in Gas and Two-Phase Flow." Paper Presented at the 25th Annual Loss Prevention Symposium, (August 18–22, 1991).
89. Leung, J. C., "Reaction Forces During Two-Phase Discharges." *J. Fluids Eng.,* 114(12), 689–692 (December 1992).

Emergency Relief System Design (General)

90. American Petroleum Institute, "Sizing, Selection and Installation of Pressure-Relieving Devices in Refineries." API Recommended Practice 520. Part I—Sizing and Selection, 6th Ed (July 1993). Part II—Installation, 3rd Ed (November 1988).
91. American Petroleum Institute, "Guide for Pressure-Relieving and Depressuring Systems." API Recommended Practice 521, 4th Ed (March 1997).
92. American Petroleum Institute, "Venting Atmospheric and Low-Pressure Storage Tanks—NonRefrigerated and Refrigerated." API Recommended Practice 2000, 4th Ed (September 1992).

Safety Relief Valve Design and Operation

93. Emerson, G. B., "Selecting Pressure Relief Valves." Chemical Engineering, 195–200 (March 18, 1985).
94. Cox, O. J., Jr. and Weirick, M. L., "Sizing Safety Valve Inlet Lines." *CEP,* 76(11) 51–54 (November 1980).
95. Van Boskirk, B. A., "Sensitivity of Relief Valves to Inlet and Outlet Line Lengths." Chemical Engineering, 77–82 (August 23, 1982).

Rupture Disks

96. Nazario, F. N., "Rupture Discs—A Primer." Chemical Engineering, 86–96 (June 20, 1988).
97. Huff, J. E., "Measurement of Flow Resistance of Rupture Disc Devices." *Plant/Operations Progress,* 11(3), 187–200 (July 1992).

Emergency Relief for Fire Exposure

98. Crozier, R. A., Jr., "Sizing Relief Valves for Fire Emergencies." Chemical Engineering, 49–54 (October 28, 1985).
99. Fisher, H. G. and Forrest, H. S., "Protection of Storage Tanks from Two-Phase Flow Due to Fire Exposure." *Process Safety Progress,* 14(3), 183–199 (July 1995).

Reaction Kinetics

100. Ahmed, M., Fisher, H. G. and Janeshek, A. M., "Reaction Kinetics from Self-Heat Data—Correction for the Depletion of Sample." In Proceedings of the CCPS "International Symposium on Runaway Reactions." AIChE/CCPS, New York (1989).
101. Ahmed, M., Goetz, D. D., Fisher, H. G. and Janeshek, A. M., "Reaction Models from Adiabatic Calorimeter Data." In Proceedings of the "International Symposium on Runaway Reactions and Pressure Relief Design." Melhem, G. A. and Fisher, H. G. (Ed), AIChE/DIERS, New York (1995).

Assurance of Safe Pilot Plant Scale-Up of Chemical Processes

L. Tuma and C. Bagner
Merck Research Laboratories
Rahway, NJ 07065

1. ABSTRACT

The assurance of a safe pilot plant scale-up of a chemical process is dependent upon the early identification of process hazards, both chemical and thermal, in a process and an evaluation of the impact of these process hazards upon the pilot plant scale-up. Thermal hazards are identified using various thermal analytical techniques, with chemical hazards being identified through a hazards analysis review. The identified thermal and chemical hazards are then evaluated for their potential initiation during pilot plant processing. Once these hazards are identified and evaluated, implications for processing are then determined via the use of a hazard analysis review provided by the process development group. The early identification of these thermal hazards facilitates process modifications which may be necessary to assure safe pilot plant scale-up.

Merck Research Laboratories (MRL) has developed a three tier evaluation program which identifies both thermal and chemical hazards and evaluates them for potential initiation during pilot plant processing. In Tiers I and II, various thermal analytical techniques are used to identify hazards which can lead to thermal runaways. Using DTA, DSC, the Mettler RC1[1] Calorimeter and FAI's Reactive System Screening Tool (RSST),[2] a process safety screening can be performed rapidly, providing information on heats of reaction and exothermic oxidation/decomposition reactions including initiation temperatures and the accompanying rates of temperature and pressure increases. Tier III provides for a review of the identified chemical and thermal hazards and heat of reaction data in the context of the process description and in conjunction with a hazard analysis review. The hazards analysis review, provided by the process developmental team, identifies chemical hazards and potential processing hazards and is used to help define an intrinsically safe process.

2. INTRODUCTION

Hazards evaluation in a multipurpose pilot plant, by its very nature, is different from hazard evaluation in a manufacturing context. The use of general purpose pilot plant equipment, short lead time and continual process development require a flexible approach to hazards evaluation. Integrating this need for flexibility, with the recent advances in testing technology and methodology of hazards testing, is a key objective of the MRL Operational Hazards Evaluation Laboratory (OHEL).

Before a process moves from the laboratory bench to the MRL pilot plant, it is the subject of a detailed operational hazard evaluation. This early identification of thermal and chemical hazards facilitates process modifications which may be necessary to assure safe pilot plant scale-up. The OHEL has developed a three tier hazards assessment program which identifies process hazards and evaluates them for potential initiation during pilot plant operations.

The Tier I and II evaluations consist of a small-scale hazards identification program, using process samples from a 200-mL-scale reaction and small-scale analytical techniques, which allow for the rapid identification of potential thermal runaways. This small scale thermal hazards identification program utilizes a number of unique small scale thermal analytical techniques developed in the OHEL. These techniques allow for the identification/determination of exothermic activity in both open and closed reaction systems, of accurate exothermic onset temperatures, rates of temperature (*dT/dt*) and pressure increases (*dP/dt*) associated with the exothermic decomposition, and heats of reaction. For subambient or high temperature reactions, a specially designed small vessel (~200 mL), adapted for use with a Mettler RC1 Calorimeter, is used to determine the heats of reactions. Samples for the Tier I and II hazards assessment are also provided from this small scale RC1 run. Using these small scale analytical techniques, safety screening for a process step (reaction, workup/isolation) can be performed in ~2 to 3 days.

There are four small scale thermal analytical techniques which are utilized in the first two Tiers of the hazards evaluation program. They include the use of specially designed DSC reusable metal crucibles[3] to identify closed system exothermic activity in process samples; the Small Scale Isothermal Age (SSIA)[4] technique to accurately determine exothermic onset temperatures; the Small Scale Adiabatic Dewar Age (SSAD)[5] technique to accurately determine exothermic onset temperatures and the accompanying rates of temperature (*dT/dt*) and pressure (*dP/dt*) increases; and Syringe Injection Calorimetry (SIC)[6] to determine heat of reactions which occur at room temperature.

The Tier III evaluation, the identification of an intrinsically safe process, consists of a review of the identified thermal hazards and heat of reaction data in the context of the process description and in conjunction with the hazard analysis review provided by the process chemists and engineers.

Safety testing for thermal hazards is required before any large scale operations are performed, with reevaluation being required when any changes are made to the existing process. All processes, including waste treatments, are subjected to a minimum of Tier I screening and Tier III evaluations, with Tier II screening being required based on the results of the Tier I screening. Samples for testing must be supplied from the exact procedure to be used for scale-up, and be provided by a chemist/engineer familiar with the process. All samples must be provided for from the same experiment and as they are generated. It is critical that samples from the complete process be subjected to the Tier I screening. These samples must include all starting materials, all combinations of the reaction mixture, extracts, concentrates, distillates, solvent switches, waste streams, by-products, filter cakes, filtrates, unwashed and washed product cakes, dried products, etc. This may appear to be an excessive number of samples. However, numerous examples exist where unexpected exothermicity in a sample could have resulted in a potential thermal runaway: a filtered by-product was found to be shock sensitive; an unwashed product decomposed exothermically while the washed product showed no activity; a distillate (later subjected to solvent recovery) showed large exothermic activity; a wet product showed an exotherm slightly above the drying temperature, but the dried product had no exothermic activity. These are but a few of the many examples which could be cited.

Any change in the process; including changes in starting materials, order of additions, reverse quenches, changes in solvents, changes in concentrations, revised isolations, etc. require that the safety testing be repeated. Changes in processing temperature, require a reevaluation of the thermal/chemical data to assure that the safety envelope has not been compromised.

3. THREE-TIER PROCESS SAFETY EVALUATION PROGRAM

3.1. Operational Hazards Evaluation Data Sheet

The process hazards evaluation starts with the use of a Mini HAZOP technique which uses a specially designed Operational Hazards Evaluation Data Sheet, consisting of two sections, which must be filled out before any safety evaluation is undertaken for a chemical process. This data sheet must be authored by the chemist or engineer submitting the process for safety testing. The definition of the chemistry through balanced equations and completion of a chemical/chemical interaction grid are stressed. Moreover, the author of the data sheet must review a checklist designed to stimulate "what if" scenarios.

Thus, the author acts both as supplier of basic process information and as a first step hazard reviewer.

The first part of the Operational Hazards Evaluation Data Sheet, Section A, provides data on the toxicity of the chemicals used in the process, information on reactants which are sensitive to water/jacket service intrusion and information on exothermic heats of reactions studies. The information provided in Section A (see Attachments 1 and 2) identifies all water reactive chemicals and exothermic reactions which require water/jacket service intrusion studies and the determination of heats of reactions. The second part of the data sheet, Section B (see Attachments 3, 4, 5 and 6), provides data that will be required to evaluate and document any thermal hazards (decomposition/oxidations/heats of reaction/water intrusion) that are identified. Section B includes a hazardous situation checklist, an adaptation of HAZOP suited to pilot plant needs. The modifications employed stress identification of the hazards associated with physical and chemical interactions and the recognition of inherent hazards associated with the principal chemical reactions, side reactions, and formation of by-products. It also includes a checklist designed to stimulate the imagination when considering undesirable scenarios. The expected and deviation reaction tables require a balanced chemical equation for all expected and deviation reactions whether they are exothermic or not. This checklist also provides data on toxic compounds, fire/explosion potential, dust or odor problems and any known significant potential hazards. The chemical and material interaction grid provides information on incompatibilities between chemicals and services in the process and materials of construction of process vessels. Once this data sheet is supplied and evaluated for completeness, the safety study is undertaken.

3.2. Tiers I and II—Thermal Hazards Assessment Program

Tiers I and II of the Thermal Hazards Assessment Program identify and characterize potential thermal hazards which can lead to a thermal runaway situation. The runaway can be initiated in several ways: through an uncontrollable heat of reaction, the initiation of an exothermic decomposition or oxidation, or a combination of these two. Differential Scanning calorimetry (DSC), Differential Thermal Analysis (DTA), Syringe Injection Calorimetry (SIC) and FAI's Reactive System Screening Tool (RSST) are thermal techniques and tools which are used in determining the exothermicity associated with process samples. By using DTA, DSC, SIC and the RSST, safety screening can be performed in less than two days with a reaction run on a 200-mL scale.

The first part of the Tier I assessment uses DSC and DTA, two thermal techniques which allow for the rapid identification of exothermicity in process samples using very small samples to identify unknown exothermic activity. These two techniques provide information on the size, rate of heat release and initiation temperatures of exothermic decompositions or oxidations. Any number of commercially available thermal analysis systems can be used for this purpose.

DSC testing is frequently used to determine quantitatively if an exotherm is present in process samples in a closed reaction system. Each sample is normally evaluated from exothermic activity from 0 to ~300°C. Typically, high pressure DSC or crimped aluminum pans are used. But the problems of low pressure containment and chemical reactivity associated with the use of crimped aluminum pans and noisy baseline and reproducibility with high pressure DSC make their use very difficult and time consuming. Merck has overcome these problems with the use of a specially designed reusable sample crucible, the Merck "closed bomb" (CB).[3] The design of this special reusable closed bomb is shown in Figure 1 and consists of a screw-on cap, with a replaceable 2- μm Teflon rupture disk with a pressure rating of ~400 psi, and a threaded bottom. The crucibles are typically made of 316L or Hastelloy B with tantalum lined and glass liners available for highly corrosive samples. The crucible has a capacity of ~60 μL and an operating range of from ~20 to 300°C. A specially designed set of wrenches is used in holding the bottom of the crucible and in tightening the cap. The crucibles calibration coefficient *E* was determined using indium and was consistent with those reported by Van Humbeeck and Bijvoet[7] and it has a high heat transfer capac-

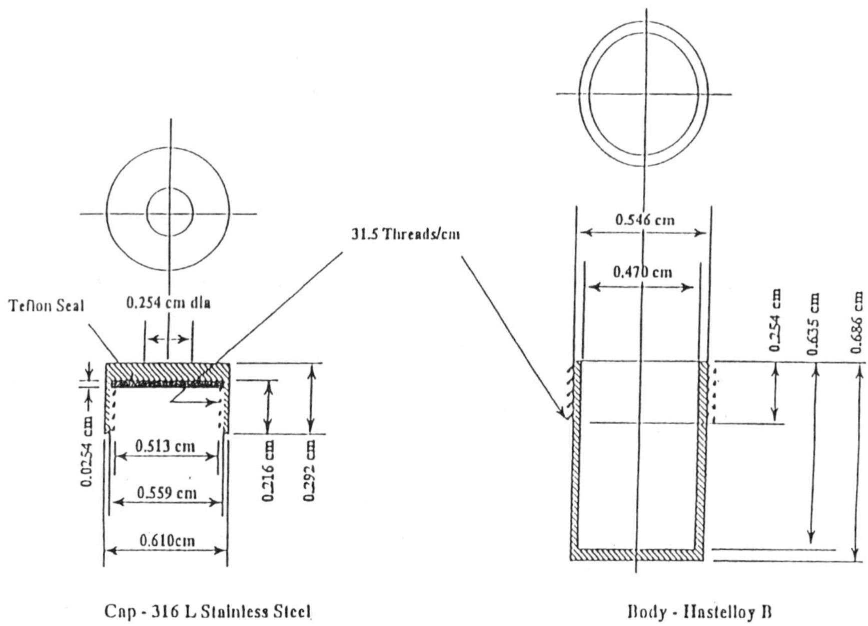

Figure 1. Reusable closed bomb.

ity. The Merck Closed Bomb crucible is capable of providing accurate and reproducible thermal data for samples which must be run in a closed system.

DTA scans can be used to evaluate the exothermicity of process samples in an open reaction system. A standard 4-mm capillary tube is used to contain the sample. The samples are evaluated from room temperature up to ~325°C. The use of the standard DTA cell, with a silver heating block, allows for evaluation of process samples in an open reaction system without the use of the DSC open pan sample container. Repeated open pan runs in a DSC cell usually result in the corrosion of the cell and a much shortened useful life of the DSC cell.

The shape of the DSC and/or DTA traces, in which exothermic activity is found, provide an indication of the rate of the heat release. DSC and DTA work are performed in an air atmosphere. Whether or not an exotherm is caused by a decomposition or oxidation is determined at a later time if there is a potential for thermal runaway is found.

If exothermicity is found in an isolated product sample, Thermogravimetric analysis (TGA) is performed. This instrument, in conjunction with a Mass Spectrometer, identifies the onset temperature for weight loss and the gaseous decomposition products. TGA has proven to be a valuable tool in the identification of low temperature decomposition onset temperatures.

The second part of the Tier I assessment is the determination of the heat of reactions for all reactions reported to be exothermic, and the quantification of heat releases/gas generations associated with the intrusion of water/jacket service fluids into sensitive chemicals/reaction mixtures which were identified in the Operational Hazards Evaluation Data Sheet.

The increased focus on safety has resulted in the requirement that heats of reaction, as well as water/jacket service liquids intrusion data, be available before any process is scaled-up. Uncontrollable heats of reaction, or the unintentional introduction of water/jacket service liquids into a batch, can result in the vaporization of the solvent present and/or the initiation of an exothermic decomposition/oxidation of a process sample which can result in the over pressurization of a process vessel and possibly cause an explosion and/or a fire.

The standard methods used to calculate heats of reactions use bond energies, heats of formation and heat of combustion data. Although the calculations are simple, only a limited data base exists, making it difficult to perform the necessary calculations in many cases. Traditional experimental procedures, such as dewar calorimetry and reaction calorimetry, require large amounts of reactants and demand considerable experimental effort. The use of Syringe Injection Calorimetry (SIC) and the Reactive System Screening Tool (RSST) as a calorimeter have been developed to overcome the drawbacks of the calculation and other experimental techniques. The SIC and RSST calorimeter techniques provide reasonably accurate data for liquid–liquid and liquid–solid reaction systems.

Syringe Injection Calorimetry (SIC) is a novel DSC technique which has been developed for use with TA[8] Instruments, Inc. thermal analytical system to overcome the drawbacks of the calculation and experimental techniques and provides reasonably accurate data for liquid–liquid and liquid–solid reactions. The reactions studied must meet the following criteria: reaction occurs rapidly at room temperature, reactants mix easily, solution is homogeneous after the reaction and there is negligible gas generation. The Syringe Injection Calorimetry technique has been calibrated using the HCl + NaOH neutralization. The plot of Q versus the experimental heat of neutralization was linear over the range of 0.4 to 3.0 Joules of total heat measured and had a positive slope which is consistent with the data published by Van Humbeeck and Bijvoet.[7] Several classical heats of reaction, solution and dilutions were studied using this SIC technique. The results are presented in Table 1.

In SIC, one of the reactants is accurately weighed into the sample crucible and held at room temperature in the DSC cell. After the system reaches equilibrium, a precise amount of the second reactant is rapidly injected into the crucible via a microsyringe. The heat flow is measured versus time and is used to calculate the heat of reaction of the system using

$$\Delta H_{\text{corr}} = \Delta H_{\text{rx}} \times E_{\text{SIC}} \tag{1}$$

TABLE 1
SIC Determined Heats of Reactions, Solutions and Dilutions

Test Reactions	Concentration (M)	$\%\Delta_{\text{th}}$ Measured
HCl + NaOH	0.5	100.2
HCl + NaOH	1.0	99.8
HCl + NaOH	2.0	99.9
HCl + NaOH	3.0	100.0
HCl + NaOH	4.0	100.1
HNO_3 + NaOH	1.0	101.3
HNO_3 + NaOH	2.0	98.1
HNO_3 + NaOH	3.0	99.0
HNO_3 + NaOH	4.0	98.2
KCl + H_2O	NA	99.0
$CaSO_4$ + H_2O	NA	97.6
H_2SO_4 + H_2O	NA	103.5

Δ_{th} = theoretical heat of reaction

where ΔH_{corr} is the corrected heat of reaction in kilocalories per mole of the limiting reactant, ΔH_{rx} is the heat of reaction in kilocalories per mole of the limiting reactant as determined from the SIC value for the heat of reaction and E_{SIC} is the calibration coefficient for the SIC system.

With the RSST calorimetric system, one reactant is accurately weighed into the 10-mL reactor flask and is held at room temperature. The heater system is left off and the RSST unit itself is left unpressurized. After the system reaches thermal equilibrium, an accurately weighed sample of the second reactant is rapidly introduced into the reactor flask via a syringe. The RSST computer system will record both the temperature change and the rate of temperature change (*dT/dt*). The RSST computer system also has the capability to reduce this data (ΔT and *dT/dt*), creating a file from which data can be plotted. From the plots of this data, or the reduced data printout, ΔH_{rx} for the reaction can be determined using

$$\Delta H_{rx} = \frac{m_1 \times H_{cap} \times \Delta T}{[LR_m \ / \ LR_{mw}] \times \Delta T} \tag{2}$$

where ΔH_{rx} is the heat of reaction in kilocalories per mole of limiting reactant as determined from the measured ΔT, m_1 is the total mass in grams of both reactants in the reaction system, H_{cap} is the heat capacity for the reaction system in cal g^{-1} deg^{-1}, ΔT is the temperature rise (°C) that accompanies the reaction as determined from the reaction run in the RSST unit, LR_m is the actual mass of the limiting reactant in grams and LR_{mw} is the molecular weight of the limiting reactant in g (g mol)$^{-1}$.

The plot of the log *dT/dt* versus 1/100 *T* (*T* in kelvin) permits the determination of the maximum rate of temperature rise that takes place during the reaction. This parameter is useful in assessing the rate of reaction and whether one or more reactions are occurring simultaneously.

SIC and RSST are capable of providing accurate, rapid and reproducible results for heats of reaction. A typical reaction can be studied by both techniques in less than four hours using milliliters and milligram quantities of reactants. Under the best of conditions, the heat of reaction determined using these techniques is within $\pm$5% of the theoretical value. For subambient or high temperature reactions, the OHEL uses a specially designed small-scale RC1 vessel (~200 mL). This RC1 also provides the required sample for hazards assessment. The RC1 calorimetry run also provides heat transfer coefficient, specific heats for reaction mixtures, adiabatic temperature rises and kinetic data.

In cases where these techniques are inappropriate or incapable of providing the heat of reaction data or where additional information is needed, use of a reaction calorimeter or dewar calorimetry is recommended. If a reaction calorimeter is not available, dewar calorimetry can be used. There are numerous literature references which detail this technique.

Water Intrusion Testing, an often overlooked hazard in a process, provides for the evaluation of the consequences of water/jacket service fluids intrusion into reactors and storage containers as well as the safety of storing chemicals in sprinkled areas or where water contamination is possible. Using the RSST unit over pressurization scenarios can be determined based on the rates of pressure increase and residual pressure after cooldown. In cases where the formation of particularly hazardous gases (hydrogen, acetylene, etc.) are possible with water/jacket service fluids, the RSST is used to confirm formation of these gasses which could result in over pressurization if water/jacket service fluids intrusion occurs.

Samples which were identified in the hazards data sheet as potentially interactive with water/jacket service fluids are subjected to this testing. These studies evaluate the potential of heat releases and gas generation upon the introduction of water/jacket service fluids into the reactor. A sealed RSST unit is used to provide the size of the heat release, the amount of gas pressure generated, whether the pressure generated is caused by a condensable or non-condensable gas and the rates of temperature (dT/dt) and pressure (dP/dt) generated by the water/jacket service liquid intrusion.

Tier II of the hazards assessment identifies potential thermal hazards which could arise from the hazards identified in Tier I testing, namely exothermic decompositions, heats of reactions and heat releases associated with water/jacket service fluids intrusion. Tests are performed to identify exothermic onset temperatures and the accompanying rates of temperature (dT/dt) and pressure (dP/dt) releases. Testing requirements are determined by evaluating: (1) the size of the decomposition exotherm, the rate of the heat release and the exothermic initiation temperature versus the proposed operating temperature; (2) the rates and size of temperature and pressure increases associated with the exothermic decomposition as well as whether non-condensable gases are formed; and (3) the impact of the heats of reaction and heats of water/jacket service fluids intrusions on any exothermic decompositions present in the reaction mixtures.

The Tier II level testing utilizes two specially developed Merck small scale testing techniques to determine exothermic onset temperatures, the Merck developed Small Scale Adiabatic Dewar (SSAD) and the Small Scale Isothermal Age technique (SSIA). The Reactive System Screening Tool (RSST), Vent Size Packaging (VSP) and Radex test units are used to determine the rates of temperature and pressure increases associated with the decomposition exotherm. The use of the VSP unit permits calculation of the appropriate vent size to handle the exothermic decomposition.

Typically, adiabatic Dewar calorimetry and/or ARC®[9] (Accelerating Rate Calorimetry) studies are used in determining exothermic onset temperatures. These tests use relatively large amounts of sample. SSAD and SSIA were developed to provide reliable information as to exothermic onset temperatures using small quantities of test material. Large quantities of materials are often unavailable during the early stages of pharmaceutical process development and thus these two techniques currently play a key role in determining exothermic decomposition onset temperatures and in safe process design.

The SSAD technique is a novel calorimetry technique which has been developed for use with the ASI Radex[10] and Fauske RSST calorimetry instruments, using specially designed Merck dewars. The Merck dewar provides reasonably accurate exothermic onset temperatures utilizing small samples (~1 gram). The SSAD technique employs a reusable glass dewar cell. The sample is held under adiabatic conditions at a preset temperature for an extended period of time, usually ~12 hours, with the aged sample being reevaluated for any change in the size of the original exotherm (ΔH). A decrease in the size of the original exotherm indicates heat was evolved during the age, and thus the exotherm initiated during the age. In addition, any auto-catalytic behavior may be detected.

The SSIA test is an isothermal DSC technique in which reusable metal crucibles are used to accurately determine exothermic onset temperatures. A typical solid sample, 50 to 75 mg, is held isothermally at a preset temperature for an extended period of time, usually 12 to 16 hours. The aged sample is then run at a standard heat-up ramp to determine any decrease in the size of the original exotherm relative to the unaged sample. The same DSC cell, metal crucible, heat-up rate and sample weight are used to determine the exotherm size in both the unaged and aged sample.

Both the SSAD[5] and SSIA[4] techniques have been calibrated using dicumyl peroxide. The SAAD technique determined the exothermic decomposition onset temperature of dicumyl peroxide to be between 103 and 105°, with the SSIA determining the exothermic decomposition onset temperature of greater than 92.5°C. The exothermic decomposition onset temperatures determined using the SSAD and SSIA techniques are in agreement with the data obtained in a DIERS Users Group Phase VII Round Robin Test[11] of 97 to 107°C (103.4 $\pm$ 3.2°C).

The SSAD and SSIA techniques are applicable to both solids and liquids and have the following advantages: (1) small sample size, (2) ability to handle compounds in which exothermic activity is accompanied by large pressure increases without damaging experimental equipment, (3) accurate determination of exothermic onset temperature, (4) ease of experimental set-up, (5) ease of data interpretation and (6) rapid experimental turn around time.

Once the exothermic decomposition onset temperature has been determined, further evaluation using RSST, Radex , ARC, VSP or other suitable technology is used to determine the potential consequences of exothermic initiation during chemical processing. These techniques provide the rates of temperature (dT/dt) and pressure (dP/dt) releases which accompany the exothermic decomposition. The rates of temperature and pressure increases can also be determined from the RSST/Radex adiabatic Dewar runs if thermal runaway occurs during the age. The VSP technique allows for the determination of vent sizes as well.

Tier II level evaluations also include the determination of potential shock sensitivity, Flash Point and the impact of the introduction of air into a chemical system. Drop weight testing and Detonation Velocity testing are used to determination potential shock sensitivity. Drop Weight testing is performed on all process samples which have exotherms greater than 100 calories/gram. Positive results are indicated by sample color change, direct and single impact, damage to sample holder, sound, sparks, etc. Each sample is evaluated for positive results and Detonation Velocity testing is done by an outside vendor for samples with potential risk. Flash Point determinations are done for chemical systems with the potential to develop flammable vapor phase gases. Air oxidation studies determine if the size and/or onset temperature for an exothermic decomposition are affected by air. All open system and closed system testing in Tier I is done under an air atmosphere, with further air oxidation studies performed if the potential for thermal runaway is determined. Air oxidation studies include running the sample under both an air and nitrogen atmosphere in both an open reaction system in which a stream of air or nitrogen is passed through the test cell and in a high pressure DSC cell in which the cell is pressurized with either air or nitrogen. In addition, SSAD studies under air and nitrogen determine if the rates of temperature and/or pressure increase are affected by the presence of air.

3.3. Tier III—Identification of an Intrinsically Safe Process

In Tier III an intrinsically safe process is identified by reviewing the identified thermal hazards, heat of reaction data, and water intrusion hazards in the context of the process description and in conjunction with a hazard analysis review, provided in the OHEL Hazards Evaluation Data Sheet, which identified the chemical and potential processing problems. In Tier III the factors involved in potential process hazards, both thermal and chemical, are evaluated in conjunction with both the known causes of thermal runaways and the implications of the known chemistry, and results in the identification of potential process hazards.

The first step in the identification of an intrinsically safe process is the identification of potential chemical hazards. Potential chemical hazards are identified in the Operational Hazards Evaluation Data Sheet which provides information on known chemistry re: potential toxicity, heat releases associated with water intrusion and heats of reactions, balanced chemical equations, chemical interactions, reaction by-products, etc. All reactions must be evaluated for the potential for vapor phase reactions re: volatile solvents, oxidizing agents (due to decomposition, pH, or addition rates) and formation of a flammable vapor phase mixture. Other areas of chemical hazard review include the incompatibility between the chemicals used in the process, the potential of the process solvents to form peroxides, the use of highly electrostatic solvents, the formation of toxic gases and the incompatibility between materials of construction of vessels and chemical used in the process. The complete review of a process from a chemical hazards point of view is critical and must be done by experienced personnel.

The second step in the identification of an intrinsically safe process is the identification of potential thermal runaways with respect to the thermal and/or chemical hazards which have been identified in the process. The factors which must be evaluated are: (1) the size and rate of a heat release and the initiation temperature versus the proposed operating temperature, (2) the size and rate of gas generation associated with the decomposition/heat of reaction and whether the gasses generated are condensable or non-condensable and (3) the impact of the heat of reaction upon any exothermic decompositions in the process.

Once the potential thermal hazard scenarios are identified and evaluated, it is then determined whether they can be controlled or if the process must be modified to eliminate the potential thermal hazard. It should be noted that this evaluation takes place early in the process development at time when changes can be easily incorporated into the process.

4. APPLICATION OF THE MERCK THREE-TIER HAZARDS EVALUATION PROGRAM

In Tier I, the size of the exotherms is quantified and the rates of heat release are estimated by evaluating the slope of the of the exothermic curves. Exotherms are considered small if the heat release calories per gram, moderate if 10 to 25 calories per gram, and large if 25 calories per gram. The value for the large category may seem low, but one should realize that a 25-calorie heat release in a solvent system is capable of raising a batch temperature by ca. 50°C in most cases. Rates of heat release are classified by the shape of the exothermic DSC or DTA traces. Figures 2, 3, 4, and 5 are examples of slow, moderate, rapid and detonation type rates of heat releases. The main exothermic decomposition criteria evaluated to determine the need for Tier II testing are (1) the exothermic initiation temperature versus the proposed operating temperature, (2) the size of the exotherm, and (3) the estimated rate of temperature release. The exotherms potential impact on processing equipment, exothermic reactions which could initiate the exotherm, and heat input to the exothermic stream are also included in the evaluation to determine the need for Tier II testing. The "Determination of Need for Tier II Level Testing" (Table 2) demonstrates how the size, initiation temperature and relative rate of temperature release of an exothermic decomposition, as seen on a DSC curve, is evaluated to determine the type of and need for Tier II testing.

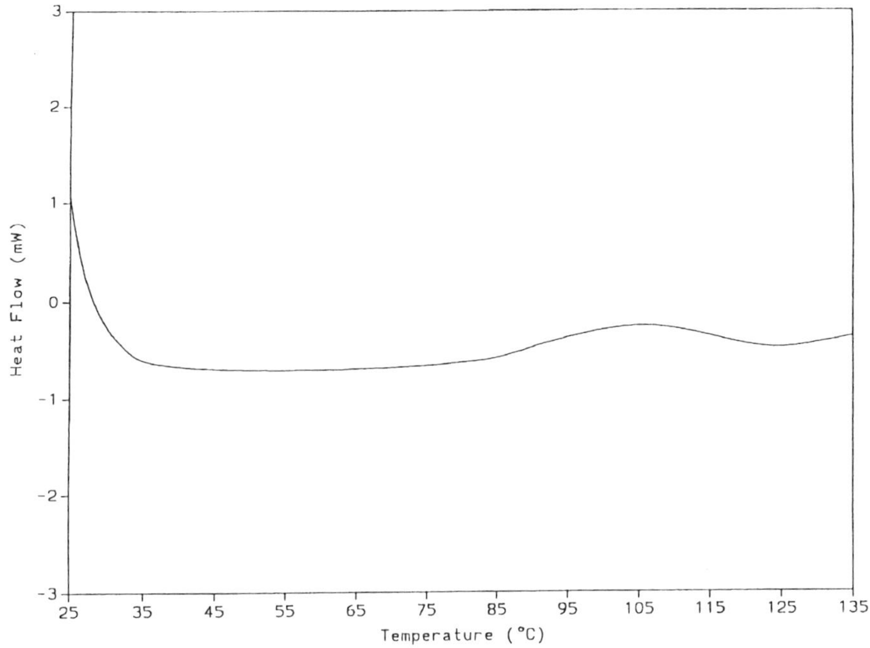

Figure 2. Slow rate of heat release.

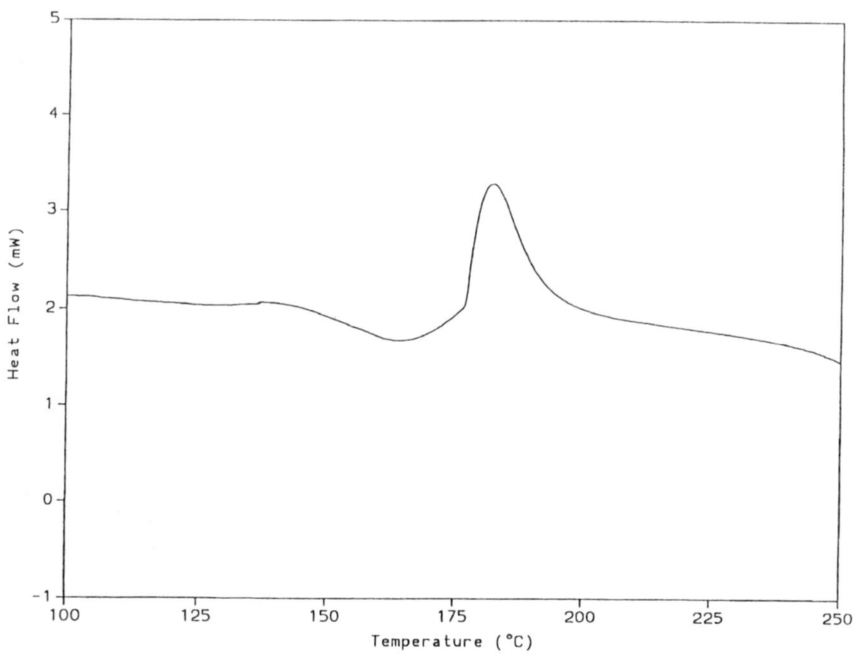

Figure 3. Moderate rate of heat release.

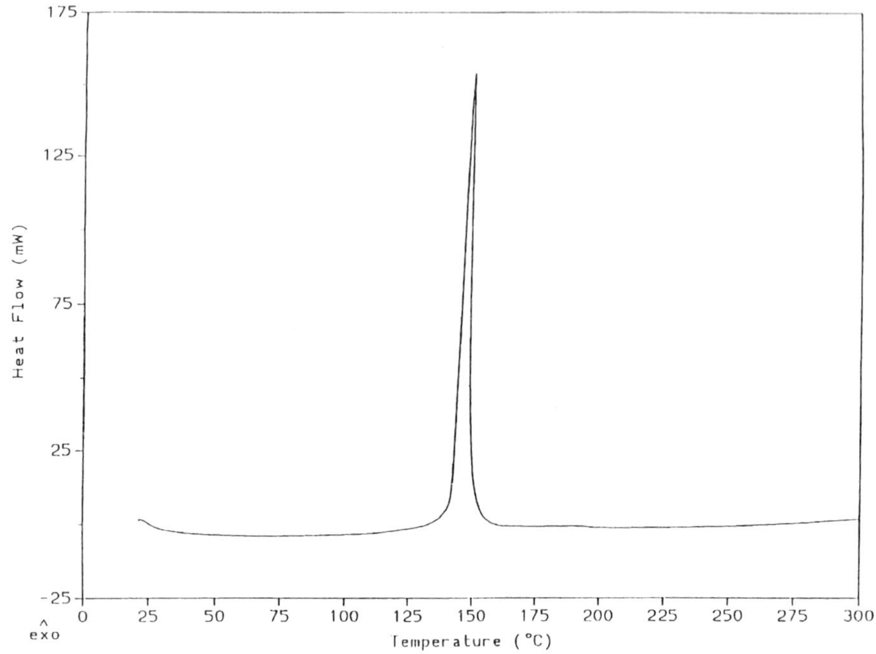

Figure 4. Rapid rate of heat release.

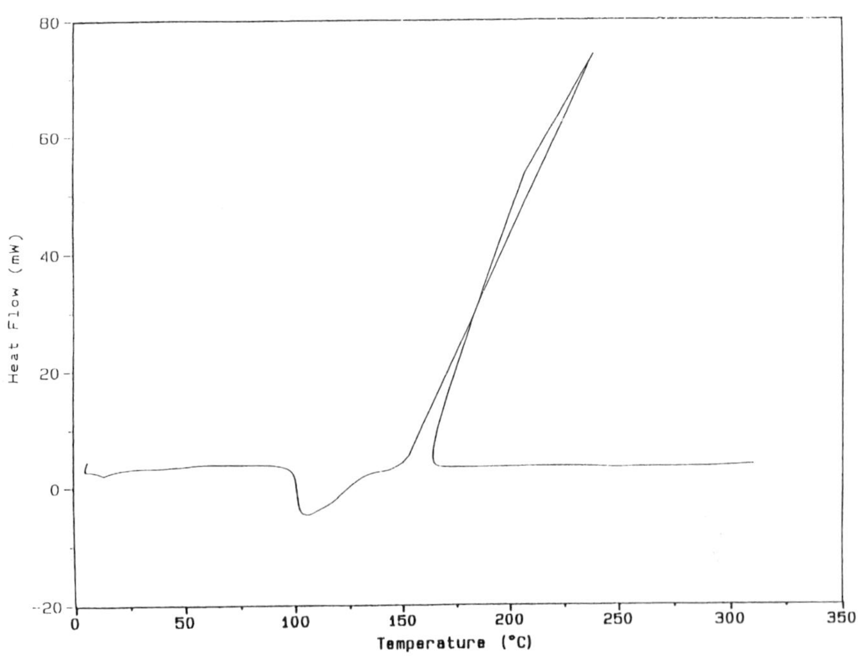

Figure 5. Detonation-type rate of heat release.

A standard rule for Tier II testing is the requirement of Drop Weight Testing for all samples in which the exotherm is 100 calories per gram. This is a very conservative number; however, it provides a safety net in assuring the identification of any potentially shock sensitive samples.

The thermal data provided from Tiers I and II in conjunction with the data provided in the OHEL Hazards Evaluation Data sheet are used during the Tier II evaluations to identify potential thermal runaways. The causes of thermal runaways evaluated in this step include: (1) highly exothermic decompositions, (2) highly exothermic reactions, and (3) loss of process control.

In the majority of cases, heats of reaction and/or exothermic decompositions/oxidations/chemical interactions have little if any impact on pilot plant scale-ups of chemical processes. Most heats of reaction can be controlled via addition rates and/or cooling. Many exothermic decompositions/oxidations/chemical interactions are either small in size or cannot be initiated due to processing temperatures and conditions. However, the potential for uncontrolled heats of decomposition and/or heats of reactions, vapor phase reactions and loss of process control need to be identified and their impact on pilot plant operations must be determined.

Heats of decomposition may lead to hazardous operating conditions when one or more of the following exists:

- an inadequate heat sink is present to contain heat releases if initiated
- the reaction is run near or at the exothermic decomposition temperature
- exothermic decomposition initiates upon the loss of the solvent barrier and the reaction is run at or near the reflux temperature
- an exothermic decomposition initiates close to a drying temperature
- an exothermic decomposition initiates below the melting temperature
- an exothermic decomposition accompanied by a rapid rate and large volume of gas generation
- a sample exhibits shock sensitivity
- exothermic decomposition is enhanced by oxygen or contaminants
- exothermic decomposition temperature is lowered by contaminants/impurities
- overconcentration results in potential loss of heat sink to control a heat release
- human errors
- mechanical failures.

Heats of reaction may play a role in process hazards when:

- there is uncontrollable and rapid generation of heat and/or gases
- the heat generated is adequate to raise the reaction temperature to the exothermic decomposition temperature

TABLE 2

Determination of the Need for Tier II Level Testing[*]

Exothermic Initiation Temperature vs Operating Temperature (•C)			Size of Exotherm (cal/gram)			Estimated Rate of Exothermic Heat Release			Size of Heat Input and Size andEase of Control of Exothermic Heats of Reaction				
0 <50	≥50 <100	≥100	<10	≥10 <25	≥25	Slow	Mod	Rapid/ Det.	3[1]	4[1]	5[1]	6[1]	7[1]
X					X			X	A/B	A/B	A/B	A/B	A/B
X				X				X	A/B	A/B	A/B	A/B	A/B
	X				X			X	B	A/B	A/B	A/B	A/B
		X			X			X	B	A/B	A/B	A/B	A/B
X					X		X		B	A/B	A/B	A/B	A/B
X				X			X		B	A/B	A/B	A/B	A/B
	X				X		X		B	A/B	A/B	A/B	A/B
	X			X				X	B	A/B	A/B	A/B	A/B
	X			X			X		B	A/B	A/B	A/B	A/B
		X			X		X		B	A/B	A/B	A/B	A/B
		X		X				X	B	A/B	A/B	A/B	A/B
X					X	X			B	A/B	A/B	A/B	A/B
		X		X			X		B	B	A/B	A/B	A/B
X				X		X			B	B	B	A/B	A/B
	X				X	X			B	B	B	A/B	A/B
	X			X		X			B	B	B	B	A/B
X			X					X	NT	B	B	B	B
	X		X					X	NT	B	B	B	B
		X			X	X			NT	B	B	B	B
X			X				X		NT	NT	B	B	B
	X		X				X		NT	NT	NT	B	B
		X		X		X			NT	NT	NT	NT	B
		X	X					X	NT	NT	NT	NT	NT
	X		X			X			NT	NT	NT	NT	NT
X			X			X			NT	NT	NT	NT	NT
		X	X				X		NT	NT	NT	NT	NT
		X	X			X			NT	NT	NT	NT	NT

[*]These guidelines should be used as guidelines in determining required tier II testing. Every sample should be individually assessed and not simply assigned to a specific category as presented in this table.

3[1] No heat input to sample re: batch temperature ambient or less and no exothermic heat of reaction.

4[1] Either low heat input or a small easily controllable heat of reaction, one or the other not both.

5[1] Low heat input and a small easily controllable heat of reaction, both are allowed.

6[1] Low heat input and a moderate easily controllable heat of reaction, both are allowed.

7[1] Any high heat input or a moderate/high difficult to control heat of reaction regardless of any other low heat input or small/moderate easily controllable heat of reaction.

NT= No Tier II testing required.

A = Rates of temperature and pressure increases (RSST/VSP/Radex testing) associated with the exothermic decomposition must be determined.

B = SSIA or SSAD testing must be done to determine the initiation temperature of the exothermic decomposition.

- a lack of reaction initiation results in an uncontrollable reaction once the reaction is initiated
- an incomplete reaction being heated to a temperature at which the reaction initiates at an uncontrollable rate.

Vapor phase reactions and loss of process control are often overlooked in the evaluation of potential thermal runaways. The hazards associated with vapor phase reactions include:

- the presence of volatile solvents
- the formation of volatile oxidizing agents through decomposition, pH or addition rates and
- flammability limits of solvents/reaction by-products.

The potential loss of process control can occur with:

- the loss of a solvent barrier
- the buildup of excess reactants
- mischarge of reactant
- service interruptions
- reverse flows and
- insufficient rate of heat removal.

Once the thermal hazards have been identified and evaluated, their implications on the pilot plant scale-up of the process are determined by reviewing the thermal and calorimetry data in the context of the process description and in conjunction with the hazard analysis section of the Operational Hazards Evaluation Data Sheet (Attachments 1–6).

The thermal data is then reviewed in the context of the process description and the information provided on the data sheet. Additional discussions with the appropriate process groups are undertaken when issues are apparent. These interactions bring to light potential hazards which are usually controllable by appropriate operating restrictions. If operating restrictions do not provide for a safe operating envelope, alternative chemistry must be identified and developed in order to provide for an intrinsically safe process.

Potential thermal hazards are controlled through safe process development and the control of operating procedures. A key factor in safe process development is the early identification of potential thermal hazards such that changes in process chemistry to eliminate the thermal hazards are accomplished without hindering the pilot plant implementation of the chemical process.

Examples of process chemistry changes include:

- revised isolation procedures or nonisolation of intermediate
- the replacement of a starting material to prevent formation of a hazardous intermediate

- the solubilization of shock sensitive products
- the replacement of a solvent system to provide a large heat sink
- the inplace destruction of a hazardous by-product and
- the development of new chemistry to avoid formation of a shock sensitive chemical.

The majority of potential thermal hazards are controllable through control of operating procedures as in the following examples:

1. **limiting reaction temperature**
   - the use of a higher reaction temperature, with controlled addition of a reactant, to assure that the reaction occurs rapidly without buildup of reactants
   - a limitation on the batch and/or jacket temperature during additions, concentrations, drying, etc.
   - the use of a split jacket to keep heat sources below batch level;

2. **limiting concentrations**
   - to prevent shock sensitivity
   - to preserve safety provided by heat sinks
   - to limit levels of exothermicity;

3. **dilution**
   - to prevent a temperature rise during a reactant addition so that the exothermic decomposition temperature cannot be reached
   - to prevent the initiation of an exothermic decomposition from raising the batch temperature to reflux;

4. **limiting the rates of addition**
   - to prevent uncontrolled or excessive rates of heatup and/or pressure generation
   - to prevent the initiation of exothermic decomposition
   - the addition of a reactant at a controlled rate at a higher temperature, rather than charging all reactants, followed by a heat up
   - to prevent initial high heat or pressure generation rates;

5. **design of a continuous reaction system**
   - where chemistry changes are not possible
   - to limit the volume of batch being processed at one time when large decomposition exotherms are present which cannot be controlled through normal process control
   - to improve and automate control
   - to minimize human contact;

6. **extraction of highly exothermic streams**
    - to prevent concentration to isolate highly exothermic streams
    - to prevent the isolation and drying of thermally unstable products;

7. **emergency quench**
    - to stop thermal runaway;

8. **equipment interlocks**
    - temperature/addition rate
    - temperature dump to quench
    - temperature jacket service
    - oxygen level/nitrogen purge
    - oxygen level/feed rate;

9. **equipment venting**

10. **facility construction.**

Examples 8, 9, and 10 are not commonly used in pilot plant facilities and are replaced with "high trained technical coverage" by experienced personnel present on the pilot plant floor during processing.

If significant hazards are identified during this Tier III evaluation which cannot be controlled through process changes or operating controls, a hazard review meeting is held. Those attending this meeting will include the Process Hazards Assessment Task Team, a group consisting of experienced chemists, engineers and safety experts; and the process development team. This group will reevaluate the process and provide insight into possible causes of the hazards and suggest alternative chemistry for producing the compound. Under no circumstances is an unsafe process taken forward into the pilot plant until all process hazards have been addressed and a safe pilot plant operating envelope provided.

After the three tier hazards evaluation is complete, a process safety memo is written which documents the results of the testing and provides the conditions under which the process must be run. The memo also contains a complete list of all samples tested, the types of tests performed and the testing results. A description of the process tested, as well as the Operational Hazards Evaluation Data Sheet, is attached. Each OHEL safety memo is given an identifying number, which must be included on the pilot plant batch sheet. Without this documentation, a process cannot be run in the pilot plant.

4.1. Examples of Safe Process Scale-ups in a Pilot Plant

The safe pilot plant scale-up of a process may involve operational controls, process chemistry changes or both. The factors taken into account when defining a safe operating envelope include the identified thermal and chemical hazards.

In one case, the use of operational controls allowed making a product which posed known operational hazards and for which alternative chemistry could not be developed. During routine Tier I analysis a large exotherm present in both an open and closed reaction system was identified in the reaction mixture. The exotherm initiated at ~80°C, with a heat release of 110 cal g^{-1} (Figure 1). A large percentage of this exotherm was attributable to a polymerization which the sample underwent at temperatures approaching 100°C, with the remainder attributable to a heat of reaction which had the potential to raise the temperature to the polymerization initiation temperature. The process procedure called for a one hour age at 75 to 85°C and then for a four hour age at 90 to 95°C.

It was concluded that this exothermic activity posed a serious potential hazard and Tier II testing was undertaken to fully characterize this exotherm. This included the following:

- *Drop Weight Testing*—negative test results indicated that the reaction mixture was not shock sensitive.
- *Adiabatic Dewar Testing*—indicated that there was a slow heat release initiating at temperatures approaching 50°C, accompanied by a ΔT of 2 to 3°C. At temperatures approaching 80°C the rate of heat release was sufficient to push the reaction mass into the exothermic polymerization with a ΔT greater than 200°C
- *Contamination Studies*—studies indicated that contact with carbon steel resulted in a decrease of the exothermic initiation temperature.
- *VSP Testing*—results indicated a violent and rapid rise in both temperature (dT/dt 100°C min^{-1}) and pressure ($dP/dt > 50$ lbf in^{-2} sec^{-1}).

A continuous reaction system was developed in which only small volumes of the reaction mixture were handled, using remote operations. Moreover, continuous introduction of a cold feed and removal of warm product provided for additional means of removing part or all of the heat generated. An emergency quench procedure was also introduced to rapidly stop any potential thermal runaway if the batch temperature reached a predetermined maximum.

Numerous examples exist where alternate safe process development enabled otherwise hazardous processes to be safely run in a pilot plant. One process produced a shock sensitive oxime solid as a by-product of the reaction. Chemical destruction of the oxime in the reaction solution before it precipitated eliminated the need for filtration of the shock sensitive material. Another process required the addition of liquid ammonia into a hot ethanol solution of the starting intermediate near the initiation temperature of a major exothermic decomposition. It was found that the reaction could be run in concentrated ammonium hydroxide at 30°C without the need for another solvent system. In another case, a completely new chemical route was devised to avoid using iodoxybenzoic acid which was found to be shock sensitive. Solubilization of a shock sensitive peracid with acetic acid eliminated the shock sensitivity.

Many operational procedure changes are recommended where chemical interactions could result in a hazardous operating condition. A major consideration is the potential for flammable mixtures to form in the vapor space above reaction mixtures. In particular, the use of strong oxidants such as hydrogen peroxide and sodium hypochlorite in reactions where organic solvents are present can present an opportunity for such flammable mixtures to form. Often, recommendations are made to use a controlled oxidant addition combined with a nitrogen sweep to help prevent the formation of a flammable mixture in the vapor space. It is especially important to apply this to waste treatment situations where use of strong oxidants is common to reduce toxicity or odors. Off gas hazards must be evaluated and the proper recommendations made for handling flammability, rates of pressure generation and venting requirements. Recommendations are made regarding alternative solvents if severe peroxide forming solvents such as ethers or highly electrostatic solvents such as hexane are used. When peroxides can be formed, recommendations are made requiring checking for peroxides before concentrations and the addition of stabilizers. Equipment incompatibilities with processing streams are brought to the attention of the process development team to prevent equipment failures. The interaction of HF or hot caustic with glass is an example. The formation of toxic gases during processing or the handling of highly hazardous materials, such as phosgene or cyanide, will require the evaluation of numerous hazardous scenarios and the determination of how to prevent them from occurring.

5. CONCLUSIONS

Merck has successfully used the Three Tier Hazards Assessment Program to identify thermal and chemical hazards in chemical processes and to evaluate the potential for thermal runaways. The Three Tier Hazards Assessment Program can be used as a guideline for determining and evaluating process hazards. However, each chemical process must be evaluated on a case by case basis. "Safe Operating Envelopes" should be individually determined, not simply assigned based on a specific example or category presented in this paper or elsewhere.

References

1. Registered Trademark of Mettler, Hightstown, NJ.
2. Fauske & Associated, Inc., 16W0709 West 83rd Street, Burr Ridge, IL.
3. Tuma, L. D. A Reusable Metal Crucible for Use in Differential Scanning Calorimetry Analysis of Closed Reaction System Samples. *Thermochim. Acta* (1992), 179–187.
4. Tuma, L. D. Small Scale Isothermal Age Technique to Determine Exothermic Onset Temperatures. *Thermochim. Acta*, 284 (1996), 135–144.
5. Tuma, L. D., and Wieczorek, A. A. Small Scale Adiabatic Dewar Technique to Determine Exothermic Onset Temperatures. *Journal of Thermal Analysis,* 49 (1997), 1653–1660.
6. Tuma, L. D., Bacher, S.,Bagner, C., and Dyke, E. Syringe Injection Calorimetry (SIC), *Proceedings of the 18th NATAS Conference* (1989), 887–891.
7. Van Humbeeck, J., and Bijvoet, A. Calibration Study of the Du Pont DSC 910 Module, *Thermochim. Acta,* 120 (1987), 55.
8. TA Instruments, New Castle, DE.
9. Registered Trademark of Arthur D. Little, Inc., Cambridge, MA
10. Astra Scientific International, Inc., Pleasanton, CA.
11. DIERS Users Group Phase VII Round-Robin Testing Results" presented at DIERS User Group "Meeting, Boston, MA, September 30, 1994.

Attachment 1
TOXICITY

Please list suspected compounds, i.e. carcinogen, tetratogen, etc.

WATER/SERVICE INTRUSION SCENERIO

1. Please list all compounds used in this process which are known or suspected to be interactive with water and/or jacket services.

A. ______________________ B. ______________________

C. ______________________ D. ______________________

2. If there are any ways (other than jacket or condenser failure) in which water might be introduced into the reaction mixture, please list them.

A. ______________________ B. ______________________

______________________ ______________________

______________________ ______________________

C. ______________________ D. ______________________

______________________ ______________________

______________________ ______________________

TEST SAMPLE REQUIREMENTS

1. 20 ml (liquid) or 2 grams (solid) sample of each known or suspected water/service jacket sensitive compound.
2. 20 ml sample of jacket service fluid, if not water.

ADDITIONAL INFORMATION

Please supply any additional information about these compounds, or experiences you have had in using them, which may be of use in evaluating the water/jacket sensitivity of the compounds listed above.

Attachment 2
HEAT OF REACTION DATA SHEET

1. Please list all known exothermic reactions: Consider all reactions in addition to the primary reaction (quenches, neutralizations, etc.).

A. ______________________ B. ______________________

C. ______________________ D. ______________________

2. For each exothermic reaction listed above, please suggest a model reaction with known thermochemistry so a heat of reaction can be calculated.

A. ______________________ B. ______________________

C. ______________________ D. ______________________

3. Observed Laboratory Heat Effects for each reaction listed in Question 1:

Reaction	Reagent Conc.	Add'n Time	Maximum ΔT	Method of Temp. Control (Measures taken to Maintain a Constant Temperature)	Difficulty of Temp. Control		
					Low None	Mod.	High
A:							
B:							
C:							
D:							

Use the back of this sheet to provide additional information.

4. If measurement of the heats of reaction is desirable, based on your knowledge, observation, experience, etc., please fill out the Reaction Calorimetry Data Chart on the following page.

5. If the reaction is run sub-ambient would a RC1 run be appropriate? Please check reactions for possible studies (*sub-ambient reactions only*).

A ❑ B ❑ C ❑ D ❑

Attachment 3
EXPECTED REACTION TABLE

PLEASE DUPLICATE THIS SECTION FOR EACH REACTION.
USE NEW PAGE FOR EACH REACTION.

A. Please fill out this table for the expected primary and all side reaction(s).

Class of Reaction (1)	Balanced Equation

Products and Byproducts of Reaction	Toxic		Fire Explosion Detonation		Corrosive to Human Tissue		Dust or Smell		Significant Potential Hazard	
	Y	N	Y	N	Y	N	Y	N	Y	N
1										
2										
3										
4										
5										

(1) e.g.: oxidation, reduction, alkylation, neutralization, etc.

B. Evaluation

Please answer the following questions when applicable. If additional space is required, please use the back of this page.

1. Elaborate on each potentially significant hazard that was indicated.
2. If there are important pieces of information not available, please list and explain.
3. Are there safety problems or unusual operating conditions not addressed in this table? If so, please discuss.

Attachment 4
DEVIATION REACTION TABLE: WHAT IF SCENARIOS

PLEASE DUPLICATE THIS SECTION FOR EACH DEVIATION REACTION.
USE NEW PAGE FOR EACH REACTION.

A. Please fill out this table for reactions which will occur for the wrong order of addition, overcharges, undercharges, reactions with services, etc.

Class of Reaction (1)	Balanced Equation

Products and Byproducts of Reaction	Toxic		Fire Explosion Detonation		Corrosive to Human Tissue		Dust or Smell		Significant Potential Hazard	
	Y	N	Y	N	Y	N	Y	N	Y	N
1										
2										
3										
4										
5										

(1) e.g.: oxidation, reduction, alkylation, neutralization, etc.

B. Evaluation

Please answer the following questions when applicable. If additional space is required, please use the back of this page.

1. Elaborate on each potentially significant hazard that was indicated.
2. If there are important pieces of information not available, please list and explain.
3. Are there safety problems or unusual operating conditions not addressed in this table? If so, please discuss.

Attachment 5
CHEMICAL AND MATERIAL INTERACTION GRID

Chemicals Services and Materials / Chemicals and Services	Chemical and Services: A	B	C	D	E	F	G	H	I	J	K	L	M	N			Materials: 316L	Hastelloy B	Hastelloy C	Glass Lined	Teflon	Polypropylene	PVC	Neoprene	Viton	Graphite	Seal Fluid			
(A)																														
(B)																														
(C)																														
(D)																														
(E)																														
(F)																														
(G)																														
(H)																														
(I)																														
(J)																														
(K)																														
(L)																														
(M)																														
(N)																														

Key		
	+	Yes
	–	No
	0	Not Known

List all chemicals and materials which are used in this process on both the vertical and horizontal grid. Then consider each intersection for interactions and fill in as per the key.

Attachment 6
HAZARDOUS SITUATIONS

A. Consider the following procedural activities/equipment problems and determine those which could result in hazardous operating conditions. Please use your best judgment.

	Results in Hazardous Operating Conditions?		
1. Procedure-Related Activities:	**Yes**	**No**	**Why?**
1a. Mischarging of reactants(*)			
1b. Wrong sequence of reactants(*)			
1c. Contamination due to residual cleaning solvents (*)			
1d. Reverse flows			
1e. Improper service conditions			
1f. Hazardous hold points and overnight storage			
1g. Labeling of materials			
1h. Interactions with other processes			
1i. Assay errors			
1j. Leaks or spills			
1k. Storage and disposal of materials			
1l. Inertion			
1m. Sampling			
1n. Physical problems such as emulsions, foaming, bumping, line pluggage, pumping characteristics or sublimation			
1o. Drying			
1p. Improper operating conditions			
1q. Cleanout and cleanup			

2. Equipment-Related Activities:	**Results in Hazardous Operating Conditions?**		
	Yes	**No**	**Why?**
2a. Power failures			
2b. Service failures			
2c. Equipment failures (jackets, condensers, vacuum source, agitation, recorders, controllers, gauges, safety valves, rupture disks, leaking valves both internal and external, rotometers and metering pumps, etc.) (*)			
2d. Lack of isolation			
2e. Facilities including grounding stations, air mask stations, scrubber, ventilation, services, antidotes, toxic chemical detection equipment, etc.			

* ***Should be described in Table II-Deviation Reactions if these scenarios lead to reactions not on Table 1.***

B. Discuss each activity that was checked yes. If additional space is required, please use the back of this sheet.

The Reactive System Screening Tool (RSST): An Easy, Inexpensive Approach to the DIERS Procedure

H. K. Fauske
Fauske & Associates, Inc.
Burr Ridge, IL

ABSTRACT

The RSST DIERS vent sizing methodology is revised to provide realistic design equations for reactive systems consistent with available large-scale experience. Using easy to obtain RSST data such as rate of temperature rise and rate of pressure rise excellent agreement is illustrated for hybrid, vapor and gassy reactive systems.

1. INTRODUCTION

Knowledge of chemical reaction rates, character, and energy release is essential to modern process design. The Design Institute for Emergency Relief Systems (DIERS) program (Fisher, 1985), sponsored by 29 companies under the auspices of the AIChE and completed in 1985, provided the chemical process industry with tools necessary to gather such data (Fauske and Leung, 1985). A primary purpose of the effort was evaluation of emergency relief vent requirements, including energy and gas release rates for systems under upset conditions and the effects of two-phase flow on the emergency discharge process. An easy, inexpensive approach to the DIERS procedure was later provided by the RSST (Creed and Fauske, 1990; Fauske, 1993). The RSST which is commercially available from Fauske & Associates, Inc. at a base price of only about $15,000 has now become a standard industry tool for chemical system characterization for relief system design data acquisition. This paper provides an update of the RSST methodology and presentation of relief system design methods consistent with available large-scale data including hybrid, vapor and gassy reactive systems.

2. RSST DESCRIPTION

The RSST (Figures 1 and 2) consists of a spherical glass reaction vessel and immersion heater (optional), its surrounding jacket heater and insulation, thermocouples and a pressure transducer, a stainless steel containment vessel that serves as both a pressure simulator and safety vessel, and, not shown, a magnetic stirrer base, a control box containing the heater power supply, temperature/pressure amplifiers, and a data acquisition and control panel. The sample cell volume is 10 mL and the containment volume is 350 mL. A key feature of the apparatus is its low effective heat capacity relative to that of the sample whose value, expressed as the capacity ratio, is approximately 1.04 (i.e., quite adiabatic). This feature allows the measured data to be directly applied to process scale.

Typically, a sample is heated at a constant rate and the sample self-heat rate is found as a function of sample temperature. This heatup rate may be varied and is controlled by feedback from the sample temperature measurement. The imposed linear ramp rate can be varied from 0.1°C/min. to ramp rates approaching those required to simulate fire exposure by using the immersion heater option. The heater compensates for losses, and the self-heating rate of an exothermic system is adjusted for heater input. Reagents may be added to a sample during a test. Data handling programs produce plots of pressure vs. temperature, temperature vs. time, pressure vs. time, and Arrhenius plots.

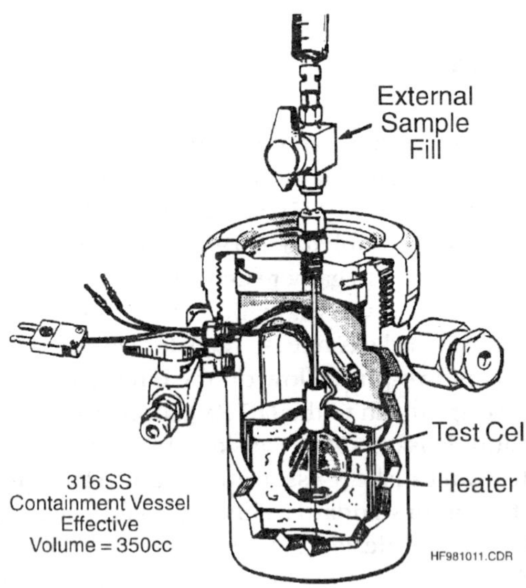

Figure 1. Reactive System Screening Tool (RSST) containment.

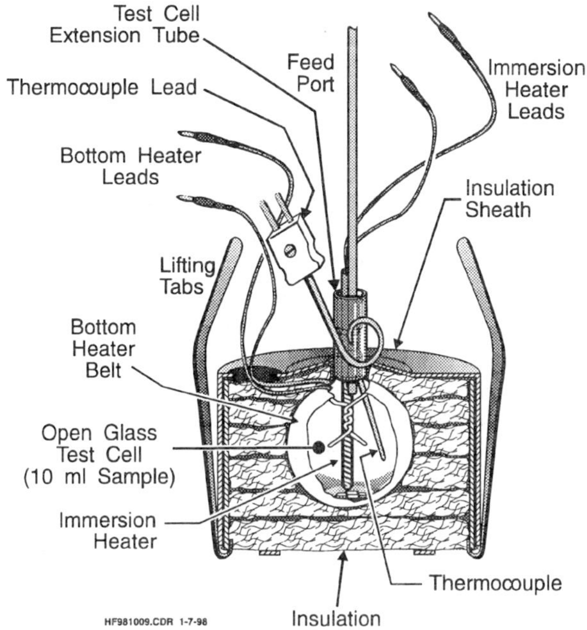

Figure 2. Reactive System Screening Tool (RSST) test cell.

The RSST quickly and safely determines the potential for runaway reactions and measures the rates of temperature and, in case of gassy reactions, pressure increases to allow reliable determinations of the energy and gas release rates. This information can be combined with analytical tools (summarized later) to assess reactor vent size requirements. This is especially useful when screening a large number of different chemicals and processes is necessary. In a typical RSST test, the user loads 10 mL of the sample into the glass cell and places the appropriate backpressure in the containment vessel. For intrusion or addition studies, the user can put part of the sample in the glass test cell and inject the remainder at the appropriate time. Next, the user enters the heating sequences and shutdown criteria into the computer. From this point, the computer records the data.

3. SYSTEM CHARACTERIZATION AND VENT SIZING

Three types of reactive systems are distinguished for their venting and mitigation character. A vaporizing, or tempered, system contains components (often a solvent) whose vapor-liquid equilibrium controls the system temperature and reaction rate for a fixed containment pressure. This occurs when one or more components boil prior to potential gaseous decomposition, so heat from exothermic reactions is removed by the latent heat of vaporization, and the system pressure is equal to the component vapor pressure. In the RSST, one compound (or more) evaporates from the glass vessel and condenses on the cold containment walls, thereby removing the reaction energy and tempering the system, as would occur in a vented reactor. So long as there is sufficient tempering liquid before the reactants are exhausted, and the vent is sized to pass the vapor as it is produced, tempering will limit the pressure rise. The principal parameter determining the vent size is the rate of temperature rise at the relief set pressure.

A gassy system has noncondensable reaction or decomposition products which continually increase the containment pressure and do not effectively temper the system, whose temperature may escalate rapidly. The total pressure is equal to the gas pressure, and the maximum rate of pressure rise determines the vent size. Note that a vapor system may evolve into a gassy one when tempering is lost by complete boiloff of solvent.

A hybrid system exhibits both tempering and gassy characteristics. Gaseous decomposition occurs prior to boiling, but the rate of reaction (gas production by decomposition) is still tempered by vaporization. The total pressure in the reactor is the sum of the gas partial pressure and the vapor pressure, and both the rates of temperature and pressure rise are needed to determine the proper vent size.

3.1. Hybrid System Vent Sizing

The unique design offered by the RSST is ideally suited to characterize vent requirements for hybrid systems. Decomposition of hydrogen peroxide ($H_2O_2 \rightarrow H_2O + \frac{1}{2}O_2$) will be used to illustrate the method since reaction kinetics and physical property data are well known and this system reflects a hybrid system where the decomposition reaction is tempered by latent heat of vaporization. Large-scale runaway data are also available with this system (Wilberforce, 1988).

Results from RSST tests (10^{-2} kg samples of 25 wt% H_2O_2 solution) are illustrated in Figures 3 and 4. The high pressure test (where vapor stripping is suppressed) clearly shows that the system produces noncondensable gas with the rate of pressure rise, $\dot{P} >> 0$, and with a maximum rate $\dot{P}$ = 180 psi/min. The low pressure RSST test also clearly illustrates that the reaction is tempered since the rate of temperature rise, $\dot{T} \rightarrow 0$, as the temperature approaches 100°C. The corresponding rates of temperature and pressure changes are obtained from the high pressure test and are 4°C/min and 1 psi/min, respectively.

The rate of gas generation, $\dot{Q}_g$ (m^3/s), at the tempering pressure, $P(\sim 10^5$ Pa), is obtained from

$$\dot{Q}_g = v\frac{\dot{P}}{P} \qquad (1)$$

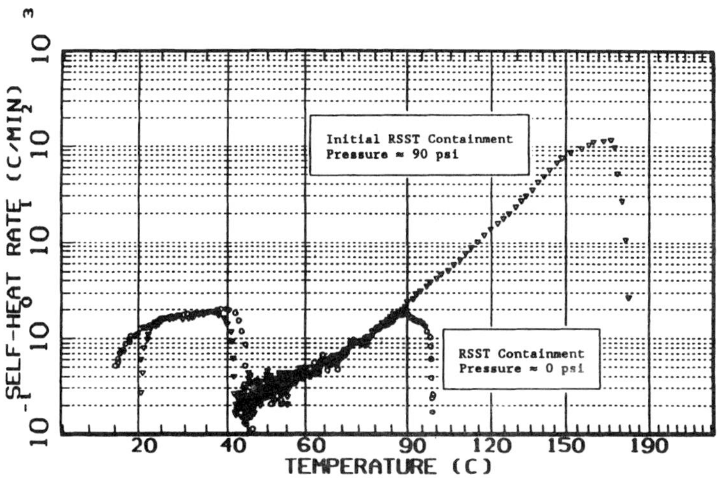

Figure 3. Self-heating rate from high- and low-pressure RSST tests with 25 wt% H_2O_2.

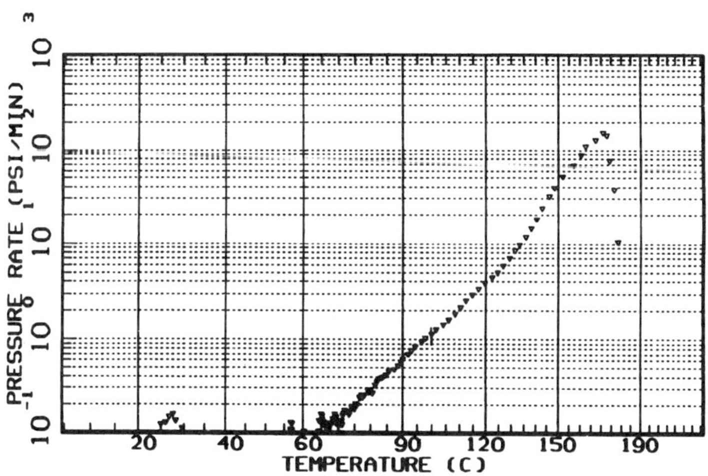

Figure 4. Rate of pressure increase from high pressure RSST test with 25 wt% H_2O_2.

where v (3.5×10^{-4} m^3) is the RSST containment volume and with = 1.15×10^2 Pa/s results in = 4×10^{-7} m^3/s. Since the H_2O_2 system is used where kinetics and physical properties are well known (Hydrogen Peroxide Physical Properties FMC Technical Bulletin 67), an independent check of the validity of the RSST gas volumetric rate measurement can be made as follows.

$$\dot{Q}_g = \frac{m_t c \dot{T} M_{w,O_2}}{\Delta H_R 2 M_{w,H_2O_2} \rho_{O_2}} \qquad (2)$$

where m_t (10^{-2} kg) is the RSST sample mass, c (3800 J/kg-K) is the reactant specific heat, $\dot{T}$ (6.7×10^{-2} K/s) is the rate of temperature rise corresponding to tempering, M_w is the molecular weight, ΔH_R (2.8×10^6 J/kg) is the heat of reaction, and ρ_O (1.03 kg/m^3) is the oxygen gas density, and results in $\dot{Q}_g$ = 4.2×10^{-7} m^3/s.

Since the RSST Phi factor is essentially 1.0, the measured information can be directly applied to process scale with the total gas–vapor volumetric rate given by

$$\dot{Q}_T = \dot{Q}_v + \dot{Q}_g = \frac{Mc\dot{T}}{\lambda\rho_v} + v\frac{\dot{P}}{P}\frac{m}{m_t} \tag{3}$$

where m (kg) is the mass of reactant, λ (J/kg) is the latent heat of vaporization, and ρ_v (kg/m^3) is the vapor density.

Considering vapor–gas venting the relief vent areas, A (m^2), are given by

$$A \approx \frac{3}{2}\frac{Q_T}{Q_D}\left(\frac{M_W}{RT}\right)^{1/2} \tag{4}$$

$$A = \frac{\dot{Q}_T}{Q_D}\left(\frac{M_W}{2(1-P_b / P)RT}\right)^{1/2} \tag{5}$$

for critical flow and highly subcritical flow $(1 - P_b/P) << 1.0$, respectively, where C_D is the discharge coefficient, P_b (Pa) is the back pressure and R (8314 Pa-m^3/K · kg-mole) is the gas constant.

The above vent sizing methodology for hybrid systems is consistent with the large-scale 200 kg 50 wt% H_2O_2 runaway reaction trials reported by Wilberforce (1988). Relevant parameters for the four reported test trials and estimated values of $\dot{Q}_T$ are provided in Table 1.

TABLE 1

Parameters for the 200 kg 50 wt% H_2O_2 Test Trials Reported by Wilberforce (1988)

Trial No.	Free Vent Diameter, inches	Self-Heat Rate at Tempering*, °C/min	$\dot{Q}_T$ at Tempering, m^3/s	Maximum Pressure, psig	Observation
1	~ 5	~ 20	~ 0.16	~ 0	No liquid ejection
2	~ 2	~ 20	~ 0.16	~ 0	No liquid ejection
3	~ 3	~ 70	~ 0.58	~ 0.9	No liquid ejection
4	~ 2	~ 200	~ 1.65	> 40	Liquid ejection following pressurization

* Corrected for vapor stripping at the tempering temperature.

As an example, considering test trial No. 3 with an observed overpressure of 0.9 psi and use of Equation (5) results in a vent diameter of about 3.2 inches which is also consistent with the observation of no liquid ejection at the tempering condition. Furthermore, the noted pressurization for test trial No. 4 is not surprising. Limiting the overpressure to 0.9 psi in this case and considering gas–vapor venting leads to a vent diameter requirement of about 5.4 inches compared to the available vent diameter of about 2 inches.

As noted by Wilberforce the absence of liquid ejection as tempering is approached for the test trials shown in Table 1 is seriously at odds with the often used DIERS two-phase flow methodology. Considering the churn-turbulent flow regime with $C_o = 1.5$, a self-heat rate at tempering in excess of about 3.4°C/min would result in liquid ejection, which is clearly not the case. In other words, consideration of two-phase flow initiation upon reaching tempering conditions would lead to a significant overestimate of the vent size requirement. In case of test trial No. 3 conditions a vent diameter of about 17 inches would be required in order to keep the overpressure to about 0.9 psi.

The absence of liquid ejection suggests that the vapor-gas release is highly nonuniform due to inherent nonequilibrium conditions with the majority of the release occurring at the free liquid surface, resulting in insignificant liquid swell in the absence of "foamy" conditions (Fauske et al., 1986). For such conditions, the free board height, h (m), which assures insignificant liquid ejection can be estimated from entrainment considerations as follows (Epstein et al., 1989)

$$h > \left(\frac{\dot{Q}_T}{2\pi u_e} \right)^{1/2} \tag{6}$$

where u_e (m/s) is the entrainment velocity given by (Kutateladze, 1972)

$$u_e = 3\left(\frac{\sigma g \rho_l}{\rho_{v,g}^2} \right)^{1/4} \tag{7}$$

where σ (kg/s^2) is the liquid surface tension, g (9.8 m/s^2) is the gravitational constant, ρ_l (kg/m^3) is the liquid density and $\rho_{v,g}$ (kg/m^3) is the hybrid vapor-gas density. The above first order estimate of the absence of significant liquid ejection is consistent with the observations noted in Table 1.

In case of "foamy" system behavior, the occurrence of gas generation and extensive liquid swelling well before reaching tempering will result in significant early mass loss of reactants again resulting in safe pressure relief based upon all vapor–gas venting from Equations (4) and (5). This is illustrated later for gassy systems and is in contrast to pure vapor systems discussed below, where "nonfoamy" and "foamy" systems must be treated separately.

3.2. Vapor System Vent Sizing

For nonviscous nonfoamy vapor systems Eqs. (4) and (5) reduce to

$$A \approx \frac{3}{2}\frac{m c \dot{T}}{C_D P \lambda}\left(\frac{RT}{M_{w,v}}\right)^{1/2} \tag{8}$$

$$A = \frac{m c \dot{T}}{C_D P[2(1 - P_b / P)]^{1/2} \lambda}\left(\frac{RT}{M_{w,v}}\right)^{1/2} \tag{9}$$

for critical and highly subcritical flow conditions, respectively.

An example of a nonviscous nonfoamy vapor system is the methanol-acetic anhydride reaction. RSST test results for this system are shown in illustrating tempering at a set pressure of 50 psig with resulting self-heat rate, $\dot{T}$, of about 40°C/min. Large-scale runaway reaction data are also available for this system. For a reaction charge of about 3770 kg, and a relief set pressure of 50 psig, Kirch et al. (1990) reported an adequate vent diameter of about 3.5 inches with no significant overpressure. Using the RSST information with = 0.67 K/s, and the following parameters for methanol, c = 1834 J/kg-K, λ = 7.7 × 10^5 J/kg, and $M_{w,v}$ = 32, and Eq. (8) results in a vent diameter of about 3.6 inches. It is noted that considering two-phase flow and a 20% overpressure (see discussion below) results in a vent size requirement of about 8 inches. The absence of two-phase flow at the tempering pressure of 50 psig is consistent with Equation (6) prediction.

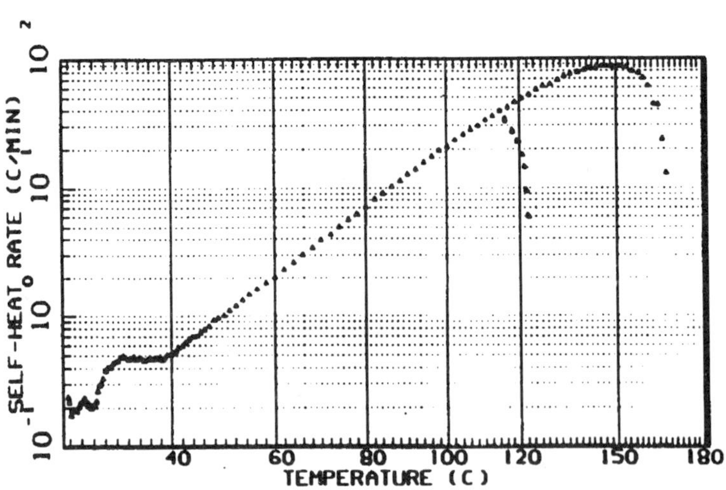

Figure 5. Illustrations of RSST tests performed on methanol and acetic anhydride with 300 and 50 psig back pressures.

Equations (8) and (9) can also be used for estimating vent size requirements in case of fire exposure. Typical range of self-heat rates associated with fire exposure can be simulated in the RSST using the available immersion heater. Again, large-scale data are available confirming the validity of Eqs. (8) and (9) under fire exposure with nonfoamy systems (Fauske et al., 1986).

In case of viscous and/or foamy-like vapor systems, an adequate vent size can be estimated from (Fauske, 1989)

$$A = \frac{5}{C_D 2} \frac{m\dot{T}}{(T/c)^{1/2} P} \tag{10}$$

Equation (10) accounts for two-phase flashing flow and an overpressure of 20% based on absolute basis.* Examples of foamy and/or viscous systems are condensation-like reactions such as the phenol–formaldehyde system and monomer polymerization reactions such as methyl metacrylate and styrene systems.

Howard (1975) recommends the following correlation for safe prediction of vent sizes (based on large-scale experience) for the runaway phenol-formaldehyde reaction

$$D = 0.3\sqrt{V} \tag{11}$$

where D is the vent diameter given in inches and V is the volume of reactant given in gallons. Equation (11) is noted to be valid for a set pressure of 15–16 psia (i.e., a horizontal vent line is mandatory to avoid pressure buildup from a static head of liquid) and a self-heat rate of the order of 6.5°C/min. For a 1000-kg reactor batch the Monzanto correlation suggests a vent diameter of about 5 inches. Using Equation (11) with the above self-heat rate and set pressure and T = 380K, c = 4000 J/kg-K and C_D = 0.8, we calculate a vent diameter of about 4.8 inches.

Kirch et al. (1990) based upon experiments reported an adequate vent diameter of about 8.5 inches for 3770 kg batch of methyl methacrylate at a relief set pressure of about 10 psig with a corresponding self-heat rate of 13.6°C/min. Using this information with c = 2500 J/kg-K and T = 393K we calculate from Equation (10) a vent diameter of about 8 inches.

Finally, we note that Equation (10) provides reasonable agreement with DIERS styrene polymerization tests (Fisher et al., 1992). Considering Test 32-9 with bottom venting and m = 21 kg, = 0.4 K/s, P = 515,000 Pa, a vent diameter equal to 14.1 mm and L/D = 430 resulted in an overpressure of about 105,000 Pa or 20% based upon absolute basis. With T = 490 K, C_D = 0.57, and c = 2500 J/kg-K, the vent diameter from Equation (10) is estimated to be 14.2 mm.

*For more detailed considerations including variations in overpressure Leung's analytical model is recommended (Leung, 1986).

Gassy System Vent Sizing

For these systems Eq. (4) reduces to

$$A \approx \frac{3}{2C_D} \frac{\rho v \dot{P}}{m_t P} \left(\frac{M_{w,g}}{RT} \right)^{1/2} \tag{12}$$

which can be restated as

$$A/V \approx \frac{3}{2C_D} \frac{\rho v \dot{P}}{P m_t} \left(\frac{M_{w,g}}{RT} \right)^{1/2} \tag{13}$$

where V is the volume of the reaction vessel and ρ is the reactant density. For gassy systems (i.e. no tempering) the maximum rate of pressure rise, $\dot{P}$, is of interest and again is provided by the RSST. An example of such data is illustrated in Figure 6 for 37.5 wt% 3,5,5 - trimethyl hexanoyl peroxide in isododecane with an external heating rate of about 0.8°C/min. The maximum rate of pressure rise measured in the RSST containment is about 1000 psi/min. Extensive data have been obtained with this system (deGroot et al., 1981 and Wakker and deGroot, 1996) as illustrated in Figure 7. Also shown are comparisons with prediction from Equation (13) using the RSST information ($\dot{P}$ = 1.15×10^5 Pa/s with $m_t = 10^{-2}$ kg), $C_D = 1.0$, $\rho = 750$ kg/m^3, $M_{w,g} = 40$ and T = 480K. The agreement based upon all gas venting is quite satisfactory over the

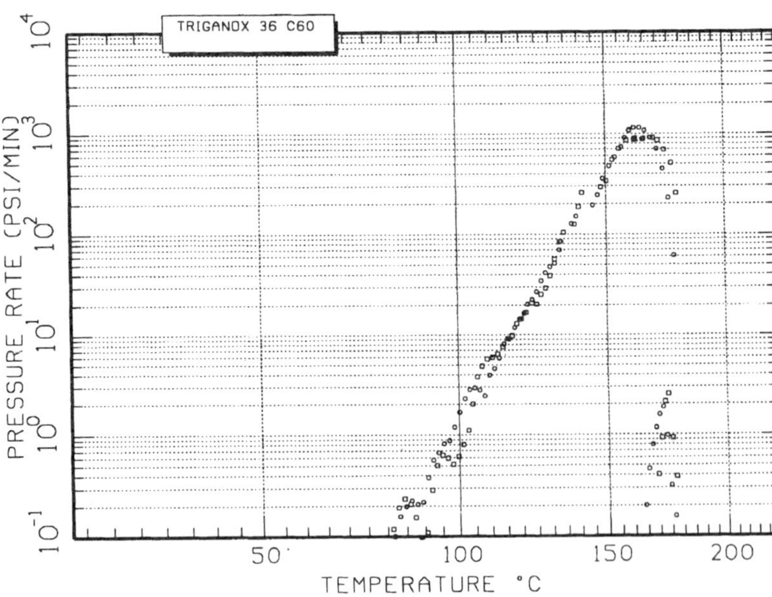

Figure 6. Duplicate RSST tests with 37.5 wt% 3,5,5 trimethyl hexanoyl peroxide in isododecane with uniform imposed heating rate of about 0.8°C/min.

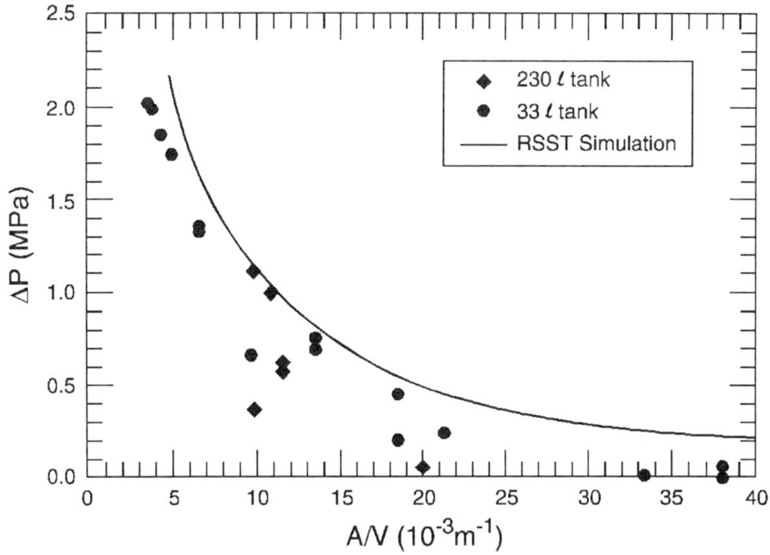

Figure 7. Runaway decomposition data of 37.5 wt% 3,5,5 trimethyl hexanoyl peroxide in isododecane with imposed heating rate of about 0.8°C/min and comparison with RSST simulation with all gas venting.

entire pressure range and essentially independent of test vessel size. Considering two-phase flow at peak reactive conditions (Creed and Fauske, 1990) on the other hand, would overestimate the vent areas by at least an order of magnitude.

Large-scale data are also available for neat dicumyl peroxide. RSST data for this system are shown in Figure 8 with a maximum rate of pressure rise of

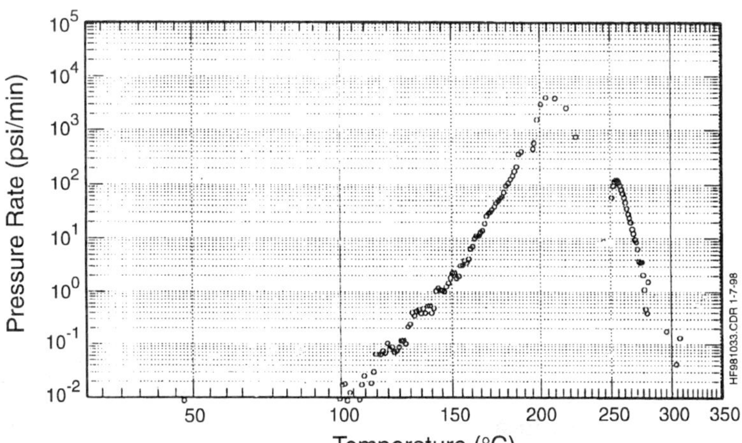

Figure 8. RSST test with 100 wt% dicumyl peroxide with imposed heating rate of about 1°C/min.

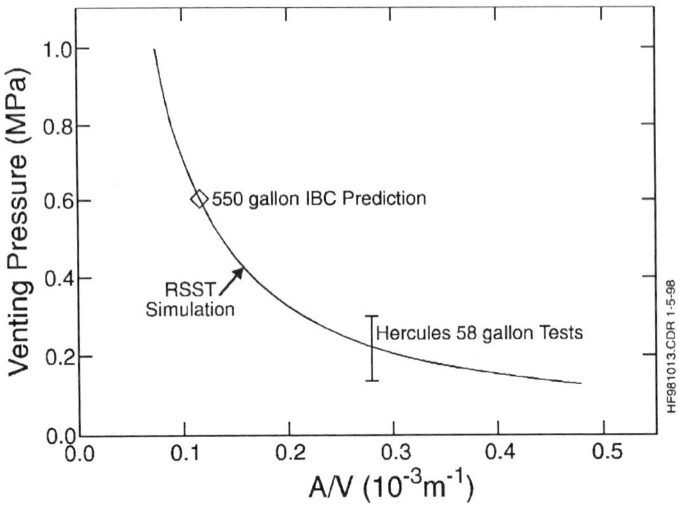

Figure 9. Runaway decomposition data of neat dicumyl peroxide with imposed heating rate of about 1°C/min and comparison with RSST simulation with all gas venting.

about 4,400 psi/min. The vent area requirement for this system estimated from Eq. (13) using the RSST information ($\dot{P}$ = 5.1 × 10^5 Pa/s with $m_t = 10^{-2}$ kg and $v = 3.5 \times 10^{-4}$ m^3), with $M_{w,g} = 40$ and $T = 450$K is illustrated in Figure 9. These predictions are consistent with the large-scale Hercules tests (Gove, 1996). Also noted in Figure 9 is the estimated venting pressure for a 550 gal Intermediate Bulk Container (IBC) with a vent diameter about 22 inches. While this pressure is substantial, safe venting is predicted. A similar pressurization event is suggested if the IBC is only partially filled because of early reactant loss prior to reaching peak reaction condition. For the latter condition, Equation (6) is satisfied for a vessel fill fraction of only about 20%. As such, considering fire exposure, it follows that a fully loaded vessel which would prevent vessel overheating would be advantageous relative to a vessel that is only partially loaded.

Finally, we note that some systems are too energetic to practically relief using top venting. An example is concentrated cumene hydroperoxide which leads to a peak rate of pressure increase of about 40,000 psi/min with a 10^{-2} kg sample in the RSST. Considering a relief pressure of 30 psia Eq. (6) leads to $A/V = 3.5$ m^{-1}. For such systems replacing top venting with a dip tube design can facilitate early removal of reactant (Wakker and de Groot, 1996). Furthermore, application of the dip tube concept merits consideration for all gassy systems including hybrid systems, with the combined result of decreasing the vent size as well as downstream containment requirement in terms of noncondensable gas production (Fauske, 1988; Walter and Edwards, 1996).

SUMMARY

Due to its ease of use as well as its low cost, the RSST is now a standard industry tool for chemical system characterization for relief system design data acquisition. The RSST data which can be scaled directly to full-size applications have been demonstrated to produce excellent agreement with a large number of large-scale venting tests including hybrid, vapor and gassy systems. Easy to use design equations are provided for these systems.

REFERENCES

Creed M. J., and Fauske, H. K., 1990. "An Easy, Inexpensive Approach to the DIERS Procedure." *Chemical Engineering Progress,* pp. 45–49, March.

deGroot, J. J., Groothuizen, T. M., and Verhoeff, J., 1981. "Safety Aspects of Organic Peroxides in Bulk Tanks," *Ind. Eng. Chem. Process Des. Dev.*, 20(1), 131–138.

Epstein, M., Fauske, H. K., and Hauser, G. M., 1989. "The Onset of Two-Phase Venting Via Entrainment in Liquid-Filled Storage Vessels Exposed to Fire," *J. Loss Prevention Process Industries,* 2(January), 45–49.

Fauske, H. K., and Leung, J. C., 1985. "New Experimental Techniques for Characterizing Runaway Chemical Reactions," *Chem. Eng. Prog.,* 81(8), 39–46.

Fauske, H. K., et al., 1986. "Emergency Relief Vent Size for Fire Emergencies Involving Liquid-Filled Atmospheric Storage Vessels," *Plant/Operations Progress,* 5(4) 205–208.

Fauske, H. K., 1988. "Proposed Safety Assessment of R-B-09 Reactor," Fauske & Associates, Inc. Report, March.

Fauske, H. K., 1989. "Emergency Relief System Design for Runaway Chemical Reaction: Extension of the DIERS Methodology," *Chem. Eng. Res. Des.,* 67(March), 199–202.

Fauske, H. K., 1993. "Reactive System Screen Tool," U.S. Patent Number 5,229,074, July 20.

Fisher, H. G., 1985. "DIERS Research Program on Emergency Relief Systems," *Chem. Eng. Prog.*, 81(8), 33–36.

Fisher, H. G. et al., 1992. *Emergency Relief System Design Using DIERS Technology,* New York: American Institute of Chemical Engineers.

Gove, S. H., 1996. "Emergency Pressure Relief for Intermediate Bulk Containers Containing Dicumyl Peroxide," DIERS User Group Meeting, February 5, Houston, Texas.

Howard, W. B., 1973, "Reactor Relief Systems for Phenolic Resins," Monsanto Company, June.

Kirch, L. S., Magee, J. W., and Stuper, W. W., 1990. "Application of the DIERS Methodology to the Study of Runaway Polymerization," *Plant/Operations Progress,* 9(1).

Kutateladze, S. S., 1972. "Elements of the Hydrodynamics of gas–Liquid Systems," *Fluid Mechanics—Soviet Research*, 1, 29.

Leung, J. C., 1986. "Simplified Vent Sizing Equations for Emergency Relief Requirements in Reactors and Storage Vessels," *AIChE Journal,* 32(10).

Wakker, J. P., and deGroot, J. J., 1996. "Venting of Decompositions of Energetic Liquids Using Bottom Vent," Proc. of Process Plant Safety Symposium, Vol. 2, pp. 55-71, Houston, Texas.

Walter, K. L. and Edwards, V. H., 1996. "Consider Bottom Venting for Pressure Relief of Hybrid and Gassy Systems," Proc. of Process Plant Safety Symposium, Vol. 2, Houston, Texas.

Wilberforce, J. K., 1988. "Emergency Venting of Hydrogen Peroxide Tanks," CEFIC Hydrogen Peroxide Safety Conference, Gotherburg, September 22.

Adiabatic Calorimeter Evaluation for Safe Process Scaleup: A Systematic Approach Illustrated with an Example

Geatesh Tampy and Loren B. Schreiber
Fragrance Chemical Development, R&D
International Flavors and Fragrances Inc.
1515 Highway 36, Union Beach, New Jersey 07735

ABSTRACT

An adiabatic calorimeter such as the Accelerating Rate Calorimeter (ARC®) can be valuable for evaluating safe process conditions. This instrument provides heat release rate and pressure data as a function of temperature from reaction samples. These data can be useful for assessing the thermal stability and the potential for thermal decomposition with gas evolution in plant reactors under various process conditions. Guidelines are provided, with an example, to show how data from an Accelerating Rate Calorimeter can be interpreted to establish the operating envelope for safe processing and storage.

1. INTRODUCTION

At International Flavors & Fragrances Inc. we implement a comprehensive safety program to assess, manage, and significantly lower the risks associated with our chemical processing operations. In our safety program, we try to incorporate the latest scientific and engineering practices to strive for safer process designs and prevent accidents [11]. An important and integral part of this program is chemical reaction hazard evaluation, which includes testing.

Most safety guidelines recommend that process hazard assessment should begin early in the development of a process and include a "desktop" screening for reactivity hazards [22]. This should be backed up by laboratory measurements to evaluate and quantify thermal stability, exothermic activity and possible gas evolution. Isothermal calorimetry is often utilized for evaluating "normal" processing under controlled conditions, and differential scanning calorimetry (DSC) or adiabatic calorimetry for simulating "accident" conditions, such as thermal runaways [33]. Care should be taken to ensure that the basis of safety and key results are clearly communicated, well understood, and effectively implemented as the process moves from research, through piloting, into the manufacturing stage.

2. OBJECTIVE

The Accelerating Rate Calorimeter [44], or ARC® (registered trademark of Arthur D. Little, Inc.) is an instrument that has been used in process development at IFF extensively since 1980 for adiabatic testing of our reaction samples. The ARC® provides valuable data pertaining to thermal stability, exothermic activity and gas evolution. However, the interpretation of ARC® data to address practical engineering concerns is not always straightforward. In this work, we have assimilated valuable pointers from the literature on process safety and hazard evaluation to show how the ARC® data may be translated to useful results, in a systematic fashion.

Furthermore, process development at IFF is conducted in close cooperation with our research and production facilities. There is thus a need to present the results from the ARC® in a format that is readily understood by other technical personnel, who are not necessarily experts in ARC® operation or terminology.

3. APPROACH

An example application is used to illustrate our typical approach on utilizing the ARC® and interpreting the data. Our intent here is to cover the more routine applications of the instrument.

This work is not directed toward the expert ARC® user. Rather, it is meant to provide a useful "how to" overview for nonexperts learning about adiabatic calorimetry techniques. Key elements of the results are explained in practical terms, with the help of an illustrative example, and without resorting to extensive mathematical derivations.

Indeed, recent publications have touched upon the more advanced approaches, such as for example, the use of ARC® data for developing complex reaction kinetic models [55,66]. These sophisticated applications, however, are beyond the scope of this work. Although kinetic analysis is briefly discussed, the emphasis is on the various experimental techniques that are applicable even when kinetic modeling is not feasible, or is complicated.

4. EXAMPLE APPLICATION

The evaluation of an oxime (oximes have the structure: R–CH=N–OH) is chosen as an example to explain our overall approach. In this case, there was concern about the thermal stability of the oxime. Since the oxime was an intermediate that went from one reaction step to the next, there was a question as to whether it could be safely stored temporarily (for ~1 day or so) under ambient conditions.

5. DESKTOP SCREENING

In the first step, desktop screening, the chemical structure is examined for clues that may suggest potential reaction hazards [7–14]. For instance, oximes can undergo dehydration to produce nitriles, as follows:

$$\underset{\text{Oxime}}{\text{R–CH=N–OH}} \rightarrow \underset{\text{Nitrile}}{\text{R–C}\equiv\text{N}} + H_2O \qquad \Delta H_{r,\ \text{estimated}} = -100 \text{ kJ/mole}$$

The oxime dehydration reaction is different from other dehydration reactions (such as alcohols dehydrating to alkenes) in that it can release 100 kJ/mol (25 kcal/mol) of heat and is therefore significantly exothermic. In other words, if this dehydration reaction is initiated, the released heat will probably have to be removed by applying cooling, to prevent a self-sustaining runaway reaction. Of course, the thermochemistry provides no indication of the *rate* of heat release.

One can use group contribution methods to estimate, from the molecular structure, the maximum heat of decomposition to lower-energy components [15,167]. In our example, the oxime can *potentially* decompose to liberate carbon, methane, nitrogen and water, along with 2.5 kJ/g (600 cal/g) of heat. Also, in this *potential* decomposition scenario, each mole of oxime will liberate 6 moles of gas/vapor (methane, nitrogen and water).

In actual practice, however, the oxime is not in pure form but contains about 33% by weight of solvent. Due to this dilution, the heat release potential reduces to 1.7 kJ/g (400 cal/g).

The CHETAH program [15,178] suggests a hazard evaluation criterion based on the enthalpy of the decomposition reaction. Total heat-release values of 2.9 kJ/g or higher are considered to indicate "high" hazard potential. The ABPI [2] also recommends a similar value to screen for possible explosion potential. On the other hand, it is noted that another literature source [189] suggests that an energy release of 0.5 kJ/g is sufficient indication that deflagration is a possibility.

By assuming a heat capacity of 2J/g.K (0.5 cal/g.K) and adiabatic conditions (i.e., no heat losses), we estimate that the released heat is sufficient to raise the reaction mixture to a temperature of ~850K. Shanley and Melhem [7] have indicated that known explosive compositions tend to have calculated adiabatic reaction temperatures (CART) that exceed 1200K (800K for peroxides).

In summary, the desktop screening results for the example oxime do indicate the potential for of both thermal and pressure hazards. However, detonation of the oxime appears to be unlikely, based on the desktop screening.

6. RESULTS FROM ARC® TESTING

Laboratory test instruments, such as the ARC®, can be used to determine how the potential reactivity suggested by the desktop screening impacts upon various processing situations.

The ARC® is an instrument that measures the temperature, temperature rate and pressure changes of a reaction sample as a function of time, under adiabatic "runaway" conditions. The ARC® provides data pertaining to the thermochemistry and kinetics, which may then be translated to engineering concerns that are related to the process. These include: assessing thermal stability (for safe storage and to detect undesirable secondary reactions); obtaining total heat release and heat release rate data (to prevent thermal runaways); and measuring pressure and pressure rate data due to gas release (to prevent vessel rupture and for the design of emergency relief and mitigation systems).

In the ARC®, the sample is held in a container (referred to as the "bomb") under adiabatic conditions, which are achieved by heating the chamber surrounding the bomb to the same temperature as the bomb. If the temperature of the bomb increases due to an exotherm, the temperature of the surrounding chamber is also correspondingly increased. The rate of temperature increase (self-heat rate) and bomb pressure are also monitored. If the self-heat rate exceeds a predetermined value (typically ~0.02°C/min), an exotherm is registered. Details of the design, operation, and performance of the ARC® are well described elsewhere [4, 19–25].

6.1. The ARC® Exotherm

Figure 1 shows the time-versus-temperature curve that was obtained from a sample of the oxime in the ARC® test. Starting at ambient temperature, the sample was heated first to 30°C, and then in 5°C increments. After each increment, the sample was held under adiabatic conditions to "search" for an exotherm by monitoring the self-heat rate (dT/dt). At 150°C, an exothermic

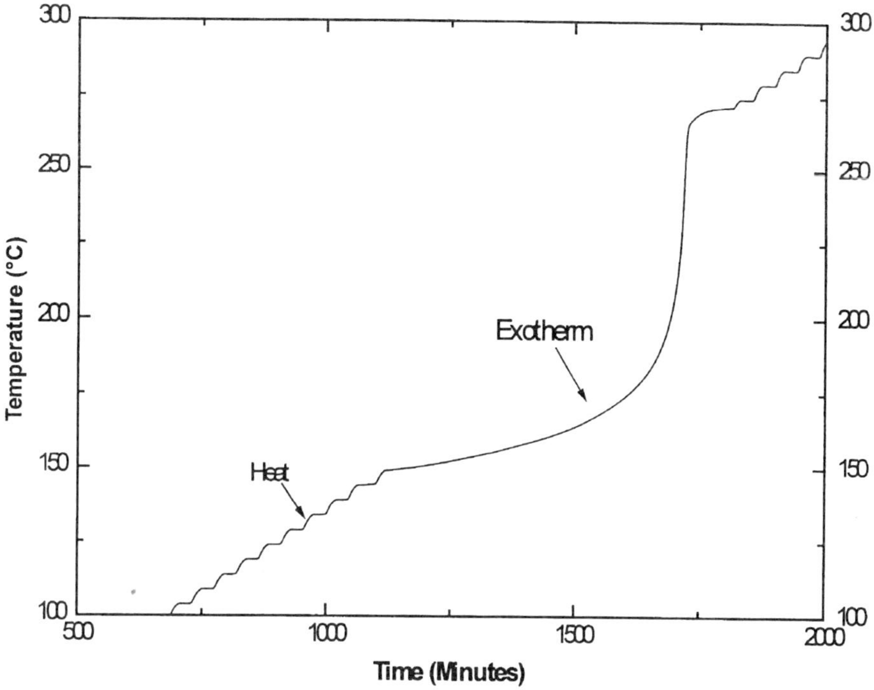

Figure 1. ARC® time-versus-temperature data for the example oxime.

reaction is seen to begin (self-heat rate exceeds 0.02°C/min) and the sample continues to "self-heat." The exotherm raises the sample temperature to a final temperature of 272°C.

Figure 2 depicts the self-heat rate (°C/min) as a function of temperature (°C). The preferred choice of scales for the axes for this plot is log (self-heat rate) on the *y*-axis and temperature plotted as $-1/T$ (in kelvin) on the *x*-axis. These scales qualitatively reflect the Arrhenius relationship: the logarithm of reaction rate constant is proportional to inverse temperature. Indeed, Figure 2 shows a linear relationship in the early portion of the curve, where the reactant concentration, being high, has only a minor influence on the rate (i.e., reaction is pseudo zero-order) [19].

From Figure 2, the onset temperature for the exothermic reaction is seen to be ~150°C. Oftentimes, a storage temperature is chosen that is about 100°C below the onset temperature measured using a Differential Scanning Calorimeter (DSC) and 20°C below the onset obtained from a Dewar calorimeter [2]. The ARC is considered to provide a better estimate of the onset temperature than the DSC, and margins of safety lower than 100°C have been suggested [21,2610]. In our example, the ARC® onset temperature is well

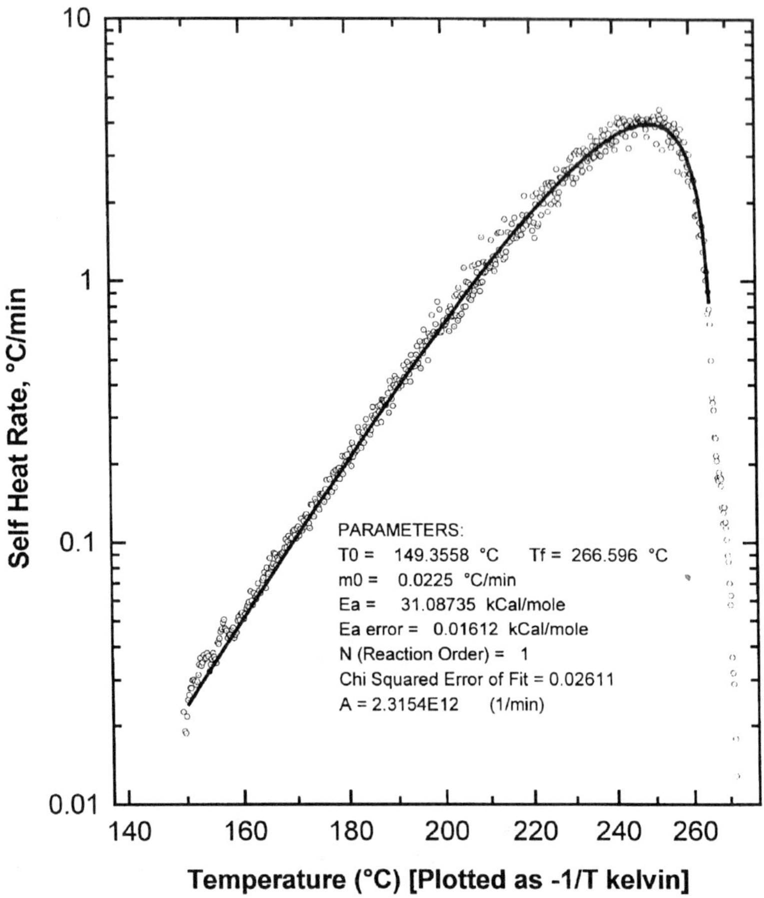

Figure 2. Self-heat rate analysis. ARC® data are shown along with a fitted model obtained by assuming the following kinetic parameters: reaction order = 1, activation energy = 31.08 kcal/mol and frequency factor = 2.31E12 min^{-1}.

over 100°C above ambient temperature. Thus we have a preliminary indication that the oxime may be safely stored temporarily at ambient temperature.

However, it must be emphasized that the observed onset temperature provides only an approximate indication of the safe storage temperature. Heuristic approaches based on the observed onset should always be used with caution. For example, the DSC 100-degree rule has been shown to have limitations [2711]. The detected onset temperature depends on the sensitivity of the calorimeter. Besides, the heat capacity of the sample container also tends to increase the observed onset temperature, typically by 5 to 10°C (see box, page 78). In any case, confidence in safety during storage may be increased using the ARC® techniques that will be discussed in Section 7.

Figure 2 depicts the maximum self-heat rate to be ~4°C/min. It is important to ascertain whether this maximum value exceeds ~10°C/min. Above 10°C/min the ARC® heaters are unable to provide sufficient heat input to "keep up" with the sample self-heat rate and conditions may not be truly adiabatic.

We see from Figure 2 that the exotherm that begins at ~150°C, ends at a final temperature of ~272°C, yielding an adiabatic temperature increase of ~122°C. This value is the uncorrected measured temperature increase for both sample and bomb. The correction involves multiplication by the term ϕ, where ϕ is the ratio of the total heat capacity of sample and container to that of sample alone.* In our experiment, the ϕ-factor had a value of 1.68, and the corrected adiabatic temperature rise was therefore 205°C. The corrected adiabatic final temperature then becomes 622K or 349°C (see box, page 78).

The effect of the ϕ-factor (also called thermal inertia) is to slow down the reaction in the ARC®. In other words, if the ϕ-factor were lower, the reaction would have been faster and would have attained a higher final temperature. It is possible that other exothermic reactions might be initiated at these higher temperatures, which could increase the final temperature to still higher values [28,29]. For critical applications (such as relief sizing) where accuracy of the high temperature, pressure and rate data are crucial, it is desirable to repeat the test in a low-f calorimeter [28,29,30,31].

By assuming a heat capacity of 2 J/g·K, we obtain a value of 400 J/g for the total heat release from the oxime sample. This value also translates to a heat of reaction of 105 kJ/mole (25 kcal/mol). This measured heat of reaction is in good agreement with the estimated value for the oxime dehydration reaction. Thus it appears that the dehydration is the dominant reaction that was occurring during the ARC® test.

Figure 2 also shows that a curve generated by assuming a "reaction order" of one and "activation energy" of 30 kcal/mol fits the experimental data reasonably well. The kinetic parameters may then be used, if necessary, to predict the rate of heat release for the same oxime decomposition reaction in a plant vessel or tank, where the ϕ-factor is close to unity (i.e., a relatively small portion of the released heat is absorbed by the container).

6.2. Pressure Data

Figure 3 shows the pressure data from the ARC® for the oxime. Again, the preferred choice of axes for this plot is log (pressure) on the y-axis, and temperature plotted as $-1/T$ kelvin on the x-axis. This choice reflects the Clausius–Clapeyron relationship, wherein the logarithm of the vapor pres-

* $\phi = (M_s \cdot C_{vs} + M_b \cdot C_{vb})/(M_s \cdot C_{vs})$, where M (g) is the mass, C_v the heat capacity (J/g·K), and the subscripts "s" and "b" refer to the sample and bomb, respectively.

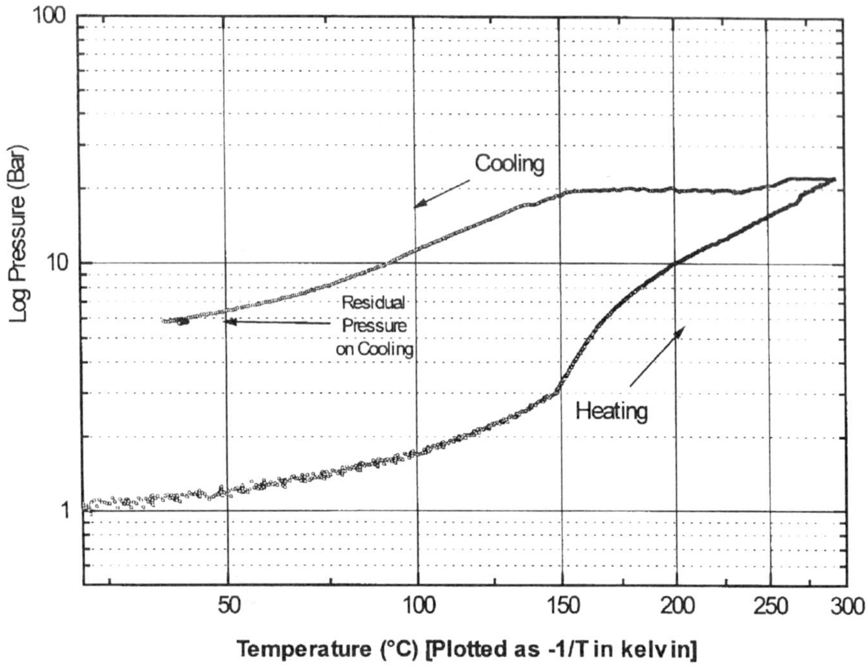

Figure 3. ARC® data showing log (pressure, bar) versus temperature, °C (plotted as $-1/T$ in kelvin).

sure for a pure material varies inversely with absolute temperature. In other words, if pure (thermally stable) material were heated in the ARC®, the pressure temperature relationship plotted in this fashion will approximately follow a straight line.

In Figure 3, we see that the plot is more-or-less linear up to a temperature of 150°C. A sharp increase is observed during the exothermic reaction temperature range of 150 to 270°C. That is, above 150°C, the increase in pressure is greater than what may be expected from vapor pressure effects alone.

We also observe that on cooling a residual pressure of ~6 bar is observed. This final cool-down residual pressure is indicative of the formation of noncondensable gas. We can estimate the quantity of gas release to be ~1 mol (25L) for 5-kg of oxime.* Thus the pressure data from the ARC® suggest that during storage the onset of the exothermic reaction must be avoided to prevent a pressure hazard due to gas release.

*The number of moles of gas is given by $\Delta n = (\Delta P \cdot V)/(RT)$, where ΔP is 5 bar, the head space volume, V, is assumed to be ~5 mL for the 10-mL bomb with 5-g oxime; R = 83.14 mL·bar/(mol·K) and T is ~300K.

6.3. Preliminary Results Summary

The results from the ARC® experiment lead to the following conclusions:

i. The observed onset temperature of 145–150°C suggests that the oxime is likely to be relatively stable at ambient temperature. This result, however, requires further confirmation.
ii. Higher temperatures, which can lead to the onset of the exothermic reaction, are to be avoided during storage. At temperatures above ~140°C, the oxime can decompose exothermically with the release of gas (a pressure hazard).

We also know from the desktop screening that deflagration or detonation of the oxime is unlikely.

7.0. STORAGE STABILITY DATA FROM THE ARC

Wilberforce [32] has shown how the ARC® may be used for storage stability testing. Subsequently, Fisher and Goetz [33] have measured self-accelerating decomposition temperatures (SADT) for the safe storage and transportation of several materials using the ARC®. They suggest that ARC® testing is much more convenient than other more elaborate tests such as the US SADT test [33].

7.1. Time to Maximum Rate Analysis

There are two approaches to study storage stability at a particular temperature by using the ARC®. One is an analytical approach based on further evaluation of the self-heat rate data, called the "time to maximum rate" (TMR) analysis.

In this analysis, as outlined by Townsend and Tou [19], the self-heat rate versus temperature relationship is integrated. This integration provides the time at any temperature that will take the sample into an exothermic decomposition reaction. The temperature at which the maximum self-heat rate was experimentally determined is chosen as a limit of integration (i.e., zero time). Although the TMR analysis involves kinetic parameters, Townsend [20] has shown that it may be applied to any dynamic system for which the logarithm of the rate of a measured parameter x [i.e., $\ln(dx/dt)$] varies in proportion to $1/x$.*

* Time (t, to exotherm) can be related to the initial temperature (T) by the expression: [$\ln(t) = (E/R)(1/T) - \ln(A)$]; E is the activation energy, A the frequency factor (see Fig. 2) and R the gas constant. Townsend [19] has demonstrated this expression to be a good approximation for several actual reactions. One requirement, however, is that the "activation energy" be relatively large (greater than 20 kcal/mole), so that the influence of the "reaction order" on the time-temperature relationship can be neglected in the early portion of the exotherm.

Figure 4 shows the log of time to maximum rate plotted against inverse temperature for the oxime. It is seen that the TMR plot is linear at the lower temperatures. Both the raw data and the data corrected for the f-factor are shown. The correction involves simply dividing the time at any temperature by the f-factor (see box) and is valid for the linear portion of the curve [19]. The linear portion may be extrapolated down to temperatures that are well below the detection limit of the ARC®. For example, the time to maximum rate for the oxime is about 10 days at 100°C and 100 days at 75°C. That is, the oxime can be stored safely at ambient conditions.

The TMR data from the ARC® provides a rational storage-stability time estimate that is based on the specific reactive system, rather than some arbitrary "factor of safety." Also, the time-temperature relationship can be useful not only for determining storage temperatures, but also for deciding on temperatures at which other action may have to be taken, such as the triggering of alarms. It is particularly useful for last-ditch backup safety measures that are best decided based on response time, such as the injection of a reaction inhibitor into a storage vessel [21,34].

It should be noted that the time estimates from the ARC® involve a significant extrapolation of data measured at relatively high temperatures. Yu and

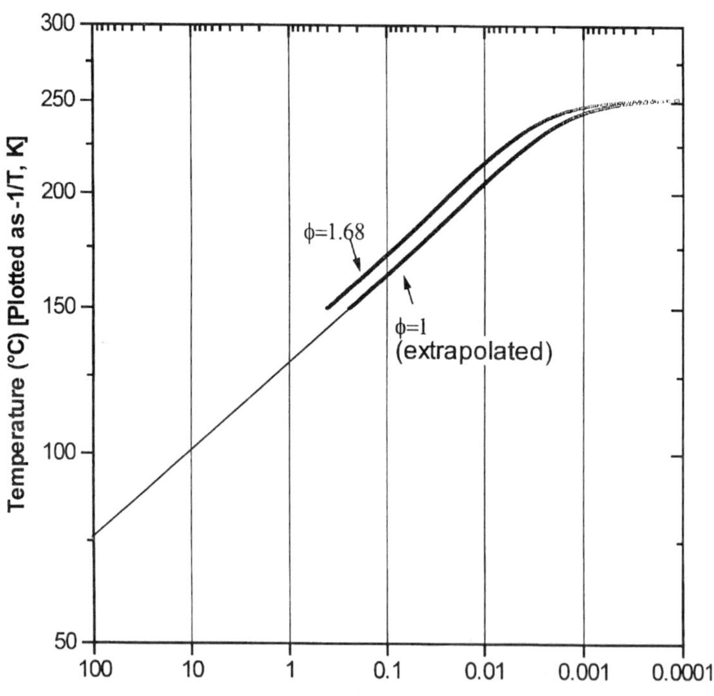

Figure 4. Time to maximum rate plot for the example oxime.

Hasegawa [35] have attempted to obtain accurate self-accelerating decomposition temperatures (SADT) for several materials and have compared the results with those obtained by Fisher and Goetz [33] from ARC® data.

The adiabatic test from the ARC® lacks the sensitivity to detect slow chemistry that may occur at lower temperatures. The analysis implicitly assumes no change in reaction mechanism at the lower temperatures. Also, the actual measured data from the ARC have to be corrected to adjust for the heat that is absorbed by the sample container during the experiment. This correction significantly lowers the time estimate, particularly at lower initial temperatures.

7.2. Isothermal Age Experiment

The ARC® also allows an "isothermal age" run, which is a convenient alternate *experimental* approach to assess storage stability. In the "isothermal age" mode, the sample is held at a constant temperature for an extended period of time. If an exothermic reaction occurs to self-heat the sample above a predetermined rate (typically 0.02°C/min), the ARC® registers an exotherm and tracks the sample temperature increase in adiabatic mode.

Townsend [36] has indicated that isothermal aging can be a useful technique for thermally stressing a material and can be particularly valuable for unearthing unique chemistry due to autocatalytic activity, parallel reactions, or trace impurities. For such materials, isothermal age tests (at temperatures 10 to 20°C below the observed onset temperature in a conventional ARC® run) will often yield positive results. Generally, the selection of temperature is based on some expected condition in the plant [36].

For our example, a sample of the oxime was held in the ARC® at a temperature of 50°C for 6 days. No exotherm was detected, nor was any pressure buildup observed. More importantly, analysis of the postrun sample (by gas–liquid chromatography) showed no nitrile (or any other component different from those in the starting material), thus providing strong evidence that a slow decomposition reaction was not occurring over this time period.

8.0. OTHER SCALEUP CONCERNS

8.1. Thermal Runaway

The ARC® TMR analysis has been extended to determine the conditions that may lead to thermal runaway in reactors or storage tanks [19,37]. This involves determining an "equipment time line" (or "time constant") and a "temperature of no return."

The equipment time line is defined by the ratio of the heat capacity of the reactor and its contents to the reactor heat transfer area and heat transfer coefficient.* The temperature corresponding to the equipment time line on the TMR plot is the temperature of no return. Above the temperature of no return, the rate of heat generation from the reaction will exceed the available cooling rate and the reaction will "run away."

Townsend and co-workers [19,37] have determined the time constants for several storage and transportation containers. They measured temperature changes with time as cooling occurred due to ambient heat loss.

For example, the time constants for a 55-gallon drum and a 10,000-gallon fiberglass insulated storage tank can be estimated to be ~1 day and 24 days respectively for storage of the oxime based on their data [37]. From Figure 4 we see that the temperature of no return for a 55-gal drum containing oxime will then be ~120°C. The temperature of no return for the storage tank is ~90°C. Above these temperatures, the rate of heat generation from the oxime decomposition reaction can be expected to be greater than the heat losses from these vessels. Consequently, the reaction may "run away" (assuming that reaction *does* occur at these temperatures at a rate suggested by the extrapolated TMR plot).

Below these temperatures of no return, however, ambient heat losses from the vessels will be sufficient to prevent a temperature increase. Thus we see that the temperature of no return for storage of the oxime in a 55-gal drum or storage tank is relatively high. Storage under ambient conditions, therefore, can be expected to be relatively safe. Safety can be greatly enhanced if, in addition, the temporary storage vessel has a temperature monitor and is equipped for jacketed cooling.

Townsend and Tou have determined the time constant for a well-insulated tank car containing organic material to be ~23 days [19]. From Figure 4 we see that the temperature of no return for a tank car containing oxime will then be ~90°C. This analysis suggests that the oxime can be safely transported by railcar provided the temperature is kept below 90°C.†

8.2. Pressure Relief

For temporary storage of the oxime, pressure relief is not a serious concern due to the fact that the onset temperature for the decomposition reaction is relatively high. The high onset temperature provides a wide margin for safety. The wide temperature margin also allows for easier detection of abnormal deviations. Furthermore, as we have seen, the high onset temperature trans-

*The equipment time line or time constant θ (sec) is given by $\theta = MC_p/UA$, where MC_p represents the heat capacity of the reactor and its contents (J/K), U is the overall heat transfer coefficient (W/m^{2}K) and A the heat transfer area (m^2).

lates to more time for emergency response. In other words, *prevention* of the onset of the exothermic decomposition of the oxime during temporary storage is a reasonable basis for safety.

There may, however, be other situations wherein the probability of a thermal runaway and pressure increase is higher and the consequences may be worse. An attractive practical solution may thus involve emergency venting to temper the thermal runaway, relieve the pressure, and thereby avoid a vessel rupture or an explosion. For such situations, the ARC can provide useful data for the design of an emergency pressure relief vent [22,24].

The approach involves correcting the time, temperature and heat rate data from the ARC for a ϕ-factor of 1 (see box) to correspond to the typical plant-scale vessel. Huff [24] has shown an approach by which the pressure data may also be corrected. The corrected closed-cell temperature and pressure rate data may then be applied using known DIERS methods [24,38,39] to design the emergency vent.

Limitations of using the ARC for pressure-relief design include having to correct for the ϕ-factor [24,30,36,39]. For complex reactions, the correction (which requires some knowledge of kinetic parameters) can fail. The pressure correction involves additional complexities. However, Huff [24] has shown that ARC evaluations can be applied even to multiple reactions for "nonoverlapping" portions of the runaway. Huff [24] points out that the corrected calorimeter data (when applicable) yields conservative designs and is simpler than alternative approaches of developing detailed computer simulation models.

Computer software programs are available today that are designed to bridge the gap between theory and experiment. For example, the SuperChems Expert software program can use ARC® data as input for developing kinetic reaction models [5]. For rigorous design, a reliable kinetic model combined with tests in a "low-ϕ" calorimeter that allows "vented" tests is desirable [39].

9.0. CONCLUSIONS

The ARC® results for this particular oxime suggest that there are no thermal stability problems under ambient conditions. That is, this oxime may be stored temporarily under ambient conditions.

The ARC® tests also show that this oxime does undergo exothermic thermal decomposition with pressure build-up at a temperature of ~150°C. This result suggests that it is desirable to avoid long term storage, as this material may present special hazards, for example, if there were a fire nearby. Also, the ARC® test does not specifically search for slow chemistry (say, over weeks) at lower temperatures.

The Effect of the Heat Capacity of the Sample Container: ϕ-Factor

The equations that govern the effect of the ϕ-factor on the measured temperatures, temperature rate, and time are presented. Derivations are available elsewhere [19,23,24].

The adiabatic temperature increase for a ϕ-factor of unity (ΔT) is telated to the measured adiabatic temperature (ΔT_m):

$$\Delta T = \phi \Delta T_m \tag{1}$$

For the oxime example, $\Delta T_m = 122\text{K}$ and $\phi = 1.68$. Thus, $\Delta T = 205\text{K}$.

The onset temperature increase for a ϕ-factor of unity (T_o) is telated to the measured onset temperature (T_{om}):

$$1/T_o = 1/T_{om} + (R/E)\ln(\phi) \tag{2}$$

The same sensitivity (0.02°C/min) is assumed at $\phi = 1$. In our example, $T_{om} = 150°\text{C}$, $E/R = 1.5 \cdot 10^4\text{K}$. Thus, $T_o = 417\text{K}$ (144°C). Note that a lower activation energy E results in a lower corrected onset temperature.

The final temperature (for $\phi = 1$) is given by

$$T_f = T_o + \Delta T \tag{3}$$

For the oxime, $T_f = 622\text{K}$ (349°C).

Any intermediate temperature (for $\phi = 1$) is given by

$$T = T_o + \phi(T_m - T_{om}) \tag{4}$$

The temperature rare for a ϕ-factor of unity (dT/dt) is telated to the measured adiabatic temperature (dT_m/dt):

$$dT/dt = \phi(dT_m/dt)\exp[(E/R)(1/T_m - 1/T) \tag{5}$$

For the oxime, maximum rate of $dT_m/dt = 4°\text{C/min}$ was measured at a temperature $T_m = 250°\text{C}$ (523K). The corresponding temperature for $\phi = 1$ is obtained from Eq. (4):

$$T = 417 + 1.68 \cdot (250 - 150) = 585\text{K or } 312°\text{C}$$

The corresponding maximum self-heat rate for a ϕ-factor of 1 is given by Eq. (5):

$$dT/dt = (1.68) \cdot (4°\text{C/min})\exp[(1.5 \cdot 10^4)(1/523 - 1/585)] = 140°\text{C/min}$$

For the special case where $T_m \sim T_{om}$ and $E > 20$ kcal/mol, $T \sim T_m$ and Eq. (5) reduces to

$$dT/dt = \phi(dT_m/dt) \tag{5b}$$

Equation (5b) may not be applied, however, for $T_m > T_{om}$, when the exponential term in Eq. (5) becomes dominant

The corrected values of the temperature and the temperature rate are used to obtain the corrected time, t (for $\phi = 1$), at each data point i, where the subscripts $i = 1, \ldots, n$ represent the measured n data values that were obtained using the ARC

$$t_{i+1} = t_i + [(T)_{i+1} - (T)_i]/(dT/dt) \tag{6}$$

These were results that we were able to convey to our plant personnel. We also did point out that ARC® tests were conducted with a few representative samples of the oxime. For storage even at ambient conditions, it is important that there is no contamination that may catalyze the reaction at lower temperatures. For storage situations, monitoring the temperature to ensure that there is no unanticipated temperature increase (due to, say, contamination) is desirable.

10. REFERENCES

1. *Guidelines for Engineering Design for Process Safety,* Center for Chemical Process Safety (CCPS) of the American Institute of Chemical Engineers: New York (1993).
2. *Guidelines for Chemical Reaction Hazard Evaluation,* Association of the British Pharmaceutical Industry, London (1989).
3. Sharkey, J. et al. Process Safety Testing Program for Reducing Risks Associated with Large Scale Manufacturing Operations. *Plant/Operations Progress,* 11(4), 238–246 (1992)
4. *ARC User Manual,* 1995.
5. Melhem, G. An Advanced Method for the Estimation of Reaction Stoichiometry and Rates from ARC® Data. In *International Symposium on Runaway Reactions and Pressure Relief Design* (G.A. Melhem and H. G. Fisher, eds.), AIChE: New York (1995), pp. 311–333.
6. Ahmed, M., Goetz, D., Fisher, H. and Janeshek, A. Reaction Models from Adiabatic Calorimeter Data, ibid , pp. 283-311.
7. Shanley, E. and Melhem, G. On the Estimation of Hazard Potential for Chemical Substances. In *International Symposium on Runaway Reactions and Pressure Relief Design* (G.A. Melhem and H. G. Fisher, eds.), AIChE: New York (1995), pp. 1–9.
8. Frurip. D. et al. *Determination of Chemical Process Heats by Experiment and by Prediction. In International Symposium on Runaway Reactions and Pressure Relief Design* (G.A. Melhem and H. G. Fisher, eds.), AIChE: New York (1995), pp. 95–149.
9. Lothrop, W. and Handrick, G. The Relationship between Performance and Constitution of Pure Organic Explosive Compounds. *Chemical Reviews,* 44, 419–445, (1949).
10. Shanley, E. and Melhem, G. The Oxygen Balance Criterion for Thermal Hazards Assessment. *Process Safety Progress,* 14(1), 29–31 (1995).
11. "Fire and Explosion Hazards of Peroxy Compounds." ASTM Special Technical Publication No. 394, American Society for Testing and Materials: PA (1965).
12. Shanley, E. Organic Peroxides: Evaluation and Management of Hazards. In *Organic Peroxides* (D. Swern, ed.), Vol. III, Wiley: New York (1972), pp. 341–364.
13. Melhem, G. A Detailed Method for Estimating Mixture Flammability Limits Using Chemical Equilibrium. *Process Safety Progress*, 16(4), 203–218 (1997).
14. Stull, D. *Fundamentals of Fire and Explosions*, AIChE Monograph Series 10, 1977.
15. *NIST Structures & Properties Software Program, Version 2.02*, US Dept. of Commerce, 1995.
16. *CHETAH, Version 7.0, Computer Program for Chemical Thermodynamic and Energy Release Evaluation*, 3rd Edition, American Society for Testing and Materials: PA (1994).
17. Shanley, E. and Melhem, G. A Review of ASTM CHETAH 7.0 Hazard Evaluation Criteria. *Journal of Loss Prevention in the Process Industries,* 8 (5), 261–264 (1995).
18. *Chemical Reaction Hazards,* (J. Barton and R. Rogers, eds.), Institution of Chemical Engineers: London (1993), p. 24.
18. Townsend, D. and Tou, J. Thermal Hazard Evaluation by an Accelerating Rate Calorimeter, *Thermochimica Acta*, 37, 1–30 (1980).

20. Townsend, D. Accelerating Rate Calorimetry. *International Chemical Engineering Symposium Series* No. 68, pp. 3Q:1–13 (1981).
21. Coates, C. The ARC® in Chemical Hazard Evaluation. *Chemistry and Industry,* 6, 212–216 (1984).
22. De Haven, E. Approximate Hazard Ratings and Venting Requirements from CSI-ARC® Data. *Plant/Operations Progress,* 2(1), 21–26 (1983).
23. Yin, F. Adiabatic Calorimeter: Fundamentals and Application in Thermal Hazard Evaluation. In *International Symposium on Runaway Reactions and Pressure Relief Design* (G.A. Melhem and H. G. Fisher, eds.), AIChE: New York (1995), pp. 58–94.
24. Huff, J. Emergency Venting Requirements, *Plant/Operations Progress,* 1(4), 211–229 (1982).
25. Tou, J. and Whiting, L. The Thermokinetic Performance of an Accelerating Rate Calorimeter, *Thermochimica Acta,* 48, 21–42 (1981).
26. Fenlon, W. A Comparison of ARC® and Other Thermal Stability Test Methods, *Plant/Operations Progress,* 3(4), 197–202 (1984).
27. Hofelich, T. and Thomas R., The Use/Misuse of the 100 Degree Rule in the Interpretation of Thermal Hazard Tests. In *International Symposium on Runaway Reactions,* AIChE: New York (1989), pp. 74–85.
28. Leung, J., Fauske, H., and Fisher, H. Thermal Runaway Reactions in a Low Thermal Inertia Apparatus. *Thermochimica Acta,* 104, 13–29 (1986).
29. Singh, J. Reliable Scale-up of Thermal Hazards Data Using the PHI-TEC II Calorimeter. *Thermochimica Acta,* 226, 211–220 (1993).
30. Ahmed, M., Fisher, H., and Janeshek, A. Reaction Kinetics from Self-Heat Data-Correction for the Depletion of Sample. In *Proceedings of the International Symposium on Runaway Reactions*. AIChE: New York (1989).
31. Young, Y. and Chippett, S. Design and Operation of an Automatic Pressure Tracking Adiabatic Calorimeter. In *International Symposium on Runaway Reactions and Pressure Relief Design* (G.A. Melhem and H. G. Fisher, eds.), AIChE: New York (1995), pp. 23–58.
32. Wilberforce, J. "The Use of the Accelerating Rate Calorimeter to determine the SADT of Organic Peroxides." ARC Technical Document (1981).
33. Fisher, H. and Goetz, D. Determination of self-accelerating Decomposition Temperatures using Accelerating Rate Calorimeter. *Journal of Loss Prevention in the Process Industries,* 4(10), 305–316 (1991); ibid. 6(3), 183 (1993).
34. Stieff, A. Emergency Cooling and Inhibiting Agents for Preventing of Chemical Reaction Runaways. *Loss Prevention and Safety Promotion in the Process Industries,* 1, 425–440 (1995).
35. Yu, Y. and Hasegawa, K. Derivation of the Self-Accelerating Decomposition Temperature for Self-reactive Substances using Isothermal Calorimetry. *Journal of Hazardous Materials,* 45, 193–205 (1996).
36. Townsend, D., Ferguson, H. and Kohlbrand, H. Application of ARC® Thermokinetic Data to the Design of Safety Schemes for Industrial Reactors. *Process Safety Progress,* 14(1), 71–76 (1995).
37. Ferguson, H., Townsend, D., Hofelich, T. and Russell, P. Reactive Chemicals Hazard Evaluation: Impact of Thermal Characteristics of Transportation/Storage Vessels. *Journal of Hazardous Materials,* 37, 285–302 (1994).
38. Fisher, H. et al. *Emergency Relief System Design Using DIERS Technology—The DIERS Project Manual.* DIERS/AIChE, New York (1992).
39. Fisher, H. An Overview of Emergency Relief System Design Practice, *Plant/Operations Progress,* 10(1), 1–12 (1991).

1

The APTAC: A High Pressure, Low Thermal Inertia, Adiabatic Calorimeter

Simon Chippett, Peter Ralbovsky, and Roger Granville
Arthur D. Little, Inc.
Acorn Park
Cambridge, MA

ABSTRACT

A large-scale, low thermal inertia, adiabatic reaction calorimeter, the APTAC, is described. The apparatus, which is fully automated, can operate from 25 to 500°C and at pressures from vacuum to 2000 psia. Reactions are carried out in low mass spherical bombs with volumes up to 130 mL and constructed of a variety of materials including titanium, stainless steel, tantalum and glass. ϕ factors down to 1.1 are attainable. The APTAC is therefore able to simulate large-scale thermal behavior. A sample bomb wall thermocouple in addition to the liquid sample thermocouple, allows the use of smaller (and relatively heavier) sample bombs for those cases where a larger ϕ factor is needed, as for instance, when it is desired to separate multiple exotherms for kinetic studies or when the adiabatic temperature rise would be excessive in a $\phi = 1$ sample bomb. An automated sample thermocouple calibration algorithm enables the detection of exotherms at the 0.04°C/min level, although lower detection levels are possible if the machine is operated at a constant pressure or under a vacuum. Reactions can be adiabatically followed up to about 400°C/min and at pressure rates to about 10–20,000 psi/min depending on the material strength of the reaction bomb. The APTAC is also able to run "open vessel" or vented tests without contaminating the containment pressure vessel and can therefore acquire the data necessary to use the simplified DIERS vent sizing technology. A heat ramp option followed by adiabatic operation upon detection of an exotherm has also been incorporated allowing for relatively rapid testing of materials. Additionally, fire exposure can be simulated by imposing a continuous forced heating rate on the reactants before, during and after an exotherm.

Construction and operation details are described and some typical examples of data obtained in the device under a variety of test conditions are shown.

1. INTRODUCTION

Adiabatic reaction calorimetry has proven to be an extremely useful tool for assessing thermal hazards of reactive chemicals and is frequently used in the iterative process of design for maximizing safety. Thermal stability of single component materials and the reactivity of multicomponent mixtures can be evaluated using adiabatic calorimetry. Adiabatic calorimetry can be used not only for screening [3] but also for obtaining kinetic and thermodynamic data [1,4,6,8] which in turn is used for constructing computer models of chemical processes which may, under some circumstances, undergo a runaway reaction. The models are then used for predicting temperature and pressure excursions which may be reached as a result of a variety of different upset conditions as well as for sizing emergency relief devices for the process.

Much of the thermal stability testing done by chemical companies is carried out in the ARC™ (Accelerating Rate Calorimeter) [1]. However, a major drawback to the ARC™ is that a relatively heavy bomb must be used to contain the reaction products with the consequence that a significant fraction of the reaction heat is absorbed by the bomb itself. Thus, the full temperature rise of the reaction is not observed and peak reaction rates are considerably lowered. Corrections to the data can be made if the reaction is simple, but for complex reactions reliable corrections are likely to be very difficult and only possible if the chemistry is known. In addition, the ARC™ is only capable of adiabatically following exotherms up to about 20°C/min so that fast reactions cannot be adequately studied without thermally diluting the system to lower the expected peak reaction rate.

One way of overcoming the problem of a relatively massive sample bomb is to use a thin walled sample container placed within a pressure vessel which is kept at the same pressure as that in the sample container. Thus, with a small pressure differential, the sample bomb will not burst even though reactions may be carried out at very high pressures. This was the principle of the Bench-Scale Apparatus proposed by the DIERS Committee [4]. In order for this principle to be used as the basis of an accurate adiabatic calorimeter, however, some considerable practical problems have to be overcome. The most formidable of these is to totally eliminate reflux heat losses from the sample and to control the temperature of the environment surrounding the reaction vessel to within a fraction of a degree of the sample. Compounding this problem is the fact that the environmental pressure will be continuously changing as the reaction proceeds thus altering the convective heat transfer from the heaters. Coupled with this, the heaters must be capable of tracking the sample temperature at rates from about 1/20th°C/min to about 400°C/min. This paper describes the APTAC calorimeter which uses thin walled reaction vessels within a pressure containment vessel yet heat losses as the sample pressure and temperature increase are minimized by careful engineering design and control software.

2. DISCUSSION

2.1. Requirements for an Adiabatic Calorimeter.

In addition to the basic requirements for a practical adiabatic calorimeter such as accuracy and repeatability, a number of other capabilities are essential or highly desirable including the ability to:

- Vary the heat capacity of the reaction vessel.
- Add reagents.
- Stir the reactants.
- Vent the reactor under controlled conditions.
- Insert probes for continuous analysis.
- Extract samples for analysis.

Some of these requirements are very difficult to build into a small scale adiabatic calorimeter because introducing relatively large diameter tubes, probes or shafts into a small volume reaction vessel also promotes heat losses by conduction and refluxing. Eliminating reflux heat losses is made especially difficult because the probes are required to exit the pressure containment vessel through the thick stainless steel head of the vessel.

One of the more important requirements for an adiabatic calorimeter is the ability to vary the ratio of the heat capacity of the sample vessel to the heat capacity of the sample.

2.2. The Importance of the ϕ Factor

When an exothermic reaction occurs adiabatically, the heat released raises the temperature of the reactants and accelerates the rate of reaction. In reality, some of the reaction heat is absorbed by the container in which the reaction is being run, so that the full adiabatic temperature rise is not observed and therefore the peak reaction rate is considerably lowered. Some situations such as large reactors and storage tanks or reactions which produce viscous products can approach the ideal adiabatic case. The mass of the containing vessels (and therefore their total heat capacities) in these cases is small in comparison with the mass of the reactants in the vessel and the rate of heat transfer to the vessel walls is also small compared to the heat release rate by the reaction. How this reduction in the reaction rate and temperature rise affects the course of the reaction depends on the complexity of the reaction being followed. One obvious consequence is that a later reaction which occurs at a higher temperature may not be observed at all if studied in a sample bomb which absorbs a significant fraction of the heat of reaction simply because the temperature at which the reaction rate becomes significant is not reached. More subtle effects are also possible where the reaction path may be altered from that which would have occurred in a true adiabatic situation.

A useful measure of the adiabaticity of the reactants is the ϕ factor, defined as the total heat capacity of the system (vessel plus sample) divided by the heat capacity of the sample [Ref.1].

$$\phi = \{M_s C_s + M_b C_b\}/M_s C_s$$

$$\phi = 1 + \{M_b C_b / M_s C_s\}$$

where M_s and M_b are the masses of the sample and bomb and C_s and C_b are the specific heats of the sample and bomb, respectively. Thus for the case where the vessel has no mass the ϕ factor is exactly unity.

A system with a ϕ factor of two means that one half the heat of reaction will be absorbed by the mass of the vessel and one half by the reactants. It is usually assumed that if a temperature rise of 100°C is observed in an apparatus with a ϕ factor of 2.0 then a temperature rise of 200°C will be observed in an adiabatic reactor where no heat is absorbed by the vessel. This will be true for simple, single reaction kinetic schemes. For more complex situations where competing reactions can occur this may not be the case and more (or less) than twice the temperature rise may be obtained. Thus, the ability to vary the ϕ factor in a series of experiments is essential to ensure that the kinetics of the reaction do not change.

To illustrate these points, calculations of the self heat rate for two hypothetical reaction schemes have been made, the reactions being carried out in two different vessels to give ϕ factors of 1.0 and 2.0 .

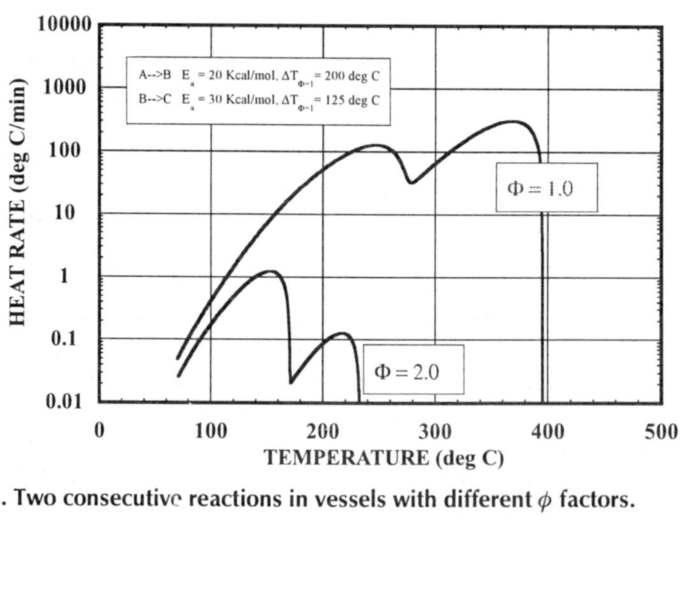

Figure 1. Two consecutive reactions in vessels with different ϕ factors.

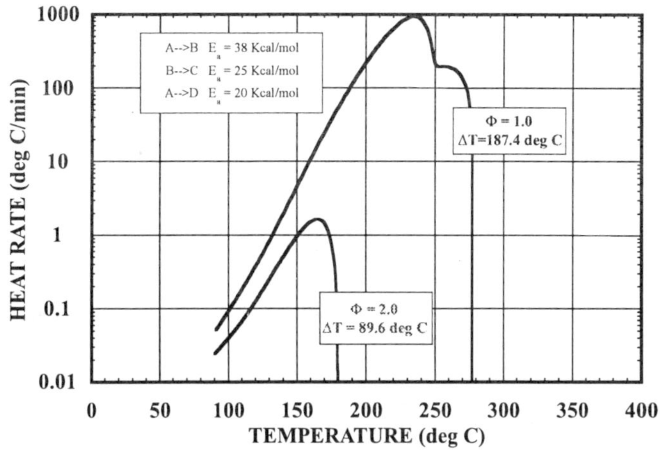

Figure 2. Competing reactions in vessels of different ϕ factor showing that $\Delta T_{\phi=1} > 2 \times \Delta T_{\phi=2}$

For the first reaction scheme (being two consecutive reactions: A→B→C) the adiabatic temperature rise is simply doubled when the ϕ factor is halved as shown in Figure 1. Both reactions are clearly seen whether conducted in the $\phi = 1.0$ vessel or the $\phi = 2.0$ vessel, although the second reaction is more prominent in the adiabatic vessel giving a maximum rate about 4500 times as great as for the $\phi = 2.0$ case. This is due to the fact that the second reaction occurs at a much higher temperature in the $\phi = 1$ reactor. It may readily be seen that a cursory or uninformed analysis of the higher ϕ factor experiment could lead to a very erroneous conclusion concerning the thermal hazard, especially if the second reaction is associated with gas production.

Figure 2 shows a similar set of calculations for a slightly more complex hypothetical scheme (A→B→C and A→D). The computed adiabatic temperature rise for the reaction carried out in the $\phi = 2.0$ vessel is approximately 89.6°C with a maximum heating rate of about 1.2 deg°C/min. However, in the "zero mass" reactor the temperature rise is greater then twice (2.092 times as great) that to be found in the $\phi = 2.0$ vessel, and the maximum heat rate is nearly three orders of magnitude greater. Because the first reaction, A→B, has a higher activation energy than the third reaction, A→D, it is preferentially followed in the low ϕ experiment where higher temperatures and rates of temperature rise are observed. It is for this reason that it is important when studying reactions in adiabatic calorimeters to obtain data in a variety of sample/bomb combinations in order to uncover any potential mechanism change as a result of a change in the ϕ factor. If the reaction mechanisms are complex and too difficult to model with the available time and resources, it is wise to have a set of data matching the ϕ factor and the fill ratio of the full-scale equipment.

If the object of the experimental study is to derive reaction kinetics, then it is essential that experiments be conducted in at least two vessels yielding different ϕ factors. Figure 3 represents the computed heat rate profile for the same consecutive reaction scheme shown in Figure 1 except that the preexponential factor for the second reaction has been increased. This acceleration of the second reaction has effectively removed the separation of the two peaks in the heat rate plot for the $\phi=1$ case but the two reactions can still be clearly distinguished for the $\phi=2$ case. If the only data available was from the low ϕ experiment, it could be wrongly inferred that only a single reaction was involved. Note also that the maximum heating rate for the low ϕ experiment is about 6000 °C/min, which is greater than can be adequately measured by an adiabatic reactor. Therefore it is often desirable to conduct experiments in a higher mass reaction vessel in order to:

- Limit the adiabatic temperature rise
- Reduce the observed peak heating rate to a measurable range and, most importantly
- Separate reactions for clarification of the reaction kinetics

which, depending on the complexity and thermokinetics of the reaction, may not be readily observable in a low ϕ experiment. Conversely, it is important to obtain experimental data in a low ϕ apparatus in order to:

- Obtain the full adiabatic temperature rise, if possible.

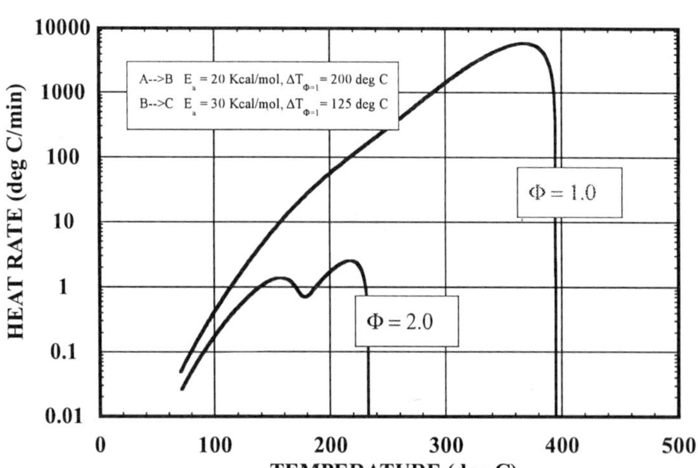

Figure 3. Consecutive reactions in vessels of different ϕ factor showing the separation of reactions by increasing the ϕ factor.

- Measure the maximum heating rate and rate of pressure rise.
- Check that additional reactions do not occur at higher temperatures.
- Confirm kinetic models derived from other experimental methods or in a higher ϕ apparatus

3. APPARATUS DESCRIPTION

The APTAC, shown schematically in Figure 4, consists of three main heaters surrounded by insulation and placed in a pressure vessel. The reaction to be studied is carried out in a thin walled spherical vessel suspended from the top heater. Pressure build up due to a reaction is balanced in the pressure vessel by introducing nitrogen gas at such a rate as to give a constant and small pressure differential (usually about 5 psi) across the wall of the reaction vessel. Mixing of the reactants is achieved by using a magnetic stirring bar. The reaction mixture is maintained in an adiabatic environment by carefully controlling the temperature of the three heaters and the temperature of a special fitting which carries the injection tubes and probes through the pressure vessel head. Control of the heaters and nitrogen flow to the vessel, data logging, sample injection and venting functions as well as most safety systems are all under the direct control of an IBM compatible PC running under Windows NT.

Samples may be injected into the reaction vessel using the single-shot injection bomb or a high pressure syringe pump. Venting from the reaction vessel is either through a computer controlled flow valve into a continuously weighed condensate pot, or directly into an external containment vessel.

There are many unique features of the APTAC which contribute to the reliability, accuracy, safety and ease of use of the machine.

- A magnadrive stirrer is attached to the bottom of the pressure vessel and the stirrer shaft passes through the bottom to a point just below the heated space. A magnet is attached to the rotating shaft and is in close proximity to the base of the reaction bomb thus giving excellent stirring by virtue of the very close coupling of the magnets. The rotation speed of the stirrer is controlled by the computer.
- Good insulation is crucial to the efficient and accurate operation of the APTAC. The insulation within the pressure vessel is made from a machinable, ceramic material. Because the insulation is nonporous, convection losses from the heaters do not change significantly as the pressure in the containment vessel increases. In addition, the insulation will not absorb chemicals which either leak from, or are dumped from, the reaction vessel due to a bomb rupture.
- The top heater and associated insulation and control thermocouple are rigidly bolted to the top head of the pressure vessel and need only

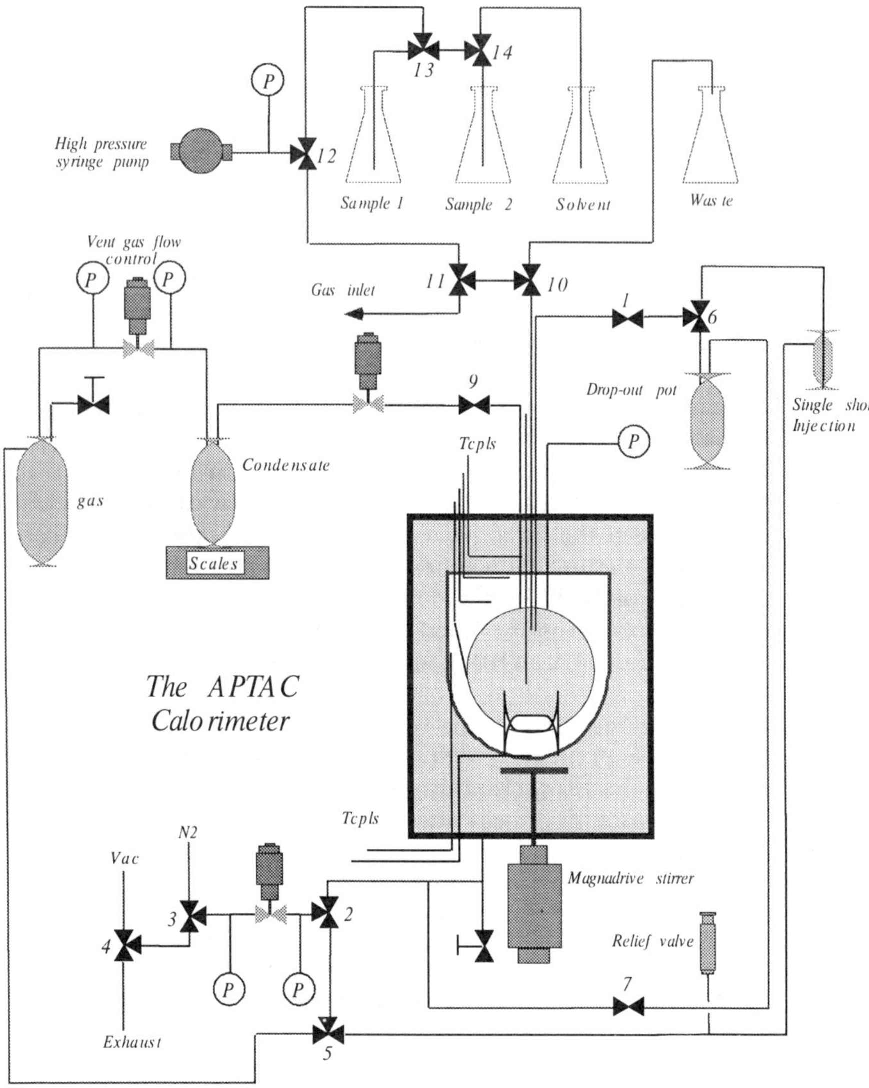

Figure 4. The APTAC Calorimeter showing the plumbing scheme.

be removed for replacement. The sample bomb is attached to the top heater and the sample thermocouple passes directly into the sample bomb through a fitting on the top heater. The sample thermocouple is reusable and may be sheathed with different materials, the choice of which will depend on the reactants being studied

- There is no internal wiring in the pressure vessel. All control thermocouples and heaters are sheathed in stainless steel and enter the vessel through split gland fittings (heaters) or multiprobe gland fittings (thermocouples). This feature allows for relatively easy cleanup in the event of leaks from the reaction vessel. The pressure vessel may simply be rinsed with a suitable solvent without the necessity for removing the internals (heaters, insulation, thermocouples). A drain valve on the bottom of the pressure vessel facilitates this operation.
- The APTAC uses spherical bombs of 2.5" diameter and 0.02" wall thickness. The bombs are available in stainless, titanium, tantalum, or glass. The necks of the vessels are 0.5" diameter thin walled tubing which allows for easy introduction of slurries, solids or viscous materials. A magnetic stirring bar may also be passed through the tube. A vessel made of titanium weighs about 30 g and has a volume of approximately 130 mL. Thus, for a typical sample size of roughly 75 g, a ϕ factor of 1.1 is readily obtainable. The bombs are, in most cases, easily reusable. If a solid product is produced, then the bombs are cheap enough that they may be discarded.
- An external pressure vessel (500 mL with a rating of 2000 psig) is attached to the sample inlet line as shown in Figure 4. This vessel expands the capabilities of the APTAC in several ways. Firstly, it acts as an emergency dropout pot for those tests where reaction rates are excessive and the computer has opened the bypass valve (#1). Secondly, it acts as a condenser/receiver when running uncontrolled open vessel or tempered tests and thus negates the need to vent directly into the pressure vessel. Thirdly, it acts as an expansion volume when testing a gas producing reaction where the rate of pressure generation would be too high in the reactant vessel alone.
- The APTAC computer control system is completely automated. Several modes of operation are possible all of which have been automated, including the heat-wait-search mode similar to the operation of the ARC™, a forced temperature ramp mode which drops into an adiabatic mode on detection of an exotherm, and a temperature ramp mode where the ramp remains constant even through an exotherm in order to simulate a fire exposure. Additionally there is the option to run any of the types of test in the open vessel configuration with or without a superimposed pressure and an option to hold the pressure at a constant preset level.
- Several safety features have been written into the control algorithms to prevent overexertion of the heaters and to minimize the possibility of the reactant vessel being either crushed by the incoming nitrogen gas or bursting from an excessive pressure generation rate.

- An automated sample thermocouple calibration mode is used to generate and store offsets between the sample and nitrogen space thermocouples. Calibration is done using a reaction vessel containing an initial pressure of nitrogen gas. Calibrations do not need to be repeated at different pressures in order to generate the table of offset values.
- The APTAC can operate at pressures up to 2000 psi over a temperature range of 25°C to 500°C. The apparatus is adiabatic up to about 400°C/min and the practical limit in following the rate of pressure rise is caused by the relatively long length of 1/16th tubing leading to the sample pressure transducer. Because of the need to limit refluxing and minimize dead volume, it is not desirable to use 1/8th tubing. Pressure rates up to about 10,000 psi/min can be accurately measured and followed, but this limit depends on the strength of the reaction vessel. For a glass bomb the pressure tracking limit may be reduced by an order of magnitude.
- A sophisticated control program operating under Windows NT allows for easy use of the apparatus. The program also has a very easy to use report generator so that the test data can be rapidly plotted and presented in a standard format.

3.1. Signal Processing

Signals from transducers and thermocouples are processed and digitized using the National Instruments SCXI series of signal conditioning units and a 16 bit A/D board. Digital inputs and outputs are also handled on the same chassis but with the appropriate modules. Analog outputs for control of heaters, flow valves and stirring are handled using a 12-bit D/A board, also from National Instruments. The high pressure syringe pump and scales are controlled via RS-232 communication with the computer.

3.2. Heaters and Temperature Control

The temperature of the heated zone surrounding the reactor is maintained by three independently controlled heaters each with full PID control directly from the computer. The surface temperatures of the heaters are measured using sheathed, 0.063" diameter, type N thermocouples. The sample temperature is measured using a sheathed thermocouple with diameter of 0.020" for following very fast reactions. The top, side and bottom heaters are constructed of a coiled stainless steel sheathed heating wire brazed to copper bases. The stainless sheaths are passed through the top head (top heater) and the bottom (side and bottom heaters) of the pressure vessel via 0.5" split gland fittings with high temperature packings. In this way no wiring connections need be made on the inside of the pressure vessel. Since it is very nearly impossible to obtain accurate surface temperature measurements using a single probe on each heater, an additional thermocouple located in the heated space is also used for control purposes. It is this thermocouple which the computer control system keeps at the same temperature as the sample by trimming the temperature of the three heaters. Thus the nitrogen surrounding the sample bomb is maintained at the same temperature as the sample. All tubes and probes exiting the sample bomb pass through the head of the pressure vessel via a special heated fitting. Reflux heat losses are thus prevented.

Figure 5 shows how the nitrogen temperature surrounding the sample bomb varies through a typical reaction. The plot is an expanded section of a test. During the sample heat step the nitrogen temperature is hotter than the sample, but as the sample temperature approaches the desired set point, the nitrogen temperature is ramped down so that at the set point it is exactly equal to the sample temperature. When an exotherm has been detected, the nitrogen temperature tracks the sample temperature by keeping the difference between the two at zero. At the peak heating rate (approximately 500°C/min) there is a small lag between the sample and nitrogen temperatures but the peak lag is small (about 12°C) and of very short duration such that heat losses from the sample are inconsequential. Moreover, a small overshoot after the peak temperature is reached compensates for the small heat loss by adding an equal amount to the sample.

The APTAC currently uses five types of heating strategies:

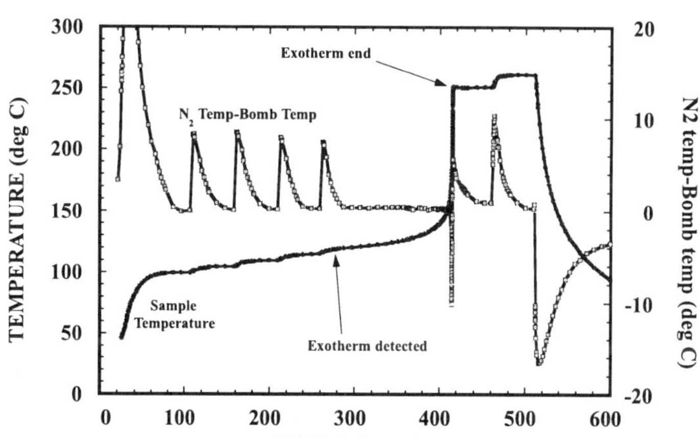

Figure 5. The temperature difference between the wall of the reaction vessel and the nitrogen space temperature during a heat-wait-search-adiabatic procedure.

- *Heat-Wait-Search* followed by adiabatic operation if an exotherm is detected. The self-heat rate for exotherm detection is a variable parameter which must be preset by the operator. Careful calibration of the sample thermocouple allows for very precise measurements of the self-heating rates. The APTAC is capable of obtaining rates at the 0.04 °C/min level or less, depending on the care with which the thermocouple calibration is carried out. The Heat-Wait-Search routine can be continued after the exotherm to check for possible heat drifts or for the presence of thermal activity at higher temperatures.
- *Straight Heat Ramp* where the heating rate is set by the operator. In this test, which is not an adiabatic mode of operation, the heaters are controlled at the set heating rate and the sample temperature and pressure are monitored. If the sample rises above the heater temperature, which it will do if the sample undergoes a significant exotherm, there will be heat losses from the sample. This test is useful for a more rapid screening of samples.
- *Heat Ramp-Adiabatic* where the apparatus will shift to an adiabatic mode of operation following a heat ramp if an exotherm is detected. An exotherm is detected when the sample temperature meets the heater temperature and the heating rate of the sample is greater than the ramp rate.
- *Isothermal Soak-Adiabatic* where the sample is held at an approximately constant temperature for a time which is specified by the operator. If the sample self-heats during the soak period the heaters track the sample temperature to keep the sample adiabatic. Every thirty minutes until the soak time has been exceeded, the self-heat rate is checked and if it is less than the apparatus detection limit, the temperature is lowered back to the soak temperature. On the other hand, if it is greater than the detection limit the apparatus ends the soak and follows the sample adiabatically.
- *Fire Exposure Simulation.* The sample is heated at a constant rate by specifying a fixed temperature difference between the heaters and the sample. The control system maintains this constant differential even through an exotherm so that a constant heat flux is superimposed on the sample thus simulating fire exposure to the walls of a tank or reactor.

3.2.1. Using the Bomb Wall Thermocouple

When the maximum self-heat rate or the adiabatic temperature rise needs to be limited, an experiment should be conducted in a relatively heavier sample bomb. For these cases, the wall temperature of the sample bomb should be used for heater control. However, care needs to be used in ensuring that temperature gradient between the sample and reaction vessel remains constant or is very small, i.e. that the rates of temperature rise of the sample and the bomb are equal. If this condition does not exist, then the sample adiabaticity will change throughout the experiment. Analysis of the data then becomes complicated by the rate of heat transfer from the sample to the bomb. Figure 6 shows how the measured wall temperature compares with the sample temperature measured in the center of the vessel for a standard 2.5" diameter titanium bomb. The vessel wall was 0.020" and the wall temperature was measured using a sheathed 0.063" diameter thermocouple. At a self-heating rate of 20°C/min, the sample and wall temperatures begin to deviate. For a thinner walled bomb and using a smaller diameter thermocouple, the self-heat rate at which this deviation becomes significant will be higher. Conversely, heavier walled bombs will impose a more severe restriction on the maximum self-heat rate which should be allowed if the system (the reaction vessel plus the sample) is to remain adiabatic. Although for this experiment the rates of sample and wall temperature do deviate, the overall effect on the data is small because the ϕ factor for the system is close to one and therefore the total heat lost to bomb is small.

3.2.2. *Importance of Minimizing Heat Losses and Gains.*

Seemingly minor heat losses from the sample or unintended heat inputs to the sample can have significant effects on the apparent thermokinetic parameters derived from the experimental data. The size of the error in the adiabatic temperature rise and the estimated activation energy for the reac-

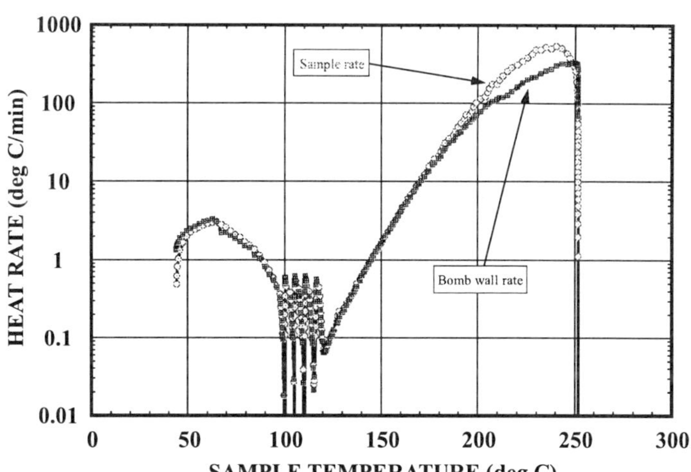

Figure 6. The outside wall temperature of the reaction vessel compared with the sample liquid temperature.

tion will depend on how the machine is operated and on the kinetics of the reaction being studied. Figure 7 illustrates the effect of a small heat loss and heat gain on the calculated self-heat rate curve for a single first order reaction with an activation energy of 38 kcal mol. The error in the calculated adiabatic temperature rise is about 15% for a small constant heat loss rate of –0.03°C/min. Most of this error accumulates at the beginning of the detected exotherm when self-heat rates are low and the sample remains at a low temperature for a long time. For a lower activation energy reaction where time spent at lower temperatures will be longer, the errors will be larger. For higher activation energy reactions, the losses will be smaller. If the heat losses are of the same order of magnitude as the exotherm detection level, the errors may become substantial. It is important therefore to control heat losses to a level substantially lower than the detection level of the machine, and to keep the machine in good calibration. Further, by heating the sample to an initial temperature where the rate of heat loss is small compared to the self-heating rate at that temperature, then total heat losses can be minimized. The disadvantage to this strategy is that reactions at the lower temperature cannot be studied.

3.3. Pressure Tracking

All on/off valves on the APTAC are air-operated ball valves with position indicators. A pneumatic flow control valve on the nitrogen/exhaust line from the

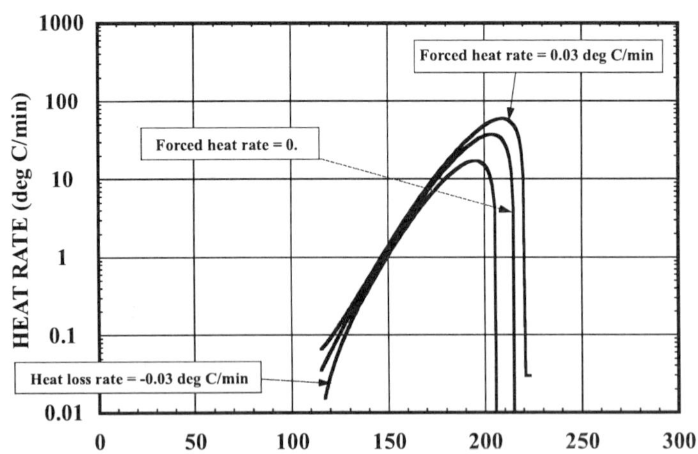

Figure 7. The effect of a small heat loss rate or a small forced heating rate on the adiabatic temperature rise and heat rate curve of a single 1st order reaction.

pressure vessel is used to regulate the flow of nitrogen into and from the pressure vessel. Pressure fluctuations in the containment vessel are therefore small and tracking is very smooth. Pressures are measured using strain gauge transducers with a range of 0–2500 psia and an accuracy of 0.1% of scale. The desired pressure differential between the sample vessel and containment vessel is a variable chosen by the operator. The computer control system is capable of tracking the pressure in the sample vessel up to several thousand pounds per minute and keeping the set pressure differential to within a 4-5 psi band. If the rate of pressure rise is too great, the pressure differential will increase. However, for most tests where the standard titanium reaction vessel is used, there is no adverse consequence because the vessels are capable of withstanding 300–400 psi differential before stretching. When using glass reaction vessels or vessels with thinner walls than the standard, some restriction on the upper bound of the rate of pressure rise is imposed.

3.4. Sample Injection

Samples may be injected into the APTAC using either the single shot injection vessel or, if the Plus system is installed, the high pressure syringe injection.

3.4.1. Single Shot Injection

A sample of liquid may be injected from the pressurized, single-shot vessel under computer control. Information required by the control program includes the volume of liquid to be injected, the volume of liquid already in the reaction vessel, the vessel volumes and the temperature at which the injection is to be made. The control program, at the appropriate point in the test sequence, pressurizes the injection vessel and automatically injects the contents at the set temperature. Adjustments to the heater control functions are made during the injection process in order to minimize wild swings in the heater temperature control due to unpredictable fluctuations in the sample temperature. This type of simple injection is uncontrolled in that the rate of injection is arbitrary and depends on the viscosity of the liquid and the pressure in the bomb. There may also be some inadvertent admission of nitrogen gas into the reaction vessel before the control system can shut off the injection valve.

3.4.2. High Pressure Syringe Injection

The high pressure syringe system allows injections to be made in a more controlled manner than the single-shot method. Two different samples may be injected at two different sample temperatures. The injection temperatures, rates of injection and the volumes to be injected are set at the beginning of the test. After the injection, the user can choose to flush the syringe and lines with a solvent as part of the test sequence and under automatic control.

During the injection process, the computer monitors the conditions in the reaction vessel to ensure that an excessive back pressure cannot be imposed on the syringe barrel. In the event that pressures become too great, the control program will close the valves leading to the pump and stop the pumping sequence. As in the single-shot injection case, the heater control functions are adjusted during the injection process.

Figure 8 shows the results of a controlled injection of a reactant into the sample bomb which contains a second reactant at 125°C. The reactant in this case is injected over a fifteen minute time period which is the maximum rate at which it could be injected in a real situation in the event of the loss of flow control. The adiabatic nature of the test also simultaneously simulates the loss of cooling to the reactor, so that the data repnresent a worst case scenario for the reaction. For this test, the reaction bomb was also vented through a 0.063" diameter line to the closed dropout pot. The data show that the heating rate is limited by the injection rate of the reactant and that the reaction pressure is controlled by the vent line up to the point where the line plugged. Note that the injection rate is not affected by the changing pressure in the reaction bomb.

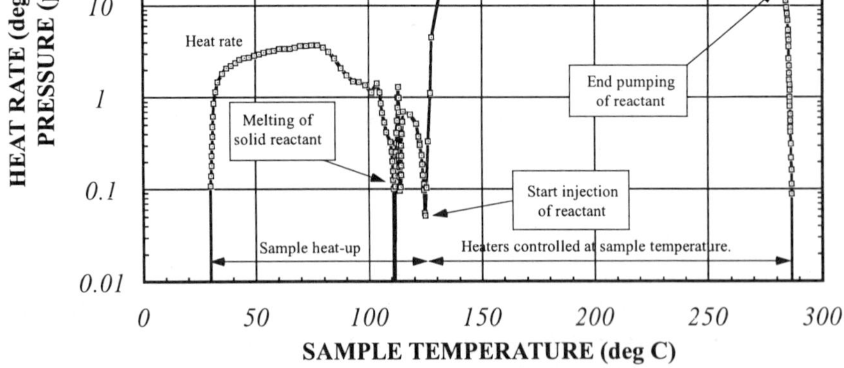

Figure 8. Injection of a reactant using a high pressure syringe pump showing that the heating rate is controlled by the pumping rate

3.4.3. Gas Injection

A gas may also be injected to a preset pressure at a predetermined sample temperature by connecting a gas cylinder to valve #10 (see Figure 4). Since there is no flow valve under the control of the computer on this inlet line, the gases must enter the reaction vessel through a small orifice in order to keep the pressurization rate to reasonable level. The containment vessel is automatically pressurized during the injection to keep the differential at the required level.

3.5. Sampling

Samples can be extracted from the APTAC during a test in one of several places. A microbore tube may be inserted into the liquid or vapor phase of the reaction vessel and small samples removed for analysis by gas chromatography and/or mass spectrometry. Extraction of samples from the vapor phase is likely to cause heat losses if too large a sample is taken. Microsampling techniques must be used if the adiabaticity of the reaction is to be preserved, which can result in plugging problems in the probes.

It is also possible and in fact much easier, to extract samples from the condensate dropout pots during vented tests and from the gas collection vessel during a controlled vent test. Alternatively, FTIR probes may be inserted directly into these vessels for a continuous analysis during the vented portion of the test. Characterization of the reaction products is important for the design of equipment for treatment of the vent stream and also for refining kinetic models of the reaction. Insertion of an FTIR probe into the reaction vessel will limit the full operating range of the apparatus.

3.6. Safety Features

3.6.1. Heaters

Because the total power available to the heaters is high (approximately 5 KW) meltdown of the heaters can occur very rapidly if an inadvertent fault were to happen such as a computer malfunction. There are therefore several different protective elements in the hardware and the software to minimize the possibility of heater destruction. Use of these features has greatly extended the life of the heaters

- The power delivered to the heaters is limited by a software control parameter which is related to the temperature and rate of heating of the sample. If the actual power demand is greater than the demand expected by the control system, the program will shut down the heaters and end the test.

- An upper temperature limit on the heaters is set by an operator controlled parameter during the test setup. The heaters will not exceed this value even if an exotherm is in progress.
- The main heater power relay is under the control of a software switch which may also be operated by hitting the 'Stop' button displayed on the CRT. This gives rapid power off under the direction of an operator when the apparatus is being manually controlled.
- There are separate, manually set, temperature cut offs for the heaters. Outputs from independent thermocouples are sent directly to the main power relay which is switched off if the temperature exceeds the upper limit for the heaters.
- If the power to the amplifiers is lost or any of the control thermocouples fail, the computer can detect this condition and immediately shuts off power to the heaters and ends the test.

3.6.2. Pressure

Automatic shutdown or other protective actions and warnings will occur if the following conditions prevail:

- If the containment vessel is not being pressurized and the pressure difference between the reaction and containment vessels exceeds a preset value.
- If the absolute pressure in the containment vessel exceeds 2000 psia.
- If the absolute pressure in the containment exceeds the maximum test pressure set by the operator.
- If the pressure difference across the reaction vessel wall exceeds a preset value and the containment vessel pressure is greater than the reaction vessel pressure, then the computer will open the bypass valve connecting the two vessels in order to prevent crushing of the thin walled reaction bomb.
- The nitrogen supply pressure is intermittently monitored and checked against the operator specified maximum test pressure.
- The vacuum pump is protected by the fact that the computer will not open the vacuum valve if a pressure greater than 20 psia exists in the line upstream of the valve.
- The pressure containment vessel is further protected by vent valve and a rupture disc.
- The single shot vessel is protected by a vent valve.

3.6.3. Spills and Leaks

All sample containers associated with the syringe pump and the controlled vent system are contained in a vented cabinet. The calorimeter assembly, high pressure tubing and valves and the single-shot injection bomb are contained in a second cabinet. Both of these cabinets are connected to a vent duct which has a pressure differential meter, the output of which is continuously monitored by the computer. Spills are thus contained in the cabinets and exposure of personnel to vapors is minimized. In addition, the area directly behind the head of the pressure vessel is vented so that when removing the sample bombs from the apparatus, exposure to vapors is again minimized.

The use of nonporous, chemically resistant, ceramic insulation in the pressure containment vessel also makes clean-up after a leak or reaction vessel breakage much easier.

3.6.4. General

A number of other features enhance the overall safety and ease of use of the apparatus, including

- A "watch-dog" timer ensures that the machine will safely shut down in the event of a computer failure. The computer control program outputs a continuous, timed, series of pulses to the watch-dog timer. If the pulse stream is interrupted, the power to the heaters and valves is automatically shut off.
- A mechanized head lift raises and lowers the heavy, top head assembly.
- Valve position indicators give positive identification of the position of each valve. The control software will not carry out some operations if valves are incorrectly positioned.
- Cabinet door locks are under computer supervision to minimize exposure to chemicals.
- An oxygen sensor on the nitrogen inlet line detects the presence of oxygen in the pressure balancing gas.

3.7. Software Features

3.7.1. General

The APTAC is operated by a custom control program operating under Windows NT. There are a number of features which enhance the ease of operation and reliability of the apparatus.

- *Experimental Set-up*. The user is guided through a series of screens to ensure that the all the input data necessary to run a test is entered. The input data may be saved as a template so that for repeat tests or tests of a similar nature, the input data need not be re-entered.
- *"Playback" of a Test.* At the completion of a test, the data file can be recalled and displayed as if the machine was conducting a test, although the hardware is nonoperational. The playback speed can be increased up to 500 times so that a long test can be viewed in a few minutes.

- *Auto Leak Checking*. At the beginning of a test, the operator can choose to do an automatic leak test on the reaction vessel as part of the test sequence. This feature has proven to be very useful in minimizing the possibility of leaks at the neck fitting of the reaction vessel. After a long test which may perhaps have lasted for a day or two, discovering that the vessel had leaked from the beginning of the test can be very frustrating. Leak checking the reaction vessel at the end of a test prior to removing the vessel from the apparatus may also be built into the test sequence.
- *Graphical Viewing of Data*. During the test, the current data is displayed on two screens and the historical data may be viewed in tabular or graphical format. The graphs available include the heat rate, pressure rate and pressure as a function of reciprocal temperature. The reciprocal temperature scale is displayed in centigrade making the plots easy to read and interpret.
- *"Manual" Operation*. The current version of the software does not allow operation of individual valves or heaters. However, operations on the machine which the user may want to conduct are programmed as a sequence. For example, rather than operating individual valves to pressurize a portion of the apparatus, the appropriate task is chosen from a menu and the operator is asked for the required pressure and the rate of pressurization. "Manual" operations include:
  —Pressurizing/Depressurizing of various portions of the apparatus.
  —Leak checking different parts of the apparatus.
  —Operating the syringe pump.
  —Operating the heaters (as a group)
  —Flushing lines with solvent if the syringe pump is installed.

3.7.1. Report Generator

An easy to use report generator is part of the operating program for the APTAC. The user may choose to print any of the ten plots available and also the data in tabular format. In addition, the input data, the thermocouple calibration data and notes about the test entered by the operator may also be printed as part of the report.

4. TYPES OF TEST

Three types of test may be conducted in the APTAC:

- Closed Vessel using any of the heating strategies described above.
- Simple Vent Test using the dropout pot downstream of valve 1.
- Controlled Vent Test.

4.1. Closed Vessel Test

Closed vessel tests give data which may be analyzed to extract heats of reaction, reaction kinetics, vapor pressure data and gas generation rates. For more complex thermochemistry it may be necessary to run a number of tests in vessels of different heat capacity and perhaps use different heating strategies such isothermal aging. In addition, analysis of intermediate compounds and reaction products may be required for understanding the kinetics.

To illustrate the operating range of the APTAC a number of experiments were run on a 25 weight % mixture of di-tert-butyl peroxide (DTBP) in toluene. The start temperature for the test, shown in Figure 9, was set at 110°C with a temperature increment of 5°C. Since an exotherm at the 0.04°C/min level was not detected at 110°C, the sample temperature was automatically incremented by 5°C. The Heat-Wait-Search strategy was continued after the reaction was completed to check for heat losses at the higher operating pressure as a result of the decomposition.

Figure 10 shows the rate of heating and the rate of pressure rise on a logarithmic scale as a function of the reciprocal temperature. A maximum self-heating rate of 453°C/min was observed, which is close to the maximum possible heating rate of the surrounding heaters and therefore the maximum self-heat rate which can be adiabatically tracked by APTAC. The sample is initially heated to the starting temperature by convective transfer from the surrounding heaters and nitrogen gas at about 3°C/min but this rate slows as the desired starting temperature is reached. The adiabatic temperature rise was

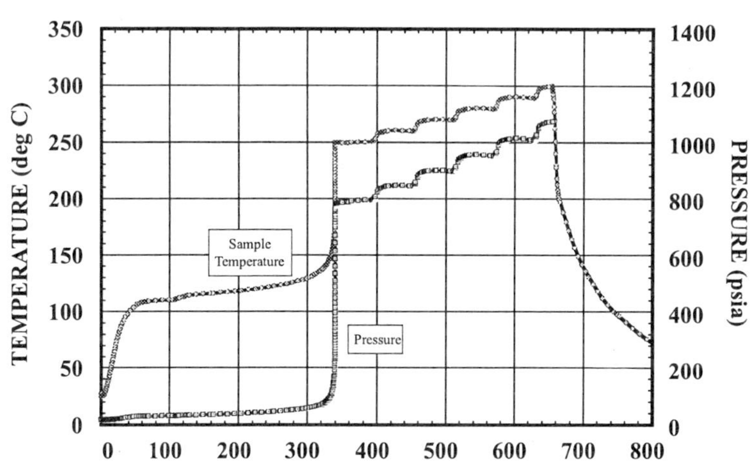

Figure 9. Temperature and pressure profiles for a closed vessel test using a 25 wt.% solution of DTBP in toluene.

measured at 136°C and the heat of decomposition was calculated to be 323.5 cals/g DTBP (or 47.23 kcal/mol) using a value of 0.14 cal/g/°C for the heat capacity of titanium and 0.505 cal/g/°C for the DTBP/toluene mixture [4].

A test on a 25% solution of DTBP in toluene was also ran in a Setaram DSC-111 calorimeter in order to compare the result with that obtained using a different method. The heat of reaction was measured at 346 cal/g DTBP at a scan rate of 2°C/min. The heat of reaction obtained in the APTAC, therefore, was a little lower than that measured in the Setaram calorimeter but higher than that measured in the ARC™ which has been quoted at 295 cal/g [1].

Low values for heats of reaction measured in the ARC™ have been observed previously [5,6]. Vynn suggested using hydrogen peroxide decomposition as the standard of choice for checking the operation of the ARC™ rather than DTBP [5]. The reason for the suggestion was that the thermodynamic properties of all the reactants and products are accurately known and the decomposition proceeds to completion without byproducts. Since the products are known, the heat release and the final pressure can be accurately computed and compared with a test result obtained in the ARC™. Vynn obtained ARC™ data and final gas composition data on hydrogen peroxide decomposition in a titanium bomb and found that the measured data were 19% lower than the calculated value. He could give no feasible explanation for the discrepancy but suggested that a study was warranted by the ARC™ Users Group. The heat rate plot for the H_2O_2 experiment gave two peaks indicating a more complex decomposition mechanism than was expected. Fur-

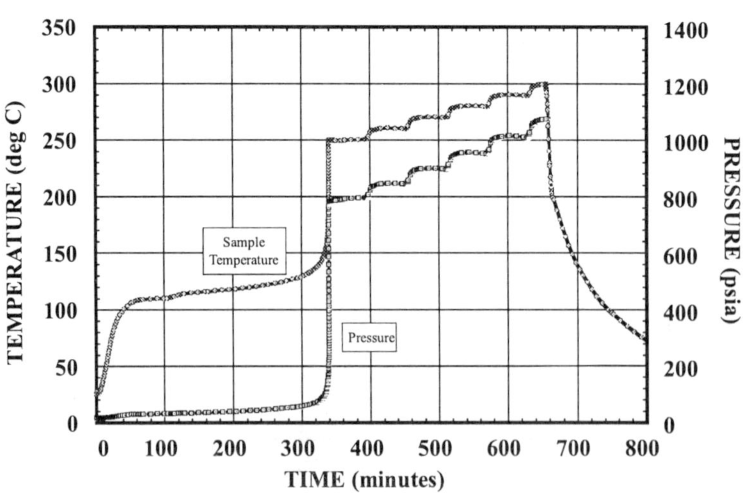

Figure 9. Temperature and pressure profiles for a closed vessel test using a 25 wt.% solution of DTBP in toluene.

ther work [7] has shown that the size and shape of these peaks are dependent on the material of the reaction bomb and the surface condition as well as the presence of any contaminants. These factors negate the usefulness of H_2O_2 as a standard for kinetic studies in the ARC™ but they should not affect the overall heat release. Although kinetic studies using H_2O_2 in the ARC™ may not be straightforward, it has been successfully studied in glass bombs in the APTAC using the simple vent method. A possible explanation for the discrepancy between the calculated and measured heat of decomposition in the ARC™ could be loss of active material outside of the heated zone by evaporation from the sample bomb and condensation in the transducer and lines leading to the transducer. An explanation invoking loss of adiabaticity would also lead to incorrect kinetics obtained by the ARC™, but the Arrhenius parameters measured for DTBP are very repeatable and agree closely with other studies [1]. Loss of sample prior to the detection of an exotherm would give a lower apparent heat of reaction but correct kinetics.

Whiting and Tou [6] studied the polymerization of styrene in the ARC™. They found that for this lower activation energy system the discrepancy between heat of reaction data measured in the ARC™ and a DSC was explainable by the presence of the radiant heater in the heated space of the ARC™. The APTAC has no uncontrolled heat sinks in the heated space surrounding the sample bomb and as such is not subject to this type of error. It is also for this reason that the magnetic stirrer is not located in the heated space where it would be a significant heat sink, although give better magnetic coupling of the stirring bar and therefore improved mixing.

Figure 11 shows the first order rate constant as a function of reciprocal temperature for duplicate tests on a 25% solution of DTBP in toluene in addition to the rates of temperature and pressure rise for a single test. The first order reaction rate, k, is obtained by

$$k = (dT/dt)/(T_f - T)$$

where T_f is the final or maximum temperature at the end of the exotherm, T is any temperature during the exotherm and dT/dt is the rate of temperature rise at the temperature T. The plot of the first order reaction rate, k, versus reciprocal temperature gives a straight line from which the activation energy, E, and preexponential factor, A, are calculated. Table 1 lists the thermokinetic parameters calculated from the experimental data for the two tests and compares them with measured data from the ARC™ [1] and the VSP apparatus [9].

4.2. Simple Vent Test

The apparatus may be operated in a vented or closed vessel mode. Closed vessel operation simply means that the reaction vessel inside the pressure vessel is closed off from the external dropoutpot by valve 1 (see Figure 5). For

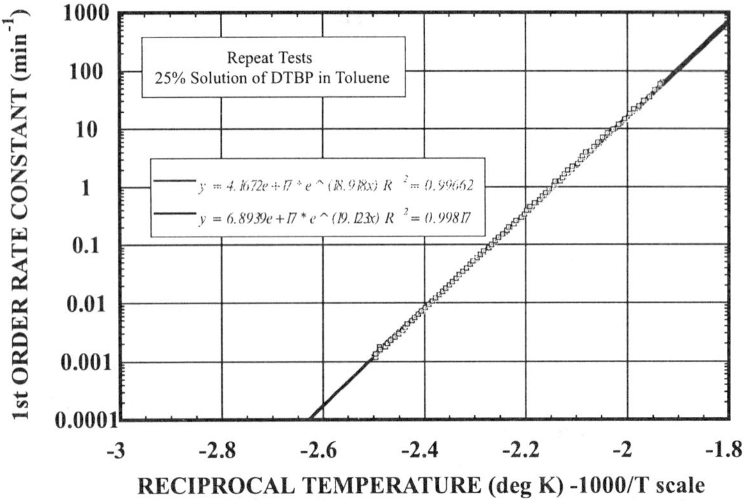

Figure 11. The 1st order reaction rate for repeat tests of the decomposition of DTBP.

TABLE 1

Thermokinetic Data for Di-Tert-Butyl Peroxide Decomposition

	E_a (kcal/mol)	$\log_{10} A$ (s^{-1})	DH_r (kcal/mol)
Test 1.	37.6	15.84	47.23
Test 2.	38.0	16.06	47.60
VSP RR [9]	36.1	15.13	47.0
ARC™ [1]	37.8	16.15	43.0

simple vent tests valve 1 is opened at some predetermined set pressure and valve 7 which is at the exit to the dropout pot can be either left open or closed. If this valve is open, then the reaction vessel is in communication with the pressure vessel and the pressure tracking system is stopped. If the pressure in the reaction vessel ceases to rise and the sample temperature remains flat or is tempered, then the reaction may be classified as a vapor system. The pressure in the reaction vessel may continue to rise, albeit at a reduced rate because of the increase in the head space of the reaction vessel, in which case the reaction system can be classified as either a gassy or hybrid system. Tempering of the sample and increasing pressure would indicate a hybrid system. At the end of the test, the contents of the dropout pot are available for analysis. If a gassy reaction is being studied the computer control system can also maintain a constant pressure in the vessels by opening valve 7 and controlling the nitrogen/exhaust flow valve or, optionally, the pressure can be allowed to build in order to quantify the gas generation rate.

Several tests were made on a 25% DTBP solution using the open vessel concept. Seventy five grams of the solution were used in a 125 mL titanium reaction vessel. Initially the dropout pot and the reaction vessel were at atmospheric pressure. As shown in Figure 12 the sample was heated to 110°C and then a further 5°C. The exotherm was detected at 115°C and followed adiabatically until the vent opened at 75 psia. During this time period, the pressure was increased in the dropout pot at the same rate as the pressure rise in the sample bomb, so that when the vent opened no pressure difference existed between the two vessels. After the vent opened, the rate of pressure rise in the sample bomb decreased due to vapor condensation in the dropout pot and because the effective volume of the reaction vessel had increased by 500 mL. as shown in Figure 13. For these tests, valve 7 was closed at the same time as valve 1 was opened. The temperature continued to rise, reaching a maximum rate of about 3.4°C/min at 168°C. The pressure rate, after the initial drop, also continued to rise reaching a maximum rate of 4 psia/min at 182°C. Because valve 7 was closed and the system was producing gas as a decomposition product, the pressure in the dropout pot increased throughout the venting period with the result that the observed temperature rise and the maximum heating rate were both higher than they would have been had the pressure been controlled at the vent set pressure.

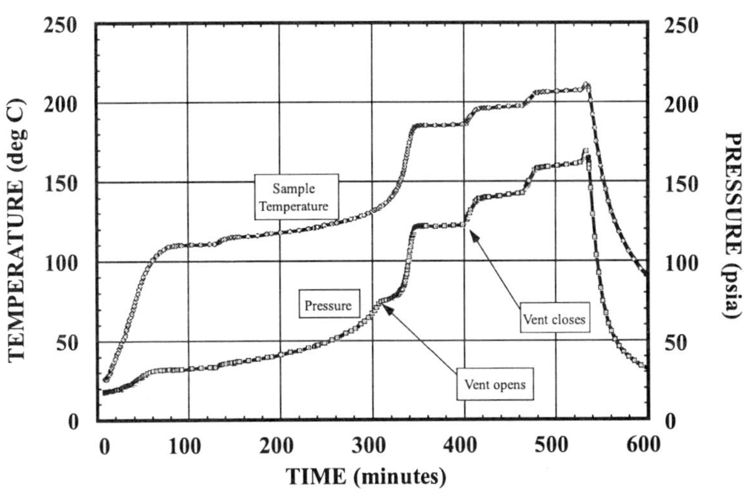

Figure 12. Temperature and pressure profiles for a simple vent test with DTBP.

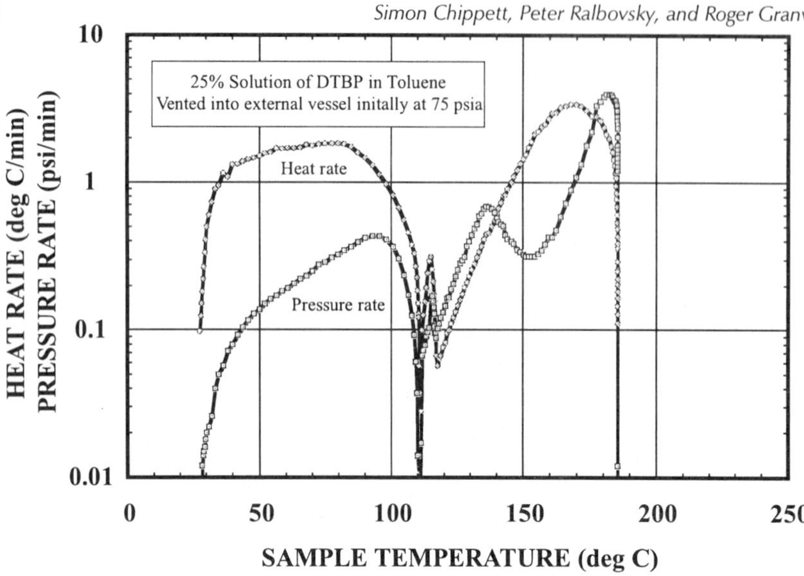

Figure 13. The rates of temperature and pressure rise for DTBP decomposition in a simple vent test.

4.3. Controlled Vent Test

Controlled venting is available on the more advanced APTAC Plus system. At the desired set pressure valve 9 is opened and the pressure in the reaction vessel is controlled at the set pressure by the downstream flow control valve. Vented gases and vapors are collected in the condensate pot which is continuously weighed. If the pressure in the pot rises above some nominal value (usually set at about 30 psia), a second flow control valve is opened and gases are allowed to flow into the preevacuated gas collection vessel. Since the volume, the temperature and the rate of pressure rise in this vessel are measured, the molar rate of gas production is known. In addition, the mass flux of condensable material vented from the reactor is also known from the condenser mass reading.

A controlled vent test was run using the same mixture as was used for the simple vent test described above. Figure 14 shows the temperature and pressure of the sample in addition to the mass vented and the gas evolved as a function of time. The exotherm was detected at 120°C and was followed adiabatically until 135°C at which point the vent valve was opened and the sample bomb was controlled at 75 psia using the flow control valve. The evolved solvent vapor and some unreacted DTBP were condensed out in the weighed dropout pot and the product gases were collected in the one gallon container. The rates of pressure rise, temperature rise and mass vented are shown in Figure 15.

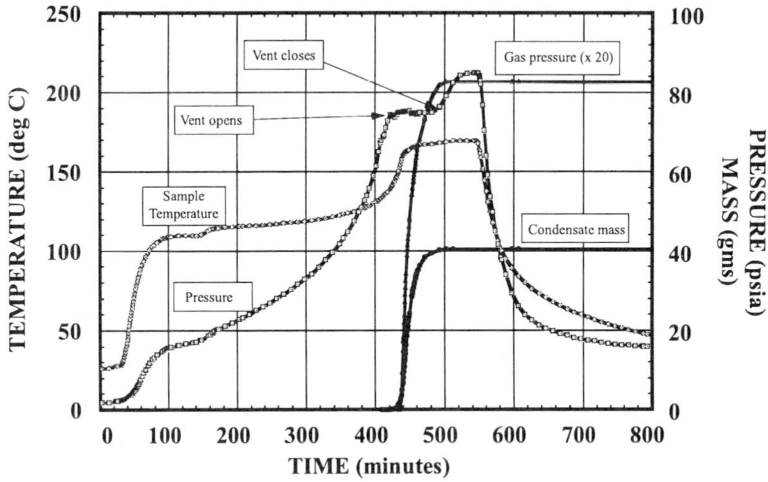

Figure 14. A controlled vent test of DTBP decomposition showing temperature and pressure in the reaction vessel and the mass of vented material and evolved gases in the vent collection vessels

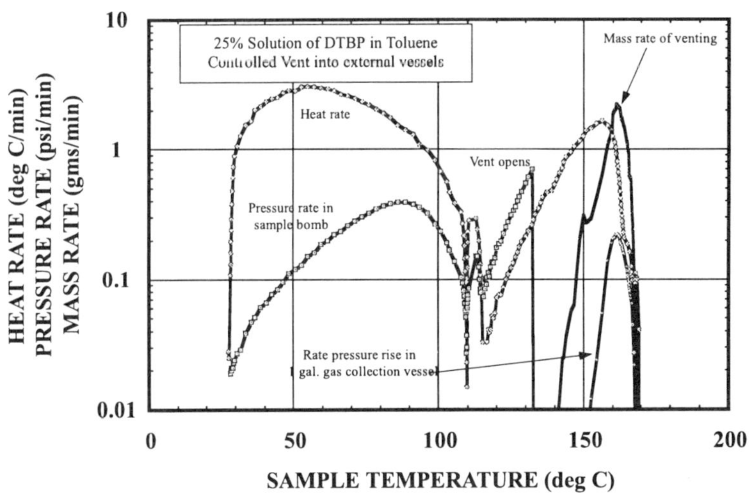

Figure 15. A controlled vent test of DTBP decomposition showing rates of temperature and pressure rise in the reaction vessel and the rate of evolution of condensable and noncondensable materials

Compared with the simple vent test, the maximum heat rate and temperature rise are smaller for the controlled vent test because the pressure in the reaction vessel was controlled at the set pressure throughout the venting period.

5. CONCLUSIONS

A fully automated, adiabatic calorimeter has been designed, built and operated. Special care has been taken in the design to ensure ease of use, maintainability and reliability. Essential requirements for a reaction calorimeter such as the ability to add reagents, stir the reactants and vent the reaction vessel are standard features of the APTAC. The instrument has the accuracy of the ARC™ but can follow much faster exotherms in very thin walled vessels, thus simulating the exotherm which would be observed in full scale equipment in the event of a runaway reaction. This gives the APTAC a powerful advantage over the ARC™ and reduces the possibility of misinterpretation of data by people inexperienced in the techniques of adiabatic calorimetry. In addition, the apparatus can simulate fire exposure to the contents of a reactor or storage vessel and is also capable of acquiring data from vented reactions for use in sizing safety relief devices using the DIERS methodology.

BIBLIOGRAPHY

1. Tou, J. C. and Whiting, L. F. The Thermokinetic Performance of an Accelerating Rate Calorimeter. *Thermochimica Acta*, 48 (1981), 21–42
2. Grolmes, M. A., Leung, J. G. and Fauske, H. K. Reactive Systems Vent Sizing. *Proceedings of the International Symposium on Runaway Reactions, Boston, MA, March 7–9, 1989.* New York: AIChE/CCPS, 1989.
3. Fauske H. K., Clare, G. H., and Creed, M. J. Laboratory Tool for Characterizing Chemical Systems. *Proceedings of the International Symposium on Runaway Reactions, Boston, MA, March 7–9, 1989.* New York: AIChE/ CCPS, 1989.
4. Leung J. C., Fauske H. K., and Fisher, H. G. Thermal Runaway Reactions in a Low Thermal Inertia Apparatus. Thermochimica Acta, 104 (1986), 13–29.
5. Vynn, W. T. "Adiabaticity of the ARC," ARCO Chemical Co. Internal Correspondence. Letter to the ARC Users Group. July 23, 1982.
6. Whiting, L. F., and Tou, J. C. Thermal Hazard Evaluation of Styrene Polymerization by Accelerating Rate Calorimetry. *J. Thermal Analysis*, 24 (1982), 111–132
7. Chippett S., Unpublished data.
8. Ahmed, M., Fisher, H. G., and Janeshek, A. M. Reaction Kinetics from Self-Heat Data—Correction for the Depletion of Sample. *Proceedings of the International Symposium on Runaway Reactions, Boston, MA, March 7–9, 1989.* New York: AIChE/CCPS, 1989.
9. Leung, J., Creed, M., and Fisher, H. G. "Round-Robin Vent Sizing Package Results." *Proceedings of the International Symposium on Runaway Reactions, Boston, MA, March 7–9, 1989.* New York: AIChE/CCPS, 1989.

Runaway Reaction Characterization: A Round-Robin Study on Three Additional Systems

Joseph C. Leung*
Leung Inc.
1521 Evergreen Lane
Darien, Illinois 60561

Harold G. Fisher
Union Carbide Corporation
P.O. Box 8361
South Charleston, WV 25303

1. INTRODUCTION

Runaway reaction characterization is a key step in the assessment of the thermal hazard and operational stability of a given chemical process. Such characterization has been increasingly carried out using DIERS and related apparatus. Round-robin testing was initiated about ten years ago as part of the DIERS Users Group Design/Testing committee activity. An earlier paper [1] discussed results from the first three systems: (1) styrene polymerization initiated with 0.5% wt. benzoyl peroxide, (2) 15% wt. di-tert-butyl peroxide decomposition, and (3) methanolysis of acetic anhydride. In this second paper, three additional systems are described,

(4) Hydrogen peroxide
(5) Benzoyl peroxide
(6) Tetraethylene glycolate

Most of the participants (see Acknowledgment section) conducted their tests in the ARC®, RSST, VSP, PHI-TEC, and APTAC. A brief description of thise equipment is in order here prior to presenting the results of our round-robin testing.

*Consultant, Fauske & Associates, Inc., 16W070 West 83rd Street, Burr Ridge, IL 60521.

2. TEST EQUIPMENT

2.1. ARC® (Accelerating Rate Calorimeter)

The ARC® was developed in the 1970s by Dow Chemical Company as an adiabatic calorimeter to measure self-heat rates and pressure rise rates during runaway conditions [2]. A standard ARC® run consists of approximately 5 g or 4 mL sample placed in a one-inch diameter metal sphere (bomb) and situated in a heated oven, the temperatures of which are controlled to follow the sample temperature, as measured by the thermocouple attached to the sample bomb (see Fig. 1). Typically a heat-wait-search mode of operation is used to detect an exotherm. The ARC® is run in a closed-system mode and hence the upper pressure limit is dictated by the strength of the bomb. Adiabatic conditions, considering the sample and the bomb together, are maintained for self-heat rates up to 10°C/min. ARC® claims to be able to a detect self-heating onset of 0.02°C/min.

2.2. VSP (Vent Sizing Package)

The VSP is the commercial version of the DIERS Bench-Scale Apparatus [3,4], which was developed at Fauske & Associates, Inc. under the sponsorship of AIChE's DIERS. The VSP and its latest version VSP2 employ a 120-mL low ϕ-factor (or thermal inertia) SS-304 or Hastelloy test cell—cylindrically

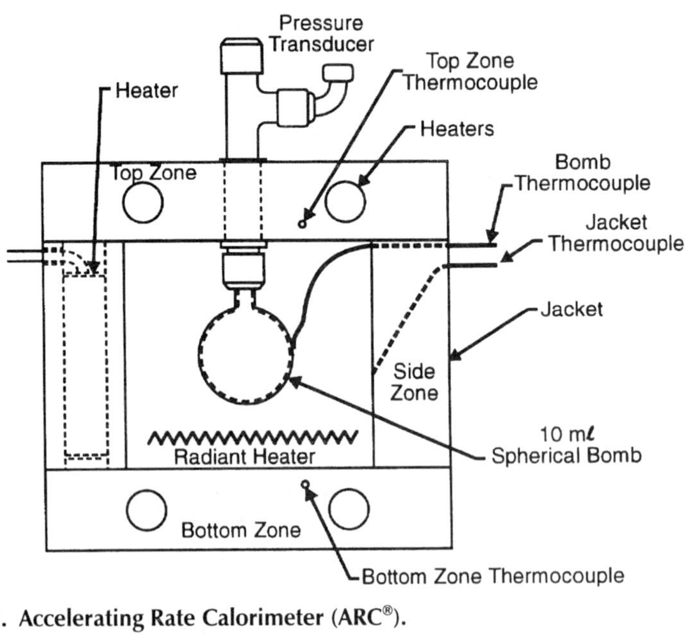

Figure 1. Accelerating Rate Calorimeter (ARC®).

shaped with a welded seam and lids—in a 4-liter high-pressure containment vessel, as shown in Fig. 2. The typical ϕ-factor is 1.05–1.08 for a test cell wall thickness of 0.005" to 0.007". The sample temperature was measured with a Type-K thermocouple T_1. The test cell is enclosed by two heater elements which are in turn enclosed by thermal insulation material. The purpose of the inner auxiliary heater is to heat the test sample to a desired temperature. During the wait, search and subsequent runaway period, the auxiliary heater remains off and the guard heater is controlled to keep an outer aluminum can at nearly the same temperature as T_1. For closed (nonvented) test cells, the pressure in the containment vessel P_2 is regulated to follow the test cell sample pressure P_1. This pressure tracking feature makes possible the utilization of the thin wall (low ϕ-factor) test cell design. Vented or open tests, where the vapor or gas generated is vented either into the containment vessel or to an external container, are unique capabilities of this apparatus. The typical onset sensitivity is 0.1°C/min for the VSP and 0.05°C/min for the VSP2.

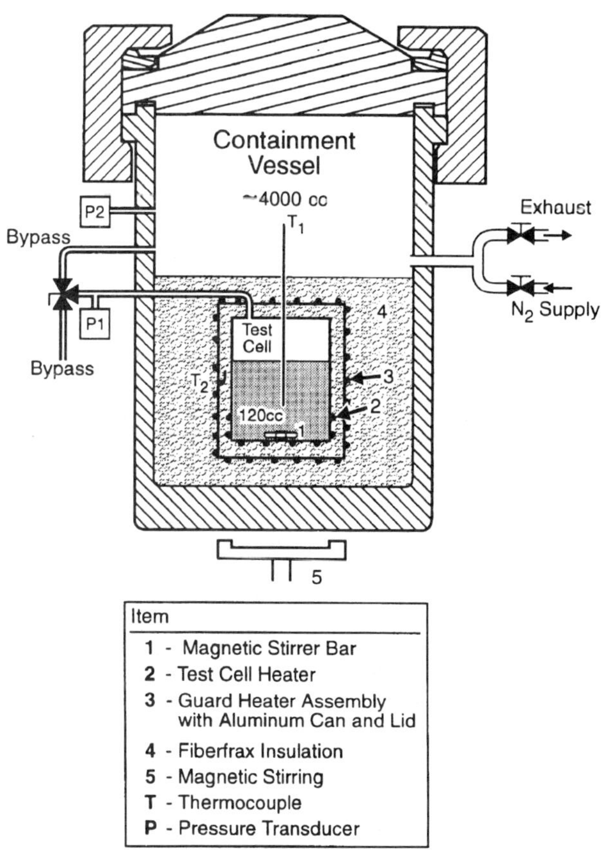

Figure 2. VSP Apparatus.

2.3. APTAC (Automated Pressure Tracking Adiabatic Calorimeter)

The APTAC is essentially an enhanced version of the DIERS Bench-Scale Apparatus developed at Union Carbide Corporation [5] and recently marketed by Arthur D. Little. The APTAC utilizes a 130-mL spherical test cell with a 0.020" wall thickness situated inside a 4-liter pressure vessel as illustrated in Fig. 3. The typical ϕ-factor is 1.15 (which is slightly higher than the VSP due to the thicker wall). Instead of using porous insulation—which is disposable—as in the VSP, the APTAC uses expensive solid ceramic insulation. The APTAC employs a high-power heater assembly which serves to heat the sample as well as provide the temperature control of the chamber surrounding the test cell. A continuous nitrogen supply is used for pressure balancing. The APTAC has an onset sensitivity of 0.05°C/min.

2.4. PHI-TEC

The PHI-TEC is essentially identical to the VSP and is marketed by Hazard Evaluation Laboratories, Ltd. [6]. It uses similar cylindrical test cell of 120 mL

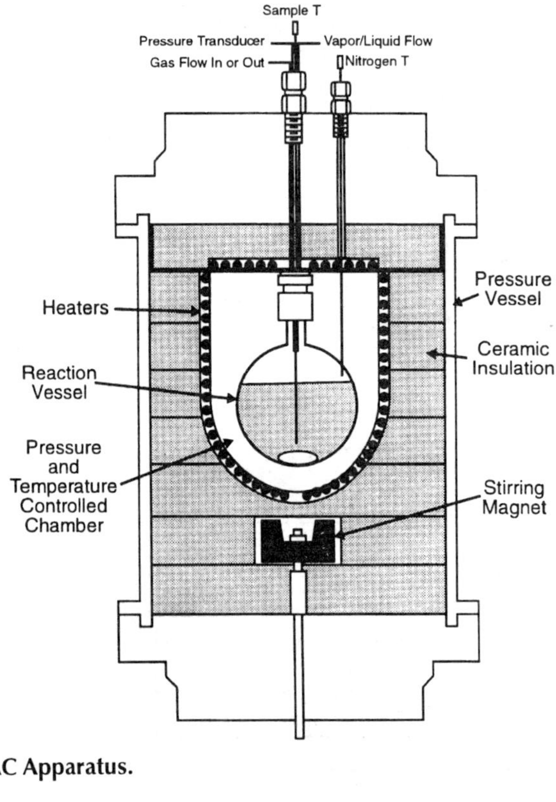

Figure 3. APTAC Apparatus.

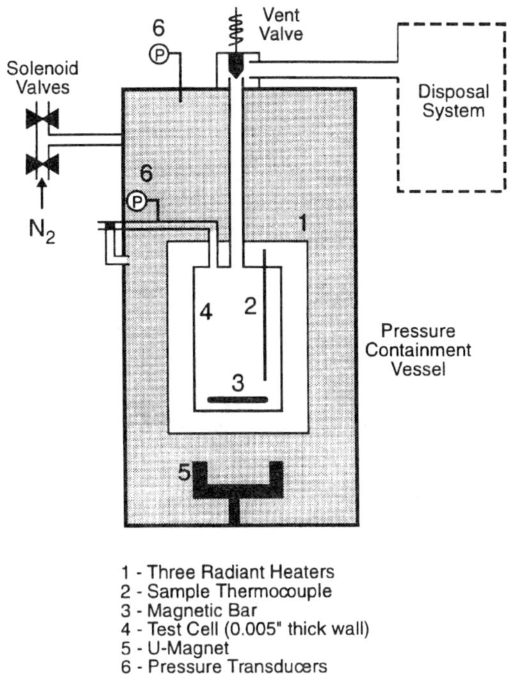

Figure 4. PHI-TEC Apparatus.

size and a 4-liter pressure vessel for containment (see Fig. 4). PHI-TEC claims to have a rather elaborate heater control and can achieve an onset sensitivity of 0.02°C/min.

2.5. RSST (Reactive System Screening Tool)

The RSST was developed by Fauske & Associates, Inc. as an inexpensive screening tool [7]. It lacks the temperature balancing technique around the test cell to be called a truly adiabatic calorimeter. However, it utilizes the ramp rate feature (similar to the DSC) where the sample is subjected to a prescribed heating rate of 0.25–2°C/min. The basic features of the RSST are illustrated in Fig. 5. This apparatus measures the sample temperature but not the sample pressure since the glass test cell is always vented to the containment vessel. The test cell is equipped with either an immersion heater or an external bottom heater which is controlled by feedback from the sample temperature measurement to assure enough power to overcome heat losses and produce a specified constant ramp rate. In case of an exotherm onset, this same ramp rate will be added to the reaction heat release.

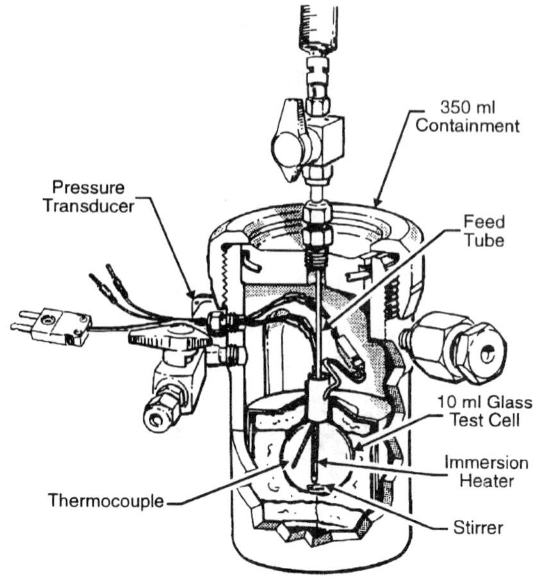

Figure 5. RSST Apparatus.

3. ROUND-ROBIN TEST DATA

3.1 Hydrogen Peroxide (H_2O_2)—4th Round-Robin System

It is well-known that H_2O_2 decomposition is susceptible to minute amounts of contaminant and wall effects of the sample container. To overcome such difficulties, an iron contaminant was purposely introduced to a 30% wt. H_2O_2 solution. In this particular round-robin testing campaign, FMC kindly provided both solutions, thus further eliminating any sample variation among suppliers:

30% wt. H_2O_2 solution (stabilized)

0.0215% wt. $Fe_2(SO_4)_3$ solution

A special test cell preparation was given to participants as a further precaution. Silver solder was not to be used in VSP type test cells. Both the test cell and thermocouple were to be oxidized with 6% nitric acid solution for two hours. Then they were rinsed with deionized water and filled with 30% H_20_2 solution for at least two days. This passivation solution was removed just prior to the test.

The testing instruction called for mixing five parts of 30% wt. H_2O_2 solution to one part Fe solution in an ice bath and heating the sample up to 50°C in the passivated test cell and allowing it to self heat or runaway. Initially a few exploratory tests were conducted in the RSST apparatus using 5:1 and 3:1 (H_2O_2 Fe solution) mix ratios. The resulting temperature rise rate data shown in Fig. 6 demonstrate identical behavior regardless of the mix ratio, implying that the reaction may follow the rate law,

$$-\frac{d[H_2O_2]}{dt} = k_1[H_2O_2][Fe] \tag{1}$$

i.e. the reaction rate is first order with respect to both H_2O_2 and Fe level.

The submitted round-robin data—VSP, APTAC, and RSST data—were compared in terms of the self-heat rate at a temperature of 100°C as shown in Fig. 7. In these tests, SS-304 test cells were used in the VSP while glass test cells were utilized in the APTAC and RSST. For the *38 data sets*, we obtained

$$\text{mean } dT/dt = 17.2°\text{C/min}$$

$$\sigma = \pm 7.1°\text{C/min}$$

and about eight data sets fell outside one standard deviation. By eliminating these 8 data sets and performing the statistical analysis on the remaining *30 data sets*, we obtained

$$\text{mean } dT/dt = 14.9°\text{C/min}$$

$$\sigma = \pm 3.1°\text{C/min}$$

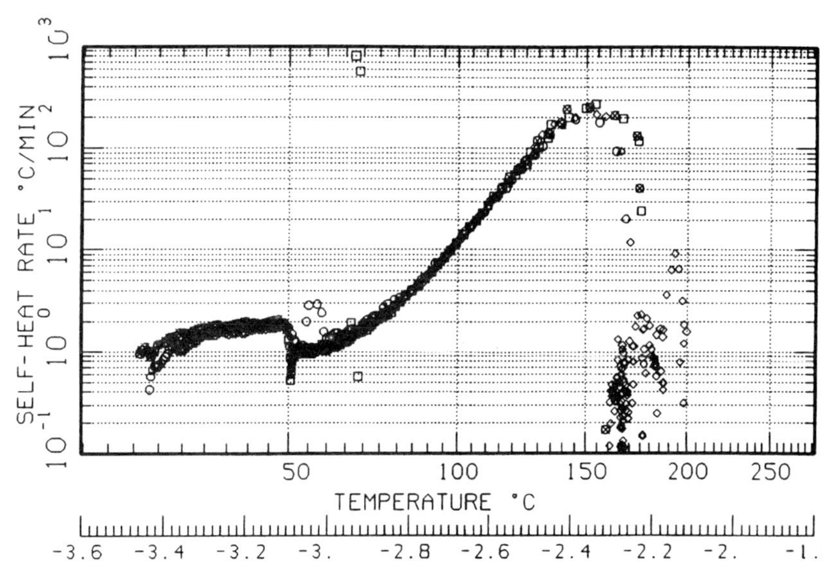

Figure 6. Four RSST runs with 5:1 and 3:1 H_2O_2 to Fe Solution Mix Ratios (100 psig Imposed Pressure).

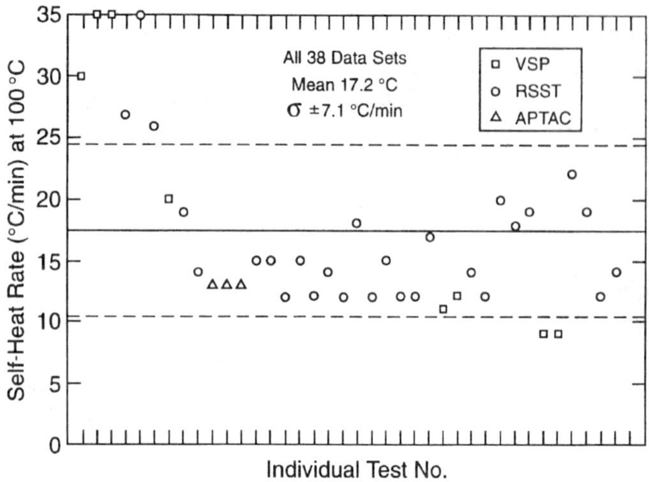

Figure 7. Statistical Analysis of *dT/dt* Data at 100°C for 38 Data Sets.

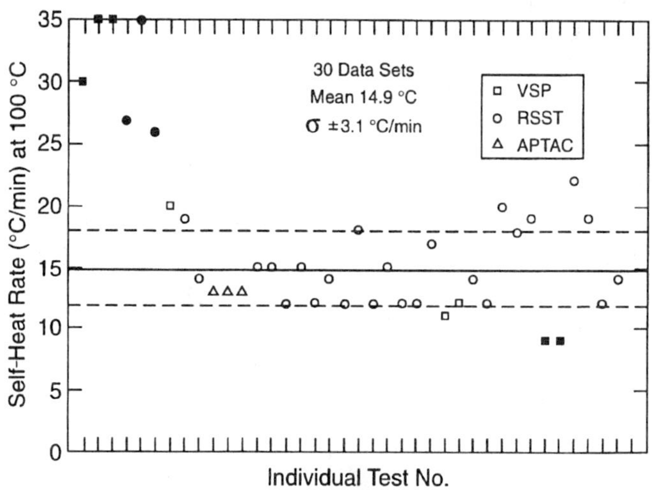

Figure 8. Statistical Analysis of *dT/dt* Data at 100°C for 30 Selected Data Sets.

and this time six data sets fell outside one standard deviation. As shown in Fig. 8, these six outlying data points are not all that far from the one standard deviation. Carrying out the analysis one last time on the 30 data sets, we arrived at the final culled data set which yields

$$\text{mean } dT/dt = 13.3°\text{C/min}$$

$$\sigma = \pm 1.6°\text{C/min}$$

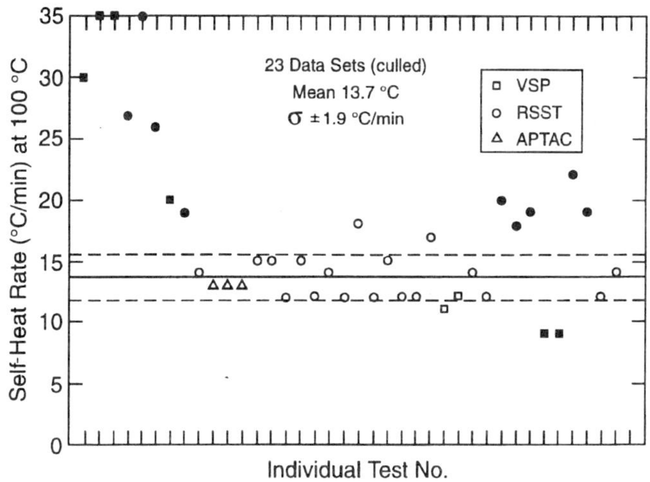

Figure 9. Statistical Analysis of *dT/dt* Data at 100°C for 23 Selected Data Sets.

This is shown in Fig. 9 where 18 sets of data fall within the one standard deviation.

Figures 10 and 11 illustrate the self-heat rate data for the culled set—consisting of VSP, APTAC, and RSST data—at the high and low back pressures, respectively. Since H_2O_2 is considered a hybrid reactive system with gas (O_2) generation and vapor (H_2O and H_2O_2) formation via latent heat, the lower

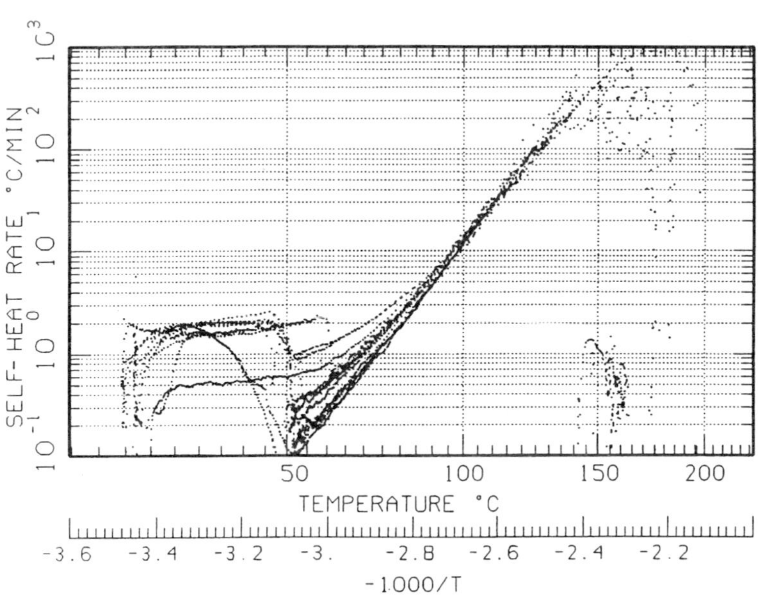

Figure 10. Self Heat Rate Data (VSP, APTAC, RSST) for H_2O_2 System in High Back Pressure (Initially 100 psig) Tests.

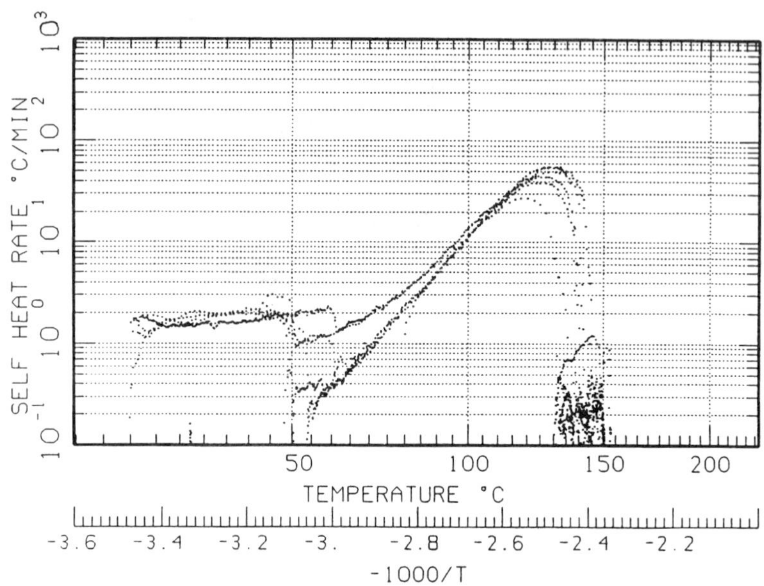

Figure 11. Self Heat Rate Data (VSP, APTAC, RSST) for H_2O_2 System in Low Back Pressure (Initially 5 psig) Tests.

back pressure runs resulted in significantly lower peak dT/dt and peak temperatures. This was due to the tempering (or evaporative cooling) effect.

Figure 12 shows the first-order rate constant (k_1) for the VSP and APTAC data where k_1 is calculated from

$$k_1 = \frac{dT/dt}{T_m - T} \tag{2}$$

Here T_m is a calculated peak temperature based on a heat of reaction of 666 cal/g for H_2O_2. For uncontaminated H_2O_2, FMC provided a k_1 equation given by

$$k_1 = 5.94 \times 10^7 \exp\left(-\frac{12{,}200}{T}\right) \tag{3}$$

Our round-robin data can be reasonably fitted by a similar expression except the preexponential frequency factor has to be increased by 5000 times, i.e.

$$k_1 = 2.97 \times 10^{11} \exp\left(-\frac{12{,}200}{T}\right) \tag{4}$$

The best-fit of our data, however, can be given by the equation

$$k_1 = 9.32 \times 10^{12} \exp\left(-\frac{13{,}400}{T}\right) \tag{5}$$

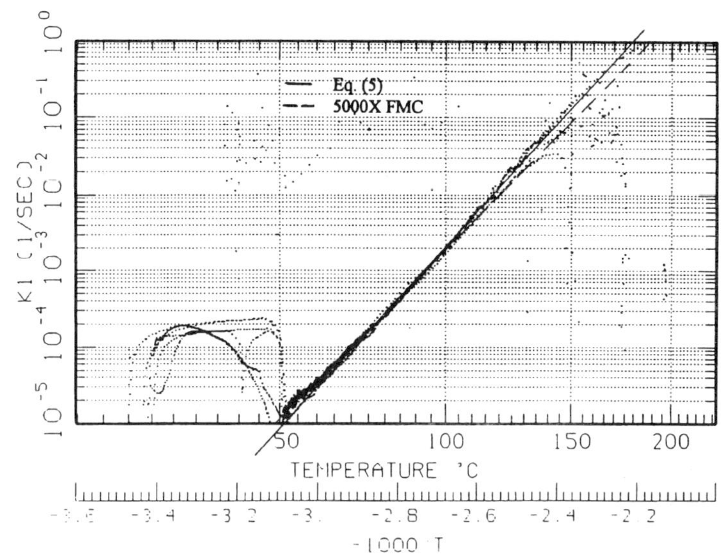

Figure 12. k_1 Rate Constant Data from VSP and APTAC for H_2O_2 System.

which gives an activation energy of 26.6 kcal/mole, slightly higher than 24.3 kcal/mole as implied by the FMC formula. Figure 13 shows that Eq. (5) can correlate the RSST data reasonably well also. Note that the positive deviation of the data below 80°C from Eq. (5) was due to external imposed heating in the RSST equipment. Thus our results indicate that a 10 ppm level of Fe con-

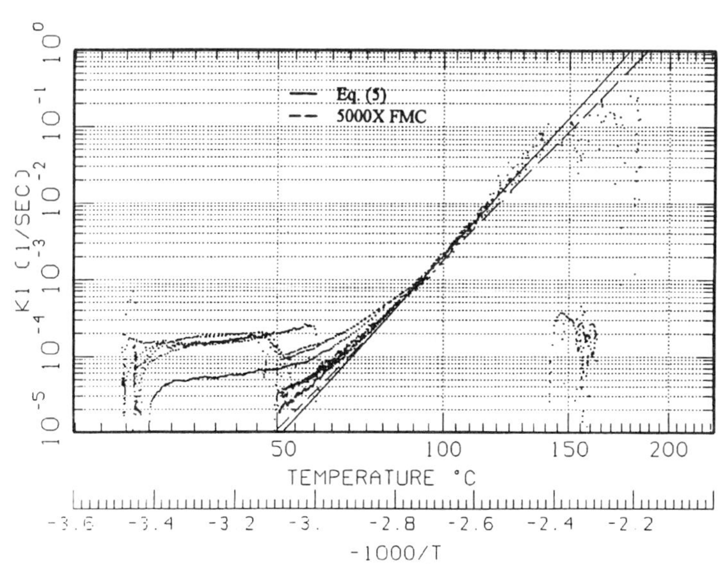

Figure 13. k_1 Rate Constant Data from RSST for H_2O_2 System.

tamination can yield a 5000 times increase in H_20_2 decomposition rate and at such a level the wall effect—ranging from glass to SS-304 to Hastelloy—seems to have been overwhelmed by the particular Fe contamination.

3.2. Benzoyl Peroxide (BPO)—5th Round-Robin System

BPO is a common free-radical initiator used in many polymerization processes. Its half-life data are well known [8]. In this round-robin testing, AZTEC Catalyst Company kindly provided a 10-lb sample of 22% wt. BPO (22.2% by gravimetric titration, 22.9% by HPLC method) in a high boiling-point solvent (dicyclohexyl phthalate). This phlegmalized BPO sample came in a powder form which had a melting point of around 55°C. The testing instruction called for heating the sample (5 g for ARC®, 7.5 g for RSST, and 50g for VSP) to 55°C before allowing it to self-heat.

Ordinarily, BPO decomposition follows the stoichiometry

$$(C_6H_5COO)_2 \rightarrow C_6H_5C_6H_5 + 2\ CO_2$$

and using heats of formation data, we estimate the heat of decomposition for BPO to be

$$\begin{aligned}\Delta H_r &= -301.6\ \text{kJ/mole BPO(s)}\\ &= -298\ \text{cal/g BPO(s)}\\ &= -306\ \text{cal/g BPO(l)}\end{aligned}$$

Here (s) and (l) denote solid and liquid phases. For the last value, a heat of fusion of 2 kcal/mole has been assumed for BPO. The ΔH_r values as obtained from four different DSC apparatus are shown in Table 1. The rather large standard deviation (±18%) may be attributed to different equipment, sample size (1–10 mg) and ramp rate (1–10°C/min). The ΔH_r values as calculated from ΔT_{ATR} (adiabatic temperature rise) data obtained from the ARC®, RSST, and VSP, i.e.

$$\Delta H_r = \frac{C_p \phi \Delta T_{ATR}}{x_{BPO}} \tag{6}$$

are summarized in Table 1 also. Here an average specific heat value of 0.5 cal/g°C has been assumed and the BPO mass fraction x_{BPO} is 0.22. These ΔH_r data are regarded as satisfactory in view of the uncertainty in the C_p value. The standard deviations for all three calorimeters are within 10% of the mean value.

The Mettler DSC also gave an analysis of the rate data. The resulting first-order rate-constant (k_1) expression is shown in Fig. 14. Comparing with the Lucidol line based on BPO half-life data, the DSC yielded a lower activation energy (E_A), of 26 kcal/mole versus 31.9 kcal/mole from Lucidol.

TABLE 1
Heat of Reaction Values from BPO Decomposition

Apparatus	No. of Data Set	Sample Wt.	ϕ Factor	T_{on}	ΔT_{ATR}	ΔH_r (cal/g BPO)
DSC[(a)]	5	1–10 mg	—	78–92°C	—	355 ± 55
ARC®	7	5 g	1.45–2.0	58–66°C	74–91°C	309 ± 20
VSP	8	50 g	1.12–1.15	65–84°C	123–134°C	343 ± 24
RSST	21	7.5 g	1.0 (approx.)	50–63°C	105–162°C	323 ± 28

(a)—Consisting of one HPDSC, two Mettler type, and two DuPont DSC.

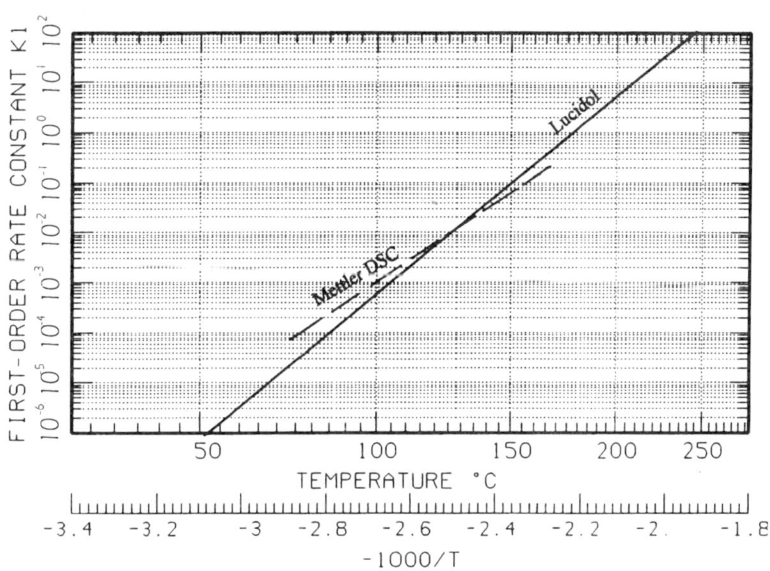

Figure 14. BPO Decomposition Rate Constant from Mettler DSC versus Lucidol Half-Life Data.

Figure 15 is a composite plot of the *dT/dt* data from all seven submitted ARC®runs. One data set was obviously an outlier as the self heat data were nearly an order of magnitude higher than the rest. Figure 16 is a composite plot of the *dT/dt* data from all seven submitted VSP runs. Note that the peak rates were in excess of 1000°C/min. Figure 17 is a composite plot of the *dT/dt* data from the 19 RSST runs. In general these peak rates were in good agreement with the VSP data, both being taken in low thermal-inertia apparatus. The *dT/dt* data below 100°C in the RSST runs were influenced somewhat by the externally applied heating rate which was at about 0.5°C/min.

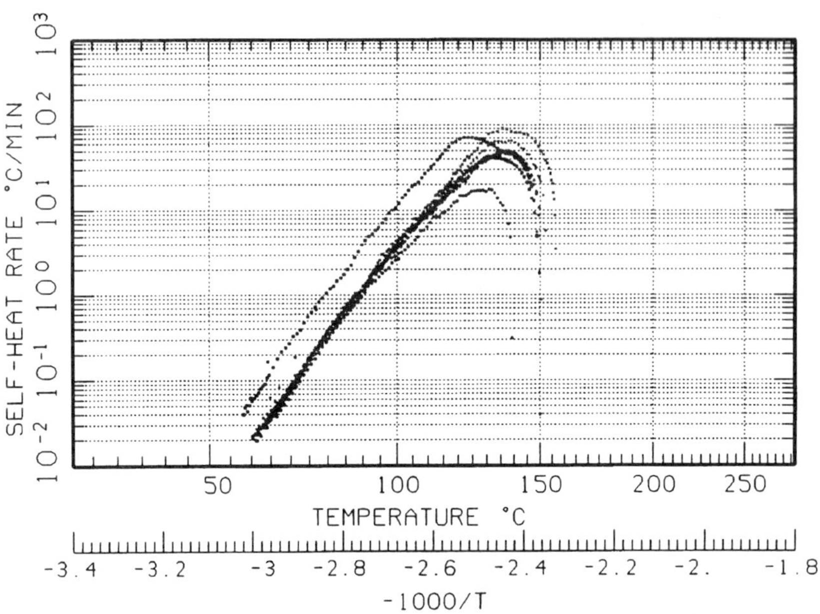

Figure 15. Self Heat Rate Data for BPO Decomposition from Seven ARC® Runs.

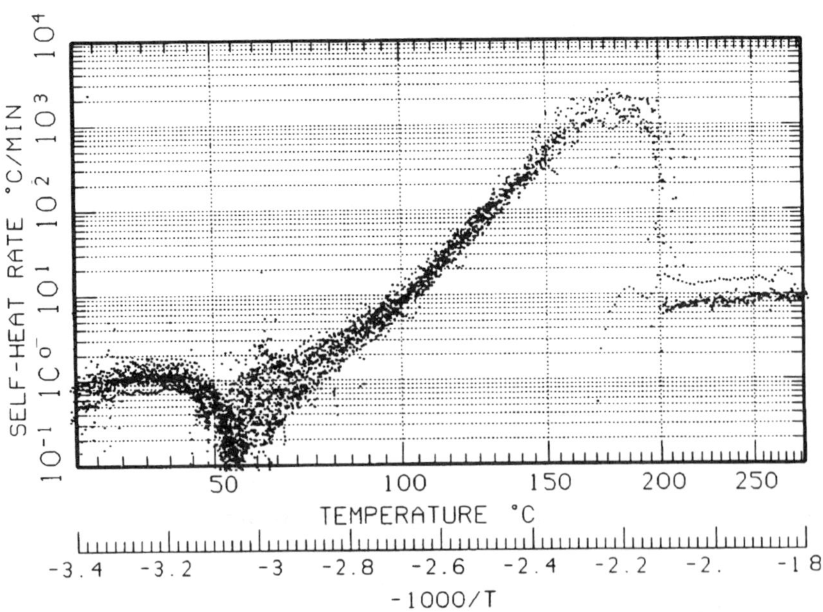

Figure 16. Self-Heat Rate Data for BPO Decomposition from Seven VSP Runs.

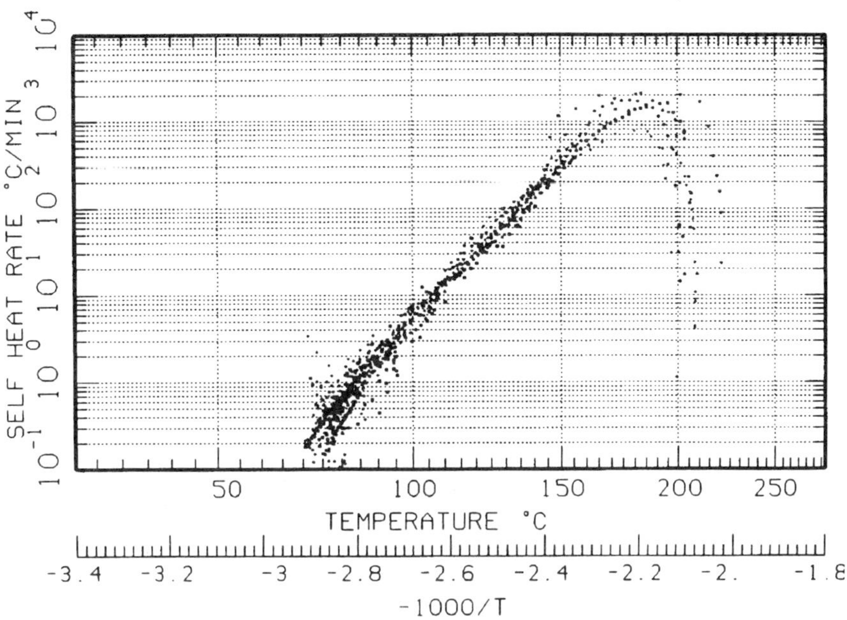

Figure 17. Temperature Rise Rate Data for BPO Decomposition from 19 RSST Runs.

Next we examine the k_1 rate constant as calculated from Eq. (2) based on the experimental dT/dt data and the observed peak temperature T_m. Figure 18 compares the ARC® k_1 data against the Lucidol curve. The outlying set naturally exhibits significantly higher k_1 values. The remaining six runs were very repeatable but their data lie significantly above the Lucidol line except near the onset temperature (T_{on}). The disparity was later found to be caused by two factors [9]. The first was due to lack of stirring in these ARC® runs. The second was due to the placement of the bomb thermocouple at the "equator" (see Fig. 1) when the bomb was less than 50% full. As a result, a significant time lag in thermocouple response to the change in sample temperature could occur. It is therefore recommended that the bomb thermocouple should be placed near the bottom which is in closer proximity with the sample [10]. An alternative would be to employ an internal thermocouple which is in direct contact with the sample inside the bomb [11]. Figure 19 demonstrates that the use of a stirred ARC® bomb and a thermocouple in direct contact with the sample for temperature measurement yields almost perfect agreement with the Lucidol line [9].

Figure 20 shows the VSP k_1 data in comparison with the Lucidol line; the agreement is quite satisfactory. The data scatter near the peak temperature may in part be due to the numerical difficulty of dividing dT/dt by a vanishingly small temperature difference as indicated by Eq. (2).

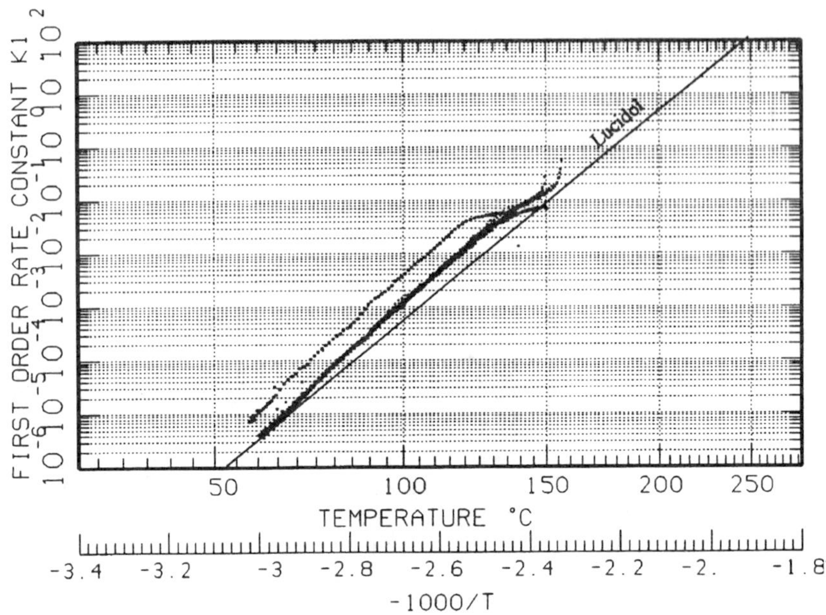

Figure 18. k_1 Rate Constant Data for BPO Decomposition from Seven ARC® Runs.

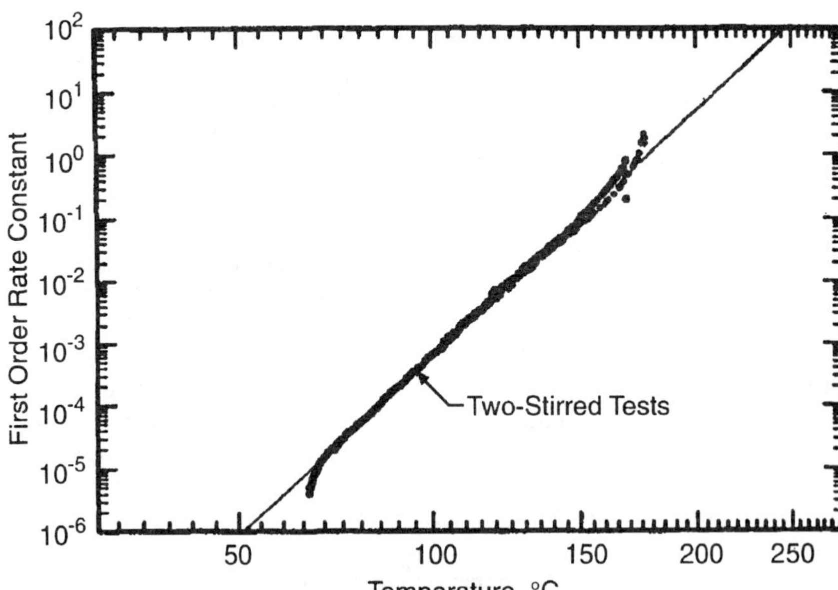

Figure 19. k_1 Rate Constant Data for BPO Decomposition from Two Stirred ARC® Runs Using Internal Thermocouple for Sample Temperature Measurement.

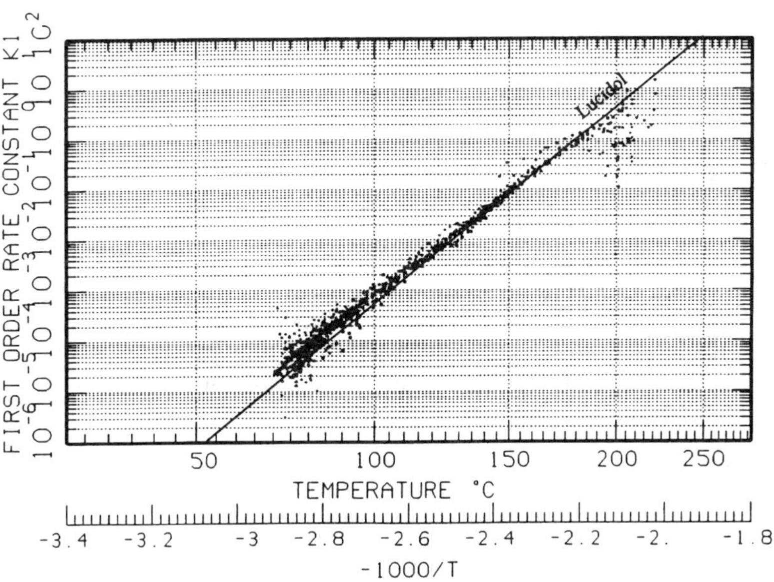

Figure 20. k_1 Rate Constant Data for BPO Decomposition from Seven VSP Runs.

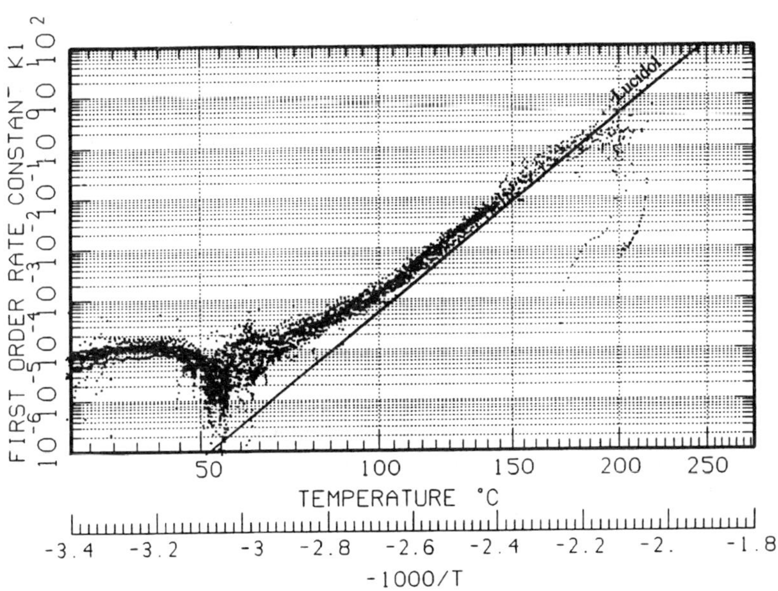

Figure 21. k_1 Rate Constant Data for BPO Decomposition from 19 RSST Runs.

Figure 21 illustrates the RSST k_1 data in comparison with the Lucidol line. The data below 50°C are not relevant since they were associated with the sample heatup procedure. In general the RSST data show a positive deviation from the Lucidol line. This result is believed to be caused by the externally supplied heating rate which became less significant in comparison with the self heating rate at higher temperature. The data tend to exhibit better agreement at the higher temperature.

3.3. Tetraethylene Glycolate—6th Round-Robin System

This round-robin system was suggested by Union Carbide Corporation. Polyethylene glycols form sodium salts when combined with sodium metal under the proper conditions. In general, neat glycols are thermally stable up to approximately 300°C in an oxygen free environment. Formation of the sodium salt, or glycolate, reduces the thermal stability. The decomposition of either species is exothermic and produces a variety of volatile and noncondensable gases:

Glycol Decomposition:

$$\text{ROH} \rightarrow \text{Res} + \text{Volatile} + \text{Gas}$$
(multiple reactions yielding one or more peaks)

Glycolate Decomposition:

$$\text{RONa} \rightarrow \text{ResNa} + \text{Volatile} + \text{Gas}$$
(one or two exothermic peaks expected)

where:

$$\text{ROH}=\text{HO–C}_2\text{H}_2\text{–O–C}_2\text{H}_2\text{–O–C}_2\text{H}_2\text{–O–C}_2\text{H}_2\text{–OH}$$
(tetraethylene glycol)

$$\text{RO-Na}^+=\text{HO–C}_2\text{H}_2\text{–O–C}_2\text{H}_2\text{–O–C}_2\text{H}_2\text{–O–C}_2\text{H}_2\text{–O–Na}^+$$
(sodium glycolate)

Res = Residue of Nonvolatile Species

ResNa = Sodium Residues of Nonvolatile Species

In this round-robin testing, Union Carbide Corporation kindly provided a 50 mole percent mixture of tetraethylene glycol and its sodium salt to all participants with specific instructions for the various equipment:

	ARC	RSST	VSP/PHI-TEC
Sample	2.2 g	10 g	80 g
Test Cell	Titanium	Glass	Type II 1/8" vent
Back Pressure	—	300 psi	300 psi
Initial Heatup	2°C/min	1°C/min	2°C/min
Heatup Cutoff	150°C	400°C	200°C

TABLE 2
Summary of Tetraethylene Glycolate Data from DSC and C-80

Apparatus	User Code	Sample	Ramp Rate (°C/min)	Onset T_{on} (°C)	Max. T_m (°C)	ΔH_r (J/g)
SETARAM DSC	1	8.1 mg	5	241	297	205.7
SETARAM DSC	1	8.1 mg	5	238	297	230.6
SETARAM DSC	2	18.8 mg	5	215	288	489.0
Perkin Elmer DSC	3	4.6 mg	10	264[(a)]	305	144.7
DSC-15	4	9.1 mg	10	235	304	255.1
DSC-15	4	5.8 mg	10	235	302	209.6
TA-DSC	5	0.82 mg	10	225	271	70.2
DSC[(b)]	6	1.03 mg	10	230	296	314.5
C-80	7	19.7 mg	0.25	195	260	615

(a)—Onset temperature not reported, shown is intercept value.
(b)—DSC manufacturer not known.

In addition to these runaway calorimeters, a number of DSC, C-80 and RADEX users also participated. The DSC and the C-80 data summary are shown in Table 2. Note that the DSC yielded significantly higher onset temperatures than C-80. The DSC data exhibited quite a bit of scatter in the ΔH_r value, 240 J/g solution with a standard deviation of 124 J/g. On the other hand, the C-80 gave a much higher ΔH_r value of 615 J/g. The three tests submitted by the RADEX—a differential thermal analyzer with pressure measurement—are summarized in Table 3. The onset temperatures were in good

TABLE 3
RADEX Data Summary for Tetraethylene Glycolate Sample

Test No. - Company ID	Sample Weight Pre/Post-test (g)	Exotherm		Pressure			Maximum	
		Onset (°C)	Max. T (°C)	Initial (psig)	Max. (psig)	Final (psig)	dT/dt (°C/min)	dP/dt (psi/min)
RAD01-2	1.36 / 1.19	228	308	0	450	56	0.45	33
RAD02-11	1.25 / 0.86	224	340	9	555	11	0.5	20
RAD03-11	1.24 / 0.96	224	340	9	574	11	0.49	21

TABLE 4
ARC® Data Summary for Tetraethylene Glycolate Sample

Test ID	T_{on} (°C)	T_{max} (°C)	@250°C		@350°C		Max. Rates	
			dT/dt (°C/min)	dP/dt (psi/min)	dT/dt (°C/min)	dP/dt (psi/min)	dT/dt (°C/min)	dP/dt (psi/min)
ARC01	212	400	0.8	1	0.7	N	3.0	30
ARC02	180	430	0.75	3.3	0.65	13	1.7	28
ARC03	190	430	0.7	3.3	0.65	14	1.7	29
ARC04	190	435	0.75	3.1	0.60	12	1.6	21
ARC05	198	400	0.6	3.7	0.52	14	0.95	27
ARC06	207	370	0.08	0.45	0.11	3.5	0.11	3.7
ARC07	190	362	0.3	1.8	N	N	0.3	2.3
ARC08	198	370	0.45	0.07	0.17	5.2	0.55	5.2
ARC09	208	340	0.40	2.4	—	—	0.42	3
ARC10	193	400	0.53	3.4	0.37	12	0.75	16
ARC11	189	395	0.6	3.7	0.5	10	0.7	15
ARC12[(a)]	221	484	0.4	2.0	0.2	4.7	0.6	4.7

N = noisy data.
(a)—PHI-TEC ARC configuration.

agreement with the DSC data in Table 2. The peak temperatures, however, were higher.

The ARC® runs were all conducted with 2.2 g sample (±10%). Table 4 summarizes these data. Figure 22 illustrates two distinct exotherms and the wide scatter in terms of self-heat rate. The ϕ-factor in these runs was about 1.7.

The VSP, APTAC and PHI-TEC data are summarized in Table 5. Note that some runs were conducted in closed test cells. The EAC (Enhanced Adiabatic Calorimeter) [12] is a hybrid version between the VSP and APTAC. The onset temperature varies from 180–220°C, which is about the same range as in the ARC®. Test 9 was run in the APTAC initially in closed system mode until the sample pressure reached 300 psia when the test cell was vented to an external catch vessel under a constant back pressure of 300 psia. In general there was good agreement among these tests with the exception of Test 10. This test, conducted in the PHI-TEC, gave an order of magnitude higher throughput rate.

To provide further analyses of these data, the first major exotherm (i.e. 200–360°C) was examined in terms of the Arrhenius rate parameters. These data plotted in terms of k_1 are shown in Figs. 23, 24, 25, and 26 for the ARC®,

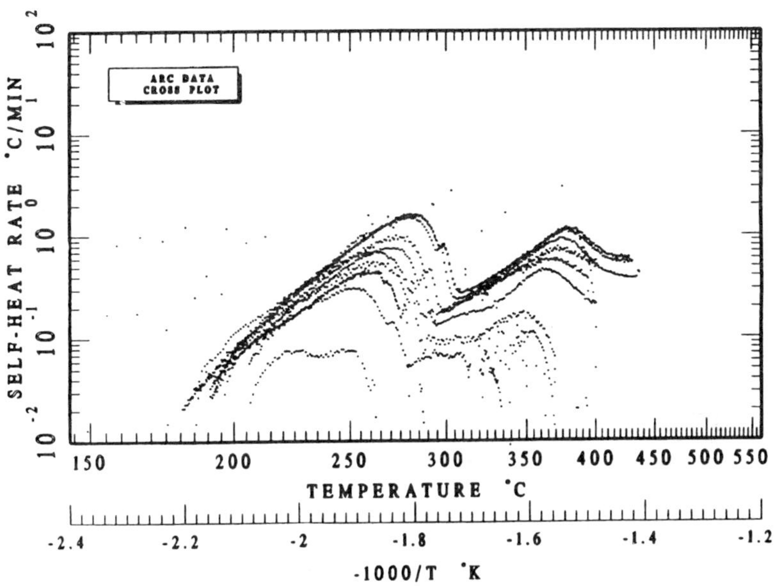

Figure 22 Self-Heat Rate Data for Tetraethylene Glycolate System from 11 ARC® Runs.

TABLE 5
VSP, APTAC, PHI-TEC Data Summary for Tetraethylene Glycolate Sample

Test ID	Test Apparatus	Test Cell Type	Sample Wt. (g)	Exotherm Onset (°C)	Exotherm Max. (°C)	@250°C *dT/dt* (°C/min)	@250°C *dP/dt* (psi/min)	@250°C *dT/dt* (°C/min)	@250°C *dP/dt* (psi/min)	Max. Rates *dT/dt* (°C/min)	Max. Rates *dP/dt* (psi/min)
1	VSP	O	82	200	398	7	N	50	3.5	150	N
2	VSP	C	25.9	—	390	4	15*	55	400*	100	400*
3	VSP	O	80.2	220	430	3.5	0.6	28	17	150	60
4	VSP	O	94.5	215	425	3.5	0.7	28	15	130	45
5	VSP	O	80.1	210	424	3.2	0.8	26	11	100	35
6	VSP	O	80.8	220	425	3.5	0.45	18	6	65	20
7	EAC	O	70.0	195	407	1.9	0.8	18	9	50	37
8	APTAC	C	50.4	190	472	2	13*	22	200*	80	650*
9	APTAC	C/O	49.2	180	450	2.7	13*	22	N	50	70
10	PHI-TEC	O	81.4	200	493	30	10	130	40	1030	4120

O = open test cell.
C = closed test cell.
* = in closed test cell.
N = noisy data.

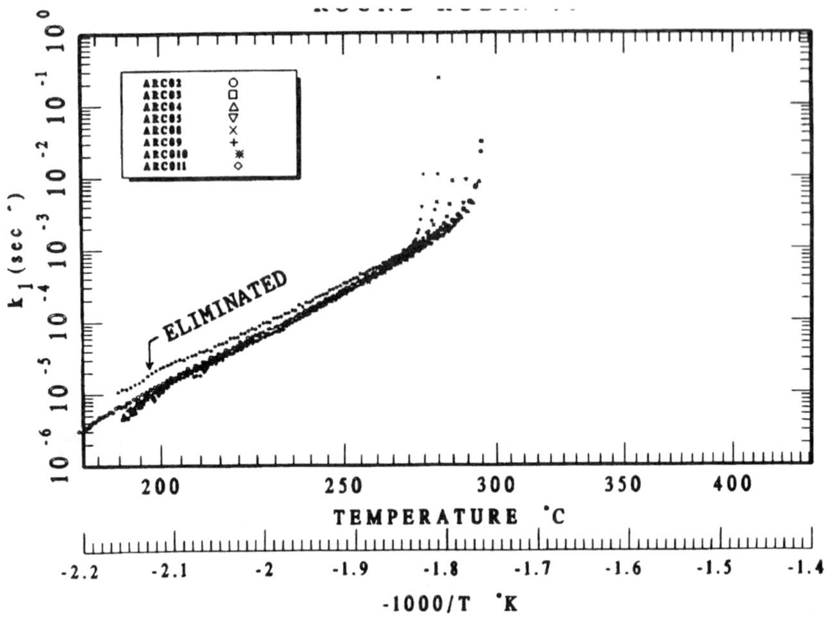

Figure 23. k_1 Rate Constant Data for Tetraethylene Glycolate System from 8 ARC Runs.

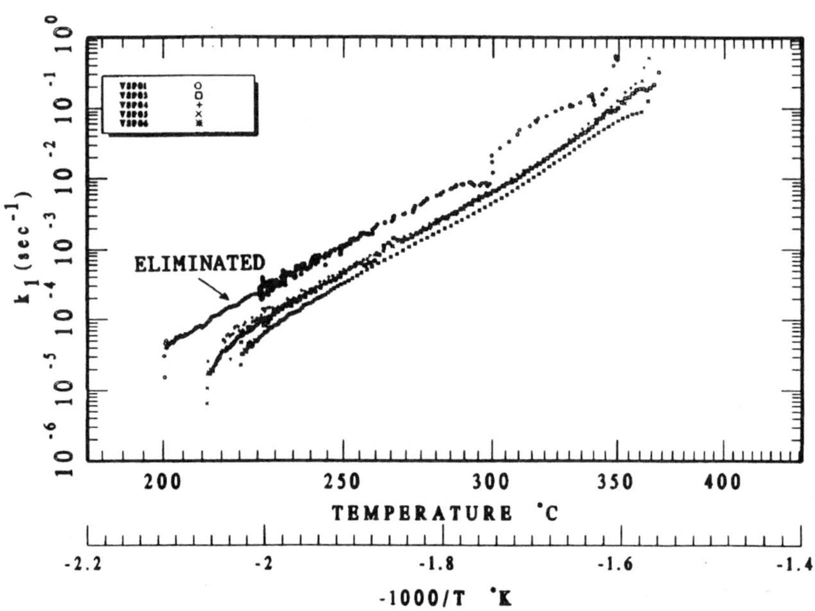

Figure 24. k_1 Rate Constant Data for Tetraethylene Glycolate System from 5 VSP Runs.

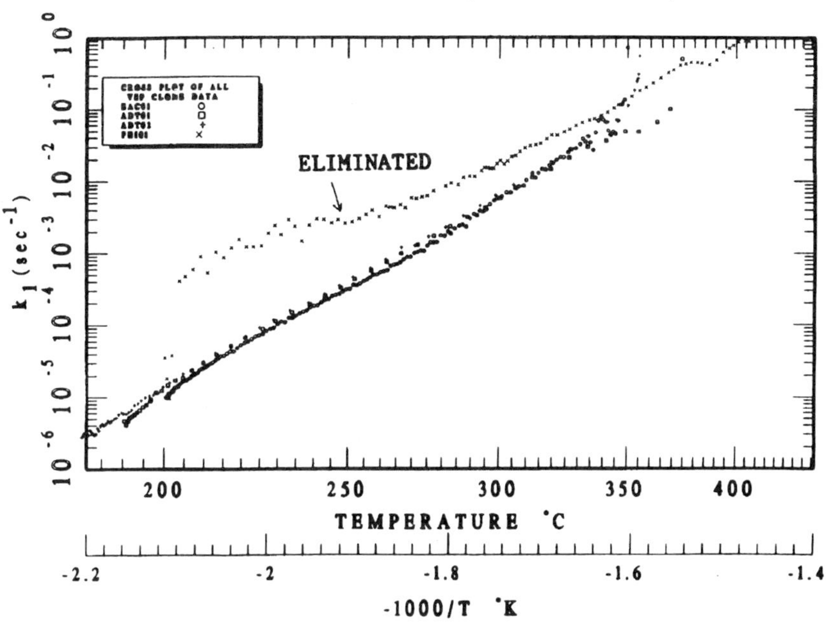

Figure 25. k_1 Rate Constant Data for Tetraethylene Glycolate System from one EAC, two APTAC, and one PHI-TEC.

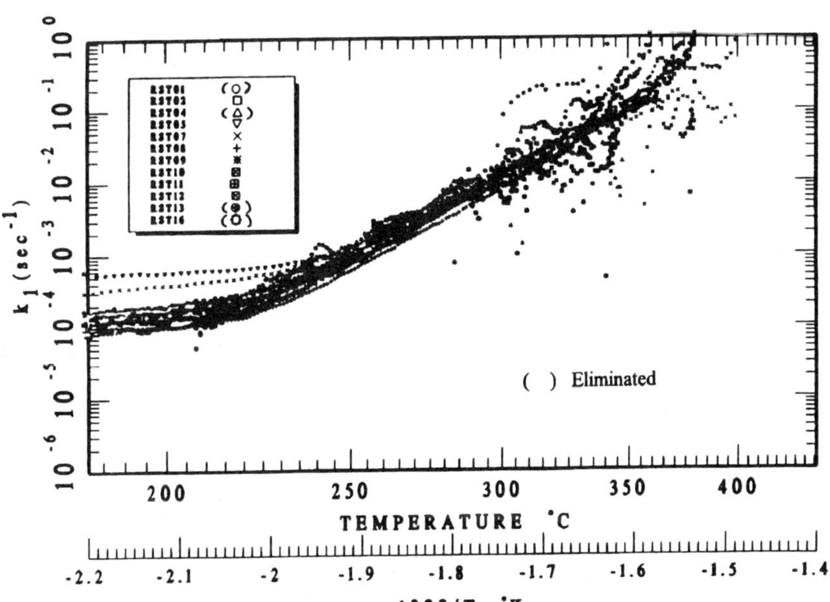

Figure 26. k_1 Rate Constant Data for Tetraethylene Glycolate System from 12 RSST Runs.

VSP, APTAC/PHI-TEC, and RSST data, respectively. The outliers were eliminated and the culled data sets were regressed with a linear fit. The results are shown in Table 6 in terms of the preexponential frequency factor and the activation energy. The RSST data gave a slightly lower activation energy because of the influence of heater power early on in the exotherm. The curve fits for the culled data sets are compared in Fig. 27 and show the excellent agreement among the VSP, APTAC, and EAC results.

TABLE 6
Arrhenius Kinetic Parameters for Tetraethylene Glycolate First Exotherm (200–360°C)

Data Set	Excluded Data	Preexponential Frequency Factor (sec^{-1})	Activation Energy E_A (kcal/mole)
ARC	#1, #6, #7, #11, #12	1.181×10^9	30.3
VSP	#1, #2	2.897×10^{10}	33.1
APTAC/EAC/PHI-TEC	#10 (PHI-TEC)	2.964×10^{10}	33.3
RSST	8 out of 16 sets	1.789×10^9	29.3

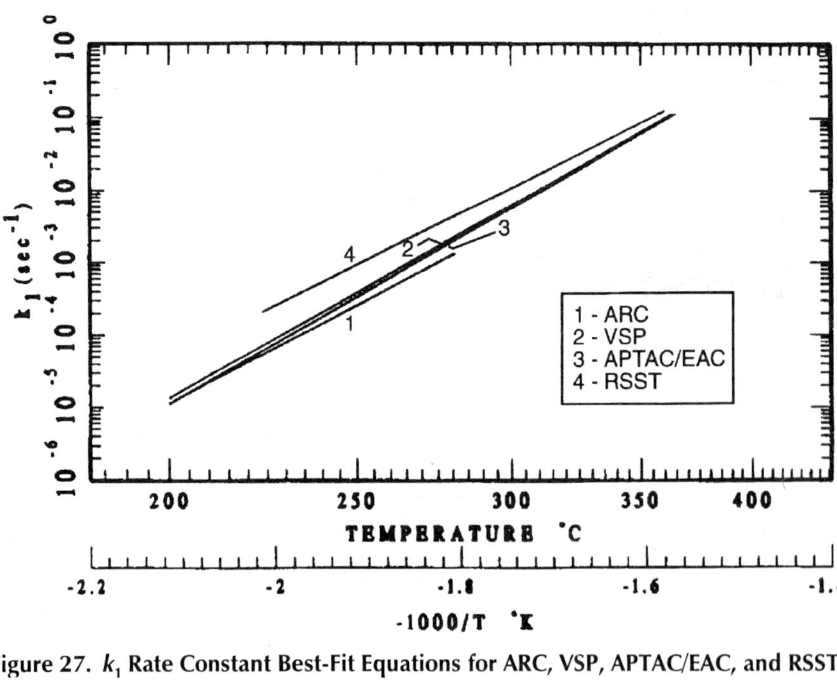

Figure 27. k_1 Rate Constant Best-Fit Equations for ARC, VSP, APTAC/EAC, and RSST.

4. CONCLUSION

About forty companies including one university participated in the round-robin testing using mostly DSC, ARC®, VSP, and RSST. The 25% wt. H_2O_2 decomposition under Fe contamination yielded a rate about 5000 times higher than the normal uncontaminated sample. The agreement among submitted data was satisfactory despite the concern regarding metallic wall effects of the test cell and thermocouple. The 22% wt. BPO decomposition rate data—with the exception of the DSC—were in good agreement with published half-life data. The discrepancy observed in the ARC® data was due to the use of an unstirred sample and external wall thermocouple. The tetraethylene glycolate system gave rise to, a large variation in heats of reaction obtained from the DSC and in self-heat rates measured in the ARC®. Rate constant data were in good agreement among the culled data set from the VSP, EAC, APTAC, and ARC®.

ACKNOWLEDGMENTS

The following individuals and companies are to be thanked for providing the chemicals used in the round-robin testing:

System	Individual	Company
H_2O_2	M. L. Pinsky	FMC Corporation
BPO	M. King	AZTEC Catalyst Co.
Glycolate	M. A. Young	Union Carbide Corp.

In addition the following companies have participated in this series of round-robin tests—listed in alphabetical order: Abbott Laboratories, American Cyanamid Co., Amoco Corp., Ashland Chemical Co., Elf Atochem (France), BASF Corp., Borden Chemical Inc., Bristol-Meyers Squibb, Buckman Labs Inc., Ciba Corp., Cytec Industries Inc., Dixie Chemical Co. Inc., Dow Chemical USA, E. I. Du Pont de Nemours & Co., Inc., Eastman Kodak Co., Exxon Chemical Co., Fauske & Associates, Inc., FMC Corp., Hazard Evaluation Lab (UK), Hercules Inc., Hoechst A.G. (Germany), Hoechst Celanese Corp., ICI Fine Chemicals (UK), ITRI (Taiwan), Mallinckrodt Specialty Chem., Merck & Co., Mitsubishi Kasei Corp. (Japan), Monsanto Co., Rhone-Poulenc Inc., Rhone-Poulenc Industrialisation (France), Rohm and Haas Co., G. D. Searle, Searle/Continental Pharma Inc. (Belgium), Sterling Organics (UK), Union Caribe Corp., Wayne State University, Zeneca Fine Chemicals (UK).

REFERENCES

1. Leung, J. C., Creed, M. J., Fisher, H. G., "Round-Robin 'Vent Sizing Package' Results." Paper presented at the International Symposium on Runaway Reactions, March 7–9, 1989, Cambridge, MA.

2. Townscend, D. I., and Tou, J. C., "Thermal Hazard Evaluation by an Accelerating Rate Calorimeter." *Thermochimica Acta,* 37 (1980), 1.
3. Fauske, H. K., and Leung, J. C., "New Experimental Technique for Characterizing Runaway Chemical Reactions." *Chem. Eng. Prog.* 81(8) (1985), 39–46.
4. Leung, J. C., Fauske, H. K., and Fisher, H. G., "Thermal Runaway Reactions in a Low Thermal Inertia Apparatus." *Thermochimica Acta,* 104(13) (1986).
5. Young, M. A., and Chippett, S., "Design and Operation of an Automatic Pressure Tracking Adiabatic Calorimeter (APTAC)." Paper presented at International Symposium on Runaway Reactions and Pressure Relief Design, August 2–4, 1995, Boston, MA.
6. Singh, J., "Design of Process Plant Handling Highly Unstable and Flammable Materials by Use of Test Data." Paper presented at Preventing Major Chemical and Related Process Accidents, May 10–12, 1988, London, UK. *Also in Inst. of Chem. Engrs. Symp. Series No. 110* (1988).
7. Creed, M. J., and Fauske, H. K., "An Easy, Inexpensive Approach to the DIERS Procedure." *Chem. Eng. Prog.* 86(3) (1990), 45.
8. Lucidol Technical Data, "Evaluation of Organic Peroxides from Half-Life Data," Bulletin No. 30.03, Lucidol formerly with Pennwalt, now within Elf Atochem (1986).
9. Yue, M. H., "Limitations of the ARC's Standard Testing Mode in Runaway Kinetics Study: Discussion Based on BPO Testing." Paper presented at 11th DIERS Users Group Meeting, Columbus, OH (May 6–8, 1992).
10. Young, M. A. of Union Carbide Corporation, South Charleston, WV, private communication (May, 1992).
11. Chakrabarti, A. of Dow Chemical USA, Midland, MI, private communication (May, 1992).
12. Yue, M. H., "An Enhanced Adiabatic Calorimeter for Thermal Hazard Analysis." *J. Hazardous Material* 38 (1994), 13.

Relief System Sizing for Exothermic Runaway: The UK HSE Strategy

J. C. Etchells, T. J. Snee†, and A. J. Wilday†*
**Directorate of Science and Technology, Health and Safety Executive, UK*
†Health and Safety Laboratory, Health and Safety Executive, UK

1. INTRODUCTION

In the UK, the Health and Safety Executive (HSE) is the government agency for enforcement of health and safety in industry. In addition to its enforcement role, other important functions of the HSE include the provision of guidance and arranging for research to be carried out on health and safety issues. Much of this research is then used as an input to further guidance.

To date, HSE involvement in guidance on chemical reaction hazards has centered on reaction hazard assessment and control. In particular, references 1–3 were developed in liaison with UK industry. Of the measures used to protect reactors, emergency relief systems are probably the most commonly used in the UK. Although much work has been done on relief system sizing, particularly in the US by DIERS, a recent survey in the UK found that there is still a need for more information, particularly by design engineers in small/medium sized companies. In response to the need for more guidance, HSE has recently prepared a Workbook[4] which concentrates on the analytical (hand calculation) methods for chemical reactor relief system sizing. This paper describes the current guidance by HSE and gives information on the survey and the Workbook.

In support of this work, HSE is also carrying out a program of research to assist in validating the DIERS methodology, particularly the hand calculation methods. This includes a number of pilot-scale runaway reaction experiments Further work is also underway, in liaison with the US and European DIERS Users Groups, to test the main computer models used in the industry against some of the experimental results. The experiments are also being used as an input to a European Community project on disposal system design. The tests and main conclusions of the work to date are also described.

2. WORK TO DATE

In the UK, exothermic runaway incidents meeting specific criteria[5] are required to be reported to the HSE. These incidents have been reviewed in a number of studies[6,7,8] which highlighted the need for further guidance, particularly in the areas of chemical reaction hazard assessment, control and relief system sizing. This led the HSE and others to develop a text book on the subject[1]. This has recently been revised[9] and now includes a number of case histories. A video has also been produced by HSE[2] and this was followed by an IChemE training package[3], which includes the video, book and incident case histories.

More recently, HSE has published a free leaflet on the hazards of chemical reactions[10]. This is particularly aimed at small to medium companies without their own "in-house" experts. This should give companies a clear picture of the hazards and where further information is available. In the longer term, an HSE Guidance Note will also be published and further action to increase awareness is being considered.

Other research contract reports which have recently been published are given in references 11 and 12.

3. SURVEY ON RELIEF SYSTEM SIZING

The design of emergency relief systems for chemical reactors in Great Britain was the subject of a recent survey by the HSE. Data on 94 vessels used for exothermic reactions at 82 different factories was collected by HSE specialist inspectors and collated onto a central data base.

The main findings were as follows:

- Emergency relief venting was provided in 73 vessels (the others relied on alternative bases of safety, including inherently safer design).
- A significant percentage of companies did not have ready access to the design standards to which their reactors were constructed.
- Less than half of the companies had clearly carried out any form of reaction hazard assessment, either to determine that overpressurization was possible or to identify the "worst case" reaction for relief system design purposes. It is possible that in some companies this work had been done at some stage, but it had not been recorded.
- Although many papers have been published on reactor relief system design, particularly by the Design Institute for Emergency Relief Systems (DIERS), and there is a DIERS Project Manual[13], only a few of the companies appear to have used the available technology. Many companies either did not know how the emergency relief systems had been

designed (and the information was unobtainable) or had used invalid techniques, such as:

—API 520 for fire venting of flammable solvents (in 2 cases the solvent was known to be thermally unstable, but this had not been taken into account);

—Factory Insurance Association (FIA) Charts (this method has been withdrawn and was never intended as a design tool[14]);

—"Professional judgment" (without any calculations or reaction runaway data to back it).

—Largest existing vessel opening.

- Most of the relief systems were specified by professional engineers, either employed by the companies themselves, or by relief system suppliers.
- No correlation between following the correct relief system sizing procedures and company size or vessel age could be found. This may be due to the limited size of the sample studied.
- Companies that were part of large multinational groups were more likely to have followed the recommended methodology.
- Although some multinational companies had sufficient expertise in-house, this expertise had not always been called upon.
- Just over half of the vessels were multipurpose, the remainder being used for one reaction only

It should be emphasized that the prime purpose of this survey was to identify what methods had been used to design emergency relief venting of chemical reactors. Of particular interest was the extent to which design techniques developed by DIERS had been adopted. Consideration of other safety measures provided for the vessel, in addition to the emergency relief vents, was not a core part of the survey.

4. WORKBOOK FOR CHEMICAL REACTOR RELIEF SYSTEM SIZING

4.1. Purpose

As a result of the need for further information, a Workbook, entitled Workbook for Chemical Reactor Relief System Sizing, has been written by HSE[4]. The purpose of the Workbook is to give information on methods available for the sizing of emergency relief systems for exothermic runaway reactions in liquid-phase chemical reactors. The Workbook summarises the main hand calculation methods available as a result of the DIERS project and their limits of applicability. Relief system sizing is by no means an easy subject, so a number of worked examples are given to help the reader understand their application. The experimental information required to size an emergency relief system properly is also discussed.

The Workbook is written mainly for chemical engineers or applied chemists with a good basic training in both chemical reaction kinetics and fluid flow. It is particularly aimed at engineers in small to medium sized companies who may not have had ready access to the DIERS meetings. However, experience in the development of appropriate physical properties from databases (or small-scale experiments if appropriate) for the reacting mixtures under consideration will be needed. In addition, such companies will need to obtain advice on the development of kinetic parameters from experimental results, if they do not have their own "in-house" expertise.

The Workbook should also be useful to others who, while they may not wish to design a relief system themselves, want to ensure that the correct procedures have been followed during relief system design of their processes.

4.2. Scope

The Workbook is intended to be self-sufficient for sizing calculations for the more straightforward applications. The emphasis of the Workbook is on the use of simple (yet adequate) equations suitable for solving with a pocket calculator, rather than on more complex computer models.

In addition to these hand calculation methods, there are also a number of computer models available for relief system sizing. The best known of these are referred to in the Workbook, but they are not dealt with in any detail. This is a highly specialized field and a potential user needs to discuss their application with the code supplier. A number of requirements for computer models are suggested. The Workbook references the forthcoming CCPS *Guidelines for Pressure Relief and Effluent Handling Systems*[15] which covers computerised design tools in more detail, but gives much less detail on hand calculation methods. Similarly, the Workbook gives a brief overview of the selection and design of disposal systems for the effluent from a pressure relief systems, but references the CCPS Guidelines for more detail. It is intended that the Workbook will be complementary to the CCPS Guidelines, with minimal duplication.

Unlike relief system sizing for nonreacting systems, a considerable amount of experimental information is normally required for the design of chemical reactor relief systems. It is necessary to assess all the credible maloperations and system failures that may occur on the process/plant to determine the reaction runaway that requires the largest relief system. The Workbook summarizes the main steps necessary to do this. Reaction hazard assessment, other than for the purposes of relief system sizing, is not dealt with. Information on this is given in reference 9.

Although there have been considerable advances in the technology over the last 20 years, it is still evolving. The information contained in the Workbook is a summary of the best available technology. Much work is still to be done and the design of relief systems for certain types of systems, e.g. viscous systems and systems containing significant levels of solids, is still complex and is outside the scope of the document. Where emergency relief system design for any particular system is outside the scope of the Workbook, the reader is referred elsewhere, e.g. to specialist computer models.

The Workbook does not deal with fire relief of vessels (except where external fire modifies the relief sizing for runaway exothermic reaction) or with the mechanical integrity of either the process vessels or relief systems. Guidance on these is available elsewhere[16].

4.3. Approach taken

The Workbook is concerned with "how to?" more than with "why?". Sizing methods are given together with conditions of applicability and some limited background information. Sources of information for theory, derivations of equations and some more unusual methods are referenced. Use is made of decision trees to guide the user to the appropriate part(s) of the Workbook. Worked examples are given for all the recommended methods.

A contents list for the Workbook is given in Table 1 and the structure is summarized by the flowchart in Figure 1 which indicates the paths to be taken through the Workbook when carrying out any particular relief system design.

TABLE 1
Contents of the Workbook

Chapters	Annexes
Foreword	
1. Introduction	A1. Basis of safety
2. Overview	A2. Experimental methods
3. Determining the worst case	A3. Level swell calculations
4. Classification of relief systems	A4. Computer codes
5. Relief system sizing	A5. Additional sizing methods
6. Vapor pressure systems	A6. Sizing for single phase relief
7. Gassy systems	A7. Safety factors
8. Hybrid Systems	A8. The Omega method
9. Calculation of two-phase flow capacity	A9. Glossary
10. Special cases	A10. Nomenclature
11. Disposal systems	A11. Index
12. Reaction forces	
13. Maintenance of hardware and software	

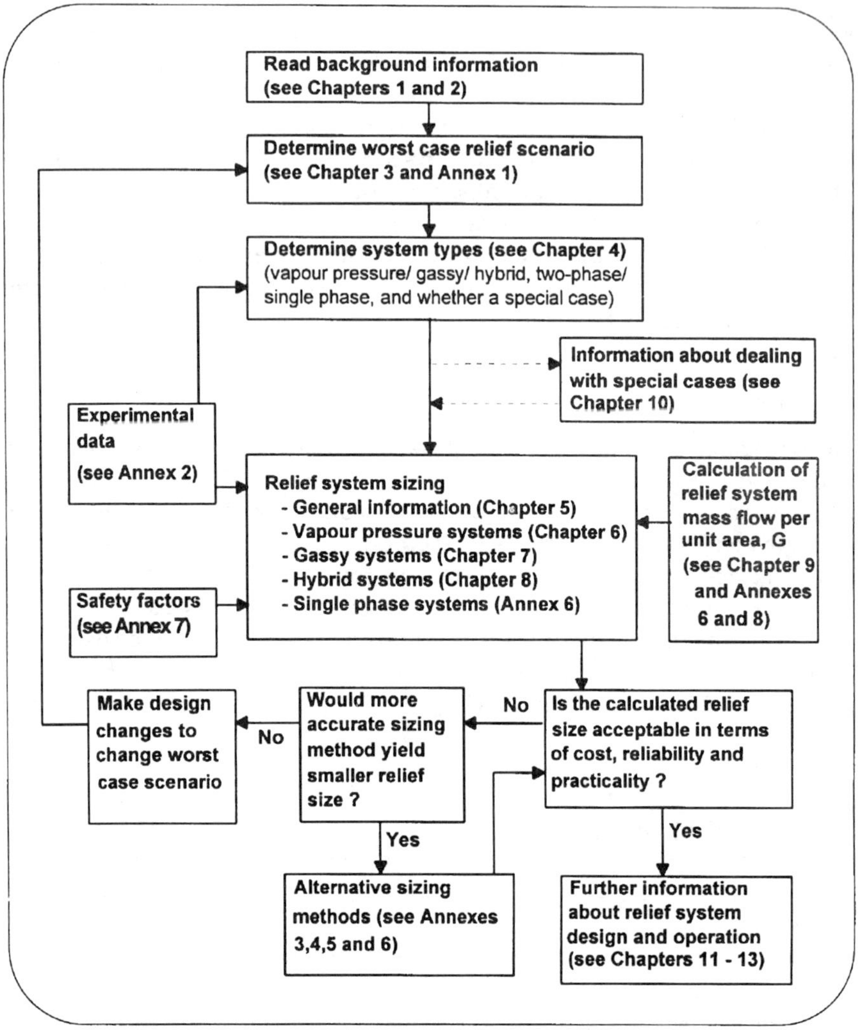

Figure 1. Flowchart for Use of the Workbook.

The first two chapters of the Workbook contain background information. In the Introduction (Chapter 1), it is emphasized that there are a number of means of achieving safe operation of chemical reactors, the best option being to avoid the hazards completely, or at least minimize them, by inherently safer design. However, in many cases this is not practical and, to maintain a viable process, other safety measures will be needed, either alone or in combination.

Information on the various options and the main advantages and disadvantages of emergency relief systems are discussed in Annex 1 of the Workbook.

The Introduction also lists the information required for the design of emergency relief systems for exothermic runaway. This includes:

a) The credible maloperations and system failures that might occur during reaction. It is very important to assess the "worst case" for relief sizing from these credible maloperations and failures.
b) The kinetics of the reaction under runaway conditions.
c) Whether the reaction pressure is from vapor or gas (or both).
d) The flow regimes, both in the vessel and relief system, during relief.
e) The design and layout of the relief system.

Unless such information is obtained and applied properly in its design, then an emergency relief system may be wrongly sized and a false sense of security placed upon it.

Chapter 2 gives an overview of the Workbook from the point of view of its use during the design process of a pressure relief system for a chemical reactor. Figure 2 illustrates the steps in the design of a pressure relief system, and indicates those steps which are covered by the Workbook.

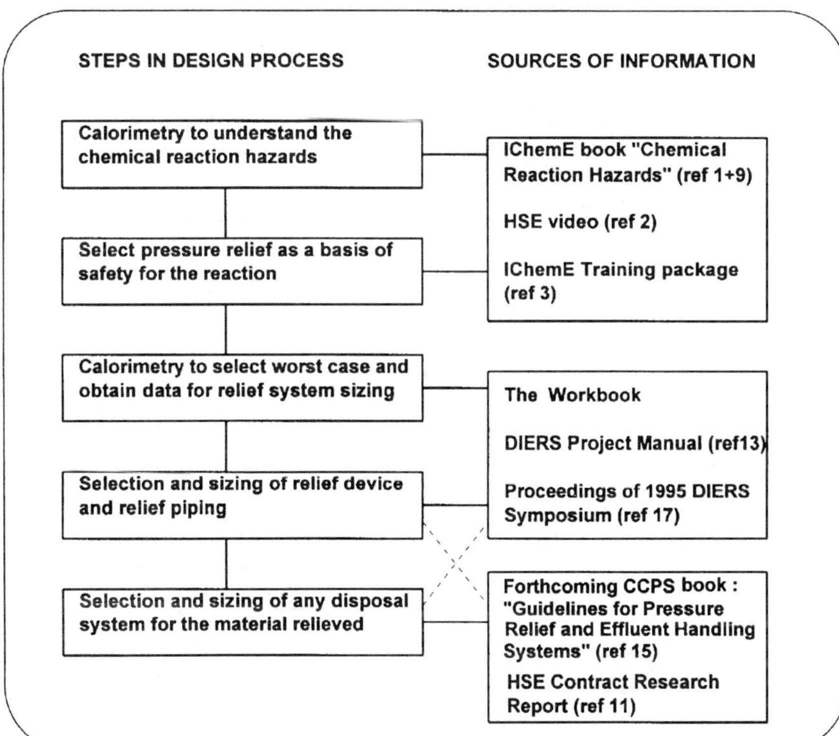

Figure 2. Steps in the Design of a Pressure Relief Systn for a Runaway Chemical Reaction.

The design process begins with Chapter 3 which explains the process of determining the worst case relief scenario on which the relief system design is to be based. This process entails determining the credible combination of failures and maloperations which gives rise to the largest required relief size. The next stage in the design process, described in Chapter 4, is to determine the system type for the purposes of relief sizing: vapor pressure, gassy or hybrid (a mixture of gassy and vapor pressure). Small-scale experiments are involved, which are described in Annex 2. Chapter 4 also deals with the determination of whether the relief flow will be two-phase or gas/vapor only as this will affect the design method selected and whether the flow in the relief system will be laminar rather than turbulent, as this will affect the calculation of the mass flow capacity per unit area (G).

The relief sizing itself can then be carried out. Chapter 5 gives important background information about relief sizing. This includes a decision tree (shown as Figure 3) for selecting the chapter or annex to be used for sizing in

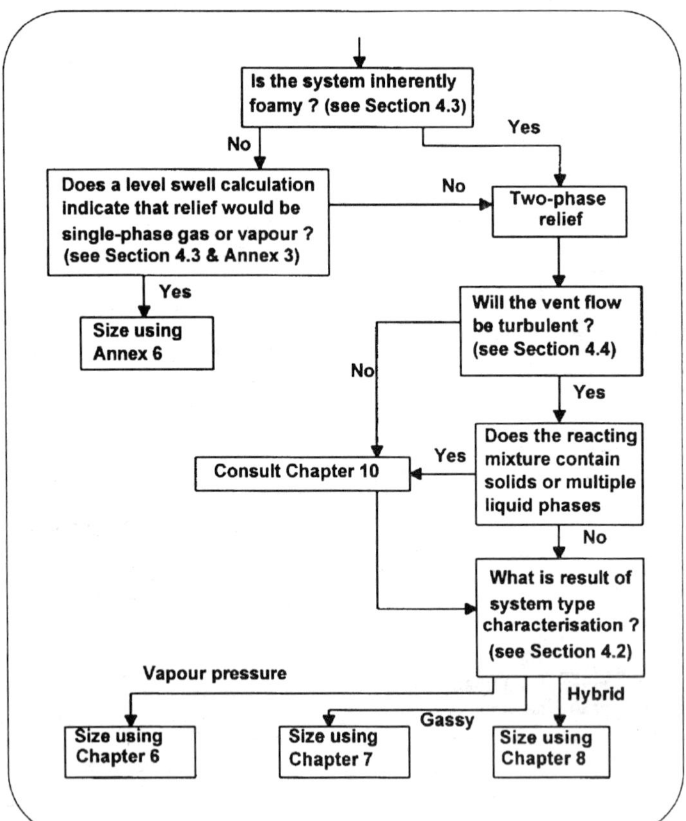

Figure 3. Decision Tree to Select Sizing Method.

a particular case. For example, the application might require a relief system for a vapor pressure system to be sized using Chapter 6.

Chapters 6, 7, and 8 give methods for calculating the relief system size for each system type. In most cases, the simplest and most usual methods are given first, followed by references to valid alternative methods (in Annexes) should the initial methods be inapplicable or likely to oversize due to their underlying assumptions.

Chapter 9 contains important background information about two-phase flow, and calculation methods for the two-phase mass flow capacity per unit area, G, which is required to obtain the relief area for most of the relief sizing methods. Some system types are special cases involving laminar flow, solids and/or multiple liquid phases. In these cases Chapter 10, rather than Chapter 9, should be used for calculation of G.

Chapters 11–13 cover the selection and sizing of downstream disposal systems, reaction forces which require piping and vessel supports, maintenance, documentation and change management. Additional material is given in Annexes 1 to 8 and is referenced from the text as required. This includes consideration of the safety factor to be applied to the calculated relief size.

The design of a relief system often involves iteration and recycle. The Flowchart in Figure 1 shows that possible recycle in the design process may involve changing the assumptions about the worst case relief scenario or changing the sizing method used.

5. RESEARCH

To further validate the DIERS hand calculation methods, the HSE's Directorate of Science and Technology has instigated a program of research which is being carried out by the HSE's Health and Safety Laboratory (HSL) at Buxton. A 350-liter pilot-scale chemical reactor plant has been built specifically to carry out vented runaway chemical reactions. The facility is supported by a range of adiabatic and isothermal calorimeters and other laboratory-scale equipment. A principal part of the exercise is to measure the experimental overpressures and compare the experimental vent sizes with vent sizes calculated to give the same overpressures.

5.1. Research Strategy

The following overall strategy is being followed in the project:

a) Initial selection of candidate reactions using published physical and chemical property data on the reagents and products;
b) Determination of the temperature and concentration dependencies of the reaction kinetics using adiabatic and isothermal calorimetry;

c) Theoretical analysis, using the calorimetric data, to predict the critical conditions which lead to a runaway exothermic reaction for the candidate reactions in laboratory and pilot-scale vessels;
d) Laboratory-scale experiments (1.6 liter) to test the reliability of the theoretical predictions of the temperature–pressure–time profiles and to obtain a qualitative appreciation of reaction venting phenomena;
e) Selection of systems for pilot-scale (350 liter) investigation and preliminary emergency pressure relief system design calculations;
f) Pilot-scale investigation of runaway reaction venting and the performance of the relief system;
g) Comparison of the pilot-scale experimental results with the simplified DIERS hand calculation methods; and some limited comparisons with predictions using computer models for reactor venting.

5.2. Pilot Reactor Facility

5.2.1. Equipment

The pilot-scale facility for investigating runaway reactions is based around a 350-liter, jacketed, glass-lined reactor equipped with three glass feed vessels and connected, via a vent line, to a 2,500-liter stainless steel catch tank. A schematic diagram of the system is shown in Figure 4. The reactor and the catch tank have a maximum working pressure of six barg. The 75-mm diameter vent line between the reactor and catch tank is fitted with an actuated valve followed by an optional restricting orifice plate or flow restriction nozzle. The valve is linked to a pressure controller and is opened automatically (within one second approximately) when the pressure reaches a preselected value. The catch tank is fitted with a 75-mm vent direct to atmosphere. Heat transfer fluid is circulated to the reactor jacket and to heating coils in the feed vessels. The temperature of each vessel can be independently controlled. Interconnecting chemical transfer pipe work is fitted with remotely operated actuated valves. A large number of temperature and pressure transducers are installed on the facility, and the catch tank is fitted with load cells to indicate the mass of the contents. A load cell is also installed below the vent line to measure changes in reactive force due to the two-phase discharge. A gamma ray densitometer is installed around the vent line above the orifice plate. Video cameras are used via site glasses to observe the contents of the reactor, vent line and catch tank. The pilot plant is controlled and monitored remotely from a control room 100 meters from the reactor building.

5.2.2. Pilot-Scale Procedure

Pilot-scale experiments were performed as batch processes with heat transfer fluid passing through the reactor jacket at a constant temperature and flow

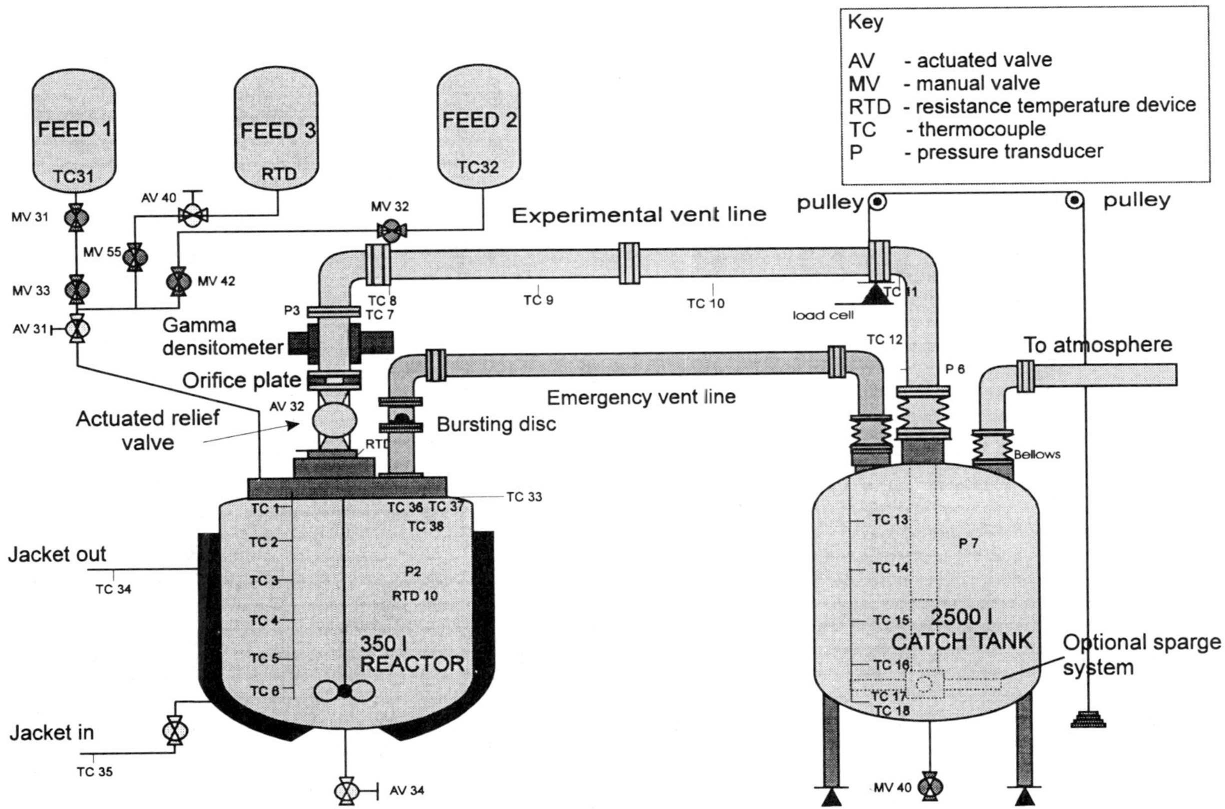

Figure 4. Schematic Diagram of Pilot-Scale Facility.

rate. In general, reagents and solvents were preheated to the same temperature as the reactor jacket and then mixed in the reactor to initiate the runaway.

When the preselected relief set pressure was reached, the vent valve between the reactor and catch tank was automatically opened. Video recording was used to observe the two-phase discharge from the reactor to the catch tank. When the pressure in the reactor had returned to atmospheric pressure, the contents were allowed to cool whilst maintaining a constant jacket temperature. At the end of an experiment the contents of the reactor and catch tank were transferred to storage vessels and weighed. The products from the reactor and catch tank, and a sample taken from a collection vessel placed below the exit of the vent to atmosphere, were chemically analyzed. The experimental method was chosen so that the conditions leading to runaway could be accurately reproduced.

5.2.3. Data Acquisition and Processing

The output of each transducer was recorded at an interval of 0.1 sec and the computer data files were summarized prior to detailed analysis. Factual reports were produced after each experiment. These results were also recorded and archived in a way that would facilitate further detailed analysis and comparison with dynamic computer simulations

At the end of a series of experiments on a particular reaction system, the data were analyzed to compare the experimental results with predictions based on the analytical methods for relief system design.

5.2.4. Conditions for Pilot-Scale Experiments

The chemical reactions for pilot scale investigation have been selected to include examples of vapor pressure and gassy systems. Model reactions with well defined chemical and physical properties have been selected so that general conclusions can be drawn from the experimental results. A test matrix is prepared for each series of pilot-scale experiments designed to systematically investigate the effect of one or more of the following key parameters:

- initial temperature
- relief set pressure
- vent area
- catalyst concentration
- batch size

Experiments were designed to give two-phase flow, but vapor and gas-only flow were also investigated. Variations in the initial temperature and catalyst concentration allowed evaluation of emergency relief system design methodology over a wide range of vapor generation and gas evolution rates.

It is not possible to include here the detailed interpretation for all the reaction systems studied to date. However, in order to illustrate the research strategy, some results for one vapor system and one gassy system are presented.

5.3. Experimental Results for a Vapor Pressure System

The esterification reaction between propionic anhydride and isopropanol, catalyzed by various concentrations of sulfuric acid, was chosen as one of the vapor pressure systems:

$$\underset{\text{isopropanol}}{CH_3CH(OH)CH_3} + \underset{\text{propionic anhydride}}{(CH_3CH_2CO)_2O} \xrightarrow{H_2SO_4} \underset{\text{isopropyl propionate}}{C_2H_5CO_2(CH_3)CHCH_3} + \underset{\text{propionic acid}}{C_2H_5CO_2H}$$

Pilot- and laboratory-scale reaction venting experiments were performed on this reaction system using five different catalyst concentrations. Adiabatic and isothermal calorimetry were used to determine the temperature and concentration dependence of the rate of heat generation for each acid concentration. The adiabatic calorimetry also provides the fundamental data for the relief system sizing calculations.

5.3.1. Adiabatic Calorimetry

Figure 5 shows the temperature dependence of the rate of heat generation determined by adiabatic calorimetry[18] for each acid concentration and Figure 6 shows the corresponding pressure/temperature relationships. The initial temperature for these adiabatic experiments was chosen to be the same as the jacket temperature used for the pilot-scale venting experiments.

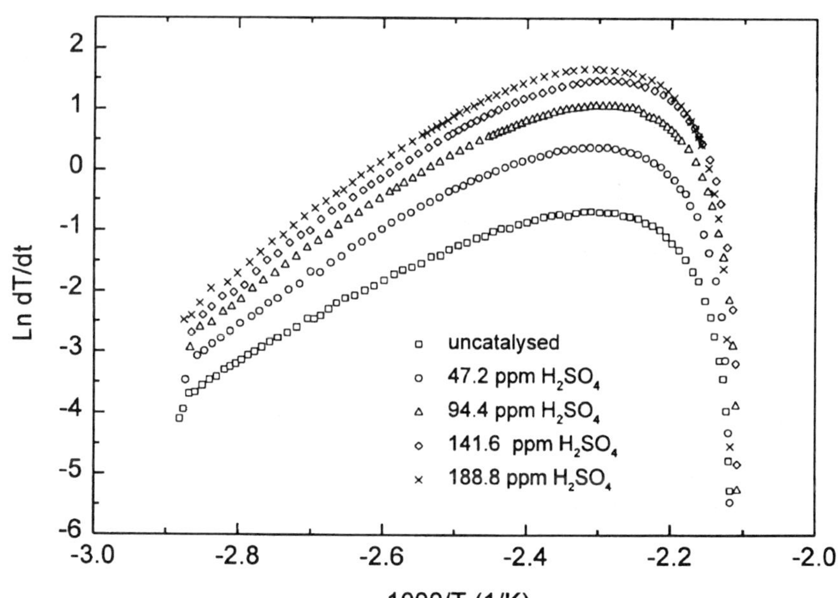

Figure 5. Phi-Tec Plots of Self-Heat Rate against Temperature for the Esterification Reaction over a Range of Catalyst Concentrations.

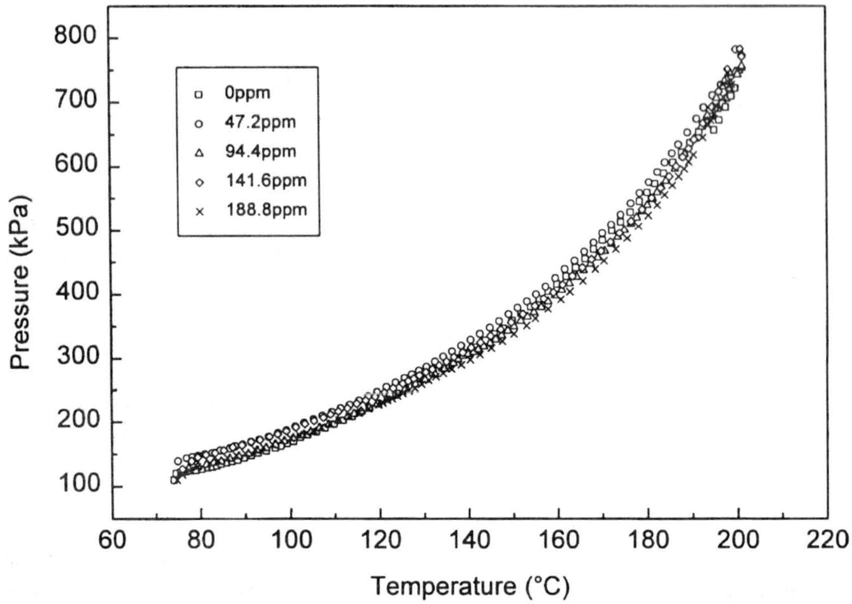

Figure 6. Pressure–Temperature Relationships from Adiabatic (Phi-Tec) Data for the Esterification Reaction.

Figure 5 shows that sulfuric acid has a strong influence on the rate of heat generation but the adiabatic temperature rise and therefore the heat of reaction is unaffected by the concentration of catalyst. The pressure–temperature relationships shown in Figure 6 are largely independent of catalyst concentration, indicating that the addition of catalyst has no strong influence on the physical properties of the reacting mixture. The calorimetric data show that, by choosing a range of catalyst concentrations, the vent-sizing methodology for vapor pressure systems can be tested systematically, over a wide range of heat and vapor generation rates.

5.3.2. Isothermal Calorimetry

Isothermal reaction calorimetry[19] was used to determine the heats of reaction for each acid concentration. These values were consistent with heats of reaction obtained from the adiabatic data.

Figure 7 shows plots of rate of heat generation against time obtained from isothermal calorimetry for a catalyst concentration of 94.4 ppm at a reaction temperature of 50°C. (This temperature was chosen to give measurable rates of heat generation without exceeding the heat transfer capacity of the calorimeter.) The maximum rate occurs some time after initiation, indicating autocatalytic behavior. Autocatalysis was evident for all except the uncatalyzed composition, but was not immediately apparent from the

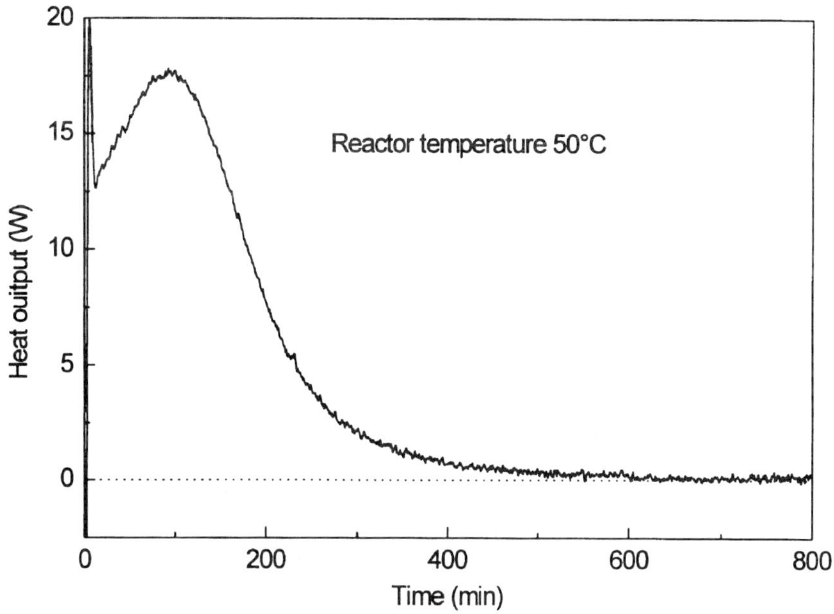

Figure 7. Isothermal Variation Rate of Heat Generation for Esterification Reaction Catalyzed by 94.4 ppm H_2SO_4 Measured Using Reaction Calorimetry.

adiabatic plots (Figure 5) which gave satisfactory agreement with normal (*n*th order) kinetic expressions. Computer simulations which model of the course of the reaction in a jacketed reactor showed that it was important to include the autocatalysis in the kinetic model for the reaction.

The adiabatic and isothermal results were used to develop kinetic expressions for each acid concentration. These expressions were used in the design of the pilot-scale experiments using computer models and physical property correlations.

5.3.3. Laboratory-Scale Tests

Closed system and vented runaway reaction experiments on the esterification reaction were performed in a 1.6-liter jacketed laboratory-scale reactor before proceeding to pilot-scale tests.

The closed system experiments were used to test predictions of the temperature–time history for the laboratory reactor using kinetic expressions derived from the adiabatic and isothermal calorimetric data. Satisfactory agreement was obtained between the experimental results and simulations using a heat transfer coefficient appropriate to the laboratory-scale reactor. These findings were used in the safety assessments for the pilot-scale experiments. However, the temperature rises for the closed system laboratory-scale experiments were always well below the adiabatic temperature rise for the esterification reaction (126°C) determined from the adiabatic results. This was due to the relatively high surface to volume ratio of the laboratory-scale reactor giving efficient heat transfer to the reactor jacket. As anticipated, the laboratory-scale experiments could not provide direct quantitative prediction of reaction venting in a large process vessel.

Qualitative features of reaction venting were established from laboratory-scale tests using an actuated valve to vent the reactor at a preselected pressure. These results were used as a preliminary test (over a limited range of vapor generation rates) of the level swell calculations to predict the degree of vapor–liquid disengagement in the vessel.

5.3.4. Test Matrix for Pilot-Scale Experiments

The test matrix for the pilot-scale experiments on the catalyzed esterification reaction was designed to achieve the following objectives:

a) Direct tests of vent sizing calculations with various degrees of vapor–liquid disengagement with the composition entering the ventline ranging from vapor-only to well developed two-phase flow.
b) Comparison of the calculated and observed overpressure where the limiting flow restriction is either due to an orifice plate or due to a reduction in pipe diameter over the length of 100 mm reported to be necessary to give vapor–liquid equilibrium over the restriction (AIChE, 1992).
c) Evaluation of flow capacity calculations assuming various degrees of vapor–liquid disengagement in the reactor.

Accordingly, seven experiments were performed on the esterification reaction using various catalyst concentrations and ventline configurations but with the same jacket temperature (75°C) and relief set pressure (1.5 bara).

5.3.5. Discussion of Pilot-Scale Results

Figure 8 shows the reactor temperature and pressure records from one of the pilot-scale experiments. Two-phase flow occurred immediately after vent opening when the temperature in the vapor space became approximately equal to the liquid temperature measured at the bottom of the vessel. The temperature continued to rise after vent opening and an overpressure of approximately 50% of the relief-set pressure was reached.

The maximum reactor pressure has been plotted against catalyst concentration in Figure 9 along with calculated values obtained using the hand calculation method due to Leung [20] and the Equilibrium Rate Method for calculating the flow capacity[13]. Figure 9 shows that the calculated values substantially exceed the experimentally observed overpressures. This finding was true for most of the analytical methods (see Table 2) provided that they were used within their stated range of applicability. Most of the methods which

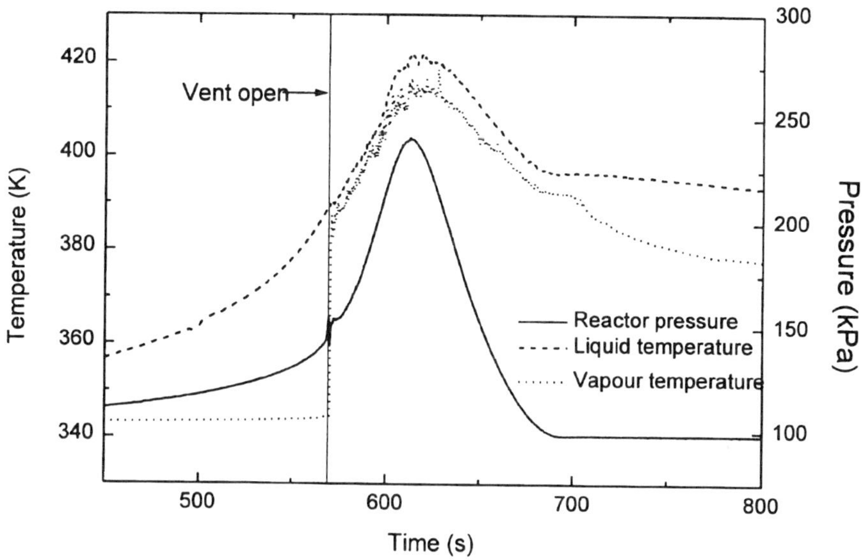

Figure 8. Reactor Temperature and Pressure Records from a Pilot-Scale Runaway Reaction Experiment on the Esterification Reaction.

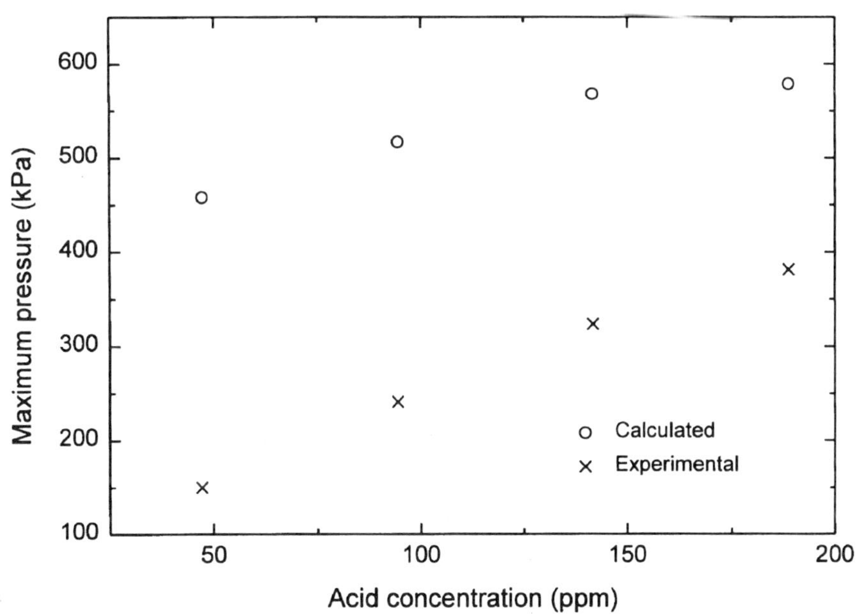

Figure 9. Comparison of Calculated and Experimentally Observed Maximum Pressures for Pilot-Scale Experiments over a Range of Catalyst Concentrations.

TABLE 2
Vent Sizing Calculation Methods Used in Comparison with Experimental Results

Method	Ref.	Applicable?	Conservative?
Leung (assuming homogeneous vessel)	20	Yes	Yes
Leung (as above but q formulated for high overpressure)	22	Yes	Yes
Fauske (with no disengagement)	23	Yes	Yes
Fauske (assuming disengagement at experimentally measured void fraction)	23	No[25]	
Huff	24	Yes	Yes
Modified Boyle	14	Yes	Yes
Wilday (Leung's method modified to account for homogeneous relief until disengagement)	25	Yes	Yes
Wilday (stepwise application of equation in Leung's method derivation)	25	Yes	Yes
Two-phase relief constant pressure (steady-state volume balance)	14	Yes	Yes

take account of overpressure and assume homogeneous flow from the reactor gave broadly similar results. Some of the methods which assume disengagement in the reactor gave nonconservative vent diameters but these methods were outside their stated range of applicability.

Vent sizing calculations for vapor pressure systems generally assume that a pseudosteady state is achieved when the rate of heat generation becomes equal to the rate of heat removal by vaporization. Some methods assume that a homogeneous two-phase mixture vents from the reactor resulting in a rapid reduction in reaction mass to a point where the rate of heat removal is sufficient to cause a temperature and pressure turnaround. Detailed analysis of this series of experiments indicates that the steady state and vessel emptying assumptions do not describe reactor venting under the conditions of the pilot-scale tests. Measurements of the mass discharge rates indicated vapor fractions at the inlet to the vent line which were much higher than the homogeneous value. Calculations using the adiabatic data and the observed discharge rates indicated that the pressure turnaround was due to reactant consumption. The reduction in maximum pressure compared to the adiabatic case could be calculated using theoretical flow capacities and the integrated vapor evolution between the set pressure and the maximum pressure. The details of this analysis will be published separately.

Two experiments were performed using identical conditions except for the flow restriction at the start of the vent line. In one of the tests an orifice plate was used while the other tests the orifice was replaced by a short pipe of equal diameter but a length of 100 mm. Comparison of the results from the two tests indicated that replacement of the orifice plate with a short pipe had only a small influence on the maximum recorded pressure. The results suggest that vapor–liquid equilibrium was attained at the choke point for both the orifice and the short pipe.

There were significant differences between the experimentally observed flow calculations and values obtained using various theoretical models. The Equilibrium Rate Model[13] and the Omega method[21] overpredicted the flow compared to the experimental values. This difference is thought to occur because the actual void fraction entering the vent line was higher than that assumed in the flow calculations. The variations in the calculated values resulted in significant variations in the recommended vent diameter.

The homogeneous vessel assumption and the emptying time principle did not adequately describe the results of the pilot-scale experiments. Further experiments are planned, using foamy systems, to establish the margins of safety when conditions are closer to the assumptions of many of the DIERS analytical methods.

5.4. Experimental Results for a Gas Generating Reaction

The decomposition of *tert*-butyl peroxy-2-ethylhexanoate (a peroxy-ester) was chosen as an example of a gas generating reaction. The reaction was performed in a hydrocarbon solvent with a high boiling range so that the reactor temperature during the runaway would remain below the boiling point of the most volatile component of the solvent.

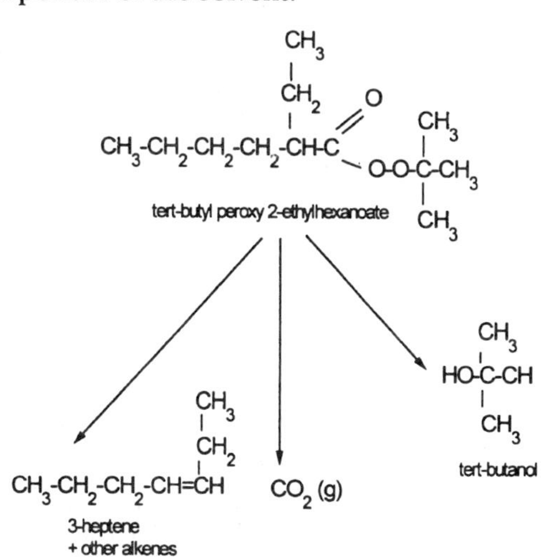

Carbon dioxide is the major gaseous decomposition product and the decomposition of one mole of the peroxy-ester is expected to produce one mole of gas. The decomposition proceeds via a free radical mechanism and some variation in the nongaseous products can be expected but the major decomposition reaction is as indicated in Equation (2). The peroxy-ester should be stored at temperatures less than 20°C. Significant rates of decomposition occur at temperatures greater than 35°C.

The decomposition of the peroxy-ester can be catalyzed by the addition of cobalt octoate accelerator.

5.4.1. Adiabatic Calorimetry

Figure 10 shows the adiabatic plots of log (self-heat rate) against reciprocal temperature for a range of accelerator concentrations with an initial temperature of 95°C. The plots show that the addition of accelerator has a strong influence on the reaction rate but the adiabatic temperature rise, and therefore the heat of reaction, is not affected by the addition of accelerator. These results were used to design pilot-scale experiments to provide a systematic evaluation of vent sizing methodology for this gassy system.

Figure 11 shows adiabatic plots of pressure against temperature for three accelerator concentrations with an initial temperature of 95°C. These plots are approximately independent of the accelerator concentration. This confirms the indications from Figure 10 that the accelerator affects the reaction

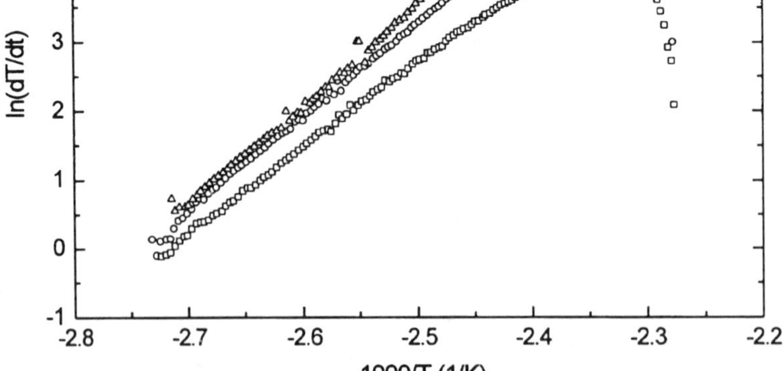

Figure 10. Phi-Tec Plots of Self-Heat Rate against Temperature for the Decomposition of the Peroxy-ester over a Range of Accelerator Concentrations.

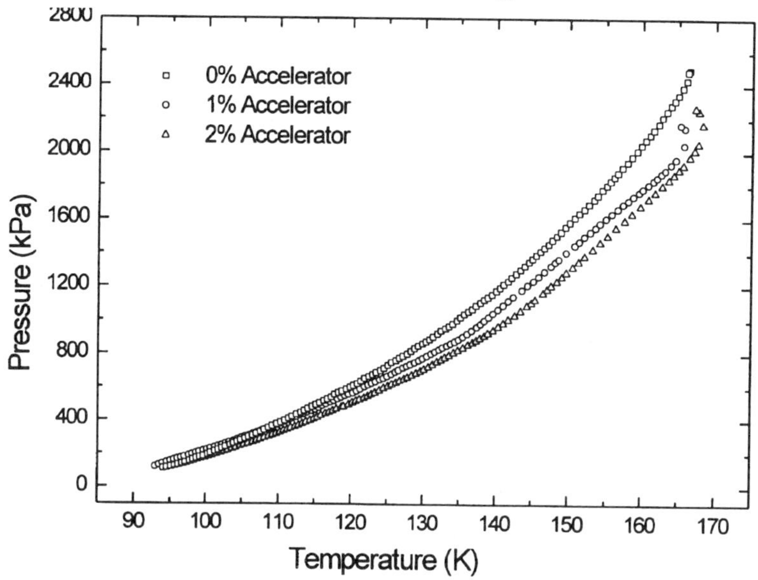

Figure 11. Pressure–temperature Relationships from Adiabatic (Phi-Tec) Data for the Decomposition of the Peroxy-ester.

rate but not the final extent of reaction or total volume of gas produced. However, the average of the maximum pressures shown in Figure 11 is substantially less than would be expected from calculations for the complete decomposition of the peroxy-ester. This was thought to be because some of the carbon dioxide produced by the decomposition of the peroxy-ester dissolves in the hydrocarbon solvent in the adiabatic test can.

The adiabatic self-heat rate plots gave good agreement with rate equations assuming first order kinetics and were used to determine the activation energy and preexponential factor for the decomposition of the peroxy-ester with various concentrations of accelerator.

5.4.2. Isothermal Calorimetry

The isothermal calorimeter[19] was used to determine the rate of heat and gas generation from the peroxy-ester with the temperature held constant.

The heats of reaction and initial rates of conversion from the isothermal tests were broadly consistent with the adiabatic results. Measurements of the total volume of gas evolved during the open isothermal tests were consistent with the assumption that one mole of peroxy-ester produces one mole of carbon dioxide.

Figure 12 shows rates of conversion as function of time from isothermal calorimetry at 90°C using three concentrations of accelerator. The variation of rate with conversion for the uncatalyzed composition gives first order

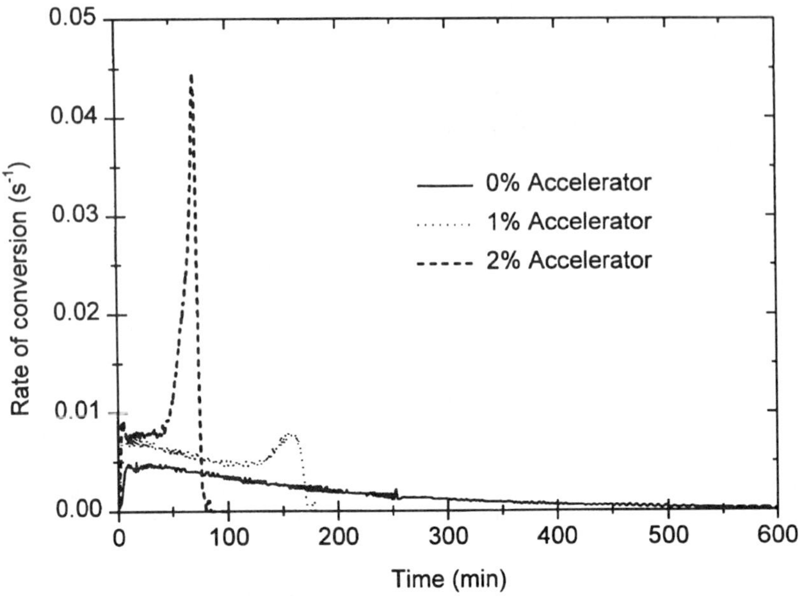

Figure 12. Isothermal Variation of the Rate of Gas Evolution for Decomposition of the Peroxy-ester in the Reaction Calorimeter over a Range of Accelerator Concentrations.

dependence in agreement with the adiabatic results. However, the isothermal results for the catalyzed compositions are not consistent with evidence of first order decomposition from the adiabatic test. Combined interpretation of the isothermal and adiabatic data was used to derive kinetic expressions for the decomposition of the peroxy-ester. These expressions were used for the design and interpretation of the pilot-scale experiments.

5.4.3. Laboratory-Scale Experiments

The decomposition of solutions of the peroxy-ester in the hydrocarbon solvent was studied over a wide range of conditions in the 1.5-liter laboratory-scale facility. The laboratory-scale results were used as part of the safety assessment for the pilot-scale experiments and to determine qualitative features of the venting of gas generating reactions so that suitable conditions for the pilot-scale tests could be established. The experiments confirmed that venting of the two-phase mixture had no significant influence on the temperature-time profile for the reactor. However, there was evidence that heat losses to the jacket and to the thermal mass of the vessel could cause a significant reduction in the maximum temperature in the reactor, particularly for the tests with no accelerator or with a low batch size or jacket temperature.

Quantitative data on the volumetric gas production from the decomposition of the peroxy-ester was obtained by installing a gas meter on the vent line from the laboratory-scale reactor. The measured volumes confirm the theoretical assumption, and the evidence from isothermal calorimetry, that decomposition of one mole of peroxy-ester produces one mole of carbon dioxide.

5.4.4. Test Matrix for Pilot-Scale Experiments

The test matrix for the gassy pilot-scale experiments was designed to achieve the following objectives:

a) Direct tests of the DIERS vent sizing calculations by systematic variation of the reaction mass and the peak gas generation rate keeping the relief set pressure and the vent line geometry constant.
b) Assessment of flow capacity calculations by measuring the two-phase mass discharge rate and void fraction in the vent line.
c) Assessment of level swell calculations to predict the onset of two-phase flow.

Accordingly, ten experiments were performed on the decomposition of the peroxy-ester with various concentrations of accelerator and vent line configurations but with the same jacket temperature (95°C) and relief set pressure (2.0 bara).

5.4.5. Discussion of Pilot-Scale Results

Figure 13 shows the reactor temperature and pressure records from one of the pilot-scale tests. In contrast to the vapor system results, the reactor pressure returns to atmospheric soon after vent opening. However the operation of the relief system does not affect the rate of temperature rise in the vessel. The reaction rate continues to rise until the rate of gas evolution is sufficient to cause two-phase flow. The onset of two-phase flow is indicated by an increase in reactor pressure and a corresponding sharp increase in the temperature at the top of the vessel.

Figure 14 shows a plot of maximum reactor pressure against batch size. The results are compared with values calculated using the equation normally used for emergency relief system design for gassy systems[13] with the flow capacities obtained by the Omega method[21]. The experimental values substantially exceed the calculated values. This difference is attributable to effects of dissolved gas. The isothermal calorimetry and laboratory-scale tests indicated that the volume of gas produced was higher than that which would correspond to the maximum pressure in the adiabatic calorimeter. This means that gas evolution rates are underestimated if dissolved gas is neglected in the interpretation of the adiabatic pressure data. Gas production rates can be calculated from reaction rates determined from the adiabatic rates of tempera-

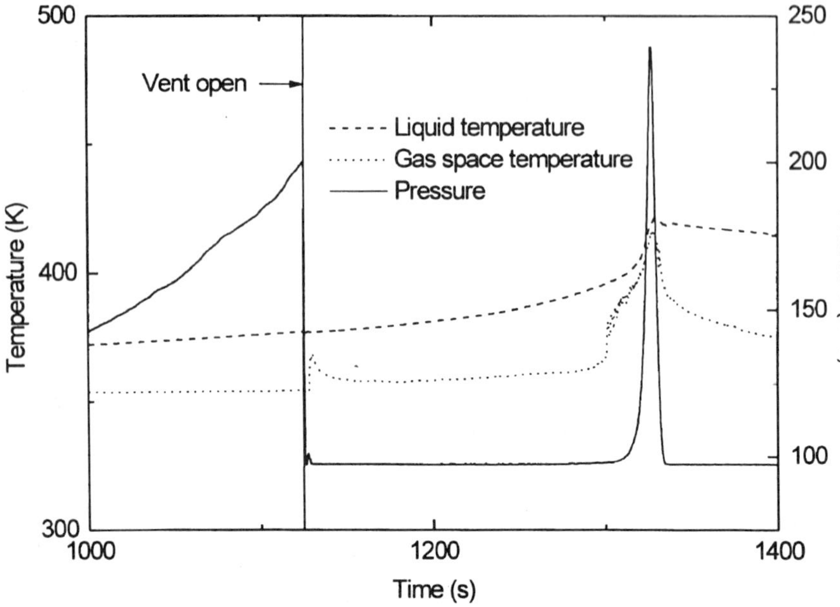

Figure 13. Reactor Temperature and Pressure Records from a Pilot-Scale Runaway Reaction Experiment on the Decomposition of the Peroxy-ester.

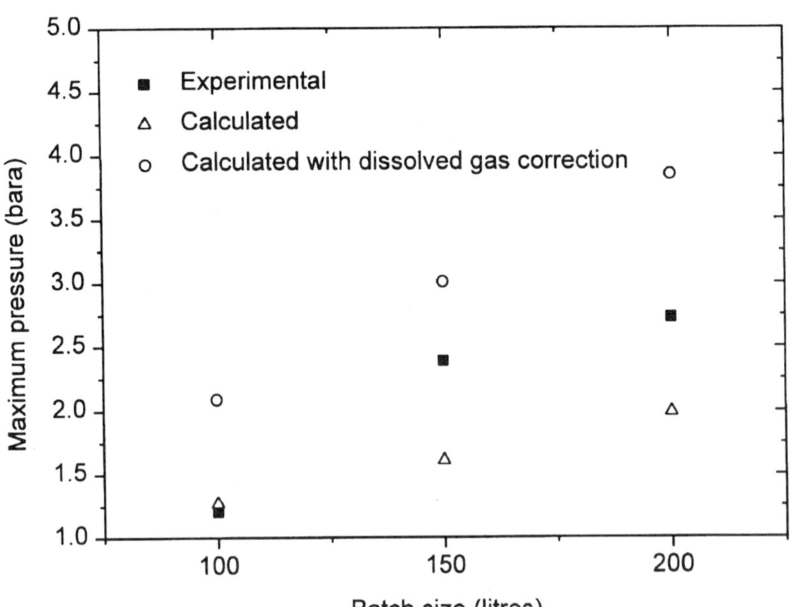

Figure 14. Comparison of Calculated and Experimentally Observed Maximum Pressures for Pilot-Scale Experiments over a Range of Accelerator Concentrations.

ture rise. This gives values which are more representative of gas evolution rates in a vented reactor where the pressure is much less than that in the closed adiabatic test. Maximum reactor pressures calculated with corrections for the effect of dissolved gas are shown in Figure 14. The corrected values lie well above the experimental values indicating that safe vent sizes can be obtained by this procedure. The adiabatic data shown in Figures 10 and 11 were obtained from closed system tests. Adiabatic calorimetry can be performed using an open test with a preimposed back pressure in the outer containment vessel. If the back pressure is carefully selected, the gas evolution rates calculated from the observed rates of pressure rise may be closer to those that would be obtained in a vented reactor.

Detailed analysis of the pilot-scale experiments on gas generating reactions will be reported separately. This will include comparison of experimental and calculated mass discharge rates and an assessment of vent sizing calculations which take account of the mass discharged from the reactor before the peak pressure is reached. Methods for calculating the degree of gas-liquid disengagement in the vessel are also examined.

5.5. Pilot-Scale Evaluation of Relief-Stream Disposal Systems

As part of the Commission of the European Community (CEC) Environmental Protection program, a joint CEC project (EV5V-CT94-0532) was set up entitled ***Chemical Hazard Evaluation and Emergency Relief Systems*** (CHEERS). The project consists of two distinct subprojects. The first is concerned with the prevention of chemical reaction hazards (CHEERS-P) and the second with the disposal of emergency relief streams (CHEERS-D). The objective of the CHEERS-D subproject was to develop guidance for the design of emergency relief and disposal systems using, in particular, jet-condenser quenching for the safe disposal of the relief stream from runaway chemical reactions. Three types of reaction system (vapor pressure, gassy and hybrid) were to be investigated. HSL's role in the project was to use two reaction systems (vapor pressure and gassy) to assess the performance of conventional quench systems for comparison with results obtained by other partners using jet condensers.

Two sets of pilot-scale experiments were performed, one set using a sparge pipe (shown in Figure 4) and quench fluid in the catch tank and, otherwise under identical conditions, the other set with no sparger or quench fluid. The results indicated that the use of a sparger causes a substantial reduction in the amount of liquid or condensable vapor vented to atmosphere for both vapor pressure and gassy systems. Under the conditions of the pilot-scale experiments, the sparger reduced the mass of liquid or condensable vapor vented to atmosphere to a level which would be hazardous only for very toxic decomposition products. The detailed results of this investigation are being submitted to the European Commission.

5.6. Round-Robin on Emergency Relief System Sizing

HSE are proposing to make available detailed information on one of the vapor pressure pilot-scale experiments for a vent design Round-Robin to be carried out by volunteer organizations within the US and European DIERS Users Groups.

The aim of the Round-Robin vent design problem is to promote consistency and good practice in emergency relief system design and to provide data for computer code validation. Participants may use any appropriate relief system sizing method, but it is hoped that a range of computer models will be used by different participants. Over 20 organizations have expressed interest in participating.

HSE will provide an information package including relevant dimensions of the pilot-scale facility, operating procedures, initial experimental conditions, adiabatic calorimetric data and kinetic expressions, physical properties data and heat transfer data. Participants will be invited to model the experiment "blind" and submit results for comparison with the experimental results. After the "blind" runs have been submitted, full experimental results will be made available to participants to facilitate understanding of deviations between predictions and experimental results.

If this exercise is successful, then a further Round-Robin on one of the gassy system tests may be put forward.

6. CONCLUSIONS.

1. Although guidance and training on chemical reaction hazard assessment and control is available in the UK, a recent industry survey on standards of relief system sizing for exothermic runaway indicates that further information on this topic is needed. A Workbook on chemical reactor relief system sizing has been completed by HSE which summarizes the main hand calculation methods available as a result of the DIERS project and their limits of applicability. The Workbook is particularly aimed at chemical engineers in small and medium sized companies, and reactor manufacturers and suppliers and is intended to be complementary to the CCPS Guidelines.
2. Research at HSE's Health and Safety Laboratory has shown, to date, that the DIERS methodology provides a sound basis for the design of emergency pressure relief systems for chemical reactors and that, in general, the analytical methods produce conservative vent diameters when used within their stated limits of applicability. There are however some important findings from the research program which indicate areas in which the methodology can be investigated further. This work is continuing and will be reported more fully at a later date.

3. A large archive of experimental data is being established which should be of substantial value to the development of DIERS methodology using both analytical and computational procedures.

REFERENCES

1. Barton, J. A., and Rogers, R. L., *Chemical Reaction Hazards,* 1st ed., IChemE, Rugby, 1995
2. "Control of Exothermic Chemical Reactions," HSE Video, available from CFL Vision, PO Box 35, Wetherby, Yorks LS23 7EX
3. IChemE Training Package, "Control of Exothermic Chemical Reactions," IChemE, 1993
4. Etchells, J. C., and Wilday J., "Workbook for Chemical Reactor Relief System Sizing," HSE Contract Research Report Number 136, ISBN 0 7176 1389 5, HSE Books, 1998
5. *A Guide to the Reporting of Injuries, Diseases and Dangerous Occurences Regulations 1995*, HSE, HSE Books, ISBN 0 7176 1012 8
6. Barton, J. A., and Nolan, P. F., "Runaway Reactions in Batch Reactors," *IChemE Symposium Series* No. 85, 1984
7. Barton, J. A., and Nolan, P. F., "Incidents in the Chemical Industry Due to Thermal Runaway Reactions," *IChemE Symposium Series* No. 115, 1989
8. Etchells, J. C., "Prevention and Control of Exothermic Runaway," IBC Conference on the Assessment and Control of Chemical Reaction Hazards, Dec. 1993.
9. Barton, J. A., and Rogers, R. L., *Chemical Reaction Hazards,* 2nd ed., IChemE, Rugby, 1997
10. HSE Leaflet "Chemical Reaction Hazards and the Risk of Thermal Runaway," HSE 1997
11. Singh, J., "Safe Disposal of Vented Reacting Fluids," HSE Contract Research Report No.100/1996, 1996
12. Jones, D. O., "Process Intensification of Batch Exothermic Reactors," HSE Contract Research Report No.105/1996, 1996
13. Fisher, H. G., et al., *Emergency Relief System Design Using DIERS Technology,* New York: DIERS/AIChE, 1992
14. Duxbury, H. A., "The Sizing of Relief Systems for Polymerisation Reactors," *The Chemical Engineer* (January 1980) 31–37.
15. CCPS, *Guidelines for Pressure Relief and Effluent Handling Systems,* New York: CCPS/AIChE, to be published 1998
16. Parry, C. F., *Relief Systems Handbook,* IChemE, Rugby 1994
17. Melham, G. A., and Fisher , H. G. (eds.), *International Symposium on Runaway Reactions and Pressure Relief Design,* New York: AIChE, 1995
18. Singh, J., In *International Symposium on Runaway Reactions,* New York: AIChE, 1989, pp. 313–330,
19. Singh, J., Waldram, S., and Appleton, M., "Loss Prevention and Safety Promotion in the Process Industries," 8th International Symposium, Antwerp, 1995.
20. Leung, J. C., "Simplified Vent-sizing Equations for Emergency Relief Requirements in Reactors and Storage Vessels," *AIChE Journal*, 32 (10), 1622–1634.
21. Leung, J. C., "The Omega Method for Discharge Rate Evaluation." In *International Symposium on Runaway Reactions and Pressure Relief*, G. A. Melhem and H. G. Fisher (eds.), New York: AIChE, 1995, pp. 367–393,
22. Leung, J. C., "Simplified Vent Sizing Methods Incorporating Two-phase Flow." In *International Symposium on Runaway Reactions and Pressure Relief Design,* G. A. Melham and H. G. Fisher (eds.), New York: AIChE, 1995, pp. 200–236
23. Fauske, H. K., "Pressure Relief and Venting: Some Practical Considerations Related to Hazard Control," *IChemE Symposium Series* No 102, 133–141, 1987

24. Huff, J. E., "Emergency Venting Requirements," *Plant/Operations Progress*, 1(4) (October 1982), 211–229
25. Duxbury H. A., and Wilday, A. J.,"Efficient Design of Reactor Relief Systems," In *International Symposium on Runaway Reactions,* New York: AIChE/CCPS, 1989, pp. 373–393

Choice of Runaway Reaction Scenarios for Vent Sizing Based on Pseudo-adiabatic Calorimetric Techniques

Jean-Louis Gustin
Rhone Poulenc Industrialisation
24, avenue Jean Jaurès F69153 Décines, FRANCE

Emergency Relief Systems to control Runaway Reactions are designed taking into account the occurrence of a two-phase release. Catastrophic accidents such as those of Bhopal, Seveso, Griesheim and others, proved that emergency relief systems should not only prevent vessels from bursting but also provide a reliable release containment system. A two-phase release of the vessel inventory sprayed over the neighborhood is not acceptable. The size and cost of containment systems is such that one cannot afford to accommodate the worst credible scenario.

Runaway reactions may occur in various pieces of equipment : reactors, distillation column boilers, pipes, high inventory storage tanks. Therefore much effort is devoted to the design of safe processes and safe process conditions. These are accomplished by the reduction of runaway scenarios using high integrity interlocks, the prevention of runaway decompositions or polymerizations, and the reduction of inventories of reactive, flammable and toxic chemicals. This results in the optimization of both the process and the size of its emergency relief system.

Runaway reaction hazards are investigated through literature searches, thermal stability screening and experimental investigation of runaway reaction scenarios for normal and abnormal process conditions.

- Thermal stability screening is achieved using DSC/DTA methods.
- The heat of reaction in normal process conditions and the possible controlling reactant accumulation are obtained from reaction calorimetry.
- The systematic investigation of runaway reaction scenarios is best achieved using low phi-factor pseudo-adiabatic techniques such as the Vent Sizing Package, where the process conditions / process deviations are reproduced. Possible decompositions are initiated under temperature scan or isothermal exposure conditions.

VSP tests provide both information on the severity of the runaway reaction and the data necessary for vent sizing. Various processes / process conditions are compared considering their possible runaway scenarios. The safer process / process conditions may be selected on this basis. A wide range of process deviations may be controlled by high integrity safety interlocks which greatly reduce the probability of occurrence. The residual runaway reaction hazard is controlled using emergency relief systems. Examples of this current approach are given based on previous work i.e. The phenol + formaldehyde runaway reaction, the vinyl acetate bulk polymerization and others.

1. INTRODUCTION

The most severe runaway reaction accidents occurred in the Chemical Industry were not caused by the reactor or vessel rupture. The pressure generated by the runaway reaction was released through an emergency relief vent, with significant liquid carry-over. The vent was large enough, though not designed for this scenario, to prevent the vessel rupture.

The severity of the accidents was only related to the release toxic properties and to the area concerned by the released material fallout.

The accidents of Bhopal (1984), Seveso (1976), and Griesheim (1993), mentioned only because they are well documented in the literature, proved that emergency relief systems should not only prevent vessels from bursting but also provide a reliable release containment. The aim of Process Safety in the field of runaway reaction hazards is not only to prevent vessels from bursting but also to design safe processes which will never cause pollution or injuries to the community.

In the three runaway reaction accidents mentioned, the relief system was not designed for the runaway reaction scenario which caused the accident and the vent design conditions were not specified. One should realize that large rupture disks or pressure safety valves on reactors or storage vessels, designed for the fire case or any other scenario without a catchment system, are potential sources of two-phase vapor-liquid release should a runaway reaction occur in that vessel. A two-phase release may also be produced by the unwanted depressurization of a liquid inventory under the pressure of soluble gases such as CO_2, H_2 and others.

The accident case histories have shown that a two-phase release of the vessel inventory, sprayed over the neighborhood, is not acceptable in densely populated countries. A catchment system is necessary where disengagement correlations show that a liquid carry-over is likely. The possible toxicity of the release may be taken into account to evaluate the need for a catchment system. This was the spirit of the Seveso Directive in the European Community, imposing hazard studies to the chemical industry on the basis of inventories of toxic and flammable products. However, this regulation would only consider the inventories of reactants and of the reaction products obtained in normal process conditions.

The products of unwanted, unexpected or unknown reactions initiated in runaway reaction accidents are not considered. The analysis of the final products of runaway reaction accidents such as those mentioned has shown that abnormal products far from the normal process products are obtained. This type of analysis is made only in accident investigations. The possible products of runaway reactions remain unknown and depend very much on the runaway scenario and release conditions. It is reasonable to consider that the release is potentially toxic. This leads to the conclusion that a catchment system must be provided in most cases.

The installation of a catchment system on emergency relief vents is considered for chemical reactors. However, runaway reactions are also possible in storage vessels, column boilers, piping systems, packaging used for the transportation of chemicals. This would imply that complex networks of emergency relief vents should be connected to a great number of catchment and disposal systems in chemical plants.

Considering the accident case histories mentioned above, it appears also that existing vents were not designed for the accident scenario and were occasionally actuated during the runaway event considered and consequently did not mitigate or increased the consequences of the runaway reaction incident. Emergency relief vents, piping and catchment systems may be adequate for a given scenario and have a reverse effect in a different circumstance.

This problem may arise in batch plants designed on a standard basis without knowing the products which will be manufactured in that plant over a long period of time. This situation is common in specialty chemicals, fine chemicals and pharmaceuticals manufacturing industry and less frequent in the petrochemicals, plastics and polymers industry where the same product will be continuously manufactured in the plant.

For many years the chemical companies have been moving from the manufacture of commodity chemicals to the manufacture of fine chemicals. This induced an increased demand for process safety studies concerning fine chemicals multistep processes with small products inventories.

In this type of chemistry, the severity of runaway reaction accidents is lower than in the heavy chemical industry. However the probability of having problems is higher due to the more complex reactions and molecules involved. Here the aim of process safety is obviously more to prevent runaway reaction incidents than to mitigate the consequences of an accident.

Considering more widespread heavy chemical processes, there is no reason why the same care should not be taken for the prevention of chemical accidents thus eliminating the well identified causes of runaway reaction accidents.

The design of safe processes where runaway reaction incident scenarios have been reduced or eliminated is a cost-effective option compared to the control/mitigation of severe runaway reactions using an emergency relief system including a release catchment system.

The study of runaway reaction scenarios to obtain the data necessary for vent sizing allows the identification of dangerous process deviations. This allows the choice of safe process versions and the prevention of dangerous process deviations using safety interlocks.

2. PRACTICAL APPROACH TO THE DESIGN OF SAFER PROCESSES

2.1. On the Interest of Literature Search

In many instances when a runaway reaction chemical accident has occurred, it is considered that the possibility of such an event was unknown and could not be foreseen. However, after severe accidents it is often recognized that the same accident with the same chemistry had already occurred once or even several times. Some processes are known to have very frequent accidents with a limited number of scenarios which are always the same. An inquiry published in the U.K. by the HSE. (1) (2) shows that in the period 1962–1987, 134 accidents involving a runaway reaction in a batch or semibatch process occurred in the U.K. of which 64 were related to polymerization reactions and 13 to the phenol + formaldehyde reaction alone. An information booklet published by the British Plastic Federation (3) on this reaction helped to eliminate this type of accident in the recent years. However accidents are still occurring everywhere with this process.

Considering the accident in Bhopal, it is known that isocyanates give violent exothermic reactions with water, producing amines, CO_2 and many byproducts. A violent pressure surge results from the production of CO_2. The most critical conditions are when the process is not onstream, during cleaning operations, where water can inadvertently come into contact with isocyanates in closed vessels.

Considering the accident in Seveso, 1976, the production of 2, 4, 5, trichlorophenol by alkaline hydrolysis of 1, 2, 4, 5 tetrachlorobenzene was known to have experienced several accidents. There are two processes, one using methanolic sodium hydroxide and the other using ethylene glycol and xylene. The Seveso process was the glycol process. As pointed out by Grewer (4) seven serious accidents had occurred with this reaction before the Seveso accident, four with the methanol process and three with the glycol process, but for different reasons.

In the Seveso accident, the batch had been dehydrated and left at a temperature of 158°C over the weekend. Unfortunately ethylene glycol and caustic soda give exothermic secondary reactions producing sodium polyglycolates, sodium oxalate, sodium carbonate and hydrogen, and exhibiting an autocatalytic behavior. These reactions caused a temperature rise allowing the production of tetrachlorodioxin and the vessel pressurization.

These reactions were studied by A. Wurtz, 1859, (5) for ethylene glycol and KOH and J. U. Neff, 1904, (6) for ethylene glycol and NaOH. After the accident in Seveso, these reactions were also investigated by P. Cardillo and A. Girelli (7).

Considering the Griesheim accident, the same accident, on the same process with the same cause, no stirring during the injection of the controlling reactant, had occurred in 1973 in Japan (8).

After every severe runaway reaction accident it is always possible to establish that this accident did not occur for the first time and that the same accident is already described in case-histories. Some specific processes appear with significant contributions in accidents statistics i.e. polymerizations, nitrations, hydrolysis (1, 2).

When unexpected secondary reactions are involved in runaway reaction accidents, the relevant chemistry is already described in the literature with warnings concerning a possible loss of control of the reaction or gas evolution (9, 10).

In some instances concerning very simple molecules i.e. ethylene glycol, polyglycols and caustic soda, the relevant literature may go back to the beginning of the century or the middle of the nineteenth century and remains undetected by computer literature search.

Therefore, the first important step in a process safety study on a new or existing process, is to proceed to a careful literature search concerning the accidents occurred with this process, their causes and the chemistry involved in the wanted reaction as well as in possible side reactions. This preliminary work is time consuming, but will provide information on possible dangerous process deviations which may be overlooked in the experimental study and in Hazop reviews. A minimum requirement is to take into account this information in the design and operating conditions of the process considered.

2.2. Experimental Investigation of Runaway Reaction Hazard

A runaway reaction accident may result from the thermal decomposition of a reactant, a product or a reaction mixture. Two reactants may be incompatible or give unwanted reactions in the process conditions. The decomposition reaction of these samples may follow n order Arrhenius kinetics or exhibit an autocatalytic decomposition occurring after an isothermal induction period.

2.2.1. DTA/DSC Tests under Temperature Scan Conditions

The thermal stability of representative samples and their decomposition kinetics may be investigated using DTA or DSC experiments as a screening method.

DSC/DTA tests are first performed under temperature scan conditions in closed cells of various materials. Stainless steel, Hastelloy, titanium, or gold plated DTA closed cells or glass ampoules are used. The test cell wall material may influence the results obtained and must be chosen according to the plant equipment construction material.

The rate of temperature scan depends on the experimental set-up thermal inertia. The most common operating rates of temperature scan are 5°C/min. for metal cells and 2°C/min. when using sealed glass ampoules.

A rate of temperature scan of 10°C/min. would be too fast and induce a significant shift in the exotherm onset temperature. 0.5°C/min. or 1°C/min. would produce excellent quality results but would be time-consuming over a temperature range between ambient temperature and 450°C.

The thermograms obtained show various reaction or decomposition exotherms and provide the following data:

- the heat of reaction or decomposition for every exotherm,
- the exotherm(s) onset temperature,
- the exotherm(s) shape,
- the exotherm(s) kinetics.

The heat of reaction or decomposition obtained by integration of the exotherm heat-flux provides a first estimate of the severity of the possible runaway reaction, should this exotherm be initiated.

The following comments on the exotherm obtained is suggested:

$\Delta H_d < 210$ J/g:	Small or medium exotherm
210 J/g $< \Delta H_d <$ 630 J/g:	Significant exotherm
$\Delta H_d > 630$ J/g:	Large exotherm

The exotherm(s) onset temperature is compared to the normal or possible process temperature to check if the exotherm considered may be initiated in normal process conditions or in case of possible process deviation.

If several exotherms are obtained, it is necessary to check if the first exotherm initiated under the process temperature can initiate the second exotherm and the latter, the next exotherm.

This is done by evaluating the adiabatic temperature rise produced by every exotherm assuming a realistic sample specific heat. The sample specific heat may be obtained from DTA/DSC experiments in temperature ranges where no exotherm is obtained. This method provides a first guess of the possible heat release when the first exotherm is initiated under process conditions.

The exotherm shape is also considered. For a given heat of decomposition, a narrow exotherm over a limited temperature range represents a faster reaction than a flat exotherm over a large temperature range. This is also shown by the consideration of the maximum specific heat-flux. An asymmetric exotherm obtained under temperature scan conditions is an evidence of a possible autocatalytic reaction or decomposition. If such evidence is obtained, or if reactive monomers are present, specific tests for autocatalysis are performed using the same DTA/DSC machine.

2.2.2. Tests for Autocatalysis using DSC/DTA Machines

In these tests, the sample is submitted to isothermal exposures in closed cell at three or more temperatures below the exotherm onset temperature obtained under temperature scan conditions.

The exotherm may appear after an isothermal induction period depending on the test temperature, thus proving the decomposition reaction chemical acceleration. The isothermal induction period may be evaluated at the exotherm onset detection or at the maximum heat flux. As shown in reference (11), the isothermal induction period in Log scale is a linear function of the reciprocal temperature with a slope $-E/R$ where E is the reaction activation energy and R the ideal gas constant.

The exotherm maximum specific heat-flux is also an Arrhenius function of the temperature. The maximum specific heat-flux in Log scale is a linear function of the reciprocal temperature with a slope $+E/R$.

The consequence for the process, of the decomposition or reaction autocatalytic behavior, is the possible initiation of the reaction after an isothermal induction period under the process conditions. The autocatalytic behavior of decomposition reactions may explain runaway reaction accidents which occurred at much lower temperature than expected after a long exposure to the process conditions.

Autocatalytic decompositions are widespread in all types of chemistry and are often found in runaway reaction accidents. Therefore their detection is of utmost importance for process safety.

2.2.3. Prediction of the Decomposition Time to Maximum Rate from DTA Experiments

In countries under German and Swiss influence, the kinetics obtained by fitting the reaction/decomposition exotherm with a simple one step Arrhenius equation are used to predict the decomposition adiabatic Time to Maximum Rate as a function of temperature.

This is done also using a formula obtained for zero order reactions (12).

$$\mathrm{TMR}_{\mathrm{ad}} = \frac{C_{\mathrm{p}} R T_0^2}{q_0 \cdot E_{\mathrm{a}}}$$

where C_p is the sample specific heat; q_0 is the maximum specific heat flux measured in an isothermal exposure at the temperature T_0 considered; E_a is the reaction activation energy; and R is the ideal gas constant.

This method gives only a very rough approximation of the Time to Maximum Rate which is very dependent on the care taken to obtain the kinetics from DTA/DSC determinations. Other experimental methods would provide a better estimate of the Time to Maximum Rate i.e. ARC, VSP.

2.2.4. Determination of Decomposition Kinetics from DSC/DTA Experiments

DTA/DSC apparatuses provide software to obtain a kinetic interpretation of decomposition exotherms using simple Arrhenius n-order kinetics. However, this type of kinetic interpretation of DTA experiments is simplistic and should not be used for reaction rate or induction period prediction. More sophisticated kinetics software is available on a commercial basis to obtain kinetics from DTA experiments and other calorimetric techniques (13, 14, 15).

The use of this software requires extensive experimental work using various calorimetric techniques to produce consistent experimental results. Also the use of this software shows that most exothermic reactions are complex combinations of Arrhenius *n*-order and autocatalytic steps. Simple Arrhenius kinetics are very seldom found in practice. Therefore predictions based on thermal kinetics must be carefully cross-checked on an experimental basis. The use of kinetic software for rate prediction is outside the scope of this paper.

2.2.5. Heat-flux Measurements Using the Setaram C80 Isothermal Calorimeter

The careful determination of the heat-flux produced under isothermal conditions by samples is necessary for the study of reactive chemicals storage or packaging. The same problem appears for the determination of the decomposition Time to Maximum Rate using the formula given in paragraph 2.2.3. above, where the maximum heat-flux q_0 released under the temperature T_0 is necessary.

Precise heat-flux determinations under isothermal conditions may be carried-out using the Setaram C80 isothermal calorimeter. The C80 is a wide spread twin-cells isothermal calorimeter using 15 cm^3 in volume tests cells. A sample size of 2–5 g is suitable. Heat-fluxes as low as 10 mW/kg sample may be measured with careful calibration of the calorimeter. This is the order of magnitude of the critical heat-flux for large storage vessels.

The experimental temperature range is between ambient temperature and 250°C. Special test-cells allow the mixing of two reactants, gas–liquid or gas–solid, and reactions to be performed under the mixing effect produced by the agitation. There is also an option which allows heat-flux and gas production measurements to be carried out simultaneously.

The most frequent determinations using the C80 are isothermal heat-flux measurements but measurements under temperature scan conditions may also be performed using slow temperature scanning rates, between 0.1 and 1°C/min. The C80 isothermal calorimeter is especially useful to obtain input data for kinetic modeling of decomposition reactions.

2.2.6. Study of Thermal Stability Using the Accelerating Rate Calorimeter

The Accelerating Rate Calorimeter (ARC) is present in most Process Safety Laboratories in Europe. However, the use of ARC for thermal stability testing is not so extensive as it is in the United States where it is considered as a standard or reference method.

The advantages of the ARC are:

- an excellent onset detection sensitivity,
- a good theoretical background for the experiment kinetic interpretation (16),
- the obtaining of pressure information.

The drawbacks of the ARC are:

- the experiment duration is much longer than DTA tests,
- the correction of complex kinetics obtained to adiabatic conditions is far from being straightforward and needs methods which are not provided with the experimental set-up (17),
- the correction of the pressure record to adiabatic conditions is a difficult problem,
- there is no stirring in the original set-up,
- the need for skilled operators and maintenance problems,
- troubleshooting new software on the PC versus the old control computer.

A new interest for this high inertia pseudo-adiabatic calorimetric technique is to allow the separation of exothermic reaction steps thus allowing kinetic interpretation of each step to be combined into a reaction model. This data provides a high quality input for kinetic models allowing the check of the consistency of the results of other methods. High inertia calorimetric techniques provide better kinetic information than low inertia pseudo-adiabatic techniques where all reaction steps are mixed in a fast runaway reaction.

2.2.7. Use of Reaction Calorimeters

It is a common practice for semibatch or batch processes to reproduce the reaction conditions in a reaction calorimeter to measure the heat of the wanted reaction under isothermal conditions. A simultaneous measurement of the reaction mixture specific heat allows the estimation of the possible reaction adiabatic temperature rise for the whole controlling reactant charge.

One should take care for large adiabatic temperature rise that the reaction mixture specific heat has been measured under the process temperature and would be quite different under high final temperature thus affecting the adiabatic temperature rise estimate. Clearly, reaction calorimeters are not the best experimental set-up to estimate large adiabatic temperature rises.

Reaction calorimeters allow the measurement of the controlling reactant accumulation during operation. The controlling reactant accumulation is thought to be the main cause of runaway reaction accidents in countries under German and Swiss influence.

In case of loss of cooling, the reaction of the accumulated quantity of the controlling reactant would raise the reaction mixture temperature, causing possible vessel pressurization or the initiation of secondary decomposition reactions.

The accumulated quantity of the controlling reactant is obtained by measuring the heat produced after interruption of the controlling reactant injection in the reaction mixture.

The controlling reactant accumulation is influenced by the choice of the process temperature. Too low, a process temperature would cause higher controlling reactant accumulation. Conversely a higher process temperature would minimize the controlling reactant accumulation. Too high a controlling reactant feed rate would also induce reactant accumulation.

Several reaction calorimeter runs may be necessary to optimize both the process temperature and the controlling reactant feed-rate.

These experiments provide information on the expected specific heat production by the reaction mixture and on the reactor cooling requirement. They also allow the optimization of the process productivity.

This type of measurement is also interesting for the study of reactions where an unstable intermediate may accumulate.

However there are other causes of the controlling reactant accumulation i.e. process deviations such as "no agitation" or "no catalyst" which may cause severe runaway reaction accidents.

2.2.8. Autoclave Experiments

Autoclave experiments are used to obtain information on pressure effects during fast reactions or decompositions. Autoclave experiments performed with small autoclaves, such as the Kühner mini-autoclave, are a valuable screening method to obtain information on the pressure evolution to be compared with DTA thermograms. Measurements are made under temperature scan conditions or under isothermal exposure. A standard data processing method for autoclave experiments is described in Ref. (11).

Autoclave experiments are performed prior to proceeding to VSP experiments to avoid having violent explosions in VSP experiments.

2.2.9. Pseudo-Adiabatic Calorimetric Methods

Pseudo adiabatic calorimetric methods are the most suitable experimental techniques to study runaway reactions. They include

- The Accelerating Rate Calorimeter (ARC) already mentioned.
- The closed Dewar experiment.
- The Vent Sizing Package (VSP).
- The Reactive System Screening Tool (RSST).

These methods are well known by DIERS users and are described elsewhere (18). The use of the ARC as a screening method has been considered above.

2.2.10. Closed Dewar Experiments

Closed Dewar experiments using stainless steel Dewar vessels are suitable for long duration experiments under moderate pressures not exceeding 25 bar. The experiment adiabaticity coefficient is typically 1.15–1.20 with some influence of vaporization.

Closed Dewar experiments using glass Dewar vessels are performed when stainless steel Dewar vessels cannot be used. The experiment adiabaticity coefficient is 1.7–1.8 due to the weight and specific heat of glass.

The most frequent Dewar experiments are isothermal exposures. A simple data processing method convenient for Dewar experiments is described in Ref. (11).

2.2.11. Use of the VSP as a Screening Method to Study Runaway Reaction Hazard

The cause of runaway reactions is not only the thermal decomposition of products or reaction mixtures due to too high a temperature or too long an exposure to the process temperature. Most accidents are caused by wrong operating conditions or process deviations which cannot be reproduced in screening methods such as DSC/DTA or ARC experiments.

It is therefore necessary to screen the wanted reaction process conditions to determine if a runaway reaction hazard is present in normal process conditions and in case of various possible process deviations. This can be done readily using VSP experiments.

VSP experiments allow the comparison of various process versions, the direct determination of the wanted reaction adiabatic temperature rise and the check of the possible initiation of secondary reactions.

If no secondary reaction is initiated at the wanted reaction adiabatic final temperature, a further temperature scan allows the determination of the temperature difference between the wanted reaction adiabatic final temperature and the subsequent decomposition reaction onset temperature.

This very important assessment is made under adiabaticity conditions representing process conditions and with the reaction mixture representative specific heat, which was not the case when such estimation was made using Reaction Calorimeter experiments.

VSP experiments are not suitable to measure the controlling reactant accumulation under normal process conditions. This is obtained using reaction calorimeters. However the consequences of the runaway reaction of the previously determined accumulated quantity of controlling reactant is best measured in a VSP experiment where this amount of controlling reactant is injected in a representative reaction mixture under normal process conditions.

VSP experiments are also suitable to assess the consequences of runaway decomposition reactions after their identification using other screening methods (DTA/DSC, ARC). The decomposition is then initiated by a temperature scan or an isothermal exposure, depending on the known kinetic behavior of the reaction.

In some instances where a reliable baseline cannot be determined on DTA thermograms, a VSP experiment allows a better and safer determination of the decomposition exotherm.

VSP experiments provide both thermal information on runaway reactions and information on pressure effects. The type of pressurization, following the DIERS methodology, i.e. vapor pressure, production of noncondensable gases or both, can be obtained from VSP experiments.

A simple straightforward data processing method is described in reference (11) providing an easy understanding of the runaway reaction characteristics and the data necessary for vent sizing.

A wide range of experimental conditions can be achieved using the VSP:

- Temperature up to 350°C–400°C under temperature scan conditions.
- Pressure up to 200 bar in our modified European version.
- Closed or open test cells. Open test cells are connected to a second containment vessel.
- Test cells of various materials including stainless steel, Hastelloy C 276 and glass test cells. Glass test cells of various size and shape are suitable for testing fine chemical and pharmaceutical products/reaction conditions, when the process is in glass vessels only or when small samples only are available, or if the samples are very expansive.
- *Mechanical stirring*:Mechanical stirring is recommended for testing polymerization reactions or viscous reaction mixtures. This requires the use of taller containment vessels than the original one to install the electric motor for the agitator.

Therefore the VSP has grown into a screening method for testing reactions and thermal stability whereas the other screening methods are suitable for testing thermal stability only.

2.2.12. The Reactive System Screening Tool (RSST)

The RSST is a simple pseudo-adiabatic calorimeter which provides some of the capabilities of the VSP. The RSST is familiar to DIERS users and has been described elsewhere (18).

- The reasons for VSP owners to purchase a RSST are as follows:
- The small glass test cell, 10 mL in volume, useful when only small samples are available or if the sample is expansive.
- The possible installation of the RSST on site when the samples cannot be transported.
- The low cost of the RSST.

The drawbacks of the RSST compared to the VSP are as follows:

- The small test cell making the study of process reaction steps more difficult.
- Only open cell tests are possible using the RSST.

3. APPLICATION OF EXPERIMENTAL METHODS TO RUNAWAY REACTION STUDIES AND VENT-SIZING

3.1. The Phenol + Formaldehyde Runaway Reaction

As pointed out earlier, the phenol + formaldehyde reaction is commonly used in the chemical industry to manufacture formo-phenolic resins and has caused many runaway reaction accidents (3).

In a previous paper (19) we considered as an example, a Resole base catalyzed reaction where the formaldehyde to phenol molar ratio was $R = 1.897$. This process includes two separate reaction steps for the methylolation reaction and the salification of the methylolated phenol.

Methylolation Step

In the methylolation step, formaldehyde is reacted with phenol to obtain a methylolated phenol in the presence of a catalytic quantity of caustic soda.

The reaction is described below.

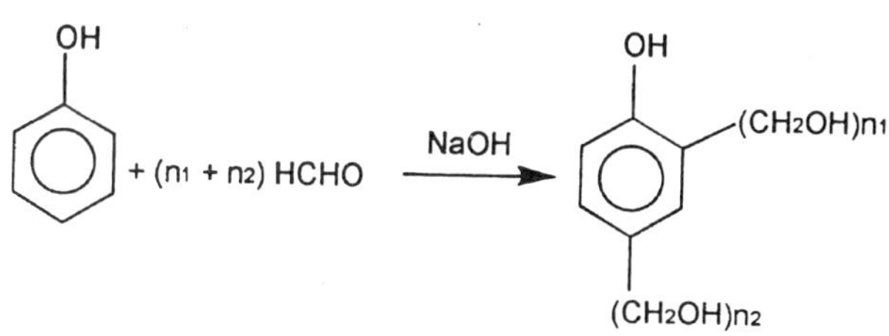

The reaction is carried out in a batch stirred reactor equipped with a condenser. The reactor can be heated or cooled through a jacket. The protection of this reactor by an emergency relief vent is considered.

In the methylolation step, phenol is charged to the reactor with the catalytic amount of caustic soda. The temperature of the mixture is set to 60°C and the aqueous solution of formaldehyde is introduced over two hours, under temperature control at 60°C with cooling through the jacket. The final solution mixture is kept at 60°C to allow the reaction to come to completion. Then the reactor is cooled to 50°C.

The recipe used is defined by Table 1, the table of the net charges.

Anhydrous phenol was used. The formaldehyde used was 37 % by weight aqueous solution containing methanol as a stabilization additive. The caustic soda was 30 % by weight solution. The concentration of water in the methylolation reaction mixture is 35.1 % by weight.

Salification Step

In the salification step, a stoichiometric amount of caustic soda is introduced over 30 minutes under temperature control at 50°C, to obtain a phenate.

The reaction is described below.

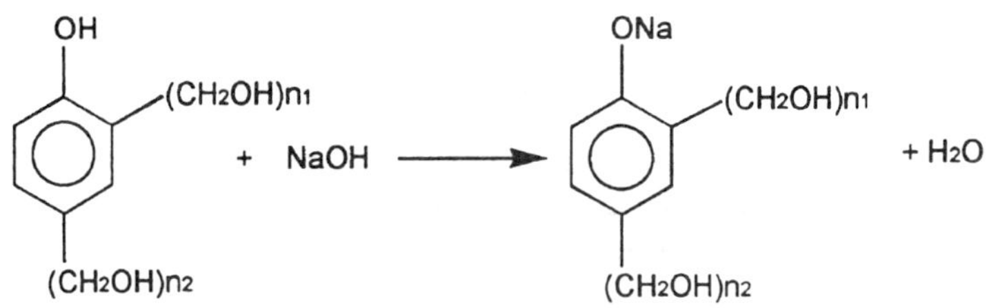

TABLE 1
Net Charges

	kg	% by weight	Mole	% by mole	Mole ratio
Phenol	2200	35.67	23400	11.58	1.000
Formaldehyde	1332	21.59	44400	21.97	1.897
Caustic soda	110.4	1.79	2760	1.36	0.118
Methanol	360	5.84	11250	5.57	0.481
Water	2165.6	35.11	120310	59.52	5.141
Total	**6168**	**100.00**	**202120**	**100.00**	

Process Deviation

There is a well known dangerous process deviation which consists of forgetting to introduce the catalytic quantity of caustic soda in the methylolation step, not detecting the mistake and starting the introduction of the stoichiometric amount of caustic soda in the salification step. Obviously, the reactant of the salification step is the catalyst of the previous methylolation step. Should this process deviation occur, the exothermic methylolation reaction is initiated when the whole inventory of formaldehyde is present and unreacted. The fast temperature rise will trigger the polymerization reaction causing the reactor to pressurize and often to explode, together with the gelification of the reactor inventory.

The gelification reaction is described below:

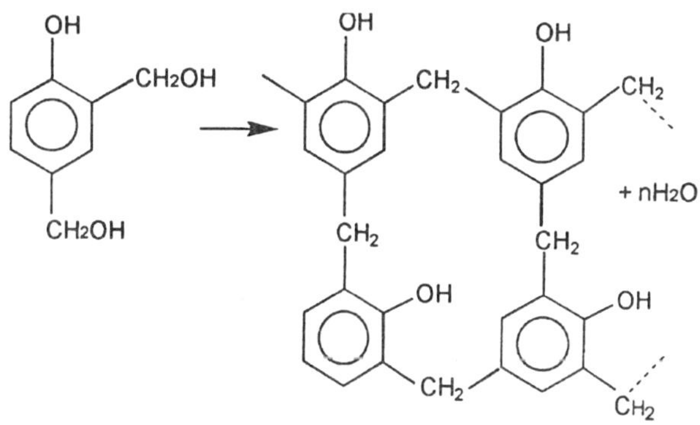

The absence of catalyst in the methylolation step may remain undetected because it has no effect on the reactor temperature which is controlled at 60°C.

To detect this mistake, it is necessary to check the reactor cooling demand during the formaldehyde introduction. The cooling demand does not appear if the catalyst is absent.

Review of the Literature

In their paper on the phenol + formaldehyde reaction, Booth et al. (20) illustrate the amount of formaldehyde reacted on gelification as a function of the formaldehyde to phenol molar ratio. According to this diagram, gelification occurs when 40 % of the overall exotherm is obtained for a reaction mixture with a formaldehyde to phenol molar ratio of 1.9.

Leung et al. (21) give the result of a VSP experiment where the abovementioned worst credible deviation is considered for a reaction mixture with a formaldehyde to phenol molar ratio of 3.2 and a water concentration of 48 % by weight. The reaction mixture was shown to be a high vapor system where pressurization was due to water vapor pressure. The reaction was initiated by a temperature scan on a sample containing the catalytic amount of caustic soda. The exotherm onset temperature was 40°C with a 0.2°C/min. temperature scan. The heat-rate curve showed a one step reaction with an experimental temperature rise of 120°C.

Booth et al. (20) gave an estimate of the heat of reaction for the linear polycondensation, i.e. 723 cal/g of formaldehyde. On this basis, assuming a reaction mixture specific heat of 0.7 cal/g/°C, Leung et al. (21) concluded that the theoretical value of the heat of reaction accounts for only 65 % of the experimental temperature rise measured in the VSP.

Their conclusion was that the polymerization is non linear. Further to our previous paper on the phenol + formaldehyde runaway reaction (19), Kumpinski (22) discussed the influence of the formaldehyde to phenol ratio and reaction mixture pH on the runaway reaction heat-rate measured in the RSST machine.

Also interesting in this paper is the discussion on the heat of reaction of the methylolation reaction and on the heat of reaction of the combined methylolation and condensation reactions.

Experimental Study of the Phenol + Formaldehyde Runaway Reaction

The runaway reaction was first studied using isothermal calorimetric methods DTA/DSC and Setaram C80 isothermal calorimeter.

A sample of formaldehyde + phenol mixture with the recipe composition and molar ratio but without caustic soda was studied in DTA stainless steel closed cell under 5°C/min. temperature scan conditions.

A single one step exotherm of 2416 J/g formaldehyde was obtained between 170°C and 255°C. See Figure 1.

The same experiment was carried out on a unreacted reaction mixture with the recipe composition, containing the catalytic amount of caustic soda.

A complex three-step exotherm of 3381.6 J/g formaldehyde was obtained between 38°C and 192°C. See Figure 2.

The methylolation reaction was carried out with the process recipe and the catalytic amount of caustic soda, under isothermal conditions at 60°C in the Setaram C80 isothermal calorimeter under the agitation provided by the thermostat continuous revolution. An exotherm of 1032.5 J/g formaldehyde is obtained over four hours. See Figure 3.

The final reaction mixture of the Setaram C80 experiment was further studied under temperature scan in DTA and in the C80 itself.

In a DTA experiment in stainless steel closed cell under 5°C/min. temperature scan conditions, a two-step exotherm of 1329.2 J/g formaldehyde is obtained between 62°C and 207°C. There is a possible further exotherm above 300°C. See Figure 4.

In the Setaram C80 calorimeter under 1°C/min temperature scan conditions, the methylolation reaction final reaction mixture presents a two-step exotherm of 1467.2 J/g formaldehyde between 67°C and 200°C. See Figure 5.

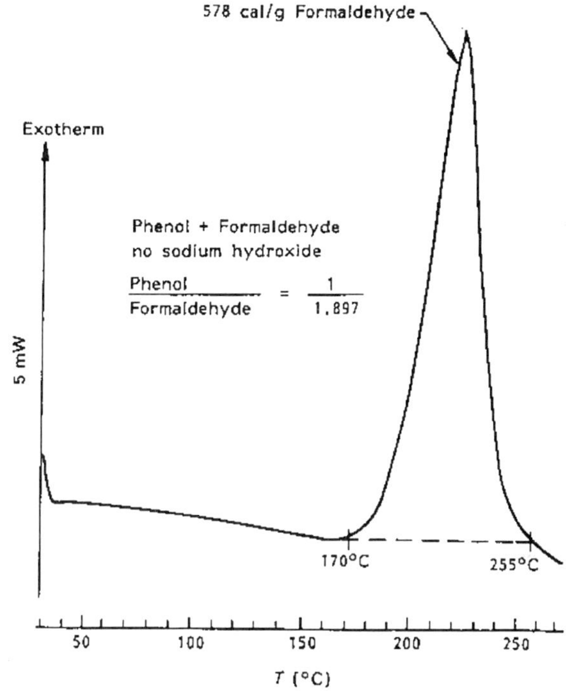

Figure 1. DTA experiment on a mixture of phenol and formaldehyde without caustic soda

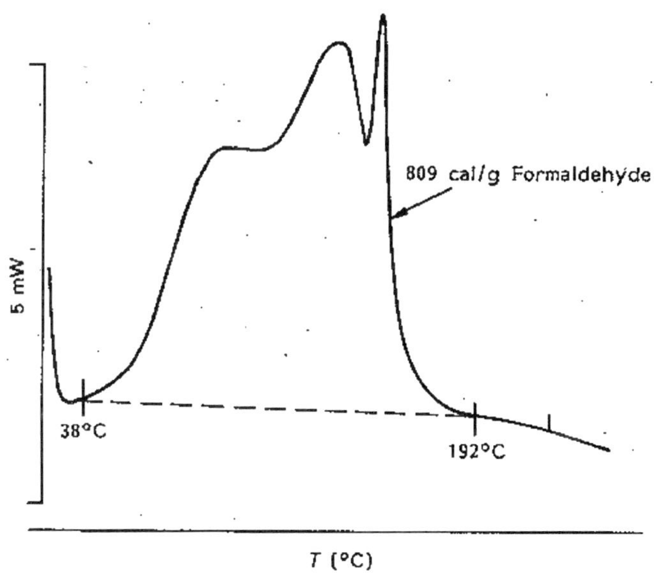

Figure 2. DTA experiment on a mixture of phenol and formaldehyde with the catalytic amount of caustic soda.

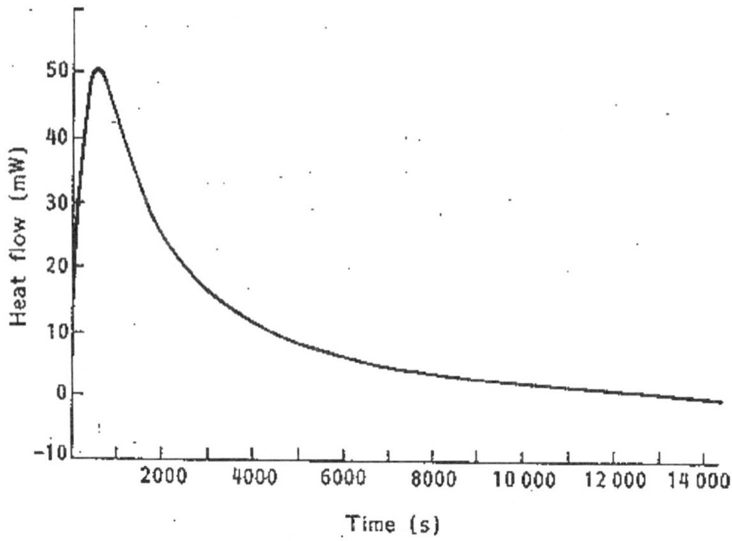

Figure 3. Methylolation reaction carried out under isothermal conditions at 60°C in the calorimeter C80.

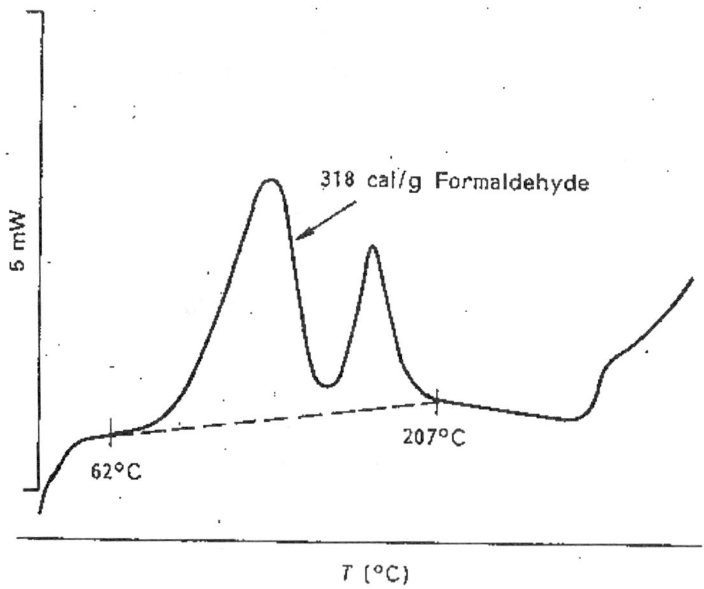

Figure 4. DTA experiment on a reaction mixture with the catalytic amount of caustic soda after methylolation carried out under isothermal conditions in the calorimeter C80.

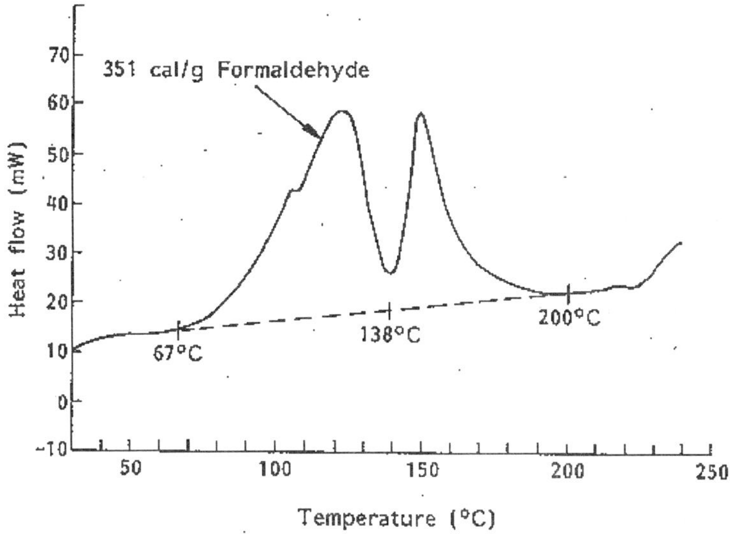

Figure 5. Reaction mixture after methylolation reaction carried out at 60°C in the calorimeter C80 submitted to a temperature scan in the same calorimeter.

The two step exotherm obtained in DTA and C80 isothermal calorimeter, on the methylolation final reaction mixture, is attributed to the methylolated phenol polymerization reaction.

The polymerization reaction is a two-step reaction. The results obtained in this simple experimental approach are summarized in Table 2.

Application of Experimental Results to Process Safety

The results obtained indicate that the methylolation reaction does not occur at 60°C if the catalyst is not present. In the presence of the catalytic amount of caustic soda, the unreacted reaction mixture presents a large exotherm initiated at a temperature well below the process temperature of 60°C or 50°C. In this exotherm, the methylolation and polymerization reactions are not separated. Consequently, there is a runaway reaction hazard, related to a process deviation where the reaction is initiated at 60°C or 50°C by introduction of the catalyst on the unreacted reaction mixture where no catalyst was present.

At 60°C, the methylolation reaction is fairly active in the presence of the catalyst. Limited controlling reactant (formaldehyde) accumulation should occur. However, if the reaction mixture temperature is allowed to decrease below 60°C where the reaction is slower or even 38°C where the reaction does not initiate, the formaldehyde feed to the reactor would accumulate because the process temperature is too low.

Should the reaction of the accumulated quantity of formaldehyde be initiated, the reaction exotherm could trigger also the methylolated phenol polymerization which is active a few degrees Celsius above 60°C.

TABLE 2
Experiments in Twin Cell Calorimeters

	C80	DTA	DTA	DTA
Experimental technique	Phenol, formaldehyde and caustic soda: normal mixture			No caustic soda
Methylolation (isothermal conditions)				
Isothermal temperature	60°C			
Exotherm/g of formaldehyde	1032.5 J/g			
Exotherm/g of sample	222.8 J/g			
ΔT_{AB} (estimated)*	76.2°C			
Polymerization				
Onset temperature	67°C	62°C		
Final temperature	200°C	207°C		
Exotherm/g of formaldehyde	1467.2 J/g	1329.2 J/g		
Exotherm/g of sample	316.8 J/g	287.2 J/g		
ΔT_{AB} (estimated)*	108.2°C	98°C		
Shape	2 steps	2 steps		
Methylolation + Polymerization				
	Total observed	Total observed	Combined reaction	
Onset temperature			38°C	170°C
Final temperature			192°C	225°C
Exotherm/g of formaldehyde	2499.6 J/g	2361.7 J/g	3381.6 J/g	2416 J/g
Exotherm/g of sample	539.2 J/g	510 J/g	729.8 J/g	
ΔT_{AB} (estimated)*	184.4°C	174°C	249.5°C	
Shape			3 steps	1 step

*C_p = 0.7 cal/g/°C = 2.926 J/g/°C

The controlling reactant can accumulate either because the reactor temperature is too low or because the formaldehyde introduction is too fast. In both cases, the reaction of the accumulated reactant will be initiated due to its increasing concentration in the reaction mixture or due to small temperature fluctuations during operation.

Another common cause of the controlling reactant accumulation is "no agitation" in which case a large amount of unreacted formaldehyde could react suddenly with violence if the agitation was restarted.

In addition, the polymerization reaction can be initiated if, during formaldehyde introduction or later, the reactor temperature is allowed to rise a few degrees above normal process temperature. The process presents a runaway reaction hazard due to too high a temperature resulting from loss of cooling or inadequate temperature control.

Concerning the ratio of the total exotherm obtained when gelification would occur, the Setaram C80 experiments show that the ratio of the methylolation heat of reaction to the combined heat of reaction for the methylolation and the polymerization is :

$$\frac{1032.5\ \text{J/g formaldehyde}}{2249.6\ \text{J/g formaldehyde}} = 0.413$$

This is very close to the prediction of Booth et al. (20) but this is true for the reaction carried out under isothermal conditions. If the combined exotherm is obtained in a DTA experiment, a higher heat of reaction of 3381.6 J/g formaldehyde is obtained and the heat of reaction of the methylolation is only 30.5 % of the heat of reaction of the combined reactions.

The above conclusions are reached without any kinetic analysis of the experimental results. A complex kinetic analysis of the experimental results obtained would be inadequate since the experiments were made with limited or no agitation. Agitation is expected to have an influence on the reaction kinetics.

Results from ARC Experiments

Two ARC Experiments were completed on the initial reaction mixture containing the catalytic amount of caustic soda. One in a stainless steel bomb and one in a titanium bomb. The characteristics and results of the ARC experiments are given in Table 3. The heat rate curves obtained in the ARC experiments are given on Figure 6.

TABLE 3
ARC Tests

TEST	ø	T_0 (°C)	ΔT_{AB} (°C)	$(dT/dt)_{max}$ (°C/min)	ΔH_R (J/g)	ΔHr/HCHO (J/g)	ΔT_A (kJ/mol)	n
208 (1) Stainless steel	1.46	31	199.1	7.07	491.6	2273.9	75.24	1
209 (2) Titanium	1.247	28	182.7	10.10	453.1	2098.4	71.1	1

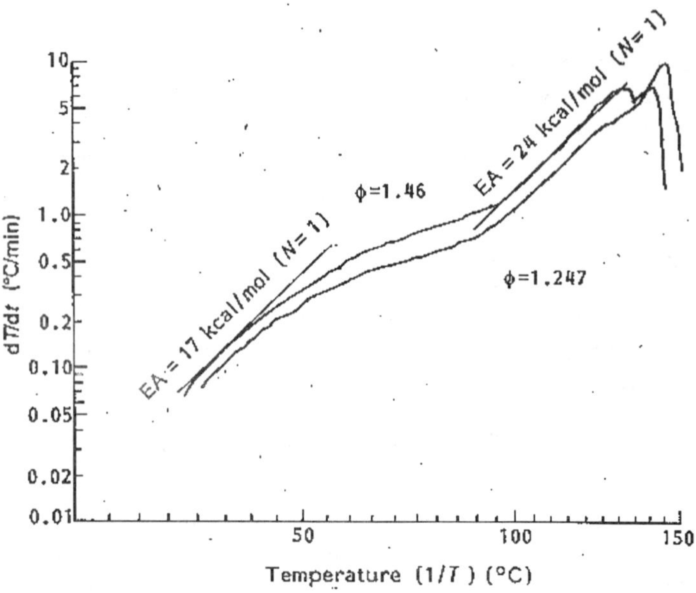

Figure 6. ARC tests. Heat rate curves

The reaction appeared to occur in three steps in the ARC experiments. The first step is the methylolation reaction. The methylolation reaction exotherm is not separated from the polymerization exotherm.

Results from Closed Dewar Experiments

A series of stainless steel closed Dewar experiments were completed to study two problems related to the worst credible process deviation i.e.. The introduction of the catalyst, caustic soda, when all the charge of formaldehyde is present in the reactor and unreacted.

Influence of the Quantity of Caustic Soda Introduced

In previous work on this subject (21) it is assumed that only the catalytic amount of caustic soda is added to the reaction mixture. This may not be the worse case.

To study this question, the reaction was initiated in a 1-liter stainless steel closed Dewar on mixtures containing:

- half the catalytic quantity of caustic soda (test 1),
- the catalytic quantity of caustic soda (test 2),
- twice the catalytic quantity of caustic soda (test 3),
- four times the catalytic quantity of caustic soda (test 4).

The table of the charges is given in Table 4.

TABLE 4
Phenol + Formaldehyde Runaway Reaction. Dewar Tests Initiation by a Temperature Scan

Test number	Charge of phenol (g)	Charge of 37% w/w formaldehyde (g)	Charge of 30% caustic soda (g)
1	71.5	116.7	6
2	71.5	117.6	12.3
3	71.7	118.0	24.6
4	71.7	120.0	49.0

The reaction was initiated by setting the furnace temperature to 60°C. The reaction onset was 35°C.

The effect of increased quantity of caustic soda is a dilution of the reaction mixture decreasing the reaction final temperature and final pressure. See Table 5. Nevertheless, the maximum heat-rate is obtained in the presence of the catalytic quantity of caustic soda specified in the process conditions (test 2). It is concluded that the most severe runaway reaction would be obtained in the presence of the catalytic quantity of caustic soda.

Influence of the Reaction Initiation Method

A fifth Dewar test was carried out where a reaction mixture was heated to 60°C before the reaction was initiated by injection of the catalytic quantity of caustic soda (test 5). The comparison with the Dewar test N°2 shows that the maximum heat-rate is increased by a factor of 2 when the reaction is initiated by the catalyst introduction. See the comparison of test 2 and test 5 in Table 6.

This result can be accounted for by the higher reaction onset temperature in Dewar test N° 5.

TABLE 5
Results of the Dewar Tests

Test number	Mass of caustic soda (g)	Charge of reaction mixture (g)	T_0 (°C)	T_{max} (°C)	$T_{max} - T_0$ (°C)	P_{max} (bar)	$(dT/dt)_{max}$ (°C/min.)
1	6	194.2	35	192.8	157.8	15.91	73.18
2	12.8	201.4	35	184.1	149.2	12.15	94.11
3	24.6	214.3	35	180.1	145.1	10.75	42.83
4	49.0	240.7	35	174.1	139.2	9.49	21.39

TABLE 6
Comparison of Dewar Tests 2 and 5. Influence of the Reaction Initiation Method in the Presence of the Catalytic Quantity of Caustic Soda

Test number	T_0 (°C)	T_{max} (°C)	$T_{max} - T_0$ (°C)	P_{max} (bar)	$(dT/dt)_{max}$ (°C/min.)
2	35	184.18	149.20	12.15	94.11
5	60	206.30	146.30	12.56	212.02

2: Initiation by a temperature scan
5: Initiation by introduction of the catalyst

The closed Dewar experiments carried out in this study could have been replaced by VSP experiments with the benefit of a better adiabaticity coefficient. The closed Dewar experiments estimated phi factor was 1.15 to be compared to a phi factor of 1.05 for VSP experiments. The heat-rate curve and pressure versus temperature curve obtained in Dewar test N°5 are shown in Figures 7 and 8.

Results from VSP Tests

Dewar tests N°2 and 5 were repeated in the VSP where the phenol + formaldehyde reaction was carried out with the catalytic quantity of caustic soda and two methods of initiation:

- initiation by a temperature scan (test 6),
- initiation by catalyst introduction at 60°C (test 7).

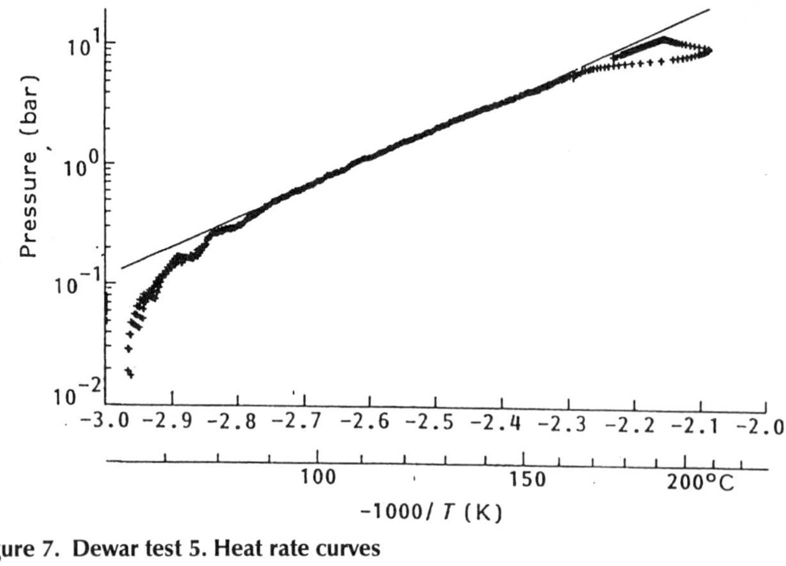

Figure 7. Dewar test 5. Heat rate curves

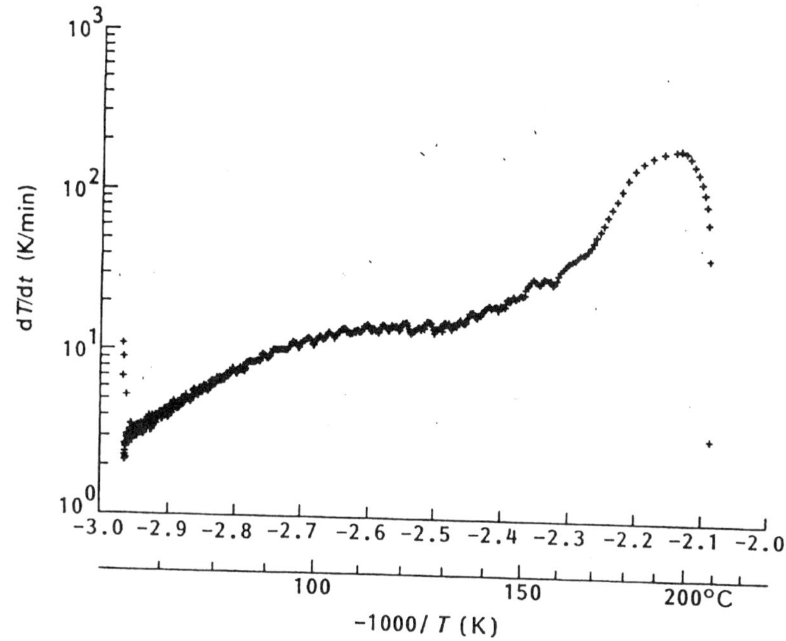

Figure 8. Dewar test 5. Pressure corrected for the nitrogen pad *versus* temperature

The charges used in these experiments are given in Table 7. The results obtained are given in Table 8. The results of the VSP test 7 are given in Figures 9 and 10.

TABLE 7
VSP Tests—Charges Used

Test	Charge of phenol (g)	Charge of 37% formaldehyde (g)	Charge of 30% caustic soda (g)
6	28.6	46.6	4.8
7	28.6	46.6	4.8

TABLE 8
VSP Tests—Experimental Results

Test	T_0 (°C)	T_{max} (°C)	$T_{max} - T_0$ (°C)	P_{max} (bar)	$(dT/dt)_{max}$ (°C/min.)
6	47.5	222.20	174.70	29.93	278.05
7	60.0	238.97	178.97	36.93	682.41

6: initiation by a temperature scan
7: initiation by introduction of the catalyst.

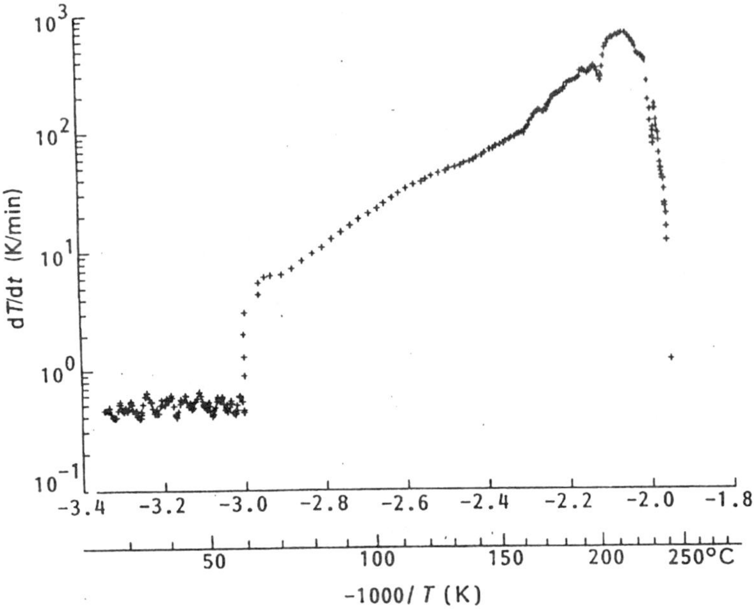

Figure 9. VSP test 7. Heat rate curve

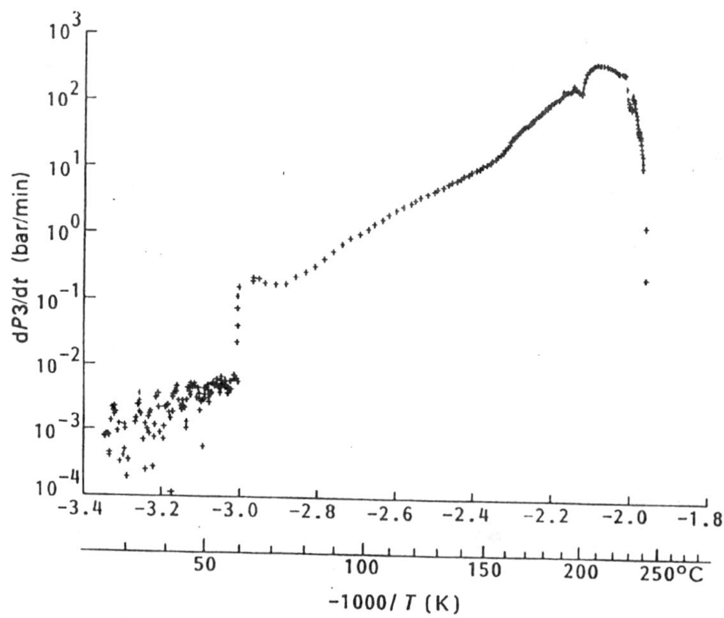

Figure 10. VSP test 7. Pressure rise rate curve

The maximum heat-rates obtained in the VSP tests are greater than the maximum heat-rates obtained in the Dewar tests by a factor 3.

The experimental temperature rise, the final temperature and the final pressure are also higher. These results explain directly why this runaway reaction is very dangerous and leads to the destruction of plants. A comparison of the results of Dewar test N°2 and 5 and VSP tests N°6 and 7 are given in Table 9.

- These results show the advantage of using VSP experiments for testing runaway reactions.
- The phi factor is very close to one and the experiment adiabaticity conditions are representative of the plant conditions if a runaway reaction is initiated.
- The process deviation considered can be reproduced in the VSP and compared to other reaction initial conditions.
- Mechanical agitation could be used but was not available when this study was made.
- The heat-rate and pressure versus temperature curves are readily obtained and provide the information necessary for vent-sizing.
- The correction to adiabatic conditions is small.

Application of Experimental Results to the Prevention of Runaway Reaction Accidents

The process considered is very critical and a number of safety interlocks are recommended to prevent runaway reaction accidents:

1. Interlocks to prevent wrong reactant introduction in the methylolation step
   - The formaldehyde introduction must not be allowed if the catalytic quantity of caustic soda has not been previously introduced.

TABLE 9

Comparison of Results of Experiments with the Same Reaction Mixture Composition in the Presence of the Catalytic Quantity of Caustic Soda

Test	Apparatus	T_0 (°C)	T_{max} (°C)	T_{max} - T_0 (°C)	ø	ΔT_{AB} (°C)
2	Dewar (1)	35.0	184.2	149.2	1.15	171.6
5	Dewar (2)	60.0	206.3	146.3	1.15	168.4
6	VSP (1)	47.5	222.2	174.7	1.05	183.4
7	VSP (2)	60.0	239.0	179.0	1.05	187.9

(1) initiation by a temperature scan

(2) initiation by introduction.

- The formaldehyde introduction must be interrupted if the cooling demand is not obtained in the reactor jacket cooling loop.
- The catalyst must not be introduced after the beginning of formaldehyde injection.
- The introduction of the controlling reactant, formaldehyde, must be prevented or interrupted if there is no agitation.

2. Interlocks to prevent the controlling reactants accumulation:
   - Low temperature interlock interrupting the formaldehyde introduction.
   - Limitation of the formaldehyde feed-rate by a diaphram.
3. Interlocks to prevent the reaction mixture polymerization by too high a temperature:
   - High temperature interlock interrupting the formaldehyde introduction and providing full cooling capacity on the reactor jacket.

In addition, the methylolation final reaction mixture must not be kept for a long period of time without temperature control in a large reactor at the process temperature or under ambient temperature because the polymerization proceeds slowly and may speed up.

Safety interlocks must be high integrity interlocks, hard wired or by high integrity software control techniques and must be redundant.

The merit of high integrity interlocks may be taken into account to reduce runaway reaction accident scenarios ; however, for the process considered only the worst credible deviation is taken into account in vent sizing calculations.

Vent Sizing to Protect a Reactor from the Phenol + Formaldehyde Runaway Reaction

The vent size is based on the results of the VSP test 7 where the highest heat-rate was obtained. In this experiment, the reaction was initiated by introduction of the catalytic quantity of caustic soda into the reaction mixture initially at 60°C. This is the worst credible deviation in this process. A temperature of 60°C was preferred to a temperature of 50°C as in normal process conditions, for the introduction of the stoichiometric quantity of caustic soda, to simplify operation and obtain conservative results.

The choice of the vent sizing method and of the two phase flow calculation method was made according to the reaction system characteristics:

- high vapor system
- flashing two-phase flow.

Vent sizing is achieved considering the adiabatic heat-rate $(dT/dt)_{TS}$ at the reaction mixture boiling point T_s under the vent set pressure P_s.

The adiabatic heat-rate is obtained from the measured heat-rate by:

$$\left(\frac{dT}{dt}\right)_{T_S\cdot\phi=1} = \phi \times \left(\frac{dT}{dt}\right)_{T_S\cdot\phi>1}$$

Gelification temperature:

As the reaction mixture gelification is effective when 40% of the reaction exotherm is obtained, we consider emptying the reactor before gelification of the reaction mixture. The gelification temperature is:

$$T_G = T_0 + 0.4 \times \Delta T_{AB} = T0 + 0.4 \times \phi \times \Delta T_{exp}$$

$$T_G = 60°C + 0.4 \times 1.05 \times 179°C = 135°C$$

Vent sizing using the Leung equation (23):

The Leung equation is derived from the mass and energy balances during venting. The required vent area is given by:

$$A = \frac{m_0 \times \frac{C_{PL}}{2} \times \left[\left(\frac{dT}{dt}\right)_{T_S} + \left(\frac{dT}{dt}\right)_{T_{max}}\right]}{G \times \left[\left(\frac{V}{m_0} \times T_s \times \frac{dP}{dt}\right)^{1/2} + \left(C_{PL} \times \Delta T\right)^{1/2}\right]^2}$$

where

$$G = \frac{dP}{dT} \times \left(\frac{T_S}{C_{PL}}\right)^{1/2}$$

The results are given in Table 10.

TABLE 10
Vent Size Requirement Using Leung Equation

P_s (atm)	T_s (°C)	$(dT/dt)_{TS}$ (°C/min.)	$ø \times (dT/dt)_{TS}$ (°C/min.)	A 6 tons (m^2)	Diameter (m)
1	100	22	23	0.18	0.48
2	120	40	42	0.123	0.39

The following data are used in this equation: C_{PL} = 2926 J/kg/K; m_0 = 6000 kg; V = 7.5 m^3.

To allow understanding of detailed numerical calculations, the following example calculations are given:

For P_S = 1 bar absolute and T_S = 100°C

$$A = \frac{6000\ \text{kg} \times \dfrac{2926\ \text{J/kg/K}}{2}\left[\dfrac{23^\circ\text{C/min}}{60} + \dfrac{23.7^\circ\text{C/min}}{60}\right]}{4743\ \text{Pa/K}\left(\dfrac{373\text{K}}{2926}\right) \times \left[\left(\dfrac{7.5\ \text{m}^2}{6000\ \text{kg}} \times 373\text{K} \times 4347\ \text{Pa/K}\right)^{1/2} + \left(2926\ \text{J/kg/K} \times 4.4^\circ\text{C}\right)^{1/2}\right]}$$

$A = 0.183\ \text{m}^2$.

For P_S = 2 bar absolute and T_S = 120°C

$$A = \frac{6000\ \text{kg} \times \dfrac{2926\ \text{J/kg/K}}{2}\left[\dfrac{42^\circ\text{C/min}}{60} + \dfrac{47.25^\circ\text{C/min}}{60}\right]}{8968\ \text{Pa/K}\left(\dfrac{393\text{K}}{2926}\right) \times \left[\left(\dfrac{7.5\ \text{m}^2}{6000\ \text{kg}} \times 393\text{K} \times 8968\ \text{Pa/K}\right)^{1/2} + \left(2926\ \text{J/kg/K} \times 4.4^\circ\text{C}\right)^{1/2}\right]}$$

$A = 0.123\ \text{m}^2$

Conclusions from vent sizing calculations:

Considering the results of vent size calculations, an ideal vent size of 50 cm in diameter is suggested, with a set pressure as close as possible to ambient pressure. This vent size does not take into account the vent line length. The mass flux reduction or vent size increase to account for a 20 m length horizontal vent line (L/D = 40) has negligible influence on the vent diameter which remains ≈50 cm.

With this vent size the reactor emptying time will be of the order of magnitude of 10-20 seconds. No prediction is made on vapor-liquid disengagement in the reactor. The reaction mixture viscosity is expected to increase and make liquid carry-over easier. Complete emptying of the liquid inventory is likely in the present case.

During atmospheric venting the temperature is controlled between 100°C and 120°C and should not reach the gelification temperature as shown by the VSP test 7 history.

3.2. The Bulk Polymerization of Vinyl Acetate

The bulk polymerization of vinyl acetate is not used in industrial operations due to the violent reaction obtained. However the unwanted bulk polymerization of vinyl acetate is involved in many industrial accidents where the polymerization occurred in storage vessels containing fresh or recycled monomers or in premix vessels where a polymerization initiator was dissolved in vinyl acetate monomer, previous to its use in the polymerization process.

Therefore, the study of vinyl acetate bulk polymerization is of great interest from a process safety point of view, even if it is not a commercial widespread process.

The bulk polymerization of vinyl acetate in storage vessels occurs spontaneously under constant temperature conditions due to a chemical acceleration phenomenon related to the free-radical nature of vinyl acetate chain polymerization.

3.2.1. Polymerization Incidents in Premix Vessels

It is a surprising practice, from a process safety point of view, that concentrated solutions of polymerization initiators are used in premix to be injected in a continuous or semibatch polymerization process.

Premix vessel polymerization accidents are known for many reactive monomers. This type of incident has occurred also in vinyl acetate polymerization processes.

1.8 % wt dilauroyl peroxide was dissolved in vinyl acetate in a premix to be injected in a VAM polymerization process. The premix was prepared in advance to be used 8 hours later. Due to the hot summer temperature, the premix happened to polymerize. The reaction was extremely violent.

A two-phase flow of polymer and vapor was ejected in the vessel vent header. The release did not ignite.

This incident has been investigated on an experimental and theoretical point of view to assess the conditions of occurrence and to evaluate the venting requirement of this scenario.

Our comment on this issue is that the practice of dissolving the polymerization initiator in the reactive monomer in a premix should be eliminated from polymerization processes for process safety reasons.

3.2.2. Cause of the Isothermal Induction Period Phenomenon

In storage vessels of monomer where some polymerization inhibitor is present, the polymerization induction period is the time necessary for the thermally generated *radicals* R• *or* M• to consume the polymerization inhibitor and produce enough radical concentration to initiate the chain polymerization reaction.

Radicals are produced by the polymerization initiator if any, or by the monomers themselves. Consequently, the isothermal induction period is influenced by the inhibitor and polymerization catalyst concentrations in the monomer.

In the case of commercial monomer storage, there should be no polymerization catalyst present, and the induction period is the time necessary for the monomer thermally generated radicals to consume the inhibitor concentration (24).

$$\tau = \frac{m[Z]}{k_0 e^{-E/RT} \cdot [M]}$$

where m is the number of monomer radicals reacting with each molecule of inhibitor Z and $k_0 e^{-E/RT} \cdot [M]$ is the rate of radical generation by the monomer.

$$\tau = K[Z]e^{-E/RT}$$

For storage vessels of commercial monomer, the slope of the Arrhenius plot of the isothermal induction period is a characteristic of the monomer. This may not be the case if a polymerization catalyst is present and if the inhibitor is eliminated first. Then the isothermal induction period is:

$$\tau = \frac{m[Z]}{k_0 e^{-E/RT} \cdot [M] + k_{d0} e^{-E_d/RT} [I]}$$

In this case, the polymerization inhibitor is consumed by the monomer thermal generation of radicals and by the decomposition of the polymerization initiator.

Under moderate temperature where the polymerization initiator is active and the thermal generation of monomer radicals negligible compared to that of the initiator, the above equation reduces to:

$$\tau = \frac{m[Z]}{k_{d0} e^{-E_d/RT} [I]}$$

In this case, the polymerization isothermal induction period is an Arrhenius function of the inhibitor and polymerization initiator concentrations with an activation energy E_d characteristic of the polymerization initiator:

$$\tau = K[Z][I]^{-1} e^{-E_d/RT}$$

The slope of the Arrhenius plot of the induction period is a characteristic of the polymerization radical initiator.

The Arrhenius plot of the polymerization induction period is a key piece of information to prevent runaway polymerization accidents. Also interesting is the experimental data on the polymerization exotherm, rate of reaction, pressure effect.

3.2.3. DTA Determinations

VAM samples containing 1.4 % wt, 1.8 % wt and 2.2 % wt dilauroyl peroxide were submitted to DTA test under 5°C/min. temperature scan conditions in a stainless steel closed cell. (Fig. 11).

The samples exhibit two separate exotherms:

- the first asymmetric exotherm, $\Delta Q1 = 639–676$ J/g is due to the monomer initiated radical polymerization,
- the second exotherm of $\Delta Q = 296–469$ J/g is due to the monomer high temperature thermal polymerization.

This is proved by the uncatalyzed thermal polymerization of VAM giving up an exotherm of 1221 J/g above 290°C.

The polymerization overall exotherm of 970–1221 J/g is of the order of magnitude of VAM heat of polymerization known in the literature (25). $\Delta H = -1036$ J/g.

3.2.4. Polymerization Induction Period of Vinyl Acetate in the Presence of 1.8 % wt Dilauroyl Peroxide

The bulk polymerization of vinyl acetate containing 1.8 % wt of radical initiator dilauroyl peroxide can be initiated by constant temperature exposure under different temperatures. The polymerization induction period was measured using various experimental techniques:

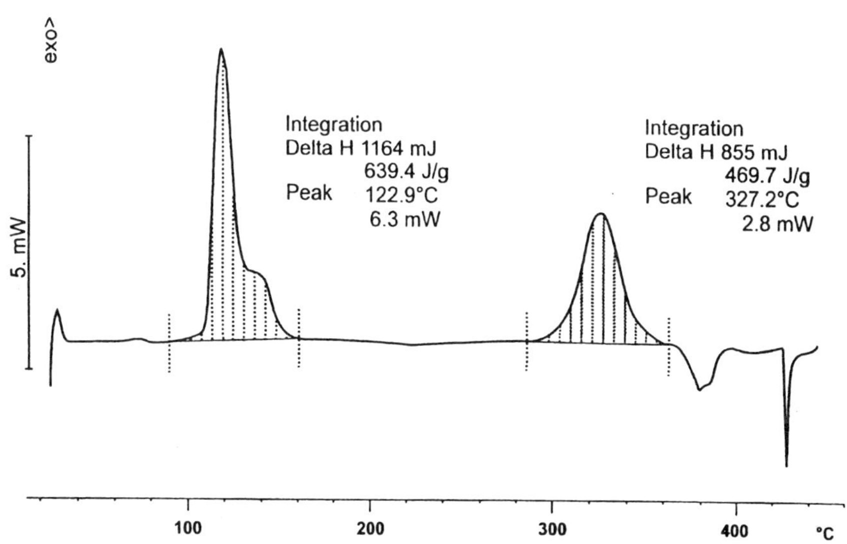

Figure 11 : Polymerization of vinyl acetate monomers containing 1,4 % w/w dilauroyl peroxide. DTA determination—Stainless steel closed cell. 5°C/min. temperature scan

- DTA for high temperature short induction periods
- closed Dewar experiments and VSP experiments for low temperature long induction periods.

A summary of results is given in Table 11.

The isothermal induction period is given for the exotherm onset detection since in some VSP experiments the test cell exploded due to the very fast pressure surge. Consequently, the maximum rate of reaction could not be observed.

The Arrhenius plot of the polymerization induction periods obtained was given in ref. (11).

A change in peroxide concentration between 1.4 % wt and 2.2 % wt does not induce significant change in the induction period. Induction periods in the range of 5–30 min. obtained in DTA are not reproducible.

3.2.5. VSP Closed Cell Experiments

The runaway polymerization of vinyl acetate containing 1.8 % wt dilauroyl peroxide was initiated on a 80 g sample under isothermal exposure conditions in a 116 cm^3 Hastelloy C VSP closed cell. The isothermal exposure temperatures were 37°C and 45°C. In one experiment at 37°C the test cell did not

TABLE 11

Polymerization Induction Period of Vinyl Acetate Containing 1.8 % wt Dilauroyl peroxide, Measured under Isothermal Exposure Conditions in Stainless Steel Closed Dewar, Hastelloy C276 VSP Closed Cell, Stainless Steel DTA Closed Cell

Experimental method	**Temperature (°C)**	Polymerization induction period (min.)
Closed Dewar	37	1100
VSP closed cell	37	790
VSP closed cell	37	820
VSP closed cell	45	425
DTA	65	35
DTA	70	34
DTA	80	8
DTA	80	19
DTA	85	11
DTA	90	5

rupture. The maximum heat-rate was 4967°C/min. The maximum rate of pressure rise was 2544 bar/min. i.e. 42.4 bar/s.

The maximum rates were obtained for a temperature of 144–174°C.

The final temperature was 268°C and the final pressure 44.5 bar.

The heat-rates obtained are given on Fig. 12.

The pressure corrected for the nitrogen pad as a function of the reciprocal temperature is given on Fig. 13 for the three experiments. The VAM vapor pressure is also represented.

After the polymerization initiation, the heat-rate curve is the same for the three experiments because the heat-rate is controlled by the test thermal inertia which was the same for the three experiments as long as the test cell did not rupture.

The pressure evolution is also the same for the three VSP experiments and similar to that observed in Dewar experiments. Let us consider the VSP test where the test cell did not rupture on Figures 14 and 15.

As the runaway polymerization proceeds, the pressure corrected for the nitrogen pad presents a negative deviation with respect to the monomer vapor pressure, showing the formation of soluble heavies.

Above 135°C, the reaction becomes very fast and the temperature measurement is late with respect to the pressure, because of the thermocouple thermal initial, causing a pressure overshoot above monomer vapor pressure.

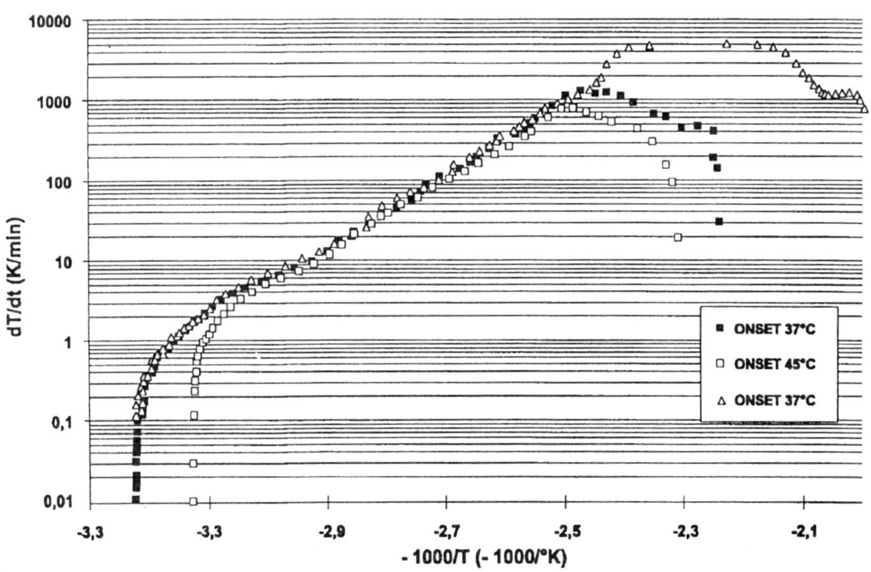

Figure 12. VSP closed cell experiment. Polymerization of vinyl acetate monomers with 1.8 % w/w dilauroyl peroxide initiated by a constant temperature exposures at 37°C and 45°C. Heat rate curves.

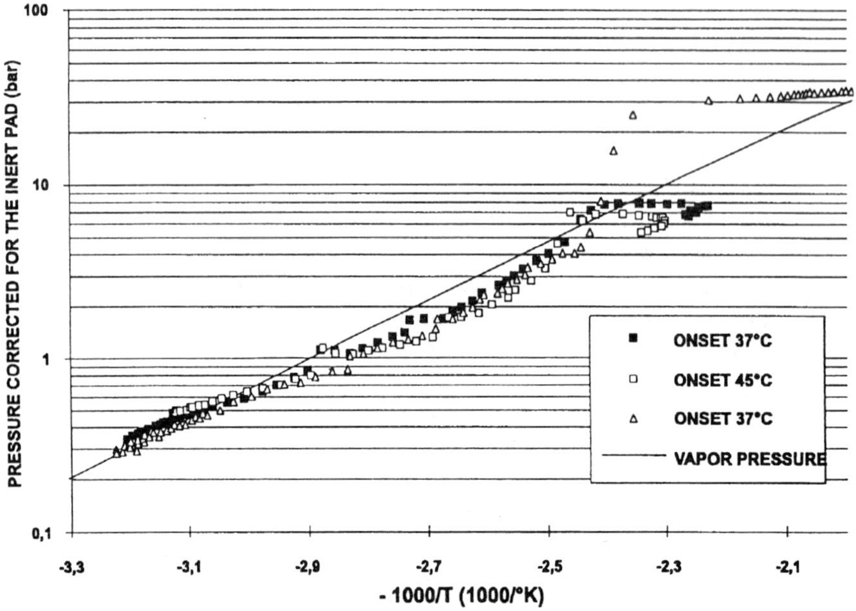

Figure 13. VSP closed cell experiment. Polymerization of vinyl acetate monomers with 1.8 % w/w dilauroyl peroxide initiated under constant temperature exposures at 37°C and 45°C. *P* vs. *T* data: Curve represents vapor pressure of monomers.

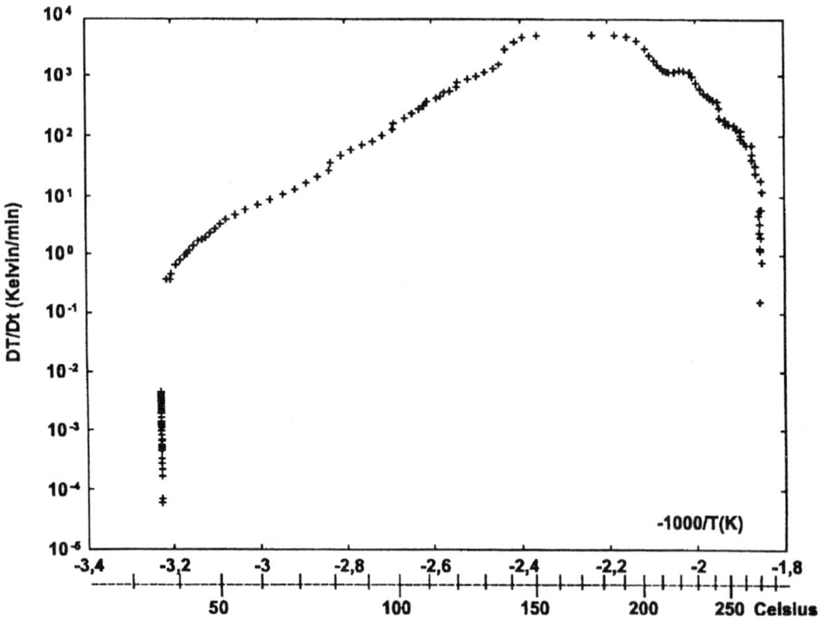

Figure 14 : VSP closed cell experiment. Polymerization of VAM containing 1.8 % w/w peroxide. Heat rate curves.

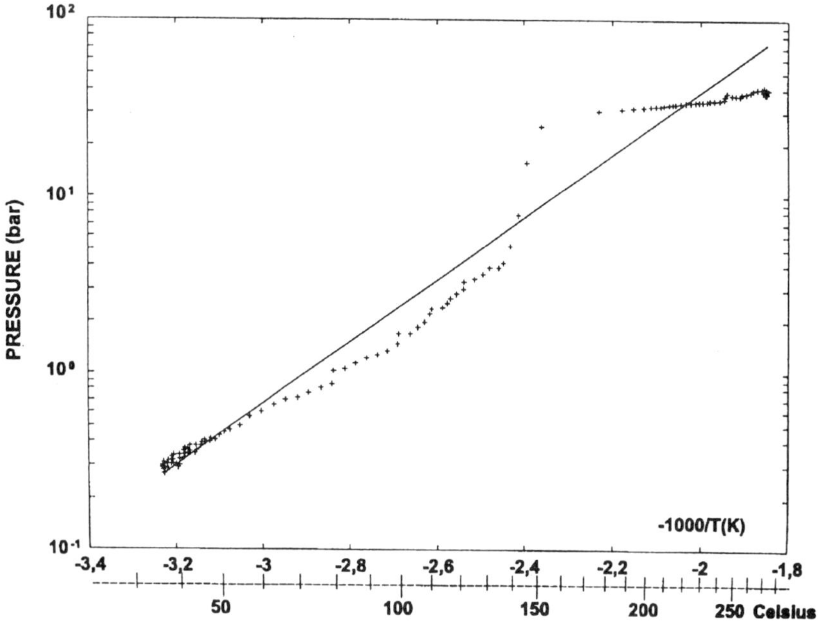

Figure 15. VSP closed cell experiment. Polymerization of VAM containing 1.8 % w/w peroxide. *P* vs *T* curve.

Above 175°C the reaction slows down and the pressure becomes lower than the extrapolated vapor pressure.

There were no decomposition gases produced.

The final temperature, 268°C, is lower than the polymer decomposition onset temperature found in autoclave experiments above 280°C.

3.2.6. Protection from the Runaway Polymerization of VAM

Considering the information obtained from runaway experiments, venting is necessary to protect vessels from the runaway bulk polymerization of vinyl acetate monomers.

In incidents or accidents reported, it was observed that on depressurization of vessels where the runaway polymerization occurred, a polymeric material was released and that the vessel was found empty after venting. In addition, the release was always violent and almost instantaneous. Long vent lines with bends must be excluded.

Following this information, it is assumed that the vessel content behaves homogeneously during venting and that all the vessel inventory is released.

Based on our experimental results, the runaway polymerization of VAM is a high vapor system as long as the polymer decomposition onset temperature is not reached (280–300°C).

3.2.7. Choice of the Vent Sizing Equation

The vented material is made of viscous polymeric materials, however the vent flow is not a viscous flow but a highly turbulent flashing flow carrying small pieces of polymers.

At the moment, there is no particular flow model for such fast transient phenomenon. Flashing flow methods are currently used. See Ref. (26). Leung's analytical vent sizing equation can be used (23).

However, in the present case due to the very fast reaction, there is no benefit to take into account the energy balance to obtain a reduced vent size. The required vent size is best discussed using Boyle's equation and considering emptying times short enough to prevent the reaction from becoming too fast before complete emptying.

$$A = \frac{m_0}{G \cdot \Delta t_p}$$

where the mass flux G is obtained from Leung's Omega method (27).

The emptying time Δt_p is chosen considering adiabatic temperature rise times obtained from VSP closed cell experiments.

The experimental temperature rise times are corrected to adiabatic conditions using the zero order correction method (16).

$$\Delta t_{\phi=1} = \frac{\Delta t_{\phi>1}}{\phi}$$

The phi factor of the VSP experiments is 1.035 (VAM sample mass = 80 g, C_{PL} = 1964.6 J/kg/°K).

3.2.8. Choice of Vent Actuation Pressure and Emptying Time

The vent actuation pressure is chosen as low as possible and the emptying time is limited to exclude high heat-rates. The minimum vent set pressure considered is 1.1 bar absolute to obtain critical mass flux and reasonable conditions for rupture disk pressure resistance. The maximum pressure during venting to 1.7 bar absolute so that heat-rates do not exceed 170°C/min.

3.2.9. Vent Size Calculation

The vent size calculations using the Boyle equation are shown in Table 12. In this table

- P_S is the vent set pressure or the vessel pressure
- T_S is the reaction mixture boiling point under PS obtained from Fig. 15
- $(dT/dt)_{Ts}$ is the experimental heat-rate obtained from Fig. 14.
- Δt_{exp} is the time necessary for the pressure to increase from 1.1 bar absolute to the current pressure. This information is obtained from the VSP experimental data file.

TABLE 12
Vent Size Calculations

P_S (bar abs)	T_S (°C)	$(dT/dt)_{Ts}$ (°C/min)	$\phi(dT/dt)_{Ts}$ (°C/min)	Δt_{exp} (s)	$\Delta t_{exp}/\phi$ (s)	ω	G^*	G_C (kg/m²/s)	A for 1000 kg (m²)
1.1	84	50	51.75	0	0	28.89	0.169	1457	
1.2	86	60	62.1	3.9	3.77	31.17	0.158	1413	
1.3	89	70	72.45	6.3	6.01	33.47	0.145	1349	0.119
1.5	95	100	103.5	9.78	9.44	37.97	0.124	1240	0.078
1.6	97	115	119	10.08	9.74	40.07	0.115	1187	0.078
1.7	100	163	168.7	11.58	11.19	42.43	0.107	1139	0.069

- $\Delta t_{exp}/\phi$ is the previous time difference, corrected using the zero order correction.
- G is the HEM mass flux.
- A is the vent area required for 1000 kg inventory.

Table 12 shows for example that if an emptying time of 11.2 sec is allowed, considering an average mass flux between 1.1 bar abs. and 1.7 bar abs. of $G = 1300$ kg/m²/s, the vent area required is :

$$A = \frac{m_0}{G \cdot \Delta t} = \frac{1000\,\text{kg}}{1300\,\text{kg}/\text{m}^2/\text{s} \times 11.2\,\text{s}} = 0.0687\,\text{m}^2$$

i.e. a 30 cm diameter vent.

3.2.10. Vent Line Considerations

The above vent size is given as a minimum requirement. Provision must be made for vent line head losses.

The vent line must be as short as possible and horizontal. At least one bend is necessary. A long vent line is likely to be plugged by polymers and would endanger the facility. The vent exit should be directed to a containment system or a stagnation plate (18) could prevent polymers being spread over the neighborhood.

The design of a quench tank to contain the two-phase release is hindered by the fact that VAM is only slightly soluble in water. The possible solvents for VAM have a low flash point and could give flammable vapors. A very large containment vessel could be necessary. A long vent line to this tank would be a major drawback for the use of a quench tank.

3.2.11. Partial Pressure Containment

A total pressure containment of this runaway reaction is not realistic because the final pressure is at least 40 bar gauge without polymer decomposition. Nevertheless, enough pressure resistance should be given to the vessel to allow for variations in the runaway behavior and differences between design and true venting scenario. A pressure resistance is useful up to the point where the reaction becomes very fast and the vent requirement very large. This is approximately obtained when $T = 130°C$ and $P = 4$ bar absolute.

We suggest that the equipment have a pressure resistance of four bar gauge. This is usually granted for the type of equipment used in such operations.

3.2.12. Conclusion on the Venting Option

The above vent size calculations as well as the runaway events observed in industrial practice suggest that the vessel protection by venting remains a reasonable option even for such very fast polymerization reactions, provided that the vessel inventory is not too large and that a minimum pressure resistance of the vessel is granted. However, we recommend considering accident prevention first by eliminating the process conditions where such a runaway polymerization can occur.

4. CONCLUSION

The two examples of runaway reaction hazard studies presented in this paper are representative of a current practical approach providing the opportunity to improve process safety.

This methodology includes:

- A review of the literature concerning the process considered,
- The identification of the process conditions or process deviations leading to a runaway reaction accident.
- An experimental study of the possible runaway reaction scenarios.
- Proposals and recommendations on possible changes of process or process conditions to eliminate hazardous conditions.
- The installation of safety interlocks to prevent hazardous process deviations.
- The control of runaway reaction scenarios by venting.

One hesitates to recommend not considering the most severe runaway reaction scenario for vent-sizing when the reliability of the safety interlocks is not specified. A reliability study would help to the selection of the runaway reaction scenarios to be considered for vent-sizing with the objective to reduce the necessary vent area. This would significantly reduce the cost and size of the release catchment system to be provided.

Literature

1. Barton, J. A., and Nolan, P. F., "Incidents in the chemical industry due to thermal runaway chemical reactions." Conference on Chemical Hazards, 5–6 December, 1989.
2. Barton, J. A., and Nolan, P. F. In *Safety of Chemical Batch Reactors and Storage Tanks,* Eurocourses—Reliability and Risk Analysis, Volume 1. Dorcrecht, The Netherlands: Kluwer Academic Publishers, 1991, pp. 1–17.
3. The British Plastic Federation Thermosetting Material Group, "Guidelines for the Safe Production of Phenolic Resins." British Plastics Federation, London, 1979.
4. Grewer, T. Thermal Hazards of Chemicals Reactions, *Industrial Safety Series* Volume 4, 1994, 324–328.
5. Wurtz, A. Mémoire sur les glycols ou alcools diatomiques, *Ann. Chim. Phys.*, 3, (1859), 400.
6. Nef, J. U., Ueber das Verhalten der Glycole und des Glycirins gegen Aetzalkalien und gegen oxydationsmittel, *Annalen*, 335, (1904), 310.
7. Cardillo, P., and Girelli A., Comportamento termico di miscele NaOH/glicol etilenico e NaOH/glicol dietilenico, *La chemia e l'industria,* 64, 12, (1982), 781.
8. Sumitomo Chemical Co Ltd. Accident case history of reactive chemicals in Japan between 1971 and 1994.
9. Bretherick, *Handbook of Chemical Reactive Hazards.*
10. *Methoden der organischen Chemie, (Houben-Weyl) vierte Auflage,* Georg Thieme Verlag, Stuttgart, 1976.
11. Gustin, J.L., "Assessment of Runaway Reaction Hazards Using the Experimental Methods of the Process Safety Laboratories." *First International Symposium on Runaway Reactions and Pressure Relief Design,* August 2–4, 1995, Boston, MA.
12. Gygax, R., In *Proceedings of the International Symposium on Runaway Reactions,* March 7–9, 1989, Cambridge, MA. New York: AIChE/CCPS (1989), pp. 52–73.
13. Benin A. I., Kossoy A .A., and Smykalov P.U., *J. Therm. Anal.* 38 (1992), 1151–1165.
14. Kossoy A. A., Benin A. I., Smykalov P. U., and Kasakov A. N., *Thermochimica Acta,* 203 (1992) , 77–92.
15. Kossoy A., Belochvostov V., andGustin J.L., *J. Loss Prev. Process Ind.*, 7(5) (1994), 397–402.
16. Townsend D. I., Tou J. C., *Thermochimica Acta,* 37 (1980), 1–30.
17. Kohlbrandt H. T., In *Proceedings of the International Symposium on Runaway Reactions,* March 7–9, 1989, Cambridge, MA. New York: AIChE/CCPS (1989), pp. 86-111.
18. Gustin, J.L., "Calorimetry for Emergency Relief System Design." In *Safety of Chemical Batch Reactors and Storage Tanks,* Euro-courses—Reliability and Risk Analysis, Volume 1. Dorcrecht, The Netherlands: Kluwer Academic Publishers, 1991, pp. 311–354.
19. Gustin, J. L., Fillion, J., Treand, G., and El Biyaali, K., "The Phenol + Formaldehyde runaway reaction. Vent sizing for reactor protection." *J. Loss Prev. Process Ind.,* 6(2) (1993), 103–113.
20. Booth, A. D., Karmarkar, M., Knight, K., and Potter, R. C. L., *Trans. I.Chem.E.,* 58 (1980), 75.
21. Leung, J. C., Fauske, H. K., and Fisher, H. G., *Thermochim. Acta,* 104 (1986), 13.
22. Kumpinsky, E., "Ph Effects on phenol-Formaldehyde Runaway Reactions." *Ind. Eng. Chem. Res.,* 34 (1995), 3096–3101.
23. Leung J. C., *AIChE. Journal*, 32(10) (1986), 1622–1634.
24. Brandrupt J., and Immergut E.H., *Polymer Handbook*, 3rd ed., Vol. II, p. 371. New York: Wiley.
25. *Encyclopedia of Polymer Science and Engineering,* 2nd ed., Vol. 17, p. 504.
26. Leung, J. C., and Fauske, H.K., *Plant/Operations Progress,* 6(2) (April 1987), 78–79.
27. Leung J. C., *J. Loss Prev. Process Ind.*, 3 (1990), 27–32.

The Effect of an External Fire on a Runaway Reaction

Michael I. Hawksworth and Caroline B. Ladlow
Allied Colloids Ltd., PO Box 38, Bradford BD12 0JZ, United Kingdom

ABSTRACT

An external fire was identified as a possible worst case scenario for one of Allied Colloids continuous processes. Therefore the effect of an external fire had to be simulated on a small scale. A batchwise hold-up in the reactor with the reactor contents initially at their steady state composition was assumed. From previous work it was known that a loss of cooling would result in exothermic work-off of the reactor contents.

Experiments have been carried out in a modified Vent Sizing Package apparatus (VSP) to simulate the effects of an external fire on this runaway reaction. The experimental procedure used has been described.

The effect of a range of external heat inputs on various parameters has also been investigated, e.g. relief set temperature; ideal vent size; any secondary runaway reaction.

Since this study was undertaken, equipment and control procedures have been introduced as part of the Basis of Safety to eliminate fire as the possible worst case. Quench / inhibition studies have been carried out to address the problems associated with the uncontrolled hydrolysis reaction. This now forms part of the Basis of Safety.

1. INTRODUCTION

During normal operation the reactors at Allied Colloids contain a steady-state mixture of acrylonitrile, acrylamide and water in the presence of a catalyst.

The reaction is as follows:

$$\underset{\text{Acrylonitrile}}{CH_2{=}CH{-}CN} + H_2O \xrightarrow{\text{cat}} \underset{\text{Acrylamide}}{CH_2{=}CH{-}CONH_2}$$

As this is a continuous process, the reactor contents are held at a constant temperature to allow the hydrolysis of acrylonitrile to acrylamide to take place safely. The acrylonitrile does not normally reach 100% conversion to acrylamide in the reactors.

In the event of a hold-up in the reactor and loss of cooling, the normal reaction, in which acrylonitrile is converted to acrylamide, can occur in an uncontrolled manner.[1] This uncontrolled reaction may result in high temperatures at which a second uncontrollable runaway reaction may occur. Upon investigating the chemical reaction hazards associated with the reactors, development of a Basis of Safety was required. It was identified that an external fire could affect the Basis of Safety. Any process equipment which can be exposed to an external fire usually requires a relief system (vent) to prevent the design pressure being exceeded.

The effect of varying external heat inputs to these runaway reactions was simulated using a modified Vent Sizing Package (VSP) apparatus.[2]

2. EXPERIMENTAL

2.1. The Apparatus

The VSP apparatus is a bench-scale test instrument which was developed for, and forms an integral part of, the Design Institute for Emergency Relief Systems (DIERS) methodology.[3] It is designed to provide system characterisation data, which along with the applicable vent sizing equations incorporating two-phase flow, leads to an emergency relief system (ERS) evaluation.

Our modified VSP is shown in Figure 1. The modified VSP consists of a thin walled test cell (approx. 120 mL in capacity), which is heated radiantly by a mica band test heater and located inside a guard heater to keep the test cell in an adiabatic environment. Fiber glass insulation is placed around the test cell and heater assemblies to further minimise heat losses to the surrounding environment. The entire assembly is housed in a 4 liter containment vessel. Agitation within the test cell is provided by a 30-mm glass-covered magnetic stirrer bar to ensure the sample remains homogeneous.

The pressure and temperature of the test cell contents are measured and recorded at varying time intervals throughout the experiment, depending on the rates of temperature and pressure rise. An absolute pressure transducer is employed to measure the sample pressure directly. The walls of the test cell are very thin (typically 0.006 in.). Therefore as the sample pressure increases, nitrogen is charged to the containment vessel to maintain a maximum of 1 bar differential across the test cell wall to prevent the test cell from rupturing. This pressure control is carried out utilizing a differential pressure transducer connected across the test cell and containment vessel.

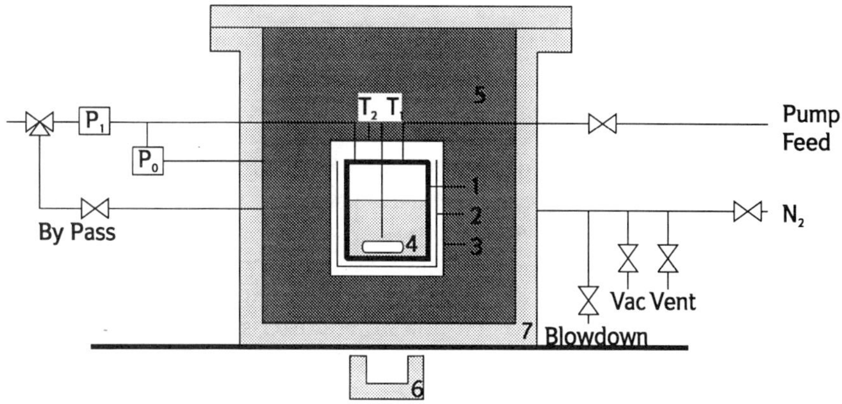

1. Test Cell
2. Test Cell Heater
3. Guard Heater Assembly
4. Magnetic Stirrer Bar
5. Fibreglass Insulation
6. Magnetic Stirrer
7. 4 Litre Containment Vessel

T_1. Sample Thermocouple
T_2. Guard Heater Thermocouple
P_1. Test Cell Pressure Transducer
P_0. Differential Pressure Transducer

Figure 1. The Modified Vent Sizing Package Apparatus

The test cell can be evacuated prior to the test commencing by opening the 3-way and bypass solenoid valves and applying a vacuum to the containment vessel. Similarly, pressure adjustments can be made to the test cell during the experiment in the same manner.

Chemicals can be charged to the test cell, either at the start of the test or at any time during the test, using a high pressure pump via a 1/16-inch feed line. Provisions can be made for additional feed lines as and when required.

Two modes of heating of the sample are commonly employed.[(4)]

(a) Constant rate—heater power adjusted to maintain a prescribed temperature rise rate.
(b) Constant power—heater power maintained at a constant value.

2.2. Evaluation and Calibration of Heat Input to Simulate External fire

The heat input to a vessel can vary due to the different fire protection factors that may be applied.[(5)] For a given mass of reactants in the reactor and a calculated heat input (see Appendix 1), a rate of temperature rise of the contents due to the external fire can be calculated.

$$Q = m \cdot C \cdot \left(\frac{dT}{dt} \right) \tag{1}$$

This heat-up rate needs to be applied to the reaction.

If the test heater was set in the constant rate mode, as the reaction (self-heating) takes place, the power being supplied to the heater would reduce to maintain the prescribed heat-up rate. Therefore the test heater was calibrated to determine the required constant power setting to give an external heat-input equivalent to that calculated in Eq. (1). To achieve this, the test heater was set in constant rate mode at the required temperature rise-rate for the external fire while the sample was being heated to the required operating temperature (i.e. no reaction was occurring). The test heater then settled down at a constant power setting to achieve the required temperature rise rate. On reaching the operating temperature, and thus the start of the simulated fire, the test heater was put into constant power mode at the observed power setting. This then had the effect of adding the required external heat input to the reaction.

2.3. Experimental Procedure

The steady state composition of the system under study was calculated.[6]

The required amounts of catalyst, water and acrylamide were charged to the test cell and heated. The acrylonitrile is not added at this stage as it would start to react during the initial heat-up and prior to the operating temperature being reached. The cut-off temperature of the heaters was adjusted to allow for the addition of cold acrylonitrile to the test cell at a later stage. The test heater was calibrated for the external fire during the initial heat-up (as described in 2.2).

On reaching the desired cutoff temperature the acrylonitrile was charged to the test cell. This had the effect of cooling the contents of the cell to the required operating temperature. At this point the sample pressure was adjusted to the prescribed operating pressure. The test heater was then put into constant power mode at the predetermined power setting and the experiment was allowed to run away.

To study the effects of the external fire on this reaction, the above experimental procedure was repeated by applying external fire heat-up rates of 1, 2, and 3 K/min.

3. EXPERIMENTAL RESULTS

In order to assess the effect of the external fire it was necessary to perform the experiment in the absence of external heat input. Figure 2 shows the sample being heated, the drop in sample temperature as the acrylonitrile is charged to the test cell followed by the subsequent uncontrolled hydrolysis reaction. This hydrolysis reaction is superseded by a period of low thermal activity followed by a second runaway reaction which is accompanied by a gassy decomposition.

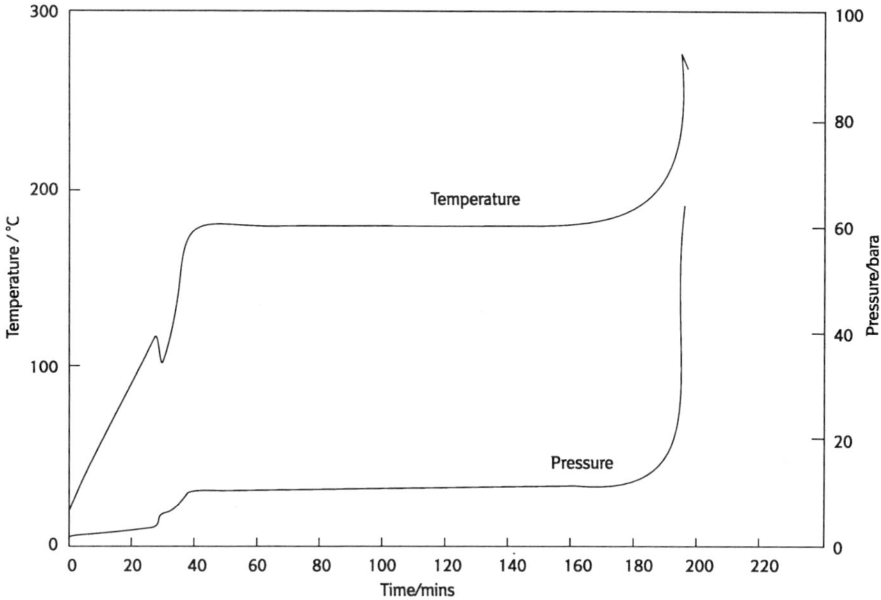

Figure 2. No Fire Case Experiment

Figure 3 shows the effect of each of the external fire heat-up rates on the temperature profiles of the reactions. The time scale has been adjusted so that each set of experimental data start at the same point to allow comparison of the traces.

The results in Table 1 show the actual experimental values and other data derived from these. All times shown are taken from the normal operating temperature (i.e. the start of the fire). ΔT is the difference between the relief set temperature (T_s) and the maximum temperature at the allowable overpressure (T_M), using a 20% overpressure allowance. $(dT/dt)_{av}$ is the measured average rate of temperature rise between T_s and T_M. Time to T_{MAX} shows the time taken to reach the maximum observed temperature during the test from the normal operating temperature. The modified Leung's method for external heating[(7)] was used to calculate the ideal vent diameter (see Appendix 2).

Table 1 clearly shows that as the rate of external heating is increased:

- the temperature at which relief will occur falls significantly.
- the time taken for relief to occur from normal operating conditions falls significantly.
- the temperature difference between T_s and T_m is reduced considerably.
- Time to T_{MAX} reduces drastically.
- T_{MAX} increases significantly.

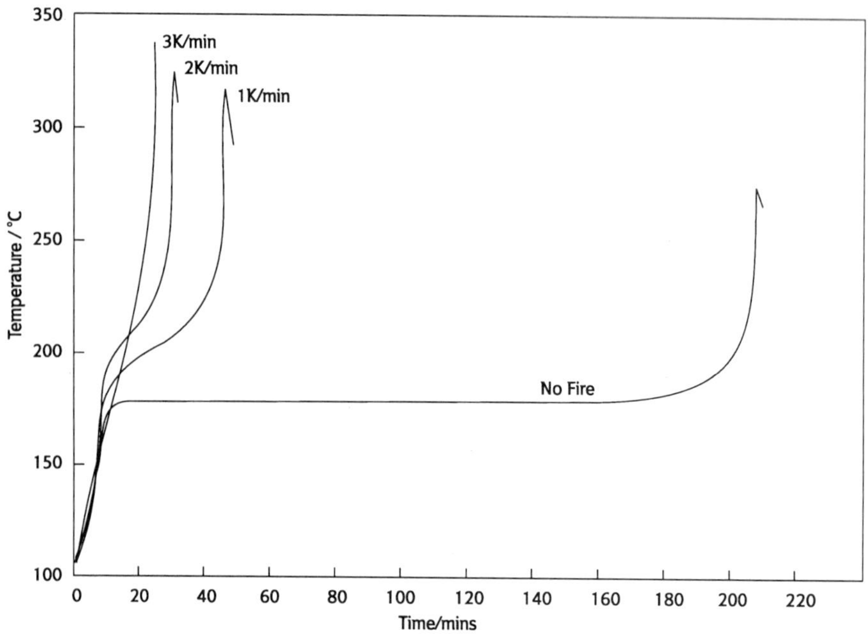

Figure 3. Effect of Varying External Heat-up Rates on the Temperature Profile

Figure 4 shows the effect that increasing the external heat input has on the pressure profile of the reactions. Again, the time scale has been adjusted to allow easy comparison of the traces.

Figure 5 shows the impact of increasing the external heat input on the ideal vent diameter.

TABLE 1

Effect of Fire Rate on Experimental Results

Fire rate (K/min)	T_s (°C)	Time to T_s (min)	ΔT (K)	$(dT/dt)_s$ (K/min)	$(dT/dt)_{av}$ (K/min)	$(dP/dt)_s$ (bar/min)	Time to T_{MAX} (min)	T_{MAX} (°C)	Ideal vent diameter (in.)
0	153	8.1	23.6	10.4	6.8	0.73	208.2	275	5.2
1	151	7.5	19.5	12.3	13.0	1.10	46.2	317	7.2
2	145	6.4	17.8	13.2	14.5	1.10	30.5	325	7.8
3	139	5.7	12.6	9	10.9	0.90	24.7	338	6.7

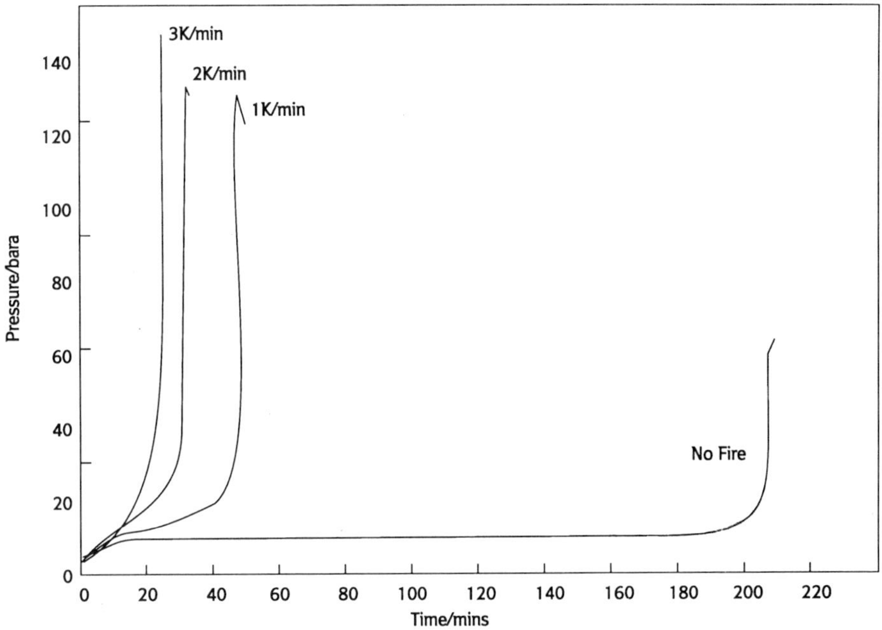

Figure 4. Effect of Varying External Heat-up Rates on the Pressure Profile

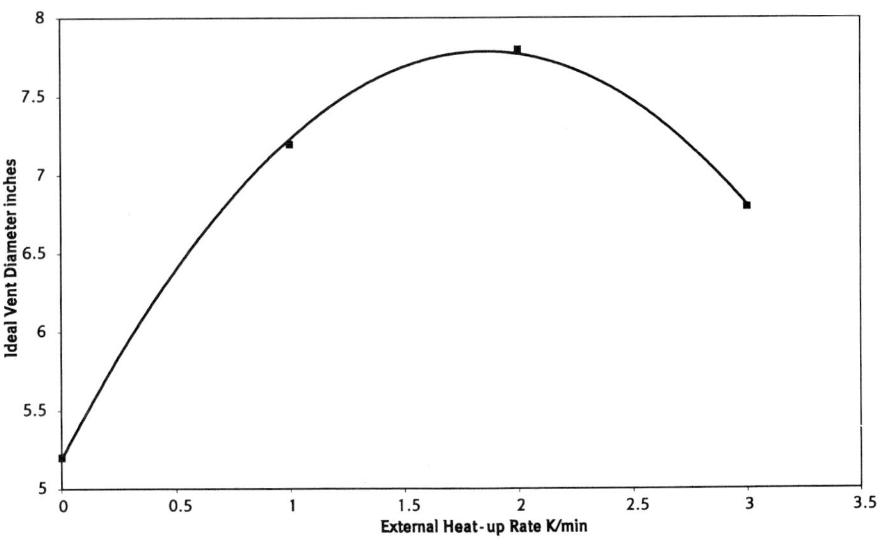

Figure 5. Relationship between External Heat-up Rate and Ideal Vent Diameter

4. DISCUSSION

As mentioned earlier, Figure 2 shows the sample being heated, the sample temperature dropping as the acrylonitrile is charged to the test cell followed by the uncontrolled hydrolysis reaction. The hydrolysis reaction reaches completion 15 minutes after the normal operating temperature and pressure have been achieved, therefore there will be no acrylonitrile present at this stage. The end of this uncontrolled reaction is followed by a temperature plateau. This period of low thermal activity lasted for 160 minutes, after which time a further accelerating , uncontrolled runaway reaction, accompanied by a gassy decomposition of the test cell contents occurred. It is thought that this secondary reaction is due to the polymerisation of the acrylamide with decomposition, and the subsequent formation of various heterocyclic compounds. A substantial residual pressure in the test cell after the experiment had ended supports the gassy decomposition hypothesis.

As the rate of external heat input is increased the relief set temperature, T_s, and the time taken to reach the relief set pressure fall significantly. This is due to the external heat input ramping the temperature of the reaction mixture, and therefore less acrylonitrile is consumed prior to the relief set pressure being reached. So as a consequence, the vapor pressure effect of acrylonitrile not being consumed is causing the relief set pressure to be reached at a lower temperature than for the no fire case. This effect is progressively more marked as the external heat input is increased.

The acrylonitrile concentration in the reaction mixture also influences the maximum temperature at the allowable overpressure, T_m. As a result the temperature difference between T_s and T_m (ΔT) is reduced considerably as the rate of external heat input is increased. Again, this is due to the vapor pressure of acrylonitrile (being the more volatile component) dominating the system.

Higher temperatures are reached at the end of the hydrolysis reaction as the rate of external heat input is increased (see Figure 3). After the completion of the hydrolysis reaction (for the 1 and 2 K/min external temperature rise rates), there is period of linear heating equal to that of the external heat input. This linear temperature rise rate is equivalent to the plateau seen when no external heat input was applied, i.e. this is a period of minimal exothermic activity. It is known from previous work[(1)] that the time to T_{MAX} is a function of the temperature at the end of the hydrolysis reaction, i.e. the higher the temperature at which the plateau is attained, the shorter the time interval to T_{MAX}. In the case of the 3 K/min external temperature rise rate, there is no period of linear heating and the hydrolysis reaction leads straight into the polymerization/decomposition reactions. In all of these cases, a significant residual pressure was observed after the experiments had ended.

On analysis of Figure 3, there appears to be little difference in the rates of temperature rise, (dT/dt), due to the hydrolysis reaction before 150°C, whether external heating is being applied or not. This phenomenon is not as expected, as an external heat input can result in a higher reactant concentration at a particular temperature which can give self-heat rates several orders of magnitude higher than that measured adiabatically.[3,7] For heterogeneous catalytic reactions, the rate of reaction is limited by the turnover number of the catalyst and its available surface area.[10] Therefore, for the system under study, the reaction rate is dependant upon the number of active sites present on the catalyst surface, the concentration of substrates (acrylonitrile and water in this case), and the temperature of the reaction mixture.

It is thought that the initial reaction rates (up to 150°C) shown in Figure 3 signifies the attainment of a maximum rate in the catalysed reaction (i.e., the catalyst is converting the maximum amount of acrylonitrile to acrylamide in unit time as function of temperature). Therefore, an increase in the external heat input rate will not increase the initial rate of reaction for this system.

The temperature rise rate at the relief set temperature, $(dT/dt)_s$, is determined by the relief set pressure, which is a function of the vapour pressures of acrylonitrile and water in the reaction mixture. The concentration of acrylonitrile at a particular temperature is in turn affected by the external heat input. Therefore, as mentioned previously, as the external heat input increases, the relief set temperature falls.

For the no fire case, the relief set temperature is 153°C, at which point the rate of temperature rise is decreasing as it nears the end of the hydrolysis reaction (i.e. nearly all the acrylonitrile has been converted to acrylamide). This results in the average self-heat rate being lower than the rate of temperature rise at the relief set temperature for the no fire case. For the 1 and 2 K/min external fire rates, the relief set temperature is reached when the temperature rise rate for the hydrolysis is near a maximum with respect to temperature. This results in the average rate of temperature rise being higher than the no fire case. For the 3 K/min fire, the relief set pressure is reached at a much lower temperature. Therefore, the hydrolysis reaction has not reached its peak rate before the maximum allowable pressure is reached. Thus the calculated average rate of temperature rise results in a smaller vent size being obtained. It appears, for this system, that with increasing external heat input, the ideal vent diameter reaches a maximum due to the vapour pressure effect of acrylonitrile.

It is possible that with increasing external heat inputs, the reaction may change and the kinetics/thermodynamics of the system could be altered significantly.

5. CONCLUSIONS

The effects of external heat input into a system can be very complex as the kinetics and thermodynamics are altered. Therefore each individual system needs to be characterised within well defined boundaries. Unfounded assumptions cannot be made and extrapolation must be avoided.

The effects of varying rates of external heat input into the acrylonitrile hydrolysis reaction system have been well studied and the following conclusions drawn.

- As the rate of external heat input is increased, the rate of the hydrolysis reaction does not increase significantly, as the catalyst is the rate limiting factor (i.e. the turnover rate of the catalyst is at its maximum).
- With increasing external heat input, the relief set temperature falls significantly due to the vapour pressure effects of the acrylonitrile. This is due to the acrylonitrile not being consumed as the external heat input ramps the temperature of the reaction mixture.
- ΔT is reduced considerably as the rate of external heat input is increased.
- Higher temperatures are reached at the end of the hydrolysis reaction as the rate of external heat input is increased.
- The time taken to reach the second runaway reaction is reduced considerably as the rate of external heat input is increased.

For this system, the calculated ideal vent diameter reaches a maximum due to the presence of a catalyst and the vapour pressure effects of the acrylonitrile.

NOMENCLATURE

A	Cross-sectional flow area of relief system (m^2)
A_{ws}	Wetted surface area of vessel (m^2)
C	Heat capacity (J/kg.K)
D	Relief system diameter (m)
F	Dimensionless fire protection factor
G	Two-phase mass flow rate per unit flow area ($kg/m^2.s$)
h_{fg}	Latent heat of vaporization (J/kg)
q	Heat release rate per unit mass of reacting mixture (W/kg)
$q_{modified}$	Heat release rate per unit mass including external heat input (W/kg)
Q	Heat input (W)
m	Mass (kg)
(dP/dt)	Rate of pressure rise (bar/s)
(dP/dT)	Rate of pressure rise with respect to temperature ($N/m^2.s$)
T	Temperature (K)
(dT/dt)	Temperature rise rate (K/s)
ΔT	Temperature difference between T_s and T_m (K)

V	Volume of vessel (m^3)
v_{fg}	Difference between vapor and liquid specific volumes (m^3/kg)

Subscripts

f	Fluid
m	Value at allowable overpressure
R	Reactor Contents
s	At relief set pressure
av	Average value between T_s and T_m
external	Due to external heating
HEAT	Due to external heating
measured	Measured in experiment
reaction	Due to runaway reaction

REFERENCES

1. Internal Technical Report, Basis of Safety for the R2000 Reactors, D.K. Bellerby and C.B.Ladlow.
2. Internal report for ISM ,The Upgrade of the Vent Sizing Package (VSP) Apparatus, Caroline Ladlow, October 1997.
3. *Emergency Relief System Design Using DIERS Technology: The Design Institute for Emergency Relief Systems (DIERS) Project Manual*, New York: AIChE, 1992.
4. *Fike Vent Sizing Package System Manual Methodology and General Operations,* Burr Ridge, IL: Fauske and Associates, Inc.
5. API 521. "Guide for Pressure-Relieving and Depressurising Systems." API, March 1997.
6. Bellerby, D. K. Allied Colloids R2000 Recipe Calculator, June 1997.
7. Wilday, J., and Daskalakis, G. "Pressure Relief Design for Chemical Reactors Exposed to External Fire." Paper Presented at AIChE Summer National Meeting, Denver, August 1994.
8. API 2000. "Venting Atmospheric and Low Pressure Storage Tanks." API, September 1992.
9. "Allied Colloids Engineering Safe Systems of Work, Process Safety Guide (Pressure Relief) Standards, Codes of Practice and Guidelines." ACL\REF\PSG Process SHE Guide No. 8.
10. Thomas, J. M., and Thomas, W. J. *Principles and Practice of Heterogeneous Catalysis.* New York: and Weinheim: VCH, 1997.

Appendix 1
Heat-up Rate Due to External Fire

When a vessel containing a liquid is exposed to external fire, the rate of heat transfer from the fire is determined by the wetted surface area in the vessel, the type and thickness of any insulation and the intensity of the fire.[8] Any enclosure of the fire zone containing the equipment and/or any effect of sloped ground near the vessel are taken into consideration. Simple formulas for calculating heat input are given in API Standard 2000,[8,9] for several ranges of vessel surface area and internal pressure.

Therefore, for a vessel with a total wetted surface area of vessel, $A_{ws} = 40$ m^2, the heat input, Q, is given by,

$$Q = 224 A_{ws}^{0.556} F \quad \text{kW}$$

The above applies for vessel wetted surface areas of between 20 and 93 m^3 with good drainage.

For a fire insulated vessel, the fire protection factor,[9] F, is equal to 0.3.

Therefore,

$$Q_{HEAT} = 224 \times 40^{0.556} \times 0.3$$

$$= 542 \text{ kW}$$

The temperature rise rate in the reactor due to the external fire is calculated using,

$$Q_{HEAT} = m_R \cdot C_f \left(\frac{dT}{dt}\right)_{external}$$

Thus,

$$\left(\frac{dT}{dt}\right)_{external} = \frac{Q_{HEAT}}{m_R \cdot C_f}$$

Mass in reactor = 9899 kg

Specific Heat Capacity of reactor contents = 3300 J/kg.K

$$\left(\frac{dT}{dt}\right)_{external} = \frac{542 \times 10^3}{3300 \times 9899} = 0.0166 \text{K/s}$$

Therefore temperature rise rate due to external fire is 1.0 K/min.

Appendix 2
The Modified Leung's Method for External Fire[7]

The modified Leung's method combines the runaway reaction for a vapor pressure system with external heating, such as external fire. This method uses the same conditions of applicability as the Leung method for modelling two-phase volatile/tempered reactions,[3] but with a modified value of the heat release rate.

The modified value of q is given by,

$$q_{modified} = q + \frac{2Q_{HEAT}}{m_R} \tag{A2.1}$$

This modification effectively divides the external heating rate, Q_{HEAT}, by the average mass in the reactor during relief, $m_R/2$, in order to get a safe estimate to add to the value of the q from the chemical reaction alone.

Therefore, substituting $q_{modified}$ for q in Leung's equation gives,

$$A = \frac{m_R \cdot q_{modified}}{G\left[\left(\frac{V}{m_R} \cdot \frac{h_{fg}}{v_{fg}}\right)^{0.5} + \left(C_f \cdot \Delta T\right)^{0.5}\right]^2} \tag{A2.2}$$

Or using the alternative version, taking the Clausius–Clapeyron relationship into account,

$$A = \frac{m_R \cdot q_{modified}}{G\left[\left(\frac{V \cdot T}{m_R} \cdot \frac{dP}{dT}\right)^{0.5} + \left(C_f \cdot \Delta T\right)^{0.5}\right]^2} \tag{A2.3}$$

The lab-scale simulation which measures dT/dt for the runaway reaction must be performed in a way which simulates the same external heating rate as for the full-scale reactor. There will be no mass loss from the small-scale test, so the initial mass of reactants, m_R, rather than $m_R/2$ is used in the calculation. The results of the test can then be analyzed as follows.

$$\left(\frac{dT}{dt}\right)_{reaction} = \left(\frac{dT}{dt}\right)_{measured} - \left(\frac{dT}{dt}\right)_{external} \tag{A2.4}$$

Example of Calculation for External heating at 1 K/min

As in Appendix 1, the reactor has a volume of 12.44 m^3 and contains 9899 kg of steady-state mixture having a density of 980 kg/m^3 and a specific heat of 3300 J/kg·K. The relief set pressure is 7.0 barg and 20% overpressure allowance is used in the calculations. The heat input due to the external fire, Q_{HEAT}, is 542 kW (equivalent to a temperature rise rate of 1 K/min).

Therefore,

$$\left(\frac{dT}{dt}\right)_{external} = \frac{1}{60} = 0.0167\ \text{K/s}$$

At 8.0 bara, T_s = 424 K = 151°C
At 9.6 bara, T_m = 444 K = 171°C

$$\Delta T = T_m - T_s = 20°\text{C}$$

The average observed heat-rate between T_s and T_m, $(dT/dt)_{av}$, was found to be 13.0 K/min.

Thus the average self-heat rate for the reaction can be evaluated using Eq. (A2.4).

$$\left(\frac{dT}{dt}\right)_{\text{reaction}} = 13.0 - 1.0$$
$$= 12.0\,\text{K/min} = 0.2\,\text{K/s}$$

Therefore the average value of q for the runaway reaction is given by,

$$q = C_{\text{f}}\left(\frac{dT}{dt}\right)_{\text{reaction}}$$
$$= 3300 \times 0.2 = 660\,\text{W/kg}$$

The modified value of q for use in Leung's method can now be evaluated.

$$q_{\text{modified}} = q + \frac{2Q_{\text{HEAT}}}{m_{\text{R}}}$$
$$= 660 + \frac{2 \times 542 \times 10^3}{9899}$$
$$= 770\,\text{W/kg}$$

Therefore, given that,

$$\left(\frac{dP}{dT}\right)_{\text{s}} = 7.836 \times 10^3\ \text{N/m}^2\,\text{K}$$

and,

$$G = 2.81 \times 10^3\ \text{kg/m}^2\ \text{s}$$

then

$$A = \frac{m_{\text{R}} \cdot q_{\text{modified}}}{G\left[\left(\frac{V \cdot T}{m_{\text{R}}} \cdot \frac{dP}{dT}\right)^{0.5} + \left(C_{\text{f}} \cdot \Delta T\right)^{0.5}\right]^2}$$

$$= \frac{9899 \times 770}{2.81 \times 10^3\left[\left(\frac{12.44 \times 424}{9899} \times 7.836 \times 10^3\right)^{0.5} = \left(3300 \times 20\right)^{0.5}\right]^2}$$

$$= 0.0262\,\text{m}^2$$

$$D = \sqrt{\frac{4A}{\pi}} = \sqrt{\frac{4 \times 0.0262}{\pi}} = 0.183\,\text{m} = 7.2''$$

Pressure Relief Requirements for Organic Peroxides and Other Related Compounds

Michael A. Grolmes
Centaurus Technology, Inc.
P.O. Box 598
Simpsonville, Kentucky 40067-0598

1. INTRODUCTION

Organic peroxides represent an extensive class of thermally unstable compounds of significant commercial utility in a variety of product synthesis reactions. Their commercial value is directly related to their predictable thermal instability. This same characteristic is also the source of transportation and bulk storage hazards. There currently exist a number of empirical hazards screening tests [1,2,3] having the objective of providing a relative ranking or classification. Purely analytical procedures have also been proposed [4,5].

The subject of emergency pressure relief protection for peroxide storage and transport vessels is an ongoing concern. Satisfactory pressure protection via either conventional top or bottom venting can be provided if the peroxide is sufficiently dilute. Difficulties arise as peroxide concentration increases. DIERS vent sizing methods accounting for both gas and liquid discharge provide conservative but in some cases impractical results.

Pressure vent sizing analysis and reaction hazards classification are closely related in fundamental characteristics—although not widely recognized as such. This paper reviews organic peroxides as a class of reactive compounds having similar characteristics and illustrates the close relation between hazards classification and pressure vent sizing. The methods developed illustrate the fundamental sensitivities of the analysis results to intrinsic and extrinsic properties of the specific compound.

The concepts presented in this paper are supported by integral analysis methods and assumptions which lead to purely algebraic relations. At the outset it is recognized that on a case by case basis the same end results could and perhaps, in some instances, should be obtained using currently available numerical integration models, such as SAFIRE or SuperChems [6,7].

However, a compelling reason for considering classification or relative hazards ranking of a similar group of compounds such as organic peroxides is to reduce the requirements for detailed evaluation of all cases. Classification analysis can quickly indicate whether any new case is more or less severe than one previously considered in detail. Organic peroxides as a class of similar compounds are of special interest since from a hydrodynamic perspective these compounds do not generally have a viscous or foaming characteristic. The decomposition reaction may often be described by first order kinetics and depending upon concentration, reaction rates may vary over many orders of magnitude.

2. FUNDAMENTAL CHARACTERISTICS OF RUNAWAY THERMAL DECOMPOSITION REACTIONS WITH ORGANIC PEROXIDES

2.1. General Background

This discussion will be principally concerned with liquid state organic peroxides. Peroxides initially in the solid state may, in certain instances, have special reaction hazards characteristics falling outside the general scope of this presentation.

The thermal decomposition of organic peroxides involves the splitting of the peroxide bond with the resulting formation of at least one free radical and perhaps low density volatile or permanent gas fragments. The decomposition is exothermic with a heat release typically in the range of 40 kcal/gmole to 60 kcal/gmole of peroxide, per peroxide bond. (Hydrogen Peroxide is an exception in this regard). Like most decomposition reactions, organic peroxide decomposition reactions follow first order kinetics rate relations with an Arrhenius (exponential) temperature dependence. This simplifying view is adequate for hazards classification and vent sizing even though the decomposition of some hydroperoxides is known to proceed autocatalytically in early stages. With the previously mentioned order of magnitude for reaction heat, the adiabatic temperature increase for a runaway decomposition reaction can vary from 100°C to values approaching 1000°C depending upon concentration. Typical rate constant activation energies are such that decomposition reaction rates tend to increase by a factor of two or more for every 10°C temperature increase. Thus the potential for an enormous increase in reaction rates should be readily appreciated.

The hazard potential of the decomposition reaction exotherm arises from the vapor pressure of volatile solvents and/or noncondensable gaseous decomposition products. Peroxides typically generate from 0.5 to 2 mole of noncondensable gas per mole of peroxide.

2.2. Intrinsic Properties

Most of the aforementioned general characteristics for a given peroxide are defined by the following intrinsic thermo kinetics properties. All of these properties may be determined using commercially available bench-scale calorimeters.

Intrinsic Property	Symbol	Customary Units
Activation Energy	E_a	[kcal/gmole]
Preexponential Factor	A	[min^{-1}]
Heat of Decomposition	ΔH_{Rx}	[kcal/gmole]
Noncondensable Gas Production Index	N_{ncg}	[mole gas/mole peroxide]

Most peroxide producers provide tabulations of activation energy and half-life temperatures which implicitly define the preexponential factor. The decomposition rate constant is represented as

$$K = a\exp\left(-E_a / RT\right) \tag{1}$$

and the isothermal half-life time is defined for a first order decomposition reaction as

$$t_{1/2} = \frac{\ln 2}{k} \tag{2}$$

In the above

k is the rate constant, [min^{-1}]
R is the gas constant (R = 1.987 cal/gmole K)
T is the absolute temperature, [K]
$t_{1/2}$ is the isothermal half-life time, [min]

Heat of decomposition and noncondensable gas production are less frequently tabulated. However, as a useful working hypothesis, it may be assumed that the heat of decomposition for an organic peroxide compound is adequately represented by 50 kcal/gmole peroxide per peroxide bond. That is the heat of decomposition for monoperoxides would be 50 kcal/gmole and for diperoxide compounds would be 100 kcal/gmole. Variations of up to ±20% may be observed in direct calorimetry tests depending upon the presence and nature of the diluents. The assumption of one mole of noncondensable decomposition gas production per mole of peroxide may also be considered as a best estimate. A selected tabulation of best estimate intrinsic properties for representative peroxide compounds is included as an appendix to this paper.

The intrinsic properties determine the fundamental decomposition characteristics. However, the actual magnitude of the maximum rate of reaction or energy release rate is also influenced by a number of extrinsic properties that can be varied for a given product and which with few exceptions have negligible impact on the intrinsic properties.

2.3. Extrinsic Properties

The following list of extrinsic properties is highly relevant to classification and further evaluation of organic peroxide thermal decomposition hazards:

(a) peroxide concentration
(b) solvent (diluent) type and volatility
(c) application of external heat
(d) container size, pressure rating, relief vent size and location

2.3.1. Peroxide Concentration

Peroxide concentration along with the heat of decomposition defines the total energy content of a product. The theoretical adiabatic temperature increase given by

$$\Delta T_a = \frac{x \Delta H_{Rx}}{C\,Mw} \tag{3}$$

is a convenient measure of the product energy content. In Eq. (3)

ΔT_a	is the theoretical adiabatic temperature increase, [°C]
x	is the peroxide concentration on a mass basis, [-]
ΔH_{Rx}	is the heat of decomposition, [cal/gmole]
C	is the liquid heat capacity of the mixture, [cal/gm °C]
Mw	is the peroxide molecular weight, [gm/gmole]

As an example, consider a 20 wt% solution of t-butyl peroxyacetate in odorless mineral spirits for which

x	=	0.2	ΔH_{Rx}	=	50×10^3 cal/gmole
Mw	=	132.2	C	=	0.55 cal/gm °C [1]

From Eq. (3) one finds $\Delta T_a = 158$°C

The adiabatic temperature increase at 75 wt% concentration would be 506°C. It will be shown that the maximum reaction rate potential increases

[1]A heat capacity value of 0.55 cal/gm °C or 2300 J/kg K is a typical value applicable to most organic liquids and is assumed valid throughout this discussion. However, heat capacity increases with increasing temperature and use of a constant value will overestimate ΔT_a at high concentration.

exponentially with increase in concentration unless offset by temperature limiting effects of solvent or peroxide boil-off.

2.3.2. Solvent Type

Inert solvent or diluent volatility can have a significant effect on the decomposition hazards potential of organic peroxides. Simply stated, if solvent or reagent boiling is not a factor, then the full adiabatic temperature increase can be achieved in a runaway decomposition. As a rule of thumb, the maximum reaction rate occurs at a temperature corresponding to ≈ 90% to 95% of the adiabatic temperature increase. Solvent or other reagent boiling can limit the maximum temperature and hence the maximum decomposition rate by dissipating reaction heat at the expense of latent heat—assuming a constant pressure process.

Under certain conditions (negligible external heat), a decomposition reaction can be fully "tempered" by solvent boil-off if the ratio $E_{F/S} < 1$, where

$$E_{F/S} = \frac{x \Delta H_{\mathrm{Rx}}}{y \lambda} \tag{4}$$

where

$E_{F/S}$	is the ratio of the total heat of decomposition of the "fuel" to total latent heat of evaporation of the diluent, [-]
x	is the peroxide "fuel" mass fraction, [-]
y	is the volatile solvent mass fraction, [-]
ΔH_{Rx}	is the heat of decomposition, [J/kg]
λ	is the latent heat of the solvent, [J/kg]

If the peroxide product mixture contains only fuel and volatile solvent, then $y = (1 - x)$. One can use Eq. (4) to assess the maximum fuel concentration, x_{T}, for which full tempering by solvent boiling, in the absence of external heating, is possible. This leads to the relation

$$x_{\mathrm{T}} = \left[\frac{\Delta H_{\mathrm{Rx}}}{\lambda} - 1 \right]^{-1} \tag{5}$$

Again, to illustrate by example, consider a typical but hypothetical peroxide and volatile solvent pair, for which $\Delta H_{\mathrm{rx}} = (1.16 \times 10^6$ J/kg) and $\lambda = (0.3 \times 10^6$ J/kg).

From Equation (5) one finds $x_{\mathrm{T}} = 0.35$

Evaluation of this ratio for real peroxide and solvent combinations typically yields values of x_{T} in the range of $0.2 < x_{\mathrm{T}} < 0.45$.

The effectiveness of evaporation heat dissipation at constant temperature is highly variable, being dependent on the relation between the solvent boiling temperature and the maximum reaction temperature along with the presence or absence of external heating effects. Low volatility, high boiling temperature solvents/diluents such as mineral spirits, mineral oil, and higher boiling isoparaffins have little practical tempering significance. On the other hand, low boiling temperature solvents may be evaporated by external heat with negligible loss of fuel. This process can lead to concentration of a peroxide under external fire emergency considerations.

The significant point to be observed is that solvent volatility, peroxide reactivity and concentration along with external heating effects are important considerations in the assessment of the energetic hazards of peroxide decomposition reactions.

It should also be noted that decomposition reaction kinetics can sometimes be altered by the choice of diluent. For many products this effect is small, but the potential should be considered.

2.3.3. External Heat

External heating effects have been introduced in context with solvent volatility. From a hazard evaluation context, external heating is a common design upset condition for storage and transportation of organic peroxides (such as fire scenario).

The bulk heat-up rate that would be associated with an external fire in which the vessel heat input due to fire is based on API-520 [8] guidelines can be represented as

$$\frac{dT}{dt_{\text{ext}}} = \frac{1.05 \times 10^7 f}{(1-\alpha)\rho C V^{0.454}} \tag{6}$$

where

dT/dt_{ext}	is the bulk heat-up rate due to an external fire heat source, [°C/min]
1.05×10^7	is a dimensional constant
f	is the dimensionless environmental factor related to fire mitigation features, [-]
α	is the vessel free volume fraction, [-]
ρ	is the liquid density, [kg/m^3]
C	is the liquid heat capacity, [J/kg K]
V	is the vessel volume, [m^3]

In developing Eq. (6) it has been assumed that the vessel surface area, S, is simply related to the vessel volume by

$$S \approx 5.5V^{2/3}$$

This approximation is valid for vessels with aspect ratios[(2)] between 0.7 and 3. The environmental factor f can take on values between f = 0.026 (minimum

[2] Aspect ratio is defined as vessel length to diameter ratio.

value for insulation allowance in Ref. [8]) and $f = 1.0$ for a bare unprotected vessel. A heat capacity value of $C = 2300$ J/kg K is appropriate

for most organic liquids. Equation (6) clearly illustrates that the bulk heat-up rate is approximately inversely proportional to the square root of vessel volume and that the environmental factor is most significant in reducing the bulk heat-up rate. Typical bulk heat-up rates due to fire are indicated in Table 1, (Other fire heat input guidelines can lead to different numerical values with the same trend).

Values in Table 1 are based on an assumed bare vessel and can be scaled downward in direct proportion to the environmental factor f. Note the approximate square root volume dependence in the table values. As the volume increases by a factor of four the bulk heat-up rate decreases by a factor of two.

The significance of the externally driven bulk heat-up rate should be placed in context with the bulk heat-up rate due to any exothermic peroxide decomposition. At low fuel burn up, the bulk heat-up rate at any temperature due only to the decomposition of peroxide can be expressed as

$$\frac{dT}{dt_{\mathrm{Rx}}} = \frac{x \Delta H_{\mathrm{Rx}}}{C\,Mw} A \exp(-E_{\mathrm{a}}/RT) \tag{7}$$

The ratio of Eqs. (7) and (6) is a measure of the significance of reaction heat to external heat input at any temperature. This ratio is illustrated in Table 2 for three peroxide solvent pairs evaluated at the normal solvent boiling temperature. The assumed peroxide concentration is $x = 0.2$ and other parameters are indicated in the table. Table 2 illustrates a wide range of outcomes depending on assumptions and property values. The implications of the table results are that if the ratio of reaction heat to external heat is low

TABLE 1
Bulk Heat-Up of Vessel Contents Under Fire Exposure for Environmental Factor f = 1.0 (bare vessel) and α = 0.1 (free volume fraction)

Vessel Volume		Bulk Heat-Up Rate $(dT/dt)_{ext}$
gallon	m^3	[°C/min]
50	0.189	13.3
100	0.379	9.7
200	0.757	7.1
500	1.893	4.7
1000	3.785	3.4
5000	18.93	1.6

TABLE 2
Ratio of Reaction Heat-Up Rate to External Heat Input Rate for Various Peroxide Solvent Pairs

Peroxide	10 hr Half-Life Temp. [°C]	Solvent	Solvent Boiling Temperature	Reaction Heat/External Heat	
				$f = 1.0$	$f = .026$
t-butyl peroxy neodecanoate	49	trichloro ethylene	87°C	4×10^{-4}	0.015
t-butyl peroxyacetate	103	decane	174°C	0.01	0.4
t-butyl peroxy neodecanoate	49	decane	174°C	1.0	42

Table assumptions: $x = 0.2$; $\alpha = 0.15$; $V = 200$ gal. (.757 m^3) $p = 950\ kg/m^3$

then at the solvent boiling temperature solvent evaporation will be due principally to external heat and the peroxide will be concentrated. This will be a characteristic of low boiling solvents combined with stable peroxides (10 hr half-life temperature is one indication of thermal stability). The converse of these observations also holds.

The important observation is that external heating can both raise bulk temperature and conserve "fuel"as well as in some instances concentrate the peroxide content. These possibilities can be related to well defined property values and boundary conditions.

2.3.4. Container Size

One instance has already been noted where bulk heat-up rate can be related to container size. Thermal inertia effects related to container/sample holder mass and heat capacity are also found to have an inverse relation to one another. Container thermal inertia is discussed briefly in Sec. 3.1.1.

2.4. Overall Significance

The overall significance of the preceding discussion is to indicate how intrinsic and extrinsic properties of a peroxide solvent combination work to influence the decomposition hazards potential. The net hazards potential or some figure of merit to gage or rank relative systems is at the heart of classification concepts. This is difficult to represent by a single index although it should be intuitively clear that related concepts such as total reaction energy content and maximum reaction rate are important. The unique specification of a hazards index for a product is made difficult by sensitivities to external boundary conditions and concentration sensitivities. These concepts are quantified in the following material.

3. ANALYTIC CLASSIFICATION AND PRESSURE VENT AREA RELATIONS

An analytic procedure can be set forth which leads to a hazards classification index based on mass specific energy and volume gas generation rate source terms. The relations are entirely algebraic in nature but have been verified by direct numerical simulation. At the outset one must be prepared to differentiate between decomposition reactions with and without external heat and with and without solvent boiling. This of necessity builds complexity into the methodology. However, the algebraic nature of the presentation should minimize conceptual difficulties.

3.1. Classification Relations for Decomposition without Solvent Boiling

For convenience in subsequent developments a reduced activation energy will be used symbolically where

$$B = E_a / R \tag{8}$$

with

B = reduced activation energy, [K]
E_a = peroxide decomposition activation energy, [cal/gmole]
R = gas constant, [1.987 cal/gmole K]

3.1.1. Onset Temperature

The reaction onset temperature is that temperature where it is assumed that significant fuel consumption begins. With external heating the onset temperature can be taken as that temperature where the bulk heat-up rate due to the decomposition reaction is equal to the bulk heat-up rate due to the external heat source.

This temperature can be represented by the relation

$$T_{onset} = \frac{B}{\ln\left[\dfrac{x_0 \Delta H_{Rx} A}{C \phi \dot{T}_{ex}}\right]} \tag{9}$$

where

T_{onset}	is the onset temperature, [K]
x_0	is the initial peroxide mass fraction, [-]
ΔH_{Rx}	is the heat of decomposition, [J/kg]
A	is the rate constant, Eq. (1) preexponential factor, [min^{-1}]
C	is the liquid heat capacity of the product, [J/kg K]
ϕ	is the dimensionless thermal inertia factor for the sample holder or product container, [-]
$T_{ex}^{\bullet}$	is the bulk heat-up rate driven by an external heat source, [°C/min]

An expression for the bulk heat-up rate due to external fire exposure has been given by Equation (6). The dimensionless factor, ϕ, is represented by the relation

$$\phi = 1 + \frac{(mC)_{vessel}}{(mC)_{product}} \tag{10a}$$

where mC is the product of mass and heat capacity for the vessel and product as indicated by subscripts. A valid approximation for a steel vessel holding an organic product is

$$\phi \approx 1 + 11.5\, W/D \tag{10b}$$

where w is the vessel wall thickness and D is the vessel diameter. A value of $\phi = 1.0$ is always a conservative assumption.

In the absence of external heating the concept of an onset temperature representing the beginning of fuel consumption is arbitrary. One of the following possibilities may be reasonably suited to specific evaluation circumstances.

3.1.2. Time-to-Maximum Rate

An onset temperature may be selected based upon an arbitrary time-to-maximum rate which can be calculated by the relation

$$t_{mr} = \frac{C\phi T^2 \exp(B/T_{onset})}{x_0 \Delta H_{Rx} AB} \tag{11}$$

Equation (11) can be inverted to yield

$$T = \frac{B}{\ln\left[\dfrac{t_{mr}\, x_0 \Delta H_{Rx} AB}{C\phi T^2}\right]} \tag{12}$$

Equation (12) yields a temperature, T, [K] corresponding to a time-to-maximum rate, t_{mr}, [min] using a simple successive substitution solution procedure. An initial guess of $T = 350$ K in the RHS of Eq. (12) will yield a solution value of T on the LHS of Eq. (12) within 1% or on an absolute basis, ±3°C. Convergence is attained within several successive substitution iterations.

The temperature corresponding to a 3-day (4320 min) time-to-maximum rate is a suggested arbitrary onset temperature in the absence of external heating since this temperature corresponds to a weekend of unattended standby. The choice is arbitrary. A 7-day (10,080 min.) Time-to-maximum rate corresponds to the SADT (self-accelerating decomposition temperature) for large containers. For a typical organic peroxide there is a 5°C temperature difference between a 7-day and a 3-day time-to-maximum rate.

3.1.3. Half-Life Temperature

An alternative choice is the temperature corresponding to an arbitrary isothermal half-life time. Half-life time for a first order decomposition is determined by the rate constant as given by Eqs. (1) and (2) which can be inverted to yield a temperature corresponding to a given half-life time as

$$T = \frac{B}{\ln\left[1.4427 A t_{1/2}\right]} \tag{13}$$

3.1.4. Maximum Reaction Temperature

Once an onset temperature is determined, the maximum reaction temperature is determined by the adiabatic temperature rise and any contribution due to external heat input. The theoretical adiabatic temperature increase was given by Eq. (3) which is restated here as

$$\Delta T_a = \frac{x_0 \Delta H_{Rx}}{C\phi} \tag{14}$$

Equation (14) differs from Eq. (3) only by including the thermal inertia factor, ϕ, previously defined by Eqs. (10a) and (10b). This factor is often significant in data evaluation and application.

It has been found through exact numerical simulation that the contribution to the overall temperature increase from external heat input can be represented as

$$\Delta T_{ex} = \frac{\dot{T}_{ex} t_{mr}}{2} \tag{15}$$

In Equation (15)

- ΔT_{ex} is the temperature increase attributed to external heating effects, [°C]
- $T_{ex}^{\bullet}$ is the bulk heat-up rate due to external heating alone such as given by Eq. (3) for external fire, [°C/min]
- t_{mr} is the time-to-maximum rate as determined by Eq. (11) with temperature T set equal to T_{onset}, [min]

With these relations one can determine the maximum reaction temperature, T_{max}, by

$$T_{max} = T_{onset} + \Delta T_a + \Delta T_{ex} \tag{16}$$

3.1.4 Maximum Reaction Rate Conditions

Having determined the maximum reaction temperature, the temperature corresponding to the maximum reaction rate for a first order reaction is

$$T_{mr} = \frac{B}{2}\left[\left(1 + 4\frac{T_{max}}{B}\right)^{1/2} - 1\right] \tag{17}$$

In Eq. (17) both T_{max} and T_{mr} are absolute temperatures. The temperature at maximum rate is typically about 95% of T_{max} on an absolute basis. Equation (17) is exact for an adiabatic decomposition reaction in the absence of external heat input. However, since external heat effects are negligible near the maximum rate, Eq. (17) is a satisfactory representation even with external heating. This has also been verified by numerical simulation. Once the temperature corresponding to the maximum rate has been defined, all other necessary reaction relations follow.

3.1.5. Concentration at Maximum Rate

The peroxide concentration at the maximum reaction rate can be represented by

$$x_{mr} = x_0\left[\frac{T_{max} - T_{mr}}{T_{max} - T_{onset}}\right] \tag{18}$$

Equation (18) is exact for adiabatic conditions, but is also a good approximation for the case with external heat input.

3.1.6. Rate Constant at Maximum Rate

The rate constant at the temperature of maximum reaction rate is

$$k_{mr} = A\exp(-B/T_{mr}) \tag{19}$$

3.1.7. Maximum Reaction Rate

The maximum reaction rate is given by

$$m^{\bullet}_{pmr} = -m_0 x_{mr} k_{mr} \tag{20}$$

where

$m^{\bullet}_{pmr}$ is the maximum rate of peroxide decomposition, [kg/min]

m_0 is the initial mass of peroxide, [kg]

3.1.8. Maximum Rate of Decomposition Gas Production

One can now express the maximum rate of generation noncondensable decomposition gas as

$$\dot{n}_{gmr} = -N_{ncg} \frac{\dot{m}_{pmr}}{Mw_p} \tag{21}$$

where

$\dot{n}_{gmr}$ is maximum rate of production of noncondensable decomposition gas product, [kg mole/min]

N_{ncg} is the intrinsic product decomposition parameter, [mole gas/mole peroxide]

$\dot{m}_{pmr}$ maximum decomposition rate defined by Equation (20), [kg/min]

Mw_p molecular weight of peroxide, [kg/kg mole]

At this point one may define several indices of decomposition reaction severity.

3.1.9. Volumetric Gas Generation Rate

A standard maximum volumetric gas generation rate can be identified as

$$SV_g = \frac{RT_{mr}}{P_s} \frac{\dot{n}_{gmr}}{m_0} \tag{22}$$

or with the aid of Eqs. (20) and (21)

$$SV_g = \frac{N_{ncg}}{Mw_p} \frac{RT_{mr}}{P_s} \frac{x_{mr}k_{mr}}{60} \tag{23}$$

In Equation (23)

SV_g is a classification index for a decomposition reaction without solvent boiling and represents a maximum standard volume generation rate of noncondensable gas having units [m^3/kg sec]

N_{ncg} is the product specific noncondensable gas production parameter

R is the gas constant (R = 8314 J/kg mole K)

T_{mr} is the absolute temperature at the maximum reaction rate, [K]

$x_{mr}k_{mr}$ are the concentration and rate constant at T_{mr} defined by Eqs. (18) and (19)

Mw_p is the molecular weight of the peroxide, [kg/kg mole]

P_s is the standard evaluation pressure, P_s = 101325 N/m^2 (P_s is arbitrarily taken at 1 atm or 14.7 psia)

60 is a unit conversion factor, [sec/min]

3.1.10. Thermal Power Factor

A thermal energy generation rate can also be identified at the maximum reaction rate as

$$E_{mr} = \frac{\Delta H_{Rx} x_{mr} k_{mr}}{60} \tag{24}$$

where

E_{mr}	is the maximum thermal energy generation rate, [watt/kg]
ΔH_{Rx}	is the heat of decomposition, [J/kg]
60	is a unit conversion factor, [60 sec/min]
$x_{mr}\ k_{mr}$	have been previously defined in Eqs. (9) through (19)

3.1.11. Thermal Explosion Vessel Test (TEVT)

In Ref. [1] an empirical classification test protocol is described for a fixed sample mass reacted in a closed sample bomb. The TEVT classification index is defined as the product of the empirically measured maximum pressure, $P_{max,}$ and the maximum rate of pressure increase, dP/dt_{max}.

Pressure—Determinations of pressure and pressure rate for a pure gas generating decomposition reaction of necessity require reference to a closed containment volume. Consider a decomposition starting from an initial pressure and temperature (onset condition) for which the pressure in a constant, unvented volume is given by

$$P = P_{onset} \frac{T}{T_{onset}} + n_g \frac{RT}{V} \tag{25}$$

The first term on the RHS of Equation (25) represents the contribution to pressure from the initial fill gas. The second term represents the contribution from the noncondensable decomposition gas. The decomposition gas, n_g, is given by

$$n_g = N_{ncg} \frac{m_0 (1 - x/x_0)}{Mw} \tag{26}$$

Equations (25) and (26) can represent any arbitrary extent of decomposition. For example at maximum reaction rate;

$$P = P_{mr}$$
$$T = T_{mr}$$
$$x = x_{mr}$$

at completion of decomposition;

$$P = P_{max}$$
$$T = T_{max}$$
$$x = 0$$

Equation (25) can be differentiated to determine the rate of pressure increase at the maximum reaction rate as

$$P^{\bullet}_{mr} = \left[\frac{P_{onset}}{T_{onset}} + \frac{N_{gmr}R}{V}\right] T^{\bullet}_{mr} + \frac{RT_{mr}}{V} n^{\bullet}_{gmr} \tag{27}$$

The only previously undefined term in Eq. (27) is the rate of temperature increase at the maximum reaction rate, $T^{\bullet}_{mr}$, which is given by

$$T^{\bullet}_{mr} = (T_{max} - T_{mr})k_{mr} \tag{28}$$

The maximum rate of pressure increase in a closed volume, Eq. (27) can also be related to the previously defined standard volume rate index by combining Eq. (22) with Eq. (27) to obtain

$$P^{\bullet}_{mr} = \left[\frac{P_{onset}}{T_{onset}} + \frac{N_{gmr}R}{V}\right] T^{\bullet}_{mr} + \frac{m_0 P_s}{V} SV_g \tag{29a}$$

The first term on the RHS of Eq. (29a) is usually small in comparison with the second such that

$$P^{\bullet}_{mr} \approx \frac{m_0 P_s}{V} SV_g \tag{29b}$$

The empirical TEVT index can be represented analytically as the product

$$I_{TEVT} = P_{max} P^{\bullet}_{mv} \tag{30}$$

where P_{max} and $P^{\bullet}_{mv}$ are defined by Eqs. (25) and (29a) respectively. One may similarly neglect the first term on the left-hand side of Eq. (25) and with (29b) one has

$$I_{TEVT} = \left(\frac{m_0}{V}\right)^2 \frac{x_0 N_{ncg} RT_{max}}{Mw_p} P_s SV_g \tag{31}$$

3.1.12. Section Summary

It should be apparent in the previous relations that all of the suggested indices for classification include the product $x_{mr} k_{mr}$. This is especially relevant for nontempered gas generating decompositions. The same product also naturally arises in pressure relief vent sizing relations.

3.2. Vent Area Relations for Gas Producing Decomposition without Tempering

The traditional DIERS vent sizing method for pure gassy systems is to provide a vent area sufficient to pass a volumetric discharge rate equal to the maximum volumetric decomposition gas generation rate [9]. It is customary to establish the latter through empirical small-scale testing [10] and scale test data to the application. For organic peroxides and other similar compounds for which the intrinsic and extrinsic properties are known one may use the appropriate solutions in Section 3.1 to show that the maximum volumetric gas generation rate is

$$V_g^{\bullet} = N_{ncg} \frac{Mw_g}{Mw_p} m_0 \frac{x_{mr}\, k_{mr}/60}{\rho_g} \tag{32}$$

The required volumetric discharge rate is

$$V_{g,ex}^{\bullet} = \frac{C_d AG}{\rho_{ex}} m_0 \tag{33}$$

Combining Eqs. (32) and (33) leads to a required effective vent area per unit mass of product as

$$\frac{C_d A}{m_0} \approx N_{ncg} \frac{Mw_g}{Mw_p} \frac{x_{mr}\, k_{mr}/60}{\left(\dfrac{\rho_g}{\rho_{ex}} G\right)} \tag{34}$$

The new terms in Eqs. (32), (33), and (34) are

C_d	vent device discharge coefficient, [-]
A	vent area, [m^2]
m_0	reaction mass, [kg]
$V_g^{\bullet}$	volumetric gas generation rate, [m^3/s]
Mw_g	molecular weight of decomposition, [kg/kg mole]
Mw_p	molecular weight of peroxide, [kg/kg mole]
ρ_g	decomposition gas density, [kg/m^3]
ρ_{ex}	vent mixture density
G	vent mass flux, [kg/m^{2}s]
60	unit conversion factor, [60 sec/min]

3.2.1. *All Gas Vent Flow*

For all gas vent flow $\rho_{ex} = \rho_g$ and

$$G = \frac{F(\gamma,\eta)P}{\sqrt{RT/Mw_g}} \tag{35}$$

for pressures, P, greater than approximately 15 psig, $F(\gamma, \eta) \approx 0.63$. For pressures, P, less than approximately 15 psig

$$F(\gamma,\eta) = \left(\frac{2\gamma}{\gamma - 1} [1 - \eta^{(\gamma-1)/\gamma}] \eta^{2/\gamma} \right)^{1/2} \tag{36}$$

where γ is the gas specific heat capacity ratio and η is the ambient pressure ratio P_{amb}/P.

This leads to a convenient vent area relation expressed as a product of vent area per unit mass times maximum venting pressure as

$$\frac{C_d A}{m_0} P = N_{ncg} \frac{Mw_g}{Mw_p} \frac{x_{mr} \, k_{mr}/60}{F(\gamma,\eta)} \sqrt{\frac{RT_{mr}}{Mw_g}} \tag{37}$$

It may be argued that the molecular weight of the decomposition gas or gases is not known with precision. However, common organic peroxide decomposition products are carbon dioxide (Mw = 44), ethane (Mw = 30) and acetone (Mw = 58). Therefore an assumed value for Mw_g in the range of 40 to 45 represents a useful estimate.

Also, note that the form of Eq. (37) is hyperbolic in the product of vent area and pressure and that all terms on RHS of Eq.(37) are based on the intrinsic and extrinsic properties of the product and need only be evaluated one time. Furthermore, it is noted that Eq. (37) is also an analytical form of the UN Pressure Vessel Test (PVT) [1,5].

3.2.2. Homogeneous Gas–Liquid Mixture Vent Flow

For this case, $\rho_{ex} = (1 - \alpha)\,\rho_l$ where α is the free volume fraction and ρ_l is the liquid density. Using methods of Ref. [11] the nonflashing two-phase mass flux for sonic flow is given by

$$G = G^* \sqrt{(1-\alpha)\rho_l P} \tag{38}$$

With the additional valid assumption that $[(1 - \alpha)/G^*]^{1/2} \leq 1.0$, Eq. (34) becomes

$$\frac{C_d A}{m_0} P = \frac{N_{ncg}}{Mw_p} \frac{x_{mr} \, k_{mr}/60}{P^{3/2}} RT_{mr} \sqrt{\rho_l} \tag{39}$$

Equation (39) is similar to Eq. (37) in the dependence on the reaction rate terms $x_{mr}\,k_{mr}$. The true relationship between the all-gas and homogeneous mixture vent flow assumptions is given by the ratio

$$\frac{\left(\dfrac{C_d A}{m_0} P\right)_{\text{hom,2-phase}}}{\left(\dfrac{C_d A}{m_0} P\right)_{\text{all gas}}} = F(\gamma,\eta)\sqrt{\frac{\rho_l}{\rho_g}} \tag{40}$$

This ratio typically has values ranging from six to twelve depending upon pressure. At this point it should be apparent that for non-tempered gas generating decompositions, classification and vent sizing are nearly identical in terms of property dependence and sensitivity. Vent area determination either empirical or analytical is another form of classification.

4. VENT SIZE DATA COMPARISON

Open literature vent area test data are quite scarce, especially data whose test protocol is sufficiently well described to permit interpretive evaluation. Vented decomposition data for 37.5 wt% 3,5,5-trimethyl hexanoyl peroxide in isodecane (boiling point 175°C) is shown in Fig. 1. These data have been reported in Refs. [12,13] and elsewhere. The adiabatic temperature increase for 37.5 wt% 3,5,5 TMHP using Appendix A data and Eq. (3) is ≈108°C. The

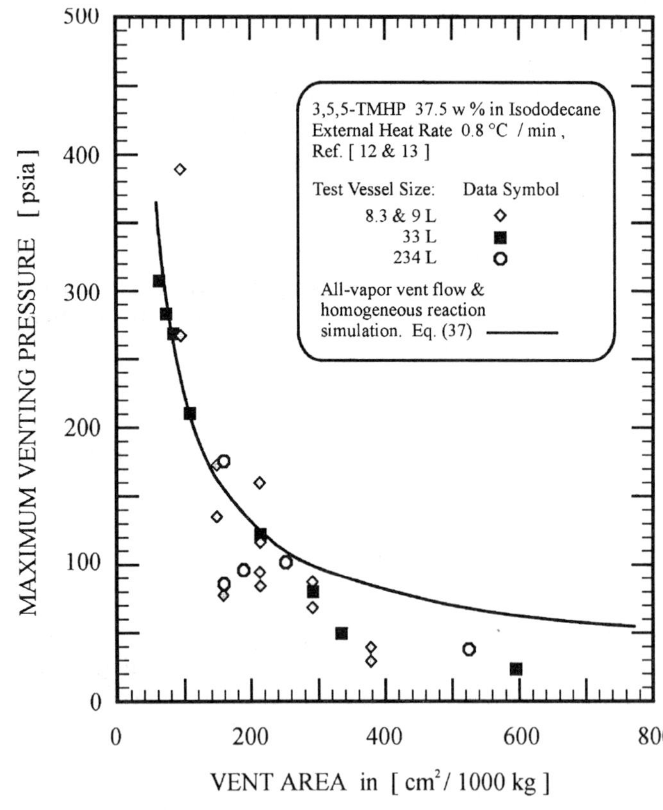

Figure 1. Runaway decomposition of 37.5 wt% 3,5,5-trimethyl hexanoyl peroxide in isodecanewith external heat input of 0.8°C/min (ni\ominal). Data compared with Eq. (37) assuming all-vapor vent flow. Data source: [12, 13].

onset temperature using an external heating rate of 0.8°C/min and Equation (9) with $\phi = 1.1$ is $\approx$75°C. With any elevation in pressure above atmospheric, solvent evaporation effects are negligible and the system may be considered as a pure gas generating decomposition reaction.

The solid line in Fig. 1 is the result obtained using Eq. (37) and intrinsic data for 3,5,5-TMHP in Appendix A. The agreement suggests that the all-gas vent flow assumption best characterizes the test data. These data have been cited as evidence that small scale 10L test data results can be scaled to larger size vessels. The fact that the vent discharge is predominantly defined by all-gas flow weakens the generality of this argument. For all-gas vent flow, the vent area versus pressure relation is always scale independent. It will be shown in the next section that 37.5 wt% 3,5,5-TMHP is a relatively weak peroxide compound. This weakness enhances the predominance of all-gas vent flow as well as the scale independence of the results.

A similar comparison can be illustrated between Eq. (37) and 20 wt% t-butyl peroxypivalate in odorless mineral spirits and Isopar-C diluents. Data shown in Fig. 2 are taken from Ref. [14] using test methods reported in Ref. [15]. These data are less well characterized by virtue of the heavy walled test vessel and uncertainty in external heat-up rate. Reliable vented test data are not known in the open literature for higher strength compounds.

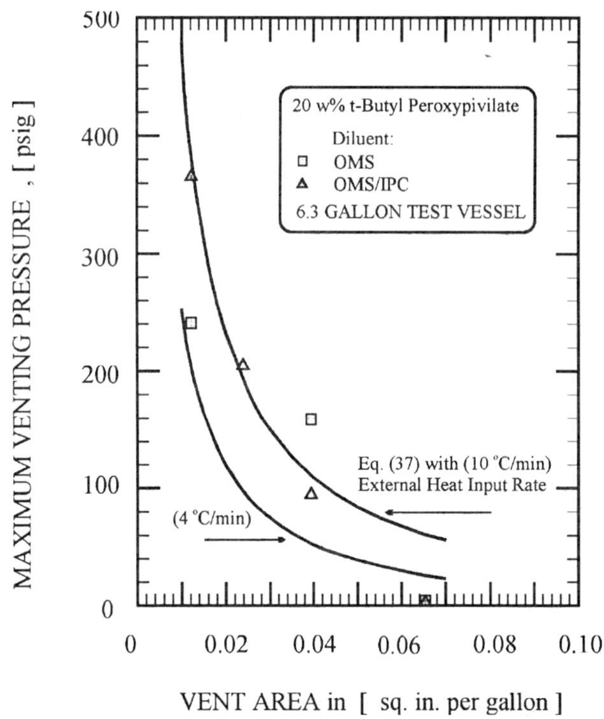

Figure 2. Comparison of Ref [14] test data and Eq. (37) using Appendix A parameter values and two bracketing external heat input rate values.

5. CLASSIFICATION COMPARISON

It is useful to compare a number of peroxide compounds as illustrated in Table 3. The comparison is based upon property values in Appendix A and the relevant equations in Section 3. An external heat input rate equivalent to 2°C/min is used. The relative ranking indices include adiabatic temperature rise and the all-gas vent area pressure product, Equation (37). It is interesting to note that for the same external heat input conditions, the adiabatic temperature increase criterion proposed by Melhem and Shanley, Ref. [4] provides an excellent and simple rule for classification. However, the magnitude of the differences between products and the effect of concentration is better captured by the vent area pressure index. From Table 3 one may obtain a better quantitative perspective on the low reaction strength of the products considered in Figs. 1 and 2.

Tables 4 and 5 show the sensitivity of the same classification indices to variations in peroxide concentration at constant heat input and to variations in external heat input at a fixed concentration for t-butyl peroxy 2-ethyl hexanoate. Table 4 clearly demonstrates the strong sensitivity to concentration. This sensitivity can be quantified by the reaction rate doubling temperature interval as follows. For this product a 10 wt% increase in concentration increases the adiabatic temperature rise by ≈38°C or 3.8 rate doubling temperature intervals. The value of $2^{3.8}$ is equal to 14 which is the approximate magnitude of the 10 wt% increase in concentration on the vent area pressure index. This is precisely the underlying reason why a ten percent variation in

TABLE 3

Relative Ranking of a Select List of Peroxide Compounds($T_{ex}^{\cdot}$ = 2°C/min)

		Mw	ΔT_a [°C]	$(C_dA/m_0)P$ [N/kg]
t-butyl peroxyacetate	75 wt%	132.2	469	4.28E6
t-butyl peroxy 2-ethylhexanoate	98 wt%	216.3	375	4.00E6
t-butyl peroxypivalate	75 wt%	174.2	356	1.16E6
dicumyl peroxide	98 wt%	270.4	300	8.95E4
t-butyl peroxyacetate	20 wt%	132.2	125	9.06
t-butyl peroxypivalate	20 wt%	174.2	95	3.55
3,5,5 trimethyl hexanoyl peroxide	37.5 wt%	314.5	98	2.94
t-butyl peroxy 2-ethylhexanoate	20 wt%	216.3	76	1.04

TABLE 4
Variation of Hazard Ranking Index for t-butyl peroxy-2-ethylhexanoate at Various Product Concentration ($T_{ex}^{\bullet}$ = 2°C/min)

Product Concentration [wt%]	ΔT_a [°C]	$(C_dA/m_0)P$ [N/kg]
20	76.5	1.04
30	114.8	16.5
40	153.0	195
50	191.3	1.76E3
60	229.6	1.24E4
70	267.8	7.12E4
80	306.1	3.42E5

estimation of the maximum reaction temperature can easily have an order of magnitude effect on the vent area pressure estimate.

It is also a fundamental concern in purely empirical methods advocating direct scale-up on vent area to mass.

The effect of external heat input illustrated in Table 5 is smaller than the concentration effect. But this influence is not reflected in the adiabatic temperature rise. Other sensitivities such as related to the decomposition gas generation parameter are typically confined to a factor of two.

TABLE 5
Variation of Hazard Ranking Index for t-butyl peroxy-2-ethylhexanoate with External Heat-Up Rate at a Fixed Concentration x = 0.40

External Heat-Up Rate $T_{ex}^{\bullet}$ [°C/min]	ΔT_a [°C]	$(C_dA/m_0)P$ [N/kg]
1	191.3	1.27E3
2	191.3	1.76E3
4	191.3	2.44E3
8	191.3	3.40E3
16	191.3	4.76E3

6. VENT SIZE CONSIDERATIONS FOR PEROXIDES AT HIGHER CONCENTRATIONS

At low peroxide concentrations vent sizing for gassy systems can be based on all-gas venting. At present there are insufficient data to provide specific guidelines as to the concentration level on classification index which requires consideration of gas and liquid vent flow. This situation is made more difficult and complex by a known tendency for higher strength products in an unstirred vessel to react with strong spatial nonuniformity. Results of vented tests with 98 wt% dicumyl peroxide [16] indicate such nonuniformity.

Such nonuniform reaction characteristics arise from temperature stratification present in volume heated natural circulation system [17,18]. Verhoff [19] has verified such behavior with peroxide decomposition reactions. The net effect of spatial nonhomogeneous decomposition reactions is to significantly reduce (by orders of magnitude) the gas generation decomposition reaction rates. In this regard extrapolation of vent sizing based upon small-scale data either by direct area to mass scaling or by kinetic modeling is likely to be conservative. In a similar manner, even sophisticated computer models such as SAFIRE and SuperChems, which treat the reaction mass as spatially uniform in temperature, will likewise yield conservative vent size results.

7. SUMMARY

It has been shown that hazards ranking and vent sizing are identical procedures. For a class of similar compounds, DIERS empirical or analytical vent sizing methods also provide an effective hazards ranking. Where sufficient data is available such as for organic peroxides, one need only establish a limit of reaction severity for a specific application. With methods similar to those proposed here, relative ranking of alternative products may be rapidly established.

This work has emphasized pure gassy systems. The list of true tempered systems is relatively small and confined to low molecular weight peroxides. In most cases solvent boiling under external heat input tends to boil-off solvent and concentrate the peroxide. Hence pure gassy systems are systems of the greatest hazard significance.

APPENDIX
Peroxide Thermo-decomposition Property Summary

Compound Code	**HP 1**	**HP 2a**	**HP 2b**	**HP 3**	**HP 4**	**DAK 1**
Chemical Formula	H_2O_2	$C_4H_{10}O_2$	$C_4H_{10}O_2$	$C_5H_{12}O_2$	$C_9H_{12}O_2$	$C_8H_{18}O_2$
Molecular Wt. [kg/kg mole]	30.015	90.12	90.12	104.15	152.2	146.23
Half-Life Temperatures at:						
10 hour $T_{½, 10 hr}$ [°C]	140	117	172	178	158	129
1 hour $T_{½, 1 hr}$ [°C]	175	140	200	210	190	149
Activation Energy, E_a [kcal/gmole]	24.2	32.1	34.4	31.5	28.55	38.8
Reduced Activation Energy B [K]	12181	16137	17321	15844	14369	19545
Preexponential Factor A [min^{-1}]	7.36E9	1.06E15	9.14E13	6.0E12	3.44E11	1.48E18
Heat of Decomposition						
ΔH_{Rx} [kcal/gmole]	22.6 [3]	38 [3]	38 [3]	40 [3]	50	50
ΔH_{Rx} [kJ/kg]	3150	1764	1764	1607	1374	1431
Heat Capacity, C [kJ/kg K]	3.3	2.3	2.3	2.3	2.3	2.3
Adiabatic Temp. Increase ΔT_a [°C]	954	767	767	698	598	622
Mole gas/mole peroxide N_{ncg}	0.5	1	1	1	1	1
Melt Temperature, T_m [°C]	0	−2.8	−2.8	−1.5	−9	−40
Boiling Temperature, T_b [°C]	115	96	96	unk	170	111

Compound Code	Compound Name	Generic Acronym
HP 1	hydrogen peroxide	H_2O_2
HP 2a	t-butyl hydroperoxide	TBHP (1)
HP 2b	t-butyl hydroperoxide	TBHP (2)
HP 3	t-amyl hydroperoxide	TAHP
HP 4	cumene hydroperoxide	CHP
DAK 1	di-t-butyl peroxide	DTBP

Notes: (1) Half-life data typical of 70 w% TBHP with 20 w% DTBP and 10 w% H_2O.
(2) Half-life data typical of 90 w% TBHP with 10 w% H_2O.
(3) Heat of decomposition based on liquid state heats of formation for reactant and products.

Compound Code	DAK 2	DAK 3	DAK 4	PE 1	PE 2	PE 3
Chemical Formula	$C_{18}H_{22}O_2$	$C_{16}H_{34}O_4$	$C_{16}H_{30}O_4$	$C_6H_{12}O_3$	$C_9H_{18}O_3$	$C_{10}H_{20}O_3$
Molecular Wt. [kg/kg mole]	270.4	290.45	286.4	132.2	174.2	188.3
Half-Life Temperatures at:						
10 hour $T_{½, 10\ hr}$ [°C]	117	120	131	102	58	55
1 hour $T_{½, 1\ hr}$ [°C]	137	140	152	123	76	74
Activation Energy, E_a [kcal/gmole]	36.6	37.2	37.4	32.4	29.4	27.4
Reduced Activation Energy B [K]	18423	18700	18840	16295	14790	13805
Preexponential Factor A [min^{-1}]	3.72E17	5.25E17	2.03E17	8.45E15	2.88E16	2.16E15
Heat of Decomposition						
ΔH_{Rx} [kcal/gmole]	50	100	100	50	50	50
ΔH_{Rx} [kJ/kg]	774	1441	1461	1583	1201	1111
Heat Capacity, C [kJ/kg K]	2.3	2.3	2.3	2.3	2.3	2.3
Adiabatic Temp. Increase ΔT_a [°C]	337	626	635	688	522	483
Mole gas/mole peroxide N_{ncg}	1	1	1	1	1	1
Melt Temperature, T_m [°C]	38	5	5	–30	–18	–42
Boiling Temperature, T_b [°C]	395	260	243	unk	unk	unk

Compound Code	Compound Name	Generic Acronym
DAK 2	dicumyl peroxide	DCP
DAK 3	2,5-di-methyl-2,5-di-t-butylperoxy hexane	25DMDTBPH
DAK 4	2,5-dimethyl-2,5-di-t-butylperoxy hexane-3	25DMDTBPH-3
PE 1	t-butyl peroxyacetate	TBPA
PE 2	t-butyl peroxypivalate	TBPIV
PE 3	t-amyl peroxypivalate	TAPIV

Compound Code	PE 4	PE 5	PE 6	PE 7	PE 8	PE 9
Chemical Formula	$C_{11}H_{14}O_3$	$C_{12}H_{16}O_3$	$C_{12}H_{24}O_3$	$C_{13}H_{26}O_3$	$C_{14}H_{28}O_3$	$C_{15}H_{30}O_3$
Molecular Wt. [kg/kg mole]	194.2	208.3	216.3	230.4	244.4	258.4
Half-Life Temperatures at:						
10 hour $T_{½, 10\ hr}$ [°C]	104	100	77	75	48	46
1 hour $T_{½, 1\ hr}$ [°C]	125	122	95	92	66	64
Activation Energy, E_a [kcal/gmole]	32.7	30.7	32.8	34.2	27.7	27.4
Reduced Activation Energy B [K]	16465	15433	16490	17219	13933	13765
Preexponential Factor A [min^{-1}]	1.05E16	1.06E15	3.28E17	3.48E18	8.02E15	6.21E15
Heat of Decomposition						
ΔH_{Rx} [kcal/gmole]	50	50	50	50	50	50
ΔH_{Rx} [kJ/kg]	1077	1004	968	910	857	810
Heat Capacity, C [kJ/kg K]	2.3	2.3	2.3	2.3	2.3	2.3
Adiabatic Temp. Increase ΔT_a [°C]	469	436	421	396	393	352
Mole gas/mole peroxide N_{ncg}	1	1	1	1	1	1
Melt Temperature, T_m [°C]	8	- 4	unk	unk	unk	unk
Boiling Temperature, T_b [°C]	unk	unk	unk	unk	unk	unk

Compound Code	Compound Name	Generic Acronym
PE 4	t-butyl Peroxy Benzoate	TBPBEN
PE 5	t-amyl Peroxy Benzoate	TAPBEN
PE 6	t-butyl peroxy-2-ethyl hexanoate	TBP2EH
PE 7	t-amyl peroxy-2-ethyl hexanoate	TAP2EH
PE 8	t-butyl peroxy neodecanoate	TBPND
PE 9	t-amyl peroxy neodecanoate	TAPND

Compound Code	PK 1	DAC 1	DAC 2	DAC 3	DAC 4
Chemical Formula	$C_{14}H_{28}O_4$	$C_{14}H_{10}O_4$	$C_{24}H_{46}O_4$	$C_4H_6O_4$	$C_{18}H_{34}O_4$
Molecular Wt. [kg/kg mole]	260.4	242.2	398.6	118.1	314.5
Half-Life Temperatures at:					
10 hour $T_{½,\ 10\ hr}$ [°C]	97	73	62	69	59
1 hour $T_{½,\ 1\ hr}$ [°C]	116	92	80	87	78
Activation Energy, E_a [kcal/gmole]	34.7	30.4	30.1	31.3	28.1
Reduced Activation Energy B [K]	17456	15318	15141	15763	14135
Preexponential Factor A [min^{-1}]	3.50E17	1.91E16	4.81E16	1.18E17	3.50E15
Heat of Decomposition					
ΔH_{Rx} [kcal/gmole]	100	50	50	50	50
ΔH_{Rx} [kJ/kg]	1607	864	525	1771	665
Heat Capacity, C [kJ/kg K]	2.3	2.3	2.3	2.3	2.3
Adiabatic Temp. Increase ΔT_a [°C]	698	376	229	770	289
Mole gas/mole peroxide N_{ncg}	1	1	1	1	1
Melt Temperature, T_m [°C]	unk	0	54	unk	unk
Boiling Temperature, T_b [°C]	unk	unk	unk	ink	unk

Compound Code	Compound Name	Generic Acronym
PK 1	1,1-bis-t-butylperoxy cyclohexane	11DTBPC
DAC 1	benzoyl peroxide	BPO
DAC 2	di lauroyl peroxide	DLP
DAC 3	acetyl peroxide	AP
DAC 4	3,5,5-trimethyl hexanoyl peroxide	355TMHP

REFERENCES

1. *Recommendations on the Transport of Dangerous Goods—Manual of Tests and Criteria,* 2nd rev. ed., Publication E.95.VII.2. New York: United Nations , 1995.

2. Yoshida, T., Wada, Y., and Foster, N., *Safety of Reactive Chemicals and Pyrotechnics,* Industrial Safety Series, Vol. 5. Amsterdam: Elsevier, 1995.

3. *Suggested Relative Hazard Classification of Organic Peroxides,* The Society of the Plastics Industry, Inc., Manual 19A OPPSD, 1975.

4. Melhem, G. A., and Shanley, E. S., "On the Estimation of Hazard Potential for Chemical Substances," *Process Safety Progress,* 15(3) (1966), 168–172.

5. Grolmes, M. A., and King, M. J., "Simulation of Vented Pressure Vessel Tests for Organic Peroxides," *J. Hazardous Materials,* 44(2–3) (1995), 253–266.

6. Shaw, D. A., "SAFIRE Program for Design of Emergency Pressure Relief System," *Chemical Engineering Progress,* 86(7) (1990), 14–17.

7. Melhem, G. A., "Advanced ERS Design Using Computer Simulation." In *Proceedings of the International Symposium on Runaway Reactions and Pressure Relief Design.* New York: AIChE, 1995, pp. 502–566.

8. API-520, "Sizing, Selection and Installation of Pressure Relieving Devices in Refineries, Part I —Sizing and Selection." American Petroleum Institute, 1220 L Street, NW, Washington, DC, 20005.

9. Fisher, H. G., et al., *Emergency Relief System Design Using DIERS Technology*, New York: AIChE, 1992.

10. Creed, M. J., and Fauske, H. K., "An Easy, Inexpensive Approach to the DIERS Procedure," *Chemical Engineering Progress,* 86(3) (1990), 45.

11. Leung, J. C., "Simplified Vent Sizing Methods Incorporating Two-Phase Flow." In *Proceedings of the International Symposium on Runaway Reactions and Pressure Relief Design.* New York: AIChE, 1995, pp. 201–236.

12. de Groot, J. J., Groothuizen, T. M., and Verhoff, J., "Safety Aspects of Organic Peroxides in Bulk Tanks," *Ind. Eng. Chem. Process Des. Dev.,* 20(1) (1981), 131–138.

13. Walker, J. P., and de Groot, J. J., "Venting of Decomposition of Energetic Liquids Using a Bottom Vent," *Proceedings of Process Plant Safety Symposium,* Vol. 2, pp. 55–71, 1996.

14. King, M. J., Gesl, R. T., and Grolmes, M. A. "Interpretation of t-Butyl Peroxypivalate Decomposition in 6.3 Gallon Vented Pressure Vessel', Unpublished.

15. Wagle, U. D., Houston, G. W., Anderson, A. J., and Mageli, O. L., "Organic Peroxide Decomposition Characteristics," *Safety in Polyethylene Plants,* Vol. 3. New York: AIChE, 1978, pp. 62–69.

16. Gove, S. H., "Emergency Relief for Intermediate Bulk Containers Containing Dicumyl Peroxide." Presentation to DIERS Users Group, Spring 1996 Meeting., Houston, TX.

17. Aydemir, N. V., et al., "Thermal Response of LPG Tanks Exposed to Fire," *J. Hazardous Materials,* 20 (1988), 239–262.

18. Yu, C. M., Aydemir, N. V., and Venart, J. E. S., "Transient Free Convection and Thermal Stratification in Uniformly Heated Partially Filled Horizontal Cylindrical and Spherical Vessels," *J. Thermal Sci.,* 1(2), (Nov. 1992), 114–122.

19. Verhoff, J., "Experimental Study of the Thermal Explosion of Liquids." Doctoral Thesis, Prins Maurits Laboratorium, TNO, The Netherlands, 1983.

Thermal Decomposition Kinetics of Cumene Hydroperoxide

Yih-Shing Duh and Chen-Shan Kao
Center for Industrial Safety and Health Technology
Industrial Technology Research Institute

Her-Huah Hwang and William W.-L. Lee
Department. of Environmental and Safety Engineering
National Yunlin University of Science and Technology

ABSTRACT

The thermal decomposition and runaway reaction of cumene hydroperoxide (CHP) in cumene were studied by thermal analysis and the VSP-II adiabatic calorimeter. Decomposition kinetics were determined in various CHP concentrations in cumene. The reaction order of 20 wt%, 35 wt%, 50 wt%, 65 wt%, and 80 wt% was determined to be 0.5, and the Arrhenius parameters were measured to be Ea ($kJmol^{-1}$) = 122.0 ± 3.0 and $\ln A$ ($min^{-1} M^{1/2}$) = 30.0 ± 1.2. This new discovery of the reaction order equal to one half implies the same mechanism of decomposition regardless of CHP concentration. We combined the chromatography and IR spectroscopy with calorimetry data to propose the thermal decomposition kinetics. The preliminary rate determining step was proposed to be the decomposition of CHP by dimer association. A reaction mechanism for CHP thermal decomposition is proposed.

1. INTRODUCTION

Most organic peroxides can be used to initiate free-radical polymerization. In fact, this application accounts for about 90% of the total consumption in Taiwan. However, many thermal runaway or explosion incidents were caused by the decomposition of organic peroxides due to their thermal instabilities or reactive incompatibilities (Table1)[1]. Hydroperoxides (such as hydrogen peroxide, and cumene hydroperoxide) are extremely sensitive or reactive to acids, metal ions and impurities.

TABLE 1
Thermal Explosion Accidents Caused by Peroxide in Taiwan

Date	Chemicals	Injuries	Fatalities	Hazard	Cause
1979 7.13	MEKP[a]	49	33	Explosion (Storage)	Thermal decomposition
1981 4.21	CHP[b]	3	1	Explosion (Condensation)	Thermal decomposition
1984 2.18	MEKP[a]	55	5	Explosion (Reactor)	Thermal decomposition
1986 5.02	CHP[b]	0	0	Explosion (Reactor)	Thermal decomposition
1987 9.05	H_2O_2	20	0	Explosion (Storage)	Incompatibility
1988 7.25	TBHP[c]	19	0	Explosion (Storage)	Cooling failure
1989 3.14	Organic Peroxide	0	0	Fire (Tank)	Cooling failure
1989 8.04	Organic Peroxide	0	0	Fire (Storage)	Cooling failure
1989 9.01	MEKP[a]	5	7	Explosion (Storage)	Thermal decomposition
1996 10.07	MEKP[a]	47	10	Explosion (Storage)	Thermal decomposition

a. Methyl Ethyl Ketone Peroxide, MEKP
b. Cumene Hydroperoxide, CHP
c. Tert-Butyl Hydroperoxide, TBHP

The process of organic hydroperoxide decomposition in a variety of solvents is usually more complicated compared with other organic peroxides catagories (e.g. Dialkyl peroxides or Ketone peroxides). Our previous report [2] indicated that thermal decomposition of cumene hydroperoxide (CHP) was dependent upon the CHP concentration. Based on the Accelerating Rate Calorimeter (ARC) test and the isothermal and dynamic differential scanning calorimeter (DSC) tests, the reaction order of 35 wt% CHP decomposition was determined to be 0.5. The Arrhenius parameters were measured to be E_a (kJmol^{-1}) = 120.6 ± 3.0 and ln A (min^{-1} $M^{1/2}$) = 30.2 ± 1.3 [2].

The aim of this research was to continue our work to verify thermal decomposition kinetics and the mechanism of CHP in nonisothermal and isothermal conditions. Both DSC and VSP-II (Vent Sizing Package II) techniques were used for thermal analysis to obtain thermal kinetics data. Identification of intermolecular and intramolecular hydrogen-bonded association of CHP was carried out using infrared (IR) spectroscopy. The decomposition kinetics of CHP, affected by its associated complex, have been confirmed by incorporating the thermokinetic data, IR spectroscopic and chromatographic method. A reaction mechanism for CHP thermal decomposition was proposed from the identification of some distinguishing decomposition products by the chromatographic method.

2. EXPERIMENT

2.1. Samples

80wt% CHP purchased directly from the supplier was measured to determine both density and concentration, then stored in a 4°C environment. Cumene was used as the dilution solvent in preparing various CHP samples.

2.2. DSC(Differential Scanning Calorimeter)

The dynamic screening and isothermal experiments were performed on a Mettler TA4000 system coupled with a DSC25 measuring cell [3]. The system was connected to an IBM PC computer in which data were evaluated and stored. Disposable high pressure crucibles (ME-26732) were used for acquiring thermograms and isothermal traces. Standard aluminum crucibles were used for heat capacity (C_P) measurements. The average heat capacity ($Jg^{-1}K^{-1}$) of various solutions of cumene hydroperoxide of 80 wt%, 35 wt%, 20 wt%,and 15 wt% were determined to be 2.219, 2.217, 2.154, and 2.074, respectively. The measuring method was used to check the heat capacity temperature function of aluminum oxide (α-Al_2O_3 single crystal) with the data provided by the National Bureau of Standards (NBS). The scanning rate chosen for the temperature programmed ramp was chosen to be 4K min^{-1}.

2.3. VSP (Vent Sizing Package)

A PC controlled adiabatic calorimeter system, the Vent Sizing Package II (VSP-II) manufactured by FAI [4], was used to measure the thermokinetic and thermal hazard data such as temperature and pressure trace in relation to time. The low heat capacity of the cell assures that essentially all the reaction heat released remains within the test sample. Thermokinetics and pressure behaviors in the small test cell (112 mL) can therefore be extrapolated directly to process scale due to the low thermal inertia of about 1.05 to 1.2. Three types of cells for closed testing, top venting or bottom dumping are available.

Detailed information on the performance of the VSP calorimeter can be found in the literature [5,6]. Basic design data for emergency relief systems can be acquired from the VSP II calorimeter and related DIERS methodology. The two-phase flow pattern such as churn-turbulent or homogeneous equilibrium flow can be determined directly from top venting experiments using the VSP II.

2.4. IR (Infrared Spectra)

Infrared spectra were measured with a Bornem MB FT-IR instrument using a solution cell made of KBr. The spectrum range from 400 cm^{-1} to 4000 cm^{-1} was scanned to seek the characteristic bands or fingerprint associated with molecular vibrations, bendings and interactions.

2.5. Chromatographic Studies

The decomposition products were identified by HPLC using a UV-970 detector employing C18 μm column. The eluting solvent was a mixture of acetonitrile and water (3:7) at a flow rate of 1.3 mL/min. The detection was carried out at a wavelength of 254 nm. A gas chromatography (HP5890 series II plus) linked with a Mass Dectector (HP5972) were also used to monitor the decomposition products using a RTX-502.2 column (60.0 m $\times$ 0.32 mm ID $\times$ 1.8 μm). The carrier gas was helium at pressure around 10.0 psig. The injection temperature was set at 225°C.

3. RESULTS AND DISCUSSION

3.1. Thermal Analysis

The onset temperature and heat of decomposition of CHP can be easily acquired by using DSC programmed scanning. Figure 1 shows the typical temperature and heat flow curves for the thermal decomposition of 50 wt%, 65 wt% and 80 wt% CHP/Cumene solution. Themal decomposition hazards of CHP with different concentrations can be characterized using a non-isothermal method. The peaks on the thermograms indicate that the higher CHP concentrations will result in potential hazard in case of thermal decomposition. Thermal analysis data are listed in Table 2. The onset temperature was determined to be 80°C. The heat of decomposition was measured to be (1810 $\pm$ 50) J per gram of CHP and in good agreement with another study [7]. However, the onset temperature was determined to be about 150°C by using a higher scanning rate of 20 $Kmin^{-1}$ in the annual reports of JRIIS [7].

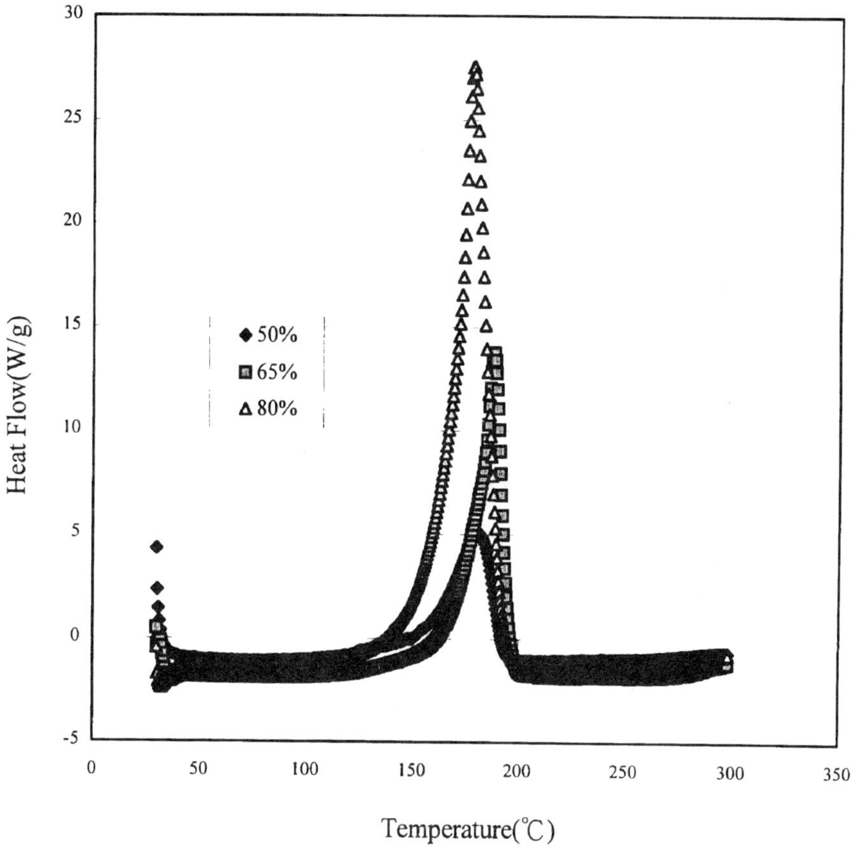

Figure 1. The heat flow versus temperature of dynamic thermal decomposition of different CHP concentrations

3.2. Hydrogen-Bonded Association

In the IR spectra, a series of concentrated solutions of CHP (5 wt%–80 wt%) in cumene showed the hydrogen-bonded effect of the –O-H stretching vibration. Two distinguishable peaks were observed and listed in Table3. The $\nu_{O\text{-}H}$ bond centered at 3430.56 cm^{-1}(band 1) was due presumably to free CHP with a bridge between the –O-O-H group and the π-electrons of the aromatic ring. This result suggests that the formation of associates by intramolecular hydrogen bonding and the$\nu_{O\text{-}H}$ is not influenced by CHP concentration.

However, another absorbance (band 2) occurred at 3489 cm^{-1} and shifted to 3502 cm^{-1} with increasing concentration, which was assigned to be the –O-H stretching mode, was due to the higher association and increased gradually with increasing concentration [8,9]. The absorption peak of –O-H stretching was originally at 3300 cm^{-1} without any perturbations of intermolecular or

TABLE 2
Heat of Decomposition and Initial Exothermic Temperature of Cumene Hydroperoxide

	Scan Rate (°C min^{-1})	METTLER DSC 25			
		T_0(°C)	ΔH (–Jg^{-1})		ΔH (–Jg^{-1})
35wt% CHP	10	110	626.7	1790(100%)	1874(100%) (JRIIS)
	4	95	618.6	1767(100%)	
	2	80	618.7	1768(100%)	
	1	80	640.0	1828(100%)	
50wt% CHP	2	90	883.0	1766(100%)	1874(100%) (JRIIS)
65wt% CHP	2	90	1172.6	1804(100%)	1874(100%) (JRIIS)
80wt% CHP	2	90	1503.2	1879(100%)	1874(100%) (JRIIS)
	4	100	1424.8	1781(100%)	
	4	100	1417.6	1772(100%)	

CHP Cumene Hydroperoxide
T_0: thermal decomposition initial exothermic temperature
ΔH: heat of decomposition
JRIIS: Japan Research Reports of Research Institute of Industrial Safety[7]

intramolecular hydrogen bonding. The shifted phenomena of the absorption peak was due to the intermolecular hydrogen boundings formed between the –O-O-H functional group(s) and the oxygen atom(s) of the cyclic dimers.

(1.1)

(1.2)

where $R=(C_6H_5)C(CH_3)_2$

Evidently one can expect that the formation of associates can retard the decomposition of organic hydroperoxides by affecting the reaction order of the decomposition pathway as well as the activation energy. The formation of a dimetric structure indicates that the rate-determinating step depends strongly on the concentration of dimer associates. It has been found that the dimeric associate of the hydroperoxide was the most stable structure and could be investigated using IR spectroscopy even at 75°C [9].

TABLE 3
Infrared Absorption Bands of Cumene Hydroperoxide in Cumene (22°C)

	Band	
Concentration	**Band center 1 cm^{-1}(transmittance)**	**Band center 2 cm^{-1}(transmittance)**
5%	3430.56 (0.829)	3502.21 (0.550)
10%	3430.56 (0.586)	3510.37 (0.329)
20%	3430.56 (0.257)	3497.37 (0.214)
40%	3430.56 (0.057)	3489.23 (0.043)
80%	3430.56 (0)	3461.52 (0)

3.3 Decomposition Kinetics

3.3.1 Isothermal method

Previous work in our laboratory has shown the correlation of runaway behaviors with respect to CHP concentration. We found that the reaction order of 35 wt% CHP decomposition was 0.5. The dynamic method described above does not lead to a full understanding of the complex decomposition mechanism. On the other hand, isothermal analysis does eliminate the thermal lag effect and maintain good thermal equilibrium using a *n*th-order model.

The rate equation for various CHP concentrations can be written as

$$-d[\mathrm{CHP}]/dt = k\,[\mathrm{CHP}]^n \quad (2)$$

Equation (2) can be integrated and becomes

$$[\mathrm{CHP}]^{1-n} = (n-1)kt + [(\mathrm{CHP})_0]^{1-n} \quad (3)$$

The isothermal tests were conducted at 140°C with a series of different concentrations, or at fixed concentration with different temperatures. As shown in Table 4, the reaction order was determined to be a constant value of 0.5 in spite of differences in temperatures or concentrations.

3.3.2. Nonisothermal Method

Interpretation of IR and thermal analysis data from dynamic and isothermal DSC verifies the formation of association complexes in various CHP/cumene mixtures. The nonisothermal approach was performed by using on adiabatic

TABLE 4
The Reaction Order Determined from DSC Isothermal Experiments

Temperature	20 wt%	35 wt%	50 wt%	65 wt%	80 wt%
120°C	—	—	—	—	0.51
130°C	—	—	—	0.48	0.50
140°C	0.49	—	0.50	0.49	0.52
150°C	0.52	0.50	0.50	0.50	0.51
160°C	0.52	—	0.47	0.52	0.50

calorimeter VSP?. A model for CHP thermal decomposition kinetics was proposed in the following equations.

$$dQ/dt = k[\mathrm{CHP}]^{1/2}\Delta H\, V\rho \tag{4}$$

$$\frac{dT}{dt} = K\frac{(T_{\mathrm{f}} - T_0)^{1/2}}{([\mathrm{CHP}]_0)^{1/2}}[T_{\mathrm{f}} - T]^{1/2} \tag{5}$$

The experimental measurements of self-heating rate and temperature behavior in 15 wt % CHP were in good agreement with the theoretical calculation data. The self-accelerating temperature to time curve was recorded in Figure 2. Figure 3 shows that the experimental data on self-heating rate were in fact bounded quite close to the proposed model with a reaction order of 0.5. VSP? experiment data of CHP 15wt% are summarized in Table 5.

3.4 The Chromatographic Data of the CHP decomposition

The detection of decomposition products of CHP at various temperatures (130°C, 140°C, 150°C) proves the presence of acetone, a-methylstyrene, and acetonphenol. But 2-phenyl, 2-propanol and phenol, not detected by gas chromatography, were clearly identified by HPLC. Interestingly, α-methylstyrene, not detected by HPLC, was shown using gas chromatography. These final products can be regarded as direct evidence for proposing the decomposition mechanism of CHP. The ratios of decomposition products at various temperatures are shown in Tables 6, 7, and 8.

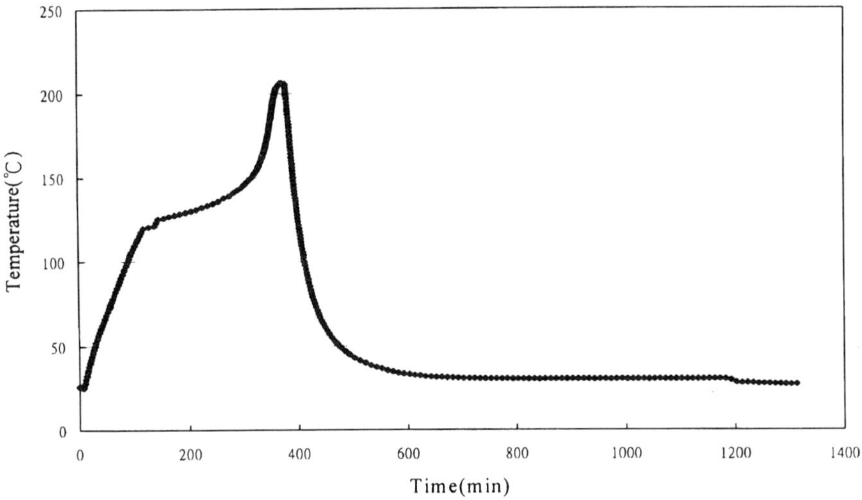

Figure 2. The time versus temperature for thermal decomposition of CHP 15 wt%

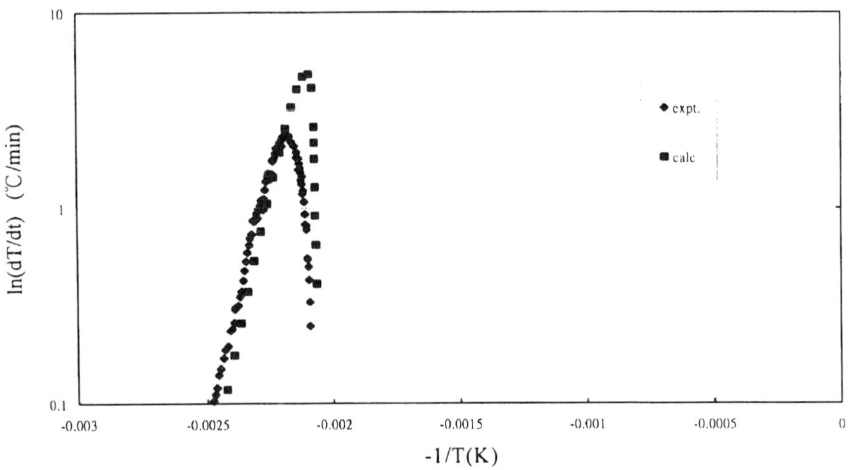

Figure 3. The self-heating rate versus temperature for thermal decomposition of CHP 15wt%

TABLE 5
VSPII Experiment Data of CHP15%

CHP wt%	15%
CHP weight(g)	8.47
Cumene weight(g)	48.0
Total sample weight(g)	56.47
Total sample volume(ml)	64.03
Can weight(g)	43.0
Can volume(mL)	112
ϕ	1.19
T_0(°C)	100
T_f(°C)	206.2
T_{cf}(°C)	210.9
$(dT/dt)_{max}$(°C min^{-1})	2.80
$(dT/dt)_{cmax}$(°C min^{-1})	4.80
C_p(Jg^{-1} K^{-1})	2.074
ΔH (J/g CHP100%)	1810
ln A (min^{-1} M$^{1/2}$)	29.71
E_a (J mol^{-1})	125000
$[CHP]_0$(M)	0.87

TABLE 6
Analyses of CHP Composition Products from Gas Chromatography

Product	130°C	140°C	150°C
Acetone	100.000[a]	4.426	79.679
α-Methylstyrene	46.600	100.000[a]	69.190
Acetophenone	5.082	18.954	100.000[a]

a. The highest intensity as base peak

TABLE 7
Analysis of CHP Decomposition Product Ratios by HPLC Diluted 1000 Times

Product	130°C	140°C	150°C
Acetone	1.306	1.178	1.344
Acetophenone	1.366	4.420	8.422
Phenol	100.000[a]	100.000[a]	100.000[a]
2-Phenyl,2-propanol	1.302	1.965	2.354

a. The highest intensity as base peak

TABLE 8
Analysis of CHP Decomposition Product Ratios by HPLC Diluted 100 Times

Product	130°C	140°C	150°C
Acetone	12.598	12.686	10.115
Acetophenone	10.057	7.125	3.050
Phenol	100.000[a]	100.000[a]	100.000[a]
2-Phenyl,2-propanol	5.166	16.847	8.823

a. The highest intensity as base peak

3.5. Mechanism of Thermal Decomposition

Evidence about the major products of CHP decomposition as verified by GC and HPLC chromatography are associated with kinetic data, on overall rate law of 0.5 order, and the mechanism of CHP thermal decomposition are proposed in this study. A free radical reaction scheme for the thermal decomposition of CHP in cumene is as follows:

$$2C_6H_5C(CH_3)_2OOH \underset{k_{-1}}{\overset{k_1}{\rightleftharpoons}} [C_6H_5C(CH_3)_2OOH]_2 \quad \text{Fast equilibrium, } k_1 >> k_{-1}$$

$$C_6H_5C(CH_3)_2OOH \xrightarrow{k_2} C_6H_5C(CH_3)_2O\cdot + \cdot OH \qquad \text{slow}$$

$$C_6H_5C(CH_3)_2O\cdot \xrightarrow{k_3} C_6H_5COCH_3 + \cdot CH_3$$

$$C_6H_5C(CH_3)_2H + \cdot CH_3 \xrightarrow{k_4} CH_4 + C_6H_5C(CH_3)_2\cdot$$

$$C_6H_5C(CH_3)_2H + \cdot OH \xrightarrow{k_5} H_2O + C_6H_5C(CH_3)_2\cdot$$

$$C_6H_5C(CH_3)_2O\cdot + C_6H_5C(CH_3)_2H \xrightarrow{k_6} C_6H_5C(CH_3)_2OH + C_6H_5C(CH_3)_2\cdot$$

$$C_6H_5C(CH_3)_2O\cdot + \cdot OH \xrightarrow{k_7} C_6H_5OH + CH_3C(=O)CH_3$$

$$C_6H_5C(CH_3)_2\cdot + \cdot OH \xrightarrow{k_8} C_6H_5CCH_3{=}CH_2 + H_2O$$

$$2C_6H_5C(CH_3)_2\cdot \xrightarrow{k_9} [C_6H_5C(CH_3)_2]_2 \qquad (6)$$

The overall rate law can be derived from the reaction scheme and rate-determining step.

$$R = \frac{-d\left[C_6H_5C(CH_3)_2OOH\right]}{dt} = k_2\left[C_6H_5C(CH_3)_2OOH\right] \qquad (7)$$

and

$$\left[C_6H_5C(CH_3)_2OOH\right] = \left[\frac{k_1}{k_2}\left(C_6H_5C(CH_3)_2OOH\right)_2\right]^{1/2} \qquad (8)$$

According to the evidence of intermolecular hydrogen-bonding from IR spectra and studies of the literature [8,9]. Most of the CHP molecules are associated into the extremely stable dimers via a –O-H—O bond. The intial and decomposition concentrations of CHP were simplified to be

$$\tfrac{1}{2}\left[C_6H_5C(CH_3)_2OOH\right]_{t=0} = \left[\left(C_6H_5C(CH_3)_2OOH\right)_2\right]_{t=0} \qquad (9)$$

and

$$\tfrac{1}{2}\left[C_6H_5C(CH_3)_2OOH\right]_{(t)} = \left[\left(C_6H_5C(CH_3)_2OOH\right)_2\right]_{(t)} \qquad (10)$$

Therefore, the overall rate law for CHP in thermal decomposition can be represented in the following equation.

$$\begin{aligned} R &= \frac{k_2 k_{-1}^{1/2}}{k_1^{1/2}}\left[\left(C_6H_5C(CH_3)_2OOH\right)_2\right]^{1/2} \\ &= \frac{k_2 k_{-1}^{1/2}}{k_1^{1/2}}\left[\tfrac{1}{2}\left(C_6H_5C(CH_3)_2OOH\right)_{(t)}\right]^{1/2} \\ &= k\left[\left(C_6H_5C(CH_3)_2OOH\right)_{(t)}\right]^{1/2} \end{aligned} \qquad (11)$$

where k is the overall rate constant

The heat of decomposition is proportional to isolated CHP molecules or dimeric forms. The thermokinetics and adiabatic temperature behavior are still the same with this model.

4. CONCLUSION

The thermal decomposition of organic alkylperoxides has been studied and are well represented by simple first-order kinetics. However, special attention has been paid to the organic hydroperoxide category. The kinetics and mechanism of the decomposition of hydroperoxides are of great importance with regard to the association of intermolecular and intramolecular hydrogen-bondings. A systematic approach to verify the decomposition pathway is addressed. The application of IR, chromatography, non-isothermal and isothermal DSC and VSP?methodologies are promising techniques for generating the data for the study of decomposition kinetics. The rate-determining step is the decomposition of the associated peroxides into radicals. The overall rate constant is independent of the solvent present. The reaction order of CHP/cumene decomposition was determined to be 0.5.

NOMENCLATURE

C_P	specific heat capacity, $Jg^{-1}K^{-1}$
K	rate constant, $sec^{-1}M^{1-n}$
N	reaction order
dQ/dt	heat generation rate, W
ΔH	heat of reation, KJ mol^{-1}
V	volume of reactant, l
ρ	density of reactant, gL–1
m	mass of reactant, g
T	temperature, K
T_0	initial exothermic temperature, K
T_f	final temperature, K
T_{cf}	calc. final temperature, K
t	time, min
$(dT/dt)_{max}$	the max self-heating rate, °C min^{-1}
$(dT/dt)_{cmax}$	the calc. max self-heating rate, °C min^{-1}
dT/dt	self-heating rate, °C min^{-1}
R	rate of reaction, $Msec^{-1}$
F	thermal inertia

Acknowledgments

The authors would like to thank the Ministry of Economic Affairs , ROC for financial support of this study under Contract No.86-EC-2-A-17-0233. W.L.L wishes to thank the National Yunlin University of Science and Technology for partial support .

The authors would also like to thank Dr. Harold G. Fisher for helpful discussions in the preparation of the paper.

APPENDIX
Derivation of Equation(5)

For an adiabatic environment,

$$-\Delta H\left(\frac{-d[\mathrm{CHP}]}{dt}\right)V = mC_\mathrm{p}\left(\frac{dT}{dt}\right) \tag{a.1}$$

$$-\frac{d[\mathrm{CHP}]}{dt} = -\frac{mC_\mathrm{p}}{\Delta HV}\left(\frac{dT}{dt}\right) \tag{a.2}$$

$$\int_{[\mathrm{CHP}]_0}^{[\mathrm{CHP}]} d[\mathrm{CHP}] = -\frac{mC_\mathrm{p}}{\Delta HV}\int_{T_0}^{T} dT \tag{a.3}$$

$$[\mathrm{CHP}] = [\mathrm{CHP}]_0 - \frac{mC_\mathrm{p}}{\Delta HV}(T - T_0) \tag{a.4}$$

For a 0.5 order reaction of CHP decomposition

$$\frac{-d[\mathrm{CHP}]}{dt} = k[\mathrm{CHP}]^{1/2} \tag{a.5}$$

$$-\frac{mC_\mathrm{p}}{\Delta HV}\left(\frac{dT}{dt}\right) = k[\mathrm{CHP}]^{1/2} = k\left[[\mathrm{CHP}]_0 + \frac{mC_\mathrm{p}}{\Delta HV}(T - T_0)\right]^{1/2} \tag{a.6}$$

The CHP concentration can be related to the change of temperature.

$$-\Delta H\frac{mC_\mathrm{p}}{[\mathrm{CHP}]_0 V}(T_\mathrm{f} - T_0) \tag{a.7}$$

Equations (a.7) and (a.6) are combined and give

$$\begin{aligned}\left(\frac{dT}{dt}\right) &= -k\frac{\Delta HV}{mC_\mathrm{p}}\left\{[\mathrm{CHP}]_0 + \frac{mC_\mathrm{p}}{\Delta HV}(T - T_0)\right\}^{1/2} \\ &= k\frac{(T_\mathrm{f} - T_0)}{[\mathrm{CHP}]_0^{1/2}}\{T_\mathrm{f} - T\}^{1/2}\end{aligned} \tag{a.8}$$

REFERENCES

1. Ho, T. C., Duh, Y.-S., and Chen, J. R. "Case Studies of Incidents in Runaway Reactions and Emergency Relief," AIChE Loss Prevention Symposium, March, 1998.
2. Duh, Y.-S., Kao, C.-S., Lee, C., and Yu, S. W. *Trans IChemE,* 75, Part B (1997), 73–78 .
3. Mettler company, TA4000 Operation Instructions, 1993.
4. Fauske & Associates, Inc., VSP II Manual and Methodology, 1996.
5. Leung, J. C., Fauske, H. K., and Fisher, H. G. "Thermal Runaway Reactions in a Low Thermal Inertia Apparatus," *Thermochimica Acta,* 104 (1986), 13–29.
6. Leung, J. C., Creed, M. J., and Fisher, H. G. "Round-Robin Vent Sizing Package Results." *Proceedings of the the International Symposium on Runaway Reactions* (Cambridge, MA, March7–9, 1989). New York: AIChE, 1989.
7. Takayuki Ando, Yasuhiro Fujumeta, and Shigeru Morisaki, "DSC Data for Reactive Chemical Substances (1)," Safety Document of the Japan Research Institute of Industrial Safety, RIIS-SD-87, 1987.
8. Shopova, N., and Ivanov, S. I. *Thermochimica Acta,* 231 (1994), 193–220 .
9. Danoczy,E., and Holly, S. *J. Phys. Chem.,* 88 (1984), 1190–1194 .

PRV Relief Systems: Codes, Standards, Practices, and Other Sources of Confusion

James E. Huff
Huff Consulting Services
3628 Wood Duck Circle, Stockton, CA 95207

1. ABSTRACT

Increasing emphasis on effluent containment has led to the evolution of pressure relief system configurations not anticipated in the wording of current codes, standards and practices. In addition, the very real condition of two-phase venting has not received adequate consideration. These documents are revised regularly, but cannot keep pace with changing system requirements. The situation is particularly troublesome for systems employing pressure relief valves. Areas of concern include the allowable flow resistances of the associated inlet and outlet piping as they affect system flow capacity and stability of operation, as well as the flow capacity of the valve itself in two-phase and subsonic gas-phase service. Good engineering judgment is required to assure conformity with the intent of current standards while adhering to the best current accepted design practices. A uniform interpretation of the provisions of both Section VIII of the American Society of Mechanical Engineers (ASME) Boiler and Pressure Vessel Code, and Recommended Practices 520 and 521 and Standard 2000 of the American Petroleum Institute (API) is proposed. Application to two-phase and subsonic gas-phase flow is included.

2. INTRODUCTION

Present Codes, standards, and practices (CSPs) for pressure relief valve (PRV) systems in the USA apply primarily to single-phase flow. Within this restricted scope of coverage, the provisions relevant to sizing of the associated piping in the emergency pressure relief system (ERS) are somewhat ambiguous and differ among the various documents. In particular, the applications of conventional valves under conditions of constant superimposed back pressure receives uneven treatment.

The designer can face an apparent dilemma when current accepted good engineering practice seems to differ from provisions of the CSPs. For example, some sections of the ASME Boiler and Pressure Vessel Code have become law in the USA and some foreign countries. Thus, the mandatory provisions of the Code are to be followed. However, some flexibility is allowed in the application of the nonmandatory provisions. The wording of current editions of such CSP provisions is included in the following sections of this paper, along with the interpretation as required for consistency with current accepted good engineering practice.

The design of PRV relief systems requires consistent design information for both single- and two-phase fluids with respect to:

- Device selection and location on the protected system.
- Flow capacity parameters.
- Limits on inlet line pressure loss.
- Limits on superimposed and built-up back pressure on valve.
- Accounting for effects of submersion and liquid accumulation in ERS.

The Emergency Relief Effluent Subcommittee of the Center for Chemical Process Safety (CCPS) has been addressing the issue of design criteria. The goal has been to:

- Develop a uniform interpretation of ambiguous or conflicting provisions of CSPs.
- Propose design criteria where needed to fill the gaps, based on collective committee experience and judgment.
- Identify areas of ignorance and suggest conservative design practices pending definitive testing work.

The results are presented in the CCPS *Guidelines for Pressure Relief and Effluent Handling Systems* ("CCPS Guidelines," in publication). See the Appendix for a listing of members of this CCPS subcommittee. This present paper is based to a large extent on the rather comprehensive treatment in the CCPS Guidelines.

3. GENERAL FEATURES OF PRV RELIEF SYSTEMS

Figure 1 is a schematic representation of a PRV relief system showing the location of the principal pressure points and the associated terminology. The PRV is shown close-coupled to the vessel, but the inlet piping may actually include additional pipe, fittings, valves, rupture disk devices, etc. The nonrecoverable pressure losses in the inlet piping must be limited to avoid chatter. Chatter is the rapid reclosing and opening of the valve due to an excessive decrease in

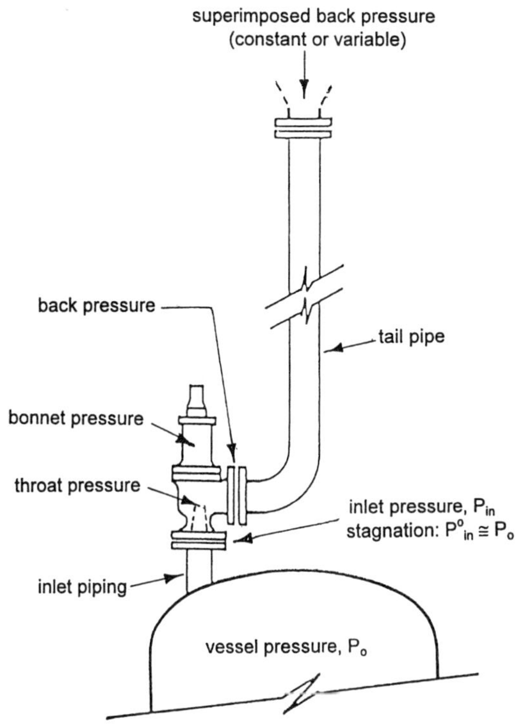

Figure 1. Pressure Relief Valve System.

the pressure at the valve inlet as flow is established, which is followed by an increase back to the popping pressure after the valve recloses.

The concept of stagnation pressure is useful both in calculating the valve flow capacity and in defining the nonrecoverable loss in the inlet piping. The stagnation pressure of a flowing stream is defined as the pressure that would exist if the stream were brought to rest along an isentropic path. Since the fluid in the vessel is flowing at a very low velocity in the vessel compared to the velocity in the inlet line, the vessel pressure P_0 can be considered to be a stagnation pressure. The stagnation pressure at the valve inlet, P_{in}^{0} , can be computed from the flowing pressure, P_{in}, and the flow rate. The loss is then the vessel pressure minus the inlet stagnation pressure. Alternatively, the loss can be estimated directly from flow formulas.

The single-phase flow equations in the CSPs are obtained by integrating the differential equation for mass flux (linear velocity times density). Integration limits for these equations are from inlet stagnation pressure to the nozzle pressure, and from zero inlet velocity to the velocity at the nozzle throat. Conventional practice for calculating the Code-certified flow capacity is to neglect the losses in the inlet piping so that the vessel pressure can be used as the stagnation pressure at the valve inlet. A safety factor is included in the computations to account for this and other approximations and uncertainties.

The bonnet pressure acts directly or indirectly on the downstream side of the sealing member of the valve and thus adds to the closing force of the spring. This bonnet pressure is either the back pressure or atmospheric pressure, depending on the valve style (see Section 4.1). When the valve is closed, the back pressure at the valve outlet is the same as the superimposed back pressure. When the valve is flowing, the back pressure is the sum of the superimposed back pressure and the pressure drop across the tail pipe. Figure 1 shows a very simple tail pipe configuration. Actual layouts will typically have two or more changes of direction; each section will be self-draining in so far as practical. There are limits on the flowing back pressure and the tail pipe pressure drop to avoid undue loss of capacity or unstable operation. The limit varies with the type of valve.

4. VALVE TYPES

The valves most commonly used in the USA process and petrochemical industries are high-lift, top-guided valves. The API standard valve orifice and pipe sizes (API 526) are generally adopted for these direct spring loaded valves. Valves certified by the National Board of Boiler and Pressure Vessel Inspectors (National Board) are used on ASME Code-stamped vessels. So-called pilot-actuated valves may also be certified by the National Board.

4.1. Direct Spring Loaded Valves

Two types of direct spring loaded valves are shown on Figure 2. In the usual installations, the bonnet of the *conventional* valve is open to the valve outlet, and thus is at the valve back pressure (note vent plug in place). The Code (ASME BPVC) makes an exception if a rupture disk device is installed on the outlet side of the valve; the bonnet must have a means to bleed off any accumulation of pressure in the valve body. For typical installations of *balanced* valves, the bonnet is isolated from the valve outlet by a bellows device and open to the atmosphere (note vent plug removed). The mean cross-sectional area of the bellows is sized to balance the area at the throat of the valve nozzle. Ideally, the bellows should provide balance against superimposed back pressure when the valve is closed as well as balance against the flowing back pressure when the valve is open. This is not possible since the effective area upon which the nozzle pressure acts is smaller when the valve is closed than when the valve is open. Compared to the conventional valve, however, the balanced

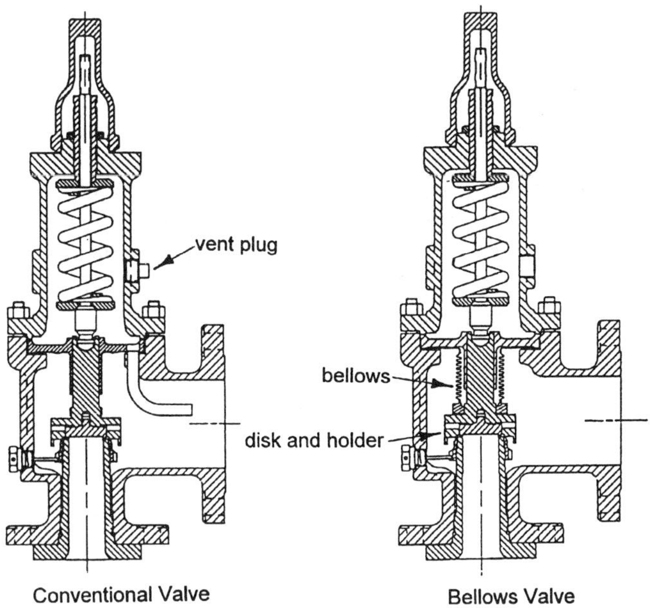

Figure 2. Typical Direct Spring Loaded Pressure Relief Valves. (After Lai 1996.)

valve is quite insensitive to back pressure with respect to both set pressure and effective flow area (disk lift). Conventional valves are suitable for use under service conditions of *constant* superimposed back pressure (but not of *variable* superimposed back pressure).

4.1.1. Opening and Reclosing Characteristics

Conventional and balanced valves have the same opening and closing characteristics. As the pressure is increased, some seepage is observed on approaching the set pressure. Then in service with compressible fluids, a rather abrupt "pop" action occurs, during which the valve will typically reach well over half the rated capacity. The capacity will continue to increase with further increase in pressure until full rated capacity is attained (at or before the pressure reaches 110% of the gauge set pressure for certified valves).

The pop action of the valve in compressible fluid service causes the valve to remain at rather high capacity as the pressure falls back to the set pressure. Typically, a valve will not "fall out of lift" until the pressure is below the set pressure. This "blowdown" below set pressure is typically 6 to 10% of the gage set pressure. Blowdown provides a margin of tolerance to inlet piping losses, since the losses must exceed the blowdown before the valve will reclose. Note that conventional and balanced valves have the same sensitivity to nonrecoverable inlet piping losses.

4.1.2. Standard vs. Liquid Trim

Pressure relief valves are available in both "standard" and "liquid" trim, as depicted on Figure 3. Liquid trim is the newer style, developed to meet the requirements for certification at the mandated pressure level (110% of the gauge set pressure; 10% "overpressure"). The close clearance between disk holder and the blowdown ring of the liquid trim generates a strong lifting force, somewhat akin to the pop action of the standard trim in compressible fluid service. Liquid trim valves can be certified for compressible fluid service as well as liquid service. Present CSPs make no distinction with regard to allowable inlet loss and back pressure, even though tests have shown large differences in the observed blowdown among the various combinations of liquid versus standard trim and compressible versus incompressible fluid service (Lai 1996).

Standard trim valves can be adjusted for liquid service by backing the threaded blowdown ring down the outside of the nozzle and away from the disk holder assembly. The performance in liquid service is typically not predictable at the 10% overpressure level and below, so valves in this configuration are not suitable for certification testing. The valves are operable above 10% overpressure, and typically become fully open at or before 25% overpressure. The flow at overpressures between 10 and 25% is estimated using the correction factor, K_p, in the API liquid flow equation (API 520-I). However, the definition of overpressure as used for obtaining values of K_p does not seem to consider service with superimposed back pressure for conventional valves. The proposed liquid flow equation in the CCPS Guidelines does account for this superimposed back pressure effect. The bonnet pressure of a balanced

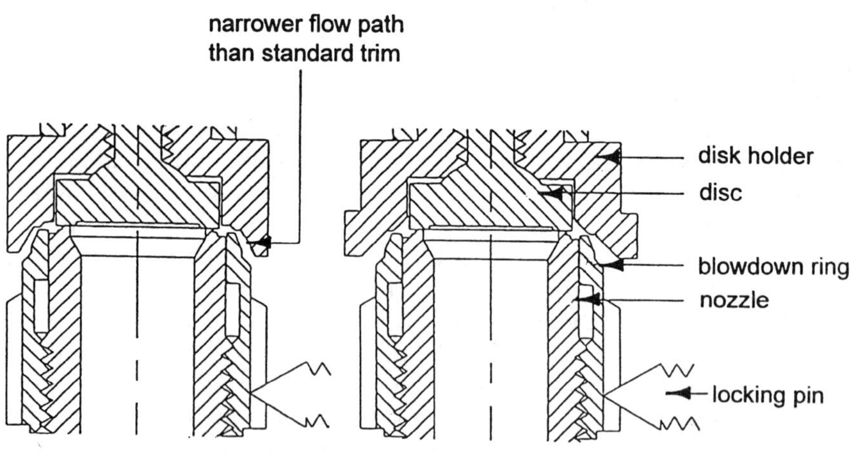

Figure 3. Standard versus Liquid Trim. (After Lai 1996.)

valve is typically atmospheric, so the K_p value is not affected by superimposed back pressure. However, some flow correction is required at higher total back pressure levels (API 520-I).

4.2. Pilot Operated Valves

Pilot operated valves consist of a main valve in the relief path, actuated by a direct spring loaded pilot valve. A pressure sensing line is normally connected from the valve inlet to the pilot. The pilot reacts to the pressure and actuates the main valve accordingly. Pilot valves may carry a small flow or be non-flowing. The main valve can be controlled to be snap-acting (similar to pop action) or to open in proportion to the overpressure.

An important option is the connection of the pressure sensing line to sense the system pressure itself, instead of the inlet pressure to the main valve. In principle, a "remote-sensing" valve is insensitive to nonrecoverable pressure losses in the inlet piping as far as operating stability is concerned. The losses do reduce the available pressure drop for flow. This reduction must be accounted for when calculating the flow capacity of the system. As will be seen, the Code (ASME BPVC) does not explicitly recognize this remote sensing provision. The API documents do include provisions for this type of valve.

5. FLOW CAPACITY

The CSPs provide equations for the flow of single-phase fluids in PRVs. In general, these equations require values of:

- Fluid physical properties.
- Nozzle ("orifice") flow area.
- Coefficient of discharge.
- Back pressure and viscosity correction factors.

The Code (ASME BPVC) provides equations for the single-phase flow of compressible gases and vapors (sonic flow only) and water. Equations are given with physical property values imbedded for dry saturated steam, air, natural gas, and water. A chart for the flashing two-phase flow of saturated water is included. The flow is essentially for test-stand conditions (per ASME PTC-25; negligible inlet line loss and back pressure).

The API documents (API 520-I and II, 2000) provide equations for both sonic and sub-sonic flow of gases and vapors, as well as an equation for the flow of incompressible fluids. Corrections for the effect of back pressure on the flow for balanced valves, as well as a correction for the effect of viscosity on the flow of liquids, are presented. Thus, the flow is for actual installations as opposed to test-stand conditions. Obsolete two-phase flow methodology of earlier editions is being replaced with current methods growing out of the AIChE Design Institute of Emergency Relief Systems (DIERS) projects (see Leung 1995 for a summary of the supporting technology). The predicted flow of saturated water will be in agreement with the Code chart (ASME BPVC).

5.1. Flow Areas and Coefficients of Discharge

The confusion in the CSPs relates mainly to the specification of flow area and coefficient of discharge. The confusion in area is due to the difference between API standard orifice areas (API-526) and the actual areas as reported in the certification reports (NB-PRDC). The confusion in coefficients arises from the fact that the Code (ASME VIII) actually defines three different values:

- K_D, the value observed in certification testing for a single test *specimen*.
- K_{av}, the average of the K_D values of the nine specimens tested to certify a valve *design*.
- K, the coefficient for that design, where $K = 0.9K_{av}$.

Note that a given K_D value can deviate as much as ±5% from the K_{av}. The *certified capacities* for that design are computed from the values of the *reported* area and K. Certified capacities are used for the selection of a valve size for a given relief flow requirement, as well as for the Code and API calculation of pressure loss and drop in the associated piping .

The API documents contain essentially the same sonic flow equations as those in the Code, and include an equation for subsonic flow. However, different flow areas are used with a given value of the coefficient of discharge. The area is the API standard area and the coefficient is representative of typical values of K_{av} for API standard valves (API 526). Since the API areas are frequently about 90% of the values in the certification reports (NB-PRDC), the resulting API flows are essentially the same as the ASME certified capacities. That is, the product of area and coefficient are essentially the same for both ASME certified flow and API flow calculations. The calculation methods in typical manufacturers' catalogs follow this API approach. As stated above, the effect of inlet losses on the flow rate are accounted for only in the case of pilot operated valves with remote sensing.

A different flow capacity value is typically used instead of the certified flow as the design flow for effluent handling systems. This "best estimate flow" is obtained using the observed flow area and K_{av}. The flow-reducing effects of the flow resistance of the associated piping may be accounted for at the discretion of the designer. If not accounted for, the best estimate flow is roughly 10% higher than the certified capacity. Some designers follow the conservative practice of using this flow in the calculation of pressure loss and drop in the associated piping (see paragraph 7). This practice is required in some non-USA standards (British for example).

5.2. Sonic vs. Subsonic Flow

The above coefficients of discharge apply to the flow of gases and vapors at sonic velocity in the throat of the valve nozzle. Subsonic flow is not covered in the Code. The API documents specify the same value of the coefficient for both sonic and subsonic compressible flow. In comparison, the coefficients for liquid flow in certified valves are typically about 75% of the values for sonic compressible flow. The lower value for liquid flow has been attributed to flow resistance in the body of the valve itself (Sherman 1993, Morris 1996). That is, the coefficient must account for nonidealities both in the nozzle and in the valve body.

The importance of the distinction between sonic and subsonic compressible flow is illustrated by the graph shown on Figure 4. This graph is a plot of the observed coefficient of discharge versus the ratio of outlet pressure to inlet stagnation pressure (Papa 1994). The ideal theoretical pressure ratio

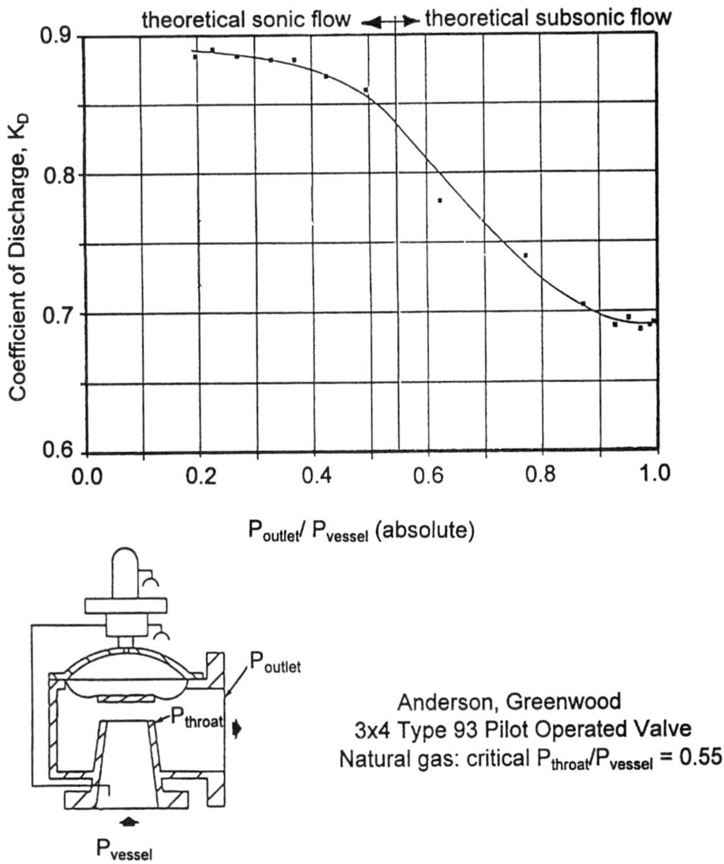

Figure 4. Coefficient of Discharge in Sonic and Subsonic Flow. (Papa 1994.)

at sonic flow (0.55 for this test fluid) is shown on the plot. This ideal ratio is computed by treating the path from the vessel to the throat of the valve as an ideal (isentropic) nozzle. For an installation with negligible inlet line losses and a K_D value of 0.89, the actual pressure ratio is closer to 0.49. The data clearly show that the K_D value decreases as the available pressure drop across the valve (inlet stagnation pressure minus the pressure at the valve outlet) decreases from the sonic to the subsonic flow regions (a phenomenon known as "sonic droop"). This droop in the *valve* K_D is not considered to be an indication of a decrease in the *nozzle* K_D. Rather, the droop results from the fact that the pressure in the nozzle for subsonic flow is not the same as the pressure at the valve outlet. The difference is due to the flow resistance in the valve body itself. The effect of this resistance is not evident in sonic flow because the flow depends only on inlet stagnation pressure (nozzle pressure is not influenced by the pressure in the valve body). In subsonic flow, however, the flow depends on the difference between inlet stagnation and the back pressure on the *nozzle*. The calculation methods for subsonic flow in the CSPs and typical manufacturer's literature use the back pressure on the valve instead of the back pressue on the *nozzle*, which yields a higher pressure drop than the actual drop across the nozzle. This results in a calculation of a correspondingly lower value of K_D to compensate for the high value of pressure drop. Thus, the observed subsonic *valve* K_D values should be used to take proper account of the flow resistance of the valve body. This is not the case in the current documents; use of the sonic-flow values of K_D is implicit.

5.3. Two-Phase Flow

The CSPs do not address the subject of calculational models for two-phase flow. The results of the AIChE Design Institute for Emergency Relief Systems (DIERS) project studies on this subject are summarized in the DIERS Project Manual (Fisher et al. 1992). Further studies are in progress under the sponsorship of the DIERS Users Group. A review of the status of flow model development is outside the scope of this present paper. However, questions as to the choice of coefficient of discharge values for two-phase flow are addressed below.

Reported values of the K_D for liquid flow in a given certified valve are typically about 75% of the values for sonic flow of gas (NB-PRDC). The question arises as to what value to use for mixtures of gas and liquid. Typical manufacturer's recommendations are to use some intermediate value based on the weight or volume ratio of vapor to liquid. The CSPs are silent on this point.

Another viewpoint is based on the observation that the K_D for subsonic gas flow on Figure 4 is about 78% of the sonic flow value at the extremes of the curve. This is quite close to the 75% figure for liquid coefficients versus the sonic gas flow values. It is reasonable to conclude that the difference between the reported coefficients for liquid and sonic gas flow is due to the relative importance of body flow resistance in sonic versus subsonic flow (paragraph 5.2), rather than due to some difference in the K_D values in the nozzle itself.

The above conclusion as to coefficient of discharge values for liquids versus gases and vapors is the basis for the CCPS Guidelines recommendation for values to use for two-phase flow. This criterion for choice of coefficient depends on whether or not the "critical" flow condition exists in the nozzle. Critical flow is used here to denote "sonic" flow of gases and vapors, and "choked" or "maximum" flow for multi-phase fluids. The recommendation is to use the K_D value for sonic gas flow if the two-phase flow is choked; otherwise use the K_D value for liquid flow.

6. SELECTION AND INSTALLATION PRACTICES

The Code (ASME BPVC) requires that "safety" valves (valves designed or set up for compressible fluid service) be used for gas and vapor service. The valves are to be connected to the vapor space. "Relief" valves (valves designed or set up for incompressible fluid service) are to be used for liquid service and be connected below the normal liquid level. Relief valves are to be used for vessels operating liquid-full. The API documents address an additional case: if vapor can be generated in a normally all-liquid system and the valve is sized based on vapor flow, be sure to locate valve so that it actually relieves vapor (API 521).

The CSPs are silent for the case of two-phase flow. Liquid trim valves can be certified for gas service and are thus predicted to be suitable for two-phase fluids of any phase ratio. However, a set of test data on vapor flow indicate a possible disadvantage due to excessive blowdown in vapor service (Lai 1996). Typical manufacturers' recommendations are to choose between standard and liquid trim based on the predominant phase (usually on a volumetric basis).

The recommendation in the CCPS Guidelines is based on the same consideration of liquid versus gas flow coefficients as presented in paragraph 5.3 above. That is, base the recommendation for the choice of standard versus liquid trim on the flow condition in the nozzle. Use standard trim if the two-phase flow is predicted to be choked; otherwise use the liquid trim.

Valves designed or set up only for incompressible fluid service are not suitable for two-phase service. Use only the certified liquid trim styles if the use of a liquid valve is indicated.

7. ASSOCIATED PIPING

The CSPs present criteria for the sizing of associated piping as developed for API-standard-type valves with standard (as opposed to liquid) trim in gas or vapor service. The detailed provisions of the Code (ASME BPVC) are presented in Nonmandatory Appendix M. The introduction to this appendix is as follows:

> **M-1 INTRODUCTION**
>
> (a) The rules in this Appendix are for general information only, because they pertain to the installation and operation of pressure vessels, which are the prerogative and responsibility of the law enforcement authorities in those states and municipalities which have made provisions for the enforcement of Section VIII.
>
> (b) It is permissible to use any departure suggested herein from provisions in the mandatory parts of this Division when granted by the authority having legal jurisdiction over the installation of pressure vessels.

It will be seen that some provisions of this appendix require interpretation to be in accord with other documents and current accepted good engineering practice.

The provisions of this appendix and other CSPs are presented in detail in the following sections. In general, the criterion for sizing the piping from vessel to valve (inlet piping) applies to conventional, balanced and pilot operated valves in compressible flow service. For conventional (nonbalanced) direct spring loaded PRVs, a tail pipe sizing criterion is presented to assure stable valve operation. The Code recognizes that other types of valves (balanced, pilot operated, etc.) exhibit various degrees of tolerance to the tail pipe flow resistance and the manufacturer's recommendations should be followed. For balanced valves in both compressible and incompressible fluid service, API (API 520-I) presents flow reduction charts that represent an average of typical manufacturer's recommendations. The CSPs are silent with respect to two-phase service.

Another question remains: should the flow resistance of the inlet and tail piping be calculated using the Code flow or the best estimate flow? The CSPs specify the use of the Code flow. However, the observed K_D for a given valve can be as much as 15% higher than the certified K for that design. This results from the ±5% allowable deviation of K_D from K_{av} for that design, with $K_{av} = K/0.9$. Actual test results are typically reported in terms of the observed value of back pressure, not some lower value as calculated from a Code-defined reduced flow rate (see Lai 1992 for example data). Conservative practice as followed by some designers is to use K_{av} instead of the K value when obtaining the flow basis for pressure loss and drop calculations.

7.1. Inlet Loss

The CSPs use the nonrecoverable frictional losses in the piping from the protected vessel to the inlet of the valve as the applicable measure of inlet piping flow resistance. The accepted limit of these losses is 3% of the gauge set pressure. However, the flow and pressure bases for calculating the losses are not clearly defined. Also, the CSPs do not explicitly deal with liquid trim valves or questions of two-phase flow.

7.1.1. CSP Wording

The Code (ASME BPVC) wording relative to inlet piping pressure losses is as follows:

> **M-7 Inlet Pressure Drop for High Lift, Top Guided Safety, Safety Relief; and Pilot Operated Pressure Relief Valves in Compressible Fluid Service:**
>
> (a) The nominal pipe size of all piping, valves and fittings, and vessel components between a pressure vessel and its safety, safety relief, or pilot operated pressure relief valves shall be at least as large as the nominal size of the device inlet, and the flow characteristics of the upstream system shall be such that the cumulative total of all nonrecoverable inlet losses shall not exceed 3% of the valve set pressure. The inlet pressure losses will be based on the valve nameplate capacity corrected for the characteristics of the flowing fluid.

Some clarification of the flow basis for the loss calculation is required in order to compute the specified flow capacity. The "nameplate capacity" is calculated from certification data for that valve design (NB-PRDC), using:

- The reported flow area (not the API standard area).
- The K value for that design (90% of the average K_D of nine specimens, K_{av}), not the K_D value used in the API documents or typical manufacturer's literature.
- The pressure as used in the certification testing (110% of the gauge set pressure).
- The physical properties of the *test fluid*.

The effect of inlet piping pressure loss on flow is covered in the 90% factor (not computed separately except for pilot operated valves with remote sensing; see the following paragraph). It follows that the "nameplate capacity corrected for the characteristics of the flowing fluid" is computed using the above area and coefficient, but using the physical properties (characteristics) of the actual relieving fluid. As noted earlier, some designers (and some non-USA standards) follow the practice of using K_{av} instead of K to obtain a conservatively higher value of flow and pressure loss. It is important to note that the criterion for sizing the inlet piping is the total nonrecoverable inlet *losses*, not the pressure *drop* from vessel to valve inlet.

The above wording of nonmandatory Appendix M would seem to require that the 3% criterion be applied to all pilot operated valves. Actually, larger losses are acceptable for pilot operated valves with remote sensing (pilot senses the vessel pressure instead of the valve inlet pressure). Such installations must meet the following mandatory provision of the Code (ASME BPVC):

> **Section UG-135(b):**
> . . . the flow characteristics of this upstream piping shall be such that the pressure drop will not reduce the relieving capacity below that required or adversely affect the proper operation of the pressure relieving device.

Implicit in this wording is the requirement that the inlet loss be accounted for in calculating the flow capacity of the relief system.

The API criterion for inlet pipe sizing (API 520-II) is in essential agreement with the above Code wording. The API wording is:

> 2.2.2 SIZE AND LENGTH OF INLET PIPING TO PRESSURE RELIEF VALVES
> . . . the total non-recoverable pressure loss between the protected equipment and the pressure relief valve should not exceed 3% of the set pressure of the valve except as permitted in 2.2.3.1 . . (see NOTE). The pressure loss should be calculated using the rated capacity of the pressure relief valve.
>
> NOTE: 2.2.3.1 deals with "pilot operated valves with remote sensing to sense the true system pressure."

This wording removes a source of confusion in earlier editions in which both "pressure loss" and "pressure drop" terminology is used as the measure of flow resistance. The definition of "rated relieving capacity" is (API 520-I):

> Rated relieving capacity is that portion of the measured relieving capacity permitted by the applicable code or regulation to be used as a basis for the application of a pressure relief device.

To be consistent with the Code (ASME BPVC), the "portion" would be 90%. The "measured relieving capacity" would be the capacity of that device (if measured on the actual relieving fluid) or the computed capacity of that design using the reported area of the particular valve, the average observed coefficient of discharge for the design, and the properties of the actual relieving fluid at relieving conditions. Not defined is the applicable relieving pressure. Conventional practice is to use the pressure as used in certification testing to establish the "nameplate" capacity (110% of the gauge set pressure).

7.1.2. Two-Phase Flow

The use of nonrecoverable loss as the measure of loss of opening force is based on the presumption that the opening force on the valve disk is determined by the stagnation pressure of the flowing fluid. This presumption is conventionally applied to two-phase as well as single-phase flow. One might question the validity of this presumption, particularly for flashing (vaporizing) two-phase streams. The use of stagnation pressure implies a conversion of momentum to pressure along a reversible adiabatic (isentropic) path.

However, the impact of a flashing fluid on the valve disk is a highly non-reversible, non-equilibrium event. This is but one of many uncertainties in the true applicability of the 3% rule. However, no test work is anticipated until such time as a problem with the application of the 3% rule is identified.

7.1.3. Liquid Trim Valves

The question of allowable inlet loss for liquid trim valves is not specifically addressed in the CSPs. The 3% loss limit is typically used by default for both compressible (gas and two-phase) flow and incompressible (liquid) flow service. Reports of any deficiency of this practice are not in evidence.

7.2. Back Pressure

The CSPs use the flowing pressure at the valve outlet as the applicable measure of the flow resistance of the tail piping. The accepted limit of this pressure varies with the type of valve and the pressure at the discharge end of the piping (superimposed back pressure; see Figure 1). However, the flow and pressure bases for calculating the flowing pressure are not clearly defined.

The following discussion of back pressure limitations applies to standard trim valves. The sensitivity of liquid-trim valve to back pressure is yet to be defined for both compressible (including two-phase fluids) and incompressible fluid service. Existing test results (Lai 1996) show a definite difference in the behavior of the two trims. Until test results become available, conservative practice is to apply the standard-trim criteria to liquid trim as well.

7.2.1. CSP Wording

The provisions of nonmandatory Appendix M of the Code (ASME BPVC) concerning discharge piping are as follows:

> **M-8 Discharge Lines from Safety Devices:**
> (c) The flow characteristics of the discharge system of high lift, top guided safety, safety relief, or pilot operated pressure relief valves in compressible fluid service shall be such that the static pressure developed at the discharge flange of a conventional direct spring loaded valve will not exceed 10% of the set pressure when flowing at stamp capacity. Other valve types exhibit various degrees of tolerance to back pressure and the manufacturer's recommendation should be followed.

The "other valve types" are presumed to be *balanced* direct spring loaded (as opposed to *conventional*) valves and pilot operated valves. Concerning the flow basis for the pressure calculations, it is presumed that "stamp" capacity is the same as the "*nameplate capacity corrected for the characteristics of the flowing fluid,*" as specified for the computation of inlet line losses.

The above Code wording of nonmandatory Appendix M would seem to prohibit use of a conventional valve if discharge is to a pressure greater than 10% of the gauge set pressure (10% superimposed back pressure). This is contrary to accepted practice; conventional valves are used for this and higher levels of *constant* superimposed back pressure.

The pertinent wording of mandatory Part UG-135 is:

> (g) . . . The size of the discharge lines shall be such that any pressure that may exist or develop will not reduce the relieving capacity of the relieving devices below that required to properly protect the vessel, or adversely affect the proper operation of the relieving devices.

A proper accounting of the effect of constant superimposed back pressure on the set pressure, flow capacity and operating stability of conventional valves is required to satisfy this Code requirement.

The general API provisions (API 521) read very much like the above mandatory Code requirements. Specific requirements (API 520-I) are as follows:

> Conventional pressure relief valves should typically not be used when the built-up back pressure is greater than 10 percent of the set pressure at 10 percent overpressure. A higher maximum allowable built-up back pressure may be used for overpressure greater than 10 percent.

This differs from the Code wording in the use of *built-up back pressure* instead of *static pressure developed at the discharge flange* as the applicable back pressure parameter. In essence, this API wording says that superimposed back pressure is not included in the computed back pressure parameter. The API definition of built-up back pressure is as follows (API 520-I):

> Built-up back pressure is the increase in pressure in the discharge header that develops as a result of flow after the pressure relief device opens

The use of the words "in the discharge header" is somewhat ambiguous here, since the pressure will vary along the length of the piping. Also, "header" refers to a collection system from several devices in some contexts and to the tail piping on a device in other contexts. In comparison to the above wording, the API definition of back pressure (API 520-I) is:

> Back pressure is the pressure that exists at the outlet of a pressure relief device as a result of the pressure in the discharge system. It is the sum of the superimposed and built-up back pressures.

This wording is explicit as to the location of the pertinent pressure point: *the outlet of a pressure device*. This interpretation in consistent with Code wording and typical current practice.

The API (API 520-II) flow rate basis for tail pipe pressure drop calculations is the same as for the inlet piping loss calculation:

> The rated relieving capacity of the pressure relief valve shall be used to size the discharge line from the pressure relief valve to the relief header.

This wording adds no confusion to the flow rate question, but does add to the ambiguity of the meaning of "relief header" and of the location of the points of measurement of the various "back pressure" parameters.

7.2.2. CCPS Guidelines

The provisions of the CCPS Guidelines are consistent with the apparent intent of Code and API provisions. The ASME definition of ***back pressure*** (Appendix I of ASME PTC-25) is accepted without modification (***"the static pressure existing at the outlet of a pressure relief device . .*** "). However, the following definition of ***built-up back pressure*** is used in the CCPS Guidelines:

> Built-up back pressure is the increase in pressure at the outlet of the pressure relief device that develops as a result of flow after the pressure relief device opens.

This definition differs from the ASME definition (Appendix I of ASME PTC-25) in the wording ***increase in pressure*** in place of ***pressure***. This change is consistent with the API definition (see paragraph 7.2.1 above). However, this CCPS definition differs from the API definition in the use of the wording ***at the outlet of the pressure relief device*** instead of ***in the discharge header***. This difference is in accord with the above ASME wording.

The CCPS definitions are applied as follows:

- ***Conventional valves:*** Adopt the API back pressure criterion (API 520-I):

  > As long as the built-up back pressure is less than the overpressure after the valve opens, the valve will remain open and perform satisfactorily under flowing conditions . . . .

  Note that the Code (ASME BPVC) uses the wording of the ***back pressure*** definition here instead of the wording of the definition of ***built-up back pressure.***

- ***Balanced valves:*** Consult manufacturer for the effect of total ***back pressure*** on flow and operability; use API average values of typical manufacturers' specifications (API 520-I) for preliminary sizing only.

Experience with back pressure calculations for conventional valves teach that the above limit becomes more difficult to meet as the relieving pressure is decreased. Thus, the calculations are performed at the lowest pressure at which the valve is assured to be at rated lift. Back pressure problems occur on the opening cycle, during which certified valves must attain full lift within 10% overpressure. Therefore, the test should be made at the 10% overpressure level. Use of the API provision (see Section 7.2.1 above for full wording from API 520-I) that "a higher maximum allowable built-up back pressure may be used for overpressure greater than 10 percent" may well be inappropriate in a given application. The overpressure for the design maximum venting pressure may exceed 10% for some design scenarios (21% for fire exposure, for example), but the tail piping must be adequate for relief of less-severe scenarios requiring only 10% overpressure for adequate relief.

7.2.3. Superimposed Back Pressure on Conventional Valves

The CSPs and typical manufacturers' literature include the required compensation in ***set pressure*** when conventional valves are adjusted for service conditions of constant superimposed back pressure. However, nothing is said concerning the possible need for a corresponding adjustment in the criterion for allowable built-up back pressure. In this application, the Code requires that the valve be marked with two "set pressures":

- Cold differential test pressure (CDTP, the pressure at which the valve actuates on cold fluid with no superimposed back pressure).
- Set pressure (actuation pressure in service; CDTP plus superimposed back pressure, corrected for relieving temperature in service).

If the valve were in service with no superimposed back pressure, the "built-up back pressure less than overpressure" rule would be applied using overpressure based on the CDTP as follows:psi allowable built-up back pressure = 0.1 × CDTP

With superimposed back pressure, the CSPs imply that the overpressure be based on set pressure as follows:

$$\text{psi allowable built-up back pressure} \cong 0.1 \times (\text{CDTP} + \text{superimposed back pressure})$$

Thus, the CSPs allow higher and higher psi built-up back pressure on a conventional valve of a given CDTP (given spring characteristics) as the superimposed back pressure is increased. This would certainly not be true if PRVs were proportional-action valves. These valves are actually pop-action at the set point, but can show some proportional action as the overpressure is increased to the level corresponding to rated lift. The higher back pressure of the CSPs would not be allowable if this region of proportional action persists as the superimposed back pressure is increased. However, as the flowing pressure increases with increasing superimposed back pressure, the pop action may increase sufficiently to achieve rated lift at higher psi back pressure levels. If so, the built-up back pressure criterion of the CSPs is reasonable as long as the flow remains sonic in the nozzle throat.

The Test Code (ASME PTC-25) prescribes testing methods for conditions of superimposed back pressure. However, data on the response of conventional valves to built-up back pressure under such conditions are not in evidence. Lacking such data, a conservative approach for applications of particular concern would be to base the overpressure and allowable built-up back pressure on the CDTP rather than on the set point in service.

7.2.4. Outlet Prediction

Experience in two-phase flow calculations of built-up back pressure in conventional valves for vaporizing liquids has shown a back pressure problem, which is not associated with tail pipe sizing. The problem arises from the fact that the vapor content increases as the fluid flashes from the pressure in the nozzle to the pressure in the valve body. The body pressure must be high enough to drive this expanded fluid through the valve outlet. If this outlet flow area is too small for a given nozzle size, the body pressure will exceed the allowable built-up back pressure limit. At some point the valve will begin to lose lift to attain a balance between nozzle and outlet flow capacities and thus become unstable.

The existence of this problem has been predicted for one set of two-phase flow models (Huff 1983) and confirmed by test (Sallet and Somers 1985). No test data are available for single-phase gas flow, though built-up back pressures are predicted to become excessive for air flow in a typical 8T10 valve as the set pressure is increased above about 150 psig. Use of a tail pipe larger than the valve outlet does not help, since the restrictive body connection remains. The recourse is to use valve(s) with a higher ratio of outlet-to-nozzle flow areas.

8. LIQUID ACCUMULATION

The accumulation of liquid in the relief piping will affect the pressure at the valve inlet and the back pressure at the valve outlet. For liquid relief, the piping is presumed to run full after the initial filling. Valve setup and flow calculations should be based on liquid-full piping. For two-phase flow, however, the piping may at least partially drain between valve cycles, making the estimate of the accumulation a difficult logic exercise. A summary of possible effects and problems has been published (Fisher 1995).

The possibility of liquid accumulation in the inlet piping brings up a question: Does an increase in static head have the same effect on valve stability as nonrecoverable pressure losses? In principal, the pressure change due to the development of the static head is recoverable if the fluid is brought back to the starting elevation. However, this is not ordinarily done in practice. On the first opening cycle, then, it would seem reasonable to add the static head buildup to the nonrecoverable losses. However, the inlet losses on the first cycle are generally of little concern with regard to valve chatter.

The test against the 3% limit for a *top-mounted* valve is properly performed on the second operating cycle (after the pressure during venting has dropped to the reclosing level). The two exteme case are:

- **Case I.** The piping does not drain back while the valve is closed (all-liquid flow, low spots in line, not enough time, etc.):

The net result of the liquid acumulation is to raise the vessel pressure at which the valve actuates. The static head effect does not change after the first cycle and is not included in the test against the 3% allowable loss.

- **Case II.** The piping drains completely while the valve is closed between cycles (two-phase flow, adequate draining time, etc.):

  The static head is established anew during each cycle and should included in the test against the 3% allowable loss.

A similar line of logic applies to the calculation of built-up back pressure in the tail piping.

The CSPs are silent on this problem of liquid accumulation in the piping. The effects calculate to be small in typical piping layouts, especially for fluids with high volumetric vapor content. However, the effects on actuating pressure and valve stability can be worrisome if large elevation changes are required in low-quality two-phase flow. Test data are not in evidence to provide guidance. If the predicted liquid accumulation is of concern, then the head should be included in the set pressre compensation and in the calculation of inlet pressure loss and tail pipe back pressure unless test results become available to indicate otherwise.

9. CONCLUSION

The codes, standards and practices provide guidance in the proper selection and installation of pressure relief devices of demonstrated relief capacities for the relief of single-phase fluids. Some commonly used system configurations are not yet considered in the CSPs, which requires that good engineering judgment be used to meet the spirit of the provisions. In particular, sizing and installation for two-phase relief is not yet covered.

An apparent dilemma can arise when certain documents are adopted and made into law by regulating bodies. Certain provisions of such documents can become obsolete and no longer seem to represent current accepted good engineering practice. It is the ERS system owner's responsibility to assure compliance with local CSPs and regulations. Consultation with the local authority having legal jurisdiction over the installation of pressure vessels may be required to identify pertinent documents. In any event, it is essential that current accepted good engineering practices be defined and followed to achieve the goal of proper relief system design. Such practices provide precedents for appropriate revisions and updating of the CSPs. The efforts of the CCPS Guidelines Subcommittee represent significant progress toward this goal.

APPENDIX
CCPS Emergency Relief Effluent Subcommittee

Chair:

Larry L. Simpson, *Union Carbide Corporation,*
succeeded Stanley S. Grossel (Hoffmann-LaRoche, Inc., retired).

CCPS Staff:

Howard E. Huckins

Members:

Matthew L. Becker, *ARCO Chemical Company (retired)*
Steve G. Coats, *E. I. DuPont de Nemours & Company, Inc.*
John DiPalma, *CYTEC Industries, Inc.*
Alan G. eiter, *Rohm & Haas Company*
Gene K. Lee, *Air Products & Chemicals, Inc.*
Al R. Muller, *Goodyear Tire & Rubber Company*
John A. Noronha, *Eastman Kodak Company*
Harvey Rosenhouse, *FMC Corporation*

REFERENCES

Codes, Standards and Practices

API 520-I. *Sizing, Selection, and Installation of Pressure-Relieving Devices in Refineries. Part I—Sizing and Selection,* API Recommended Practice RP 520, Sixth Edition, 1993. American Petroleum Institute, Washington, D.C.

API 520-II. *Sizing, Selection, and Installation of Pressure-Relieving Devices in Refineries. Part II— Installation,* API Recommended Practice RP 520, Fourth Edition, 1994. American Petroleum Institute, Washington, D.C.

API 521. *Guide for Pressure-Relieving and Depressuring Systems* (1997). American Petroleum Institute, Washington, DC.

API 526. *Flanged Steel Safety-Relief Valves.* API Standard 526, Third Edition, 1984 (reaffirmed 1989). American Petroleum Institute, Washington, D.C.

API 2000. *Venting Atmospheric and Low-Pressure Storage Tanks,* API Standard 2000, Fourth Edition, 1992. American Petroleum Institute, Washington, D.C.

NB-PRDC. *Pressure Relief Device Certifications.* The National Board of Boiler and Pressure Vessel Inspectors, Columbus, OH (1995; revisions/additions through 1997).

ASME BPVC. *Boiler and Pressure Vessel Code,* Section VIII, Division 1 (1995), American Society of Mechanical Engineers, New York, NY.

ASME PTC-25. *Pressure Relief Devices,* Performance Test Codes (1994). American Society of Mechanical Engineers, New York, NY.

Technical Papers

Fisher, H. G., "Emergency Relief System Design: The DIERS Users Group Safety Relief Valve Example Problem," *International Symposium on Runaway Reactions and Pressure Relief Design*, AIChE, New York, pp. 418-449 (1995).

Fisher, H. G., et al., *Emergency Relief System Design Using DIERS Technology: The DIERS Project Manual*. American Institute of Chemical Engineers, New York, 1992.

Huff, J. E., "Intrinsic Back pressure in Safety Relief Valves," *1983 Proceedings—Refining Department, Volume 62* (48th Midyear Meeting), American Petroleum Institute, pp. 105–111 (1983).

Lai, Y. S., "Safety Relief Valve Performance Under Back Pressure Conditions,"*J. Loss Prev. Process Ind.*, 5(1) (1992), 55–59.

Lai, Y. S., "Safety Relief Valves on Compressible and Incompressible Fluids", 1996 *Process Plant Safety Symposium, South Texas Section AIChE,* Volume 1, pp. 47–53 (1996).

Leung, J. C."The Omega Method for Discharge Rate Evaluation," *International Symposium on Runaway Reactors and Pressure Relief Design,* New York: AIChE, 1995, pp. 367–393 .

Morris, S. D., "Liquid Flow Through Safety Valves: Diameter Ratio Effects on Discharge Coefficient, Sizing and Stability,"*J. Loss Prev. Process Ind.*, 9(3) (1996), 217–224.

Papa, D. M., "Subcritical Flow Through Pressure Relief Valves," Paper presented at API Midyear Refining Meeting, Dallas, TX, (1994).

Sallet, D. W., and Somers, G. W., "Flow Capacity and Response of Safety Relief Valves to Saturated Water Flow," *Plant/Operations Progress,* 4(4) (1985), 207–216.

Sherman, D. H., "The Effect on the Flow Rate and Operational Characteristics of Spring Loaded Pressure Relief Valves on Flashing Subcooled and Saturated Water Service," Presentation to the ISO 185 Committee (1993).

The Reliability of Safety Valves and Relief Valves

Ian M. Parr

Eutech Engineering Solutions Ltd., Brunner House, PO Box 43, Winnington, Northwich, Cheshire, CW8 4FN, UNITED KINGDOM
(A wholly owned subsidiary of ICI)

1. INTRODUCTION

How reliable is a relief valve?[1] Does this matter? What is the importance relative to the integrity of pressure systems as a whole?

Questions such as these have been asked before. Conclusions drawn about the reliability of relief valves and what the effects may be on the safety of pressure systems and the importance compared to other issues of plant safety have varied.

Nevertheless, our first question remains: will a relief valve perform in service as required? There are now many published sources of relief valve reliability data and many companies have their own internal records. Whether this information is more or less important than other safety issues is a matter for debate that will be touched on towards the end of this paper. But, putting aside the question of relative safety priorities we need to address the value of the data available upon which we draw our judgments. This paper will seek to do that in the context of the design and operation of relief systems.

So what aspects of the performance of relief valves should give us cause for concern?

Clearly there is a question of safety. That a relief valve operates on demand providing protection against excessive pressure is a prerequisite. Any malfunction jeopardizing safety is of concern. There are recommendations and requirements in national codes and transnational codes, even in legislation in some countries, specifying what should and should not be done to ensure the integrity of pressure systems. However, we know things still go wrong even if we comply with all the essential requirements.

[1] The terms "relief valve," "safety valve," and "safety relief valve" are widely used in industry and have specific meanings in US standards; however, only "safety valve" is defined in most European standards and in International Standards (ISO). No distinction is made between these terms in this paper. "Relief valve" is used for consistency and because it is most widely accepted colloquially in the industry.

A second issue, less well considered in our experience, is the environmental performance of relief valves. Those factors, which affect safety are not necessarily the same as those affecting environmental performance. However, this paper will show that they are sometimes connected and that an understanding of the causes of failure of relief valves is important. This will be our third topic.

We will also look at the cost implications of safety and environmental performance. In the drive to extend plant run times and reduce maintenance costs, value engineering, risk assessment and hazards review will be among the determining factors. The balance between risk, cost and adequacy of design has been an important part of at least one previous DIERS paper (1) and we will be taking up some of the points raised in that paper.

A major concern for the author is the tendency for inquiries into the behavior of pressure relief systems to focus on single topics—relief valve reliability, rupture disk performance, process design etc. with little reference to the relationship between components and design intent and the manner in which plants are operated. This can be illustrated by the failure of many design organizations to integrate teams involved in pressure relief system design. We will discuss how this affects relief valve reliability at various stages in this paper.

Relief valve reliability statistics are an important part of the way an increasing number of companies assess the suitability of relief valves and decide the operational intervals between inspections. They are discussed in References (2), (3), and (4), which also provide the first part (Section 2) of this paper, namely the measurement of set pressure[2] of relief valves.

These data are used to help make plant operational decisions on the period between relief valve inspections and may affect the judgment of project engineers and works support personnel in deciding upon the specifications for relief valves.

It should be emphasized that data collection alone is not enough to tell us about the reliability of relief valves. An understanding of the system requirements (not just the "top" relief event) and the interaction between required and delivered valve performance, specification of the total relief system, details of relief valve installation and in-service maintenance and inspection are also essential. If failures are to be understood and avoided, these issues need to be considered. They are dealt with in the second part of the paper, also in the third where we look more closely at failure modes.

[2] The term "set pressure" as used in this paper means the pressure at which the valve is set or pressure is measured on the workshop test bench to ensure the value corresponds to the required set pressure when installed on the plant. There are very precise definitions for this term in codes and standards.

Finally, the paper will pull some of the common threads together suggesting priorities and issues that need to be addressed by plant designers and operators.

This paper was prompted by and draws upon material collected by members of the Eutech Engineering Solutions Ltd. Pressure Relief Team and members of ICI Central Maintenance Workshops, Merseyside for over 20 years. It incorporates plant observations and anecdotal experience and it references what seem to the author to be helpful published papers that provide further information on this topic. In 1987 a series of student projects were initiated by ICI Engineering Department. They aimed to study plant records and operations relating to relief valve service history. Workshop reports of relief valve set pressure performance are used in ICI to enable plant and in-service inspection personnel to monitor relief valve reliability and adjust inspection intervals to ensure operators and inspection engineers have confidence in valve reliability. The key factor is set pressure as removed from the plant compared with the required set pressure. This general process has been well documented elsewhere (5, 6, 7).

Closer examination of the reports, observations on plants and increasing knowledge of relief valve performance suggested to us that while the wisdom of using as removed data as the basis of inspection interval determination is not in doubt, any statistical conclusions were likely to be unsound. while statistics showed a steady decline in valves lifting unacceptably high,[3] a number of other factors needed more explanation:

- Changes in the specification of relief valves did not seem to improve overall reliability statistics
- Workshop records showed that components such as bellows had poor reliability and this was not reflected in the data.
- There was no change in the statistics for relief valves that lifted below set pressure.

The main reports that were commissioned from students are indicated in References (8), (9), and (10).

In summary it was decided that the value of relief valve reliability records is important but interpretation of the data is far from straightforward and a more technically aware approach is necessary.

[3] In this paper we describe relief valves lifting "heavy" meaning their set pressure measured in the workshops on removal from the plant (or as received from the manufacturer or overhaul

2. RECORDS

2.1. The Value of Records

The period between each relief valve inspection and/or overhaul is usually described as the "inspection or examination interval". The term "inspection frequency" is also used. Records of relief valve overhaul and set pressure reports are maintained by many companies because they can be used as the basis for assessing inspection frequencies. Frequencies are proposed in a number of publications [e.g. (5) and (7) give some, but they do not take account of installation conditions. Further advice is also given by the Safety Assessment Federation, SAFed (11).

Recently the possibility of using risk analysis techniques in conjunction with reliability data has been considered. This is not yet a common approach but it is claimed to offer a more rational method of determining safe operating periods than merely reviewing the statistics and operating history of relief valves. It also offers supporting arguments for extending plant run times.

The determination of inspection frequencies is discussed in "A Guide to Assessing the Operating Periods of Relief Valves," published by the Engineering Equipment and Materials Users Association (EEMUA) (12).

In the last thirty years there have been concerns about the reliability of relief valves, as defined by the numbers lifting "heavy", expressed in the public domain—some of which are noted in (3) and (4). Parry in *Relief Systems Handbook* (13), devotes two chapters and an appendix to an analysis of data from several sources. The author's private discussions with colleagues in his own and other companies have often concluded that reliability of relief valves based upon analysis of set pressure records may give reason for concern, but no consensual industry view has emerged.

We describe relief valves lifting "light" as when their set pressure is less than that determined as being acceptable.

Factors common to plant operators' overhaul and test centers and those of relief valve manufacturers and service organizations can be summarized as follows:

- Companies reviewing the performance of their relief valves use set pressure data to determine the length of time between inspections and valve overhauls. Other methods may sometimes be used.
- Inspection frequencies sometimes impact on plant run times and there is a requirement to extend them in many parts of the Chemical Industry. For these companies relief valve reliability is also a plant operations cost issue.
- A substantial percentage of relief valves do not lift on test following removal from service or do so at a pressure regarded by the operating companies as unacceptable (i.e. "heavy"); typically this value is 110% of required set pressure.

- There are no major relief valve manufacturers whose products are known to be more or less reliable than others.

However, reports commissioned in ICI suggest that much recorded data for relief valve reliability should be treated with caution. The main reasons follow:

- Test bench measured values do not reflect in-situ performance, especially if process fluids can cause blockage on exposure to air.
- The set pressure of relief valves can be affected by careless handling upon removal and transfer from the plant to the testing station; see 2.4.2.
- Immediate removal following relief is sometimes adopted. If applied consistently, this can produce "good" data for that plant but they are rarely if ever transferable.
- The performance of a relief valve can be affected by the connecting piping. This may not be apparent on the test bench.
- The set pressure of a relief valve may be within acceptable limits but if a pressure balancing bellows fitted within the valve has failed then the raw data could be misleading. See 2.2.
- Different methods of measuring set pressure of relief valves can produce different results. See 2.4.1.

2.2. Relief Valves Lifting "Heavy"

Values for relief valves lifting at more than 110% of set pressure taken from ICI Central Maintenance Workshops, Merseyside between 1986 and the present are shown in Fig. 1.

The causes of relief valves lifting "heavy" are discussed in References 2, 5 and 13. This may reflect a wider problem with pressure systems. As MacMahon shows (5), if data on relief valve reliability are based upon testing of relief valves and measurement of set pressure, it may not provide all the information needed to improve reliability. Parry (13) seems to imply (erroneously in our view) that "primary reliability" can be judged by the number of relief valves lifting "heavy" or "light."

One issue which needs early clarification is that of the impact on statistics of the performance of pressure balancing bellows in relief valves. If a bellows in a relief valve is used for pressure balancing (so that the valve set pressure in service will not be affected by back pressure), then its integrity may be critical to the valve set pressure, and by association, the safety of the pressure system. This is particularly the case if there is no compensating pressure balancing piston fitted. All assessments of relief valve data need to consider the presence of a bellows (as well as other aspects of the specification and installation). In the author's experience about half of all bellows fitted in relief valves

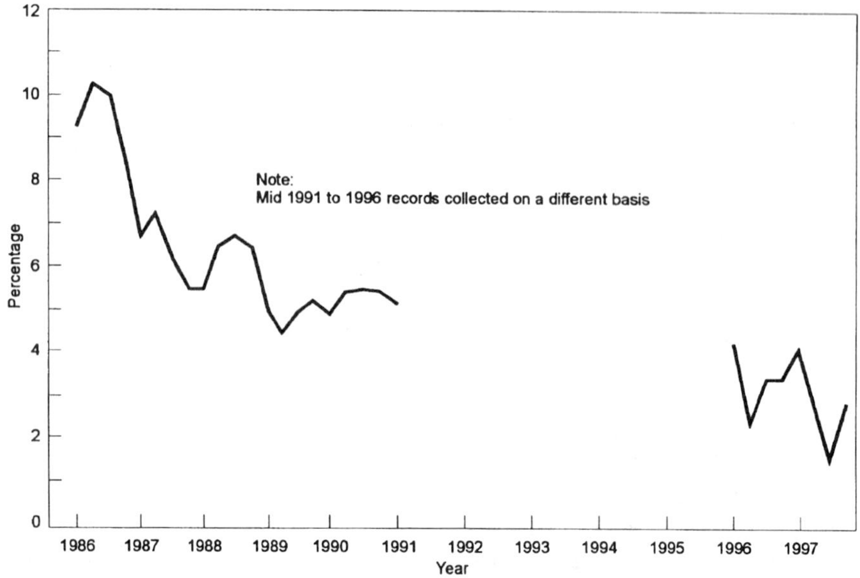

Figure 1. Valves Lifted Above 110% of Set Pressure

are essential to maintaining the relieving pressure in a pressure system within code compliance. Data which ignore the condition of pressure balancing bellows in relief valves (such as Fig. 1) is not likely to be valid if used for purposes of performance analysis.

Reliability data for pressure balanced relief valves gathered by the author and members of the Eutech Pressure Relief Team with the help and cooperation of ICI Central Maintenance Group Workshops for some ICI sites over a ten-year period indicated that about 30% of bellows were in a failed condition when returned for inspection. Figure 2 shows a typical bellows failure.

2.3. Relief Valves Lifting "Light"

2.3.1. Statistics

Most studies of relief valve statistics have tended to concentrate on valves lifting "heavy" because these are regarded as impacting upon safety and plant run times. Figure 3 shows data obtained in ICI (on the same basis as Fig. 1) for relief valves lifting at 90% or less of set pressure.

Although values for valves lifting more than 110% of set pressure have declined (it is thought, due to changes in specification and inspection intervals), there has been no similar improvement in the data for relief valves lifting "light." Between 1992 and late 1995 methods of collection changed so the

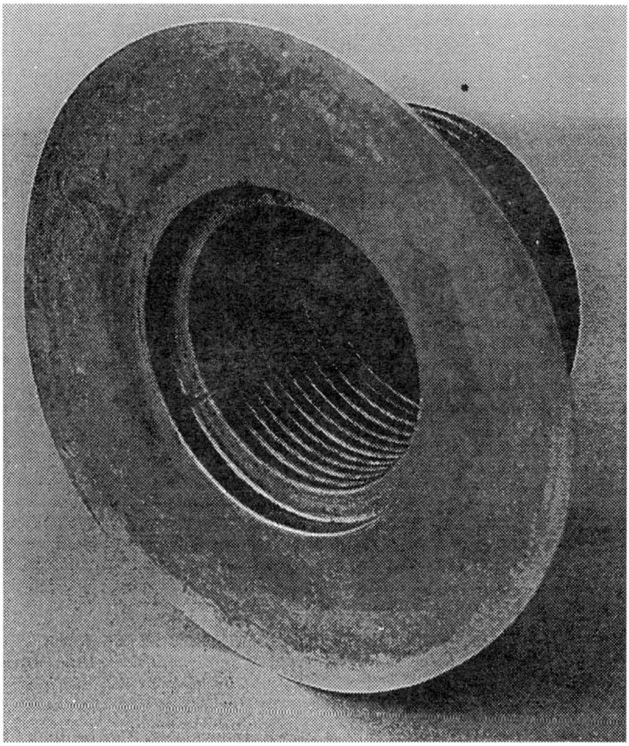

Figure 2. Example of a Failed Stainless Steel Pressure Balancing Bellows. Note failure along seam (pinholing is also common)

data have been omitted. The rapid increase in the last year has yet to be fully explained.

In 1987 a student project was initiated to investigate the observed differences in Figs. 1 and 3. The aim was to look at a selected application known to produce valves which consistently lifted "light." Relief valves on steam applications whose set pressure was less than 150 psig (10 barg) were selected. The report showed that the performance of many of the lower priced, mass-produced relief valves from manufacturers who had not obtained independent certification (British Standard, AD Merkblatt, ASME, etc.) for their valves was substantially worse than valves with appropriate certification. As a result a policy of improvement was begun and the recommendations of the report were followed. New valves with identifiably better specification replaced those shown to have performed badly. One of the main drivers for change was that the student report had shown the costs of steam leakage to be substantial—in the order of £50k for the ICI site per year in 1987 values—and that extrapolating these values for other duties gave cause for concern on grounds of cost of lost materials and potentially environmental effects.

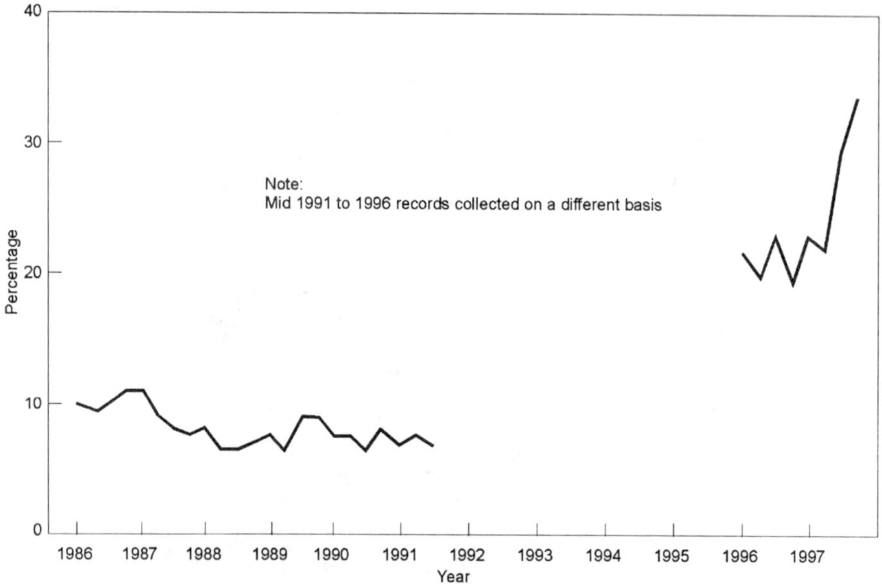

FIGURE 3. Valves Lifted Below 90% of Set Pressure

However, no long-term statistical improvement was observed. In 1996 a further student project was commissioned from Manchester Metropolitan University by Eutech Engineering Solutions Ltd. (a subsidiary of ICI) and at the same time a comprehensive relief valve reliability database was developed. The first activity was aimed exclusively at revisiting the original survey. The second was part of an ongoing improvement of data collection.

The data have not yet been fully processed. However, it is thought that failures in service fall into two categories: seat and guide damage; or spring corrosion and relaxation. Consequently these features are looked at in more detail in this paper. These preliminary findings fit well with observations made in the field and discussions with valve overhaul personnel within ICI and in other companies. For further correlation a sample of 12 relief valves that had lifted "light" was selected:

- Valves were of all sizes and manufacture from a small screwed valve to the largest 4″ × 6″ and set pressures up to 52 barg.
- Six were process and six were steam duties.
- All the steam valves required machining and lapping of seat faces; three of them had seats in poor condition.
- One of the steam valves required a new spring.
- Of the process valves all required machining and lapping of seat faces and three of them had seats in poor condition. However, one valve (hydrogen duty) had seats in good condition and no sign of why it had lifted "light."

- Only one process duty valve had a bellows fitted and this was in good condition
- This same valve had a rupture disk fitted upstream which had burst and the valve had been sent for overhaul and resetting. The valve is known to have lifted in response to a genuine demand. The duty is dirty but the valve may have chattered as a result of oversizing or restriction in the piping system. The valve seat was damaged but the bellows and spring were in good condition. The likely cause of lifting 'light' is recorded as product contamination and this accords with the service record prior to the valve being refurbished but chatter effects require further investigation.

2.3.2. Seat Damage

Seat damage can occur, either through:

- corrosion of the valve disk or nozzle which usually affects the full area of the seat
- local damage likely to be caused by poor valve handling or assembly, wire drawing or chatter (oscillation of the valve disk involving impact with the valve seat during operation).

Relief valve manufacturers have demonstrated that seat damage cannot only cause mismeasurement of set pressure but also failure of a relief valve to develop the lift required to achieve certified capacity. Thus failure of the seating capability of a relief valve (measured as set pressure below the requirement) may represent a potentially unsafe situation.

2.3.3. Spring Failures

Spring failures are pointed out as a feature of poor relief valve performance by Waters (9). Figure 4 shows typical damage. Likely causes may be environmental conditions or process fluid diffusing into the bonnet of the relief valve.

Figure 4. Corrosion of Relief Valve Spring

Some relief valves have plated springs. Our experience is that this may not improve reliability and can, under some circumstances, lead to premature spring failure through embrittlement or stress corrosion cracking caused by porosity of the coating.

The effect of deterioration in spring performance is that set pressure is reduced. This can lead to the relief valve operating inadvertently in service, followed by chatter, seat damage, etc. Tests were undertaken some years ago by the author for ICI Engineering at the United Kingdom National Engineering Laboratory in Scotland in conjunction with a major plant item package vendor. The aim was to confirm the performance of relief valves installed on a proprietary piece of equipment. A particular relief valve was shown to lift on test and chatter under full flow conditions leading to a reduction in the order of 50% of set pressure as the spring relaxed due to rapid oscillation. It can be concluded that short-term failure of a spring, or its incorrect specification, may have environmental implications leading to longer term failure which may be significant for system safety.

Spring performance in general has been a concern within ICI since the 1970s. Failures have been due to quench cracking, hydrogen embrittlement and hydrogen induced stress corrosion cracking. Nelson (6), points out similar modes of failure. More rigorous specifications and an insistence upon not using materials such as tungsten steels have helped minimize spring failures attributable to these causes in ICI in recent years.

Spring performance is critical in many relief valve process applications yet manufacturers continue to supply springs whose design is known to result in service failures. Current and proposed national and continental standards (for instance, European Standard EN 1268) remain inadequate in protecting users from a well understood and potentially hazardous condition.

2.3.4. Improvements in Specification

It is thought the reasons why improvements in specification of relief valves have not led to improvements in statistics in ICI for valves lifting below required set pressure are the following:

- Overall relief system design remains a cause of unreliability and replacing the relief valve for one of improved specification has not provided a total solution to the problem
- Increasing the frequency of inspection of a relief valve may have moved it from an apparently unsafe performance to one in which the valve set pressure is reduced during its service life. This improves the apparent safety of the system but the environmental performance may have thus become an issue requiring attention.

Further investigation is continuing in order to provide more understanding of these observations.

2.4. Some Comments on Set Pressure Measurement and Relief Valve Performance in Service

2.4.1. Measurement

Weston (8), examined, in a modest way, the measurement of set pressure of relief valves and how performance changed depending upon the method used. A description of this report and others that followed is required here because it is necessary to put set pressure measurement into context.

It is well known in the relief valve industry that different methods of measuring set pressure produce different values. Indeed, ANSI B95.1 has different definitions of how set pressure can be measured. Although European and International Standards definitions are more precise, in practice manufacturers and relief valve overhaul and repairers tend to use methods not described in the codes. This is probably because the definition of set pressure is considered applicable only for capacity certification and is not always regarded as practicable for production purposes.

Weston measured set pressure by different methods and reviewed the results. Private observation of relief valve set pressure procedures by the author leads to the conclusion that the variations described by Weston (plus or minus 11.5%) are modest compared with those that may be produced. Scully (14), also supports this view.

If a consistent method of setting relief valves is used, then this disparity does not cause problems in interpreting relief valve reliability data. However, if different suppliers and overhaul organizations are used by an operator and different methods of obtaining or measuring set pressure are employed, variations can be significant and can invalidate any statistical basis for determining inspection intervals.

Furthermore, unless relief valves are tested prior to installation by a consistent method, then later, in-service reliability data will not be useful. Weston points out that from a sample of 35 relief valves received new from various manufacturers in a specified period over 50% lifted outside code requirements. Zirps et al. (7), report that 15% of all new relief valves received at a site did not meet code requirements.

2.4.2. Checking of Set Pressure

It should be noted that many relief valve manufacturers, in recent years, have improved the way valves are packaged for transport. However, mishandling on site, especially in construction stores and during installation, can affect set pressure.

Some relief valve manufacturers do not recommend that valves are adjusted or checked at all and claim that any interference invalidates the valve warranty. The author's experience is that some of these valves cannot be reliably adjusted once disturbed and gathering further data is useless.

2.4.3. In-situ Set Pressure Measurement

Over a number of years increasing confidence in relief valve reliability data from measurement while a plant is on-line has been obtained through the use of "Trevitest." This method has been used in ICI in the UK for some time and a company procedure has been developed. Weston showed that such methods, in which an additional load applied to the relief valve is measured and translated into a value for set pressure, are potentially useful as a tool in obtaining reliable, as-installed measurement of set pressure. It does not, however, give assurance of valve lift nor that major parts are free from damage.

An important extension of "Trevitest," and similar methods now being recommended, is "fingerprinting" a valve prior to it being brought into service. It is particularly useful when fluids are likely to cause gumming of the relief valve when removed for testing and overhaul thus giving a false understanding of the valve's condition.

2.4.4. Full Lift

How essential it is that relief valves are fully lifted on test as part of the set pressure set up and measurement activity is a matter for debate. It is argued by some relief valve manufacturers that it is not possible to have full confidence in a valve unless it has been "exercised" and that the set pressure can only be determined by achieving full lift.

Woodfolk and Sanders (2) and Scully (14) discuss this. They take the view that demonstrating "popping" of the valve ensures the set pressure is measured realistically and shakedown of components takes place.

Weston shows that it is a consistent method that is needed in assessing reliability. Setting aside the practicalities of popping large relief valves at high set pressure—the issue of valve lift is a separate one from set pressure—it is reasonable to assume that properly assembled, installed and maintained relief valves will perform as intended, although there is evidence that relief valve overhaul organizations do not always properly maintain valves.[4]

2.4.5. Cost of Relief Valve Overhauls on Plant Operation

Inspection and overhaul frequencies of relief valves may impact severely on plant run times. For example, relief valves protecting a hot oil system were repeatedly found blocked when inspected and frequencies based upon these data were set at three-month intervals, necessitating a plant shutdown each time. The estimated cost to the business was in excess of £100k per annum. The systems were redesigned and valves respecified and in situ set pressure

[4] This subject needs debate in another forum and could be considered as a matter of code compliance and quality control of manufacturers and relief valve overhaul organizations. Therefore the scope of this paper has been limited to those aspects within the direct control of the design engineer and his or her plant operations equivalent.

measurement was instituted. Inspection intervals are now acceptable to the business and the costs of the modifications repaid many times over.

This question of system design in relation to relief valve reliability is dealt with in more detail in Section 3. However, cases like this show how inspection intervals, cost, and safety considerations are related to the way records are gathered, interpreted, and used.

2.4.6. Primary Reliability

Parry (13) defines "primary reliability" as relief valves which

- fail to open at the correct pressure
- fail to open fully at the relieving pressure.

These conditions are compared with other kinds of relief valve failure.

Primary reliability is used as a performance measure and thus implies a priority that may not address all issues of plant availability and run times or environmental considerations. Nevertheless, Parry recognizes the problems of measurement of set pressure but links them to direct failures in the field which we would describe as extreme—failure of valve pressure retaining parts—and which represent only 9 examples of a sample of 608 (3 helium, 5 carbon dioxide, 1 air/nitrogen). He summarizes his findings, thus:

> A complete analysis of reliability would require comprehensive data on each application and each failure.

We do not concur with this view. In particular, in Sections 3 and 4 we will try to show that better data collection and interpretation and a systems approach to plant design and maintenance would address many of the concerns Parry, rightly, emphasizes.

2.5. Summary of the Issues of Measurement of Set Pressure

1. Measurement of set pressure of relief valves is important. Methods of measurement of set pressure need to be consistent
2. Understanding of relief valve reliability requires more than interpretation of measured values of set pressure taken on removal of the valve from service.
3. Preinstallation checks of relief valve set pressure are necessary.
4. The condition of a relief valve when removed from service needs to inform any interpretation of data. See MacMahon (5).
5. Handle relief valves removed from a plant prior to set pressure testing as if "new."
6. Improvements in specification may mask more underlying problems of system design and inherent unsuitability of a relief valve for its application.

7. Few methods of measuring set pressure give assurance of the valve achieving the required lift. In some cases it is impossible to do this because of size, set pressure, etc.

3. SYSTEM DESIGN

3.1. Failure Modes

For a relief valve to become unreliable there must be a failure or a demonstrable risk of failure. Measurement of set pressure alone is insufficient. Any failure will have an initiating event or events and evidence of these events may be "heavy" or "light" set pressure or damage to, or complete failure of, valve components or relief system.

Most contributions on the topic of relief valve reliability deal with statistics, risk and the effects of service conditions. Perry (13) and Scully (14), however, indicate a number of system problems and their solutions. In this section some of these problems and solutions will be followed further and the performance of relief valves in service, in particular their interactions with the systems in which they function, will be reviewed in terms of events which may lead to unreliability.

Observations made in collecting information on failure of relief valves suggests they are affected in two ways, singly or in combination.

1. Environmental or process conditions can lead to corrosion or damage to valve components
2. Relief events, failure to install valves correctly or changes in operating conditions (transient or extended) can cause the valve to simmer or to lift leading to local failure or setting up conditions that will lead in time to failure.

It is important that potential failure events are properly understood if reliability of pressure relief systems is to be improved and plants made safer and more environmentally predictable.

Mechanisms for potential or actual failure may be summarized as:

- those which are to do with operating conditions and which therefore affect what may be termed as the "static" condition
- those which affect the relief valve once a relief event has occurred and which may be termed the "dynamic" condition.

Many of the references cited in this paper discuss these mechanisms. However, there is a general failure to see the way they impact on components in a pressure system thus leading to a misunderstanding of how to eliminate them.

3.1.1. Static Conditions Leading to Relief Valve Failure

The first category of failure mechanisms include:

- *High operating pressure as a percentage of relief valve set pressure. Codes tend to recommend values in the order of 85 to 90% which is often too high for conventional relief valves.*
  The highest reliable value will vary depending upon the type of fluid and the characteristics of the relief valve. In particular, the difference in blowdown between a relief valve on liquid duty and one on gas or vapor needs to be considered.
  The relative closeness of set and operating pressures may mean there is potential for the valve to simmer and leak which can lead to seat damage and possible corrosion of guides and spring. Once the spring has weakened or the seat is damaged, the set pressure will be affected and the valve will appear to have a "low" set pressure. In time this may lead to failure with the set pressure "high" as valve components gall or gum up.

- *Position of the relief valve relative to flow.*
  Variations in flow pattern, local turbulence, variations in temperature, slugs of liquid etc. can cause the relief valve to flutter, cause seat damage and subsequent valve failure which may follow the pattern described above. Emerson (15) mentions the effects of flutter of the bellows in relief valves, but fails to indicate their potential for failure due to fatigue (bellows are designed for a finite number of operating cycles).
  Baldwin and Simmons (16) describe vibration effects on relief valves in steam service. However, this form of failure may not be limited to high velocity steam systems.
  Relief valves protecting or installed in complex piping systems and containing liquid or liquefied gas have been particularly prone to failure as a result of surge, vapor cavity formation, or vibration. Significant environmental consequences have followed from some relief valve failures caused by such events.
  Parry (13)identifies loose adjustment as a cause of change in set pressure. Our experience is that this can be loss of adjustment of the relief valve spring load or the blowdown ring. Both arise from vibration in the valve or dynamically induced chatter.

- *Changes in the mechanical stability of the piping system.*
  The most significant external effect upon a relief valve is the transmittal of piping loads to the valve. These come into three broad categories; deadweight, thermal effects and dynamic effects. Some of these are also

significant when the relief valve is flowing or has operated; see 3.1.2. All these forms of loading can affect the mechanical integrity of the relief valve. Figure 5 shows a relief valve removed from an air compressor. The valve had been installed on its side thereby aggravating the effects of vibration and the out of balance mass.

- *Effects of dead legs.*
  All relief valves are fitted into deadleg positions. "Deadlegs" are familiar to piping design engineers and operations maintenance personnel because they introduce many reliability problems into piping systems. But the lessons of branch size and length, supporting, fabrication standards etc. seem rarely to be applied to relief valves and their position in a piping system. This is probably because their location is often dictated by provisions other than those normally applied to, for instance, instrument connections. The conditions in the piping immediately

Figure 5. Failed Relief Valve from Compressor. Note: Spring has vibrated through the bonnet and the return spring for the lift lever is displaced.

upstream, and sometimes downstream, of the relief valve are almost always stagnant and can be different from those in the rest of the piping system.

Such a situation may produce gumming up of the piping system or exposed valve components or the creation of deposits that may have the same effect.

Many of the problems observed with pressure balancing bellows used in relief valves have been attributed to the special environment experienced by relief valve components. In ICI it has led to avoidance of the use of stainless steels in these situations. Statistics have indicated a significant improvement in reliability as a result.

3.1.2. Dynamic Conditions Leading to Relief Valve Failure

The second category of mechanisms for failure, described as "dynamic," occur when the relief valve is open at some time during the relief event and there is fluid flow across the valve seat. This takes place when the forces opposing valve closure have been overcome by the pressure energy in the equipment upstream of the relief valve combined with the kinetic energy of the flowing fluid. These conditions are not the same as those in which flow occurs due to leakage across the valve seat and the valve cannot retain the fluid or provide an effective seal.

Instability of a relief valve caused by inability to deal with certain dynamic conditions is a well appreciated mode of failure which requires more discussion. It is considered in a number of the references and in the following paragraphs we try to bring them into a recognizable set of topics that may help explain certain relief valve failures.

Many aspects of the dynamic conditions affecting relief valve performance are covered by codes and design guides. Some of these are mandatory, others are not. Relief valve manufacturers also make recommendations. In total they appear an incoherent set of design principles which may be empirically sound but not in all cases. Some do not improve system safety or are expensive and compliance should not be necessary.

A number of factors contribute to relief valve performance. They tend to overlap, but for simplicity each is dealt with separately:

1. Pressure drops in piping affecting stability, sometimes leading to chatter
2. Sizing of relief valves
3. Piping routing and design generally

1. Pressure drops in inlet and exit piping

Chokes (changes in piping geometry causing sonic flow) upstream of a relief valve are a major cause of relief valve instability. Other than that the sensitivity of relief valve performance in relation to pressure drops varies and depends upon a number of factors including:

- valve flow area and flow characteristics
- relief valve spring characteristics
- set pressure
- fluid properties
- relief valve trim

Valve flow area and flow characteristics: These affect lift and dynamic stability of the valve. Knowledge of the flow capacity of a relief valve is essential to sizing inlet and outlet piping. Some designs of relief valve are more sensitive to upstream or downstream pressure drops than others. Data from manufacturers are not always helpful; although most manufacturers have knowledge of the flow characteristics of their valves for steam and air, but the correlation with other fluids, particularly mixtures and mixed phases and nonideal gases may be a matter of guesswork or at least involve the use of large safety factors.

Spring characteristics: Blowdown performance of the valve depends upon the flow characteristics. Pressure drops in piping, particularly upstream of the relief valve, can impact upon stability through the tendency for the valve to try to reseat (or blowdown).

Also the spring range selected, the spring characteristics and changes in temperature of the spring may have a significant effect upon performance. Spring materials, often specified by manufacturers, do not take into account either prolonged changes in temperature during relief or relaxation (fatigue) as a result of chatter caused by valve instability. Spring relaxation and serious failure of springs is not unknown.

In some cases a bellows in the relief valve may modify the spring characteristics and blowdown ring adjustment may be necessary.

Set pressure: Sonic flow of vapors and gases through the valve is important in evaluating this effect upon relief valve performance. If choking occurs, then the momentum effects of the fluid may be more significant in terms of the valve stability than pressure effects. Where velocities through the relief valve are subsonic, the performance will be different and more related to changes in pressure. In these circumstances, code recommendations or requirements need careful consideration to avoid conditions that may adversely affect valve reliability.

Fluid properties: Higher mass flows relative to the flow area tend to improve valve stability through increased momentum forces compared with fluids with lower molecular mass. The combination of fluid properties, set pressure and the effects of back pressure in the piping system are important when considering the performance of a given relief valve. However, fluid properties during relief events may vary or be different for different events and in some circumstances, therefore, lead to instability.

Relief valve trim: Relief valve blowdown characteristics vary not only with fluid properties but also with valve trim characteristics. These are usually determined by the manufacturers who may not have full knowledge of the various relieving conditions (see notes on flow characteristics). Designers should inform manufacturers of all identified relief events and the nature of the fluid; gas, liquid or multiphase. Such a compromise may lead to impairment of relief valve performance under some flow conditions compared with others.

All these factors have an impact upon relief valve stability during relief. They tend to lead to valve seat damage, galling of guides, spring relaxation (or total failure) and on occasions failure of the bellows within the relief valve. Weighell (17) expresses concern about these findings, see also (2, 6, 7, 13, 14).

2. Relief valve sizing

Conventional wisdom is that relief valves should not be oversized by factors greater than four or five times and that when this occurs valve instability is a consequence. Certainly there is evidence that gross oversizing can result in severe relief valve chatter and catastrophic failure of piping.

Severe chatter has been known to lead to total failure of piping systems and loss of containment. Parry, Reference 13, has some examples.

It is not the purpose of this paper to debate this matter in detail but it is unlikely that relief valve failure through chattering can be attributed solely to oversizing unless there are serious errors in design. The major factors which need to be taken into account include relieving pressure, fluid properties and period for which the relief valve needs to be open. (See also comments relating to stability of relief valves in the first part of this Section.)

Often the designer will have considered a demand frequency that is relatively high but of low mass flow compared with a demand frequency that is low (such as fire engulfment) that has a high mass flow requirement. The dilemma is then either to festoon plants with many relief valves to satisfy both sets of conditions separately, as recommended by some design codes, or to accept that chatter may occur and take account of the risks in the final selection of the protective system. Alternatively, other methods of relief or inherently safe solutions might be considered. This judgment is not always made consciously but it is an important one for relief valve stability and thus plant safety and the environment. Increasing the number of relief valves increases the complexity of systems, especially if isolation of relieving devices is needed to keep plants on stream while relief valves are maintained. The consequences of permitting chatter also have to be considered. An example of a misjudgment in such a situation is related in Section 3.2.

Another problem is that of multiphase flow conditions. The required flow area of a relief valve for liquid flow—in mass flow terms—will always be lower than for gas or vapor conditions. Two-phase flow is yet another matter! Fundamentally the relief valve trim ought to be different for both sets of conditions. The compromise can lead to chatter. The possibility of valve instability should be considered and the in-service inspection requirements specified accordingly.

An emerging issue for relief valve reliability is the increasing tendency to specify relief valves with limited lift. ASME VIII and other codes recognize this as a characteristic needed to satisfy some relieving conditions. This may be a problem being stored up for the future. A proliferation of different relief valve flow area specifications introduces unknown elements for the future for relief valve reliability. It is essential for plant engineers and overhaul organizations to maintain records of valves and their specifications. No one should rely upon the information plate attached to the valve. This might be painted over or broken off due to corrosion etc. in time.

In summary, relief valve over-sizing will frequently be a cause of instability and a possible failure mechanism. Relief valves are rarely specified to account for chatter which is an inevitable consequence of the compromises sometimes necessary in design. Codes and enforcement organizations should reflect the need for engineers to make judgments on potential for chatter, which may include some form of risk assessment.

3. Pipe routing and general design

There is much advice on pipe routing and the issues involved for relief systems design in standards and industry guides such as ANSI/API RP 520. It is not uncommon for these guidelines to be ignored in the wider interest of cost or space considerations. There are, however, some essential requirements which should be followed. In some respects system failures are outside the scope of this paper; however, where they can be attributed to inadequacies of relief valve specification, the impact of the relief valve on the system performance—safety and environmental reliability—cannot be ignored

There are four interrelated factors in piping design which can make an important contribution to relief valve reliability:

- position of the relief valve branch
- drainage
- piping loads and displacements
- luting

Position of the branch carrying the relief valve and relief system: The proximity of other piping fittings to the relief valve branch is undoubtedly a major cause of relief valve failure and is described in a number of the references. It is only necessary to walk around any chemical plant and look for a steam leak to see that often a relief valve is the apparent cause. Observe the relationship between the relief valve and any pipe fittings, isolation or control valves and the true reason is likely to be as described in 3.1.1.

Drainage: It is obvious, but not always easy to arrange, that drainage be directed away from a relief valve. Very rarely should the ability to drain the piping adjacent to the valve or from the body drain be relied upon. Failure of relief valve vent lines to drain can cause corrosion in the valve or lead to blockage in the valve or piping. Other than an initial error in design, the most likely cause of poor drainage is reassembly of the relief system following inspection or plant overhaul.

Piping loads and displacements: Reference has been made in 3.1.1 to the effects on relief valves of piping loads. During relief there are often changes in temperature in the relief system and corresponding changes to loads upon equipment—including the relief valve. Reaction forces during relief may impose further loads on piping. There have been many examples in which loads have been transmitted to vessels or piping systems through either insufficient supports (too few to restrain the piping) or too rigid a system (inadequate flexibility to take account of thermal displacements in piping or vessels) or inadequate strength (including the supporting structure). An example is quoted in 3.2.3 Case 1.

A common failing is to allow a relief valve to discharge at right angles to the main (usually horizontal) axis of a piping system. Displacement of the piping, exerting loads upon the valve or changes in the piping geometry, may not be obvious, but relief valve guides and other internals may be affected and such damage may not be observable on the test bench so there are no statistical data on the extent of this feature.

Luting: The existence of lutes unintentionally designed into piping systems does not often affect the reliability of relief valves directly.

A lute in the inlet to a relief valve may give rise to a slug of liquid hitting the valve and causing surge damage. Variations in piping size can also result in physical damage to a valve, again most commonly at the inlet. The main problem is likely to be if a slug of liquid continually impacts the relief valve disk causing a form of chatter.

While lutes in piping downstream of relief valves have resulted in failures of pressure systems—freezing or blockage—they rarely affect the reliability of the valve itself. However, it is not unusual to find piping systems designed with the relief valve at the bottom of a lute formed by the inlet and outlet piping. The additional pressure drop may lead to chatter of the valve, but the most likely result is corrosion of the relief valve internals caused by lack of drainage, condensation of product etc. Also, surge when the relief valve discharges can result in severe piping damage and may affect the relief valve and its installation.

3.2. Some Examples of Systems Causing Relief Valve Failure

3.2.1. Relief Systems

As we have shown in the preceding Section, if relief valve reliability is to be understood, it is necessary to look into the wider issue of the "relief system."

Perhaps "relief valve reliability" should be substituted for "relief system integrity." Improving relief system integrity has led us to examine the failure mechanisms in order to identify the improvements needed.

In this Section we will provide further examples to those in the cited references to show what failures occur and how they can be corrected. It should be emphasized that failure of a relief valve is usually due to a combination of factors, some of which will be dynamic and some static. In the author's experience it is rare for a failure to be attributable to only one fault. This would imply a gross error, usually in design. while many reporters tend to simplify the problems of relief valve installation, our investigation takes us into more complex areas.

Dynamic conditions (relief valve operation) are not common on process plants and if systems perform as intended usually they are not investigated. Who asks, "Why did that relief valve work so well?" Even if they do ask, "What caused the relief valve to operate?"

So it is difficult to obtain good operational data on the performance of relief systems. Most frequently problems are only manifest by catastrophic failure, which is very rare, or through "near miss" situations reported only within operating organizations.

In threading some examples together we wish to show how relief valve reliability is linked to safety, environmental performance and operational costs.

3.2.2. Design

The design of relief systems, statistically, seems quite good—there are not many catastrophic failures. But we have already shown that statistics of relief valve set pressure while not so useful for determining inspection intervals do indicate trends very well. One conclusion we can draw is that although designs rarely contain gross errors there are many smaller ones which lead to relief valve unreliability, thus threatening safety, environmental performance and plant operation costs.

Recently, the author's own company undertook design verification of relief systems on behalf of a customer. The verification process included:

- adequacy of documentation
- code compliance
- suitability of relief systems for the conditions described.

The design had been undertaken by a major contractor and comprised more than 200 relief systems.

- Over 25% had inadequate documentation—the purpose of the design was unclear.
- More than 10% did not comply with the relevant codes. In particular a substantial number of these had relief valves set at pressures greater than design.

- About 15% of relief streams had to be referred back to the designer because suitability for the duty was in doubt.

Typical faults included

- relief valves under/oversized, likely to have inadequate capacity or would chatter
- piping undersized
- methods of calculating required relief rate unsafe
- relief valve specification not adequate for the duty
- aspects of piping design, including safe discharge, not considered.

Frequently there was more than one fault in a design. The similarity with other reports is striking.

Design verification and site audits covering an average of 2000 relief systems a year for very many years have shown figures consistent with those reported by Kreder (18). The author has recorded less statistical data with respect to errors in sizing and the general performance of relief hardware, the emphasis being on individual cases. However, the overall impression is of the accuracy of Kreder's report. Figure 6 shows a typical failure of design found at one installation.

FIGURE 6. Example of Bad Design. Note: Relief valves mounted horizontally (liquefied gas duty) and long inlet piping

In a presentation in August 1996 to a meeting of the Engineering Equipment Materials and Users Association Dr Paul Murphy, of the UK Government agency the Health and Safety Executive, Pressure Systems Group, information based upon HSE records of investigated reportable incidents was provided.

Between 1986 and 1996 the following causes of failures were identified:

Incorrect set pressure of relief/safety valve	32.5%
No relief/safety valve fitted	38.5%
Blocked or stuck relief/safety valve (shut or open)	14.9%
Badly positioned outlet from relief/safety valve	8.8%
Undersized relief/safety valve for flow rate	5.3%

Taken from different angles, the above sets of records support the view that current design methods are not good enough and that changes in methods ought to be important in improving relief valve integrity. The cases that follow suggest areas where these improvements could take place.

3.2.3. *Installation*

There is inevitably an overlap between design and installation and relief system management. Supporting of relief systems, vibration and corrosion or product blockage are the most common problem areas. In many cases, and in nearly all those where there are serious problems, there are complicating factors such as those described here.

Case 1

A relief system was required as part of a major contract. It was designed to protect equipment downstream of a steam pressure control valve. The relief valve was set at about 20 bar.

Initially, there were to be two relief valves operating as a composite system to provide the required relief capacity. Piping design, including the stress analysis and supporting of the discharge piping from the relief valves, was completed.

At an advanced stage it was decided to fit only one, larger, relief valve, 4″ × 6.″ There was no review of the initial design. The system was still fundamentally safe. It was

- code compliant
- had adequate relief capacity
- could not be isolated from the protected equipment
- relief valve specification was acceptable
- pressure drops in the piping were within acceptable limits
- pipe supporting would have been adequate for the design as intended.

During commissioning the relief valve operated and the reaction force displaced the relief valve discharge piping causing failure at the relief valve branch. It was subsequently found that:

- Rreaction forces were substantially greater than the original design.
- Supports were not installed as the original design.
- The piping loads imposed by the original design were outside code limits. However, they would not have led to complete failure.

While the supporting was not as designed and calculations showed there would not have been failure of the piping, loads on the relief valve were greater than recommended. Valve guides have seized or other failures have been observed in similar circumstances.

Case 2

A plant engineer established that the relief valve protecting a compressor was undersized (a common problem!). It was decided to replace it with a larger one. However, the larger capacity valve was heavier and the effect of the increased mass was not recognized in the desire to make an unsafe system "safer." Very soon after fitting the relief valve "fell off" when the connection at the aftercooler failed.

- There were no supports for the new system
- The effect of the change of center of gravity caused by fitting the larger valve and discharge piping was not taken into account
- The significance of the change was not appreciated.

In this acute example there are telltales of more common and chronic ones. Vibration is a significant contribution to unreliability of relief valves and it is very hard to anticipate its full effects at the design stage. Figure 5 shows one example of a relief valve on compressor service. More commonly, failures occur around the spindle and guides of relief valves where they absorb the vibration energy. Such failures are taken for granted because it is accepted that vibration produces these results. A thorough assessment may indicate the need for a complete redesign. Compressor manufacturers, and similar vendors of machinery, do not see this as their business and plant engineers rarely have the time or resources to do such work.

Cases 1 and 2 indicate the need for control of change on plants and in construction management. The consequences may not always be as dramatic as these and unreliability of relief valves not appear to be the main issue, but is consequential of others; the true nature of the problem is revealed if the circumstances relating to relief valve or relief system failure are examined.

Case 3

In 2.4.5 we described how relief valves on a hot oil system were modified and there were benefits in safety and substantial savings in plant costs.

Figure7 shows the arrangement of a typical system and how it was modified.

Complex valve arrangements to provide routing to various plant items meant there was potential for pressure system failure if isolation valves were

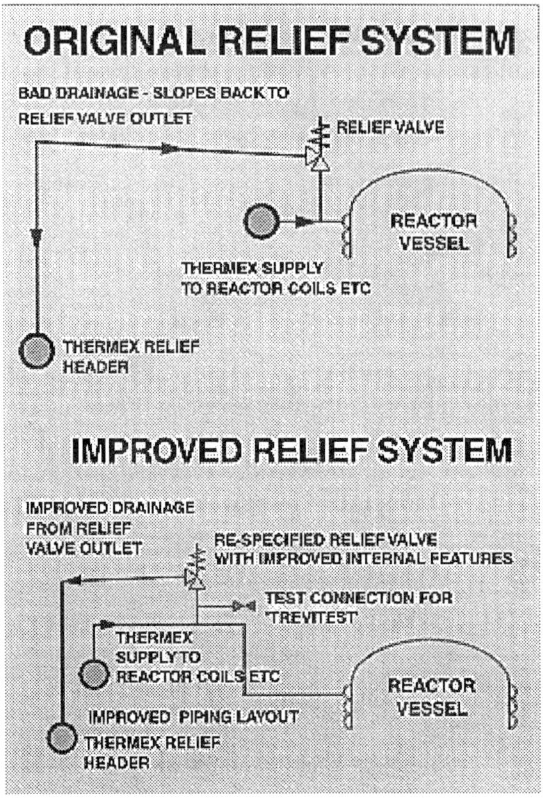

Figure 7. Hot Oil System (diagramatic). Note: Isolation valves not shown for clarity

shut and the system full of oil with heat continuing to be applied. Relief valves set at approximately 10 bar were installed at a dozen or so points in the system to provide protection against excessive pressure caused by the locked in thermal conditions which could rise to 200°C.

Unreliable operation of the relief valves, including blockage had caused inspection intervals to be steadily reduced over several years to a three month period with consequent plant shut downs.

Attempts were made to "design-out" the relief systems, but this was not possible without severe operational penalties. The relief valve specification was found to be unsuitable. There was potential for fluid build-up in the valve body that would flood the guides and valve seat. Site inspection also revealed that the relief valve discharge piping into the vent header would not drain because of thermal movements in the vent header.

The whole piping system was redesigned including connections to the relief valve inlet. The relief valves were replaced by a different design. "Trevitest" fingerprinting and monitoring were also instituted.

Failure to appreciate that piping "moves" and that it can affect relief valve reliability is common. Also, gumming up or blockage or corrosion due to "dead leg" conditions (and sometimes to changes in environment when a relief valve is removed) are often important factors in reliability.

Case 4

A steam relief system vented to atmosphere (see Fig. 8). The piping route included a horizontal section. To ensure any condensate could drain away a hole was made in the piping in the straight section of pipe. The tail pipe was anchored as shown. Thermal expansion caused by leaking condensate changed the piping geometry and prevented drainage. When there was a relief demand a slug of liquid was propelled through the piping system and the forces were such that failure of the relief valve branch at the steam main occurred.

Mention has already been made of the tendency for steam relief valves to leak. There is frequent evidence of this in relief valves taken from service. Inspection of piping will often show signs of significant change in piping configuration.

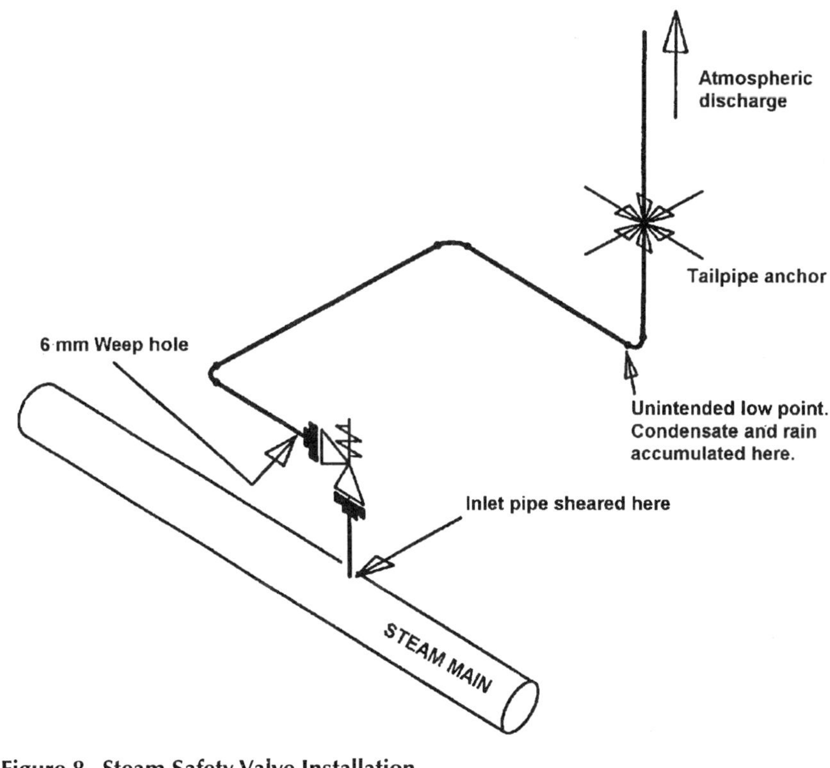

Figure 8. Steam Safety Valve Installation

3.2.4. Operation

Freezing up of relief valves when relieving has caused a number of serious accidents in which there has been loss of containment from pressure systems. Parry (13) records three events in which icing was a significant event. As with vibration, design of relief systems and specification of relief valves to avoid icing is difficult with some fluids.

Designs need to minimize the possibility of leakage across the seat of the relief valve and ensure plant conditions do not interfere with the valve's performance. An important safeguard maybe to install a changeover system to ensure a standby relief valve is available.

Case 5

Reference (19) describes the failure of a carbon dioxide storage vessel in 1988. The actual causes of vessel failure are not our concern here. However, the relief valve did not function correctly. It is believed the circumstances were as follows:

- The relief valve position on the vessel was modified late in the design process.
- Continuous heat input placed a demand on the relief system.
- The relief valve failed to operate.
- There was an internal dip pipe which probably led to freezing of the relief valve seat.
- It was known that in the past, plugs of ice had formed within relief valves on similar service due to inadequate drainage.

The lessons drawn in the report are:

- Relief valves should never be connected to a standpipe that goes through the liquid CO_2.
- Relief valves should be installed so that seats or springs cannot be rendered inoperative due to freezing.

Case 6

A relief valve was called upon to operate due to a thermal relief demand in a liquid ethylene piping system. The valve had a three quarter inch inlet and one inch outlet. Set pressure was approximately 30 bar.

The relief valve froze in the open position and ethylene was vented locally to the atmosphere. A major incident was prevented because an operator recognizing the danger changed the relief valve over to a standby system.

Cases 5 and 6 raise a number of points.

1. Drainage of piping is important and reference has already been made to it. It can be an issue in both hot and cold situations.

2. Seat leakage is also a common mode of failure and where there is potential for auto-refrigeration, there will always be a tendency for relief valves to freeze.
3. The maintenance of seat integrity is essential and potential failure of relief valve seats should be considered when determining the integrity of relief systems.
4. Ideally a relief valve should have a minimum 'simmer' period if freezing is possible.

Case 7

A distillation section that was part of an organics processing unit was protected from excessive pressure by a relief valve that discharged into an absorption system. The relief valve was 6″ × 8″ and set at about 25bar. The system operated at about 170°C.

Authority was given for pressure drops upstream of the relief valve of 10% of the set pressure. (Later recalculated at 8%.) The following reasons were given for authorization:

- There was little scope for substantial redesign of the piping.
- The frequency of demand on the relief system was considered low, being based upon control system malfunction which had a low probability of failure.
- Pressure drops downstream of the relief valve were calculated to be within acceptable limits and the valve had a pressure balanced bellows.
- Some chatter was considered acceptable and it was thought there was sufficient opportunity for operator intervention.
- Relief valve lift had been restricted in order to reduce relief valve capacity and minimize pressure drops in the system.

On two occasions the relief valve was called upon to operate not long after the plant was commissioned. The second time chatter was severe and the relief valve badly damaged and there was concern for the integrity of the pressure system.

Following the second event, it was further found that:

- The demand on the relief system was much more frequent than expected because the control system was less predictable than first thought.
- The relief valve had been open for the equivalent in capacity terms of 45 seconds.
- The fluid flowing during relief did not have ideal gas physical properties; therefore flows, pressure drops etc. were not as originally calculated and there were several chokes in the downstream piping not previously recognized.

- The relief valve selected had no surface treatment to the guide and seat

Ignoring that of plant control, some issues require further consideration:

- Kreder (18) shows that relief systems are not always designed for two-phase relief when it is necessary to do so.
- Physical properties of process fluids are not always predictable.
- Surface treatment of guides, seats and valve disk holders etc. varies greatly between relief valve manufacturers. Is cost an issue? Since chatter is an unavoidable part of pressure relief, should not relief valves be designed to cope with it?

A number of points have been made in this paper regarding the stability of relief valves. It is worth recalling those that relate to this case:

1. Some process fluids have characteristics much like steam. The fluid in Case 7 is one of them in terms of its effects on the tendency for the relief valve to chatter.
2. Galling and "pick-up" from bearing surfaces tends to increase at higher temperatures so the damage done during chatter may be greater at higher temperatures.
3. Where chatter is anticipated, blowdown can be extended to reduce its effects either by increasing the "parachute" shape around the relief valve nozzle and seat (the "huddling chamber"), fitting a softer spring so that the popping action of the valve is less violent, or ensuring the relief valve trim is suitable for the duty.

3.3. Process versus Mechanical Design

Grolmes et al. (1), consider the relative priority of providing process data, required relief rate, fluid properties etc. compared with the risk of hardware failure. Kreder (18), in 1997 provides statistical information supporting this approach and across a wider spectrum of industry than either Grolmes or Marrs et al. (20), whose paper was published in 1989.

In our case studies, and other examples given here, we have tried to show how process and mechanical engineering data are closely linked. The failures we see in relief valves lifting "heavy" or "light" are often evidence of inadequate process design as well as mechanical design. Indeed it is dangerous to discriminate between disciplines in a branch of engineering that should integrate these skills. Unless we take this view, the risk is that there will always be significant unreliability of relief valves and there will always be failures of pressure systems.

It is not a valid conclusion, in our view, to draw from the data in these references that there is some priority in one aspect of design over another. It is evident that some mechanical failures have process engineering origins. It is also obvious that process design without mechanical competence will end in failure. All activities require knowledgeable and consistent process and engineering management when a plant is operational.

3.4. Summary of System Design

The interaction of components in relief systems is a complex matter. There are considerable safety factors intentionally or inadvertently built into systems which assist relief valve reliability and ensure that pressure systems are not frequently subjected to periods of excessive pressure. However, even though designs may be in accordance with recognized good practice and are well maintained, failures occur. Overdesign (too high a factor of safety) may be a contributory factor. Failure of system integrity can be seen in the histories of individual relief valves and in the general trends revealed by thoughtfully compiled statistics. Many of the problems arise from a combination of factors. In general we can say failure of relief valves to provide the necessary integrity arises from:

- the nature of the process and general plant conditions (corrosion etc.)
- inherent weaknesses in the detail design of relief valves
- failure of designers (engineers of all disciplines) to understand the technology of pressure relief, including safety factor effects
- inability of many plant operators to recognize the link between overall relief valve performance and the consequences for safety, environment and plant running costs.

All these features can be improved. Individual records do not need to be analyzed in detail, as Parry fears, but a combination of actions appropriate to each one is necessary. In case studies and examples used in this paper we have tried to show what some of these actions might be.

4. THE RELIABILITY OF RELIEF VALVES—SOME CONCLUSIONS

When a problem with a relief system occurs, the most frequent request we receive is, "We have a relief valve that's failed. Can you please help?"

Commentators, including Bordelon (21) and Scully, have noted that relief valve reliability and relief system performance are closely related. But it is not often in the minds of engineers.

The starting point for many relief valve reliability problems is design.

Relief valves, from their service histories, seem to be designed and models certified by codes without consideration for their operational life, which might include variations in relief rate. Springs, bellows, seats and guides are all known to be prone to failure. Yet the design of these components is generally outside the control of the users. Relief valve manufacturers who troubleshoot relief valves do not seem to reflect their experience in the product.

Engineering design of relief systems is not without fault. Many engineering design contractors and plant operations' engineering departments undertake relief system design in a series of functional steps involving process engineers, vessel and piping engineers and instrument engineers. Hazard and reliability study specialists may be involved, but few organizations have Pressure Relief specialists. There is rarely the opportunity to develop them. while many acknowledge the importance of the topic, the impact on plant safety, environmental performance and running costs has not been fully appreciated. It is thus easier to have in place elaborate design management and plant management systems. Who has conducted an audit of pressure relief costs and returns in terms of safety, environment, design and plant operation? Without these data, no sensible value can be accorded to the technology.

In this paper we have tried to show that the performance and reliability of relief valves needs to be considered in the context of their part in the pressure system in which they function.

If extended run times for plants and more cost effective operation of process plants is to be reconciled with greater safety and environmental performance, then relief valve reliability monitoring needs to be directed towards identifying and resolving inherent problems in the design and selection of relief systems, rather than perpetuating them through activities such as changing inspection intervals.

We regard the present collection of statistics in industry with mistrust and have tried to show why in this paper. Policies of companies, code organizations and legislatures should not be influenced by current raw data. Indeed, our impression is that too much information on relief system operation is anecdotal and inadequately researched.

If the collection of relief valve reliability data is important, Industry needs to agree why and how it should be best used. Since data are being collected and used to revise inspection intervals, should it not also be directed, systematically, towards a proper understanding of pressure relief technology as a whole?

References

1. M. A. Grolmes, J. Tabor, M. Kenton, and R. Wachowiak. A risk perspective for emergency pressure relief system design. *Journal of Hazardous Materials,* 44 (1995).
2. W. Woolfolk, and R. Sanders. Process Safety Relief Valve Testing. *Chemical Engineering Processing,* March 1984.

3. R. J. Aird. Reliability Assessment of Safety/Relief Valves. *Trans I Chem E*, 60, 1982.
4. A. B. Smith. Safety Valve Reliability. Paper presented at Loss Prevention and Safety Promotion in the Process Industries, 8th International Symposium, 1995.
5. R. D. MacMahon. Safety Valve Reliability. Paper presented at the Pressure Vessel Technology Conference, Sydney, 1985.
6. W. E. Nelson. Improve relief valve reliability. *Hydrocarbon Processing*, January 1993.
7. G. T. Zirps, K. R. Jones, R. B. Gleason, and S. J. Overton. Improving plant availability and reliability with pressure relief devices. Paper presented at the ASME Joint International Power Generation Conference; Phoenix, AZ; October1994.
8. G. L Weston. "The Design and Assembly of a High Pressure Test Rig for the Investigation of the Set Press.ure of Safety Relief Valves." Manchester Metropolitan University, student thesis; 1991.
9. M. Waters. "Investigation into Specification for Helical Compression Springs." Liverpool Polytechnic, student engineering project report; 1989.
10. M. Waters. "Reliability Study—Pressure Relief Valves on Low Pressure Steam Duty." ICI Engineering—internal report; 1989.
11. SAFed. *Pressure Systems; Guidelines on Periodicity of Examinations*. Safety Assessment Federation; 1997.
12. *EEMUA Guide to Operating Periods for Safety Relief Valves*. Engineering Equipment and Materials Users Association.
13. C. E. Parry. *Relief Systems Handbook*. Institution of Chemical Engineers, pp. 94–109 and 193–196
14. W. A Scully. Safety-Relief Valve Malfunctions: Symptoms, Causes and Cures. *Chemical Engineering*, August 10, 1981.
15. G. Emerson. Improve Safety with Pilot Operated Relief Valves. *Hydrocarbon Processing*, October 1996.
16. R. M. Baldwin and H. R. Simmons. Flow-Induced Vibration in Safety Relief Valves. *Journal of Pressure Vessel Technology*, 108, August 1986
17. S. C. Weighell. The Performance of Relief Valves—An Equipment Users View. *I Chem E Symposium Series*. No 60, 1980.
18. R. Kreder. "Summary and Analysis of Findings from Pressure Relief System Verification Efforts." Paper presented at the Design Institute for Emergency Relief Systems Users Group meeting, 14–16 April 1997.
19. W. E. Clayton and M. L. Griffin. Catastrophic Failure of a Liquid Carbon Dioxide Storage Vessel. *Loss Prevention Bulletin* 125.
20. G. P. Marrs, F. P. Lees, J. Barton, and N. Scully. Overpressure Protection of Batch Chemical Reactors. *Chemical Engineering Res Des*, 67, July 1989.
21. T. R. Bordelon. Ten Steps to Upgrade Safety-Valve Choice, Maintenance. *Power*, October 1983.

Sizing of Pressure Relief Devices and Associated Piping for Multiphase Flows

*Joseph C. Leung**

Leung Inc., 1521 Evergreen Lane, Darien, Illinois 60561

1. INTRODUCTION

This paper provides a review of a unified methodology for sizing relief devices—both rupture disks (RDs) and safety relief valves (SRVs)—and associated relief piping for two-phase flow. The particular treatment has been described in a number of earlier publications [1,2,3]. For this review, it suffices to summarize a few key developments. Examples are used to illustrate the use of this method in subcooled inlet condition, multi-segment relief piping and reaction pipe force estimation.

2. SIZING CRITERION

Sizing of a relief device to protect a process equipment depends on the upset contingency chosen [4]. Frequently a required vent rate (W_{req}) is calculated that will prevent the system pressure from rising above an allowable accumulation, typically 10% of the MAWP of the equipment. At the peak accumulated pressure, the vent sizing criterion invariably reduces to a balance between the volumetric generation rate Q_{gen} in the process equipment and the volumetric discharge rate. In equation form, the required vent rate is given by

$$W_{req} = \frac{Q_{gen}}{v} \tag{1}$$

where v is the specific volume of the discharge fluid, while noting $W_{req}v$ is the volumetric discharge rate. Knowing the conditions of the discharge fluid will allow us to calculate the fluid density (ρ) and the specific volume ($v = 1/\rho$) as

*Consultant, Fauske & Associates, Inc., 16W070 West 83rd Street, Burr Ridge, IL 60521.

well as the discharge mass flux (G). Then the minimum vent area (A_{min}) can be evaluated as:

$$A_{min} = \frac{W_{req}}{G} \tag{2}$$

A method for evaluating G through a nozzle is hence required. As discussed in an earlier paper [5],

$$G = \frac{\left[\int_{P_0}^{P} -v\,dP \right]^{1/2}}{v} \tag{3}$$

where the integration is taken from the stagnation pressure P_0 to the throat pressure P. Using an expansion law for a general two-phase system, Eq. (3) can be solved analytically. The proposed expansion law takes the form of [5]

$$\frac{v}{v_0} = \omega\left(\frac{P_0}{P} - 1\right) + 1 \tag{4}$$

where the A parameter in the most updated form is given by [3,6]

$$\omega = \alpha_0\left(1 - 2\frac{P_0 v_{fgo}}{h_{fgo}}\right) + \frac{C_p T_0 P_0}{v_0}\left(\frac{v_{fgo}}{h_{fgo}}\right)^2 \tag{5}$$

for flashing flow system, and

$$\omega = \frac{\alpha_0}{k} \tag{6}$$

for nonflashing flow system.

For flashing two-phase flow systems, the second term in Eq. (5) dominates until α_0 approaches unity (all-vapor inlet). For nonflashing two-phase flow systems (such as air–water or nitrogen–oil), the second term vanishes (no phase change) and ω can be approximated by α_0/k, i.e. Eq. (6). Thus depending on the A value, we can have the following flow characterizations [7]:

Flashing flow	$\omega > 1$
Gas/vapor flow	$\omega \approx 1$
Non-flashing flow	$0 < \omega < 1$
Liquid flow	$\omega = 0 \quad (\alpha_0 = 0)$

The omega parameter as defined by Eqs. (5) (for flashing flow) is applicable when the thermodynamic reduced temperature is less than 0.9 or reduced pressure less than 0.5 [5]. Above this temperature and pressure, the omega correlation would tend to underestimate the mass flux and hence overpredict the required vent area. Reference [5] extends the omega correlations to near the thermodynamic critical point region. An alternative approach would be to carry out an enthalpic flash calculation to 70% of the inlet pressure (0.7 P_o) and determine the omega parameter via,

$$\omega = \frac{\dfrac{v_1}{v_0} - 1}{\dfrac{P_0}{P_1} - 1} \tag{7}$$

where v_1 is the specific volume at P_1 (= 0.7 P_0). Then for the nozzle case, Fig. 1 can be used to determine both the choking mass flux as well as the choked flow pressure. This method can be used also for supercritical fluids as well, again via an isenthalpic flash calculation.

3. OMEGA METHOD NOZZLE FLOW EQUATIONS

Upon substituting Eq. (4) into Eq. (3) and carrying out the integration, a generalized expression for flow through a perfect nozzle can be given by

$$\frac{G}{\sqrt{P_0/v_0}} = \frac{G}{\sqrt{P_0/\rho_0}} = \frac{\left\{-2\left[\omega \ln \dfrac{P}{P_0} + (\omega - 1)\left(1 - \dfrac{P}{P_0}\right)\right]\right\}^{0.5}}{\omega\left(\dfrac{P_0}{P} - 1\right) + 1} \tag{8}$$

Choking or critical flow condition as defined by the maximum value of G of the above equation yields the following (transcendental) equation for the critical pressure ratio P_c/P_0

$$\left(\frac{P_c}{P_0}\right)^2 + \left(\omega^2 - 2\omega\right)\left(1 - \frac{P_c}{P_0}\right) + 2\omega^2 \ln\frac{P_c}{P_0} + 2\omega^2\left(1 - \frac{P_c}{P_0}\right) = 0 \tag{9}$$

When P_c/P_0 is substituted for P/P_0 in Eq. (8), the critical mass flux G_c is obtained; but a much simpler formula for G_c is given by

$$\frac{G}{\sqrt{P_0/v_0}} = \frac{P_c/P_0}{\sqrt{\omega}} \tag{10}$$

Note that Eq. (8) is a more general formula, valid also in the subcritical flow condition. Thus if $P_c > P_b$, flow is choked and Eqs. (9) and (10) apply. Otherwise ($P_c < P_b$) flow is subcritical (or unchoked), and P is equated to P_b in Eq. (8) to obtain the unchoked mass flux G.

The above choked flow solutions can be conveniently represented by Fig. 1, covering the entire two-phase flow spectrum—flashing flow lies to the right of $\omega = 1$ and nonflashing flow lies to the left of $\omega = 1$. At $\omega = 1$, the present graph recovers the classical gas nozzle flow ($k = 1.001$) result. It can also be demonstrated by setting $\omega = 1$ in Eq. (8),

$$\frac{G}{\sqrt{P_0 \rho_{g0}}} = \frac{P}{P_0}\left(-2\ln\frac{P}{P_0}\right)^{0.5} \tag{11}$$

which yields a maximum at

$$\frac{P_c}{P_0} = 0.606 \tag{12}$$

$$\frac{G}{\sqrt{P_0 \rho_{g0}}} = 0.606 \tag{13}$$

which agrees with API formula [4] for $k = 1.001$. At the other limit when $\omega = 0$, i.e. $\alpha_0 = 0$ or nonflashing liquid flow, Eq. (8) reduces to

$$\frac{G}{\sqrt{P_0 \rho_{g0}}} = \left[2\left(1 - \frac{P}{P_0}\right)\right]^{0.5} \tag{14}$$

which is the classical Bernoulli's equation for incompressible flow and is also the underlying formula for the API SRV sizing equation for liquid service.

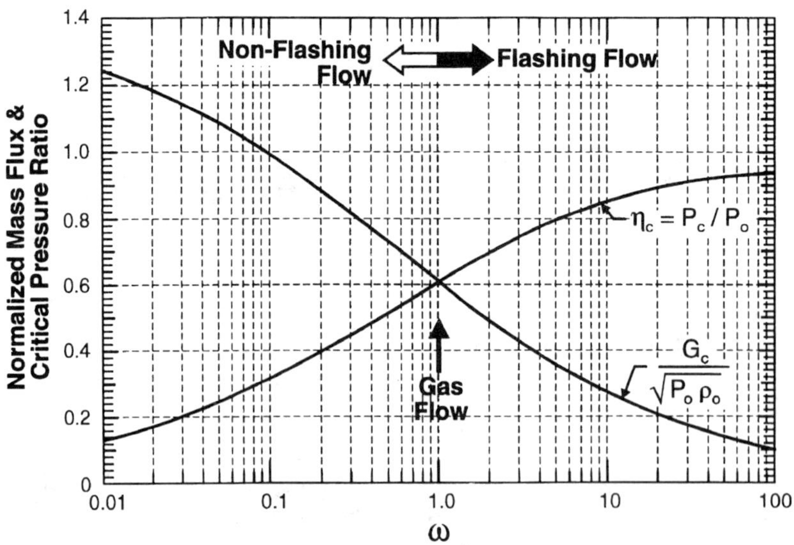

Figure 1. Generalized Chart for Two-Phase Flow Through a Nozzle.

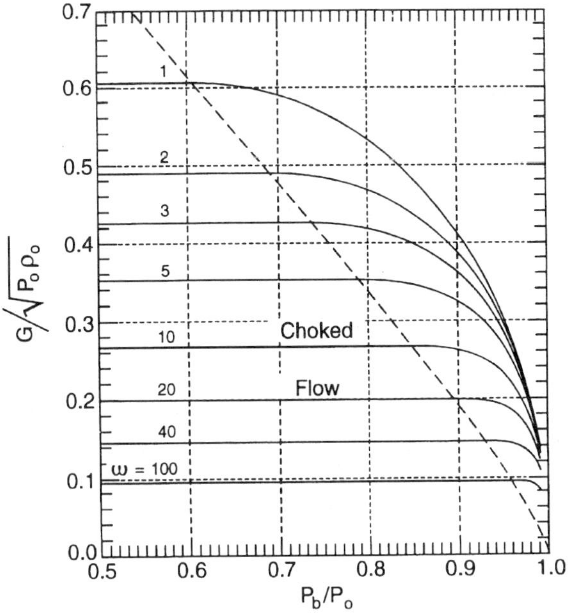

Figure 2. Effect of Back Pressure on Flashing Flow Through a Nozzle.

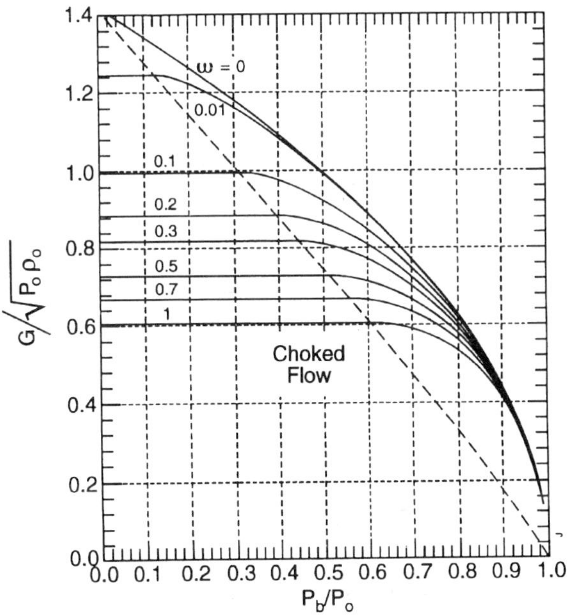

Figure 3. Effect of Back Pressure on Nonflashing Flow Through a Nozzle.

Figures 2 and 3 illustrate the solutions of Eqs. (8) and (9) for flashing flow and nonflashing flow, respectively, and in particular, show the effect of back pressure. The locus of the choking conditions is illustrated in these figures as the dashed lines.

4. OMEGA METHOD PIPE FLOW EQUATIONS

Pipe flow is, as expected, more complex than nozzle flow. Here the pressure drop is due to acceleration (sometimes also known as momentum), friction in the pipe, and gravity (elevation change). The last two components are not encountered in nozzle flow. The following discussion will be limited to the turbulent regime where the friction factor f can be assumed to be constant or some suitable average value. For two-phase flow application, a Fanning turbulent friction factor of 0.005 is adequate for most cases. The ω method reveals three dimensionless parameters governing such turbulent two-phase pipe flow [8,9];

ω = same definition as for nozzle flow

$$4f\frac{L}{D} = \left(4f\frac{L}{D}\right)_{\substack{\text{straight}\\\text{pipe}}} + \sum K_i \tag{15}$$

$$Fi = \frac{\rho_0 g L\cos\theta}{\left(4f\dfrac{L}{D}\right)P_0} = \frac{\rho_0 g H}{\left(4f\dfrac{L}{D}\right)P_0} \tag{16}$$

The fitting loss coefficients K_i are to be included in the total piping resistance factor $4f(L/D)$ as shown. Here a so-called "flow inclination" number Fi is introduced for two-phase flow with H (or $L\cos I$) denoting the elevation rise relative to the pipe inlet. Horizontal pipe flow yields $Fi = 0$ and upflow gives a positive Fi number. So, in pipe discharge the mass flux can be uniquely characterized by the following:

$$\frac{G}{\sqrt{P_0\rho_0}} = \text{function}\left(\omega, 4f\,L/D, Fi, P_b/P_0\right) \tag{17}$$

5. HORIZONTAL PIPE DISCHARGE

For convenience the following dimensionless variables are defined:

$$\eta_1 = P_1/P_0, \qquad \eta_2 = P_2/P_0, \qquad \eta_b = P_b/P_0, \qquad G^* \equiv G/\sqrt{P_0\rho_0}$$

Thus for a given $4fL/D$ and η_b, we have three unknown parameters G^*, η_1, and η_2. Hence we need three equations which are [8]

$$G = \frac{\{-2[\omega \ln \eta_1 + (\omega - 1)(1 - \eta_1)]\}^{1/2}}{\omega\left(\frac{1}{\eta_1} - 1\right) + 1} \tag{18}$$

$$4f\frac{L}{D} = \frac{2}{G^{*2}}\left[\frac{\eta_1 - \eta_2}{1-\omega} + \frac{\omega}{(1-\omega)^2}\ln\frac{(1-\omega)\eta_2 + \omega}{(1-\omega)\eta_1 + \omega}\right] - 2\ln\left[\frac{(1-\omega)\eta_2 + \omega}{(1-\omega)\eta_1 + \omega}\left(\frac{\eta_1}{\eta_2}\right)\right] \tag{19}$$

$$G_c^* = \frac{\eta_{2c}}{\sqrt{\omega}} \text{ (for the exit choking condition)} \tag{20}$$

Note that the integrated momentum equation, Eq. (19), yields the pressure drop $(P_1 - P_2)$ in the pipe implicitly as a function of $4fL/D$ and mass flux G. Also, relating the inlet pressure drop, Eq. (18) is the same formula as the nozzle flow equation, Eq. (8). Essentially, the inlet pipe section is treated as an ideal nozzle and any irreversible loss due to the sharp entrance effect is then accounted for in the overall $4fL/D$ term. Finally, the exit choking criterion, Eq. (20), is similar to Eq. (10) for the nozzle flow, except that the pressure in Eq. (20) corresponds to the pipe exit pressure. A subscript c is used to denote choking at the pipe exit.

Choking Exit: $\eta_{2c} > \eta_b$ or $P_{2c} > P_b$
Choking is implied and Eq. (20) is valid. Eqs. (18), (19), and (20) are used to solve for G_c^*, η_1, η_2 (likewise G_c, P_1, P_2) for a given 4fL/D and A.

Unchoked Exit: $\eta_{2c} > \eta_b$ or $P_{2c} > P_b$
No choking occurs, Eq. (20) is not valid. Only Eqs. (18) and (19) are needed to solve for G^* and η_1, since $\eta_2 = \eta_b$ (or $P_2 = P_b$) for unchoked exit. Under most situations it suffices to set $\eta_1 = 1$ and solve for G^* immediately by rearranging Eq. (19).

6. INCLINED PIPE DISCHARGE

The inclined pipe discharge case has identical inlet and exit relationships, i.e. Eqs. (18) and (20). However, the integral momentum equation before integration is written as

$$4f\frac{L}{D} = -\int_{\eta_2}^{\eta_1} \frac{[(1-\omega)\eta^2 + \omega\eta]\left(1 - G^{*2}\frac{\omega}{\eta^2}\right)}{\frac{G^{*2}}{2}[(1-\omega)\eta + \omega]^2 + \eta^2 Fi} \tag{21}$$

Admittedly, this formulation is too unwieldy to present here as the closed-form solution of this integral; the interested reader can consult Refs. [1] or [9].

7. DESIGN CHARTS FOR PIPE DISCHARGE

The exit choking solutions can be most conveniently presented in design-chart format, similar to Shapiro's chart for ideal gas flow [10]. Here the mass flux relative to that of a perfect nozzle as denoted by G_c/G_{0c} is plotted versus $4fL/D$ for ω ranging from 0 to 100, that is, covering both flashing and non-flashing flow. Figures 4, 5, and 6 present such design charts for *Fi* values of 0, 0.1, and 0.2, respectively. Figure 7 presents the choking exit pressure ratio P_{2c}/P_0 as a function of G_c/G_{0c} and A. However, it can be shown that critical flow pressure ratios between the pipe flow and the nozzle flow are related via

$$\frac{P_{2c}}{P_0} = \left(\frac{P_c}{P_0}\right)\left(\frac{G_c}{G_{0c}}\right) \tag{22}$$

For unchoked flow, Fig. 8 provides a design chart for estimating the effect of back pressure on the mass flux. Here, the ratio of unchoked exit mass flux to the critical mass flux G/G_c is shown as a function of $(P_0 - P_b)/(P_0 - P_{2c})$ where P_{2c} is the choking exit pressure. Although this chart was developed for the nozzle case, it is found generally to be adequate for pipe discharges also.

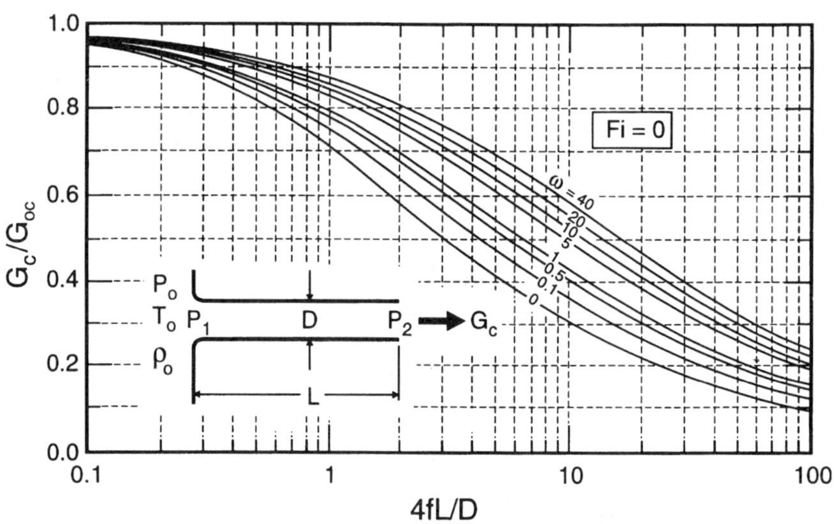

Figure 4. Critical Flow Discharge from a Pipe in the Turbulent Regime with *Fi* = 0 (Horizontal).

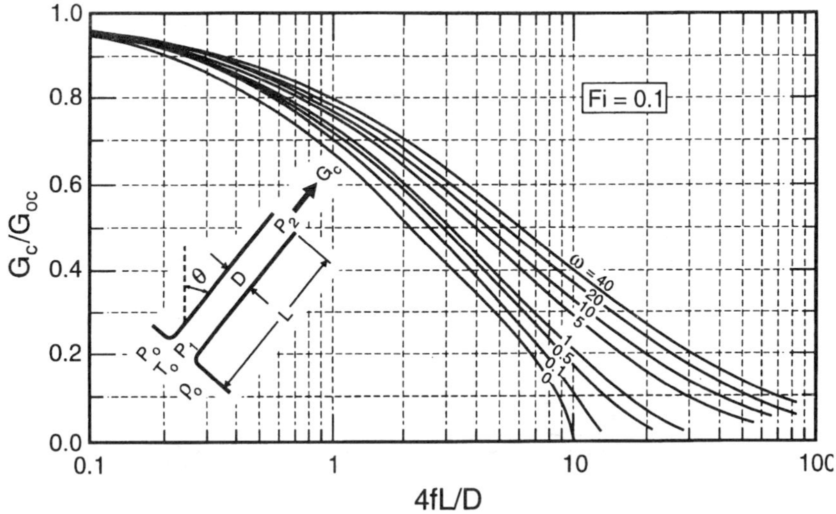

Figure 5. Critical Flow Discharge from a Pipe in the Turbulent Regime with *Fi* = 0.1.

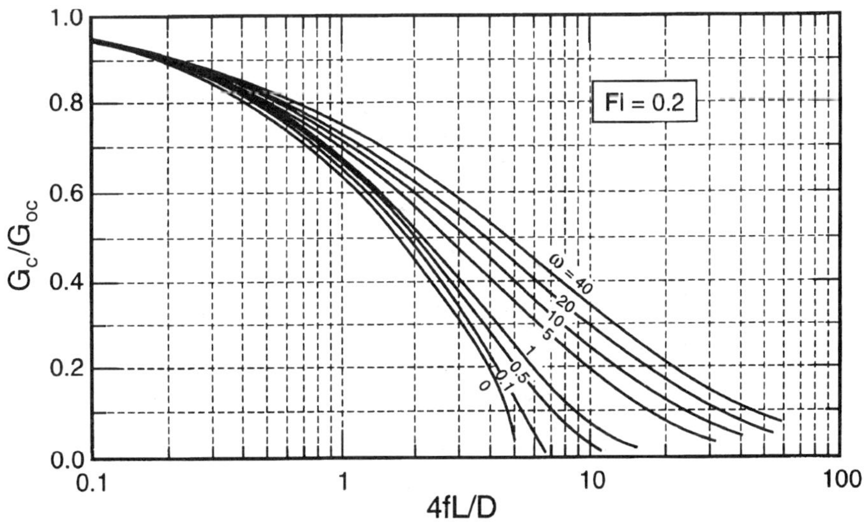

Figure 6. Critical Flow Discharge from a Pipe in the Turbulent Regime with *Fi* = 0.2.

Since these charts were developed for turbulent two-phase pipe flow, they should not be used for viscous flow situations where the friction factor is no longer invariant with the pipe flow Reynolds number.

The solution given by $\omega = 1$ in Fig. 4 is in perfect agreement with the traditional "isothermal" gas flow approximation [11]. This can be demon-

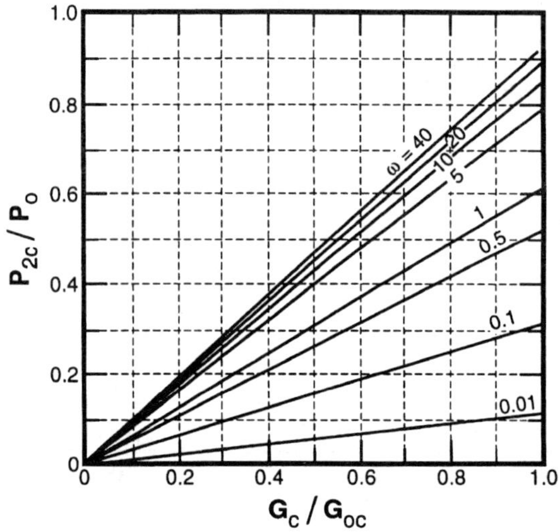

Figure 7. Exit Choking Pressure Ratio in Turbulent Flow Pipe Discharge.

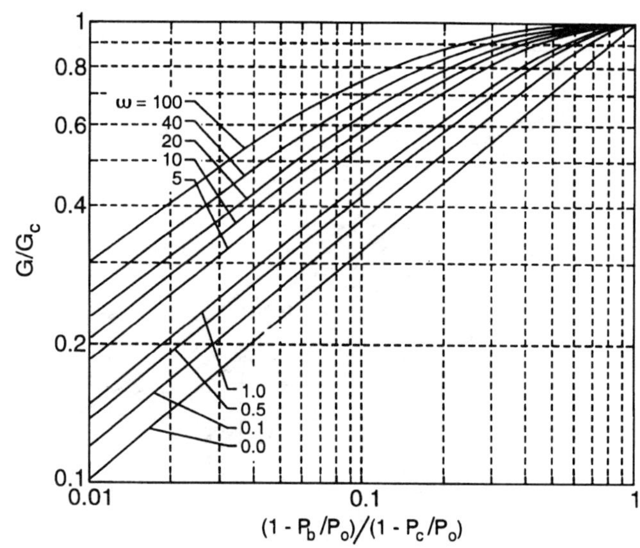

Figure 8. Effect of Back Pressure on Nozzle and Pipe Discharges.

strated also by setting $\omega = 1.001$ in Eq. (19) and expanding the first logarithmic term using Taylor's series, resulting in

$$4f\frac{L}{D} = \frac{P_0\rho_0}{G^2}\left[\left(\frac{P_1}{P_0}\right)^2 - \left(\frac{P_2}{P_0}\right)^2\right] - 2\ln\left(\frac{P_1}{P_2}\right) \tag{23}$$

while noting $P_0/\rho_0 = P_1/\rho_1$. It is further noted that at $\omega = 0$, the solutions are in agreement with the single-phase incompressible pipe flow results

$$4f\frac{L}{D} = \frac{2\rho_{f0}}{G^2}(P_1 - P_2) \tag{24}$$

8. SUBCOOLED LIQUID INLET

Flashing discharge of an inlet subcooled liquid requires special treatment [12,13]. The resulting flow rate can be substantially higher than the saturated liquid inlet case at the same inlet pressure. In this treatment, the first term in the A parameter vanishes and a "saturated" ω parameter can be defined as

$$\omega_s = \rho_{f0} C_p T_0 P_s \left(\frac{v_{fg0}}{h_{fg0}}\right)^2 \tag{25}$$

where P_s is the saturation vapor pressure corresponding to inlet temperature T_0. The two-phase expansion law now takes the form

$$\frac{v}{v_0} = \omega_s\left(\frac{P_s}{P} - 1\right) + 1 \tag{26}$$

The generalized solutions for nozzle discharge [12] are divided into low and high subcooling regions, delineated by a "transition" saturation pressure ratio (P_{st}/P_0), i.e.,

$$\eta_{st} = \frac{2\omega_s}{1 + 2\omega_s} \tag{27}$$

If the inlet subcooling is high such that $\eta_s < \eta_{st}$ (or $P_s < \eta_{st} P_0$), no vapor is formed until the exit is reached, and the above methodology simply yields a critical mass flux given by

$$\frac{G}{\sqrt{P_0 \rho_{f0}}} = [2(1 - \eta_s)]^{0.5} \tag{28a}$$

or the familiar form

$$G_c = [2\rho_{f0}(P_0 - P_s)]^{0.5} \tag{28b}$$

and the critical pressure ratio is simply

$$\eta_c = P_s / P_0 \tag{29}$$

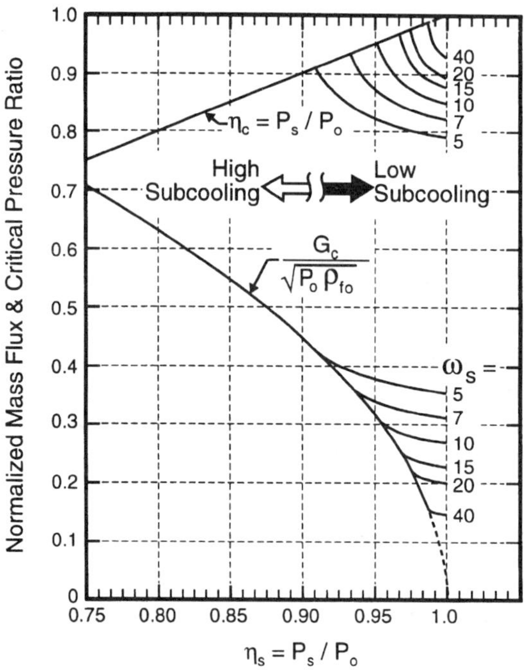

Figure 9. Nozzle Critical Flow Discharge of Inlet Subcooled Liquid.

or choking occurs at the saturation pressure corresponding to inlet temperature T_0. Figure 9 is a design chart for the nozzle flow. Consult Refs. [12] and [13] for development of additional equations for pipe discharge.

9. FURTHER CONSIDERATION

So far, the ω method has been developed primarily for one-component flashing fluids and two-component nonflashing mixtures. Limited success has been reported for multicomponent mixtures exhibiting minor vapor-phase composition change during depressurization and flashing [14,15]. Pseudo one-component physical properties can be estimated based on liquid-phase composition, except for heat of vaporization and vapor specific volume, which should be based on vapor-phase composition. An immiscible liquid system has been demonstrated to be successfully predicted by the ω method as well [16].

In a recent study [17] a total of 15 multicomponent mixtures of industrial interest were evaluated at conditions varying from saturated liquid to two-phase to almost all-vapor. Overall, the ω correlation appears adequate for reasonably narrow boiling mixtures (those with a difference in nominal atmospheric-pressure boiling point between the highest and heaviest component of less than 80°C). But for wide-boiling-range mixtures and for nonideal VLE systems, such as those containing hydrogen in particular, the ω correlation underpredicts the flow significantly. In fact, for most of the hydrogen containing mixtures, the mass flux can be closely predicted based on the purely nonflashing flow correlation ($\omega = \alpha_0$). However, even for the most nonideal systems examined, the simple expansion law Eq. (4) can describe reasonably well the fluid behavior upon depressurization. Hence, Ref. [17] suggested that the ω parameter be evaluated based on the actual flash behavior to 90% of the inlet pressure ($0.9\,P_0$), i.e. via Eq. (7) for nozzle flow (SRV) application.

It is quite common for relief devices to discharge to a common vent header and so the calculation of pressure drop in the relief piping can get tedious when multiple reliefs occur simultaneously. Essentially one has to break down the relief configuration into constant diameter segments and assign the corresponding flow rates and stream compositions. Typically one would start at the exit end of the vent header. By carrying out two flash calculations involving all the streams to two pressures—one to ambient (back) pressure and one to a higher than ambient (back) pressure—a representative ω value can be obtained, i.e.

$$\omega = \frac{\left(\dfrac{v_{22}}{v_{11}} - 1\right)}{\left(\dfrac{P_{11}}{P_{22}} - 1\right)} \tag{30}$$

Here one can use Eq. (20) to calculate the exit choking pressure ratio if indeed it is choked where $P_0 = P_{11}$ and $\rho_0 = \rho_{11}$ (or $v_0 = v_{11}$). Having established the vent header exit pressure, one can then employ Eq. (19) to calculate the pressure at the beginning of the pipe segment. Note that

$$G^* = \left(W/A_\mathrm{p}\right)/\sqrt{P_0\rho_0} \tag{31}$$

is defined based on the total flow W in the segment and the pipe flow area A_p.

The above-mentioned method should be adequate for a low pressure drop situation (about 10% of set pressure), in particular when the relief devices are SRVs. For rupture disk installations (without a SRV as the flow controlling device), the above method generally yields a satisfactory result. However, for some high pressure situations and especially for complex mixtures such as hybrid systems (i.e. containing both flashing components and noncondensible gases), the two-parameter expansion law may be necessary:

$$\frac{v}{v_0} = 1 + a\left(\frac{P_0}{P} - 1\right) + b\left(\frac{P_0}{P} - 1\right)^2 \tag{32}$$

By carrying out two flash calculations at, say, $0.9\,P_0$ and $0.5\,P_0$, the a and b parameters of Eq. (32) can be found easily. The treatment of nozzle flow and pipe flow follows closely the development presented here although the actual numerical calculation is most expediently performed by the computer. Rather than seeking an analytical closed-form solution for the integral momentum equation given by

$$4f\frac{L}{D}=\int\frac{-v\,dP-G^2v\,dv}{\frac{1}{2}G^2v^2+\frac{gD}{4f}\cos\theta} \tag{33}$$

the integration can be efficiently carried out numerically.

10. MULTISEGMENT DISCHARGE PIPE FLOW

In many industrial installations more than one diameter pipe segments are common practice. The one-segment pipe discharge model discussed above can be extended to a multisegment pipe model using the simple procedure outlined below. A four-segment pipe model is illustrated here, refer to Fig. 10 for notation:

Step Action:

1. Calculate W and P_2 with $P_3 = P_a$ (only for initial guess), i.e. based on 1st segment pipe only.
2. With W known, calculate exit pressure P_8 for the 4th (last) segment pipe.
3. Calculate inlet pressure P_7 for 4th segment.
4. Calculate exit pressure P_6 for 3rd segment.
5. Calculate inlet pressure P_5 for 3rd segment.
6. Calculate exit pressure P_4 for 2nd segment.
7. Calculate inlet pressure P_3 for 2nd segment.
8. Compare P_2 with P_3. If $P_2 > P_3$, the flow is choked. If $P_2 < P_3$, the flow is unchoked at the 1st segment pipe outlet. Use P_3 just calculated as next guess to recalculate W. Repeat Steps 1–8 until W does not change.

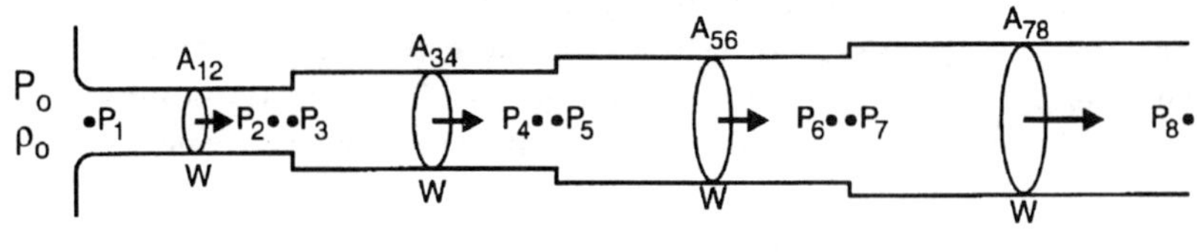

Figure 10. Notation for a Four-Segment Pipe Flow Discharge Model.

11. RD/TAILPIPE DESIGN

Here we will use an example that appears in a recent paper by Westman [18] in our illustration of a two-segment pipe calculation. Suppose a pressure vessel is to be protected by a rupture disk (RD) through which 24,000 kg/h of HCl gas must be relieved in the event of a process upset. The vessel's MAWP is 55.1 psig (3.8 bar-g) and a 10% accumulation is permitted. A 4-inch diameter RD is installed 4 ft downstream of the vessel outlet nozzle. The system is vented downstream of the RD by a 127 ft pipe length containing three short-radius elbows. The total equivalent pipe is 197 ft and the total equivalent K or $4fL/D$ value is 9.57. Table 1 is an output of the RDPIPE program [19] which utilizes the methodology discussed in this paper. The maximum allowable equivalent pipe run is 0.3 ft (excluding the entrance loss and the RD resistance) and the maximum allowable K value is 2.4. Obviously the proposed configuration will not deliver the required vent rate. Hence the pipe section downstream of the RD will have to be enlarged. In MOD A, a 6-inch diameter tailpipe is tried, giving rise to the following two-segment piping configuration [18]:

Segment	Pipe Diameter	Equiv. K
I	4"ø	2.42
II	6"ø	5.65

Our two-segment pipe flow calculation was carried out using the PIPEHDR program [19] which utilizes the methodology discussed in this paper. In this case it yielded the following pressure ratios:

$$
\begin{aligned}
P_1/P_0 &= 0.91 \quad \text{(Segment I inlet)} \\
P_2/P_0 &= 0.53 \quad \text{(Segment I outlet)} \\
P_3/P_0 &= 0.53 \quad \text{(Segment II inlet)} \\
P_4/P_0 &= 0.20 \quad \text{(Segment II outlet)} \\
P_b/P_0 &= 0.20 \quad \text{(ambient back pressure)}
\end{aligned}
$$

The equivalent discharge coefficient C_d is 0.654 and the estimated vent rate W is 23,510 kg/h which is less than the required rate. Hence it is concluded that MOD A configuration is inadequate; this conclusion is in agreement with Westman's finding. But his result was based on solving a set of algebraic equations from classical compressible gas flow theory.

In MOD B, an 8-inch diameter tailpipe is tried:

Segment	Pipe Diameter	Equiv. K
I	4"ø	2.42
II	8"ø	4.55

TABLE 1
RDPIPE Output for RD/Tailpipe Design Example

This is a single license copy for Fauske & Assoc.
—Fauske & Associates VSSP software © 1/96
RDPIPE (1.0)—RUPTURE DISK PIPING DESIGN PROGRAM
assume turbulent (nonviscous) flow regime
calc. max allowable pipe length RD tailpipe
equiv. length excludes entrance & RD losses

CEP WESTMAN ARTICLE RD/TAILPIPE SIZING

omega	= 1.00100 omega compressible flow parameter
RHOG	= 0.48000 vapor density(#/ft^3)
RHOF	= 60.00000 liquid density(#/ft^3)
PS	= 60.00000 RD set pressure(psig)
PB	= 0.00000 back pressure(psig)
XE	= 1.00000 vessel exit quality
HEL	= 0.00000 elevation above reactor(ft)
Area	= 8.50000 in^2 ideal vent area required

K-value for entrance loss = 1.000
K-value for rupture disk = 1.220

Std pipe sizes (Crane Table) selected
Fanning friction factor = 0.041

Ideal area/ Pipe area	Pipe Dia (in.)	Flow Area (in.^{2})	*Fi* no.	*4fL/D*	Eq. Length (ft)
0.85971	3.500	9.887	0.000	0.55	−3.0
0.66771	4.000	12.730	0.000	2.38	0.3
0.42479	5.000	20.010	0.000	10.39	20.8
0.29422	6.000	28.890	0.000	25.96	72.4
0.16990	8.000	50.030	0.000	86.30	341.8
0.10780	10.000	78.850	0.000	220.67	1110.0

Our PIPEHDR program gave the following results:

$P_1/P_0 = 0.91$ (Segment I inlet)
$P_2/P_0 = 0.40$ (Segment I outlet, choked)
$P_3/P_0 = 0.31$ (Segment II inlet)
$P_4/P_0 = 0.20$ (Segment II outlet)
$P_b/P_0 = 0.20$

The equivalent discharge coefficient C_d is 0.666 and the estimated vent rate W is 24,000 kg/h which just matches the requirement. Hence the MOD B configuration is adequate which is the conclusion reached by Westman.

12. SIZING OF SRV ORIFICE

For a required relief vent rate W_{req} and SRV inlet conditions, the minimum nozzle (or orifice) flow area is calculated by

$$A_{min} = \frac{W_{req}}{K_d K_b G_0} \tag{34}$$

where G_0 is the discharge mass flux for a perfect nozzle geometry. Since the valve should be fully open at 10% overpressure, it is customary to evaluate G_0 at this overpressure. In practice, the opening pressure of a full-lift SRV in gas/vapor service ranges from 1% to 5% above the set pressure. Furthermore the capacity correction factor K_b due to back pressure is given for overpressure of 10% and higher. A coefficient of discharge of 0.975 can be used for gas service SRV when the particular valve does not have a certified value. A SRV with the next larger orifice area than A_{min} will have to be selected.

13. SRV INLET PIPING DESIGN

A SRV is to be installed with as short an inlet piping as possible to minimize the inlet pressure drop. API recommends limiting the inlet nonrecoverable losses to 3% or less of the (gauge) set pressure [4]. The inlet pressure loss can be estimated by

$$\Delta P_{in} = \frac{1}{2} K_{in} \left(\frac{W}{A_{in}} \right)^2 v_{in} \tag{35}$$

where K_{in} is the combined entrance, rupture disk, and straight pipe resistance; A_{in} is the inlet pipe flow area; and v_{in} is the average (two-phase) fluid specific volume in the inlet pipe section. A design chart has been proposed as shown in Fig. 11 [20] where the maximum allowable K_{in} is given as a function of the ω parameter, the orifice diameter to inlet pipe diameter ratio, and the ambient pressure ratio P_b/P_s.

In another publication [21], Westman presented an elaborate scheme for calculating the inlet pressure drop for the case of gas/vapor flow, again based on classical compressible gas flow theory. His example is used here to illustrate our design methodology and can be restated as follows. A pressure vessel is to be protected by a SRV through which 26,000 kg/h of nitrogen gas must be relieved in the event of a process upset. The vessel's MAWP is 50 psig (3.448 bar-g), and a 10% vessel accumulation is permitted. The SRV is installed 82 ft downstream of the vessel outlet nozzle; its set pressure is at the MAWP. The original design configuration is shown as Case A in Fig. 12. Other suggestions—Cases B, C, and D—from Ref. [21] are illustrated in Fig. 12 as well.

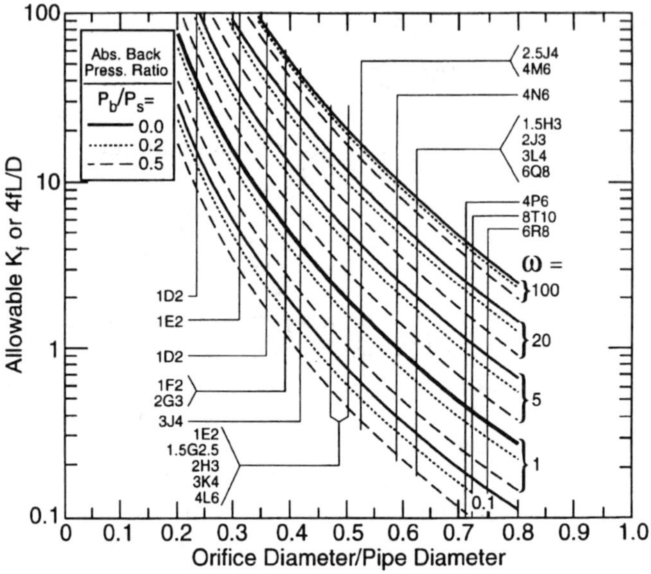

Figure 11. Allowable Inlet Piping Resistance Factor for Satisfying the API 3% Rule.

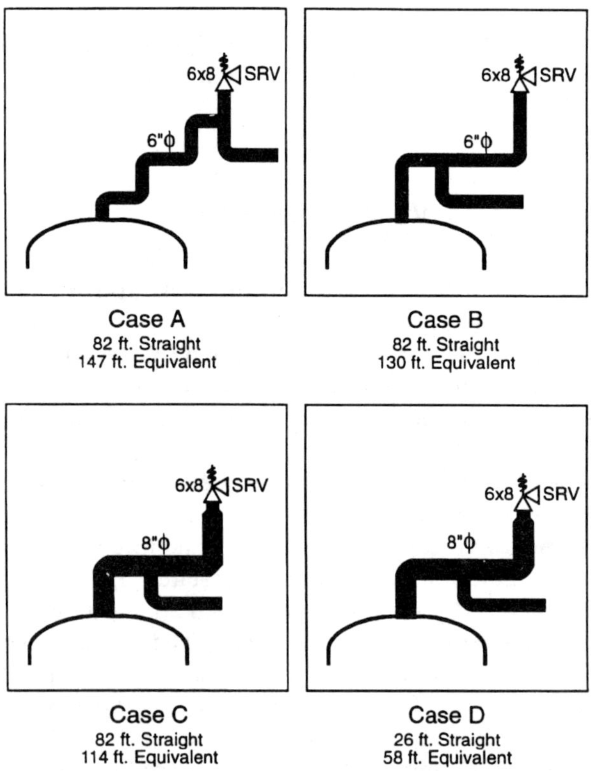

Figure 12. Four Suggested Configurations for SRV Inlet Piping Design Example.

Our calculation shows that a "Q" orifice with a flow area of 9.95 in^2 is required, and such a SRV comes with an inlet flange of 6" in diameter. Using Fig. 11, at $\omega = 1$ and $P_b/P_s = 0.2$, the allowable K_{in} is about 0.7–0.8. The Case A configuration in Fig. 12 yields a K_{in} of 4.85 which is obviously too excessive. Table 2 is an output of the SRVPIPE program [19] which utilizes the methodology discussed in this paper. The maximum allowable inlet pipe run is restated below:

Inlet Pipe Diameter	Max. Allowable Pipe Run
6"ø	12 ft
8"ø	94 ft

Comparing the four configurations shown in Fig. 12, one can quickly conclude that Case D is the only feasible design. This is in agreement with Westman's finding.

14. SRV OUTLET PIPING DESIGN

The limitation here is the length of the outlet pipe run given the maximum allowable back pressure for the particular valve. The latter is specified by the valve manufacturer and is typically 10% of the set pressure for conventional (unbalanced) valves. Higher built-up back pressure would affect the valve lift causing rapid deterioration of relief capacity. The longest equivalent pipe length that can pass the actual flow at a given pipe diameter is derived from the momentum equation as given below [similar to Eq. (19)]:

$$4f\frac{L}{D} = 2\frac{P_0\rho_0}{(W/A_2)^2}\left\{\frac{\eta_1-\eta_2}{(1-\omega)} + \frac{\omega}{(1-\omega)^2}\ln\left[\frac{(1-\omega)\eta_2+\omega}{(1-\omega)\eta_1+\omega}\right]\right\}$$
$$-2\ln\left[\frac{(1-\omega)\eta_2+\omega}{(1-\omega)\eta_1+\omega}\left(\frac{\eta_1}{\eta_2}\right)\right] \qquad (36)$$

Recall $\rho_0 = 1/v_0$ and A_2 is the outlet pipe flow area. Referring to Fig. 13, $\eta_1 \equiv P_1/P_0$ and $\eta_2 \equiv P_2/P_0$ in Eq. (36) denote the pressure ratios at the front end and at the exit end of the pipe, respectively. Note that this equation is applicable to horizontal pipe flow but should be adequate for flashing flow even with an elevation rise in the tailpipe since the fluid density is significantly reduced in the outlet pipe, resulting in negligible hydrostatic head. For the nonflashing flow situation, elevation rise in the outlet pipe can be detrimental and should be avoided. In estimating the allowable $4fL/D$ from Eq. (36), P_1 would assume

TABLE 2
SRVPIPE Output for SRV Inlet Piping Design Example

This is a single license copy for Fauske & Assoc.
— Fauske & Associates VSSP software © 1/96
SRVPIPE (Ver 1.1)- SRV PIPING DESIGN PROGRAM
assume turbulent (nonviscous) flow regime
calc. max allowable inlet pipe length tailpipe
equiv. length excludes entrance & RD losses

CEP WESTMAN ARTICLE SRV INLET PIPE SIZING

omega	= 1.00100 omega compressible flow parameter
RHOG	= 0.35000 vapor density(#/ft3)
RHOF	= 60.00000 liquid density(#/ft3)
PRS	= 55.00000 SRV relieving pressure(psig)
DSP	= 50.00000 differential set pressure(psi)*
PB	= 0.00000 back pressure(psig)
PA	= 0.00000 ambient pressure(psig)
XE	= 1.00000 vessel exit quality
HEL	= 10.00000 elevation above reactor(ft)
Area	= 9.95000 in2 ideal vent area required

note: * DSP is "back out" from input
BALANCED TYPE RELIEF VALVE CHOSEN
K-value for entrance loss = 0.500
Std pipe sizes (Crane Table) selected
Fanning friction factor = 0.004
SRV DISCHARGE COEFF. (Kd)= 0.9500
API ORIFICE SIZE TABLE USED
SRV ORIFICE LETTER IS ="Q"
SRV ORIFICE AREA = 11.050 in^2

INLET PIPE I.D. = 6.000in
SIZING OF INLET PIPING ACCORDING TO API 3% RULE
MAX ALLOWABLE INLET PIPE RESISTANCE RESULT:
note: equiv. length values excludes entrance & RD(if present) losses

MAX. EQUIVALENT LENGTH	= 11.64ft
INLET PRESSURE LOSS	= 0.893 psi
PRESS LOSS w.r.t. SRV set press	= 1.79%

INLET PIPE I.D. = 8.000in
SIZING OF INLET PIPING ACCORDING TO API 3% RULE
MAX ALLOWABLE INLET PIPE RESISTANCE RESULT:
note: equiv. length values excludes entrance & RD(if present) losses

MAX. EQUIVALENT LENGTH	= 93.63ft
INLET PRESSURE LOSS	= 0.289 psi
PRESS LOSS w.r.t. SRV set press	= 0.58 %

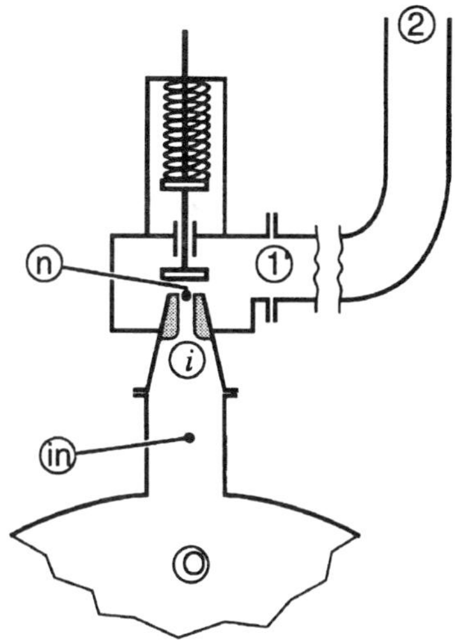

Figure 13. SRV schematic: Use of Subscripts.

the maximum allowable built-up back pressure value while P_2 is assigned the higher of the back pressure P_b or the critical exit choking pressure:

$$P_{2c} = \frac{W}{A_2}\left(\frac{P_0\omega}{\rho_0}\right)^{1/2} \tag{37}$$

Note that this equation is another form of Eq. (20).

For conventional-type SRVs (in which the bonnet is vented to the discharge side of the valve), the built-up back pressure is restricted to typically 10% of the set pressure when operating at an overpressure of 10%. Higher back pressure at this overpressure may reduce the valve capacity and induce the valve to chatter. The performance against back pressure can be improved by operating at higher overpressure. The valve manufacturers should be consulted.

Balanced-type SRVs are designed so that the opening and closing forces due to back pressure acting on the seated disc are balanced—behave better against built-up back pressure. At 10% overpressure the sum of built-up and superimposed back pressure should not exceed 30–50%; again the valve manufacturers should be consulted. Likewise the valve performance against back pressure can be improved by operating at high overpressure but at present available data is limited to a maximum value of 20% overpressure.

Example. Suppose a high-pressure vessel is to be protected by a SRV with a "D" orifice (0.11 in^2 with K_d = 0.95) set at 705 psig (48.6 bar-g), which is also the MAWP of the vessel. The vessel is liquid filled with water at 227°C and normally operates at a pressure of 550 psig (37.9 bar-g). The water is hence regarded as subcooled (i.e. below its saturation temperature corresponding to the imposed pressure) both under normal operation and during relief condition. The tailpipe consists of a 2-inch diameter pipe of 100 ft length and contains a 90° elbow.

First we estimate the flow rate through the SRV. Using the subcooled nozzle discharge model:

$$\begin{aligned}
P_0 &= 790 \text{ psia}(5.45 \text{ MPa, } 110\% \text{ MAWP}) \\
T_0 &= 227°\text{C} = 500\text{K} \\
P_s &= 384 \text{ psia} \qquad (2.65 \text{ MPa}) \\
\rho_f &= 52.1 \text{ lb/ft}^3 \qquad (833 \text{ kg/m}^3) \\
\omega_s &= 7.7 \qquad (\text{from Eq. 25}) \\
\eta_{st} &= \frac{2\omega_s}{1+2\omega_s} = 0.94 \qquad (\text{from Eq. 27}) \\
\eta_s &= \frac{P_s}{P_0} = 0.49
\end{aligned}$$

Since $\eta_s < \eta_{st}$, the discharge is in the high subcooling regime where

$$G = \sqrt{2\rho_f(P_0 - P_s)} = 68{,}300 \text{ kg/m}^2\text{s}$$

With $A_n = 0.11 \text{ in}^2$ and $K_d = 0.95$, the vent rate is

$$W = K_d A_n G = 10.4 \text{ lb/s} \quad (4.73 \text{ kg/s})$$

The discharge through the SRV is essentially all liquid flow. But the flow in the tailpipe (i.e. downstream of the SRV) is not just liquid alone because the water will flash to form a two-phase mixture. If we choose a SRV with a 1" × 2" body, then the assumption of liquid flow in the 2-inch tailpipe will yield a pressure drop of only 5 psi, hardly any built-up back pressure. This would suggest a conventional SRV is adequate. To estimate the pressure drop in the tailpipe taking into account flashing flow, a new reference condition based on the water inlet temperature and its corresponding saturation pressure (P_s) is suggested:

$$\begin{aligned}
P_0 &= P_s = 384 \text{ psia} \\
T_0 &= T_s = 227°\text{C} = 500\text{K} \\
\omega &= \omega_s = 7.7 \quad (\text{same as before})
\end{aligned}$$

Solving Eq. (36) for the pressure drop in the 2-inch tailpipe with 4fL/D being 12.3, our PIPEHDR program gave:

$$\begin{aligned}
P_2 &= 34.7 \text{ psig} \qquad (\text{exiting pressure}) \\
P_1 &= 157.6 \text{ psig} \qquad (\text{immediately downstream of SRV})
\end{aligned}$$

Thus the discharge is also choked at the tailpipe exit and the pressure drop in a flashing flow situation is seen to be significantly higher than the liquid flow case. The built-up back pressure just downstream of the SRV is 158 psig which is more than the 10% allowable for a conventional relief valve. A balanced-type SRV is hence recommended if the given tailpipe configuration is followed.

15. SONIC VELOCITY AND CRITICAL FLOW

A critical flow condition is reached at a nozzle throat or pipe exit where a decompressive disturbance cannot travel upstream. No matter how much the downstream pressure drops (even to vacuum), the flow is unchanged. Physically the exit fluid (homogeneous) velocity has attained the local speed of sound or sonic velocity. For a homogeneous fluid, the sonic velocity c can be given by the thermodynamic definition

$$c = v\left[-\left(\frac{\partial v}{\partial P}\right)_s\right]^{1/2} \tag{38}$$

where the property derivative is to be evaluated along a constant entropy path. Furthermore for a decompressive disturbance, thermodynamic equilibrium between the two phases is closely approached [22]. The HEM sonic velocity according to Eq. (38) can be simply correlated by the expression [6]

$$c^* = \frac{C}{\sqrt{Pv}} = \frac{1}{\sqrt{\omega}} \tag{39}$$

where ω takes on Eq. (5) for a flashing system and Eq. (6) for a nonflashing system. However one major distinction is that the physical properties are to be evaluated locally.

For a nonflashing system (absence of any phase change), Eq. (39) would yield

$$c = \sqrt{\frac{Pv}{(\alpha/k)}} = \left(\frac{kP}{\alpha\rho}\right)^{1/2} \tag{40}$$

By ratioing this sonic velocity in a two-phase mixture to that of the gas alone, and making the usual approximation that $\rho \doteq (1-\alpha)\rho_f$ we arrive at the well-known results [23],

$$\frac{c}{c_g} = \left[\frac{\rho_g/\rho_f}{\alpha(1-\alpha)}\right]^{1/2} \tag{41}$$

where the sonic velocity for a two-phase system exhibits a minimum at a void fraction of 0.5, and is much less than the sonic velocity in the gas phase.

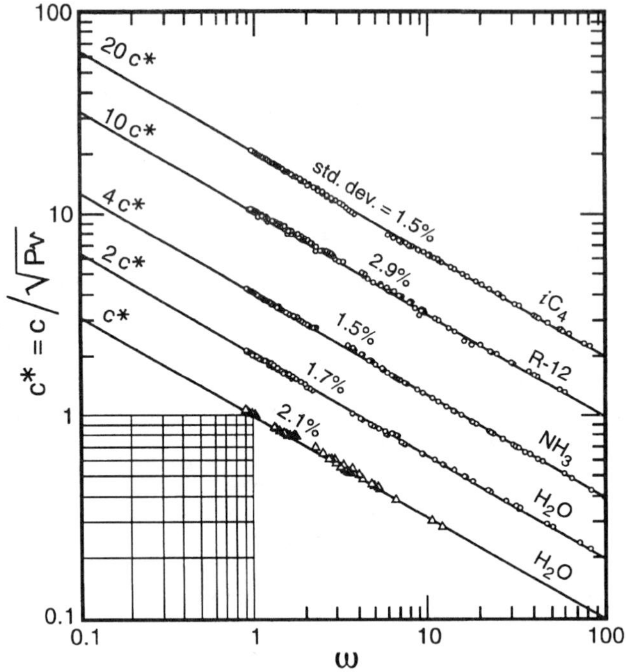

Figure 14. Comparison of Theoretical Two-Phase Sonic Velocity Data Against Present ω Correlation.

For a flashing system, the sonic velocity data—theoretically derived values—covering a wide range of thermodynamic conditions and spanning the entire two-phase range for isobutane, refrigerant 12, ammonia, and water can be correlated by the A parameter with excellent result as demonstrated in Fig. 14 [6]. As a further comparison, the sonic velocity data for saturated steam taken from the ASME Steam Table [24] are shown in Fig. 15 as a function of pressure. The present A correlation predicts the data trend very well and only underpredicts the sonic velocity by 1–2% up to 2200 psia (150 bar).

16. PIPING REACTION FORCES

Reaction (thrust) forces are produced from the discharge of a SRV or a RD. The reaction force can cause the piping and the vessel to recoil and the moment caused by the reaction force can result in excessive stress on the equipment nozzle and discharge piping. Referring to Fig. 16, the thrust developed from a *nozzle discharge* derived from consideration of momentum is,

$$T_R = Wu_e + A_e\,(P_e - P_b) \qquad (42)$$

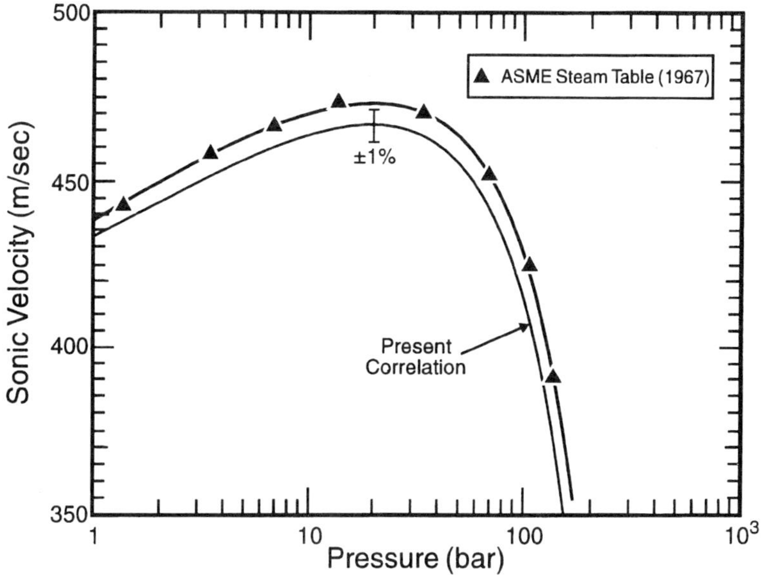

Figure 15. Comparison of Saturated Steam Sonic Velocity Data Against Present ω Correlation.

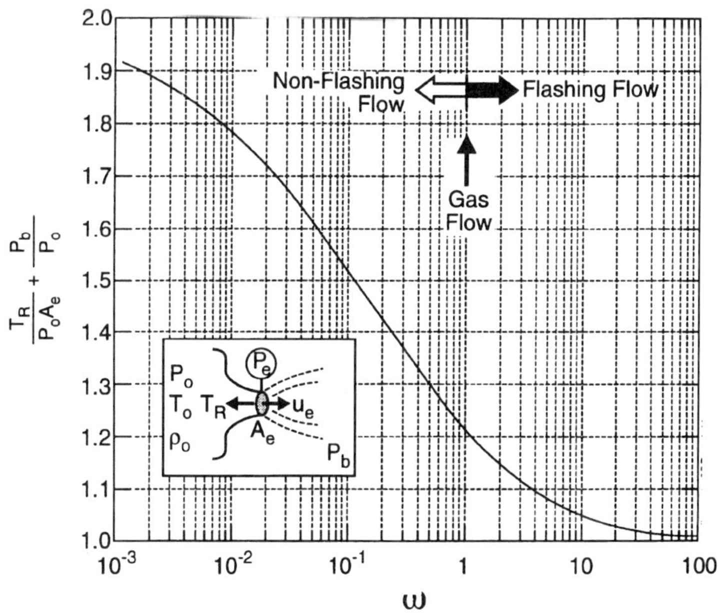

Figure 16. Generalized Chart for Choked-Flow Nozzle Discharge Thrust Forces.

Here the thrust is composed to two components: the first term represents the fluid momentum expulsion rate and the second term represents the exit plane pressure (imbalance) force. For a perfect converging nozzle the compressible (ideal) gas flow solution is given by [10]

$$\frac{T_R}{P_0 A_e} = 2\left(\frac{2}{k+1}\right)^{\frac{1}{k-1}} - \frac{P_b}{P_0} \tag{43}$$

Here T_R/P_0A_e is termed the normalized thrust. For convenience, we shall denote the quantity $[T_R/(P_0A_e) + P_b/P_0]$ as the "thrust coefficient" and according to Eq. (43) it has a numerical value of 1.21 for $k = 1.001$ (large molecular weight gas limit) and 1.27 for $k = 1.4$ (for diatomic gases such as nitrogen); thus it is nearly invariant with k. For the case of incompressible liquid discharge from a nozzle, it can be easily shown that

$$\frac{T_R}{P_0 A_e} = 2\left(1 - \frac{P_b}{P_0}\right) \tag{44}$$

or the thrust coefficient is simply $2.0 - P_b/P_0$. To arrive at a generalized thrust expression, Eq. (42) can be rearranged to give the normalized thrust as,

$$\frac{T_R}{P_0 A_e} = \frac{G^2}{P_0 \rho_0}\left(\frac{v_e}{v_0}\right) + \left(\frac{P_e}{P_0} - \frac{P_b}{P_0}\right) \tag{45}$$

which is in terms of the exit mass flux and exit specific volume v_e. For the case of critical flow from a nozzle, substitution of Eqs. (4) and (10) yields (while noting that $P_e = P_c$),

$$\frac{T_R}{P_0 A_e} = \frac{P_c}{P_0}\left[2 - \frac{P_c}{P_0}\left(1 - \frac{1}{\omega}\right)\right] - \frac{P_b}{P_0} \tag{46}$$

As has already been shown the critical pressure ratio P_c/P_0 for the nozzle case is only dependent on ω via Eq. (9), hence the thrust coefficient can be uniquely determined solely by the ω parameter as presented in Fig. 16. Here the thrust coefficient is bounded by a value of 2 at ω approaching 0 and a value of 1 at an ω of about 100. Thus at the limit of nonflashing (incompressible) liquid discharge the thrust coefficient attains the maximum asymptotic value of 2 which is in agreement with the classical solution represented by Eq. (44). At the other limit with flashing liquid discharge ($\omega > 1$), a minimum value of 1.0 is approached for the thrust coefficient. At $\omega = 1$, the thrust coefficient has a value of 1.21 which is in agreement with the classical solution (Eq. 43 with $k = 1.001$).

For the case of an unchoked exit, the exit pressure imbalance term in Eq. (42) vanishes and the thrust is only dependent on the exit momentum rate. Substituting Eqs. (4) and (8) into Eq. (45) gives

$$\frac{T_R}{P_0 A_e} = \frac{2\left[-\omega \ln\left(\frac{P_b}{P_0}\right) - (\omega - 1)\left(1 - \frac{P_b}{P_0}\right)\right]}{\omega\left(\frac{P_0}{P_b} - 1\right) + 1} \tag{47}$$

The exit thrust for a *pipe discharge* situation can be given by Eq. (45) also. The exit choking pressure ratio in this case is given by Eq. (20) and therefore is related to that for the nozzle discharge via Eq (22). Here a useful quantity for representation of the thrust result is to evaluate the following thrust coefficient ratio:

$$\frac{\left[\frac{T_R}{P_0 A_e} + \frac{P_b}{P_0}\right]_{\text{pipe}}}{\left[\frac{T_R}{P_0 A_e} + \frac{P_b}{P_0}\right]_{\text{nozzle}}} = \frac{\left(\frac{G_c}{G_{0c}}\right)\left[2 - \frac{P_c}{P_0}\left(\frac{G_c}{G_{0c}}\right)\left(1 - \frac{1}{\omega}\right)\right]}{2 - \frac{P_c}{P_0}\left(1 - \frac{1}{\omega}\right)} \tag{48}$$

This result as shown in Fig. 17 covers both flashing flow and nonflashing flow regimes. So for a given pipe flow problem with A, 4fL/D and Fi defined, the corresponding G_c/G_{0c} ratio is obtained from design charts such as Figs.

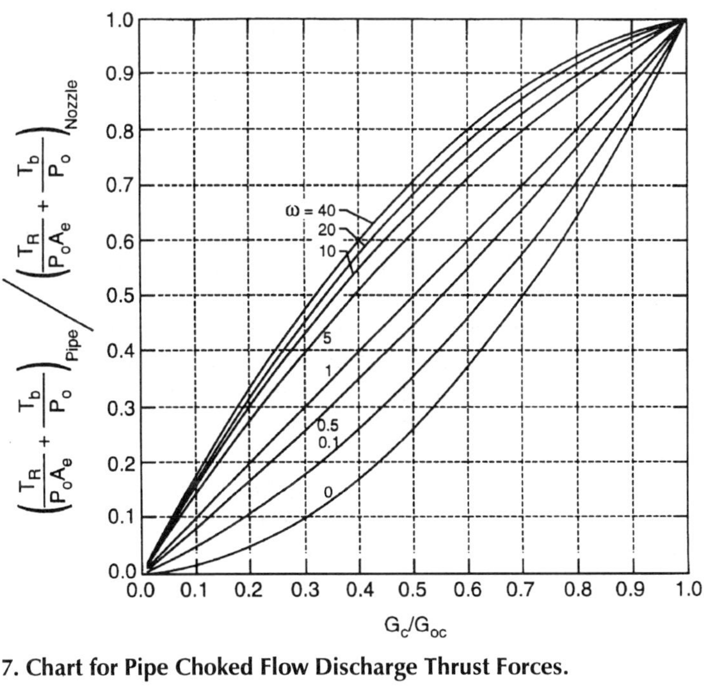

Figure 17. Chart for Pipe Choked Flow Discharge Thrust Forces.

TABLE 3
Comparison of Thrust Coefficient in Gas Flow Through Constant-Diameter Pipe

Total Velocity Head Loss $4fL/D$	Mass Flux Reduction G_c/G_{0c}	$T_R/(P_0A_p) + P_b/P_0$	
		Present Method[(a)]	DIERS Design Manual[(b)]
0.1	0.96	1.16	1.17
1	0.78	0.95	0.96
10	0.43	0.52	0.52
100	0.15	0.18	0.19

(a)—Simply obtained by 1.21 (G_c/G_{0c}).
(b)—For $k = 1.0$, Fig. V-13 of Ref. [25].

4-6. Figure 17 then offers a most expedient way of estimating the resulting exit thrust. It is interesting to note that for the gas flow ($\omega = 1$) case, Eq. (48) yields a thrust coefficient ratio equal to G_c/G_{0c}, see also Fig. 17. Table 3 demonstrates that for gas discharge from a constant-diameter pipe, the present method yields essentially identical results as the more rigorous gas dynamic analysis [25].

For a *SRV discharge*, the reaction force is determined by the conditions at the exit of the valve body or the tailpipe—thrust from the SRV nozzle is balanced by the spring force on the valve disk. In this configuration it is convenient to present the results in terms of the thrust ratio defined as

$$\text{Thrust Ratio} = \frac{\text{Thrust from Tailpipe Exit}}{\text{Thrust from SRV Nozzle}} \tag{49}$$

Note that the thrust from a SRV nozzle can be easily found from the nozzle chart of Fig. 16. The results are presented in Fig. 18 for a back pressure to set pressure ratio P_b/P_s of 0 (consult Ref. [25] for charts at 0.1, 0.2, and 0.5 ratios). This chart should provide a conservative design. Since the nozzle and the tailpipe carry the same mass flow, i.e. $W = A_nG_0 = A_pG$, the thrust ratio is shown as a function of either A_n/A_p (nozzle area to tailpipe area) or G/G_0 (tailpipe mass flux to nozzle mass flux). Note that the thrust at the tailpipe exit is higher than the nozzle case for the flashing flow ($\omega > 1$) even though the mass flow is the same. The increase is due to a higher exit velocity as a result of flashing. However, the maximum augmentation in thrust is only a factor of two. For vapor or gas discharge, the thrust ratio is identically unity—no augmentation is possible unless the flow becomes supersonic. For a nonflashing discharge, the thrust ratio is less than unity, implying that the thrust at the tailpipe exit is less than that at the nozzle discharge.

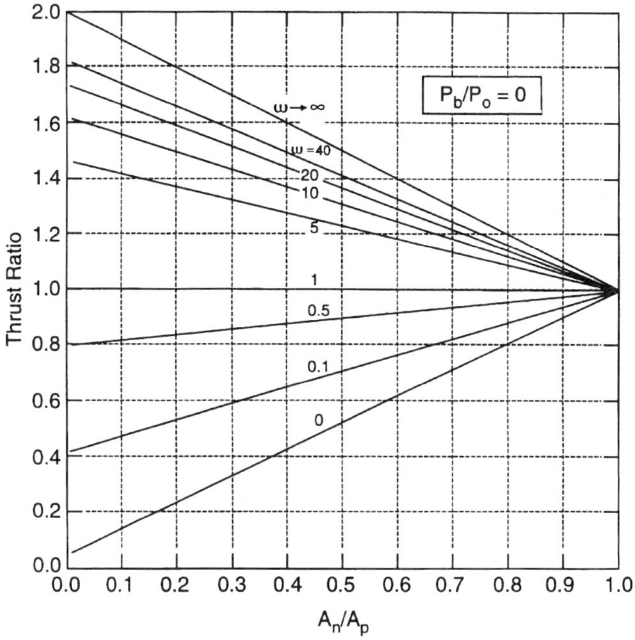

Figure 18. Tailpipe-Exit T_R to SRV-Nozzle T_R versus Nozzle to Tailpipe Flow Area.

Example. Consider a 4P6 SRV set at 600 psig and discharging anhydrous ammonia at a relieving pressure of 660 psig or 675 psia. Calculate the reaction forces at the 6-inch outlet pipe exit for both vapor discharge and saturated liquid discharge situations.

For vapor discharge, the thrust ratio from Fig. 18 for a A_n/A_p = 6.38/28.9 = 0.22 is 1.0 (i.e. independent of A_n/A_p for ω = 1). The thrust developed for a nozzle discharge is from Fig. 16 where the thrust coefficient is 1.21. Thus

$$T_R = \left(1.21 - \frac{14.7}{675}\right)(675)(6.38) = 5117 \text{ lb}_f$$

(Note that this is one of a few cases where working in the US engineering unit is recommended.) The thrust developed at the outlet pipe is hence also 5117 lb_f. This result is in close agreement with the DIERS Design Manual example (see p. 341 of Ref. [25]) where a value of 4990 lb_f was obtained for a gas having a *k* value of 1.3.

For saturated liquid discharge, an A value of about 4.0 is obtained. The thrust ratio in Fig. 18 is 1.3 in this case. First from Fig. 16 for the nozzle case,

$$T_R = \left(1.1 - \frac{14.7}{675}\right)(675)(6.38) = 4642 \text{ lb}_f$$

For the SRV pipe exit

$$T_R = 1.3\ (4642) = 6036\ \text{lb}_f$$

Despite the liquid (saturated) discharge which yields a mass flow rate about 2.5 times higher, the resulting pipe force is only 20% higher than the vapor discharge case.

17. REFERENCES

1. Leung, J. C., "The Omega Method for Discharge Rate Evaluation," in Melhem, G. A., and Fisher, H. G. (eds.), *International Symposium on Runaway Reactions and Pressure Relief Design* (Boston, August 2–4, 1995), New York: AIChE/DIERS, 1995.
2. Leung, J. C., "Size Safety Relief Valves for Flashing Liquids," *Chem. Eng. Progress,* 88(2) (February 1992), 70.
3. Leung, J. C., "Easily Size Relief Devices and Piping for Two-Phase Flow," *Chem. Eng. Progress,* 92(12) (December 1997), 28.
4. American Petroleum Institute, "Sizing, Selection, and Installation of Pressure-Relieving Devices in Refineries, Parts I and II," API RP-520, 6th ed. (March, 1993).
5. Leung, J. C., "A Generalized Correlation for One-Component Homogeneous Equilibrium Flashing Choked Flow," *AIChE Journal,* 32(10) (1986), 1743.
6. Leung, J. C., "On the Application of the Method of Landau and Lifshitz to Sonic Velocities in Homogeneous Two-Phase Mixtures," *ASME Trans. J. of Fluids Engineering,* 118 (1996), 186.
7. Leung, J. C., "Similarity Between Flashing and Non-Flashing Two-Phase Flows," *AIChE Journal,* 36(5) (1990), 797.
8. Leung, J. C. and Grolmes, M. A., "The Discharge of Two-Phase Flashing Flow in a Horizontal Duct," *AIChE Journal,* 33(3) (1987), 524; also errata 34(6) (1988), 1030.
9. Leung, J. C. and Epstein, M., "The Discharge of Two-Phase Flashing Flow from an Inclined Duct," *ASME Trans. J. of Heat Transfer,* 112(2) (May 1990), p. 524.
10. Shapiro, A., *Compressible Fluid Flow*, Vol. I, Ch. 6, Ronald Press, New York (1953).
11. Lapple, C. E., "Isothermal and Adiabatic Flow of Compressible Fluids," *Transactions of the American Institute of Chemical Engineers,* 39 (1943), 385.
12. Leung, J. C. and Grolmes, M. A., "A Generalized Correlation for Flashing Choked Flow of Initially Subcooled Liquid," *AIChE Journal,* 34 (1988), 688–691.
13. Leung, J. C., and Ciolek, W. H., "Flashing Flow Discharge of Initially Subcooled Liquid in Pipes," *ASME Trans. J. of Fluids Engineering,* 116(3) (September 1994).
14. Leung, J. C., and Fisher, H. G., "Two-Phase Flow Venting of Reactor Vessels," *J. Loss Prev. Process Industries,* 2(2) (1989), 78.
15. Leung, J. C. and Nazario, F. N., "Two-Phase Flashing Flow Methods and Comparisons," *J. Loss Prev. Process Industries,* 3(2) (1990), 253.
16. Forrest, H. G., "Calculation of Immiscible Liquid Flashing Perfect Nozzle Flow Rates for Heptane–Water System." Paper presented at 17th DIERS Users Group Meeting, Orlando, FL, (January 31–February 2, 1995).
17. Nazario, F. N. and Leung, J. C., "Sizing Pressure Relief Valves in Flashing and Two-Phase Service: An Alternative Procedure," *J. Loss Prev. Process Industries,* 5(5)(1992), 263 .
18. Westman, M. A., "Easily Design Safety Relief Vent Header Systems," *Chem. Eng. Progress,* 93(10) (October 1997), 70.
19. Fauske & Associates, Inc., "VSSP (Vent Sizing Software Program) Methodology and User's Manual," commercial PC compatible software, Report No. FAI/95-21 (April, 1995).

20. Leung, J. C. and Fisher, H. G., "A Method for Sizing Safety Valve Inlet Lines," Paper accepted for publication in *Process Plant Safety* (1998).
21. Westman, M. A., "Rapidly Design Safety Relief Valve Inlet Piping Systems," *Chem. Eng. Progress,* 93(3) (March, 1997), 80.
22. Moody, F. J., *Introduction to Unsteady Thermofluid Mechanics*, Chapter 2, pp. 72–75. New York: John Wiley & Sons, 1990.
23. Wallis, G. B., *One-Dimensional Two-Phase Flow,* Chapt. 2, p. 25, Chapt. 6, p. 144, New York: McGraw-Hill, 1969.
24. ASME 1967 Steam Tables, "Thermodynamic and Transport Properties of Steam," American Society of Mechanical Engineers, 2nd ed. (1967).
25. Fisher, H. G. et al., *Emergency Relief System Design Using DIERS Technology: DIERS Project Manual,* New York: AIChE, 1992.

18. NOTATION

A_{in} = inlet pipe flow area
A_n = nozzle area
c = sonic velocity
C_p = (liquid) specific heat at constant pressure
f = Fanning friction factor
g = gravitational acceleration
G = mass flux or velocity
G_0 = mass flux corresponding to nozzle
h = enthalpy
h_{fg} = latent heat of vaporization
k = isentropic exponent
K_b = capacity correction factor due to back pressure for gas service valve
K_d = discharge coefficient of SRV nozzle
K_i = individual fitting loss coefficient
L = pipe length
P = absolute pressure
Q = volumetric rate
T = absolute temperature
T_R = thrust force
u = velocity
v = specific volume
W = mass flow rate
x = quality or vapor mass fraction
α = void fraction
ρ = fluid density
η = P/P_0 pressure ratio
θ = angle of inclination with the vertical

Subscripts

b	=	refer to back pressure
c	=	critical or choked
e	=	exit
f	=	liquid
fg	=	difference between vapor and liquid properties
g	=	vapor or gas
min	=	minimum
n	=	nozzle
s	=	saturated condition, also set condition
v	=	vapor
0	=	stagnation inlet condition
1	=	pipe inlet
2	=	pipe exit

19. TERMINOLOGY

Design Institute for Emergency Relief System (DIERS): Institute under the auspices of the American Institute of Chemical Engineers, founded to investigate requirements for vent lines in case of two-phase venting.

Set Pressure—the inlet pressure at which the relief device is set to open (burst).

Overpressure—a pressure increase over the set pressure of the relief device usually expressed as a percentage of gage set pressure.

Stagnation Pressure—is the pressure that would be observed if a flowing fluid were brought to rest along an isentropic path.

Back Pressure—the static pressure existing at the outlet of a pressure relief device as a result of the pressure in the discharge system. It is the sum of the superimposed and built-up back pressures.

Built-up Back Pressure—pressure existing at the outlet of a pressure relief device caused by flow through that particular device into a discharge system.

Superimposed Back Pressure—the static pressure existing at the outlet of a pressure relief device at the time the device is required to operate. It is the result of pressure in the discharge system from other sources.

20. UNITS AND CONVERSION

One difficulty often encountered in relief sizing is units and unit conversion. In this article SI units are recommended. Here a quick conversion table is provided:

Parameter	US Engineering Unit		SI Unit
P	1 psi	=	6897 N/m^2 or Pa
V	1 ft^3	=	2.8317×10^{-2} m^3
m_0	1 lb	=	0.4536 kg
C_p	1 Btu/lb°F	=	4187 J/kg K
ρ	1 lb/ft^3	=	16.02 kg/m^3
h_{fg}	1 Btu/lb	=	2326.1 J/kg
W	1 lb/s	=	0.4536 kg/s
G	1 lb/s ft^2	=	4.878 kg/m^{2}s
A	1 in^2	=	1/1550 m^2
v	1 ft^3/lb	=	0.06243 m^3/kg
L	1 ft	=	0.305 m

In addition the following constant is noted:

Gravitational Acceleration g = 9.81 m/s^2

Also worth noting are:

1 Pa = 1 N/m^2 = 1 kg/m s^2
1 J = 1 N m

Since $G\sqrt{P_0\rho_0}$ and the grouping

$$\rho_0 C_p T_0 P_0 \left(\frac{v_{fgo}}{h_{fgo}}\right)^2$$

in the ω parameter are used frequently, it is worth demonstrating that they are dimensionless.

$$\frac{G}{\sqrt{P_0\rho_0}} [=] \frac{\dfrac{\text{kg}}{\text{m}^2\text{s}}}{\left[\text{Pa}\dfrac{1\ \text{kg}}{\text{Pa m s}^2}\cdot\dfrac{\text{kg}}{\text{m}^3}\right]^{1/2}}$$

$$\rho_0 C_p T_0 P_0 \left(\frac{v_{fgo}}{h_{fgo}}\right)^2 [=] \frac{\text{kg}}{\text{m}^3}\cdot\frac{\text{J}}{\text{kgK}}\cdot \text{K}\cdot\frac{\text{N}}{\text{m}^2}\left(\frac{\text{m}^3/\text{kg}}{\text{J/kg}}\right)\cdot\frac{1\ \text{J}}{\text{N m}}$$

An Easy to Use Two-Phase Flow Model Including Subcooling, Nonequilibrium, and Viscous Effects

H. K. Fauske
Fauske & Associates, Inc., Burr Ridge, Illinois

ABSTRACT

An easy to use and extensively verified two-phase flow model is outlined. Given information about stagnation conditions and flow geometry the model can handle gas–liquid, vapor–liquid and hybrid flows and accounts for subcooling, nonequilibrium and viscous effects in case of flashing flows.

INTRODUCTION

Practical and easy to use guidelines for estimating two-phase and flashing flows are provided below. These guidelines are useful in assessing various emergency releases including requirements for pressure relief and accidental breaches in vessels and pipeworks containing toxic and flammable gas–vapor–liquid materials.

While single-phase flows through relief devices such as safety valves for all liquid and all gas flows are generally specified by the valve manufacturers, i.e., discharge coefficients for all liquid and all gas–vapor flows are provided, discharge coefficients applicable to two-phase flows are generally not available. Here we extend the use of the single-phase information to predict two-phase flows as follows

$$G_{\mathrm{TP}} = \left[\frac{1 - x_0}{G^2_{x_0=0}} + \frac{x_0}{G^2_{x_0=1.0}} \right] \tag{1}$$

where G_{TP} (kg/m^2-s) is the two-phase flow rate and x_0 is the stagnation quality. Furthermore, the expression for $G^2_{x_0=0}$ (kg/m^2-s) is extended to include flashing flows (Fauske, 1985) and viscous effects as follows

$$G^2_{x_0=0} = \frac{C^2_{D,l}\left[2P_{g,0}\rho_{l,0} + \lambda^2_0\rho^2_{v,0} / T_0 C_0 N_{NE}\right]}{1+N} \tag{2}$$

with

$$N_{NE} = \frac{\lambda^2_0\rho^2_{l,0}}{2(P_0 - P_b)\rho_{l,0} T_0 C_0} + \frac{L}{L_E}; \qquad L \le L_E = 0.1\,\text{m}$$

$$N = 4f\,L/D; \qquad f = 0.08[G_{x_0=0}\,D/\mu_{l,0}]^{-1/4}$$

where $C_{D,l}$ is the liquid discharge coefficient, P_0 (Pa) is the stagnation pressure, $P_{g,0}$ (Pa) is the stagnation gas partial pressure, P_b (Pa) is the back or ambient pressure, λ_0 (J/kg), $\rho_{v,0}$ (kg/m^3), $\rho_{l,0}$ (kg/m^3), C_0 (J/kg-K) and $\mu_{l,0}$ (kg/m-s) are the latent heat of vaporization, vapor density, liquid density, liquid specific heat and liquid viscosity, respectively, corresponding to the stagnation temperature T_0 (K); L (m) is the duct length and D (m) is duct diameter.

Finally, $G^2_{x_0=1.0}$ (kg/m^2-s) is the all gas–vapor flow rate and is evaluated using the well-known expressions for either critical or subcritical flows, as required.

Evaluations based on Equation (1) are compared below to available two-phase and flashing flow data through safety valves and piping including subcooling, nonequilibrium and viscous effects.

TWO-PHASE FLOWS THROUGH SAFETY VALVES

For two-component two-phase flows such as air–water flows, is determined from

$$G_{x_0=0} = C_{D.l}\sqrt{2(P_0 - P_b)\rho_{l,0}} \tag{3}$$

where $C_{D,l}$ is the all liquid discharge coefficient usually specified by the valve manufacturer and for critical flows is given by

$$G_{x_0=1.0} = C_{D,g}P_0\left(\frac{M_w}{RT_0}\right)^{1/2}\left[k\left(\frac{2}{k+1}\right)^{(k+1)/(k-1)}\right]^{1/2} \tag{4}$$

where $C_{D,g}$ is the all gas discharge coefficient usually specified by the valve manufacturer, M_w is the molecular weight, R (8314 Pa-m^3/K-kg-mole) is the gas constant, and k is the isentropic coefficient.

Examples of all-liquid and all-gas discharge coefficients for full lift safety valves are shown below (Lenzing and Friedel, 1977).

Safety Valve	$C_{D,l}$	$C_{D,g}$
Leser	0.50	0.77
ARI	0.59	0.81
Bopp & Reuther	0.66	0.86

Air–water data obtained with these valves at three different stagnation pressures for a wide range of stagnation qualities are compared to Equation (1) predictions in Figures 1–3.

For one-component two-phase flows such as steam–water, flows can be estimated from

$$G_{x_0=0} = \rho_{g,0}\lambda_0 (T_0 - C_0)^{-1/2} \tag{5}$$

It is noted that in case of flashing flows the flow rate given by Equation (5) usually requires only a small fraction of the total pressure drop (Fauske, 1985). It follows that losses in the valve body that occur with all liquid flow conditions ($C_{D,l} < 1.0$) will not effect the flashing flows ($C_{D,l} \approx 1.0$). Similar to two-component two-phase flows $G_{x_0=1.0}$ is given by Equation (4). Steam–water flow data obtained with a Leser Safety valve at three different stagnation pressures over a wide range of stagnation qualities (Lenzing and Friedel, 1977) are compared to Equation (1) predictions in Figure 4.

In case of hybrid gas–vapor–liquid equilibrium flows Equation (2) reduces to

$$G_{x_0=0} = C_{D,l}\left[2P_{g,0}\rho_{l,0} + \lambda^2\rho_{v,0}^2 / T_0 C_0\right]^{1/2} \tag{6}$$

and is again given by Equation (4) with the molecular weight M_w given by

$$M_w = M_{w,g}\left(P_{g,0}/P_0\right) + M_{w,v}\left(P_{v,0}/P_0\right) \tag{7}$$

where $P_{v,0}$ (Pa) is the vapor pressure corresponding to stagnation temperature T_0.

ALL-LIQUID FLASHING FLOWS

For all-liquid initial conditions Equation (1) reduces to $G_{TP} = G_{x_0=0}$. In case of equilibrium flashing flows ($L < 0.1$ m) G_{TP} is determined from

$$G_{TP}^2 = \frac{C_{D,l}^2(\lambda_0^2\rho_{v,0}^2 / T_0 C_0)}{1+N} \tag{8}$$

and predictions from Equation (9) with $C_{D,l} = 1.0$ are compared to flashing Freon-11 data in Figure 5.

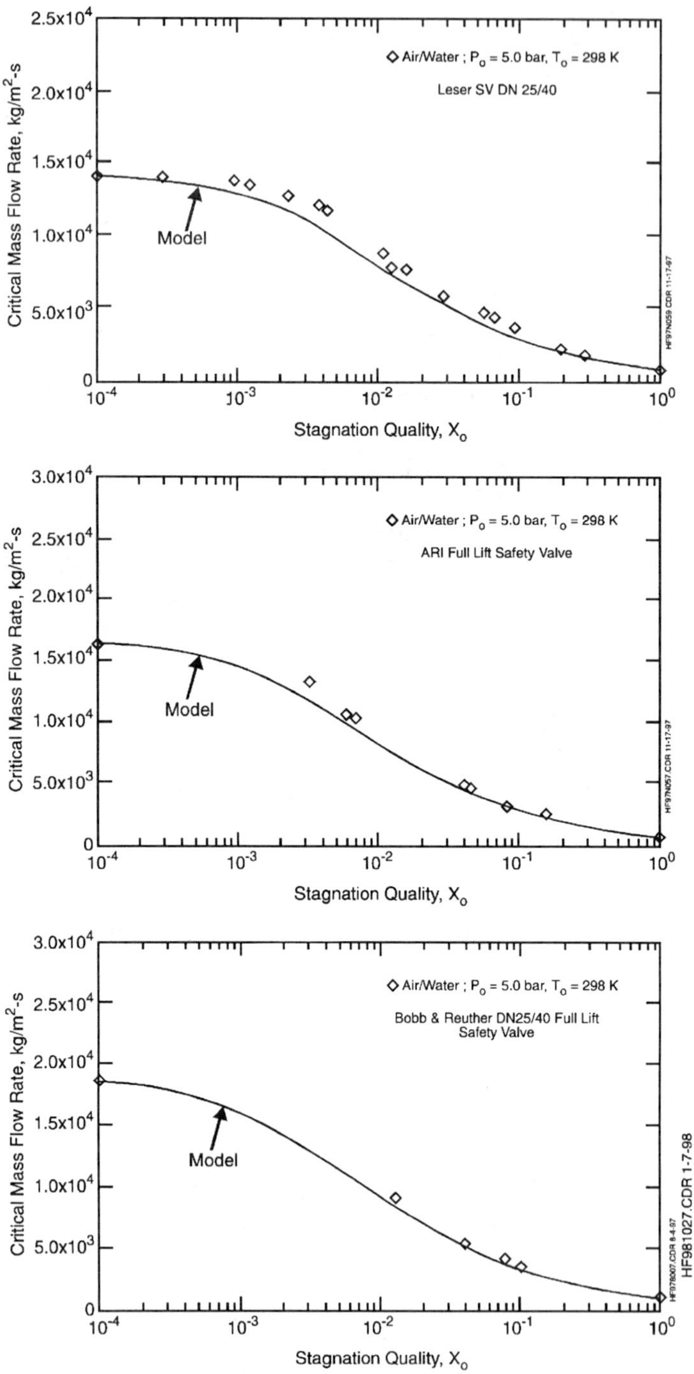

Figure 1. Comparison between measured air–water data (Lenzing and Friedel, 1997) and Equation (1) at P_0 = 5 bar.

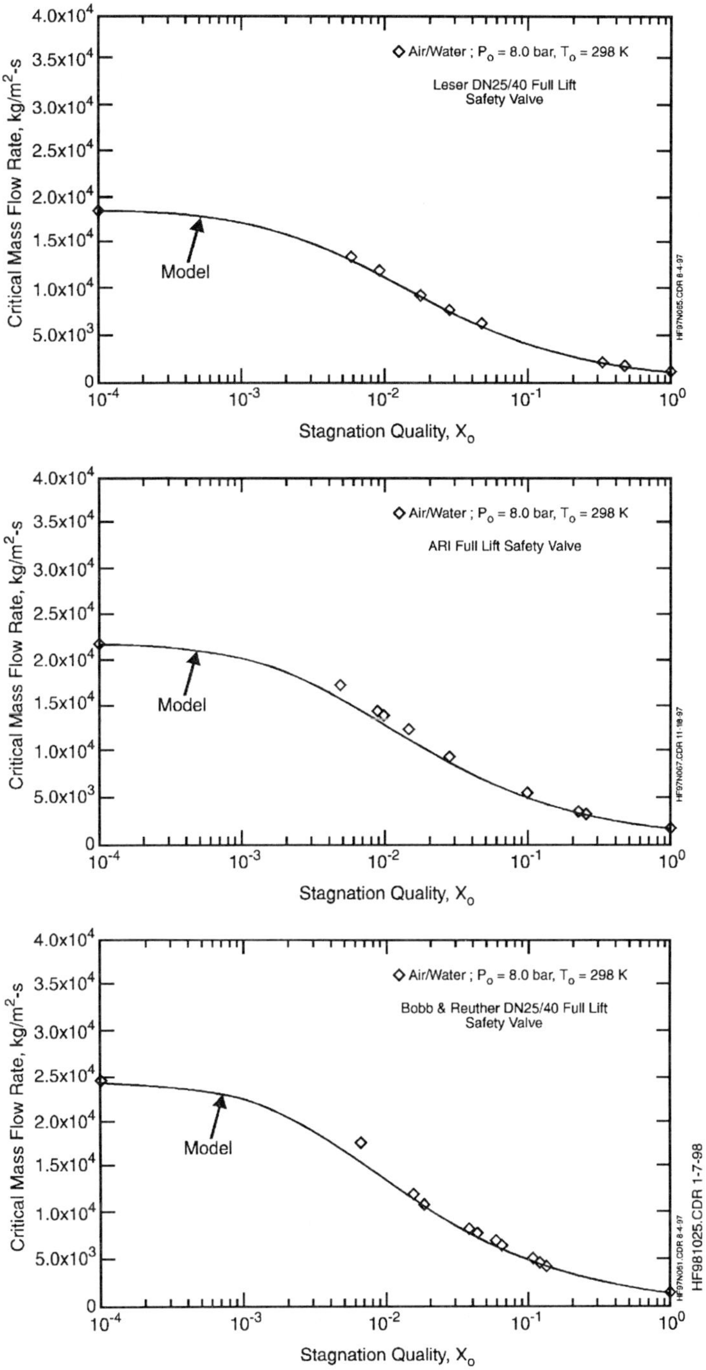

Figure 2. Comparison between measured air–water data (Lenzing and Friedel, 1997) and Equation (1) at P_0 = 8 bar.

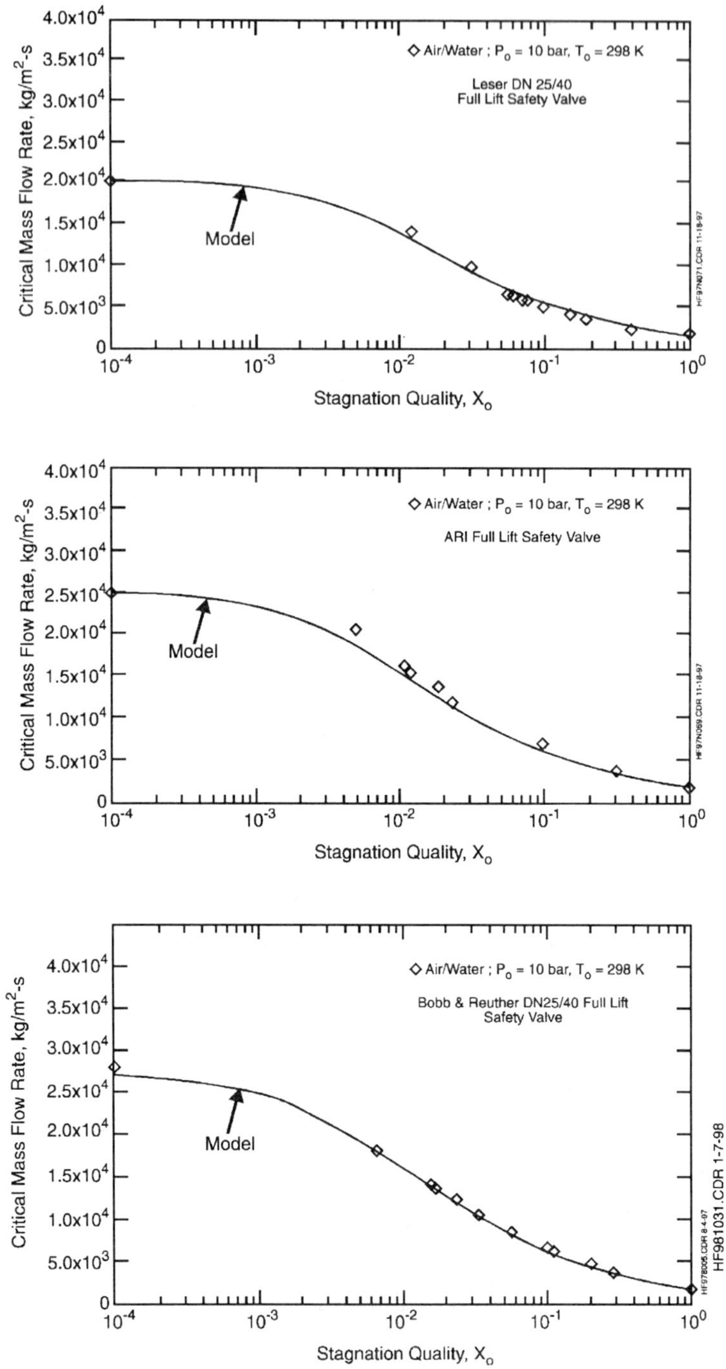

Figure 3. Comparison between mesrud air–water data (Lenzing and Friedel, 1997) and Equation (1) at $P_0 = 10$ bar.

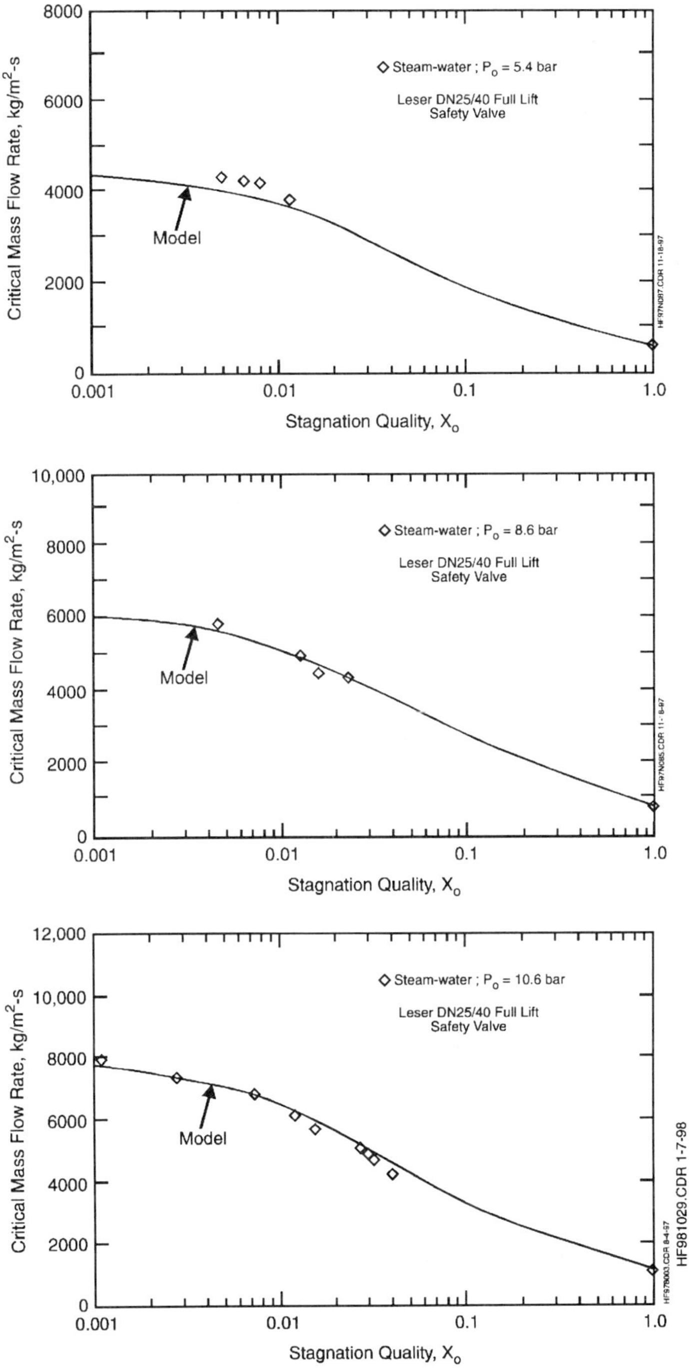

Figure 4. Comparison between measured steam–water data (Lenzing and Friedel, 1997) and Equation (1) at different stagnation pressures.

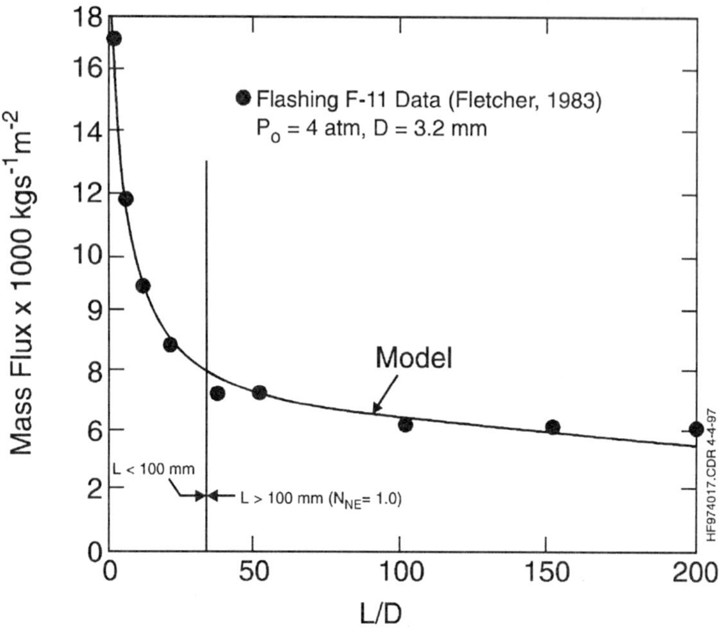

Figure 5. Comparison with flashing F-11 data (Fletcher, 1983).

Considering nonequilibrium flashing flow ($L < 0.1$ m) G_{TP} is given by

$$G_{TP}^2 = C_{D,l}^2 \lambda_0^2 \rho_{v,0}^2 / T_0 C_0 N_{NE} \tag{9}$$

and predictions from Equation (9) with $C_{D,l} = 1.0$ are compared to flashing Freon-11 data in Figure 5.

In case of subcooled equilibrium flashing flows ($L > 0.1$ m) the expression is the same as that for hybrid vapor–gas–liquid flows with $x_0 = 0$

$$G_{TP}^2 = \frac{C_{D,l}^2 \left[2P_{g,0}\rho_{l,0} + \lambda_0^2 \rho_{v,0}^2 / T_0 C_0 \right]}{1+N} \tag{10}$$

and predictions from Equation (10) with $C_{D,l} = 1.0$ are compared to flashing water data in Figure 6.

Considering subcooled nonequilibrium flashing flows ($L < 0.1$ m) G_{TP} is given by

$$G_{TP}^2 = C_{D,l}^2 \left[2P_{g,0}\rho_{l,0} + \lambda_0^2 \rho_{v,0}^2 / T_0 C_0 N_{NE} \right] \tag{11}$$

and predictions from Equation (11) are compared with flashing water data through a Crosby safety valve in Figure 7. The strong effect of nonequilibrium is clearly illustrated.

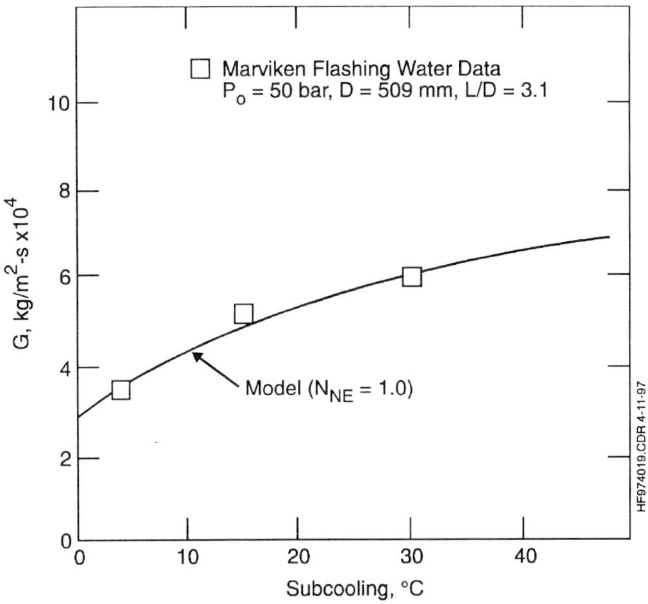

Figure 6. Comparison with flashing subcooled water data.

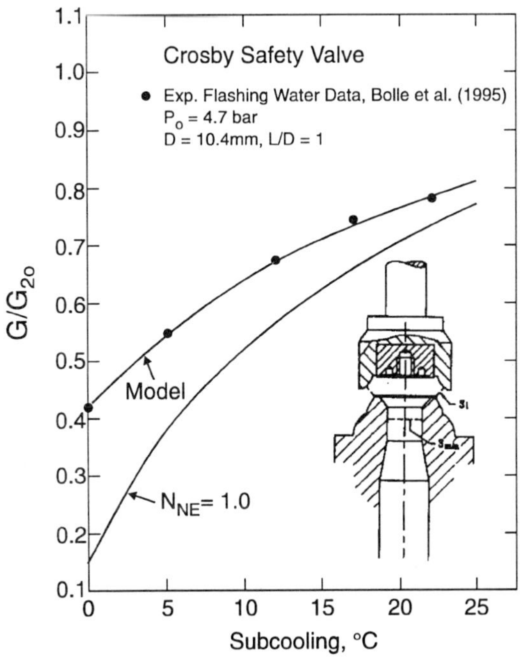

Figure 7. Comparison with flashing water data (Bolle et al., 1985).

Finally, flashing viscous flows can be assessed from Equation (8). Considering that flashing flows promote turbulence, the usually turbulent flow regime for non-flashing flows is extended well into the laminar flow regime (see dotted line in Figure 8). Predictions from Equation (8) are compared to viscous flashing flow data in Figure 9. This model also provides good agreement with the DIERS large-scale PS-EB High Viscosity Flashing Flow Tests (Fisher et al., 1992).

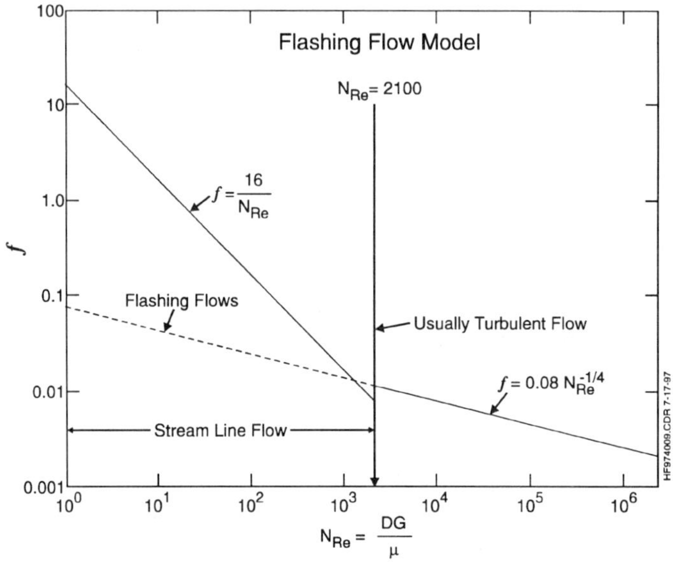

Figure 8. Illustration of flashing viscous flow model.

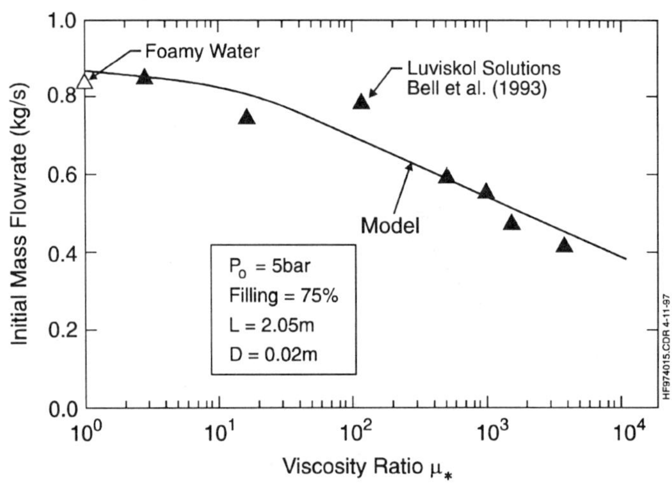

Figure 9. Comparison to viscous flashing flow data.

CONCLUSION

A generalized very easy to use two-phase flow model is presented. Given information about stagnation conditions and flow geometry the model can handle gas–liquid, vapor–liquid and hybrid flows and accounts for flashing equilibrium and non-equilibrium flows including subcooling and viscous effects. Extensive benchmarking with available data is illustrated including two-phase flows through safety valves and ducts.

REFERENCES

Bell, K. et al., *J. Loss Prev. Process Ind.* 6(1).

Bolle, J. et al. (1995), *J. Loss Prev. Process Ind.*, 8(2).

Fauske, H. K. (1985), "Flashing Flows: Some Practical Guidelines for Emergency Releases," *Plant/Operations Progress*, 4(3), 132–134.

Fisher, H. G. et al. (1992), *Emergency Relief System Design Using DIERS Technology,* New York: American Institute of Chemical Engineering.

Fletcher (1983), "Flashing Flow Through Orifices and Pipes." Paper presented at the AIChE 17th Loss Prevention Symposium, Denver, CO, August 28–31.

Lenzing, T. and Friedel, L., 1997, European DIERS User Group Meeting, May 6–7, Hamburg, Germany.

Marviken Data, 1982, "Critical Flow Data Review and Analysis," EPRI/NP-2192.

Perspectives on Relief Valve Sizing for Two-Phase Flow

Ron Darby
Department of Chemical Engineering
Texas A&M University
College Station, Texas 77843-3122

ABSTRACT

The calculations required for sizing a relief valve require a basic theoretical model for flow in the valve nozzle, a knowledge or data base for relevant fluid physical and thermal properties, and a knowledge of a suitable discharge coefficient which accounts for the deviation between the theoretical calculated nozzle flow and the actual flow in the valve. For single phase liquid and critical (choked) gas flows, the models are straightforward and information on the value of the appropriate discharge coefficient is readily available. However, if the conditions are such that two-phase (gas–liquid) flow occurs in the valve, the situation is much more complex. Under these conditions, there are a number of factors that must be considered with respect to the nature of the fluids, flow patterns, phase distribution, flow conditions, and fluid properties in order to select the appropriate model. There are a wide variety of theoretical models which apply to two-phase flow, and there is no single "correct" model which will always give the "right answer" under all conditions. Each model has limitations, and a given model (or variation thereof) may work well under certain conditions but not others. Furthermore, there is virtually no experimental information available for the proper discharge coefficient to use for two-phase flows.

This paper will provide a perspective on what is known and what has not been firmly established with respect to both single and two-phase flow in both nozzles and valves. The basic principles governing two-phase flow in nozzles and valves assuming homogeneous ("pseudo-single fluid") flow behavior will be reviewed, with an overview of those models and variations thereof which provide reasonable engineering predictions for various two-phase flow conditions such as highly subcooled liquid, slightly subcooled liquid, saturated liquid, two-phase flashing, two-phase "frozen" (e.g. constant quality), etc. Considerations of nonequilibrium flashing flow, gas-liquid slip effects, and high viscosity fluids are included, as well as a consideration of the discharge coefficient for such situations.

1. VALVE NOZZLE SIZING

The proper size for a relief valve nozzle is calculated by first determining the required relief capacity (e.g. , lb_m/s) for a given relief scenario (e.g. a runaway reaction, heating from an external fire, loss of cooling, etc., and dividing this by the mass flux capacity of the relief valve (G_{valve}) as determined from design relief pressure conditions:

$$A_{nozzle} = \frac{\dot{m}}{G_{valve}} \tag{1}$$

The mass flux capacity of the valve (G_{max}) is determined by calculating the mass flux through an ideal isentropic (frictionless) nozzle and multiplying this by an appropriate "discharge coefficient" (K_D):

$$G_{valve} = K_D G_{nozzle} \tag{2}$$

1.1. Nozzle versus Valve Flow

The discharge coefficient accounts for any discrepancy between the calculated nozzle mass flux, G_{nozzle}, and the actual mass flux in the valve, G_{valve}. This discrepancy can actually arise from two possible sources:

1. any difference between the flow in the *nozzle* and that in the actual *valve,* and;
2. any deficiency in the theoretical model used to describe flow in an *ideal nozzle*, compared to flow in a *real nozzle* for the given conditions.

If the model used for the ideal nozzle accurately represents the flow in the actual nozzle and the flow capacity of the valve is identical to that of the nozzle, the value of K_D would be unity. Hence the extent to which K_D deviates from unity may be an indication of either the presence of significant effects which influence flow in the valve which are not accounted for by the nozzle alone or a deficiency in the theoretical model for the nozzle, or both.

Although the nozzle is an integral part of the valve and is usually assumed to limit the flow capacity of the valve, there are obvious differences between flow in an actual valve and that an ideal nozzle as illustrated in Figure 1. The key reference pressures are also shown in Figure 1: P_0 is the upstream pressure in the containment vessel (assumed to be the stagnation pressure), P_n is the pressure at the "throat" or exit plane of the nozzle and P_b is the back pressure or valve discharge pressure. If the velocity in the nozzle reaches the sonic velocity, the flow will be choked (or critical) and the nozzle throat pressure at which this condition occurs is designated P_c.

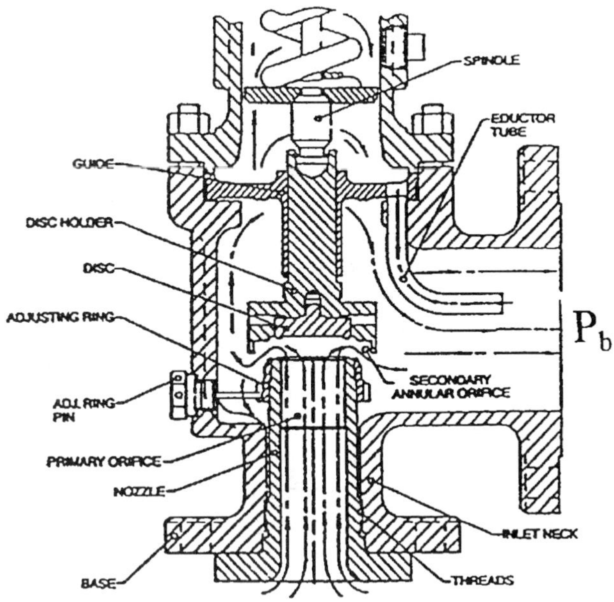

(a) Actual Safety Relief Valve

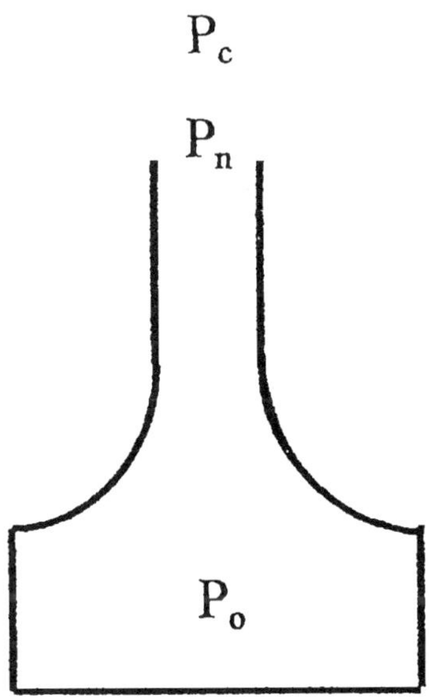

(b) Ideal Nozzle

Figure 1

1.2. Test Certification Conditions

Valve certification measurements (ASME,1986) are normally made with the valve mounted as close as possible to a reservoir vessel, with a smooth rounded entrance between the vessel and the inlet pipe. Measurements include the fluid mass flow rate, the pressure inside the reservoir, and the discharge pressure at the valve exit (which is normally atmospheric). The resulting measured values of K_D for every manufacturer's valves are reported in the "Red Book" (1997) which is updated annually.

The reservoir is large enough that the kinetic energy of the fluid in the vessel can be neglected so that the measured vessel pressure is assumed to be the stagnation pressure (P_0). The rounded entrance and the short inlet piping upstream of the valve minimize the irreversible pressure loss due to friction between the vessel and the valve, but the reversible pressure drop resulting from the increased kinetic energy as the fluid passes from the vessel to the inlet piping is not negligible. The combination of the local pressure and kinetic energy is the "total pressure," or "stagnation pressure". This pressure is the same at all points within a flow stream in which friction loss and potential energy variations due to elevation changes are negligible.

Since the nozzle discharge pressure (P_n) is not known nor measured, the measured valve discharge pressure (P_b) is taken to be the exit pressure for both the valve and the nozzle flow model.

2. NOZZLE FLOW EQUATION

The theoretical model for the mass flux through a nozzle based on the assumption of isentropic (frictionless) flow follows from applying the frictionless Bernoulli equation across the nozzle. This results in the following equation:

$$G_{\text{nozzle}} = \rho_n \left[2\int_{P_0}^{P_n} -\frac{dP}{\rho} \right]^{1/2} \tag{3}$$

This assumes that the fluid enters the nozzle from the stagnation state at pressure P_0 and leaves the nozzle at pressure P_n. However, the actual pressure entering the nozzle (P_1) differs from P_0 by the (irreversible) pressure drop in the upstream piping from the vessel to the nozzle as well as the (reversible) pressure drop due to the increase in kinetic energy. Also the nozzle exit pressure (P_n), is not normally known in a valve, and differs from the valve exit or back pressure (P_b) by any (irreversible) pressure loss or (reversible) pressure recovery from the nozzle throat to the valve exit. Thus, any effect on the valve mass flux resulting from the use of P_b instead of P_n, as well as any effect of using P_0 instead of P_1, is reflected in the value of the discharge coefficient, K_D.

Evaluation of the integral in Eq. (3) requires a knowledge of the compressible nature of the fluid, i.e. the $(P\text{–}\rho)$ (or $P\text{–}v$) fluid properties, and the manner in which the fluid density may vary as the pressure changes through the nozzle (and the valve). We will consider liquid, gas (vapor), and two-phase (gas–liquid) fluids under a variety of conditions and what is and is not known with regard to both the appropriate nozzle model(s) for various conditions encountered, as well as what is and isn't known about the value of K_D for each of these cases. The compressible nature of gases/vapors and two-phase fluids (as opposed to the constant density of liquids) results in a major difference in how well the isentropic nozzle model represents the actual valve behavior, and consequently the value of K_D. This is due to the fact that compressible fluids will "choke" when the local velocity reaches the speed of sound (i.e. the sonic velocity or the "critical" state). When this occurs, the mass flux is a maximum and any pressure drop downstream of the "choke point" will have no influence on the flow capacity. Since liquids do not choke (as the speed of sound is extremely high in a liquid), all factors which contribute to pressure changes between the vessel and the valve discharge point will influence the valve flow capacity. This applies to compressible flows under subcritical (nonchoked) conditions as well as to liquids. Thus the evaluation of the correct valve mass flux is dependent on a knowledge of:

1. the physical properties of the fluid that determine the $(P\text{–}\rho)$ relation;
2. the appropriate model that represents the nozzle mass flux for this fluid, for both critical (choked) and subcritical flow;
3. how well the nozzle model represents actual nozzle flow behavior;
4. how well the nozzle flow behavior represents flow in the actual valve.

We will discuss the extent to which these factors are known and have been verified, for various single and two-phase flow conditions.

3. SINGLE-PHASE FLOW

3.1. Liquid Flow

Since liquids are assumed to be incompressible fluids under normal circumstances, the $(P\text{–}\rho)$ relation reduces simply to ρ = constant, giving rise to the simple incompressible "Bernoulli" equation for nozzle flow:

$$G_{\text{valve}} = K_D G_{\text{nozzle}} = K_D \sqrt{2\rho(P_0 - P_b)} \tag{4}$$

where the valve discharge pressure, P_b has been substituted for the nozzle discharge pressure, P_n.

3.1.1 Low Viscosity (High Reynolds Number) Liquid Flow

Equation (4) without the K_D term follows directly from Eq. (3) with ρ = constant (assuming $P_n = P_b$), and applies only to a reversible (frictionless) system. For liquids at sufficiently high Reynolds numbers (e.g. $N_{Re} \geq 50{,}000$), the influence of viscosity on nozzle flow is negligible and the "frictionless" assumption is valid. Values of K_D reported by valve manufacturers in the "Red Book" for their liquid service valves are determined from tests with water at high Reynolds numbers where this assumption should be valid. These values are generally in the range of 0.6 to 0.8, however, which indicates a significant deviation of the predicted isentropic nozzle mass flux from the actual mass flux through the valve.

The primary reason for this discrepancy relates directly to the substitution of the valve discharge pressure P_b for the nozzle discharge pressure P_n in Eq. (4). A variety of measurements of the flow of water through nozzles indicate a close agreement with Eq. (4) when using a nozzle coefficient near unity and P_n as the discharge pressure (Alvi et al., 1978; Miller, 1983). Therefore, we may conclude that for high Reynolds number flows through valves:

1. the theoretical Bernoulli equation for liquid flow provides a reasonably good representation of flow in a nozzle and
2. there appears to be a significant effect of the flow in the valve body in addition to that in the nozzle on the flow capacity of the valve.

In fact, Morris (1996a) has examined the values of the discharge coefficients reported in the "Red Book" for various size valves, and concluded that the effect of the flow resistance in the valve body can account for as much as 40% of the total flow capacity of the valve. However, since actual measured values of discharge coefficients are available for virtually all valves (in the "Red Book"), it is not necessary to be able to predict this additional valve body pressure loss if these measured values of K_D are used with Eq. (4).

3.1.2 High Viscosity (Low Reynolds Number) Liquid Flow

If the Reynolds number through the nozzle is low (e.g., significantly below, say, 50,000), the effect of viscous dissipation on the flow cannot be neglected. In this case, the mass flux will be lower than that predicted for inviscid flow, which can be accounted for by application of a multiplier (K_v) correction factor to the high Reynolds number (inviscid) discharge coefficient. Such a correction factor is presented in API Recommended Practice 520 (1993), which is based on data from control valves having an uncertainty of about ±20%. Darby and Molavi (1997) have shown that the API curve for the K_v function can be represented by the equation:

$$K_v = \frac{0.975}{\sqrt{\frac{170}{N_{Re}} + 0.98}} \tag{5}$$

The viscous correction factor has also been determined from CFD computations by Molavi (1996) for a typical valve geometry with two different size nozzles. The results have been shown by Darby and Molavi (1997) to be represented by the equation:

$$K_v = \frac{0.975\beta^{0.05}}{\sqrt{\frac{950(1-\beta)^{1.4}}{N_{Re}} + 0.9}} \tag{6}$$

where $\beta = d_n / D_1$. The prediction of Eq. (6) is consistent with the values given by API 520 and Eq. (5), and the values predicted for F and T size nozzles bracket the uncertainty range of the control valve data.

3.2. Gas Flow

The theoretical equations for the isentropic flow of an ideal gas through an ideal nozzle are well known and are available from a variety of sources (e.g. API 520, 1993) but will be reviewed here for reference, and uncertainties in these expressions will be discussed.

3.2.1 Critical (Choked) Flow

As the pressure drops through the system the gas expands and accelerates until the velocity reaches the speed of sound. In a uniform duct this occurs at the downstream end. When the sonic velocity (c) is reached, the mass flux is a maximum for given upstream conditions and will be independent of any further decrease in the downstream pressure. This condition is referred to as critical or choked flow, and the corresponding mass flux is given by:

$$G_c = c\rho = \rho\sqrt{\left(\frac{\partial P}{\partial \rho}\right)_s} \tag{7}$$

The pressure at which critical flow occurs is P_c. If $P_c > P_n$ (or, in practice, if $P_c > P_b$), the flow in the nozzle (or the valve) will be choked, and the upper limit of the integral in Eq. (3) should be P_c. In practice, the critical mass flux and the corresponding value of P_c, can be found by evaluating the mass flux from Eq. (3) at successively lower values of P_n until a maximum in G_{nozzle} is reached

For an ideal gas (which represents air flow quite well over a wide range of conditions), the equation of state for an isentropic ideal gas can be used to evaluate the speed of sound from Eq. (7), and the critical mass flux from Eq. (3), to give:

$$G_c^2 = P_0\rho_0\left(\frac{2k}{k-1}\right)\eta_c^{2/k}(1-\eta_c)^{\frac{k-1}{k}} \tag{8}$$

where the critical pressure ratio is

$$\eta_c = \frac{P_c}{P_0} = \left[\frac{2}{k+1}\right]^{k/(k-1)} \qquad \text{where} \quad k = \frac{C_p}{C_v} \tag{9}$$

Values of the measured discharge coefficient (K_D) for air flow under critical (choked) conditions are documented for valves from all manufacturers in the "Red Book" (1997). Typical values are on the order of 0.96–0.98, which indicates that:

1. The ideal gas model gives a good representation of critical flow of air through a nozzle and
2. The nozzle model provides a reasonable representation for flow through a typical valve.

This is because the conditions downstream of the point where the flow becomes choked (i.e., the nozzle exit) have no effect on the mass flux in the nozzle/valve.

Even though this model appears to provide acceptable results for critical air flow under normal conditions, there are several effects that are not accounted for by the model which could be significant. One factor is the possible deviation from ideal gas behavior, which occurs for all gases as the critical temperature and pressure of the gas are approached. The most direct way of accounting for this is to include a compressibility factor (z) for the gas in the basic equation, which is equivalent to dividing the gas density by z.

Other concerns arise from the assumptions that friction in the nozzle is negligible and the use of the stagnation pressure as the upstream reference which implies that the nozzle area is negligibly small relative to the upstream area. Both of these assumptions can affect the actual critical pressure ratio in the nozzle throat, as shown below.

As has been shown by Morris (1996b), the friction loss in the nozzle (as represented by the deviation of the nozzle flow coefficient, C_D, from unity) influences the critical pressure ratio according to:

$$\eta_c = C_D (\eta_c)_{\text{ideal}} \qquad \text{where} \quad (\eta_c)_{\text{ideal}} = \left(\frac{2}{k+1}\right)^{k/(k-1)} \tag{10}$$

Morris (1996b) states that for actual safety relief valves, the reported "Red Book" values for gas flow show that 73% of the valves have $0.90 < C_D < 0.975$ and 27% have $0.671 < C_D < 0.90$, which means that the actual choke pressure may be significantly less than predicted for an ideal nozzle. It is normally assumed that general chemical service safety relief valves having the higher coefficients are well-designed and suitable for two-phase flow service. However, the measured discharge coefficient is defined relative to the "ideal nozzle" model which neglects any effect of friction on the choke pressure (i.e.

the right side of Eq. (10)). Since Eq. (10) shows that friction can result in the actual choke pressure being lower than that predicted for the ideal (frictionless) nozzle, the actual pressure driving force across the nozzle may be higher than the theoretical "frictionless" value resulting in a flow rate for the actual nozzle being higher than that predicted by the "ideal" model and a correspondingly larger discharge coefficient. In fact, it is not unusual to encounter a "measured" discharge coefficient which is greater than unity for choked gas flow because the calculated mass flux used as a basis for defining the discharge coefficient determined by the frictionless relations, Eq.s (8) and (9), is less than that which would be predicted if friction loss were considered.

Another possible source of error in Eq. (9) arises from the assumption that the nozzle (throat) area at which choking occurs is negligibly small compared with the upstream area. If this assumption is not valid, the critical pressure ratio is dependent upon the area ratio as follows (e.g. Darby, 1996):

$$\eta_c = \left[\frac{1}{1+\left(1-\beta^4\eta_c^{2/k}\right)\dfrac{k+1}{2}} \right]^{k/(k-1)} \tag{11}$$

where $\beta = d_n/D_1$. For $k = 1.4$ and $0 < \beta < 0.9$, this predicts $0.528 < \eta_c < 0.656$. The effect is thus opposite to that of friction [Eq. (10)], that is, the choke pressurc from Eq. (11) is higher than that predicted by the "ideal nozzle model," resulting in a lower mass flux than predicted by the ideal model and a corresponding lower discharge coefficient. It is therefore possible that some of the effects that are normally neglected may not indeed be negligible, but nevertheless tend to cancel out.

3.2.2 Subcritical Gas Flow

If the nozzle (or valve) discharge pressure is greater than P_c, the flow through the nozzle will be subcritical, i.e. choking will not occur. Under this condition, the theoretical mass flux for isentropic flow of an ideal gas in a nozzle is given by Eq.s (8) and (9), with P_c replaced by P_n, the nozzle throat or discharge pressure. Since this pressure is internal within the valve, it is not known a priori so that in practice the valve discharge pressure, P_b, is used instead. If the pressure drop (or pressure recovery) from the nozzle exit to the valve discharge is negligible, then P_n will not differ significantly from P_b.

However, measured values of the discharge coefficient for actual valves are not readily available for subcritical conditions. The API 520 Recommended Practice (1993) recommends using a discharge coefficient value of 0.975, implying a negligible effect on the valve flow capacity of the conditions from the nozzle exit to the valve discharge. However, data shown by AGCO (1996) for a typical relief valve operating under subcritical conditions indicates that the discharge coefficient is a function of the total pressure ratio across the valve, and may vary from 0.85 to 0.65.

Thus, although there are insufficient data available to draw general conclusions for subcritical gas flow in valves, it does appear that:

1. The theoretical model for subcritical flow of an ideal gas through a nozzle under isentropic conditions probably gives a good representation of actual nozzle flow (this has been confirmed by a variety of measurements on air flow through well-designed nozzles) and
2. There appears to be a significant influence of the conditions in the valve body downstream of the nozzle on the flow capacity of the valve which is not accounted for by the nozzle model and depends on the actual valve design and the pressure ratio through the valve body, but has not been quantified in a general sense.

4. TWO-PHASE FLOW

Two-phase (gas–liquid or vapor–liquid) flow is much more complex than single phase flow because of the wide variety of possible conditions that can be encountered and which must be accounted for in order to arrive at an appropriate isentropic nozzle model to predict the mass flux. In addition to uncertainties with regard to the proper model to use, uncertainties in the proper discharge coefficient are even greater since there are no published data for two-phase discharge coefficients in safety relief valves.

In order to analyze any two-phase flow problem and select an appropriate model to use, assumptions must be made with regard to a variety of factors, such as the distribution of the phases, the physical state of both liquid and gas phases relative to their flash/dew points, the degree of thermodynamic and mechanical equilibrium that exists between the phases, the composition of the fluids, etc. These considerations are addressed in detail below.

4.1. Phase Distribution

There are a variety of flow regimes which are possible in two-phase gas–liquid flow, which relate directly to the distribution of each of the two phases in the flow section. These have been described quantitatively by Lockhart and Martinelli (1949), Martinelli and Nelson (1948), and others. These regimes depend upon the relative flow rates and relative amounts of the two phases (typically described by the "quality" or mass fraction of gas/vapor in the mixture), and can range from continuous liquid with a distributed gas phase to continuous gas with a distributed liquid phase, including both uniform and nonuniform distribution of the phases. For flow in a vertical pipe, the regimes are typically described as bubbly, slug, churn, wispy-annular, and annular.

Although it is conceivable that any or all of these regimes could occur in a safety relief valve, the typical relief scenario involves the evolution of gas or vapor due to rapid depressurization of a volatile liquid (e.g. dissolution of a gas or flashing of the liquid). Under these conditions, the most likely initial flow regime is usually assumed to be churn flow with relatively low quality in which the gas exists as uniformly distributed bubbles or foam (not unlike the flow from a champagne bottle when the cork is "popped"). However, there are situations where this assumption may not be valid, such as a runaway reactor containing a highly viscous fluid, or the case in which a high degree of flashing and/or disengagement occurs in the vessel before entering the relief device.

The churn flow regime may be described by either a separated flow model or a homogeneous flow model. The separated flow model (e.g. Lockhart and Martinelli, 1949) assumes that each phase occupies a distinct but unspecified region of the flow cross section, and that the two-phase pressure drop can be described by a "two-phase multiplier," which is used with the corresponding pressure drop for either of the single phases flowing alone in the conduit. However, for the flow conditions which prevail in the typical relief device, for example, a high pressure gradient and high flow rate, the more common assumption is that the two-phase mixture can be described as ***homogeneous,*** i.e. as "a pseudo-homogeneous single phase fluid" with physical properties which are some appropriate weighted average of those of each of the separate phases.

4.2. Physical State

This refers to the thermodynamic state of the gas and liquid phases relative to the boiling/dew point of the mixture. The possible states can be described as "frozen," "flashing," saturated," or "slightly subcooled," although the "saturated" and "slightly subcooled" states could be considered as special cases of the "flashing" state.

If the fluid entering the valve is a two-phase mixture of gas and liquid under conditions which are far removed from the boiling point of the liquid and the dew point of the gas such that no vaporization of the liquid nor condensation of the gas would occur as the pressure changes through the valve, there will be no change in the mass of either phase in the valve. An example might be a two-phase mixture of cold water and air. Such a condition is referred to as "frozen flow," and is characterized by a constant quality.

If a liquid entering the valve sufficiently close to its boiling point such that saturation conditions would be encountered within the nozzle or the valve, then flashing of the liquid would be expected to occur when the pressure drops to the saturation pressure. If the entering fluid is saturated liquid, then flashing would be initiated as the liquid enters the valve/nozzle and additional flashing will occur as the pressure drops through the nozzle. If the entering fluid is a two-phase vapor/liquid mixture at equilibrium, then additional liquid will flash to vapor as the pressure drops through the nozzle. If the entering liquid is slightly subcooled (i.e. the pressure is above the saturation pressure), no flashing would be expected until the pressure drops to the saturation pressure (this assumes that the two phases are in local thermodynamic equilibrium, which is not generally true—see Section 4.3 below).

In each of these flashing cases, the relative amount of vapor and liquid will change as the pressure changes through the valve/nozzle, i.e. the quality (x) will increase as flashing progresses. For an isentropic nozzle, the quality (x) at any point can be determined by an isentropic condition applied between the upstream (stagnation) state and any local state. In practice it is usually more convenient to assume a "constant enthalpy" condition for this determination (neglecting any change in kinetic energy), which gives very nearly the same result in most cases:

$$x = \frac{S_{LO} - S_{LO}}{S_{LG}} = \frac{S_{LO} + x_0 S_{LGO} - S_L}{S_{LG}} \doteq \frac{h_{LO} + x_0 h_{LGO} - h_L}{h_{LG}} \tag{12}$$

4.3. Thermodynamic Equilibrium

If the fluid is assumed to be locally in thermodynamic equilibrium at all points, then the local composition and the relative amounts of vapor and liquid in a flashing mixture are determined by the energy balance, the local temperature and pressure and the VLE (vapor–liquid equilibrium) properties of the mixture. In this case, the local quality is determined by Eq. (12) and the densities of the vapor and liquid are determined by the local temperature and pressure. The corresponding local two-phase density is given by:

$$\frac{1}{\rho} = \frac{x}{\rho_G} + \frac{1-x}{\rho_L} \quad \text{or} \quad v = v_G + v_L(1-x) \tag{13}$$

Once flashing has been initiated and a significant amount of vapor has been generated, the assumption of local thermodynamic equilibrium appears to be valid. However the initiation and growth of vapor bubbles are rate processes which involve nucleation, heat and mass transfer, etc. These non-equilibrium processes result in an effective delay in the generation of a significant amount of vapor after the pressure reaches the saturation pressure. The result is that for initially subcooled liquid entering the valve/nozzle, flashing will be delayed beyond the point at which the saturation pressure is reached. Thus flashing could occur either within the nozzle or at the exit (or not at all) depending on the length of the nozzle relative to the distance traveled by the liquid during the "delayed flashing" period. This effect must obviously be considered when selecting the appropriate model for initially subcooled liquids.

4.4. Slip

As the pressure drops through a nozzle or valve, the gas/vapor phase will expand whereas the liquid phase density remains constant. As a result, the gas phase accelerates relative to the liquid phase, resulting in a relative velocity or "slip" between the two phases (this is sometimes referred to as "mechanical nonequilibrium"). The slip ratio is defined as

$$S = \frac{V_G}{V_L} \tag{14}$$

The effect of slip is to alter the "holdup" or local volume fraction of gas/vapor as follows:

$$\alpha = \left[1 + S\left(\frac{1-x}{x}\right)\frac{\rho_G}{\rho_L}\right]^{-1} \tag{15}$$

It is evident that the "holdup" volume fraction of gas with slip is less than the holdup with no slip.

Based on the kinetic energy flux equation, Wallis (1969) determined the effective two-phase density in the presence of slip to be

$$\frac{1}{\rho_{ke}} = \left[\frac{x}{\rho_G} + \frac{S(1-x)}{\rho_L}\right]\left[x + \frac{1-x}{S^2}\right]^{1/2} \tag{16}$$

However, if the momentum equation is used instead, the corresponding expression is

$$\frac{1}{\rho_{mf}} = \left[\frac{x}{\rho_G} + \frac{S(1-x)}{\rho_L}\right]\left[x + \frac{1-x}{S}\right] \tag{17}$$

The average or "holdup"mixture density, as determined from

$$\rho_h = \alpha\rho_G + (1-\alpha)\rho_L \tag{18}$$

is (e.g. Chisholm, 1983 b):

$$\frac{1}{\rho_h} = \left[\frac{x}{\rho_G} + \frac{S(1-x)}{\rho_L}\right]\left[\frac{1}{x + S(1-x)}\right] \tag{19}$$

Morris (1991) presented an analysis which applies to nozzles for nonflashing homogeneous two-phase flow with slip and liquid entrainment. The expression he used for the two-phase density with slip is:

$$\frac{1}{\rho_m} = \left\{\frac{x}{\rho_G} + \frac{S(1-x)}{\rho_L}\right\}\left\{x + \frac{1-x}{S}\left[1 + \frac{(S-1)^2}{(\rho_L/\rho_G)^{1/2} - 1}\right]\right\} \tag{20}$$

An appropriate expression for the two-phase density can be used in Eq. (3) to evaluate the mass flux. If Eq. (16) is used, the expression for the nozzle mass flux for a given slip ratio is (Simpson, 1991):

$$G^2 = \frac{\int_{P_0}^{P} -2[x/\rho_G + (1-x)/\rho_L]\, dP}{\{[x/(\rho_G S) \ + (1-x)/\rho_L]^2 [xS^2 + 1 - x]\}} \tag{21}$$

If the slip ratio (S) and the local gas and liquid densities (ρ_G, ρ_L) are known as a function of pressure, Eq. (21) can be integrated to give the theoretical mass flux for the ideal nozzle. Various expressions for the slip ratio have been derived based on different assumptions, all of which show that $S = fn\ (\rho_L/\rho_G)$ but result in widely varying expressions (which will be discussed later). The criterion for determining when slip effects are important is not always evident, but in general they should be most significant for relatively low quality flows (e.g. $x < 1\%$), for flashing from slightly subcooled liquids, and situations involving large pressure gradients (i.e., approaching the choke pressure, where theoretically the pressure gradient approaches infinity).

4.5. Fluid Composition

Apart from being either single- or two-phase, frozen or flashing, equilibrium or nonequilibrium, slip or uniform velocity, the fluid entering the nozzle/valve could also be a "hybrid" mixture of liquid, vapor, and noncondensable gas (e.g., steam, air and water). It could be a one-component system (e.g., flashing water, ammonia, propane or ethylene) or a multicomponent mixture. The liquid phase could be a low viscosity or high viscosity fluid or it could contain suspended solids, and it could be either Newtonian or non-Newtonian. If the liquid contains high polymers, it could also be viscoelastic as well as non-Newtonian. Each of these factors must be considered in order to properly determine the fluid density as a function of pressure, which is required to evaluate the integral in Eq. (3), as well as the effects of viscous/viscoelastic properties that are not accounted for by this equation. Many of these conditions are beyond the scope of this report.

4.6. Choking Conditions

As previously remarked, the mass flux of a compressible fluid reaches a maximum when the velocity reaches the speed of sound (sonic or critical "choked" flow), as given by Eq. (7). For an ideal gas in an isentropic nozzle, this expression is readily evaluated to give the critical pressure ratio at which choking occurs [Eq. (9)]. However, the speed of sound is not as clearly defined for a two-phase mixture, and the critical or choke condition is determined by evaluating the mass flux as a function of $\eta_n = P_n / P_0$ (from e.g. Eq. 3) and evaluating the mass flux as a function of $\eta_n = P_n / P_0$ (from e.g. Eq. 3) and thereby determine the value of $\eta_n = \eta_c$ at which the mass flux is maximum. Many flashing flows choke at values of η_c of 0.9 or higher (as compared with a value of 0.5–0.6 for an ideal gas). However, frozen two-phase flows may choke over a much wider range of pressure ratios, but generally at lower values than flashing flows.

5. TWO-PHASE NOZZLE FLOW METHODS

There are a wide variety of computational methods and/or models that may be applied to determine the mass flux for a two-phase mixture in an isentropic nozzle, many of which apply only to one or more of the special cases discusses in Section 4. For our purpose, we would like to have the simplest and easiest to apply method that gives acceptable results for engineering design purposes. This precludes consideration of the so-called "two-fluid" models, which are based upon the microscopic conservation equations for mass, energy, and momentum as they apply locally to each phase, with appropriate models for interphase transport. There are many such models in the literature which are very detailed, are inevitably highly computationally intensive, and are more suited for investigating the detailed mechanisms than for engineering predictions. It is also not evident that any of these methods give better results than the "homogeneous" models for safety relief valve sizing when the proper model is utilized under the appropriate conditions. We will, therefore, limit this discussion to the various "homogeneous" models and methods. Thus, the basic problem is the evaluation of G from Eq. (3) (or equivalent). We will review the various computer and analytical methods presently available for this purpose, with comments on those special cases for which the various methods apply.

5.1. Computational Methods

The determination of the two-phase flow capacity of a given valve (or nozzle or pipe) for given upstream conditions is not in general a simple or direct process, even for single component compressible systems. This is because the mass flux depends on the pressure drop across the device and the exit (choked) pressure depends on the mass flux. Thus an iterative process is normally required to determine both the conditions (pressure) under which choking will occur and the corresponding maximum mass flux. For sufficiently simple compressible equations of state, such as the ideal gas law for single-phase gas flow, or for a linearized equation of state for pseudo-one-component homogeneous-equilibrium two-phase flow, analytical solutions (e.g. the Omega Method of Leung, 1995, 1996) are available and will be reviewed. However, for nonideal gases, two-phase mixtures which exhibit a nonlinear P–ρ relation, or multicomponent mixtures in which the components are not close-boiling compounds, numerical methods are necessary.

The SAFIRE computer code, available from the AIChE through DIERS, and the SuperChems and TPHEM codes, which will be available soon, are designed to address some or all of these situations. Proper application of any of the analytical or computer methods requires the user to understand and specify the nature of the flow encountered during the relief process (as described above for two-phase flow) and to identify the property data that are necessary to perform the required calculations. It is not possible to program this understanding or the required judgement into a program for automatic execution, so these programs can only be used properly by those who understand the systems they are dealing with.

5.1.1 SAFIRE

The initial code developed for DIERS (e.g. Grolmes and Leung, 1985, Shaw, 1990) is called SAFIRE and is described in some detail in the DIERS project manual "Emergency Relief System Design Using DIERS Technology" (DIERS, 1992). The program is designed primarily as a "rating" code to evaluate the pressure/temperature/time history for various emergency scenarios with a specified relief system. It has limited design capabilities to determine vent size for simple geometries, but is not a suitable tool for unsophisticated users as it relies considerably upon the judgment of the user to select the proper model and a set of user-controlled options which best describes his or her application.

5.1.2 SuperChems for DIERS

This program is being developed by Arthur D. Little under contract with the DIERS Users Group as a replacement for SAFIRE (e.g. Melhem et al., 1995, Melhem and Fisher, 1997). It is intended to be a dynamic simulator, capable of emergency relief system and effluent handling designs for complex geometries and multiphase reaction systems, and includes a sophisticated multicomponent physical property package based on equations of state. Effluent handling systems coupled with the vent calculations for systems involving supercritical reactions, polymerization reactions, nonideal solution effects and phase splitting may be simulated. Slip effects (i.e. either the Moody or Fauske slip models—see below) in two-phase flow, and high viscosity liquid properties can be handles, but the program does not accommodate nonequilibrium (slightly subcooled) flows. It is intended to be considerably more "user friendly" than SAFIRE, but is still undergoing evaluation at this time so the degree of "user friendliness" is yet to be determined. The program should be available from the AIChE and DIERS in early 1998.

5.1.3 TPHEM and COMFLOW

COMFLOW for single-phase compressible flow and TPHEM for two-phase homogeneous flow are presently in the latter stage of refinement. These programs will be available on a disk which accompanies a forthcoming CCPS book "Guidelines for Pressure Relief and Effluent Handling Systems"which is scheduled for release early in 1998. The TPHEM program is based upon the homogeneous equilibrium model (HEM) and incorporates direct numerical integration of Eq. (3) for the adiabatic mass flux of a two-phase mixture through an isentropic nozzle. Various two-phase pressure-volume models can be used with one, two, or three parameters, which require the input of the phase densities and vapor-liquid ratio of the two-phase mixture at up to three pressures (e.g. Simpson, 1995), as shown in Table 1. The program handles all phase conditions, from all liquid to all gas/vapor (although the COMFLOW program is preferred for the latter), and requires the user to provide the entering phase density and vapor-liquid ratio before the flow rate can be computed. The most recent version of TPHEM has been revised to include slip and nonequilibrium effects.

The constants in these models can be determined from single- or multi-component flash calculations at several different pressures, as indicated in the table. For model A, the single constant can be determined by invoking the Clausius-Clapeyron equation at the stagnation pressure to circumvent the flash calculation, in which case this model is identical to the so-called Omega Method (e.g., Leung, 1995, 1996) which is discussed below. (It should be noted that the parameters in the Omega Method model can also be evaluated

TABLE 1
Two-Phase Volume (Density) Models (Simpson, 1995)

MODEL	POINTS	EQUATION
A	2	$(\rho_0/\rho - 1) = a(P_0/P - 1)$
B	3	$(\rho_0/\rho - 1) = a(P_0/P - 1)^b$
C	3	$(\rho_0/\rho - 1) = a[(P_0/P - 1)^b - 1]$
D	2	$(\rho_0/\rho - 1) = a(P_0/P - 1) + b(P_0/P - 1)^2$
E	2	$x = a_0 + a_1 P$ $\rho_G = b_0 P^{b_1}$ $\rho_L = c_0 + c_1 P$
F	3	$x = a_{0+} + a_1 P + a_2 P^2$ $(\rho_{G0}/\rho_G - 1) = b_0[(P_0/P)^{b_1} - 1]$ $\rho_L = c_0 + P + c_2 + P^2$

from the knowledge of the two-phase density at two different pressures, in which case the Omega Method becomes equivalent to model A above).

5.2. Two-Phase Models

We will summarize here the various models which apply to homogeneous two-phase flow through an isentropic nozzle, and discuss the advantages and limitations of each. Although it might be more satisfying to organize this information according to the various upstream conditions that could be encountered, with a discussion of which model(s) apply to each of these conditions, this could be more confusing since some models may be applied in different forms to various flow conditions. After discussing the models, we will attempt to summarize the results in tabular form according to application.

5.2.1 Homogeneous Frozen Flow Model (HFZ)

This model applies to an initial two-phase mixture of a nonvolatile (or "highly subccooled") liquid and a noncondensable gas (e.g. air and "cold" water). Under these conditions, the mass fraction of the gas phase (i.e. the quality, x) does not change as the pressure varies through the flow region. If the gas phase is assumed to behave as an ideal gas, Eq. (3) can be evaluated to yield (Henry and Fauske, 1971):

$$G_n^2 = \left(G_0 \frac{\nu_0}{\nu_n}\right) + 2\frac{P_0}{\nu_n^2}\left\{(1-x)\nu_{L_0}\left(1-\frac{P_n}{P_0}\right) + x\nu_{G_0}\left(\frac{1}{k-1}\right)\left[1-\left(\frac{P_n}{P_0}\right)^{(k-1)/k}\right]\right\} \tag{22}$$

where

$$\nu_n = (1-x)\nu_L + x\nu_{G_0}\left(\frac{P_n}{P_0}\right)^{-(1/k)} \tag{23}$$

G_0 is the mass flux at state "0" (which is normally zero if "0" represents the stagnation state).

The Omega Method (e.g. Leung, 1996) can also be applied to frozen (nonflashing) flow as a special case of the more general model. This model is an analytical solution of Eq. (3) which employs a linearized two-phase equation of state relating the mixture density (p) to the pressure by:

$$\frac{\rho_0}{\rho} = \omega\left(\frac{P_0}{P} - 1\right) + 1 \tag{24}$$

For a nonflashing ("frozen") system (Leung, 1996b):

$$\omega = \frac{\alpha_0}{k}$$

where α_0 is the volume fraction of gas at the stagnation state. The A parameter can also be determined from a knowledge of the mixture density at two pressures (e.g. at P_0 and P_1, e.g. $P_1 = 0.9\,P_0$) from

$$\omega = \frac{(\rho_0/\rho_1)-1}{(P_0/P_1)-1} = 0.9\left(\frac{\rho_0}{\rho_9} - 1\right)$$

where ρ_9 is the mixture density at $P_1 = 0.9\,P_0$. The solution for the mass flux through the nozzle is:

$$G^* = \frac{G}{\sqrt{P_0\rho_{L_0}}} = \frac{\{-2[\omega \ln \eta = (\omega-1)(1-\eta)]\}^{1/2}}{\omega(1-\eta)/\eta + 1} \tag{27}$$

where $\eta = P/P_0$. The maximum value of $G^* = G_c^*$ occurs when the flow is choked at a pressure ratio $\eta_c = P_c/P_0$, given by:

$$\eta_c^2 + \omega(\omega-2)(1-\eta_c)^2 + 2\omega^2 \ln \eta_c + 2\omega^2(1-\eta_c) = 0 \tag{28}$$

This is implicit for η_c, for a given value of ω. The corresponding equation for the choked mass flux is:

$$G_c^* = \frac{\eta_c}{\sqrt{\omega}} \tag{29}$$

This also follows from the expression for the choked flow criterion [e.g., Eq. (7)], which occurs when the fluid velocity reaches the velocity of sound (c):

$$c = \sqrt{\left(\frac{\partial P}{\partial \rho}\right)_s} \quad \text{or} \quad G_c = \sqrt{\rho^2\left(\frac{\partial P}{\partial \rho}\right)_s} = \sqrt{\left(-\frac{\partial P}{\partial \rho}\right)_s} \tag{30}$$

By setting $k = 1.001$, these expressions reduce to the classical isentropic nozzle equations for an ideal gas. For $\omega = 0$ (i.e., $\alpha_0 = 0$, or all-liquid flow) the equations reduce to the incompressible Bernoulli equation, Eq. (4), which can be expressed in the above dimensionless variables as:

$$(G^*)^2 = 2(1-\eta_b) \tag{31}$$

where $\eta_b = P_b/P_0$. Other special cases of the Omega Model will be discussed below.

Any of the computer programs SAFIRE, SuperChems, or TPHEM can be used for frozen flow calculations.

One possible deficiency of this model is related to the "no slip" assumption. Slip effects are likely to be important where the pressure gradient is large (e.g., at or near choking conditions), although there is insufficient evidence to establish any firm quantitative guidelines for when slip effects must be considered. Also, choking may occur over a much wider range of pressure ratio than for single phase gas flow, depending upon the quality and other properties of the mixture.

There are no reliable data or firm guidelines for the appropriate value of the discharge coefficient to use for frozen two-phase flow. Because a relatively small quality can represent a relatively large volume fraction of gas, the discharge coefficient for critical (choked) flows with significant quality will probably be closer to that for gas than for liquid. It is therefore likely that the coefficient for critical flow would be in the range of 0.8 to 0.9, whereas for subcritical flow it would be more likely to be on the order of 0.6 to 0.7.

5.2.2 Homogeneous Equilibrium Model (HEM)

HEM is a "generic" term for any model that applies under conditions of homogeneous flow and local thermodynamic equilibrium. That is, the possibility of slip or nonequilibrium (delayed flashing) effects is not considered. Although the HEM conditions also apply to the frozen flow case described above, common usage of the HEM term usually implies flashing flow. The assumption of equilibrium between the liquid and vapor phases is typically valid for cases in which a two-phase mixture of liquid and vapor with quality of 10% or more enters the valve. It is also valid when flashing is initiated within the valve/nozzle, provided there is sufficient flow length following the initiation point for equilibrium to be established. Studies have shown (e.g. Fauske, 1985) that this "delay length" is typically about four inches (10 cm). Thus, the HEM model would not be expected to apply for nozzles with lengths less than 10 cm long for subcooled entering conditions.

Subject to the above restrictions, the HEM can take on a variety of forms depending upon how the density of the two-phase mixture is determined (as a function of pressure) and how the result is used to evaluate Eq. (3). For either single- or multicomponent fluids, the vapor–liquid ratio, and hence the quality and mixture density, can be determined by an energy (e.g. entropy or enthalpy) balance and a series of flash calculations at various pressures from P_0 to successively lower pressures, using appropriate VLE data for the mixture. The local pressure (P_n) is reduced until either the nozzle exit pressure is reached or the mass flux reaches a maximum (corresponding to choked flow), whichever comes first. This can be done using any of the computer programs (e.g. SAFIRE, SuperChems, or TPHEM) as well. Using TPHEM, the mixture density is determined at several pressures from the flash calculations. These values are provided as input data to the program along with a specification of which of several models (see Table 1) is to be used to fit the data. This information is then used by the program to evaluate Eq. (3).

The Omega Method, Eqs. (27)–(29) can also be applied to flashing flows, if the appropriate value of ω is used. This value can be determined from (ρ, P) data at two pressures [e.g. Eq. (26)], evaluated from flash calculations as described above. If the Clausius-Clapeyron equation is employed to estimate the saturation properties of the liquid, the value of ω can be determined from fluid properties at the stagnation state only according to (Leung, 1996):

$$\omega = \alpha_0 \left(1 - 2\frac{P_0 v_{LG0}}{h_{LG0}}\right) + C_p T_0 P_0 \rho_0 \left(\frac{v_{LG0}}{h_{LG0}}\right)^2 \tag{32}$$

where

$$\alpha_0 = \frac{x_0 \rho_0}{k \rho_{G0}} \tag{33}$$

is the volume fraction of gas/vapor in the mixture at the upstream (stagnation) state. The parameter ω represents the compressibility of the mixture—the first term in Eq. (32) applies to the initial (entering) gas/liquid mixture and the second term represents the effect of the phase change upon flashing. The assumptions inherent in Eq. (32) are: (a) isenthalpic expansion; (b) $P v_{GL} = P_0 v_{Gl0}$, with $v_{GL} \approx v_G$, and; (c) constant properties, that is, $T = T_0$, $h_{GL} = h_{GL0}$, $C_{pL} = C_{pL0}$, and $v_L = v_{L0}$.

Thus, the value of ω is directly related to the initial phase state of the mixture as follows:

Flashing flow:	$\omega > 1$	
All gas/vapor flow:	$\omega \approx 1$	
nonflashing ("frozen")flow:	$0 < \omega < 1$	$(\omega = \alpha_0/k)$
Liquid Flow:	$\omega = 0$	$(\alpha_0 = 0)$

The Omega Method was developed primarily for one-component flashing fluids and two-component nonflashing systems subject to the homogeneous-equilibrium conditions. It may also be applied to multicomponent mixtures which exhibit a relatively minor composition change during flashing;. For such mixtures, average physical properties should be estimated based on the liquid phase composition, except for the heat of vaporization and vapor specific volume, which should be based on the vapor phase composition. The model has been shown to give good results for mixtures with a relatively narrow boiling range (e.g. a range of 80°C or less) over a wide range of quality values (Nazzario and Leung, 1992). For wide-boiling-range mixtures and nonideal VLE systems, especially those containing hydrogen, the method underpredicts the flow rate significantly. For most hydrogen-containing mixtures, better results are obtained if the nonflashing version is used, with $\omega = \alpha_0$. The method is also limited to conditions which are not close to the thermodynamic critical point, e.g. at reduced temperature less than 0.9 and reduced pressure less than 0.5 (Leung, 1996).

Critical (choked) flow generally occurs at considerably higher pressure ratios for flashing systems than for nonflashing two-phase or single phase flows. It is not unusual for choked flow to occur at $\eta_c = 0.9$ or higher, so that choking is the usual condition for flashing flows.

Again, there are no accepted data from which reliable values of the discharge coefficient for flashing flows can be obtained or estimated. However, for $x > 0.01$ and critical conditions, the nozzle/valve flow behavior would be expected to be similar to that for gas only flow, which would imply a discharge coefficient of about 0.8 to 0.9. For lower qualities, the liquid contributes a greater resistance to the flow, so that the coefficient would be more likely to be in the range 0.6–0.7.

5.2.3 Equilibrium Rate Model (ERM)

If the liquid entering the nozzle is saturated and thermodynamic equilibrium applies, flashing will occur when the pressure reaches the saturation value. This will occur only if the flow channel is long enough, for example, $L > 10$ cm. This model assumes that no flashing occurs until the pressure reaches the choke point at the nozzle exit. By utilizing the critical mass flux condition, Eq. (7), a constant enthalpy criterion to determine the amount of liquid flashed to vapor, and the Clausius-Clapeyron equation, the critical (choked) mass flux can be found to be (Fauske, 1985):

$$G_c = \frac{h_{LG}}{v_{LG}\sqrt{T_s C_{p_L}}} \tag{34}$$

This applies to initially saturated (zero quality) liquid. If the entering fluid is a two-phase mixture with an initial quality x_0, Eq. (34) can be modified accordingly to give (FAI, 1983):

$$G_c^{-2} = \frac{x_0 v_G}{kP_c} + (C_{p_L} T - x_0 h_{LG})\left(\frac{v_{LG}}{h_{LG}}\right)^2 \tag{35}$$

The predicted mass flux from this model is reported to be about 10% higher than that from the HEM model for flashing flow in long channels.

5.2.4 Slip-Equilibrium Model (SEM)

As vapor is generated by flashing, or as the pressure drops rapidly through the nozzle/valve, the gas/vapor phase expands and accelerates relative to the liquid phase. This effect is enhanced markedly for a flashing liquid because of the rapid generation of vapor due to flashing, and is even more pronounced as the critical velocity is approached since the (negative) pressure gradient becomes ever steeper as the speed of sound is approached. Consequently, the gas phase moves faster than the liquid resulting in "slip" or a relative velocity between the phases, as discussed in Section 4.4. The HEM model may significantly underpredict the mass flux under these conditions. This model may be applied to either frozen or flashing flow, and assumes thermodynamic equilibrium between the phases. However, some sources also refer to this as a "nonequilibrium" model, since slip implies that the two phases are not in "mechanical equilibrium."

The slip ratio, defined in Eq. (14), affects the gas phase holdup and hence the density of the two-phase mixture as shown in Eqs. (15)–(21), which illustrate the variety of expressions for the mixture density that can result from various considerations involving slip. However, under typical conditions there is usually very little difference in the two-phase density calculated by these various expressions.

Slip is a function of the relative densities of the two phases, and various expressions for the slip ratio S have been derived by various methods. Moody (1965) found

$$S = \left(\frac{\rho_L}{\rho_G}\right)^{1/3} \tag{36}$$

but determined that the value which maximizes the mass flux is

$$S = \left(\frac{\rho_L}{\rho_G}\right)^{1/2} \tag{37}$$

which was originally derived theoretically by Fauske (1962). The Moody model is reported to consistently overpredict the mass flux (e.g., Riester and Bajura, 1990). Darwish (1978) recommended experimentally derived exponents of 1.4 for horizontal pipe, and 1.9 for vertical pipes in place of the exponents of 1/3 or 1/3. However, Chisholm (1973) determined

$$S = \left(\frac{\rho_L}{\rho_G}\right)^{1/2} = \left[1 + x\left(\frac{\rho_L}{\rho_G} - 1\right)\right]^{1/2} \tag{38}$$

where

$$\frac{1}{\rho_{TP}} = \frac{x}{\rho_G} + \frac{1-x}{\rho_L}$$

For orifice flow, Chisholm (1977) concluded that Eq. (38) applies for $X > 1$, whereas for $X < 1$:

$$S = \left(\frac{\rho_L}{\rho_G}\right)^{1/4} \tag{39}$$

where X is the Lockhart-Martinelli (1949) two-phase parameter:

$$X = \left(\frac{\Delta P_L}{\Delta P_G}\right) = \frac{1-x}{x}\left(\frac{\rho_G}{\rho_L}\right)^{1/2} \tag{41}$$

Grolmes and Coats (1997) summarized many of the slip models in terms of a general equation of the form (which was adapted from Butterworth, 1975):

$$S = a_0 \left(\frac{1-x}{x}\right)^{(a_1-1)} \left(\frac{\rho_L}{\rho_G}\right)^{(a_2-1)} \left(\frac{\mu_L}{\mu_G}\right)^{a_3}$$

where the values of the parameters are shown in Table 2.

From a general expression for the momentum flux for separated flow, Chisholm (1983b) deduced the following expression for the two-phase density in slip flow:

$$\begin{aligned}\frac{1}{\rho_{TP}} &= \left(\frac{x}{\rho_G} + \frac{S(1-x)}{\rho_L}\right)\left(x + \frac{1-x}{S}\right) \\ &= v_L + (v_G - v_L)[Bx(1-x) + x^2] \end{aligned} \tag{42}$$

where B is given by

$$B = \frac{\frac{1}{S}\left(\frac{\rho_L}{\rho_G}\right) + S - 2}{\frac{\rho_L}{\rho_G} - 1} \tag{43}$$

If $\rho_L / \rho_G >> 1$, this reduces to

$$B = \frac{1}{S} \tag{44}$$

As previously mentioned, there are no strict guidelines for determining when slip must be considered, although it appears to be important in high

TABLE 2

Parameters for Slip Model Equation (Grolmes and Coats 1997)

Model	a_0	a_1	a_2	a_3
Homogeneous	1	1	–1	0
$S = (\rho_L/\rho_G)^{1/2}$ (Fauske, 1962)	1	1	–1/2	0
$S = (\rho_L/\rho_G)^{1/3}$ (Moody, 1965)	1	1	–2/3	0
Thom (1964)	1	1	–0.89	0.18
Baroczy (1966)	1	0.74	–0.65	0.13
Lockhart-Martinelli (t-t) (1949)	0.28	0.64	0.36	0.07
Constant $S = S_0$	S_0	1	–1	0

pressure gradient regions for any two-phase flow as well as for low quality flashing flows. In fact, the net effect of slip is to increase the mass flux, which is also the result of thermodynamic nonequilibrium in flashing flows, and some data for slightly subcooled flashing flows can be reproduced by either an appropriate slip model or a nonequilibrium model. However, there are no guidelines which would indicate which of the many slip models is appropriate under given conditions, and there is no evidence to indicate whether or not the presence has any effect on the valve discharge coefficient.

5.2.5 Homogeneous Nonequilibrium Model (HNE)

As discussed in Section 5.2.2, thermodynamic equilibrium implies that vaporization of the liquid will occur instantaneously at the point where the pressure drops to the saturation pressure. However, if the entering fluid is saturated or slightly subcooled, a finite time is required for nucleation to occur and bubbles to grow due to the heat and mass transfer rate processes involved. This results in what is sometimes called a "metastable" or superheated liquid state. The effect of this nonequilibrium condition can be described in terms of a finite superheat required to initiate nucleation, or an equivalent delay time, or an equivalent relaxation distance that the liquid must travel during this delay time under critical (choked) flow conditions (Fauske's approach). Lackme' (1979) (see also Feburie et al., 1993) assumed that nucleation takes place when the pressure drops to 95 to 97% of the saturation pressure.

Henry and Fauske (1971) accounted for the nonequilibrium delay length in their "homogeneous nonequilibrium (HNE) model" by including a nonequilibrium parameter N in the basic ERM model, which assumes that no flashing occurs between the stagnation and choke point. The parameter N in the HNE model represents the fraction of equilibrium flashing that takes place at the choke point. There is evidence that there is a delay time for nucleation of the order of 1 ms (Khajehnajafi and Shinde, 1994, Schrock, et al., 1977), which corresponds to a length of travel of about four inches (10 cm) at a typical critical (sonic) velocity of the order of 400 ft/s. This is in agreement with the value of the "relaxation length" as determined by Fauske (1985) from comparison of the model with a variety of data from the literature. This "relaxation time" or "relaxation length" can also be interpreted in terms of a corresponding degree of superheat, which is on the order of 2–3°C.

Various forms or the HNE model have been presented over the years. The most recent version was presented by Fauske (1997) in the following form, which also includes a friction loss term to account for viscous effects in low Reynolds number flow:

$$G_c = \left[\frac{1 - x_0}{G_{x_0}^2} + \frac{x_0}{G_{x_1}^2} \right]^{-1/2} \tag{45}$$

where:

$$G_{x_0}^2 = \frac{K_{DL}^2\left[G_0^2 + G_1^2/N_{NE}\right]}{1+K_f} \tag{46}$$

$$G_{x_1}^2 = \frac{K_{DG}^2 \dfrac{P_0^2 M_w}{RT_0}\left(\dfrac{1}{k+1}\right)^{\frac{k+1}{k-1}}}{1+K_f} \tag{47}$$

$$N_{NE} = \left(\frac{G_1}{G_3}\right)^2 + \frac{L}{L_e} \quad \text{for} \quad L \le L_e = 10\text{cm}$$

$$N_{NE} = 1 \quad \text{for} \quad L > L_e \tag{48}$$

$$G_0 = \sqrt{2\rho_{LO}(P_0 - P_s)} \tag{49}$$

$$G_1 = \frac{h_{GLO}\rho_{GO}}{\sqrt{T_0 C_{P_L}}} = G_{ERM} \tag{50}$$

$$G_3 = \sqrt{2\rho_{LO}(P_0 - P_2)} \tag{51}$$

discharge pressure. The terms K_{DL} and K_{DG} are the (measured) discharge coefficients for the specific Here, x_0 is the quality of the entering mixture, L_e = 10 cm is the length required for equilibrium to be established, P_s is the saturation pressure corresponding to T_0 , and P_2 is the valve for liquid and gas flow, respectively (e.g. the "Redbook" values). The term K_f accounts for friction loss in the nozzle/pipe, which would be significant for high viscosity (low Reynolds number) flow. For flow in pipes (or nozzles) with large values of L/D, the friction loss coefficient K_f is determined by the (Fanning) friction factor by the definition:

$$(K_f)_{\text{pipe}} = \frac{4fL}{D} \tag{52}$$

For a subcooled flashing fluid, the nonequilibrium model should be applicable for nozzles of 10 cm or less in length. However, the assumption of local equilibrium between the vapor and liquid phases (the HEM model) appears to be reasonable for an inlet flow with a quality of 10% or higher regardless of length (Schrock et al., 1977), or if the phases have coexisted for a sufficient period of time (e.g., nozzles longer than about 10 cm). According to DIERS (1992), the critical (choke) pressure ratio predicted by the HNE model is in the range of 0.85–0.95 for an initial quality below 10%.

Simpson (1991) suggested that the effect of delayed flashing could be accounted for by replacing the quality (x) in the integral form of the model [e.g., Eq. (3), or Eq. (21) if slip is present] by a "nonequilibrium" quality, x_{ne}. Simpson's expression has subsequently been modified to include an additional nonequilibrium parameter, K_{NE}, where

$$x_{ne} = x_0 + K_{NE}(x^2 - x_0^2) \qquad \text{where} \quad 0 \le x_{ne} \le x \tag{53}$$

and x_0 is the quality at the upstream saturation condition. The model is said to yield results comparable to the Henry-Fauske HNE model for $K_{NE} = 1$. This capability has been incorporated into the latest version of the TPHEM computer model.

An equivalence between nonequilibrium and slip was also determined by Chisholm (1983a) who defined a nonequilibrium factor N as:

$$N = \frac{x_N}{x} \tag{54}$$

which is the ratio of the actual quality (mass fraction vapor/gas), x_N, to that which would exist under equilibrium conditions. He showed the following relation between this parameter and the slip ratio:

$$N = BS = \frac{(\rho_L/\rho_G) + S^2 - 2S}{(\rho_L/\rho_G) - 1} \tag{55}$$

where B is the parameter given in Eqs. (42)–(43). This indicates that slip can be described as a nonequilibrium process, and is most significant at low values of x (e.g. below about 0.01). The Chisholm nonequilibrium parameter N given by Eq. (54) or (55) is equivalent to the reciprocal of the Henry-Fauske nonequilibrium parameter N_{NE} of Eqs. (47) and (48) (e.g., Henry and Fauske, 1970, 1971).

6. RECOMMENDATIONS

Based on the findings of this study, a summary of the models currently recommended for engineering design calculations for the purpose of predicting the mass flux through a safety relief valve under various entering flow conditions is presented in Table 3. Because of the scarcity of direct experimental verification of these models under the appropriate conditions, and the resulting tentative nature of the conclusions, no guarantee of the validity of any these recommendations or relations is expressed or implied.

TABLE 3
Recommended Guidelines

Entering Fluid State	G	G_c	K_D
		SINGLE-PHASE	
Liquid $N_{Re} > 10{,}000$ $N_{Re} < 5{,}000$	Bernoulli Eq. (4)		"Red Book" value Use K_v, Eq. (8)
Gas—Critical		Ideal Gas Eqs. (8), (9) ($P=P_c$)	"Red Book" value
Gas—Subcritical	Ideal Gas Eq. (8) ($P = P_b$)		0.65– 0.85
		TWO-PHASE	
Frozen	Eq. (22) or (25,27) $P_n = P_b$ Slip (?)	Eq. (22) or (25,29) $P_n = P_c$, Eq. (28) Slip (?)	Critical: 0.8 - 0.9 Sub-Crit: 0.6 - 0.7
Flashing—Single Component or "Pseudo Single-Single Component" Mixture			
Two-Phase $x > 0.1$ $x < 0.01$	HEM Computer[1] Eqs. (27) , (26) or (32) $P_n = P_b$ Slip (?)	HEM Computer[1] Eqs. (27) , (26) or (32) $P_n = P_c$ Eq. (28) Slip (?)	 Critical: 0.7–0.8 Sub-Crit: 0.6–0.7 0.6 – 0.7
Saturated		HNE, HEM, or ERM Computer[1] Eqs. (27) , (32) or (34) $P_n = P_c$, Eq. (28)	0.5 – 0.7
Subcooled $L \leq 10$ cm: $N_{NE} = 1$ $L > 10$ cm, N_{NE}		Computer[1] HNE, Eqs. (44)–(51) HEM or HNE Eqs. (27) , (26) or (32) $P_n = P_c$ Eq. (28)	0.5–0.7 or "Red Book gas and liquid K_D's with HNE Eqs.; liquid K_D with HEM model
Hybrid		Computer[1], HEM Eqs. (27), (28), (25)	0.5–0.7

[1] If available, the preferred method for all two-phase conditions would be the use of one of the computer programs having the desired capability (e.g. TPHEM, SAFIRE, SuperChems for DIERS). Due care must be given to identifying the proper input parameters to these programs which correspond to the conditions of interest with regard to the entering fluid state, nozzle length, etc. Also, a suitable database of fluid thermodynamic properties and/or a multi-component flash routine is required

TABLE 3
Recommended Guidelines (continued)

Entering Fluid State	G	G_c	K_D
Flashing—Nonideal Multicomponent or "Near-Critical" State			
Initially Two-Phase	HEM, HNE Computer Eq. (3) (+VLE data) or Eqs.(26), (27), or (45)–(51) $P_n = P_b$	HEM, HNE Computer[1] Eq. (3) (+VLE data) or Eqs.(26), (27), or (45)–(51) $P_n = P_c$ Eq. (28)	Critical: 0.8–0.9 Sub-Crit: 0.6–0.7
Saturated		HEM, HNE Computer[1] Eq. (3) (+VLE data) or Eqs.(26), (27), or (45)–(51) $P_n = P_c$ Eq. (28)	0.6–0.7
Subcooled $L \leq 10$ cm: $N_{NE} = 1$ $L > 10$ cm, N_{NE}		HNE Computer[1] Eqs. (45)–(51) HEM, HNE Computer[1] Eq. (3) (+VLE data) or Eqs. (26), (27) $P_n = P_c$ Eq. (28) or Eqs. (45)–(51)	0.6–0.7 or "Red Book" K_D's with HNE 0.6–0.8 or liquid K_D with HEM
Hybrid		HEM, Computer[1] Omega	0.6–0.8
Vapor + Liquid + Hydrogen		HEM, Computer[1] Frozen $\omega = \alpha_0$	0.6–0.8

[1] If available, the preferred method for all two-phase conditions would be the use of one of the computer programs having the desired capability (e.g. TPHEM, SAFIRE, SuperChems for DIERS). Due care must be given to identifying the proper input parameters to these programs which correspond to the conditions of interest with regard to the entering fluid state, nozzle length, etc. Also, a suitable database of fluid thermodynamic properties and/or a multicomponent flash routine is required

The above recommendations apply for high Reynolds number flows, e.g. $N_{Re} \geq 10{,}000$ (about). The Reynolds number is calculated using the nozzle diameter and the volume-weighted average density and viscosity of the two-phase mixture, Eqs. (13) and

$$\mu = \alpha\mu_G + (1-\alpha)\mu_L \tag{56}$$

where α is given by Eq. (15) with $S = 1$. For $N_{Re} < 10{,}000$, the discharge coefficient should be multiplied by the "viscosity correction factor," given in Eq. (6).

ACKNOWLEDGMENTS

The author wishes to thank the DIERS Users Group of the AIChE and the Mary Kay O'Connor Process Safety Center of the Texas A&M Engineering Experiment Station for support of this work. Valuable assistance and specific comments have been provided by H. G. Fisher, S. G. Coats, J. E. Huff, H. S. Forest, and F. E. Self of the DIERS Users Group, as well as the members of the Gulf Coast Relief Valve Consortium, composed of the Chemical and Mechanical Engineering Departments of Texas A&M University; the Chemical Engineering Department of Prairie View A&M University; Anderson, Greenwood and Company; Bechtel Engineers and Constructors; Dow Chemical Company; Dresser Valve and Controls Division; Kvaerner/John Brown Engineers and Constructors; and Teledyne Fluid Systems, Farris Engineering.

NOTATION

A area (normal to flow)
c speed of sound
C_D nozzle flow coefficient
D pipe diameter
f Fanning friction factor
G mass flux
G^* dimensionless mass flux = $\sqrt{G/(P_0\, \rho_{L0})}$
h enthalpy per unit mass
k C_p/C_v (ratio of specific heats for gas/vapor)
K_D valve discharge coefficient
K_f loss coefficient
K_v viscous correction factor for K_D
L length (pipe or nozzle)
L_e relaxation length for nonequilibrium flashing flow
$\dot{m}$ mass flow rate
N nonequilibrium model parameter
N_{NE} nonequilibrium model parameter
N_{Re} Reynolds number (based on nozzle conditions)
P pressure
s entropy per unit mass
S slip ratio = V_G/V_L
T absolute temperature
u internal energy per unit mass
V velocity
x quality (mass fraction gas/vapor)
X Lockhart-Martinelli two-phase pressure drop parameter

Subscripts

b valve exit
c critical (choked) value
e equilibrium
G gas/vapor
L liquid
LG change in property from liquid to gas phase
n nozzle
s saturation conditions
t nozzle throat
TP two-phase
0 vessel (stagnation) conditions
1 entrance of inlet piping upstream of valve
1,2 arbitrary reference points (usually, 1 = valve inlet, 2 = valve outlet)

Greek

α volume fraction of gas/vapor
β d/D_1
η P/P_0
ν specific volume = $1/\rho$
μ viscosity
ρ density = $1/\nu$
ω two-phase compressibility parameter

REFERENCES

AGCO (Anderson, Greenwood & Co.) (1996), Company Literature on Valve Sizing.

Alvi, S.H., K. Sridharan, and N.S. Lakshmana Rao (1978), "Loss Characteristics of Orifices and Nozzles," *J. Fluids Eng.*, 100, 299–307.

American Petroleum Institute (1993), "Sizing, Selection, and Installation of Pressure-Relieving Devices in Refineries, Part I—Sizing and Selection," API Recommended Practice 520, Sixth Edition, March.

ASME (1986), Boiler and Pressure Vessel Code, Appendix 11: Capacity Conversions for Safety Valves, p. 610. New York: American Society of Mechanical Engineers

Baroczy, C. J. (1966), "A Systematic Correlation for Two-Phase Pressure Drop," *Chemical Engineering Progress Symposium Series,* 62(44), 232–249,

Butterworth, D. A. (1975), "A Comparison of Some Void Fraction Relationships for Co-current Gas–Liquid Flow," *Int. J. Multiphase Flow,* 1, 845.

CCPS (Center for Chemical Process Safety) (1998), *Guidelines for Pressure Relief and Effluent Handling Systems,* New York: American Institute of Chemical Engineers. Forthcoming.

Chisholm, D. (1973), "Research Note: Void Fraction During Two-Phase Flow," *J. Mech. Eng. Sci.*, 15(3, June), 235–236.

Chisholm, D. (1977), "Research Note: Two-Phase Flow Through Sharp-Edged Orifices," *J. Mech. Eng. Sci.,* 19(3, June), 128–130.

Chisholm, D. (1983a), "Gas–Liquid Flow in Pipeline Systems," in N.P. Cheremisinoff and R. Gupta (Eds), *Handbook of Fluids in Motion,"* pp.483–513, Ann Arbor: Ann Arbor Science.

Chisholm, D. (1983b), *Two-Phase Flow in Pipelines and Heat Exchangers,* London: Geroge Godwin (New York: Longman).

Darby, R. (1996), *Chemical Engineering Fluid Mechanics,* New York: Marcel Dekker.

Darby, R., and K. Molavi (1997), "Viscosity Correction Factor for Emergency Relief Valves," *Process Safety Progress,* 16(2), 80–82.

Darwish, M. A. (1978), "Two-Phase Flashing Flow in Conduits," *Desalination,* 26, 285–297.

DIERS, *Emergency Relief System Design Using DIERS Technology* (1992), H.G. Fisher, H.S. Forrest, S.S. Grossel, J.E. Huff, A.R. Muller, J.A. Noronha, D.A. Shaw, and B.J. Tilley, New York: The Design Institute for Emergency Relief Systems of the American Institute of Chemical Engineers

Fauske, H. K. (1962), "Contribution to the Theory of Two-Phase, One-Component Critical Flow," Argonne National Laboratory Report ANL-6673, October.

Fauske and Associates, Inc. (1983), "Emergency Relief Systems for Runaway Chemical Reactions and Storage Vessels: A Summary of Multiphase Flow Methods," Report FAI/83-27, October.

Fauske, H. K. (1985), "Flashing Flows or: Some Practical Guidelines for Emergency Releases," *Plant/Operations Progress,* 4(3), 132–134.

Fauske, H. K. (1997), "General Guidelines for Emergency Relief Systems," Presented to the 21st DIERS Users Group Meeting, Pittsburgh, PA, October 29.

Feburie, V., M. Giot, S. Granger, and J.M. Seynhaeve (1993), "A Model for Choked Flow Through Cracks with Inlet Sub-Cooling," *Int. J. Multiphase Flow,* 19(4), 541–562.

Grolmes, M. A.. and J. C. Leung (1985), "Code Method for Evaluating Integrated Relief Phenomena," *Chem. Eng Prog.,* 47–52, August.

Grolmes, M. A.. and S. G. Coats (1997), "Progress Report on High viscosity Two-Phase Flow Modeling, Part II," Presentation for 20th DIERS Users Group Meeting, Phoenix, AZ, April 1–15

Henry, R. E., and Fauske, H. K. (1970), "Two-Phase Critical flow at Low qualities Part II: Analysis," *Nuc. Sci. and Eng,* 41, 92–98.

Henry, R.E. and H.K. Fauske (1971), "The Two-Phase Critical Flow of One-Component Mixtures in Nozzles, Orifices, and Short Tubes," *J. Heat Transfer, Trans. of the ASME,* 179–187, May.

Khajehnajafi, S., and A. Shinde (1995), "Prediction of Discharge Rate From Pressureized Vessel Blowdown Through Sheared Pipe," *Process Safety Progress,* 14(1), 22–25.

Lackme', C. (1979), "Incompleteness of the Flashing of Superheated Liquid and Sonic Ejection of the Produced Phases," *Int. J. Multiphase Flow,* 5, 131–141.

Leung, J. C. (1995), "Simplified Vent Sizing Methods Incorporating Two-Phase Flow," in G. A. Melhem and H. G. Fisher (Eds.), *Proceedings of the International Symposium on Runaway Reactions and Pressure Relief Design,* Boston, MA, August 2–4, pp. 200–236, New York: AIChE.

Leung, J. C. (1996), "Easily Size Relief Devices and Piping for Two-Phase Flow," *Chem. Eng Prog.,* pp. 28–50, December.

Lockhart, R. W., and R. C. Martinelli (1949), "Proposed Correlation of Data for Isothermal Two-Phase Two-Component Flow in Pipes," *Chem. Eng. Prog,* 45(1), 39–48

Martinelli, R. C. and D. B. Nelson (1948), "Prediction of Pressure Drop During Forced-Convection Boiling of Water," *Trans. ASME,* 70, 695–702

Melhem, G. A., H. G. Fisher, and D. A. Shaw (1995), "An Advanced Method for the Estimation of Reaction Kinetics, Scaleup, and Pressure Relief Design," *Process Safety Progress,* 14(1), 1–21

Melhem, G. A. and H. G. Fisher (1997), "An Overview of SuperChems for DIERS: A Program for Emergency Relief System and Effluent Handling Designs," *Process Safety Progress,* 16(3), 185–197.

Miller, R. W. (1983), *Flow Measurement Engineering Handbook,* New York: McGraw-Hill.

Molavi, K. (1996), "Viscosity Effects on Discharge Coefficients of Safety Relief Valves," Presentation to the 18th DIERS Users Group Meeting, Houston, TX, February 6.

Moody, F. J. (1965), "Maximum Flow Rate of a Single Component, Two-Phase Mixture," *Trans. ASME, J. Heat Trans.*, 134–142, February.

Morris, S. D. (1991), "Compressible Gas–Liquid Flow Through Pipeline Restrictions," *Chem. Eng. Prog.*, 30, 39–44.

Morris, S. D. (1996a), "Liquid Flow Through Safety Valves: Diameter Ratio Effects on Discharge Coefficients, Sizing and Stability," *J. Loss Prev. Process Ind.*, 9(3), 217–224.

Morris, S.D. (1996b), "Choke Pressure in Pipeline Restrictions," *J. Hazardous Materials,* 50, 65–69

Nazario, F. N., and J. C. Leung (1992), "Sizing Pressure Relief Valves in Flashing and Two-Phase Service: An Alternative Procedure," *J. Loss Prev. Process Ind.,* 5, 263–269.

The "Red Book" (1997) The National Board of Boiler and Pressure Vessel Inspectors, "Pressure Relief Device Certifications," 1055 Crupper Avenue, Columbus, OH 43229-1183

Riester, J. B. and R. A. Bajura (1990), "Two-Phase Flashing Flow Through Valves," in "Advances in Gas–Liquid Flows—1990," Kim, Rohatgi, and Hashemi (Eds), ASME, FED v. 99, HTD-vol. 155

Schrock, V. E., E. S. Starkman, and R. A. Brown (1977), "Flashing Flow of Initially Subcooled Water in Convergent-Divergent Nozzles," Trans. *ASME, J. Heat Trans,* 99, 263–268, May.

Shaw, D.A. (1990), "SAFIRE Program for Design of emergency Pressure Relief systems," *Chem. Eng. Prog.,* 86(7), 14–17.

Simpson, L. L. (1991), "Estimate Two-Phase Flow in Safety Devices" (with corr.), *Chem. Eng,* 98–102, August.

Simpson, L. L. (1995), "Navigating the Two-Phase Maze," in G. A. Melhem and H. G. Fisher (Eds.), *Proceedings of International Symposium on Runaway Reactions and Pressure Relief Design,* Boston, MA, August 2–4, pp. 394–417, New York: AIChE

Thom, J.R.S. (1964), "Prediction of Pressure Drop During Forced Circulation Boiling of Water," *Int. J. Heat Mass Transfer,* 7(7), 709–724

Wallis, G. B. (1969), *One-Dimensional Two-Phase Flow,*" New York: McGraw-Hill.

Systematic Evaluation of Chemical Reaction Hazards

G. A. Melhem
Arthur D. Little, Inc.
20 Acorn Park
Cambridge, Massachusetts 02140

ABSTRACT

Effective thermal hazard evaluation requires an integrated approach to assessment of chemical reactivity. A variety of theoretical and experimental methods can be used to determine the thermal stability of specific compounds or reactive systems. Notwithstanding this, recurrent problems occur with materials and/or reactions which are known to be inherently unstable. Correct application of physical tests and interpretation of output data is essential for effective thermal hazard evaluation.

This paper provides an integrated approach for the systematic evaluation of thermal hazards and effective scaleup of data for emergency relief system design.

THERMAL HAZARDS ASSESSMENT

In order to develop an effective approach, we need to consider the essential elements which comprise thermal hazard evaluation for pressure relief design purposes. This includes:

1. Screening Checks
2. Data Interpretation
3. Physical Testing

Screening Checks

Before initiating any thermal stability testing program, it is prudent to carry out preliminary screening checks. Complex or unfamiliar materials may exhibit detonation or deflagration properties. In such cases, specialized testing is required. For simple or familiar materials, much information can be confidently predicted by theoretical means, thus reducing the need for time consuming and costly testing. Optimization of experimental parameters for subsequent physical tests can increase cost effectiveness of testing programs.

Typical screening checks might include the following three elements, structural analysis, specialized testing (detonability, etc.) and thermochemical evaluation.

Physical Testing

For thermal analysis in connection with relief system design, we are primarily concerned with the point of incipient instability and with characterization of the runaway process. A variety of calorimetric devices exist which can be used to measure and quantify runaway and/or decomposition phenomena. Other larger-scale devices (e.g., reaction calorimeters) are intended for study of reaction heat release under controlled conditions. Our discussions here will concentrate mainly on the first category above.

A number of practical constraints affect our choice of test methods. Runaway or decomposition phenomena are extremely hazardous, other than on a relatively small scale. Notwithstanding this, generally more representative data can be obtained with increasing experimental sample size. Rapid preliminary screening is often required to target further resources for detailed analysis. Experience indicates that most test methods have inherent advantages and disadvantages.

For purposes of classification, we have defined four generic types of test apparatus.

1. Microthermal Calorimeters
2. Vent Sizing Packages
3. Macrothermal Calorimeters
4. In House Calorimeters

A variety of parameters must be quantified to characterize a runaway or decomposition process for relief system design purposes. These parameters include:

- *System Thermal Stability.* Nature and the relative magnitude of exothermic/ endothermic effects
- *Initial decomposition* temperature under specified experimental conditions
- *Total Heat Release:* Reaction heat release per (mass or molar) unit of material
- *Rate of Heat Release:* Characterized by global kinetic parameters and heat generation as a function of temperature
- *Rate of Gas Generation:* Volatilization of liquid components, vapor expansion effects, and gaseous decomposition products
- *Scaleup/Loading Factors:* Physical characteristics of full-scale plant

- *Vent Flow Characteristics:* Multicomponent/multiphase system properties, continuing chemical reactions, etc.

All of the test methods of interest involve heating a sample of material in a controlled manner and monitoring its thermal behavior, either in absolute terms or with respect to a nonreactive reference (control) sample. Accurate calibration of control and measurement instrumentation is essential to obtain meaningful data. The selected experimental parameters (sample size, containment type, heating rate) can greatly influence results. Understanding of the influence of these parameters is essential for effective experimental implementation and interpretation.

There are four fundamental heating modes which can be employed in thermal analysis experiments of this type. These are illustrated in Figure 1 and include:

1. Scanning
2. Isothermal
3. Isoperibolic
4. Adiabatic

Another important characteristic of any experimental system is thermal inertia. The inertial (or phi) factor tends to unity with increasing reacting mass, for example as encountered in bulk materials. The adiabatic sample self-heating rate (either absolute or relative to an inert reference) can be used

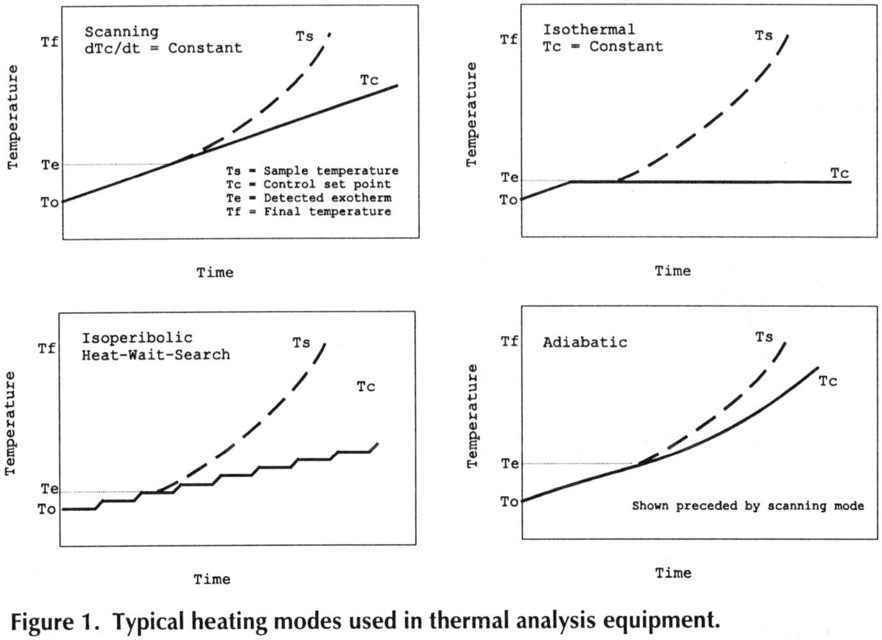

Figure 1. Typical heating modes used in thermal analysis equipment.

to generate pseudo rate data, which in turn yield global kinetic parameters describing the reaction/decomposition process.

Microthermal calorimeters are commercial instruments, primarily developed for other (nonhazardous) thermal analysis purposes. These calorimeters typically employ milligram scale samples (e.g., 1–20 mg). They provide highly accurate temperature measurement/control, but relatively low thermal detection sensitivity due to inertial effects. Experimental options include open and closed sample containers, system pressurization, noble metal containers (to avoid catalysis/inhibition effects). Heating options include scanning, isothermal, and isoperibolic.

There are three principal types of microthermal calorimetric methods. The operational principle of these devices involves monitoring thermal or physical transitions within the sample as it undergoes a predetermined heating cycle.

- *Differential Thermal Analysis (DTA):* DTA measures any sample temperature deviations from the programmed temperature profile or with respect to an inert reference (control) sample.
- *Differential Scanning Calorimetry (DSC):* DSC measures any decrease in heat input required to maintain a specified temperature profile within the sample, usually compared to an inert reference.
- *Thermogravimetric Analysis/Differential:* TGA measures weight changes in the thermogravimetry (TGA)sample as a function of temperature, either in absolute terms or with respect to a control reference.

Microthermal methods have a number of inherent advantages and disadvantages.

Advantages

1. Rapid automated screening
2. Low hazard potential due to small sample size
3. Variety of heating modes and heating rates possible
4. Limited sample requirements
5. Useful for characterizing thermal signatures
6. DSC can provide accurate heat of reaction measurement

Disadvantages

1. Representative samples only possible for highly homogenous materials
2. Inertial effects reduce detection sensitivity and can inhibit reaction
3. Rapid loss of sensitivity at high scanning rates
4. No pressure generation measurement possible
5. Kinetic evaluations possible, but frequently misleading for hazard evaluation purposes
6. Heats of reaction misleading if sample containment falls

Macrothermal calorimeters are commercial instruments developed specifically for the study of thermal decomposition processes. These calorimeters typically employ gram scale samples (e.g., 1–100g). Improved thermal detection sensitivity is possible due to reduced inertial effects. Experimental options include open and closed, noble metal or glass sample containers, and direct measurement of reaction pressurization effects. Heating options include scanning, isothermal, isoperibolic, adiabatic and combinations. A variety of proprietary instruments of this type have been marketed; for these instruments the operational principle involves programmed heating in conjunction with accurate measurement of sample temperature and/or pressure.

1. *Accelerating Rate Calorimeter (ARC):* ARC measures sample containment "bomb" temperature with respect to air jacket temperature
2. *Automatic Pressure Tracking Adiabatic Calorimeter (APTAC):* APTAC measures the temperature and pressure of the sample containment, while allowing for reagent addition and controller/simple sample venting.
3. *Reactive System Screening Tool (RSST):* RSST measures sample temperature and self-heating rate during a programmed temperature ramp
4. *SEDEX Calorimeter:* SEDEX measures sample temperature and/or pressure with respect to surrounding external oven temperature
5. *SIKAREX Calorimeter:* SIKAREX measures sample temperature with respect to external air jacket temperature

Macrothermal calorimeters have a number of inherent advantages and disadvantages.

Advantages

1. Composition of large sample is more representative in most cases
2. Reduced inertial effects improve relative thermal detection sensitivity
3. Temperature and pressure monitoring possible
4. Variety of heating modes and containment types possible

Disadvantages

1. More onerous and less reproducible experimental parameters (calibration, set up, interpretation, clean up)
2. Increased operator and equipment hazard potential due to increased sample size
3. Robust sample and/or equipment containment necessary to contain pressurization
4. Loss of sample containment probable (potential occupational exposure and disposal liability
5. Can have "placebo effect" on hazard analysts/management

A number of companies and industry groups have developed their own noncommercial apparatus for thermal hazard evaluation. These calorimeters typically employ gram scale samples (e.g., 1–100 g). They often use common laboratory apparatus such as Dewar vessels, drying or old GC ovens, and standard thermocouples in conjunction with customized sample containers. Published examples include Union Carbide, ICI organics, Sterling Organics, ABPI, and Hoechst.

In house methods have a number of inherent advantages and disadvantages.

Advantages

1. Highly cost effective vs. commercial instruments
2. Can be tailored to suit company products, testing philosophy and data requirements
3. Encourages operator to confront practical limitations on testing and results
4. Has "researcher appeal" and stimulates continuous improvement

Disadvantages

1. Potentially dangerous (e.g., explosion risk, occupational exposure risk)
2. Potentially idiosyncratic, nonuser friendly and reliant on frequent calibration/ maintenance
3. Requires skilled operator for safety/operability/interpretative purposes
4. Data output nonstandard; difficult for comparative interpretation or demonstrating regulatory compliance

Emergency relief system sizing packages are intended to characterize runaway processes (including vent flow phenomena) directly by means of physical simulation of plant conditions. These tools typically employ 10–100 gram reactant. Sample stirring and reactant addition are possible to simulate realistic chemical reaction processes. Inertial effects are reduced by large relative sample size and light (hence mechanically weak) containment construction. Containment integrity is maintained by equalization of external/internal pressure using inert gas padding. A good example of an emergency relief system sizing package is marketed by Arthur D. Little (APTAC) and Fauske & Associates (VSP).

Direct experimental determination of venting behavior and vent sizing parameters has a number of advantages and disadvantages.

Advantages

1. Low thermal inertia (typically circa 1.05) experimental set up, which is comparable to real plant conditions

2. Simulated liquid phase reaction possible
3. Flow regime characterization possible

Disadvantages

1. Equipment complex and relatively expensive
2. Difficult to establish and reproduce experimental conditions/measurements
3. Bulk effects (e.g. transport phenomena, thermal conductivity) may still change the nature of reaction in the full-scale plant

In summary, a number of critical experimental parameters which might constrain or influence testing with a specific instrument.

- Containment Type
- Heating Rate
- Thermal Inertia
- Heating Mode
- Data Interpretation

The selection of instrument type, sample containment type and experimental conditions can strongly influence results. Tempering effects can inhibit reaction in open containers. Potentially dangerous exothermic behavior can be masked by inertial effects and rapid temperature scanning, particularly for samples of limited size. Autocatalytic effects, and catalysis/inhibition by sample containment material, can lead to complex or misleading experimental results. Instrument or software characteristics can produce anomalous results.

This has been demonstrated by comparative testing of controlled samples of a range of reactive materials, each exhibiting different thermal behavior histories prior to decomposition. One such material tested is tertiary butyl peroxybenzoate (TBPB).

This has the advantage of being readily available as technical grade product with high purity specification and also the thermal stability and kinetic data from large-scale tests are available in open literature.

This provides a good comparative basis for instrument comparison with respect to exotherm detection sensitivity and extrapolation of global kinetic parameters (see Table 1).

When developing a rigorous testing strategy for thermal hazard evaluation, similar considerations and constraints exist to those which influence our choice of specific instruments and/or methodologies. The requirements of an effective test strategy include:

- Rapid preliminary screening and effective targeting of experimental resources
- Optimal significance of results with minimal operator risk
- Recognition of inherent strengths and weaknesses of specific methods

TABLE 1
Data from Various Thermal Analysis Instruments

Test Method	Experimental Parameters	Sample Mass	Detected Onset (°C)
DTA	10°C/min	8.1 mg	125
DSC	10°C/min	3.4 mg	122
	5°C/min	4.4 mg	101
	1°C/min	6.4 mg	93
ARC	Start: 50°C Heat Step: 10 Wait: 15 min PHI = 2.38	3.5 g	82
SEDEX	0.5°C/min	5.8 g	84
	"Scanning" Experiment	2.0 g	88
SIKAREX	0.125°C/min Scanning	5.0 g	72

In view of this, we propose an integrated, noncommercially aligned approach which incorporates the identified principal requirements: Theoretical/Experimental, Screening Tests, Advanced Numerical and Fluid Dynamic Modeling Techniques, and Detailed Experimental Measurements.

Figure 2 summarizes several popular experimental and theoretical tools that are often used in the characterization of thermal stability, process design data, and runaway reactions. These techniques will be described briefly in the following sections.

THEORETICAL AND COMPUTATIONAL SCREENING

Literature Searches

The first step in analyzing the hazards of a chemical under consideration involves reviewing the literature for any known and available information. Typical literature sources are provided below:

1. *The Encyclopedia of Chemical Technology* (Kirk-Othmer, Eds.), Wiley, New York
2. Handbook of Reactive Chemicals Hazards, 4th Ed., L. Bretherick, Butterworth-Heinemann, Stoneham, MA (1990)
3. National Fire Protection Association, "Manual of Hazardous Chemical Reactions," NFPA, 491M, Quincy, MA (1991)
4. Tokyo Fire Department Compilation: Handbook on Hypergolic Ignition, T. Yoshida (Ed.), Nikkan Kogyo Shinbunsha, Tokyo, Japan (1980).

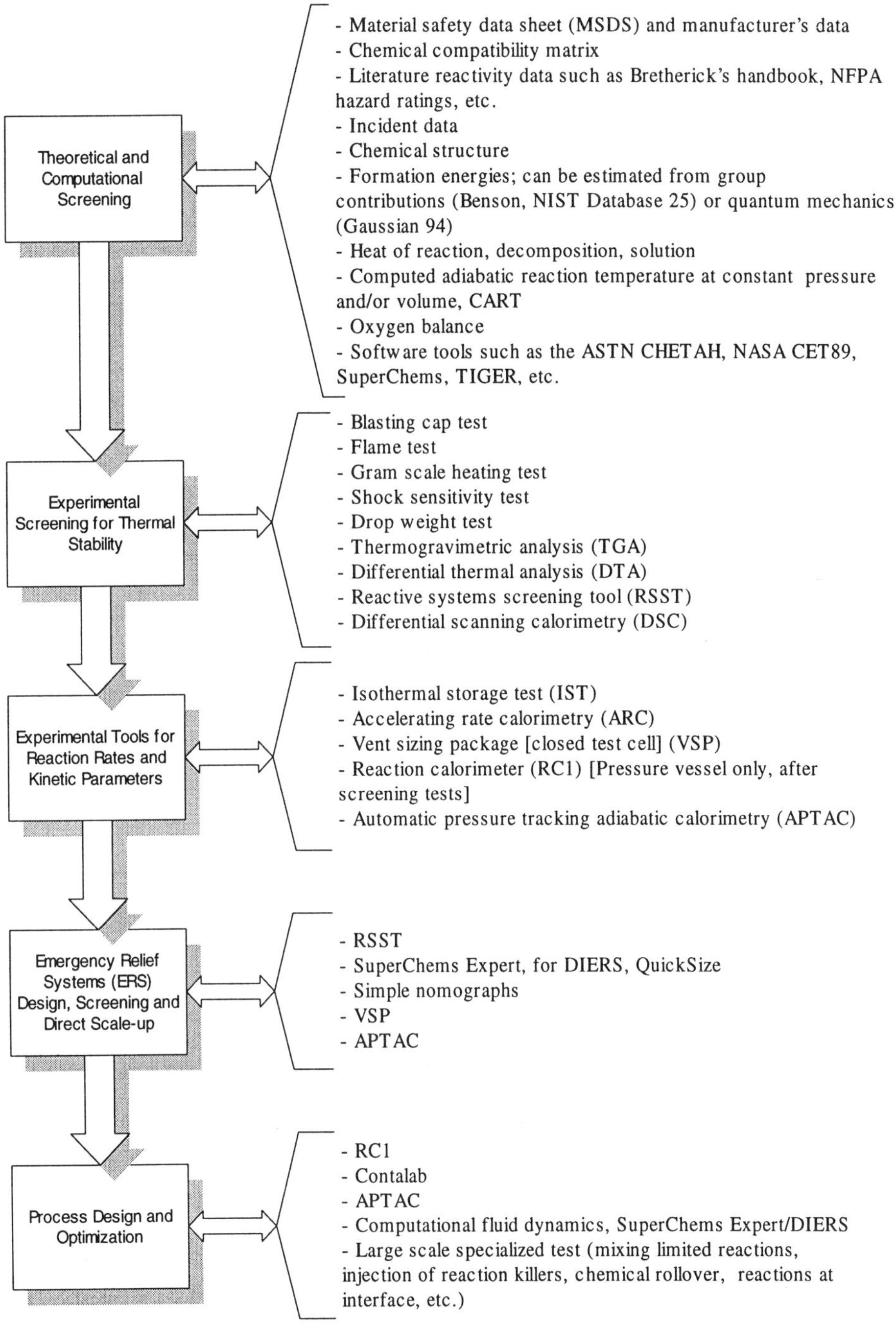

Figure 2. Available experimental and theoretical tools for reactive systems characterization

5. Weast, R. C. (ed.), *Handbook of Chemistry and Physics,* 66th Edition, CRC Press, Inc. Boca Raton, FL, 1985

Chemical Matrix

Chemical matrices (incompatibility charts) can help in organizing available data on the incompatibilities existing between expected process mixtures. Reference [1] gives one procedure for developing a chemical compatibility chart while describing some of the tools available. Reference [2] also provides a table of known incompatibility hazards.

Data can also be gathered experimentally on the compatibility of materials. Finally, incompatibility charts have been published by both the U.S. Coast Guard[1] and ASTM[2] as well as other industry groups around the world. Again, see Reference [1] for a description of experimental tests and published compatibility charts.

Structural Analysis

From a structural perspective, a number of molecular groupings are recognized as being inherently unstable such as molecules containing $-ONO_2$, –CC–, $N—NO_2$, etc. Table 2 outlines several molecular structures known to be high energy structures.

Oxygen Balance Method

In a 1949 Chemical Reviews article, W. C. Lothrop and G. R. Handrick documented a strong correlation between "oxygen balance" and the explosive performance of organic nitrates and nitro compounds. In this context, an oxygen balance of zero implies an oxygen content just sufficient to convert all carbon to CO_2 and all hydrogen to H_2O. Although valuable and universally used in

1 "Compatibility of Cargoes," published in 46 CFR Ch. 1 (10-1-93 Edition)—Part 150 with changes that appeared in the Federal Register dated April 11, 1994 as a Final Rule by Coast Guard (CGD 92-100) for Bulk Hazardous Materials. See also: "Compatibility Guide for Adjacent Loading of Bulk Liquid Cargoes (Final Report)", Report No. USCG-D-156-75.

2 ASTM (American Society of Testing and Materials) Committee D-34 , Proposal-168, "Proposed Guide for Estimating the Incompatibility of Selected Hazardous Wastes Based on Binary Chemical Reactions," and included Minority Report, March 1986. See also: "A Method for Determining the Compatibility of Hazardous Wastes," Hatayama, H.K.; Chen, J.J.; De Vera, E.R.; Stephens, R.D.; Storm, D.L.; CA Dept. Health Serv., Berkeley, (1980) 165 pp. NTIS, SPRINGFIELD VA, PB80-221005. Also published as Report No: EPA-600/2-80-076, April 1980, 165 pages.

TABLE 2
Typical High Energy Molecular Structures

Definition	*Bond Grouping*	*Definition*	*Bond Grouping*
acetylenic compounds	$-C\equiv C-$	tetrazoles; high nitrogen-containing compounds	$-N=N-N=N-$
haloacetylene derivatives	$-C\equiv C-X$	triazines (R=H, –CN, –OH, –NO)	$-C-N=N-\underset{R}{\underset{\vert}{N}}-C-$
metal acetylides	$-C\equiv C-M$	alkyl hydroperoxides; peroxyacids	$-C-O-O-H$
azo compounds	$-C-N=N-C-$	peroxides (cyclic, diacyl, dialkyl); peroxyesters	$-C-O-O-C-$
diazo compounds	$-C=N^{+}=N^{-}$	metal peroxides; peroxoacid salts	$-O-O-M$
diazeno compounds	$-C-N=NH$	amine chromium peroxocomplexes	$N\rightarrow Cr-O_2$
nitroso compounds	$-C-N=O$	azides	$-N_3$
nitroalkanes	$-C-NO_2$	halogen azides; N–halogen compounds; N–haloimides	$-N-X$
polynitro alkyl compounds; polynitro aryl compounds	$-C-NO_2$ $\llcorner NO_2$	diazonium sulfides and derivatives; "Xanthates"	$-C-N=N^{+}\ S^{-}$
acyl or alkyl nitrates	$-C-O-NO_2$	diazonium carboxylates and salts	$-C-N=N^{+}\ Z^{-}$
acyl or alkyl nitrites	$-C-O-N=O$	amine metal oxo salts	$(N\rightarrow M)^{+}\ Z^{-}$
1, 2–epoxides	$-C-C-$ (with O bridging the two C)	N–metal derivatives	$-N-M$
metal fulminates	$-C=N-O-M$	halo-aryl metal compounds	$Ar-M-X$
aci-nitro salts	$HO-(O=)N=$	hydroxyammonium salts	$-N+-OH\ Z^{-}$
N-nitroso compounds	$-N-N=O$	arenediazoates	$-C-N=N-O-C-$
N-nitro compounds	$-N-NO_2$	arenediazo aryl sulfides	$-C-N=N-S-C-$
fluoro dinitromethyl compounds	$F-C(NO_2)-NO_2$	*bis*-arenediazo oxides	$-C-N=N-O-N=N-C-$
difluoro amino compounds; N,N,N-trifluoroalkylimidines	$-N-F_2$	*bis*-arenediazo sulfides	$-C-N=N-S-N=N-C-$
N-azolium nitroimidates	$-\overset{\vert}{\underset{\Vert}{N}}{}^{+}-N^{-}-NO_2$		

Taken from *Guidelines for Chemical Reactivity Evaluation and Application to Process Design*. New York: AIChE/CCPS, 1995.

explosives practice, it has been shown that there is no necessary connection between oxygen balance and self-reactivity in general [3].

Indiscriminate use of the oxygen balance criterion can produce misleading hazard assignments.

Enthalpy of Formation

Enthalpy of formation data provide information on thermochemical stability with respect to the elements which make up the composition in question. In fact, most explosives are stable with respect to their elements, but are highly unstable with respect to some set of (usually) simpler reaction products. A positive value of the enthalpy of formation provides an early warning of hazard potential. A negative heat of formation provides no hazard potential guidance.

Table 3 summarizes data pertinent to several chemical structures with positive heat of formation and high energy release potential.

TABLE 3
Typical High Energy Molecular Structures With Positive Heat of Formation

Compound	*Structure*	ΔH_f°	
		kJ/mol	*kJ/G*
cyanogen	N≡C—C≡N	+308	+5.9
benzotriazole	$C_6H_4NH{=}N$ (ring)	+250	+2.1
nitrogen trichloride	NCl_3	+230	+1.9
acetylene	HC≡CH	+227	+8.7
allene	$H_2C{=}C{=}CH_2$	+192	+4.8
diazomethane	$H_2C{=}N^+{=}N^-$	+192	+4.6
hydrogen cyanide	HC≡N	+130	+4.8
1,3-butadiene	$H_2C{=}CHCH{=}CH_2$	+112	+2.1

Note: the numbers in the fourth column give a better relative indication of potential energy release than those in the third column.

Taken from *Guidelines for Chemical Reactivity Evaluation and Application to Process Design.* New York: AIChE/CCPS, 1995.

There are several techniques for the estimation of heat of formation data based on molecular structure. The techniques include group contribution methods, quantum mechanical estimates, and heat of combustion data.

Heat of Reaction

The difference between enthalpy of formation of the composition in question and its reaction products represents the enthalpy of reaction. This parameter provides at least a rough guide to the amount of stored chemical energy and, to this extent, to the hazard potential. Note that the specific enthalpy change, as in kJ/gram, is the relevant quantity.

ASTM "CHETAH" Hazard Criteria

The well-known CHETAH program provides a convenient, computerized means for estimating the enthalpy of formation for any composition that can be represented in the Benson group additivity system. CHETAH then uses a linear programming technique to select, from its own data bank, the set of room-temperature reaction products yielding the greatest enthalpy change. The resulting value of $\Delta H(r)$ is reported as CHETAH hazard evaluation Criterion 1. Values of −2.9 kJ/gram or more negative are taken as indicating "high" hazard potential.

The five additional hazard criteria provided by CHETAH appear to have limited value. A critical review has been published in Reference [4].

Computed Adiabatic Reaction Temperature (CART)

As noted above, chemical reactivity hazards originate with stored chemical potential energy. Without stored energy in the form of metastable compositions there would be no self-reactivity hazards. However, it is the rate of energy conversion that distinguishes detonations and fast deflagrations from ordinary chemical reactions. Reaction rates are dominated by temperature, as temperature occurs in the exponent of the rate equation. This effect is well represented by Figure 3. Figure 3 illustrates the impact of temperature on time required for ammonium nitrate to self-heat itself to explosion under adiabatic conditions. For this reason it is to be expected that adiabatic reaction temperature might be useful in hazard prediction. Study of this possibility has been facilitated by the development of computer codes designed to carry out multiphase Gibbs free energy minimization and adiabatic reaction temperature calculations, and by the proliferation of PCs capable of using these codes.

Melhem and Shanley [5] have shown through comparison with experimental data for a wide variety of chemical systems that hazard potential is

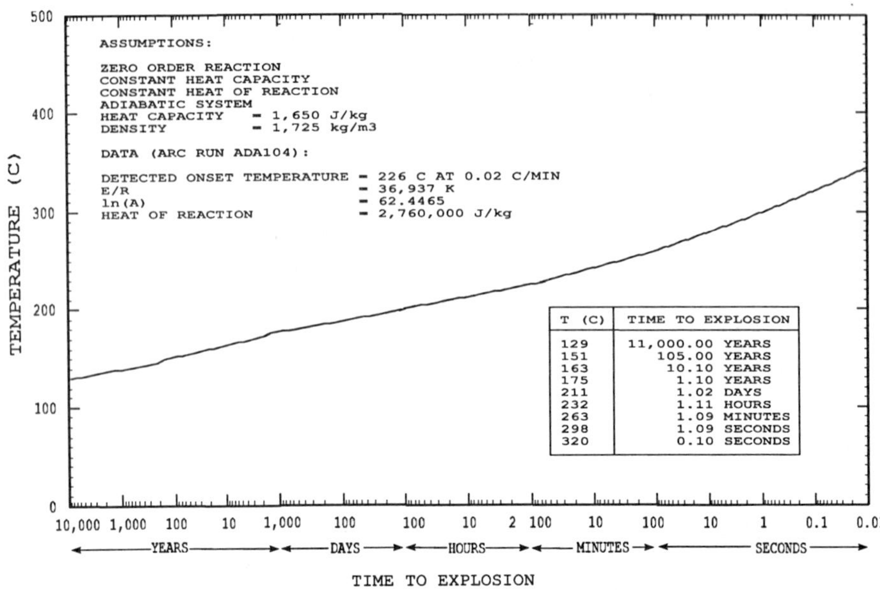

Figure 3. Impact of temperature on time to explosion for ammonium nitrate

better predicted with the use of an adiabatic temperature rather than heat of reaction or decomposition. Melhem and Shanley observe that for many materials of interest (including gases, solids and condensed phase) an adiabatic temperature value of 1200 K or more correlated with experimentally established explosive potential. Peroxide and peroxide containing systems are an exception to this criteria. Their explosive potential correlated well with temperature values greater or equal to 800 K.

NIST SP Database 25 Software

The NIST SP Database 25 [6] software is a convenient and user friendly method of applying the Benson group contribution system for the estimation of heats of formation. This tool allows the user to search an existing database or to input data by drawing the particular molecular structure under study. Using this drawing, the program estimates thermodynamic properties, including the heat of formation. The heat of reaction for a balanced chemical reaction can also be estimated.

NIST Chemical Kinetics Database Software

The NIST Chemical Kinetics Database 17 is a database of gas-phase kinetic data. It is a compilation of published kinetic data that can be searched by reactants, products, reactions and authors. It can be obtained by contacting NIST under Standard Reference Data Program, NIST, Bldg. 820, Rm.113. Gaithersburg, MD 20899-0001.

NIST Chemistry WebBook

The NIST Chemistry WebBook provides on-line access to the full array of data compiled and distributed by NIST under the Standard Reference Data Program. Located at www: webbook: NIST: gov=chemistry, this web site provides on-line access to a large compilation of chemical data, including structural, thermochemical, thermophysical and heat of reaction for over 3000 reactions. Currently NIST is not charging a fee for use of this service. The data can be searched by chemical name, formula, CAS number, molecular weight, ionization energy or proton affinity.

ASTM CHETAH Software

The CHETAH 7.0[3] program is also a computerized means of applying the Benson system. It is more cumbersome to use than NIST SP; however, CHETAH 7.0 contains many more Benson groups permitting evaluation of many compounds that cannot be modeled in SP. To estimate thermochemical properties using CHETAH, you need to toggle through the various menus of Benson groups, choosing those that apply to the compound of interest. Please see Reference [4] for a detailed review of CHETAH 7.

NASA CET89 Software

The NASA-Lewis computer program, CET89, provides convenient means for estimating the adiabatic decomposition temperature (maximum attainable temperature) for most compositions, given the elemental composition and the enthalpy of formation. This program also reports the equilibrium decomposition products at the adiabatic decomposition temperature.

TIGER Software

The TIGER[4] program, developed by the Stanford Research Institute, is capable of calculating the thermodynamic state attained in a heterogeneous system of known atomic composition. The equilibrium state is generated by minimizing the Gibbs free energy of the system at a given temperature and pressure.

3 CHETAH 7.0, The Chemical Thermodynamic and Energy Hazard Appraisal Program, Version 7

4 Cowperthwaite, M. and W. H. Zwisler, "TIGER: Computer Program Documentation," Stanford Research Institute, Menlo Park, CA (1973)

SuperChems

SuperChems [7, 8, 9, 10] is an advanced tool for thermal hazards assessment, pressure relief design, and consequence analysis. Developed by Arthur D. Little, Inc. SuperChems™ is available in two editions: Professional and Expert. The Professional Edition consists of a family of rigorous models designed to conduct consequence analysis and hazard modeling. The Expert Edition has been used as an effective design tool with capabilities in multicomponent analysis, transient quasi-three-dimensional heavy gas dispersion, multiphase reaction kinetics, transient gas explosion dynamics with obstructions, transient pipeline flows, and simultaneous phase and chemical equilibrium flash routines.

SuperChems™ is linked to a databank containing over 1200 chemicals with full database functionality, such as the option to add, delete, and modify databank compounds. In addition, several utilities are included to facilitate the manipulation of user-specific data. SuperChems™ Expert handles multicomponent mixtures in every model, using true multicomponent phase equilibria, and can perform Vapor/Liquid Equilibrium (VLE), Liquid/Liquid Equilibrium (LLE), Vapor/Liquid/Liquid Equilibrium (VLLE) with or without solids. It can also perform multiphase reversible chemical reaction calculations in vessels and piping.

Two of SuperChems™ capabilities are useful here; they are its properties estimation feature and its ability to reduce and scaleup reactivity data from several different types of calorimeters. This includes detailed methods for extracting stoichiometry data from chemical reactions and decompositions.

Gaussian 94

Gaussian 94[5] is a connected system of programs for performing a variety of semiempirical and ab initio molecular orbital (MO) calculations. Gaussian 94 is capable of predicting many properties of molecules and reactions, including:

- Molecular energies and structures
- Energies and structures of transition states
- Vibrational frequencies
- Thermochemical properties
- Bond and reaction energies
- Reaction pathways
- Molecular orbitals
- Atomic charges
- Multipole moments

5 © Copyright Gaussian, Inc. Carnegie Office Park, Bldg. 6 Pittsburgh, PA 15106 USA

Computations can be carried out on systems in the gas phase or in solution, and in their ground state or in an excited state. Thus, Gaussian 94 can serve as a powerful tool for exploring areas of chemical interest like substituent effects, reaction mechanisms, potential energy surfaces, and excitation energies.

HAZARD RANKING

Using the theoretical screening methods discussed in this paper and other recent experimental studies (see reference [11]) we can assign most chemical compositions (condensed phase and gases) to one of three self-reactivity hazard groups, as follows:

1. **NO to LOW self-reactivity hazard group:** Heat of reaction no more negative than –1.2 kJ/gram, and CART no more than 700 K.
2. **INTERMEDIATE group:** Heat of reaction between –1.2 and –3.0 kJ/gram, or CART no higher than 1600 K.
3. **HIGH self-reactivity hazard group:** Heat of reaction more negative than -3.0 kJ/gram, or CART higher than 1600 K.

Compositions falling in the NO to LOW hazard group are usually considered to be suitable for full-scale manufacture and use. Compositions falling in the HIGH hazard category may be regarded as explosive to a practical certainty. Such compositions are usually regarded as unsuitable for manufacture, handling or use under ordinary commercial circumstances.

The intermediate group includes many compositions which are in large-scale commercial use but which may require special care and design to control potential reactivity hazards.

EXPERIMENTAL SCREENING FOR THERMAL STABILITY

Flame Tests

Two small-scale heating tests can provide much insight into the explosive potential of a pure material or reaction/product mixture. The first involves the heating of approximately a 1/10 g sample on a metallic spatula, for example, by holding the spatula above a small flame. Behavior is rated when the sample is removed from the flame (i.e., flame on spatula slowly heats up sample) and when the sample is directly exposed to the flame. The second test involves the dropping of the same size sample onto a preheated metallic surface (e.g., hot plate). By observing the sample behavior under the three conditions above (slowly heated, immersed in flame, exposed to high heat) judgments can be made as to the explosive potential of the material.

Observations should be made regarding the following behaviors:

- melting
- boiling
- spontaneous ignition
- rapid reaction with sound
- carbonization

Observed burning behavior provides a guide to reactivity and fire hazard potential. Reaction accompanied by noise suggests explosive potential.

Gram-Scale Heating Test

A 3-gram sample of liquid or solid composition is placed in a standard 10 mL glass test tube. The test tube is placed in a cavity about 2.5 inches deep bored into the end of a 2-inch-diameter length of copper rod about 3 inches long. The copper block also includes a small diameter hole near and parallel to the test tube cavity. This hole accommodates an unjacketed thermocouple. The copper block is held, cavity end up, in a simple electric furnace which will raise the temperature of the block and the test tube to 300°C in about a half hour. Boiling and decomposition behavior can be viewed through an explosion shield. Heating should be continued until the sample decomposes violently, is completely volatilized, or charred. If no violent decomposition is observed, the test should be repeated by inserting a fresh test tube and sample into the preheated block at about 300°C. Violent decomposition accompanied by noise or vigorous eruption indicates a self-reactivity hazard.

The copper block will contain the worst case violent decomposition to be expected in the case of 3-gram samples of Class 2 materials as described above. Nevertheless, glass fragments may be propelled upward with considerable force. Means for intercepting such fragments should be provided, as well as an explosion shield to protect the observer. In no case should explosives or suspected explosives (class 3 compositions) be subjected to this test in a conventional laboratory setting.

Explosivity and Mechanical Sensitivity Tests

In the processing of chemical compounds it is very important to understand the mechanical sensitivity of a particular substance. Sensitivity tests can be divided into two categories which represent the types of mechanical energy expected to be seen in the process plant. These are mechanical friction (due to materials being wedged between surfaces) and mechanical shock (or impact). Several tests are available for measuring both the sensitivity to friction and impact. These are described in References [2, 12]. Two tests for identifying impact sensitive materials are described here. They are the hammer test and the drop weight test. Information concerning additional tests can also be found in References [2, 12]. These tests can provide valuable insight into the behavior of a process mixture under extreme process conditions.

Hammer Mechanical Sensitivity Test

In conducting the hammer test, a small amount of sample (approximately 0.1 g) is placed on an anvil. The sample is then hit with a hammer and the behavior observed. A comparison of the behavior of a known standard explosive vs. the behavior of a sample can provide insight into the explosive potential of the sample. Hammer tests do not provide a quantitative estimate of the impact sensitivity of the sample; however, they can provide a qualitative assessment of whether or not a particular material will explode under normal operating conditions.

Drop Weight Mechanical Sensitivity Test

Another test for determining mechanical shock sensitivity is the drop weight test. Although, difficult to replicate, behavior in the drop weight test can furnish valuable comparative information about the relative mechanical shock susceptibility of the composition in question as compared with that of better known compounds. The experimental test apparatus is rather straightforward. The sample is placed in a holder designed for the physical state and type of material being tested. A known weight is then allowed to drop from a specified height. After impact the sample is examined for signs of detonation. The weight can be dropped from higher heights if the lower impacts produce no detonation. Typical regulatory agencies require 10 nondetonates out of 10 attempts at a height of 10 inches. A typical testing apparatus has the ability to go to 25 inches in height. Drop weight tests and testing devices are described in further detail in Reference [2]. Care must be taken with these tests, as small changes in sample preparation or test setup can greatly affect the results; therefore, drop weight testing is best left to a testing laboratory with experience and equipment to run these tests. Note that both tests described above are conducted in the open and the results may or may not apply to the behavior of the material in a closed system.

Blasting Cap Test

The blasting cap test provides means for assessing the sensitivity of a new composition to relatively intense impact. It may be more appropriate for use in large scaleup problems. However, in selected cases, including the presence of highly energetic groups in the compound, it may prove useful. Advantages to carrying out these type of explosivity tests are that they are inexpensive, fairly easy to carry out and use a small sample size. Despite the fact that they are easy to carry out, current regulations regarding acquisition and use of blasting caps are so restrictive as to discourage in-house use of blasting cap tests.

In the Arthur D. Little Inc. version of this test, a No. 6 blasting cap is detonated while immersed in a column of the composition to be tested. A witness tube provides the means for determining whether or not a propagating explosion was initiated and also provides a means for comparing the energy release behavior of the composition being tested with that of better known compositions. In this manner, it can be determined whether or not a certain sample is expected to propagate a reaction.

Qualitative agreement with CART hazard rankings was reported by Shanley and Myers [11] in a recent experimental blasting cap study conducted at Arthur D. Little Inc.

More than 35 tests were conducted on a wide variety of chemicals ranging from inert substances to known high explosives. The tested materials included water, dodecane, toluene, mononitrotoluene, tetranitromethane (TNM)-toluene mixtures (50/50 and balanced), cumene hydroperoxide 80%, Di t-butyl peroxide, benzoyl peroxide, hydrogen peroxide–ethanol mixtures (40/60 and balanced), ammonium nitrate (pure and with dodecane), dinitrotoluene, and urea nitrate.

The test series demonstrated qualitative correlation between CART values and extent of deformation of the blasting cap witness tubes. For example, nine of the ten compositions which produced minimum deformation had CART values below 1700 K. Six of the seven compositions which produced significant or severe deformation of the witness tubes had CART values greater than 1700. All three compositions, which produced severe deformation, had CART values greater than 2100 K. The deformations observed with the two fueled hydrogen peroxide compositions were in good agreement with published findings. H_2O_2–40% ethanol is outside the published explosive limits, while H_2O_2–50% ethanol is on the boundary of the explosive region.

The three organic peroxides tested produced minimum deformation of the witness tubes in keeping with their relatively low CART values. The test tube containers are undamaged by these explosions, which suggest a low-rate deflagration rather than a detonation.

In tests conducted as described above, any substantial deformation of the witness tube over and above that produced by firing the cap in an inert composition provides a strong indication of reactivity hazard. Energetic but insensitive compositions such as straight ammonium nitrate and dinitrotoluene provide weak signals (no propagation), while more energetic compositions such as fueled AN and urea nitrate provide strong signals (partial or full propagation).[6]

6 Note that larger diameter charges or different packing densities (in the case of solid specimens) or different initiators for example could produce different effects.

These experiments demonstrated that the blasting cap can produce valid comparisons between novel compositions and better known materials whose hazardous properties are already established. Such comparisons are useful supplements to evaluations based upon computational methods.

Differential Scanning Calorimeter (DSC)

Differential Scanning Calorimetry, or DSC, is a microcalorimetric technique that measures either exothermic or endothermic reactions/phase transitions. Samples on the order of 10 mg are typical for this technique. The difference in energy required to keep a pan with sample and an empty pan (reference pan) on a specified heating ramp is measured and recorded. The differential technique is used to remove contributions from the pan and to eliminate noise from heater instabilities.

The primary use of the DSC is to obtain information on the thermal behavior of reactive materials. Typical information obtained include sample heat flux as a function of temperature.

A small amount of sample (5–25 mg) is placed in a sample pan which is placed in the calorimeter's cell. The sample is heated or cooled at a specified temperature ramp rate in a temperature range that varies from –170°C to 500°C and the heat absorption/generation characteristics are compared with an empty reference pan to determine endothermic/exothermic behavior. A variety of gases can be used to pad the sample although the most common gases are nitrogen, air, or oxygen. High pressure conditions up to 1000 psig can also be employed.

Exothermic or endothermic phenomena are evidenced when the DSC trace undergoes a positive or negative excursion, respectively, from the baseline. This type of phenomena includes exothermic or endothermic reactions, melting, evaporation, or phase transitions. Heat capacities of materials can be determined. This is a relatively rapid test which many companies use as the primary screening tool in evaluating the exothermicity of materials.

Differential Thermal Analysis (DTA)

Differential Thermal Analysis (DTA) is another microcalorimetric technique which closely resembles DSC. Both DSC and DTA measure exotherms and endothermic reactions and phase transitions. Data from both instruments is often very similar. Unlike the DSC, the DTA measures events by monitoring the difference in temperature between a sample pan and a reference pan while both are supplied the same amount of heat energy. DSC is more commonly used and available of the two techniques, however, many high temperature units are DTA based.

Thermogravimetric Analysis (TGA)

Thermogravimetric Analysis (TGA) is the oldest and the simplest of the micro-calorimetric techniques used today. Samples up to about 50 mg are placed in an inert pan, usually platinum. The samples are heated at a specified rate and the weight loss of the sample is recorded. A low flow of user selectable gas is used to purge the sample during heating. Inert gases are used to keep the headspace clear of gaseous and sometimes corrosive materials. Other gases, like oxygen, can be used to study the oxidation of materials. Data from this technique is usually plotted as weight loss as a percent or in mg versus temperature.

Reactive Systems Screening Tool (RSST)

The basic operational principle of the RSST[7] involves programmed heating of a sample at a preselected rate (nominally 0–2°C/minute) through a selected temperature range. Exothermic reaction in the sample is identified by positive deviation of the sample temperature rise rate (dT/dt) from the programmed set value. The sample is held in an open glass test cell, although it is possible to operate the test under inert gas pressure. Figure 4 illustrates a recommended test sequence for identification of reactive system type and vent sizing.

The selected heating rate is attained by software controlled switching of a constant voltage supply to the sample heater. The necessary heater switching rate is based on software resident calibration data for a typical organic sample (toluene), presumably under inert gas pressure to suppress boiling. As such, there is no direct control of true sample heating rate or heater power input.

A rudimentary form of on line temperature control/heating drift correction is provided in the form of the heater offset function. This increases the heater voltage pulse rate when negative deviation from the set heating rate is observed. However, the offset function does not correct positive drift, as this could lead to heater cut off during exothermic sample behavior, leading to a rapid loss of heat from the reacting sample. A major problem with this control configuration is that it is not possible to maintain linear temperature ramping for materials which undergo endothermic transitions, during which the heater input power will increase to maintain the set heating rate. When endothermic behavior ceases, the heating rate will remain at an increased level (as negative offset is not possible) unless the heater program is manually interrupted and reset. For materials which exhibit endothermic transition followed rapidly by exothermic reaction (a phenomenon which is not uncommon) it is difficult to envisage how sample thermal behavior can be interpreted.

7 The RSST is a trademark of Fauske & Associates Inc.

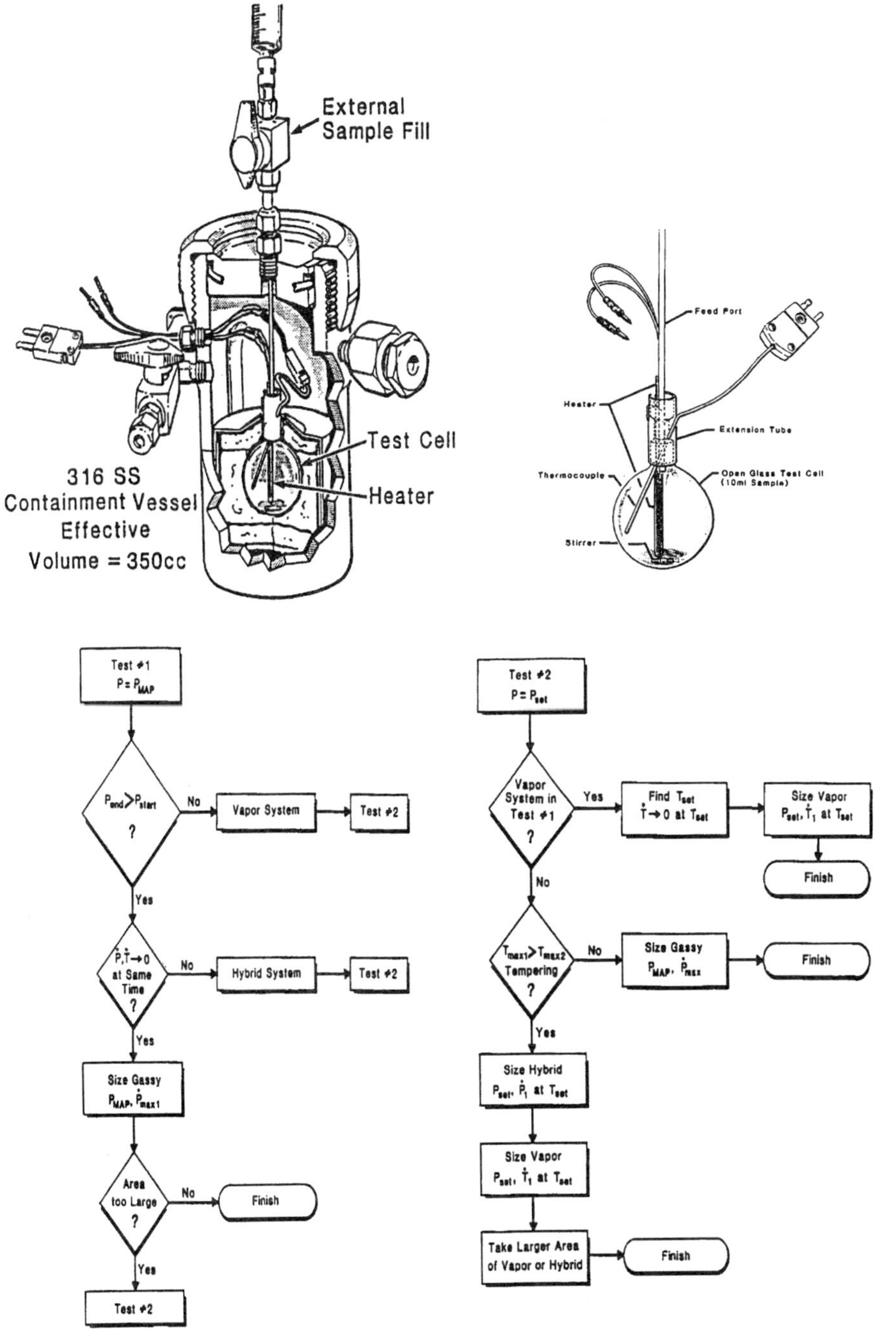

Figure 4: Reactive System Screening Tool

The slow sampling rate and significant signal disturbance for thermocouple outputs mean that the temperature resolution is estimated to be no better than ±0.25°C. Furthermore, the heating rate disturbance at 1°C/minute appears to be approximately ±0.2°C/minute. For a sample of a typical organic material with a heat capacity 2.0 J/gm/C and system phi factor 1.05, this corresponds to a detection sensitivity (q_s) given by:

$$q_s = 2 \times \left[\frac{0.2}{60}\right] \times 1.05 = 7 \times 10^{-3} \text{ W/g} \quad \text{or} \quad 7 \text{ W/kg} \tag{1}$$

This sensitivity is comparable to DSC using moderate scanning rates, but is higher than the heat generation levels where spontaneous self heating will occur in an industrial scale plant.

The RSST is very useful for rapid reactivity screening. The performance of the RSST could be improved significantly by externalizing either one or more of the heater power control, temperature program control, or sample temperature monitoring functions.

EXPERIMENTAL TOOLS FOR REACTION RATES AND KINETIC INFORMATION

Accelerating Rate Calorimeter (ARC)

The Accelerating Rate Calorimeter (ARC)[8] is an instrument that can provide adiabatic pressure and temperature time data required for establishing reaction data. The ARC can be used to obtain information on the thermal behavior of reactions and exothermic onset temperatures. The ARC is primarily used for liquid-phase reactive systems. It is also used for safety/performance evaluation of explosives and propellants. Exothermic onset temperatures of 0.01°C/min can be detected. This instrument (see Figure 5) which was developed by Townsend and Tou is known to provide ample thermokinetic data that is applicable to the design and safety/performance evaluation of reactors and storage vessels. Such thermokinetic data includes:

- adiabatic rate of self-heating,
- adiabatic time to explosion,
- rate of pressure rise,
- maximum rate of reaction,
- kinetic data such as activation energy, reaction order and preexponential factor, and
- heat of reaction.

8 The ARC is a trademark of Arthur D. Little Inc. and Columbia Scientific Industries.

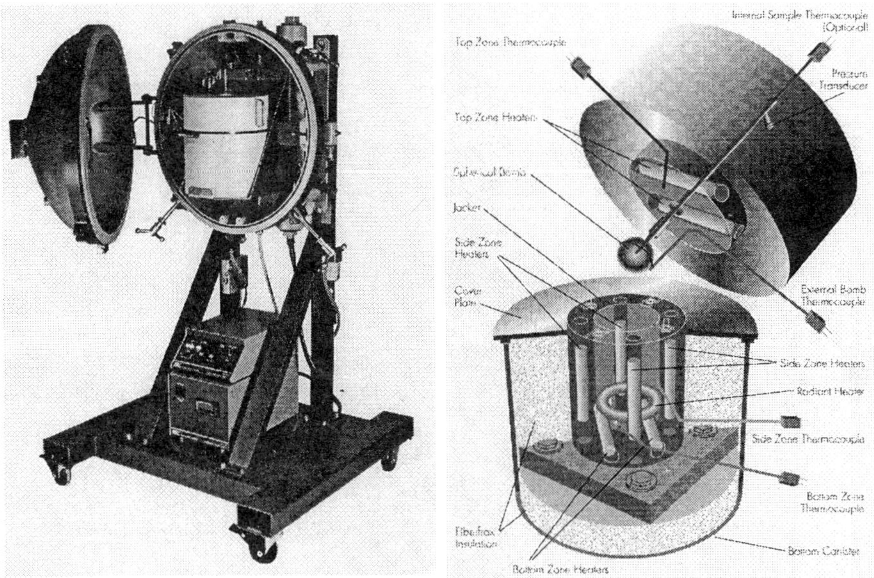

Figure 5. The Accelerating Rate Calorimeter (ARC)

The reaction mixture to be examined is introduced into a spherical cell whose volume is approximately 10 mL. The cell is equipped with a thermocouple mounted on the external wall. The pressure is recorded by means of a pressure transducer. Various cells can be used which have varying pressure ratings from 4,500 psi to 15,000 psi. As the pressure rating increases, the cell phi factor increases. The ARC is a high thermal inertia instrument whose phi factor is greater than or equal 1.4. The high phi factor requires that the temperature and pressure be corrected by established techniques which requires knowledge of the heat capacity of the reaction mixture and the cell. The significant advantage that the ARC offers over other similar techniques is exothermic onset delectability at 0.01°C/min.

The tracking of temperature and pressure rates is limited to approximately 10°C/min. Temperature drift is not a problem with proper calibration. Operation does not require significant operator supervision. Operator skill is required for proper calibration. The high phi factor can potentially mask exotherms which may occur at higher temperatures.

Vent Sizing Package (VSP)

The primary use of the VSP[9] (see Figure 6) is to obtain information on the thermal behavior of reactions (instrument data and descriptions are taken

9 The VSP is manufactured by Fauske & Associates Inc.

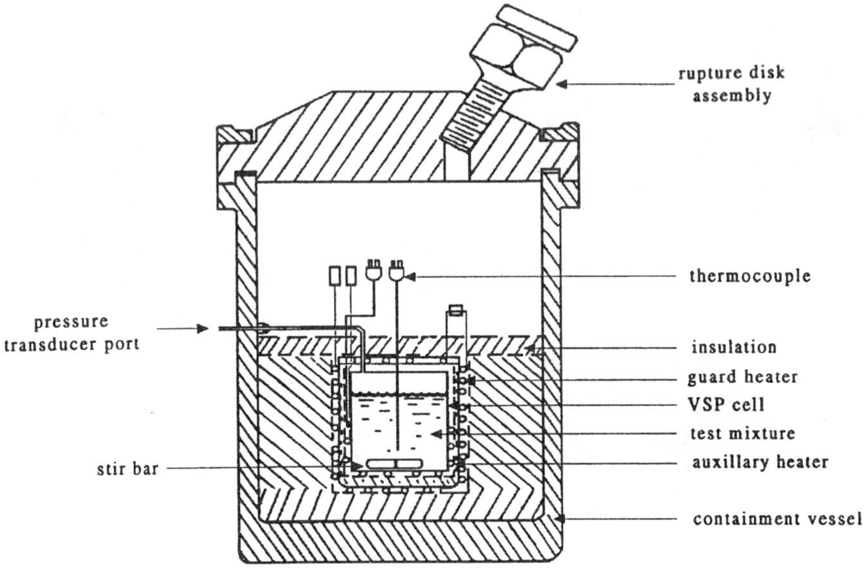

Figure 6. Vent Sizing Package

from reference [13]). It is primarily used on liquid-phase reactive systems. Information obtained includes temperature and pressure profiles. The basic data is manipulated to give self-heat rate, pressure rate, and vapor pressure.

The reaction mixture to be examined is introduced into a thin-walled cylindrical cell whose volume is approximately 110 mL. The cell is equipped with a thermocouple which directly measures the temperature of the reaction mixture. The pressure is measured by means of a pressure transducer. The cell is placed in a high pressure containment vessel. Around the cell there are two sets of heating elements called the auxiliary heater and the guard heater. The auxiliary heater raises the sample to a prescribed set temperature. The guard heater maintains the temperature of the environment around the cell at the same temperature as the cell contents to make the system adiabatic. As pressure develops in the cell, compensating nitrogen is introduced into the containment at a matching rate to prevent cell rupture. The VSP is a low thermal inertia instrument whose phi factor is approximately 1.1. The low phi factor ensures that the experimentally observed temperature and pressure profiles are the same as those which would be experienced in plant process equipment. Temperature and pressure rates as high as 100 to 300°C/min and 20,000 psi/min can be tracked.

The major use of the VSP is in the design of relief vents for runaway exothermic reactions. The vapor pressure behavior of the system can be used to distinguish between tempered, gassy and hybrid systems. The self-heat rate, or rate of adiabatic temperature rise, when coupled with the vapor pressure behavior of the system is used with the DIERS methodology to size emergency relief vents. Heats of reaction can be obtained from the adiabatic temperature rise if the heat capacity of the material is known or can be estimated. Kinetic parameters such as the rate constant and Activation energy can be obtained from the data but apply only for the reaction mixture examined. Blowdown tests can be conducted to characterize the disengagement dynamics of the systems. Rates of inert gas generation can be quantified and with the DIERS methodology, ERS can be designed for gassy systems.

The detection of the exothermic onset temperature is limited to approximately 0.2 to 0.5°C/min. Temperature drift can be a problem unless careful attention is paid to the installation of the insulation. Results can depend upon experimental technique. Operation requires significant operator supervision and skill. Application of the DIERS methodology using data obtained from the VSP or ARC is not recommended for the uninitiated. The kinetic information obtained is applicable only for the particular system tested.

Automatic Pressure Tracking Adiabatic Calorimeter (APTAC)

The primary use of the APTAC[10] (see Figure 7) is to obtain information on the thermal behavior of reactions and exothermic onset temperatures. Information obtained includes adiabatic temperature and pressure profiles for reactions. The basic data are manipulated to give self-heat rate, pressure rate, and vapor pressure. Exothermic onset temperatures of 0.04°C/min can be detected.

The reaction mixture to be examined is introduced into a spherical cell whose volume is approximately 130 mL . The cell is equipped with a thermocouple which directly measures the temperature of the reaction mixture. The pressure is measured by means of a pressure transducer. The cell is placed in a 4-liter high-pressure containment vessel rated to 1900 psig. The pressure is recorded by means of a pressure transducer. Nonisobaric pressure tracking by means of flow control valves prevents the cell from rupturing. Four independent heaters with PID cascaded control maintain adiabatic conditions. Temperature and pressure rates of 400°C/min and 10,000 psi/min can be tracked. The APTAC is a low phi factor instrument (less than 1.15) with exotherm detection capabilities of 0.04°C/min.

The APTAC combines the best features of the VSP (low phi factor and high tracking capabilities) and the ARC (low exotherm onset detection and low operator attention) into a single instrument. For more information on the APTAC please refer to the paper authored by Chippett et al. in the symposium proceedings.

10 APTAC is a trademark of Arthur D. Little Inc.

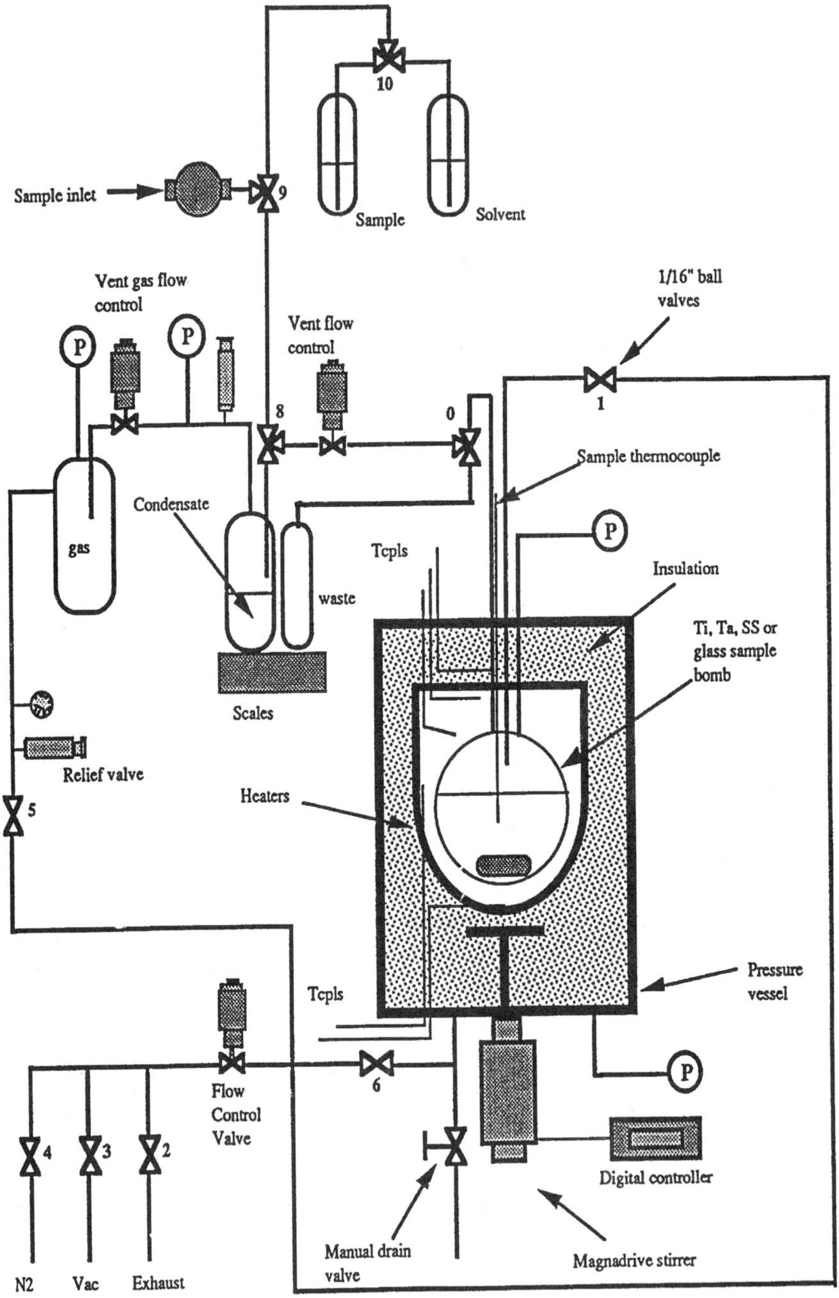

Figure 7. Automatic Pressure Tracking Adiabatic Calorimeter (APTAC)

TNO Isothermal Storage Test (IST)

In the Prins Maurits Laboratory (TNO) Isothermal Storage Test (IST) the heat generated by reacting or decomposing substances at a constant temperature is measured as a function of time. Performance of these measurements at a series of temperatures leads to a quantitative understanding of the relation between temperature and the heat generation of the substance under investigation. The IST is applicable to solids, liquids, pastes and dispersions.

The IST shown in Figure 8 consists of a large heat sink (an aluminum block) which is kept at a constant temperature. In the block are two holes with a very sensitive heat flow meter (e.g., a Peatier element) at the bottom of each hole. Holders are placed on both heat flow meters. These holders are identical. One holder contains the sample, the other an inert substance.

The heat generated by the sample results in a voltage signal from the heat flow meter which is proportional to the heat flow. Random fluctuations in the heat flow are avoided by monitoring the voltage differences between both heat flow meters.

The stainless steel sample holder has a volume of 70 cm^3 . The sample mass which depends on the heat generated by the sample, amounts to a maximum of about 20 gram. Measurements can be performed in the temperature range of 250 K to 420 K. Heat generation can be measured between the lower limit of 5 MW/kg and the upper limit of 5 W/kg with an accuracy of at least 30% in the lower range to 5% in the higher ranges.

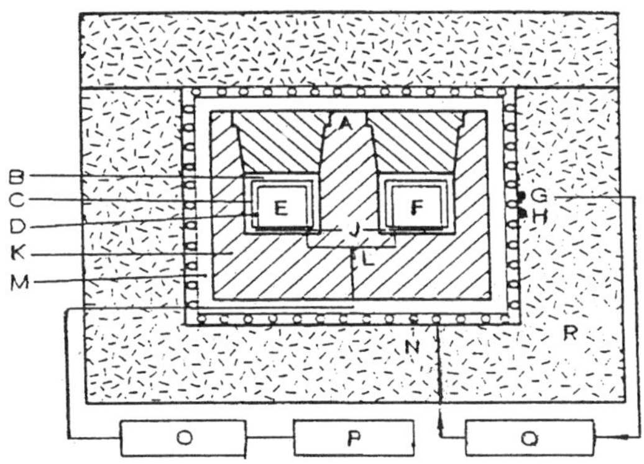

(A) PT thermometer
(B) Sample vessel
(C) Cylindrical holder
(D) Air spaces
(E) Sample
(F) Inert Material
(G) PT Sensor for temperature control
(H) PT Sensor for safety control
(J) Peltier elements
(K) Aluminum block
(L) Electric circuit
(M) Air space
(N) Heating wires
(O) Amplifier
(P) Data registration
(Q) Temperature controller
(R) Glass wool

Figure 8. Isothermal Storage Test

EMERGENCY RELIEF SYSTEMS DESIGN

There are currently two established practices for the design of emergency relief systems:

1. Simplified and direct scaleup methods
2. Detailed computer simulation methods

The simplified and direct scaleup methods are methods that are most applicable to nonreactive systems. These methods use direct scaleup techniques to obtain a vent size. The advantages of such techniques are summarized below:

- Simple and easy to use
- Can be used by non-ERS design specialists
- Calculations can be performed using a spreadsheet or charts

The main disadvantages are:

- Scaleup data are only valid for system composition and thermal inertia represented by the small-scale experiments
- Results are often conservative, especially for gassy systems. This can be costly if postrelease mitigation/effluent handling is required.
- Does not consider the presence of long inlet lines or the impact of downstream equipment on performance of the ERS system
- Little useful information can be obtained for reaction data such as onset temperatures, activation energy, and decomposition stoichiometry
- Fire exposure is difficult to simulate
- Requires an additional experiment per what if scenario for iterative designs

The detailed approach advocates a more fundamental approach to pressure relief design, especially for reactive systems. First, the reaction chemistry is qualified using small-scale experiments and then these data are coupled with fluid dynamics to design the emergency relief system. The main advantages of this approach are:

1. High accuracy
2. Can handle complex systems and systems that interact with other equipment
3. Can be used for iterative design and what-if analyses such as the direct evaluation of the impact of temperature, pressure, composition, fill level, solvent boiling point, and reduced charge
4. In addition to vent sizing information this approach provides insights into process modification and refinement as well as the establishment of a safe operating envelope.

5. Provides necessary flow data for vent containment design (if required), structural support, etc.

The main disadvantages of the detailed methods are:

1. Requires expert skills for establishing kinetic information from test data
2. Requires the use of computer simulation

Figure 9 outlines a recommended general flow chart for selecting appropriate ERS design methods ranging from simple solutions to detailed computer simulations.

FIA Chart

A simple method that can be used for screening is the FIA chart [14]. This chart has been withdrawn as an official document of Industrial Risk Insurers and is not recommended for general use since it has been shown to be nonconservative for some systems [15] while producing overly conservative results for many others.

The main advantage of the FIA chart is its ease of use. However, the FIA chart is only applicable to tempered reactions with a low relief viscosity in 100–125-psig vessels. The chart (see Figure 10) consists of four straight lines drawn on log paper. Reactor volume is plotted on the abscissa, and the required vent area is shown on the ordinate. Each line represents a different class of reaction,

The chart divides reactions into four classes:

1. *Class A:* Endothermic or very low exothermic reactions.
2. *Class B:* Reactions with low heat release/volume such as suspension polymerization of vinyl chloride, etc.
3. *Class C:* Moderately exothermic reactions, such as methyl methacrylate and styrene bulk polymerizations, mono nitration of benzene, toluene, etc., and Friedel-Crafts reactions.
4. *Class D:* Very high heat release reactions, such as oxidations using nitric acid, bulk polymerization of ethyl acrylate or methyl acrylate, caustic catalyzed phenol-formaldehyde condensation, etc.

Once the reaction class has been selected, the vent area is determined by drawing a vertical line from the vessel charge through the appropriate reaction class line, then extending a horizontal line from the intersection to the vent area axis. For example, a 1000-gallon vessel running Class C reaction requires 56 square inches of venting capacity (an 8-inch diameter vent) according to the FIA chart.

If measured energy release rate data are available for a reaction, it is possible to draw a ew line on the FIA chart to allow interpolation between the reac-

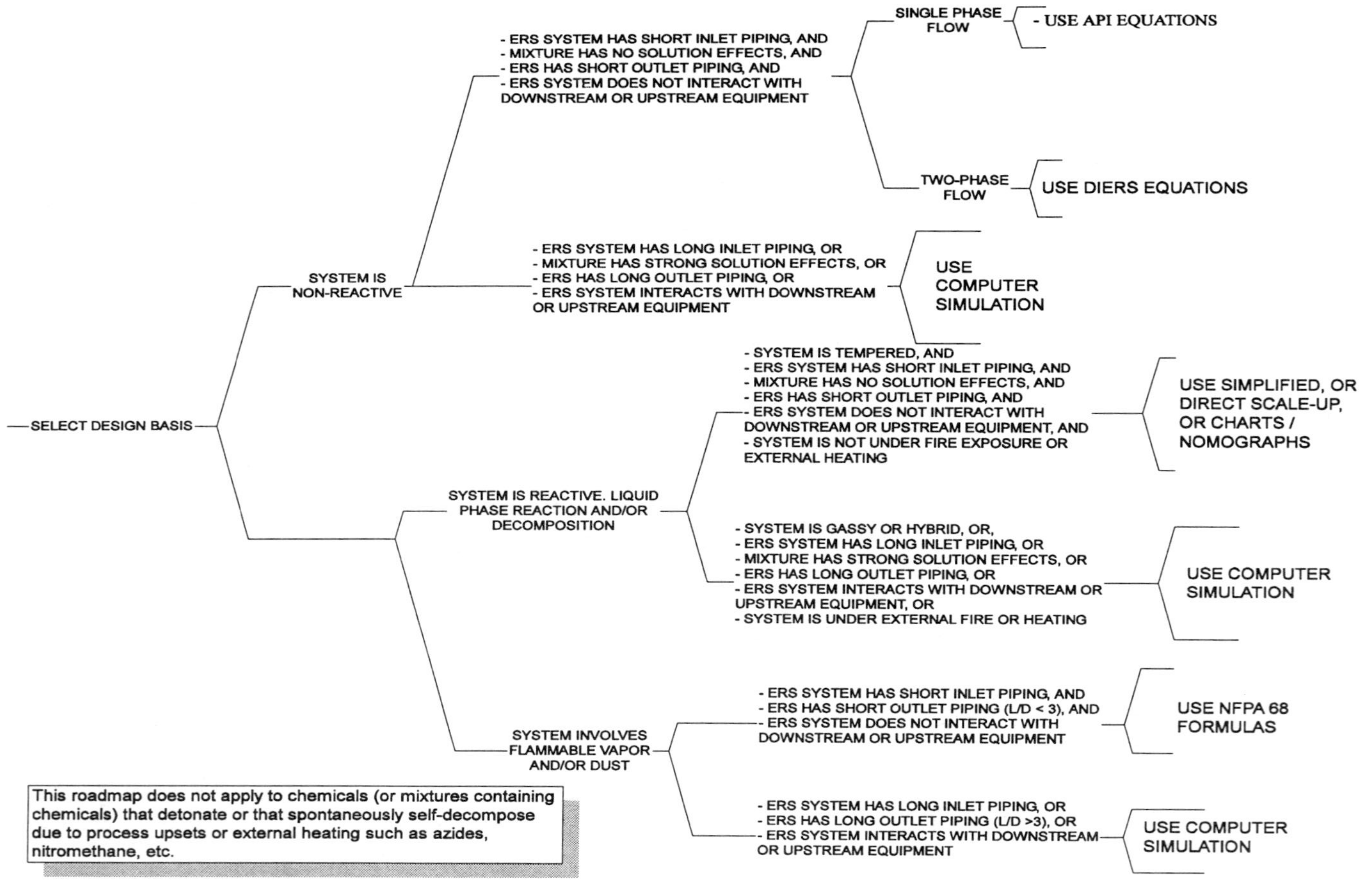

Figure 9. A selection chart for ERS design methods

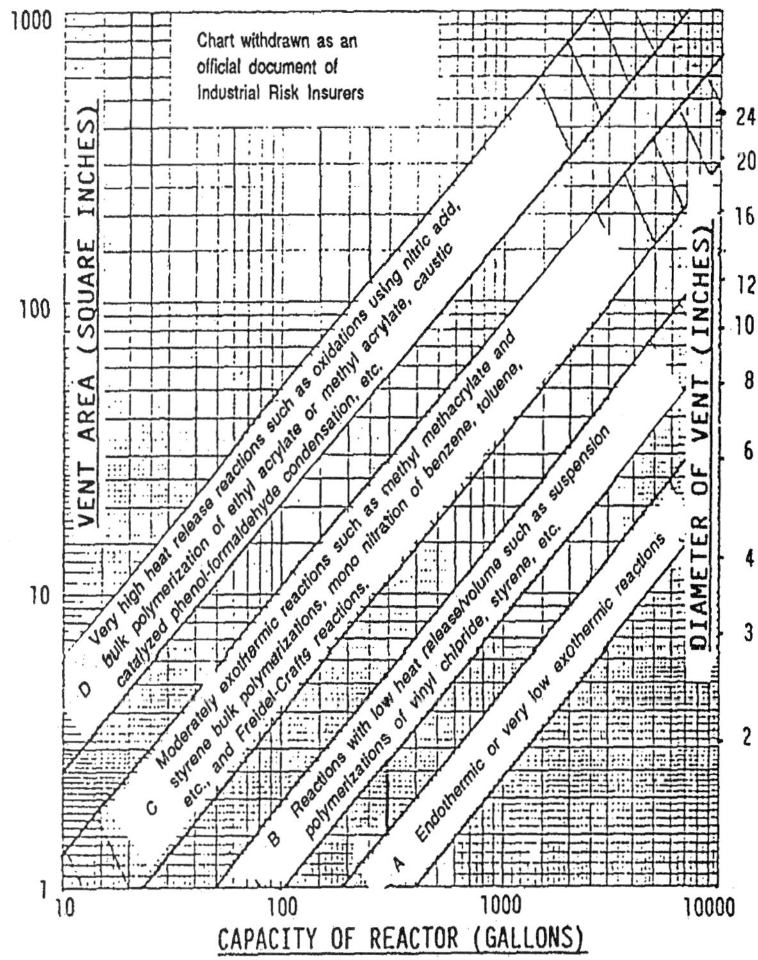

BASIS: TEMPERED REACTIONS
100-125 PSIG VESSEL PRESSURE
$\leq$ 15 FEET RELIEF LINE LENGTH*
LOW RELIEF VISCOSITY

WARNING: CHART RECOMMENDATIONS ARE NON-CONSERVATIVE FOR SEVERAL OF THE NAMED REACTIONS

*NOTE: RELIEF AREA CAN BE ADJUSTED FOR THE EFFECT OF LINE L/D ON FLOW

Figure 10. FIA Chart

tion categories. The lines for the four categories correspond to these energy levels:

1. Class A: 150 BTU/hr-lb
2. Class B: 608 BTU/hr-lb
3. Class C: 3040 BTU/hr-lb
4. Class D: 15,200 BTU/hr-1b

The chart can be reduced to a formula that can be easily coded into an electronic spreadsheet, making the vent-sizing process even easier than reading a chart. The formula is:

$$A = (0.0000442)E^{0.952}\,V^{0.93} \tag{2}$$

$$D = 1.128\sqrt{A} \tag{3}$$

where A is the required vent square area in in^2, D is the vent diameter in inches, E is the rate of evolution of energy BTU/lb-hr, and V is the vessel volume in gallons.

Simplified Nomograph Method

Boyle [16] developed a simple method for quick estimation/screening of vent area for reactive systems. He defined the required vent area as that size which would empty the reactor before the pressure could rise above some allowable overpressure for a given vessel.

$$A = \frac{V\rho}{G_{2p}\delta t_v} \tag{4}$$

where A is the relief vent area, V is the reactor volume, ρ is the density of the reactants, G_{2p} is the mass flux through the relief device, and δt_v is is the venting time which can be approximated using the following equation:

$$\delta t_v = \frac{\Delta T C_p V\rho}{q_s} \tag{5}$$

where ΔT is the temperature rise corresponding to the overpressure ΔP, C_p is the heat capacity, and q_s is the energy release rate per unit mass at the set pressure of the relief system.

Combining the above two equations yields an expression for A which depends on the reactor mass, the reaction energy release rate, allowable overpressure, two-phase mass flux and heat capacity:

$$A = \frac{V\rho q_s}{G_{2p}\Delta T C_p} \tag{6}$$

The two-phase mass flux can be estimated using various methods as illustrated later in this paper. A simple method that is often used is that developed by Fauske which assumes homogeneous equilibrium between the vapor and liquid phases (no slip) and should provide conservative estimates for relief areas (underestimates the actual mass flux which can be higher due to slip, subcooling, etc.):

$$G_{2p} = (0.9)C_d \frac{\lambda}{V_v - V_l}\sqrt{\frac{1}{C_{p,l}T_s}} \tag{7}$$

where C_d is the discharge coefficient, λ is the latent heat of vaporization, V_v is the specific vapor volume, V_l is the specific liquid volume, and T_s is the reaction temperature at the set point of the relief device.

The ratio of the latent heat over the specific volume difference can be redefined using the Clausius-Clapeyron equation:

$$\frac{\lambda}{V_v - V_l} = T_s \frac{dP}{dT_s} \tag{8}$$

Substitution of Eqs. (8) and (7) in Eq. (6) yields the following simple equation for estimation of relief area:

$$\frac{A}{M} = \frac{q_s}{0.9C_d\,\Delta P\sqrt{T_s C_{p,l}}} \tag{9}$$

where M is the vessel mass in kilograms, ΔP is the vessel overpressure over the set point in Pascals, T_s is the relief set pressure in Kelvins, q_s is the reaction energy release rate in J/kg/s, and $C_{p,l}$ is the liquid heat capacity in J/kg/K.

Equation (9) provides a conservative estimate of the vent area required. By considering the case of 20% absolute overpressure, assuming a typical liquid heat capacity of 2,510 J/kg/K for most organics, and assuming a saturated water relationship, the following equation was obtained by Fauske [17]:

$$\text{A in } (\text{m}^2/1000 \text{ kg}) = 0.00208\frac{dT/dt}{P_s} \tag{10}$$

where dT/dt is the reaction energy release rate in (°C/min) and P_s is the set pressure in bar. A simple nomograph of the results can be plotted and is shown in Figure 11. The required vent area is determined simply from the heating rate, the set pressure and the mass of reactants.

The Fauske nomograph is useful for performing quick estimates and checking the results of more rigorous estimates. The nomograph data of Figure 11 applies to a discharge coefficient of 0.5 representing a discharge (L/D) of 400. The nomograph can be used for smaller $L = D$ by multiplying by the ratio (less than 1) of 0.5 over the actual C_d value.

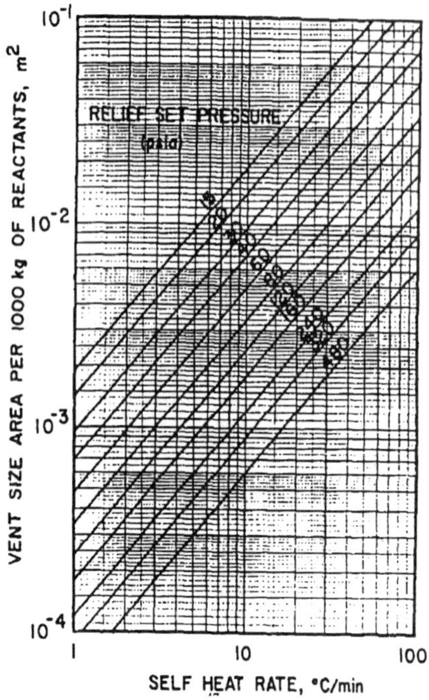

Figure 11. Nomograph for sizing two-phase reactor reliefs. Taken from Reference [18].

DIERS Methods

Two phase mass flux can be estimated by maximizing an expression involving an overall enthalpy change (stagnation – outlet) as a function of exit pressure and a selected thermodynamic path for determining the outlet vapor quality and temperature:

$$\text{Maximize } G = \rho_m \sqrt{2(H_0 - H)} \tag{11}$$

where G is the two-phase mass flux, ρ_m is the mixture density, H_0 is a two-phase stagnation enthalpy and H is the enthalpy at exit conditions. Both H_0 and H are mixture properties that depend on vapor quality. The vapor quality is typically calculated by assuming an isentropic expansion.

By invoking the thermodynamic relation between entropy and enthalpy under isentropic conditions, nozzle flow can be estimated via an integral equation.

$$TdS = 0 = dH - V\,dP \tag{12}$$

which leads to:

$$G = \sqrt{2\rho_m \int_P^{P_0} \frac{dP}{\rho}} \tag{13}$$

This expression is integrated from the exit pressure to the stagnation pressure.

Instead of using a detailed equation of state to represent the PVT and phase equilibrium behavior of pure components and/or mixture, DIERS proposed a simple two-parameter equation of state with two adjustable parameters, *a* and *b*, that are determined from either isentropic or isenthalpic equilibrium flash calculations:

$$\frac{\rho_0}{\rho} = a\left(\frac{P_0}{P} - 1\right) + b\left(\frac{P_0}{P} - 1\right)^2 \tag{14}$$

The parameters *a* and *b*are calculated by fitting the volumetric expansion behavior using two or more flash equilibrium calculations at pressures lower than stagnation. This approach may not work well for mixture with wide boiling point difference such as hydrocarbon mixtures containing hexane and hydrogen, for example. The same concept was extended to pipe flow using homogeneous equilibrium (no slip between the liquid and vapor phases) by DIERS.

Leung's simplification of the DIERS two-parameter equation, the omega method, was introduced in several publications, with an overall summary paper published in reference [19].

Leung uses the same DIERS form with $b = 0$and $a = \omega$:

$$\frac{\rho_0}{\rho} = \omega\left(\frac{P_0}{P} - 1\right) \tag{15}$$

This simplification leads to a generalized form for mass flux for nozzle flow using the ω pressure/volume equation in Eq. (14):

$$\frac{G}{\sqrt{P_0\rho_0}} = \frac{\sqrt{-2\{\omega \ln(P/P_0) + (\omega - 1)[1 - (P/P_0)]\}}}{1 + \omega[(P/P_0) - 1]} \tag{16}$$

This equation can be evaluated for different ratios of $P = P_0$ to find the point at which *G* is maximum where *P* ranges from the stagnation conditions to the back pressure imposed on the system. The critical pressure ratio can also be solved for by setting the first derivative of *G* with respect to pressure to 0. This is illustrated in Figure 12.

The value of ω ranges from 0 for all liquid flow, 1 for all vapor/gas flow, greater than 1 for flashing flow and between 0 and 1 for nonflashing flow.

The ω method is simple to evaluate for systems involving pure components and is of limited use for systems involving mixtures with wide boiling point difference or where composition changes are significant due to chemical reaction or large pressure drops. In addition, recent experimental flow data indicate that slip should be considered in estimation of the mass flux, especially for viscous systems. This becomes an issue when designing down-

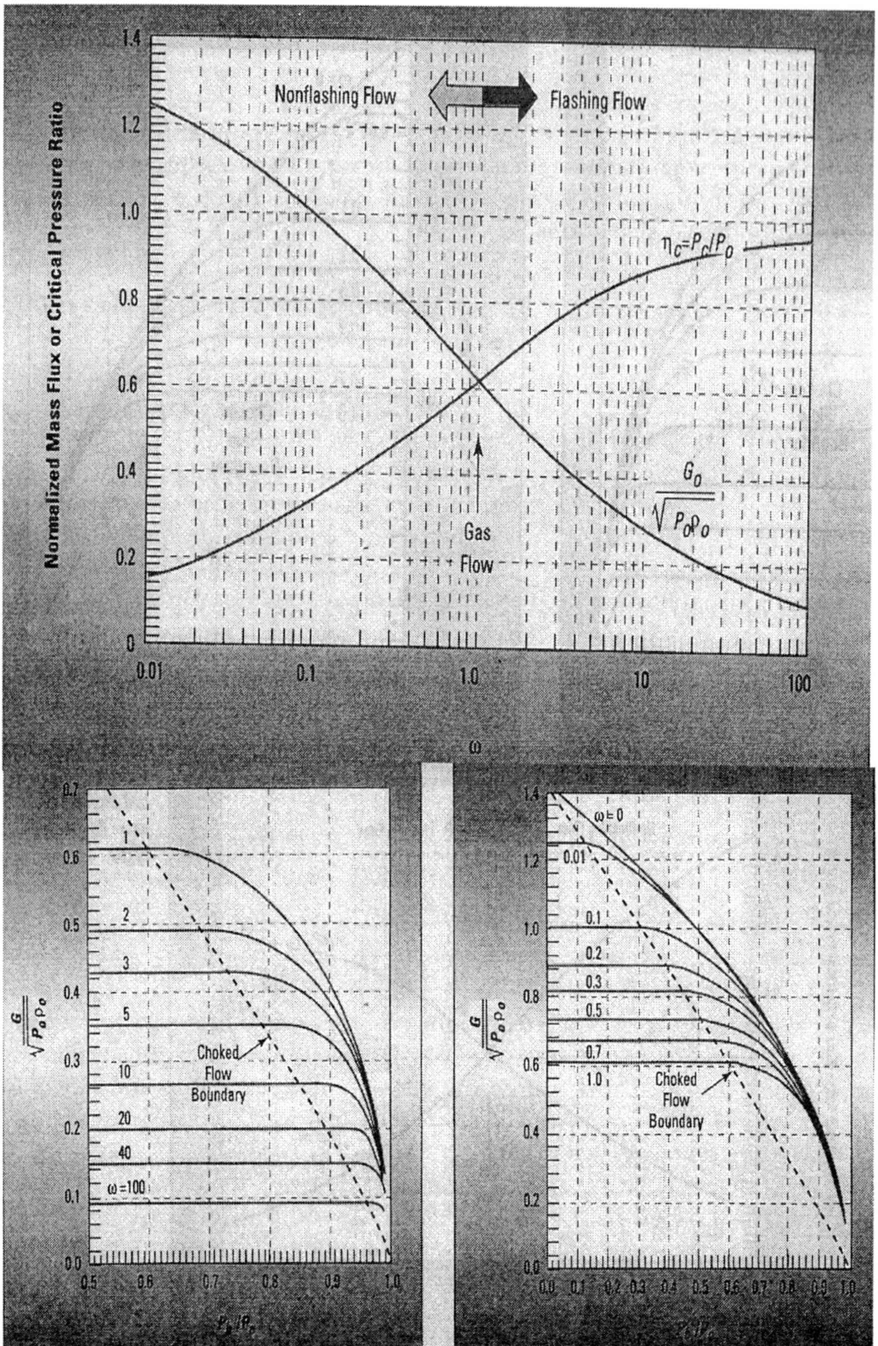

Figure 12. The omega method for estimation of two-phase mass flux and critical pressure ratio. Taken from *Chemical Engineering Progress*.

stream effluent handling equipment since in many situations the assumption of no slip between the liquid and the vapor phases leads to under estimation of the two-phase mass flux and as a result to a conservative estimate of the required relief area.

Leung provides the following expressions for ω which depend on stagnation properties only.

For flashing systems:

$$\omega = \alpha_0 + (1 - \alpha_0)\rho_{m,0} C_p T_0 P_0 \left(\frac{V_{v,0} - V_{l,0}}{H_{v,0} - H_{l,0}} \right)^2 \tag{17}$$

where α_0 is the void fraction at stagnation conditions.

For nonflashing systems:

$$\omega = \alpha_0 \tag{18}$$

Leung published many variations on the omega method extending its use to pipe flow, subcooled flow and for the calculation for reaction forces. We find these extensions less useful since they cannot handle inlet piping and outlet piping configurations with multiple segments and elbows, etc. The geometry of the inlet and outlet piping becomes very important for flashing flow, since pressure drop will lead to vapor quality change and will lead to choking at lower mass flow rates. In addition, the omega method is not capable of handling the impact of continuing reactions in piping and vent containment.

Relief Sizing for Gassy Systems Using Direct Scaleup

The volume balance equations represented in the earlier section are most applicable to reaction systems with tempering characteristics, i.e. "vapor" systems. For gassy systems, the reaction temperature can continue to increase after relief due to negligible tempering effects of noncondensable gas generation. The vent size should be able to handle the maximum rate of gas generation.

The required relief area can be estimated by equating the rate of volume generation due to reaction and the rate of volume discharge from the relief device:

$$\left(\frac{M_0}{M_{cell}} \right) \left(\frac{V_{cell}}{P_0} \right) \left[\frac{DP}{dt} \right]_{max} = \frac{C_d G A_h}{\rho_0} \tag{19}$$

where G is a two-phase mass flux to be computed using the omega method for nonflashing systems:

$$\omega = \alpha_0 \tag{20}$$

The pressure rise rate data is typically derived from a suitable open-system calorimeter. P_0 is the maximum allowable working pressure of the vessel (MAWP), M_0 is the initial vessel charge, V_{cell} and M_{cell} represent the void volume in the test cell and the mass in the test cell.

Solving directly for the relief area, we obtain:

$$A_h = \rho_0 \left(\frac{1}{C_d G} \right) \left(\frac{M_0}{M_{cell}} \right) \left(\frac{V_{cell}}{P_0} \right) \left[\frac{DP}{dt} \right]_{max} \tag{21}$$

The estimates of relief areas produced using this equation have been shown to be a factor of five to ten times larger than what is needed [see, for example, paper by A. Chakrabarti in *Proceedings of International Symposium on Runaway Reactions and Pressure Relief Design,* Boston, 1995 (New York: AIChE)].

Relief Sizing for Hybrid Systems

Sizing for hybrid systems requires the use of detailed computer simulation methods. However, an estimate of relief area requirement can be obtained by performing the sizing as a vapor system and as a gassy system and selecting the larger of the resulting relief areas.

Direct Scaleup of RSST data

The vent sizing equations used to scaleup RSST data account for two-phase flow. For a gassy system the required vent area is calculated from:

$$A_h = 3 \times 10^{-6} \frac{1}{F} \frac{M}{m} \frac{\dot{P}}{P_{mawp}^{1.5}} \tag{22}$$

where m is the sample mass, M is the plant vessel contents mass, $\dot{P}$ is the maximum measured pressure rise rate in (psia/min), and F is a flow reduction factor which compensates for vent lines. F can be estimated from:

$$\ln(F) = -0.0408 \left[\frac{L}{d} \right]^{0.554} \tag{23}$$

For a vapor system, the required vent area is calculated from:

$$A_h = 1.5 \times 10^{-5} \frac{M\dot{T}}{FP_{set}} \tag{24}$$

where $\dot{T}$ is the measured temperature rise rate (°C/min) at the vent set pressure, and F is given by:

$$\ln(F) = -0.0125 \left[\frac{L}{d} \right]^{0.671} \tag{25}$$

For hybrid systems the equation is similar to that for a gassy system:

$$A_h = 3 \times 10^{-6} \frac{1}{F} \frac{M}{m} \frac{\dot{P}}{P_{mawp}\sqrt{P_{mawp} - P_{set}}} \tag{26}$$

where F is a flow reduction factor which compensates for vent lines. F can be estimated from:

$$\ln(F) = -0.0408\left[\frac{L}{d}\right]^{0.554} \tag{27}$$

SuperChems for DIERS

The Design Institute for Emergency Relief Systems (DIERS) Users Group awarded Arthur D. Little, Inc. a contract to provide the next generation computer program for emergency relief system and effluent handling designs.

The new computer program, SuperChems for DIERS [8], is a dynamic simulator, capable of performing emergency relief system and effluent handling designs for complex geometries and multiphase reaction systems. In addition, SuperChems for DIERS is an equation-of-state based program which provides several benefits over existing non–equation-of-state based methods for systems involving supercritical reactions like polymerizations of butadiene and acrylonitrile, solution effects such as HCl/water, and a priori determination of phase-splitting.

This new computer program allows the user to dynamically simulate several common configurations for vent containment design. For example, the user is able to simulate a vessel discharging a two-phase mixture into a quench/vent tank where the catch/vent tank will vent to a stack or a scrubber. Effluent handling equipment available includes separators (horizontal and vertical), cyclones, etc. The impact of back pressure and continuing reaction in the vent containment system is accounted for in the dynamic simulations.[11]

SuperChems for DIERS Functionality

Highlights of the new computer program as outlined in the request for proposal are summarized below:

1. Provide a general purpose thermodynamic and transport properties generator with implicit corrections for nonideal behavior in both the liquid and vapor phase. The properties generator should also be able

11 This paper appeared in the *AIChE Loss Prevention Symposium Proceedings*, March 1997, Houston

to provide temperature and pressure dependent derivatives for all properties of interest. The generator should be detailed enough so that heat of solution effects and vapor–liquid nonideal equilibrium are implicit. This generator should also be able to provide all properties and all derivatives required by the computer models. The generator should be equation of state (EOS) based.

2. Create/design an interface allowing the properties generators to access a thermophysical properties database so that properties do not have to be manually added. A databank manager should be provided in order to allow the users to input/modify their own compounds, if needed.
3. Revise all flow models such that the new computer program will address the following items:
   (a) inclined flow
   (b) subcooled flow
   (c) sudden expansion/contraction
   (d) piping segments with varying diameter and orientation
   (e) viscous two-phase flow through safety relief valves and pipes
   (f) continuing chemical reaction in piping and vent containment systems
   (g) detailed energy balances for vessels and piping
   (h) detailed momentum balance for piping
   (i) implicit vapor–liquid equilibrium relations
   (j) the flow models and the vessel balances should be equation oriented.
4. Provide a suitable stiff differential/algebraic (DAE) equation-solver. Proposed schemes should include Gear's or Michelsen's methods.
5. Provide a user-friendly menu-driven interface including graphics plotting capabilities and report generation.
6. Provide required documentation including an Operations Guide, a User's Guide, and a Reference Manual.
7. Validate all new models using experimental data, where available.

Arthur D. Little Inc. (ADD) had already developed a computer program which contained all the above specifications. This program took about five years to develop and is known as SuperChems Expert Version 3. The emergency relief system design portion of SuperChems was customized for the DIERS Users Group and is available for sale from the American Institute of Chemical Engineers (AIChE).

SuperChems for DIERS is an equation-of-state based program which provides several benefits over existing non–equation-of-state based methods for emergency relief system design for systems involving:

1. Reactions with supercritical components such as polymerizations of butadiene, acrylonitrile, etc.

2. Solution effects such as HCl /water, etc.
3. A priori determination of phase-splitting

SuperChems for DIERS contains a databank of more than 3000 binary systems with equation-of-state binary interaction parameters derived from experimental vapor-liquid and liquid-liquid data. There is also a VLE/VLLE data package. This versatile utility allows the estimation of binary interaction parameters for the equation of state composition dependent mixing rule. The source of data can be one of six types:

1. Azeotropic data
2. TPXY measured data (SuperChems for DIERS also includes three thermodynamic consistency utilities for X and Y calculations)
3. Mutual solubility data
4. Henry's law constants (mostly used for gas solubility in liquids)
5. Activity coefficient model parameters
6. Infinite dilution activity coefficients

SuperChems for DIERS has an extensive database with over 1200 chemicals. The database contains 39 thermophysical properties with temperature dependent properties and data quality parameters. The program also has a detailed regression package (linear/nonlinear) which allows the reduction of tabular data to equation forms supported by the databanks.

SuperChems for DIERS Computer Implementation

Emergency relief system design does not stop at the estimation of the size of the relief device. The effluent must be treated if it is toxic and/or flammable or if it presents an environmental impact. While homogeneous-equilibrium flow (no slip) is typically used for sizing the relief device, slip-equilibrium flow should be used to establish correct pressure drops and safety/environmental impacts.

SuperChems for DIERS is a computer program which allows the integral evaluation of relief dynamics and downstream effects. For example, using SuperChems for DIERS, we can evaluate the time dependent history of pressure, temperature and composition in a reactor vessel as the relief occurs. Simultaneously, the effluent is discharged, and handled to meet established (regulatory or internal) criteria. Typically, many options are evaluated before a final design is selected. This includes separation equipment, flares, stacks, etc.

The most useful aspect of SuperChems for DIERS is its scenario/object driven architecture. Once an object (such as a vessel or piping configuration) is defined, it can be used by one or more scenarios. Once a scenario is defined, it can be duplicated and used to perform what-if or sensitivity analysis.

The detailed algorithms for SuperChems for DIERS are published in references [9], [10], and [8].

CONCLUSIONS

Our discussions have provided an overview of the range of techniques available for thermal hazard evaluation in connection with relief system design. It is clear that, in practice, a number of (possibly conflicting) factors might influence the precise approach chosen. This includes existing testing regimes, equipment familiarity/availability, regulatory drivers, time pressure, and financial resources.

Ultimately, the main requirement is for a coherent and consistent approach to thermal hazards evaluation using the best means available. All instruments and methodologies have inherent strengths and weaknesses. Diversity of approach stimulates active interpretation and understanding, and avoids common mode errors. Informed interpretation is ultimately more important than quality of hardware.

References

1. D. J. Frurip, T. C. Hofelich, D. J. Leggett, J. J. Kurland, and J. K. Niemeier. A review of chemical compatibility issues. In *31st Loss Prevention Symposium*, pages 43c–1. New York: AIChE, 1997.
2. AIChE/CCPS. *Guidelines for Chemical Reactivity Evaluation and Application to Process Design*. New York: AIChE/CCPS, 1995.
3. E. S. Shanley and G. A. Melhem. The oxygen balance criterion for thermal hazards assessment. *Process Safety Progress*, 14(1):29–31, 1995.
4. E. S. Shanley and G. A. Melhem. A review of ASTM CHETAH 7.0 hazard evaluation criteria. *Journal of Loss Prevention in the Process Industries*, 1995.
5. G. A. Melhem and E. S. Shanley. On the estimation of hazard potential for chemical substances. *Process Safety Progress*, 15(3):168–172, 1996.
6. S. E. Stein, J.M. Rukkers, and R.L. Brown. NIST standard reference database25, the NIST structures and properties database and estimation program, version 1.2. Technical report, NIST, 1991.
7. G. A. Melhem. SuperChems Expert version 3: An advanced tool for consequence analysis, thermal hazards assessment and pressure relief design, 1996.
8. G. A. Melhem and H. G. Fisher. An overview of SuperChems for DIERS: A program for emergency relief system and effluent handling designs. *Process Safety Progress*, 16(3), 1997.
9. G. A. Melhem. Advanced ERS design using computer simulation. In *International Symposium on Runaway Reactions and Pressure Relief Design*, pages 502–566. New York: AIChE, 1995.
10. G. A. Melhem, H. G. Fisher, and D. A. Shaw. An advanced method for the estimation of reaction kinetics, scale-up and pressure relief design. *Process Safety Progress*, 14(1):1–21, 1995.
11. E. Shanley and S. Myers. Blasting cap test report. Experimental Dec. 18, Arthur D. Little Inc., 1997.

12. CCPS. *Guidelines for Safe Storage and Handling of Reactive Materials*. New York: AIChE, 1995.
13. G. W. Boicourt. Experimental safety: What you need for effective process safety evaluation. In *30th Loss Prevention Symposium*. New York: Safety and Health Division, AIChE, 1996.
14. E. J. Sestak. Venting of chemical plant equipment, 1965.
15. H. K. Fauske. Venting of runaway chemical reactions and nonequilibrium effects. In *4th Miami International Symposium on Multiphase Transport and Particulate Phenomena*, 1986.
16. W. J. Boyle Jr. Sizing relief area for polymerization reactors. *Chemical Engineering Progress*, 63(8):61, 1967.
17. Hans K. Fauske. A quick approach to reactor vent sizing. *Plant/Operations Progress*, 3(3), 1984.
18. Hans K. Fauske. Generalized vent sizing monogram for runaway chemical reactions. *Plant/Operations Progress*, 3(4), 1984.
19. J. C. Leung. Easily size relief devices and piping for two-phase flow. *Chemical Engineering Progress*, 92(12):28–49, 1996.

1,3-Butadiene: Thermal Dimerization/ Trimerization, Free-Radical Polymerization, and Polymer Decomposition Reactions

Harold G. Fisher
Union Carbide Corporation, P. O. Box 8361, South Charleston, WV 25303

Georges A. Melhem
Arthur D. Little, Inc., 20 Acorn Park, Cambridge, MA 02140

Marc E. Levin
Shell Oil Company, P. O. Box 1380, Houston, TX 77252

Joseph C. Leung
Leung, Inc., 1521 Evergreen Lane, Darien, IL 60561

ABSTRACT

1,3-butadiene monomer undergoes thermally initiated, reversible dimerization/trimerization reactions with essentially the same kinetics in both the gaseous and liquid phases. Kinetics for formation of the dimer and the trimer are available from the open literature. The rate of reaction becomes significant (0.02°C/min = 29°C/day) at temperatures above 70-80°C. Inhibitors (t-butylcatechol, et al.) are used to prevent/minimize free-radical polymerization reactions in the liquid phase and to maintain product quality at ambient or subambient temperatures. These inhibitors do not prevent the dimerization/trimerization reactions.

Unless adequate emergency relief is provided, the adiabatic temperature rise from the dimerization/trimerization reactions can lead to both a free-radical polymerization, initiated by adventitious peroxides, and a thermal decomposition of the resultant polymer to produce residues, volatiles, and noncondensable gases. Temperatures of 600°C and pressures of over 2000 psig are possible. Heat rates of 10,000°C/min and pressure rise rates of 10,000 psig/min are also possible in unvented/undervented vessels.

Emergency relief devices protecting vessels containing high concentrations of 1,3-butadiene should be reviewed to identify potentially reactive cases. This review is recommended to ensure that current emergency relief system designs are adequate and that equipment is being operated with an adequate margin of safety.

1.0. BACKGROUND

Following an incident involving a butadiene still,[1] thermal stability tests of 1,3-butadiene were conducted using a then standard (but now obsolete) thermal stability apparatus. Mixtures of 1,3-butadiene and several process components were also tested. Unfortunately, the test apparatus had a high thermal inertia (heavy weight process vessel) and high heat losses. Heat rates of 30 to 40°C/min at temperatures of 200 to 340°C and pressures of 1450 to 1800 psig were found to be necessary to initiate an explosive decomposition of 1,3-butadiene. These results are due only to shortcomings of the apparatus and are not an accurate representation of the thermal stability of the chemical under truly adiabatic runaway reaction conditions.

A Material Safety Data Sheet (MSDS)[2] states that explosive decomposition (the above-described reactions) can only occur if the liquid is heated at 30 to 40°C/min at temperatures of 200 to 340°C and pressures of 1450 to 1800 psig. This statement is false and results from inadequacies of the test apparatus in use 27 years ago.

2.0. KINETIC SUMMARY

1,3-butadiene monomer undergoes thermally initiated, exothermic, reversible dimerization/trimerization reactions with essentially the same kinetics in both the gaseous and liquid phases. The forward kinetics are second order, while the reverse kinetics are first order. The dimerization product is predominately 4-vinylcyclohexene (VCH) with some *cis*, *cis*-1,5-cyclooctadiene (COD) and *trans*-1,2-divinylcyclobutane (DVCB) produced. The trimerization product is octahydro diphenyl.

1,3-butadiene readily reacts with oxygen to form polymeric peroxides, which are not very soluble in liquid butadiene and tend to settle to the bottom of a container because of their higher density. These peroxides are explosive and shock-sensitive when concentrated. The addition of antioxidants (t-butylcatechol, et al.) removes free radicals that can cause a rapid, exothermic polymerization. The polymerization inhibitors lose effectiveness at 85–100 °C. Since these inhibitors are nonvolatile, they are only effective in the liquid phase. Nitrogen blanketing is used to exclude oxygen from 1,3-butadiene storage tanks.

1,3-butadiene will undergo a thermal or "adventitious" peroxide-initiated free radical polymerization at higher temperatures to form polybutadiene polymer. At even higher temperatures, the polybutadiene polymer will decompose to form residues, dimers and noncondensable gases such as ethylene and ethane, propylene and propane, butenes and butanes.

3.0. REACTION KINETICS (Dimerization/Trimerization and Reverse Reactions)

A literature review provided kinetics for the forward dimerization/trimerization of 1,3-butadiene and the reverse reactions of the dimers/trimer. In most cases the reaction kinetic constants were measured in the vapor phase at high temperatures. Several of the studies, however, compared reaction kinetics in the vapor phase with those obtained at lower temperatures in the liquid phase. The comparisons are sufficient to indicate reasonable agreement, if not identical kinetics, for both the liquid and vapor phase reactions.

1,3-butadiene produces three dimerization products to include 4-vinylcyclohexene (VCH), 1,5-cyclooctodiene (COD) and 1,2-divinylcyclobutane (DVCB). A comparison of relative reaction rates, stoichiometric information from the literature and NMR values obtained by one of the authors indicate that the production of VCH is the dominant reaction for emergency relief design. The combined production of COD, DVCB and butadiene trimer is low compared to that of VCH.

3.1. Butadiene Dimerization

3.1.1. 4-vinylcyclohexene (VCH) Formation

The predominant product of the dimerization reaction of 1,3-butadiene (BD) is 4-vinylcyclohexene (VCH).[3,4] NMR studies conducted on a sample obtained to support this paper showed a VCH:COD ratio of 19:1. Kinetics for the Diels-Alder dimerization reaction of 1,3-butadiene to form 4-vinylcyclohexene are reported in the open literature.[4-11] An exothermic, second-order kinetic model for the dimerization of BD to form VCH is as follows:[4,6,11]

$$k \text{ (ft}^3\text{/lb mole hr)} = 5.2917\text{E}{+}11 \exp(-23690/(1.9872{*}\text{TK}))\ \text{CBD}^2$$

$$\Delta H_R \text{ (BTU/lb)} = {\sim}307.5$$

$$2\text{BD} \rightarrow \text{VCH}$$

where TK (K) is temperature and CBD (lb mole/ft^3) is the concentration of butadiene.

These kinetics reportedly[9] include the formation of COD and DVCB in addition to VCH.

An alternate second-order kinetic model for the dimerization of BD to form VCH is as follows:[7,8,10,11]

$$k \text{ (ft}^3\text{/lb mole hr)} = 5.1390\text{E}+11 \exp(-24530/(1.9872*\text{TK})) \text{ CBD2}$$

This model is slightly slower than the first model. The two models, however, bracket the data discussed in this paper.

A best estimate second-order kinetic model for the dimerization of BD to form VCH is as follows:

$$k \text{ (ft}^3\text{/lb mole hr)} = 5.6270\text{E}+11 \exp(-24110/(1.9872*\text{TK})) \text{ CBD}^2$$

The reaction rate from this model is intermediate to the primary and alternate models.

Kinetics for the higher temperature reverse reaction of VCH to form BD are also available.[4,5 8-11]

3.1.2. 1,5-cyclooctodiene (COD) Formation

Kinetics for the dimerization reaction of 1,3-butadiene (BD) to form 1,5-cyclooctodiene (COD) are reported in the open literature.[9] An exothermic, second order kinetic model for the dimerization of BD to form COD is as follows:

$$k \text{ (ft}^3\text{/lb mole hr)} = 2.5756\text{E}+12 \exp(-28440/(1.9872*\text{TK})) \text{ CBD}^2$$

$$\Delta H_R \text{ (BTU/lb)} = \sim 615$$

$$2\text{BD} \rightarrow \text{COD}$$

where TK (K) is temperature and CBD (lb mole/ft^3) is the concentration of Butadiene

Kinetics for the higher temperature reverse reaction of COD to form BD are also available.[4,9,10]

Kinetics are also available for the reaction of VCH to form COD[9] and the reverse reaction of COD to form VCH.[4,9,10,12]

3.1.3. 1,2-divinylcyclobutane (DVCB) Formation

1,2-divinylcyclobutane is formed by the dimerization of 1,3-butadiene[10] and in turn decomposes to form BD, VCH, and COD.[4,10,13]

3.2. Butadiene Trimerization (Δ3, 3′-octahydro diphenyl)

Kinetics for the reaction of 1,3-butadiene (BD) and 4-vinylcyclohexene (VCH) to form butadiene trimer (Δ3, 3'-octahydro diphenyl) are reported in the open literature.[5] An exothermic, second order kinetic model for the reaction of BD and VCH to form butadiene trimer is as follows:

$$k\ (\text{ft}^3/\text{lb mole hr}) = 7.4959\text{E}{+}15\ \exp(-38000/(1.9872{*}\text{TK}))\ \text{CBD}{*}\text{CVCH}$$

$$\Delta H_R\ (\text{BTU/lb}) = \sim 922.5$$

$$\text{BD} + \text{VCH} \rightarrow \text{Butadiene Trimer}$$

where TK (K) is temperature, CBD (lb mole/ft^3) is the concentration of butadiene, and CVCH (lb mole/ft^3) is the concentration of vinylcyclohexene

4.0. POLYMERIZATION INHIBITORS

1,3-butadiene reacts with oxygen or air to form butadiene polyperoxide. This peroxide has the explosive power of TNT and is heat and shock sensitive. The peroxide is only slightly soluble in 1,3-butadiene at high concentrations and forms layers in a process because of its greater density.[14]

Inhibitors such as t-butylcatechol (TBC) are used to prevent/minimize the formation of butadiene peroxide and "popcorn" polymer. These inhibitors react with free radicals to form more stable free radicals, which are then terminated. Oxygen in ppm quantities is required to activate most inhibitors. TBC is consumed as it reacts and must therefore be periodically replaced. The efficiency of free-radical inhibitors varies with temperature. The consumption of inhibitor is adequately described by first order kinetics when sufficient oxygen is present.

Above a certain temperature TBC is rapidly consumed by the generation of free radicals and inhibitor effectiveness is lost. At still higher temperatures the butadiene polyperoxides decompose to produce additional free radicals. The thermally-induced, free-radical polymerization of 1,3-butadiene monomer then occurs.

5.0. "POPCORN" POLYMER FORMATION

1,3-butadiene monomer also forms rubbery "popcorn" polymers caused by polymerization initiators such as free radicals or oxygen. Addition of antioxidants such as t-butylcatechol (TBC) and storage at low temperatures can substantially reduce fouling caused by these polymers. Polymer seeds (free radicals), oxygen, and rust promote the formation of "popcorn" polymers. Butadiene "popcorn" polymer nucleation occurs as a heterogeneous surface reaction. "Popcorn" polymers can grow rapidly and generate tremendous pressure resulting in plugging and/or rupture of pipes and vessels. Rigorous exclusion of oxygen, passivation of metal surfaces, and removal of "popcorn" polymer seeds can mitigate most of the problem.[6,15–17] A recent review has been published.[17]

6.0. REACTION KINETICS (Free-Radical Polymerization)

The dimerization/trimerization reactions of 1,3-butadiene are not affected by the presence or absence of free-radical initiators or polymerization inhibitors. Low temperature storage is required to prevent/limit these reactions.

t-butylcatechol (TBC) inhibitor is used to prevent/limit the thermal and/or "adventitious" peroxide initiated, free-radical polymerization of 1,3-butadiene during storage and low temperature industrial processing. TBC, however, is not an effective inhibitor at the high temperatures resulting from the exothermic dimerization/trimerization reactions. The exothermic, free-radical polymerization of 1,3-butadiene thus occurs in parallel to the dimerization/trimerization reactions at high temperature.

The "adventitious" polybutadiene peroxide catalyzed, free-radical polymerization of 1,3-butadiene monomer is proportional to the square root of the peroxide concentration. This reaction has an activation energy similar to that of the dimerization reaction. A concentration of 1200 ppm active oxygen (one active oxygen atom per molecule of peroxide—about one-half the soluble concentration) is reported to catalyze the free-radical polymerization of 1,3-butadiene at a rate 10–20 times faster than the dimerization reactions. Fortunately, exclusion of oxygen from process vessels minimizes the concentration of polybutadiene peroxide and thus the catalyzed, free-radical polymerization reaction. In the absence of peroxides the free-radical polymerization reaction is insignificant compared to the dimerization reaction.[6]

7.0. REACTION KINETICS (Polymer Decomposition)

Polybutadiene polymer is reported[18] to completely decompose over the temperature range 350–477 °C to produce residue (85.6 wt %; MW = 739), volatiles (9.95 wt %; assumed to be dimers) and lights/noncondensable gas (4.45 wt %; MW ≈ 51).[19] Fifty percent of the polymer decomposes in 30 minutes at approximately 407°C.[20] Using the literature half-life information, we derived exothermic first-order kinetics for the decomposition of polybutadiene polymer as follows:

$$k\ (\text{pph}) = 9.4672\ \text{E}{+}19\ \exp(-62000/(1.9872 * \text{TK}))$$

$$\Delta H_R\ (\text{BTU/lb}) = 632.5\ (\text{Ref. } 6)$$

$$\begin{aligned}\text{PBD} = {} & 0.856\ \text{RESIDUE (MW} = 739) + 0.0995\ \text{DIMER (VCH)} \\ & + 0.0445\ \text{LIGHTS/NONCONDENSABLE GAS (MW} \approx 51)\end{aligned}$$

8.0. VALIDATION OF KINETIC MODELS

Experimental data for the thermally initiated reactions of 1,3-butadiene are available from the Accelerating Rate Calorimeter (ARC)[21,22] (Figures 1–4), Vent Sizing Package (VSP2)[23,24] (Figures 5–8), and Automatic Pressure Tracking Adiabatic Calorimeter (APTAC)[25] (Figures 9–12). We used these data to confirm our kinetics, heats of reaction, stoichiometry, physical property, and vapor–liquid models, which were assembled from open literature sources.

The experimental data presented in this paper are primarily the dimerization/trimerization reactions of 1,3-butadiene. The reverse dimerization/trimerization reactions require high temperatures to achieve a substantial rate. Since oxygen was excluded from the reaction mixture, we would not expect the free-radical polymerization reactions of 1,3-butadiene to be significant. Since we therefore expect a low rate of polybutadiene polymer formation, we also expect a minimum pressure effect due to the decomposition reaction of polymer. Each of these expectations is supported by the available data as discussed below.

In many experiments, however, the test vessel ruptured due to the high partial pressure of 1,3-butadiene at the high reaction temperatures or due to using an initial reactant fill ratio which exceeded the volume of the test cell at the critical volume of the reaction mixture (i.e., the test cells become liquid-full and burst due to the hydrostatic pressure caused by liquid expansion). At 25°C the hydrostatic pressure of 1,3-butadiene increases by 98 psi/°C in a liquid-filled vessel due to thermal expansion. The critical volume of 1,3-butadiene is 0.245 g/mL. Vessels with higher fill ratios, which become liquid full below the critical volume, can burst due to hydrostatic pressure.

8.1. Accelerating Rate Calorimeter (ARC) Data

ARC run 4845 was available. A total of 3.5 g (initial 0.55 fill ratio) of uninhibited 1,3-butadiene were charged to a 11.5 mL lightweight, spherical, titanium bomb. Following several heat–wait–search periods to 104°C, the ARC attained a peak heat rate of 10.5°C/min, a peak temperature of 200°C and a peak pressure of 1025 psia. The test was terminated at 200°C.

Our best estimate kinetic model fit the ARC 4845 heat rate versus temperature (Figure 1), pressure rate versus temperature (Figure 2), pressure versus temperature (Figure 3) and temperature versus time (Figure 4) plots quite well.

Note: Inclusion of the 1.5 mL transducer volume and tubing was required to obtain the proper burst pressure.

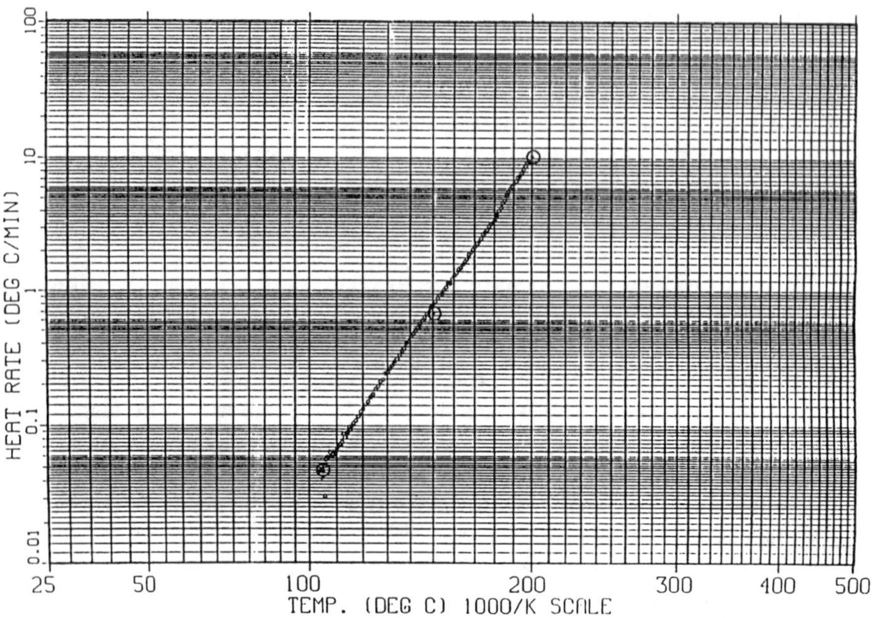

Figure 1. Uninhibited 1,3-butadiene (ARC 4845)

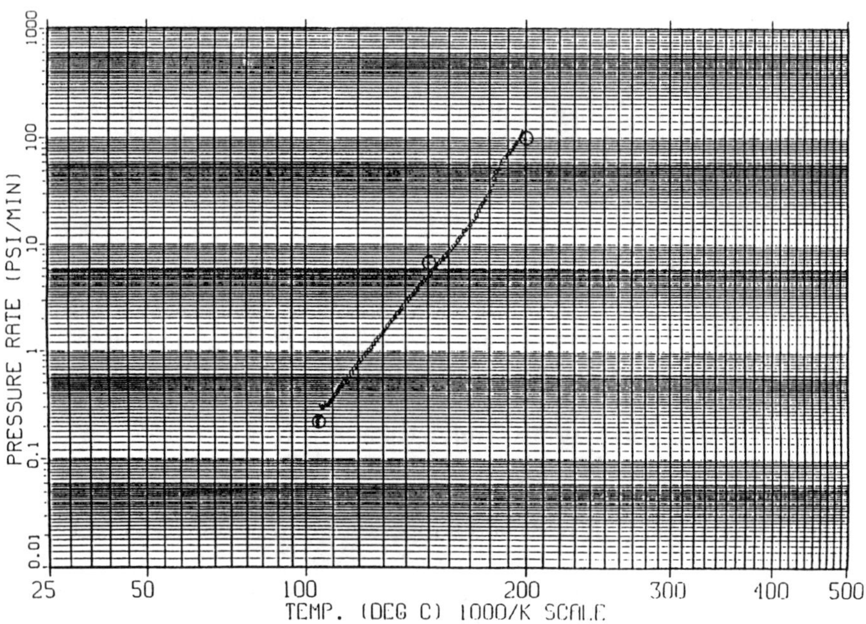

Figure 2. Uninhibited 1,3-butadiene (ARC 4845)

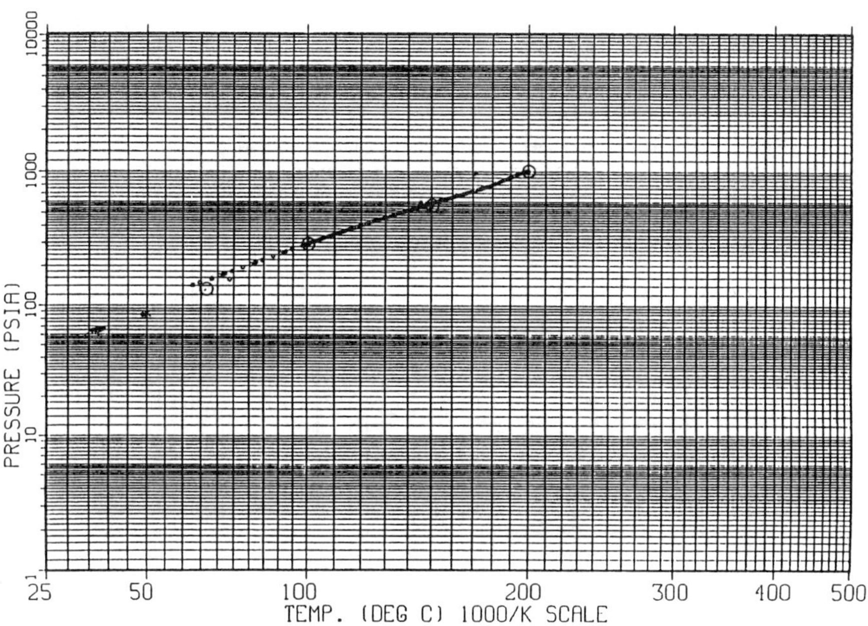

Figure 3. Uninhibited 1,3-butadiene (ARC 4845)

8.2 Vent Sizing Package (VSP2) Data

A total of 62.9 g (0.58 initial fill ratio) of 37.2 wt% 1,3-butadiene and 62.8 wt% of tridecane diluent were charged to an approximate 115 mL cylindrical test cell, which is expected to expand to a volume of approximately 140 mL at rupture. Following a ramped heating period to 120°C, the VSP2 4715 sample appeared to heat by reaction and a positive drift of approximately 0.1°C/min to a temperature of 140°C. The VSP2 then attained a peak heating rate of approximately 6°C/min, a peak pressure of 300 psia and a peak temperature of 295°C. The test cell did not rupture, which allowed a complete check of the kinetic model and the heat of reaction.

The primary kinetic model fit the VSP2 4715 heat rate versus temperature (Figure 4), pressure rate versus temperature (Figure 5), pressure versus temperature (Figure 6), and temperature versus time (Figure 7) plots quite well.

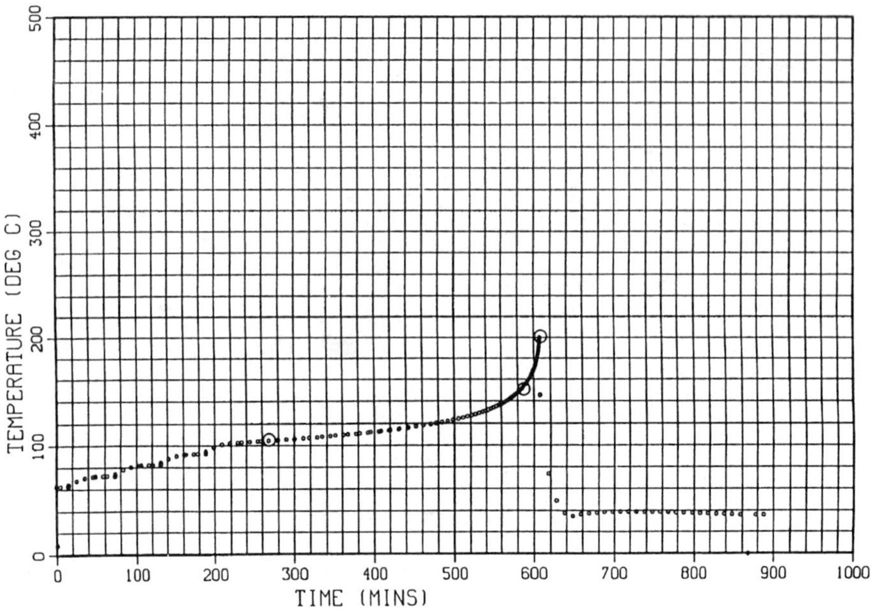

Figure 4. Uninhibited 1,3-butadiene (ARC 4845)

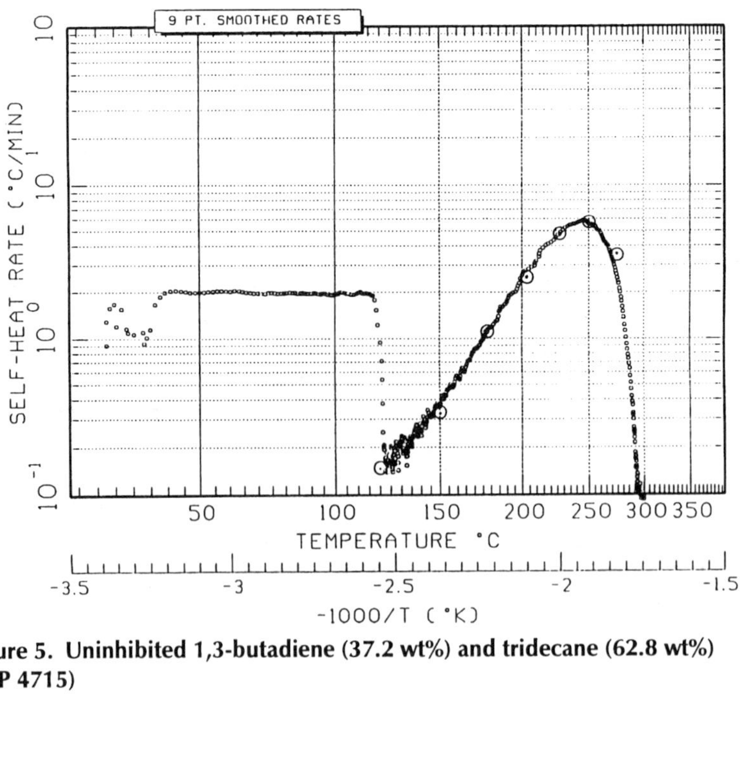

Figure 5. Uninhibited 1,3-butadiene (37.2 wt%) and tridecane (62.8 wt%) (VSP 4715)

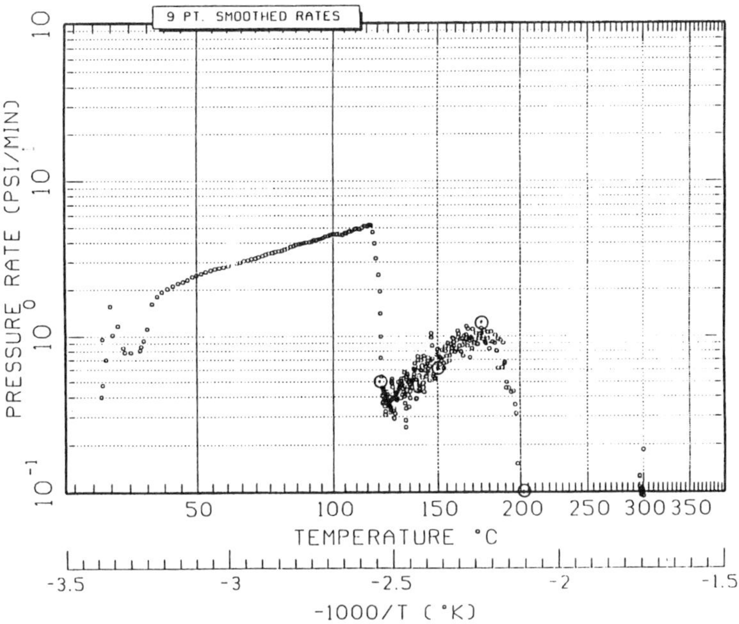

Figure 6. Uninhibited 1,3-butadiene (37.2 wt%) and tridecane (62.8 wt%) (VSP 4715)

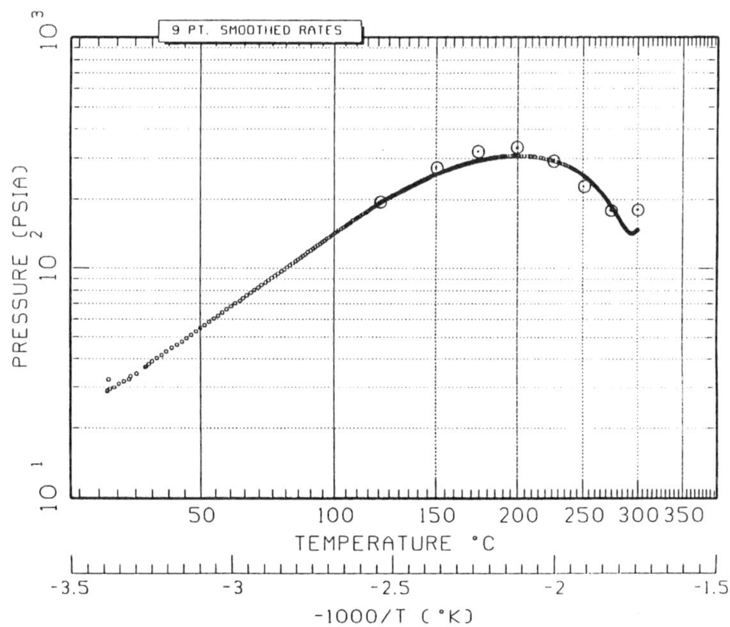

Figure 7. Uninhibited 1,3-butadiene (37.2 wt%) and tridecane (62.8 wt%) (VSP 4715)

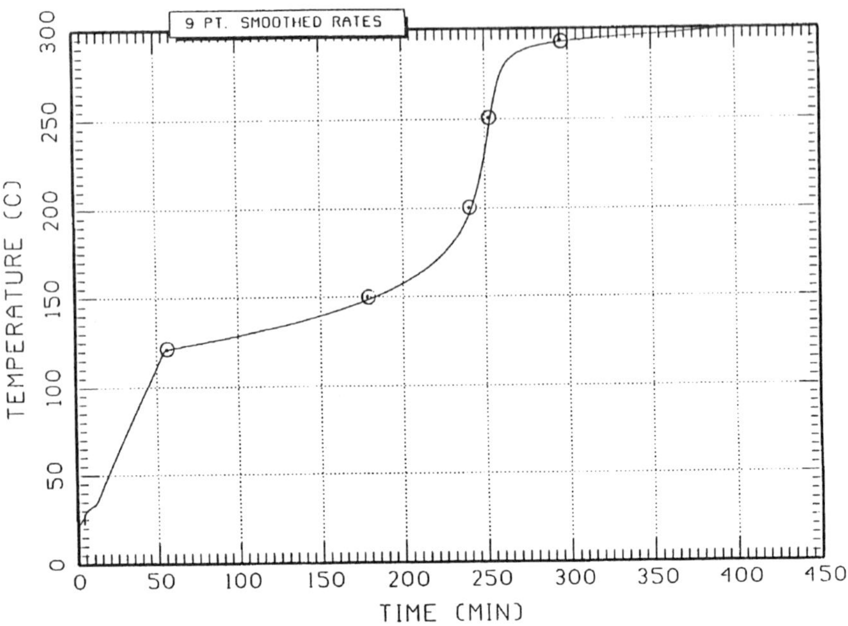

Figure 6. Uninhibited 1,3-butadiene (37.2 wt%) and tridecane (62.8 wt%) (VSP 4715)

8.3. Automatic Pressure Tracking Adiabatic Calorimeter (APTAC) Data

APTAC run 2162 was also available. A total of 32.5 g (initial 0.25 fill ratio) of 1,3-butadiene were charged to a 130 mL spherical, titanium bomb. Following several heat–wait–search periods to 115°C, the APTAC attained a peak heating rate of 835°C/min, a peak temperature of 394°C and a peak pressure of 830 psia. The bomb ruptured.

The upper limit of the heat rate compensation (adiabaticity) of the APTAC instrument is approximately 400 °C/min. The alternate kinetic model exhibits a positive bias on the APTAC 2162 heat rate versus temperature (Figure 9) plot well below this value. The positive bias is also clearly noticeable on the pressure rate versus temperature (Figure 10) plot. The vapor–liquid equilibria model causes an inflection in the pressure versus temperature (Figure 11) plot. This inflection is reflected in both pressure rate increases and decreases, which are clearly not evident on the pressure rate versus temperature (Figure 10) plot. The agreement of the model with the temperature versus time (Figure 12) plot is acceptable.

Note: Inclusion of the 15.72 g weight of the pressure transducer nut and ferrule in addition to the 31.95 g weight of the spherical titanium bomb was required for the model to approximately fit the data.

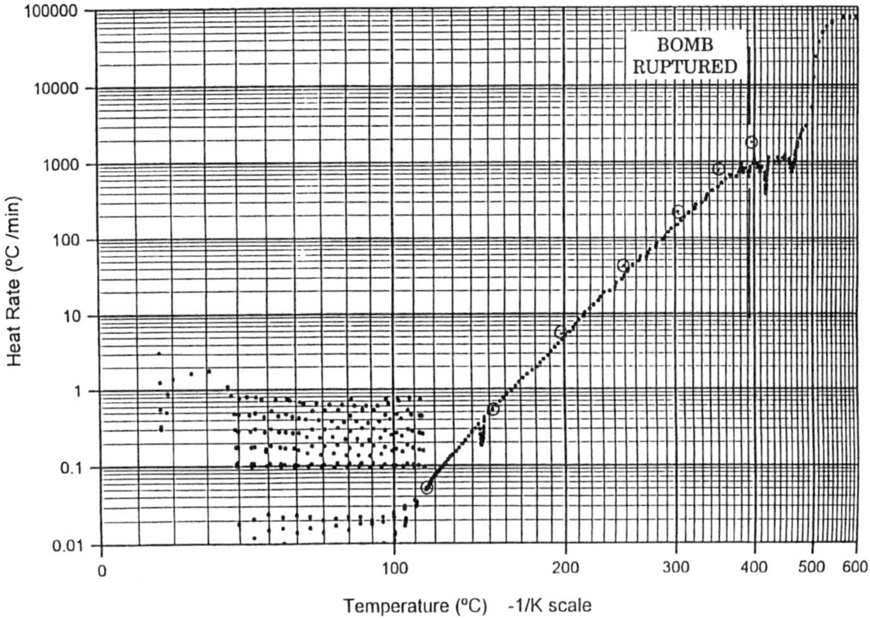

Figure 9. Uninhibited 1,3-butadiene (APTAC 2162)

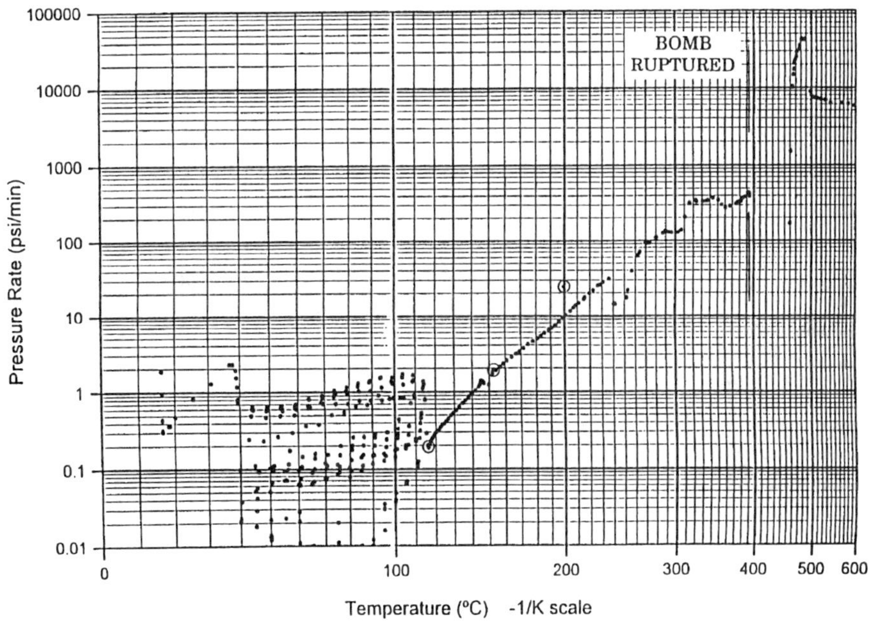

Figure 10. Uninhibited 1,3-butadiene (APTAC 2162)

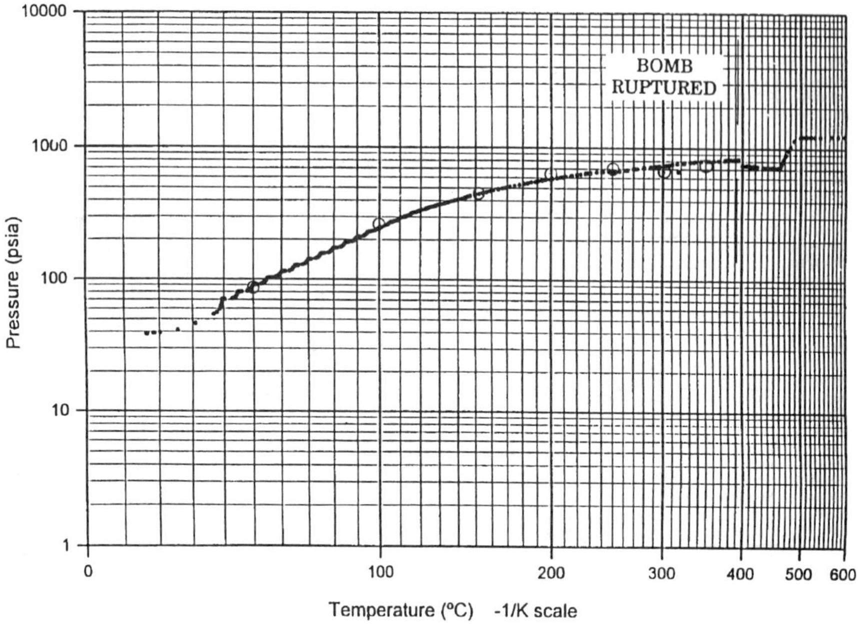

Figure 11. Uninhibited 1,3-butadiene (APTAC 2162)

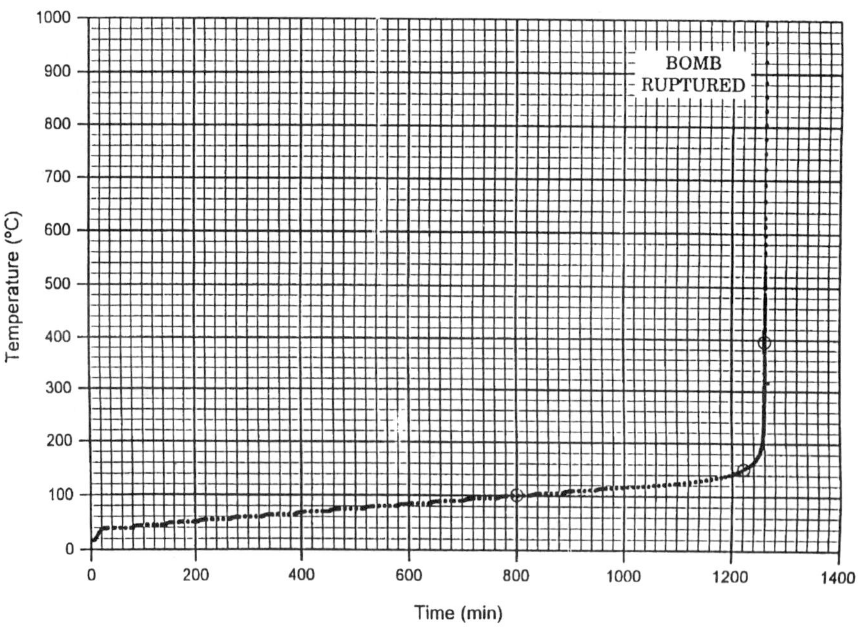

Figure 12. Uninhibited 1,3-butadiene (APTAC 2162)

9.0. CLOSURE

A recent paper[26] contains some of the information discussed in this paper. We chose to concentrate, herein, on the most likely, uncatalyzed, thermally initiated reactions of 1,3-butadiene as a basis for sizing emergency relief devices. The validity and usefulness of our model increases when the relief flow is tempered such as is possible with adequate mlysized emergency relief devices with low set pressures. The reverse dimerization/trimerization reactions become significant at higher temperatures (>300°C) and cause a volatility (pressure) increase. Since the critical pressure and density of 1,3-butadiene are 620 psig and 0.245 g/mL at the critical temperature of 152°C, our model should be valid for emergency relief devices with set pressures at or below 400 psig.

10.0. ACKNOWLEDGMENTS

C. L. Ivey (UCC) and C. L. Rhodes (UCC) provided the physical properties used in our reaction model. B. C. McDavid (CDI-UCC) assembled the physical property package. J. J. Kurland (UCC), C. F. Askonas (FAI) and S. Chippett (UCC) provided the ARC, VSP2 and APTAC experimental data, respectively. D. Lin-Vien and P. N. Tutunjian (Shell Oil) provided the compositional analysis of the 1,3-butadiene–tridecane mixture. A recent paper (Reference 26) provides additional data and information beyond the scope of this paper.

11.0. REFERENCES

1. Keister, R. G., Pesetsky, B. I. and Clark, S. W., "Butadiene Explosion at Texas City-3," Loss Prevention Symposium (5) (1971), 67–75, AIChE.
2. Anon., "1,3-Butadiene," MSDS, Sigma-Aldrich Corporation (1994).
3. Stephenson, L. M., Gemmer, R. V. and Current, S., "Dimerization of Butadiene," *JACS*, 92(20) (October 1, 1975), 5909–5910.
4. Gajewski, J. J., *Hydrocarbon Thermal Isomerizations*, New York : Academic Press, 1981.
5. Kistiakowsky, G. B., and Ranson, W. W., "The Polymerization of Gaseous Butadiene," *J. Chem. Phys.*, 7(9) (September 1939), 725–735.
6. Robey, R. F., Wiese, H. K., and Morrell, C. E., "Stability of Butadiene," *IEC*, 36(1) (January 1944), 3–7.
7. Rowley, D., and Steiner, H., "Kinetics of Diene Reactions at High Temperatures," *Discussions of the Faraday Society*, 10 (1951), 198–213.
8. Duncan, N. E. and Janz, G. J., "The Thermal Dimerization of Butadiene and the Equilibrium Between Butadiene and Vinylcyclohexene," *J. Chem. Phys.*, 20 (1952), 1644–1645.
9. Huybrechts, G., et al., "Thermal Dimerization of 1,3-Butadiene: Kinetics of the Formation of *cis, cis*-cycloocta-1,5-diene," *Int. J. Chem. Kinetics*, 9 (1977), 283–293.
10. Doering, W. von E., et al., "On the Mechanism of a Diels–Alder Reaction. Butadiene and Its Dimers," *JACS*, 94 (May 31, 1972), 3833–3844.

11. Tsang, W., "Decyclization of Cyclohexene, 4-Methylcyclohexene, and 4-Vinylcyclohexene in a Single-Pulse Shock Tube," *J. Chem. Phys.*, 43(5) (March 1965), 1805–1809.
12. Srinivasan, R., and Levi, A. A., "Thermal Decomposition Processes in the System Tricyclo [3.3.0.0^{2,6}] octane –1,5– Cyclooctadiene," *JACS*, 86 (September 20, 1964), 3756–3759.
13. Hammond, G. S., and De Boer, C. D., "Multiplicity of Mechanisms in the Cope Rearrangement," JACS, 86 (March 5, 1964), 899–902.
14. Sun, H. N. and Wristers, J. P., "Butadiene," in *Kirk–Othmer Enclyclopedia of Chemical Technology,* 4th ed. (Vol. 4, pp. 663–690), New York: Wiley, 1992.
15. Welch, L. M., et al., "Initiation and Growth of Butadiene Resinous Polymers," *IEC*, 39(7) (July 1947), 826–829.
16. Miller, G. H., et al., "Initiation of Butadiene "Popcorn" Polymerization," *JPS:* Part C, No. 4 (1963), 1109–1115.
17. Sun, H. N., "Popcorn Polymers—New Solutions to an Old Problem," in Process Plant Safety Symposium, Houston, TX (April 1–2, 1996), New York: AIChE.
18. Madorsky, S. L., *Thermal Degradation of Organic Polymers,* Huntington, NY: Robert E. Krieger, 1975.
19. Jellinek, H. H. G., *Degradation of Vinyl Polymers,* New York: Academic Press, 1955.
20. Hawkins, W. L., *Polymer Stabilization,* New York: Wiley–Interscience, 1972.
21. Townsend, D. I., and Tou, J. C., "Thermal Hazard Evaluation By an Accelerating Rate Calorimeter," *Thermochimica Acta,* 37 (1980), 1–30.
22. Granville, R. and Wallace, D., "Accelerating Rate Calorimeter: Instrumentation and Application," in *International Symposium on Runaway Reactions and Pressure Relief Design,* Melhem, G. A., and Fisher, H. G (Eds.), pp. 10–22, New York: AIChE/DIERS, 1995.
23. Leung, J. C., Fauske, H. K., and Fisher, H. G., "Thermal Runaway Reactions in a Low Thermal Inertia Apparatus," *Thermochimica Acta,* 104 (1986), 13–29 .
24. Leung, J. C., and Fisher, H. G., "Runaway Reaction Characterization— A Round-Robin Study on Three Selected Systems," in *2nd International Symposium on Runaway Reactions, Pressure Relief Design and Effluent Handling,* Melhem, G. A. and Fisher, H. G. (Eds.), New York: AIChE/DIERS, 1998.
25. Young, M. A., and Chippett, S., "Design and Operation of an Automatic Pressure Tracking Adiabatic Calorimeter (APTAC)," in "International Symposium on Runaway Reactions and Pressure Relief Design," Melhem, G. A. and Fisher, H. G. (Eds.), pp. 23–57, New York: AIChE/DIERS, 1995.
26. Fried, H. E., et al., "A Study of Free-Radical and Thermally-Initiated Butadiene Polymerization," Process Plant Safety Symposium, AIChE, Houston, TX (April 1–2, 1996).

Validation of the SAFIRE Computer Program for ERS Design

D. Arthur Shaw
Monsanto Company

1. INTRODUCTION

Computer programs used for emergency relief system (ERS) design are complex. Since these computer programs are used to design safety systems, it is critical to have accuracy where possible and conservatism where unknowns are significant. It is very important that the design engineer know the capabilities and limitations of any computer program used for ERS design.

The capabilities and limitations can be found from a structured validation process. There are two main parts to the validation process.

1. The program must be validated to insure that the complex equations were programmed properly and that the basic theory fits reality.
2. Each time the program is used it must also be fit to the specific chemical system used for the design.

This paper presents examples used in validation of a computer program for ERS design. The examples presented here are meant to insure that the complex equations were programmed properly and, where possible, compares calculations with data. Validation for a specific chemical system will not be covered. Examples of some of the complex calculations required by an ERS design program will be presented. The computer program developed by the Design Institute for Emergency Relief Systems (DIERS) was used as an example. The name of the program is SAFIRE (Systems Analysis For Integrated Relief Evaluation).

This type of validation is an ongoing effort by the Mathematical Modeling of ERS Committee of the DIERS Users Group. The DIERS Users Group periodically proposes round-robin design problems that allow checking of the SAFIRE code and validate it against experimental data and other computer programs. The Users Group has also proposed experimental research programs to help better understand the phenomena occurring during the venting of a runaway. Results from the research will allow increasing the fundamental basis for SAFIRE and other computer programs used for ERS design.

2. VALIDATION PROCESS

The validation process can be divided into several categories. The first three categories listed below are completed for the computer program itself. The fourth category is completed by the user for each chemical system studied.

I. General Checks
   A. Input
   B. Output
II. Math
   A. Integration
   B. Mass balance
   C. Energy balance
III. Complex Calculations
   A. Flash, nonideal VLE
   B. Fluid flow
      1. Incompressible
      2. Compressible vapor
      3. Two-phase flow
   C. Vapor-liquid disengagement
IV. Specific Chemical System
   A. Physical properties
   B. Mixture
   C. Pressure–temperature model
   D. Reaction model

It is best to compare the output from the computer program directly to data. Where data do not exist, an attempt is made to base the calculations on conservative assumptions.

Some aspects of the calculations are so complex that it is very difficult to determine if the computer gives the correct answers. If experimental data or simple theoretical calculations are available for comparison, then the complex program can be validated directly. Unfortunately, there are many aspects of ERS design that have no data. To check the codes for these complex cases, a "Modified Delphi" technique is being used by the DIERS Users Group. Specific problems are proposed, solved with several different computer programs (authored by different people) and the answers compared. Participants are encouraged to modify their programs until a "consensus" result is achieved.

2.1. General Checks

It may seem trivial to check the input and output, but *everything* must be checked in a complex program used for engineering design. SAFIRE has an option to print back all the input items, just as they were entered. It is a good idea to use the option to check that the program is working properly and to ensure that all the input data were entered correctly.

2.2. Math

The integration, mass balance and energy balance can be checked by rather simple examples. SAFIRE has several input files set up to check the integration routine, mass and energy balances. None of the examples are presented here, but the examples entail simple heat-up and/or reaction simulations where the correct solutions can be determined easily and independently of the computer program. The computer results are compared to the known, correct solutions to insure that the program is working correctly.

The integration algorithm used in SAFIRE is a Runge–Kutta–Gill [1] algorithm. The Runge–Kutta–Gill is a fourth-order Runge–Kutta algorithm modified to be more efficient for stiff systems. Because of the way SAFIRE is structured, an integration algorithm specially designed for stiff systems, such as GEARS, could not be used. The output should be checked carefully to insure that the integration is correct. The usual way of checking is to decrease the integration step size and rerun the case. Decreasing the integration step size should not cause any significant change in the computer program output.

2.3. Complex Calculations

2.3. 1. Vapor–Liquid Equilibrium

Vapor–liquid equilibrium (VLE) calculations are checked against data. The VLE calculations are checked for each chemical system studied. An example was presented by in a previous paper [2] and these checks are not discussed further here.

2.3.2. Fluid Flow

Simple fluid flow calculations can be compared directly to theoretical calculations. For complex fluids, such as two-phase flow, or for complex geometries the "Modified Delphi" method has been employed. Most of the DIERS Users Group Round-Robin design problems have been designed to test the two-phase flow portions of the code. Some of the Round-Robin examples are presented below.

2.3.3. Vapor–Liquid Disengagement

It is very difficult to separate the validation of the two-phase fluid flow and the disengagement parts of the computer program. Two approaches have been taken with SAFIRE. The fluid flow models are validated against data and against other programs through the DIERS Users Group round-robin design problems and the overall program has been tested against small and large scale blowdown experiments, both reactive and non-reactive. Some of the round-robin examples are presented below.

2.4. Specific Chemical Systems

Another level of validation is performed by the user each time a new chemical system is simulated. This consists of checking the reaction rate and pressure calculations against laboratory data. This is a very important validation step. Details were presented in another paper [2] and are not discussed further here.

3. COMPUTER PROGRAM

The basic requirements for a computer program to aid in the design of emergency pressure relief systems for runaway reactions were described by Huff [3]. The SAFIRE computer program incorporates these requirements into a program that can be run on any computer that has a Fortran compiler. SAFIRE was used as an example to show validation.

SAFIRE was described in detail by Tilley and Shaw [4] and by Shaw [5]. The computer program is used to help design or rate emergency pressure relief systems and/or calculate flow through pipes or relief valves. SAFIRE uses basic energy and mass balances to compute temperature, pressure and composition as a function of time in a vessel containing a reacting material or exposed to an external heat source such as a fire. The purpose of the program is to help determine what size vent is necessary to protect a vessel from serious overpressure.

The fluid flow calculations include single-phase (vapor or liquid) and two-phase vapor–liquid flow. Several vent configurations are available that allow calculations for rupture disks or relief valves. The calculations for valves include options for calculating losses in entrance and exit piping.

If the foaming behavior of the material is known, the program can be used to calculate vent sizes taking into account the disengagement of vapor and liquid in the vessel.

SAFIRE was developed for the Design Institute of Emergency Relief Systems (DIERS). The code was first released to the DIERS member companies in 1984. Since then, the program has been used in many designs. SAFIRE contains the latest methods for calculating two-phase flow in relief systems and includes the DIERS methods for calculating how much liquid is carried over into the vent (partial disengagement). The computer program is available from the publications department of AIChE in New York.

4. ROUND-ROBIN EXAMPLES

4. 1. High Viscosity "Polywater"

One of the first round-robin design problems was the Polywater Sample Problem. The problem was to compute flow for a viscous fluid under flashing and nonflashing conditions. The problem conditions were presented by Fisher [6].

A flashing liquid, with zero entrance quality, flows from a vessel with no hydrostatic head through a square-cut nozzle and 100 feet of horizontal, schedule 40, standard steel (roughness = 0.00015 feet), 4-inch diameter pipe to a surrounding pressure of one atmosphere. The fluid has the physical properties of water, except for viscosity which is given by the following:

$$\mu_L = \exp(4.49717655 + 1758.71708/T)$$

where μ_L is viscosity in cP and T is temperature in K. Table 1 presents the results of the calculations for the four cases.

The geometry for this problem was rather simple and the flows were either totally non-flashing or all flashing. There were no cases that had subcooled flow mixed with flashing flow. The logic to handle flow that is partly subcooled and partly flashing is very complex and requires special programming. Some of the later round-robin problems were designed to check code for handling subcooled flow.

4.2. Luviskol

Fisher [7] presented a round-robin design problem based on some blowdown tests at the JRC in Ispra, Italy [8]. Those experiments were simulated with SAFIRE. The calculated pressure (shown in Figures 1, 2, and 3) and final inventory (Table 2) agree with the measured values for the low viscosity tests. The high viscosity results did not agree as well (Figures 4, 5, and 6).

TABLE 1
SAFIRE Calculations for the Viscous (Polywater) Sample Problem

Case No.	Flow Model	Viscosity (cP)	Flow (pph)
1	All Liquid, no Flashing	Water	834,794
2	HEM	Water	197,613
3	All Liquid, no Flashing	As specified	73,878
4	HEM	As specified	41,206

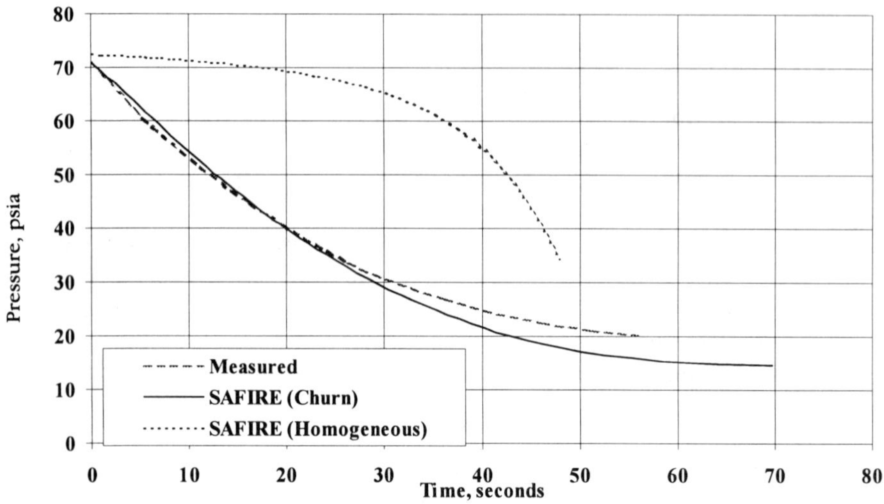

Figure 1. JRC Blowdown Test with Water, 75% Full

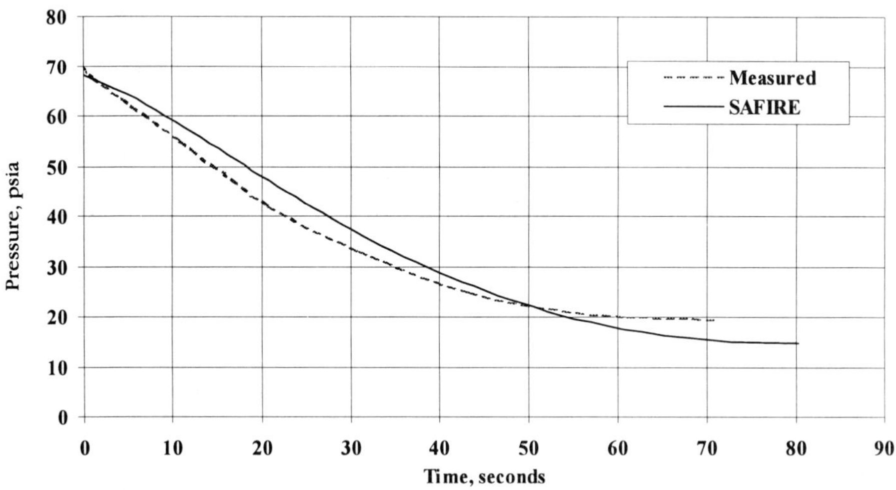

Figure 2. JRC Blowdown Test With Water, 85% Full

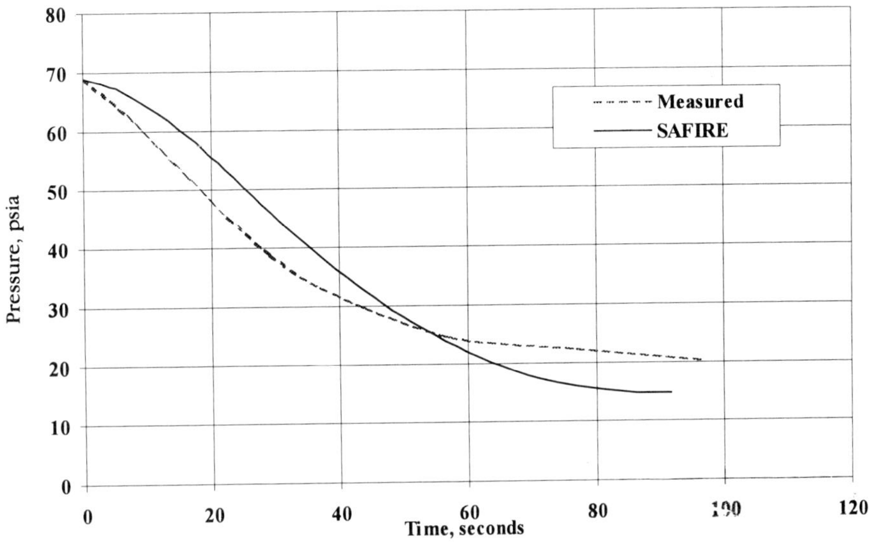

Figure 3. JRC Blowdown Test with Water, 95% Full

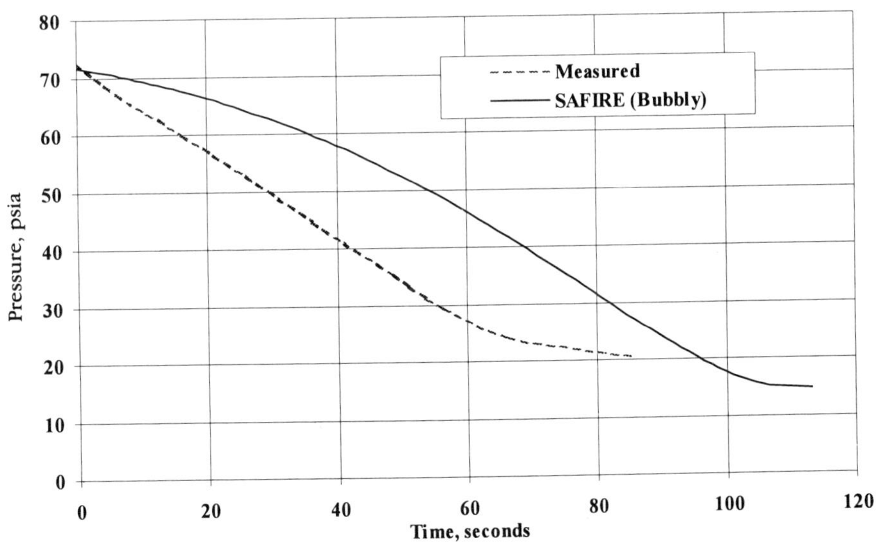

Figure 4. JRC Blowdown Test With 10% Luviskol Solution

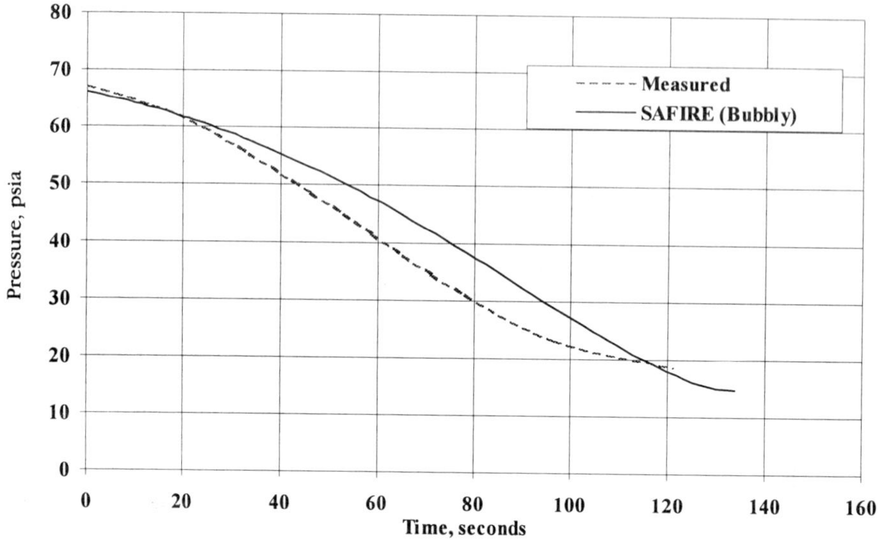

Figure 5. JRC Blowdown Test with 25% Luviskol Solution

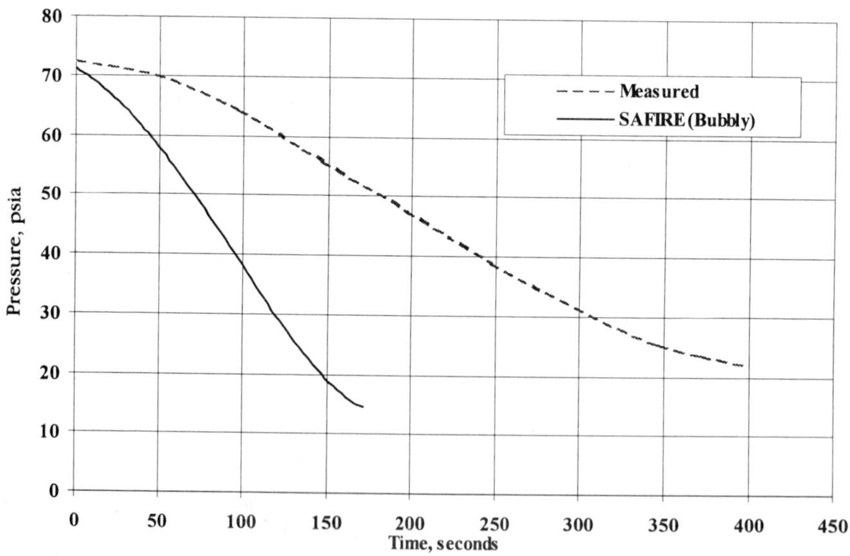

Figure 6. JRC Blowdown Test with 35% Luviskol Solution

TABLE 2
Luviskol Round-Robin Blowdown Tests

Test No.	Percent Luviskol in Water	Percent Full	Fluid Type	Final Mass (lb)		Initial Flow (lb/hr)	
				Measured	Calculated	Measured	Calculated
1	0	95	CT	65.8	64	6318	5468
2	0	85	CT	63.9	64	4730	3730
3	0	75	CT	61.6	62	3262	2616
7	0	75	H		2.7		6846
4	10	74	B	34.5	20	6262	3786
5	25	73	B	28.5	22	4468	3287
6	35	71	B	16.2	24	3333	2414

CT = Churn-turbulent, H = Homogeneous, B = Bubbly

In the low viscosity tests, the calculated initial flow was consistently lower than the measured (see Table 2). It is felt that this was because the vent pipe went straight up and the initial flow would not see any static head effects. The steady-state flow conditions were not achieved until a later time.

The calculations assumed that the pipe was instantly full of the two-phase flowing mixture. The churn-turbulent model was chosen for behavior of the fluid in the vessel and the HEM model was chosen for flow in the vent.

The high viscosity tests were simulated using the bubbly model for behavior in the vessel and the HEM model for calculating vent flow. Comparison of the calculations with the measurements show that there is more work needed before we will fully understand the high viscosity two-phase fluid behavior. A research project has been undertaken by the DIERS Users Group to further study high viscosity two-phase flow.

41.3. Safety Valve Round-Robin

Design of safety valves is fairly complex because of the necessity of calculating the pressure losses in the inlet and outlet pipes. There are straight-forward methods to calculate the losses for gas flow, but the calculations can be complex and tedious for two-phase vapor-liquid flow.

SAFIRE will do the pressure loss calculations for vapor or two-phase flow and has the logic built-in for conventional and bellows valves. If the pressure loss criteria are exceeded, SAFIRE prints an error message warning the user that the valve may not work properly. These calculations have been checked by comparing several different computer programs. Fisher [9] reported the

TABLE 3
Definition of Safety Valve Round-Robin Design Cases

Case Number	Flow Type	Onset/ Disengagement Model	Stagnation Temperature (°C)	Inlet Quality	Liquid Viscosity (cP)
1	Vapor	—	Saturated Steam at 72.5 psia	1.0	Water
2	Flashing Liquid	—	Saturated Water at 72.5 psia	0.01	Water
3	Flashing Liquid with 5 psi N_2	—	Saturated Water at 67.5 psia	0.01	Water
4	Flashing Liquid	—	Saturated Water at 72.5 psia	0.01	See Note 1
5	Flashing Liquid with 5 psi N_2	—	Saturated Water at 67.5 psia	0.01	See Note 1
6	Flashing Liquid	CT	Saturated Water at 72.5 psia	Calc.	Water
7	Flashing Liquid	HOM	Saturated Water at 72.5 psia	Calc.	Water
8	Water with 72.5 psia N_2	LIQ (BOT)	40	0.0	1
9	Water with 72.5 psia N_2	LIQ (BOT)	40	0.0	1000
10	Frozen with 72.5 psia N_2	—	151.85	0.01	1
11	Frozen with 72.5 psia N_2	—	151.85	0.01	1000
12	Flashing Liquid	—	151.85	0.01	1000
13	Flashing Liquid	—	151.85	0.01	5000
14	Flashing Liquid with 5 psia N_2	—	149.2	0.01	1000
15	Flashing Liquid with 5 psia N_2	—	149.2	0.01	5000

Note 1. Viscosity given by $\mu = \exp(-1.74 + 2959.3/T)$, where T = Temperature in K and μ = viscosity in cP.

results of the safety valve round-robin design problem from several different programs.

Calculations were performed for fifteen cases for flows through a 4P6 valve with a set pressure of 52.55 psig. The valve was mounted on a vertical, 1595 gallon, 5.25 feet ID vessel. The inlet line was six feet of vertical, schedule 40, standard steel pipe, 4-inch diameter, with a square cut entrance nozzle and a full-port gate valve. The discharge line was a 6-inch diameter line, schedule 40, standard steel pipe, having a horizontal length of one foot and a vertical height of seven feet. The discharge pipe exits to a surroundings pressure of one atmosphere. The individual cases are defined in Table 3. The results from SAFIRE are presented in Table 4.

TABLE 4
SAFIRE Calculations for the Safety Valve Round-Robin Design Cases

Case No.	Mass Flow (pph)	Choke Pressure (psia)	Choke Temp. (°C)	Choke Quality	Inlet Pressure Loss ($\%P_{set}$)DP/%	Bellows Discharge Back Press ($\%P_{set}$)DP/%	Conventional Discharge Back Press ($\%P_{set}$)DP/%
1	21,810	37.7	93.5	1.0	3.1/5.8	1.3/2.5	1.3/2.5
2	90,505	58.7	144.1	0.0254	0.8/1.5	5.6/10.7	5.6/10.7
3	97,068	59.3	143.4	0.0217	0.9/1.6	6.5/12.4	6.5/12.4
4	87,445	58.2	143.8	0.0260	1.2/2.3	6.2/11.9	6.2/11.9
5	93,490	58.7	143.1	0.0223	1.3/2.5	7.0/13.3	7.0/13.3
6	62,580	50.0	138.5	0.0857	1.7/3.3	3.0/5.7	3.0/5.7
7	102,771	62.7	146.5	0.0005 0.0113	0.4/0.8	7.3/13.9	7.3/13.9
8	715,350	—	—	—	5.9/11.2	5.6/10.7	2.7/5.2
9	577,363	—	—	—	14.9/28.4	7.9/15.0	5.0/9.6
10	199,337	42.4	151.85	0.01	3.3/6.3	2.1/2.9	1.9/3.6
11	182,111	41.0	151.85	0.01	5.7/10.9	3.5/6.7	3.5/6.7
12	80,325	57.2	143.2	0.0272	2.2/4.2	6.5/12.3	6.4/12.3
13	56,627	52.9	140.5	0.033	6.1/11.5	6.2/11.7	6.1/11.7
14	84,921	57.6	142.5	0.0235	2.4/4.5	7.0/13.4	7.0/13.4
15	57,589	53.2	139.8	0.0286	6.4/12.1	6.3/11.9	6.3/11.9

4.4. Subcooled Liquid Valve Round Robin

Sherman [10] presented measured flow rates of hot water flowing through a relief valve. The water entering the valve had various degrees of subcooling such that there was some flashing in the valve.

Calculating flows for this situation is a challenge for computer programs. The most common way to calculate flashing two-phase flow is to assume flashing throughout the nozzle. If that is done, the calculation will be incorrect and will not agree with measurements. SAFIRE version 9.3 calculates subcooled flow using the method presented by Leung [11] with a two-constant equation of state. The results of the calculations are compared to the measured flows in Figure 7. The HEM model is most commonly used for design and the data show that it either predicts very closely or underpredicts the measured flows. For purposes of vent sizing, this would be conservative.

4.5. Benchmarks

Nine pipe and nozzle flow exercises were proposed by Fisher [12]. These have been developed and adapted as benchmarks for the membership of the DIERS Users Group. The line lengths in the examples have been exaggerated to clearly illustrate programming flaws. Pipe orientation, subcooled conditions and high viscosities were included to illustrate the difficulty in calculating these flows. Results from SAFIRE for some of the benchmarks are presented below.

Since the latest version of SAFIRE cannot handle the transition from frozen flow to flashing flow in pipes, some of the benchmarks could not be

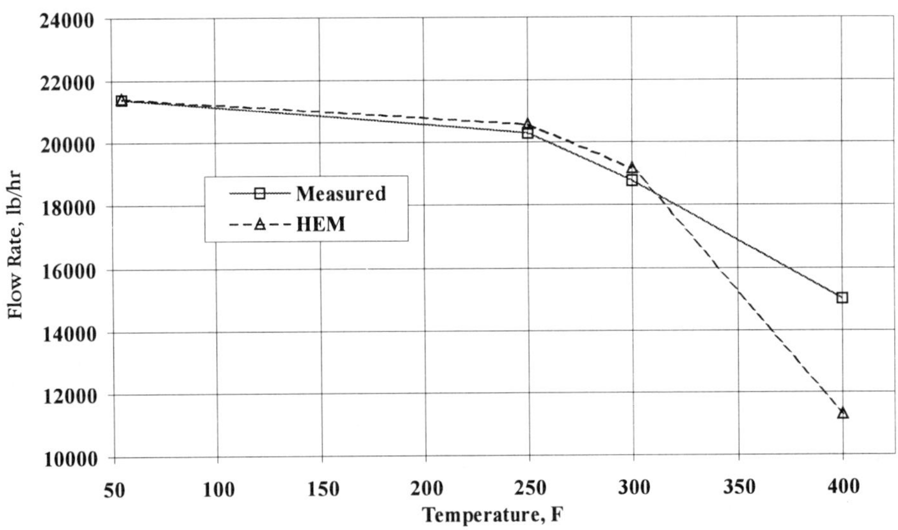

Figure 7. Hot Water Flow Through a Relief Valve

completed. SAFIRE can handle subcooled flow in pipes if there is no noncondensable gas, but can not handle subcooled flow with a noncondensable gas (frozen flow). SAFIRE can compute flows for subcooled nozzles, whether there is a noncondensable gas present or not. Modifications are being made to SAFIRE so that it can compute subcooled flow through pipes with a noncondensable gas present. The next issue of SAFIRE will include this feature.

4.5.1. Benchmark 1, Water Flow

A 100-gallon, vertical, cylindrical vessel with a diameter of 2 feet contains 633.8 lb of water at a temperature of 25°C. The total pressure (nitrogen + water vapor + hydrostatic) in the vessel is 29 psia. Bottom-mounted pipes with square-cut entrances, 2-inch and 8-inch internal diameters and lengths of 16.4, 106.4, and 1600.4 feet drain the vessel. These pipes have horizontal, vertical (downward) or sloped (downward) orientations. Assuming liquid (non-flashing) flow, calculate the initial flow rate from the vessel for each of the 18 configurations. The calculated flows using SAFIRE are presented in Table 5.

TABLE 5
SAFIRE Results for Benchmark 1, Water Flow

Orientation	ID (in)	Length (feet)	Flow Rate (pph)
Horizontal	2	16.4	119,001
		160.4	48,159
		1600.4	14,841
	8	16.4	2,570,670
		160.4	1,584,650
		1600.4	573,900
Vertical Down	2	16.4	145,736
		160.4	118,944
		1600.4	113,270
	8	16.4	3,143,540
		160.4	3,762,000
		1600.4	4,203,090
Sloped Down	2	16.4	138,436
		160.4	103,231
		1600.4	95,297
	8	16.4	2,987,130
		160.4	3,357,090
		1600.4	3,542,970

4.5.2. Benchmark 2, Saturated, Flashing Water Flow

A 169.5-gallon, vertical, cylindrical vessel with a diameter of 100 feet (no hydrostatic head) contains 1000 lbs of water at a temperature of 151.9 °C. The total pressure (water vapor) in the vessel is 72.6 psia. Bottom-mounted pipes with square-cut entrances, 2-inch internal diameters and lengths of 16.4, 160.4 and 1600.4 feet drain the vessel. The pipe roughness is 0.00015 feet. These pipes have horizontal, vertical (downward) or sloped (downward) orientations. Assuming homogeneous-equilibrium (HEM) flow with an initial quality of 0.001, calculate the initial flow rate from the vessel for each of the nine configurations.

Since the flow is saturated liquid without any noncondensable gas, SAFIRE can be used. The results are presented in Table 6.

4.5.3. Benchmark 3, Saturated, Flashing Water in a Combined Vertical/Horizontal Flow

A 169.5-gallon, vertical, cylindrical vessel with a diameter of 100 feet (no hydrostatic head) contains 1000 lbs of water at a temperature of 151.9°C. The total pressure (water vapor) in the vessel is 72.6 psia. Bottom-mounted pipes with square-cut entrances, 2-inch internal diameters, one 90 ° elbow and lengths of 16.4, 160.4 and 1600.4 feet drain the vessel. The pipe roughness is 0.00015 feet. These pipes have combined horizontal and vertical (upward) orientations with length ratios of 50:50. Assuming homogeneous-equilibrium (HEM) flow with an initial quality of 0.001, calculate the initial flow rate from the vessel for each of the six configurations.

The results computed by SAFIRE are presented in Table 7. Note that the flows calculated for the HOR/VER orientation were greater than the flows for

TABLE 6
SAFIRE Results for Benchmark 2, Saturated, Flashing Water Flow

Orientation	ID (in)	Length (feet)	Flow Rate (pph)
Horizontal	2	16.4	45,031
		160.4	25,493
		1600.4	9,457
Vertical Down	2	16.4	47,237
		160.4	47,300
		1600.4	49,181
Sloped Down	2	16.4	47,786
		160.4	43,613
		1600.4	43,468

the VER/HOR orientation. That result was expected because of the affects of hydrostatic pressure on flashing. For the HOR/VER orientation, the hydrostatic head in the vertical pipe suppressed flashing in the horizontal section. Because the flashing was suppressed, the specific volume of the flow going into the second section of pipe was less for the HOR/VER orientation than it was for the VER/HOR orientation. The lower specific volume into the second section of pipe resulted in a higher overall mass flow.

The results for short pipes agreed with computations from other programs, but for long pipes, SAFIRE calculated flows as much as twice those calculated by another program. At this time it is not known why the answers were so different. Work will continue to arrive at a consensus result.

4.5.4. Benchmark 4, Initially Subcooled/Flashing Water Flow

A 169.5-gallon, vertical, cylindrical vessel with a diameter of two feet contains 1000 lbs of water at a temperature of 151.9°C. The total pressure (nitrogen + water vapor + hydrostatic) in the vessel is 101.6 psia. Bottom-mounted pipes with square-cut entrances, 2-inch internal diameters and lengths of 16.4, 160.4 and 1600.4 feet drain the vessel. The pipe roughness is 0.00015 feet. These pipes have horizontal, vertical (downward) and sloped (downward) orientations. Assuming Frozen/homogeneous-equilibrium (HEM) flow with an initial quality of 0.001, calculate the initial flow rate from the vessel for each of the nine configurations.

Since the flow is saturated liquid with a noncondensable gas in a pipe, the current version of SAFIRE should not give correct answers. Therefore, the results of these calculations were not presented.

4.4.5. Benchmark 5, Huff Initially Subcooled/Flashing Water Nozzle Flow

This sample problem was originally proposed by Huff [13]. Saturated water at a temperature of 147.77°C discharges from a vessel through an ideal 2.009-

TABLE 7
SAFIRE Results for Benchmark 3, Saturated, Flashing Water in a Combined Vertical/Horizontal Flow

Orientation	ID (in)	Length (feet)	Flow Rate (pph)
HOR/VER	2	8.2/8.2	43,218
		80.2/80.2	2,5057
		800.2/800.2	9,287
VER/HOR	2	8.2/8.2	41,784
		80.2/80.2	23,950
		800.2/800.2	8,833

inch diameter nozzle. The quality entering the nozzle is zero. The nitrogen pressure in the vessel is 0, 1 and 2 psi resulting in total pressures of 65.0, 66.0 and 67.0 psia. Calculate the initial flow rate from the vessel for each of the three pressures assuming subcooled/flashing homogeneous-equilibrium (HEM) flow.

Table 8 presents the results computed with SAFIRE.

4.5.6. Benchmark 6, Nielsen Initially Subcooled/Flashing Water Flow

A 523-gallon, vertical, cylindrical vessel having a diameter of 3.54 feet and two ASME 2:1 elliptical heads contains 2525 lb of water at a temperature of 176.1°C. The total (water vapor + hydrostatic) pressure in the vessel is 134.8 psia. Horizontal pipes having square-cut entrances and the internal diameters, lengths and pipe roughness values shown in Table 9 drain the vessel. These drain pipes were side-mounted at 3.94 inches above the tangent line of the bottom head. The 2525 lb of water above the drain pipe result in the initial hydrostatic head of 1.9 psi. Assuming both homogeneous-equilibrium (HEM) and an appropriate slip-equilibrium flow model, calculate the initial flow rate from the vessel for each of the three pipe configurations and flow models.

TABLE 8
SAFIRE Results for Benchmark 5, Huff Initially Subcooled/Flashing Water Nozzle Flow

Pressure (psig)	Flow Rate (pph)
65	55,421
66	59,780
67	83,157

TABLE 9
Specifications for the Piping for Benchmark 6

Pipe Number	Diameter (inches)	Length (feet)	Roughness (feet)
1	3.146	6.037	0.0000783
2	1.291	6.562	0.0000425
3	0.417	6.037	0.0000425

TABLE 10
SAFIRE Results for Benchmark 6, Nielsen Initially Subcooled/Flashing Water Flow

Pipe Number	Flow Model	Calculated pph	Experimental @ t = 8 sec. pph
1	HEM	237,148	262,700
2	HEM	36,165	41,557
3	HEM	3,065	3,495

The Nielsen [14] pipe flow problem is supported by data. Table 10 shows the flows calculated by SAFIRE along with the measured flows. An inlet quality of zero was assumed for all the calculations. The results for only the homogeneous-equilibrium (HEM) model were shown in the table. The calculated flows were a little lower than the measured flows.

The slip models in the current version of SAFIRE do not handle initially subcooled flow so only the HEM results were presented in the table.

4.5.7. Benchmark 7, Saturated, Flashing, Constant Viscosity, Viscous Water Flow

Repeat benchmark 2, with a constant liquid viscosity of 5000 cP for all temperatures. The flows calculated using SAFIRE are presented in Table 11.

TABLE 11
SAFIRE Results for Benchmark 7, Saturated, Flashing, Constant Viscosity, Viscous Water Flow

Orientation	ID (in)	Length (feet)	Flow Rate (pph)
Horizontal	2	16.4	13,722
		160.4	1,938
		1600.4	195
Vertical Down	2	16.4	13,908
		160.4	2,270
		1600.4	1,677
Sloped Down	2	16.4	13,852
		160.4	2,135
		1600.4	1,187

Downward flow calculations are very difficult to do and there have been very few, if any, submissions of answers from other computer programs. Those answers that have been submitted agree with the horizontal flows, but vary widely with downward vertical or sloping flows. More work is necessary to ensure that the computer programs are working properly.

4.5.8. Benchmark 8, Saturated, Flashing, Variable Viscosity, Viscous Water Flow
Repeat benchmark 2, with a liquid viscosity that varies with temperature according to the following equation:

$$\ln(\mu) = -9.647346 + 7036.745867/T$$

where μ is the viscosity in cP and T is the temperature in K. The flows calculated using SAFIRE are presented in Table 12.

4.5.9. Benchmark 9, Saturated, Flashing, Variable Viscosity, Viscous Water Flow
Repeat benchmark 2, with a liquid viscosity that varies with temperature according to the following equation:

$$\ln(\mu) = -28.5050331 + 14073.49174/T$$

where μ is the viscosity in cP and T is the temperature in K. The flows calculated using SAFIRE are presented in Table 13.

TABLE 12
SAFIRE Results for Benchmark 8, Saturated, Flashing, Variable Viscosity, Viscous Water Flow

Orientation	ID (in)	Length (feet)	Flow Rate (pph)
Horizontal	2	16.4	26,245
		160.4	4,680
		1600.4	478
Vertical Down	2	16.4	27,327
		160.4	8,828
		1600.4	8,295
Sloped Down	2	16.4	26,984
		160.4	6,795
		1600.4	5,874

TABLE 13
SAFIRE Results for Benchmark 8, Saturated, Flashing, Variable Viscosity, Viscous Water Flow

Orientation	ID (in)	Length (feet)	Flow Rate (pph)
Horizontal	2	16.4	42,804
		160.4	19,573
		1600.4	2,743
Vertical Down	2	16.4	60,791
		160.4	60,742
		1600.4	60,738
Sloped Down	2	16.4	55,007
		160.4	54,339
		1600.4	54,281

2. REFERENCES

1. Korn, G. A., and T. M. Korn, *Mathematical Handbook for Scientists and Engineers,* p. 779, New York: McGraw-Hill, 1968.
2. Shaw, D. A., and M. E. Levin, "Application of SAFIRE in the Design of an Emergency Relief Vent for a Storage Vessel," in *International Symposium on Runaway Reactions and Pressure Relief Design,* H. G. Fisher and G. A. Melhem(Eds.), New York: AIChE/DIERS, 1995.
3. Huff, J. E., "Computer Programming for Runaway Reaction Venting," *I. Chem. E. Symposium Series* No. 85 (1984).
4. Tilley, B. J., and D. A. Shaw, "SAFIRE Computer Program for Emergency Relief Sizing," in *Emergency Relief System Design Using DIERS Technology, the Design Institute for Emergency Relief Systems (DIERS) Project Manual*, Chapter 7. New York: AIChE/DIERS, 1992.
5. Shaw, D. A., "SAFIRE Program for Design of Emergency Pressure Relief Systems," *Chemical Engineering Progress,* 86(7) (1990), 14.
6. Fisher, H. G.., "High Viscosity 'Polywater' Sample Problem." Memo to the DIERS Users Group Members (July 6, 1992).
7. Fisher, H. G., "JRC Pipe Flow Sample Problem," Presented at the 14th DIERS Users Group Meeting, Chicago, IL, September 27–30, 1993.
8. Bell, K., S. D. Morris, and R. Oster., "Vent Line Void Fractions and Mass Flow Rates During Top Venting of High Viscosity Fluids," *J. Loss Prev. Proc. Ind.*, 6(1) (1993), 31–35.
9. Fisher, H. G., "Emergency Relief System Design: The DIERS Users Group Safety Relief Valve Example Problem," in *International Symposium on Runaway Reactions and Pressure Relief Design,* H. G. Fisher and G. A. Melhem (Eds.), New York: AIChE/DIERS, 1995.
10. Sherman, D. H., "The Effects on the Flow Rate and Operational Characteristics of Spring Loaded Pressure Relief Valves on Flashing Subcooled and Saturated Water Service," Presented at the DIERS Users Group Meeting, (1993).

11. Leung, J. C., "The Omega Method for Discharge Rate Evaluation," in *International Symposium on Runaway Reactions and Pressure Relief Design,* H. G. Fisher and G. A. Melhem (Eds.), New York: AIChE/DIERS, 1995.
12. Fisher, H. G., "DIERS Benchmark Relief Device Sizing Problems," Memo to the DIERS Users Group Members (April 23, 1996).
13. Private communication
14. Nielsen, D. S., "Validation of Two-Phase Outflow," *J. Loss Prev. Process Ind.*, 4 (July 1991), 236–241.

Sizing Emergency Relief Systems Using SuperChems™ for DIERS

Daniel J. Muzzio and John J. Sharkey
Process Safety Laboratory, Merck & Co., Inc., Rahway, NJ 07065

ABSTRACT

There exist various methods for sizing emergency relief systems (ERS) based on two-phase flow models. A distinction between the methods is the amount of information and the complexity of the equations used to model the system. Considering this difference these methods can be classified as simplified (shortcut) or detailed. The shortcut methods require minimal data and are easy to use. The results, however, apply only to specific cases and the ERS design is frequently very conservative. The detailed approach involves fewer assumptions, but requires more extensive information and requires a computer to perform the calculations. Smaller ERS designs plus accurate representation of the entire venting process is, however, usually obtained. In this paper a detailed method to generate ERS design models is examined and necessary data for the modeling reviewed. Actual industrial application of the modeling is also demonstrated. Specifically, vent sizing for three scenarios is used to highlight the detailed calculation procedures utilizing the computer software program SuperChems for DIERS. This program is being offered through the AIChE's Design Institute of Emergency Relief Systems Users Group. The scenarios presented in this paper move from the relatively simple to the complex. Each case demonstrates how findings can vary based on available information and the models chosen. Solutions are also compared to results from experimental tests to validate the importance of the assumptions made and models chosen.

INTRODUCTION

The purpose of this paper it is to illustrate how upsets which can occur in a chemical plant can be modeled using SuperChems for DIERS. SuperChems

SuperChems is a trademark of Authur D. Little, Inc., Cambridge, MA for DIERS is a new simulation program being offered by the Design Institute for Emergency Relief Systems (DIERS) Users Group for modeling releases through emergency relief systems. It is designed to conduct detailed calculations to simulate venting from equipment normally used in the chemical industry. It is based on fundamental thermodynamic and transport phenomena principles and accounts for the transient effects occurring during a release in the venting system. It has the ability to simulate events leading to, occurring during, and following a release.

SuperChems for DIERS is a stand alone program which runs on a standard personal computer. It has a wide range of modeling capabilities. The venting of reactive as well as nonreactive systems can be handled by the program. The venting material can be all vapor, all liquid, or two-phase flow. Varying equipment designs including releases from pipe lines to systems fitted with collection systems can be modeled. Multicomponent nonideal behavior can also be simulated using SuperChems for DIERS.

Because of the flexibility of SuperChems for DIERS, it is difficult at first to understand how to navigate the various options available to obtain the required results. The purpose of this paper is to take the reader step-by-step through the procedures, which are needed to be performed, so that an accurate SuperChems for DIERS model simulation can be obtained. This paper is meant to be tutorial. It is hoped the reader using SuperChems for DIERS software and following the steps in this paper will get the results shown in the paper. Once the tutorial has been mastered, the knowledge can then be used to solve other related problems. The main objective of this work is to give the reader the required tools to simulate venting scenarios involving transient effects for all vapor, or two-phase flow using a detailed computer method such as in SuperChems for DIERS.

PROGRAM STRUCTURE

SuperChems for DIERS is a DOS based program which runs under MS-DOS, Windows 95, or Windows NT. The program possesses a complex computer-based file structure. Normally all modeling related to one project is saved together under a single core image. This core image is essentially a binary DOS file connecting related files in the proper manner. The related files under the core image are hidden and thus can not be observed in the standard file manager directory. All operations in SuperChems for DIERS are menu-driven.

Scenario Definition

For conducting venting simulations in SuperChems for DIERS of all vapor, all liquid, or two-phase flow, only the *Define*, *Preload*, and *Execute* pulldown menus in the main screen need to be employed. To begin a simulation it is necessary that a new scenario file be established. This is accomplished by duplicating the default scenario and renaming it. Scenario is one of the DOS files saved under the core image which contains the various files used to definc the upset under investigation. By starting in this way all parameters and settings needed by SuperChems for DIERS to run whether or not required by the scenario are defined. At the same time, the original default settings are preserved for future use. After a new scenario is established, the various elements that make up a scenario need to be specified. Each of these elements is also a DOS file saved under the core image. These elements are specified under the *Define* pulldown menu and involved the following:

- *Site:* contains information related to the location of the equipment which effects the venting process
- *Vessel:* includes equipment data
- *Piping:* describes the vent system piping
- *Reaction:* stores kinetics and stoichiometry information if a reaction is occurring during the venting process
- *Mixture:* specifies the number of components in the ventingsystem and the location of the physical properties even if only one compound is involved
- *Stream:* holds stream details if one is being added

Once the above required elements are completed, they are entered into the scenario file which is also under the *Define* pulldown menu. Next the scenario is preloaded by using the *Preload* pulldown menu and finally it is executed under the *Execute* pulldown menu.

Results/Output

Output from SuperChems for DIERS is typically viewed via prescribed plots in the program. The prescribed plots in SuperChems for DIERS range from change of mass in the vessel with time to the pressure rise rate as a function of temperature. A file listing all data in tabular form can also be obtained. This file can then be imported into other programs such as Excel and the data manipulated and plotted as desired.

Lessons in Tutorial

In this paper three tutorials involving upsets are presented. Each tutorial is meant to build on the knowledge gained from previous ones so by the end of all three tutorials the reader will be able to simulate typical upsets which result in venting through a relief device based on all vapor or two-phase flow models. Tutorial 1 involves verifying the design of a pressure relief device for an HCl vaporizer which overheats. It teaches how, using SuperChems for DIERS, to define a vessel, piping that contains an ideal relief device with a coefficient of discharge (CD), a single component mixture, and a scenario. It then goes on to show how to execute the scenario based on two-phase flow. Next it shows how the actual piping layout can be specified in SuperChems for DIERS to take transit effects into account during the venting process. Tutoria 2 linvolves accidentally pumping water into concentrated sulfuric acid in a closed vessel. It teaches how to define a multicomponent vessel mixture, a multicomponent addition stream, and the use of binary interaction coefficients. It also shows how SuperChems for DIERS varies the vessel mixture composition to obtain the desired temperature and pressure plus how to add a stream into a vessel. Tutorial 3 simulates water intrusion into a batch containing a water reactive material which is thionyl chloride. It teaches how to the define a reaction. The examples involve simple situations but their extension to more complicated problems involving complex reactions and piping is straightforward.

TUTORIAL 1: OVERHEATING A HYDROGEN CHLORIDE VAPORIZER

A continuous hydrogen chloride (HCl) vaporizer equipped with a steam jacket (see Figure 1) can overheat if the inlet and outlet valves are suddenly shut, resulting in the unit overpressuring. The vaporizer is a vertical cylinder with a length of 2.5 meters; its inside diameter is 0.387 meters. The jacket has an overall heat transfer coefficient of 405 watts per square meter per Kelvin. The area of the jacket is 2.44 square meters and the heating fluid is saturated steam at 445 K. As the vaporizer heats up the pressure will eventually build and it will exceed the set pressure of the safety value on the unit. The pressure relief device is 3 × 4 safety valve with an H (0.798 square inches) orifice. It is desired to know if the existing safety valve can prevent the pressure from exceeding 10% of the safety valve set pressure especially if two-phase flow occurs The set pressure of the safety value is at the MAWP of the unit, which is 2.17 MPa. As a result of this SuperChems for DIERS was used to simulate the venting process.

Simulation 1a. Overheating Hydrogen Chloride Vaporizer (Ideal Vent)

For simulation 1a the release from the safety valve on the HCl vaporizer will be assumed to be two-phase homogeneous flow. To simplify the calculations, the vent will be assumed to be ideal with the pressure loss taken into account with a coefficient of discharge (CD). This method is used to quickly determine

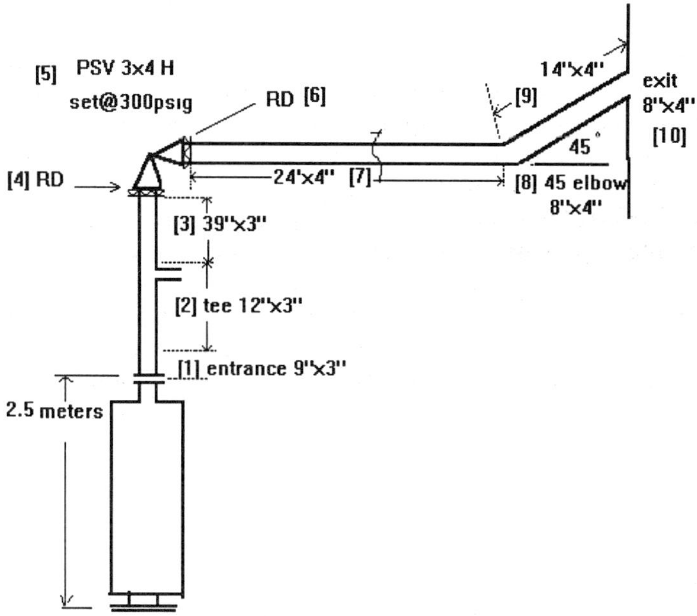

Figure 1. Hydrogen Chloride Vaporizer Schematic

if the specified relief device is suitable to prevent overpressure before a much longer determination involving actual piping is conducted.

The required steps for SuperChems for DIERS to model the sudden overheating of the HCl vaporizer with an ideal vent having a CD are presented below.

1. Create New Scenario by Duplicating Default

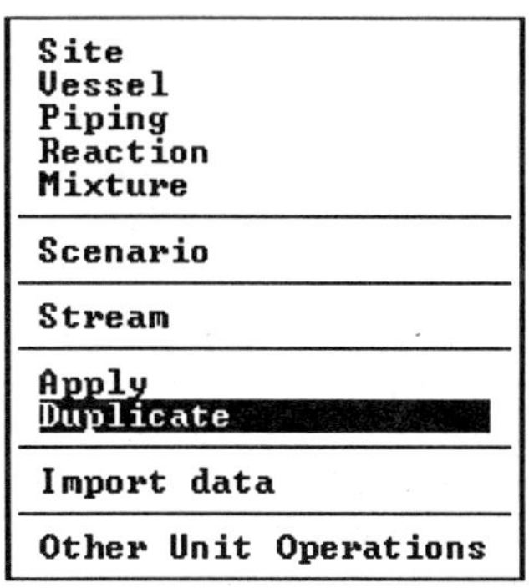

Screen 1. In the main menu select *Define* to obtain the this screen, then highlight *Duplicate* and Press F10.

```
Hazop
Fault tree
Site
Vessel
Piping
Reaction
Thermal
Toxicity
Overpressure
Mixture
Scenario
 Press F10 to select
```

Screen 2. Highlight *Scenario*. Then select *From*, which is located above this part of the screen, by using the right arrow key or the mouse

```
DEFAULT.SC
   Scenario
```

Screen 3. This screen lists all defined scenarios. Since none have been specified only the default one appears. Select

2. Define Vessel

```
Names of entries to be created    MZHCNPV
Separate using a space.
DO NOT USE EXTENSIONS !
```

Screen 4. Type MZHCNPV or any appropriate name in capitals on the line *Names of entries to be created*. The scenario name is a DOS file name and therefore must follow the DOS conventions (8 letters/numbers maximum). Press F10 and then the ESC key twice to get back to the main menu.

```
Site
Vessel
Piping
Reaction
Mixture
-------------------
Scenario
-------------------
Stream
-------------------
Apply
Duplicate
-------------------
Import data
-------------------
Other Unit Operations
```

Screen 5. In the main menu select *Define* to view this screen then highlight *Vessel* and press F10.

```
Current
Edit
  DEFAULT
```

Screen 6. Highlight *Edit* by using the down arrow key or the mouse. After selecting *Edit* press F10. If *Current* is selected, values for the vessel will be displayed but they cannot be changed. Once *Current* is selected, one needs to keep pressing F10 until the above screen reappears.

```
Vessel name [s]                    HE102
Description [s]
           >
           >
Material of construction [s]       IRON
```

Screen 7. Overwrite the default values as shown. Again the vessel name is a DOS File name and therefore must follow the DOS conventions (8 letters/numbers maximum). Comments can be placed in the 3 description lines. Select *Material of construction*, press F9, then select N, a pulldown menu with values to select will appear. Pressing F1 rather than F9 will bring up a help screen. Most input blocks have this feature.

```
Rectangular
Vertical Cylindrical
Spherical
Horizontal Cylindrical (Flat heads)
Horizontal Cylindrical (hemispherical heads)
Horizontal Cylindrical (Dished heads)
                                   Vessel Type
```

Screen 8. Select *Vertical Cylindrical* to represent the HCl vaporizer. Press F10

```
Length (m)                                                            2.5
Inside diameter (m)                                                   0.387
Vessel metal mass (kg)                                                647.141
Wall metal thickness (m)                                              0.0254
Base elevation with respect to scenario ground level (z=0) (m)
Maximum pressure limit for vessel in simulation (Pa)                  1E+38
Maximum temperature limit for vessel in simulation (K)                1000
```

Screen 9. Input the above values for the vaporizer. Press F10

```
Calculated vessel metal weight    = 647.1      kg
User defined vessel metal weight  = 647.1      kg

Do you wish to use the calculated value instead ?

                                                   NO
                                                   YES
```

Screen 10. Since the mass of the vessel is already known, select *No*.

```
   Water spray
   Insulation
   External fire
   ARC/APTAC with heat/wait/search
   Vapor eductor
   Cooling jacket
√  Heating jacket
   Solar flux
   Vessel is a pipeline
   User defined heating/cooling load
T                Toggle vessel options
```

Screen 11. In this screen more than one option can be chosen by highlighting the selection and then pressing the space bar. For the HCl vaporizer select *Heating jacket*, press space bar and then press F10.

```
Jacket overall heat transfer coefficient (W/m2/K)    405
Jacket heating area (m2)                             2.44
Jacket heating temperature trigger (K)               1000
Jacket heating fluid temperature (K)                 445
```

Screen 12. The vessel jacket information is loaded through this screen. Use the above values. The temperature trigger is the value at which heating will be shut off. Press F10

```
Vessel already exists. Overwrite ?
                                        NO
                                        YES
```

Screen 13. SuperChems will now run a subroutine to build the vessel. The above screen will only appear if the file name previously existed. If it does appear select *Yes* and press enter

```
Current
Edit
  DEFAULT
```

Screen 14. From this screen one can return to the main menu by pressing ESC key twice.

3. Define Piping with Ideal Vent Having a CD

```
Site
Vessel
Piping
Reaction
Mixture
--------------------------
Scenario
--------------------------
Stream
--------------------------
Apply
Duplicate
--------------------------
Import data
--------------------------
Other Unit Operations
```

Screen 15. In the main menu select *Define* to obtain above screen, then highlight *Piping*. Press F10.

```
Current
Edit
DEFAULT
```

Screen 16. Highlight *Edit*. Press F10.

```
Piping layout name [s]                  HCLPSV
Piping layout description [s]           IDEAL PSV 3X4 H
Number of segments []                   1
Elevation of first segment (m) []       2.499
```

Screen 17. Overwrite the default values with the above information. Elevation of the first segment is set at 2.499 m, SuperChems requires that the vent piping start below the top of the vessel which in this case is 2.500 m. In this simulation only one element will comprise the venting system which will be a relief device. Press F10.

```
1. Rupture disk. Pset(Pa/Psig) = 101325/-3.97403E-06, D(m/in) = 0.0508/2.000
Return to previous menu
                                                                      HCLPSV
```

Screen 18. Select the first line. Press F10. Notice the default screen above shows a *Rupture disk*.

```
Straight pipe / valves / fittings
Relief device
Rupture disk
Thin sharp edge orifice
Pipe expander (rounded)
Pipe reducer (rounded)
Pump
                    Segment 1/1 Type
```

Screen 19. Select *Relief device*. Press F10.

```
Relief device type                       3
Orifice/outlet flow area (m2) []         0.0005065
Set pressure (Pa)                        2169752.771
K1 value []                              0
Kinf value []                            0
```

Screen 20. Enter 3 for relief device type which is the code for a relief valve that is fully opened at 10% over set pressure and is fully closed at -7%. For the area, use the value for an H orifice. Enter the set pressure of the relief device in Pascal. Press F10.

```
Files      Inputs      Options     Other
 1. Relief device. Pset(Pa/Psig) = 2169753/300.001, Ao(m2/in2) = 0.0005/0.785
 Return to previous menu
                                                                        HCLPSU
```

Screen 21. *Select Return to previous menu*. Press F10. A screen will then appear requesting a discharge coefficient be specified. Press enter.

```
Discharge coefficient []     0.62
```

Screen 22. The screen above appears when a single element which is relief device is specified for the piping layout. The pressure losses in the piping layout can then be specified as a discharge coefficient. For the HCl vaporizer use the default value of 0.62. Press F10.

```
Current
Edit
 DEFAULT
```

Screen 23. Press the ESC key twice to return to the main menu.

4. Define Mixture

```
Mixture name [s]                              MZHCL
Description [s]                               HCL
Number of compounds []                        1
Number of solids []                           0
Is mixture soluble in water (Y/N) [s]         Y
```

Sreen 24. Select *Define, Mixture* and then *Edit* as under "Define Vessel" to obtain the above screen. Then overwrite the default values as shown. Press F10.

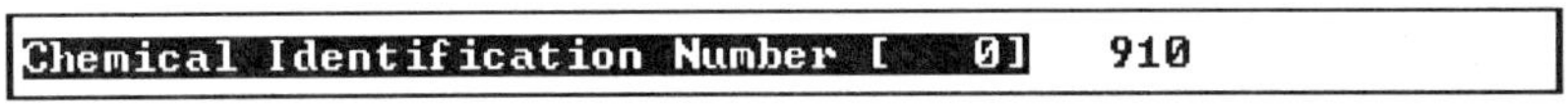

Screen 25. SuperChems uses its own chemical identification number unique to its data base. The designation for HCl in this database is 910. Enter this value and press F10.

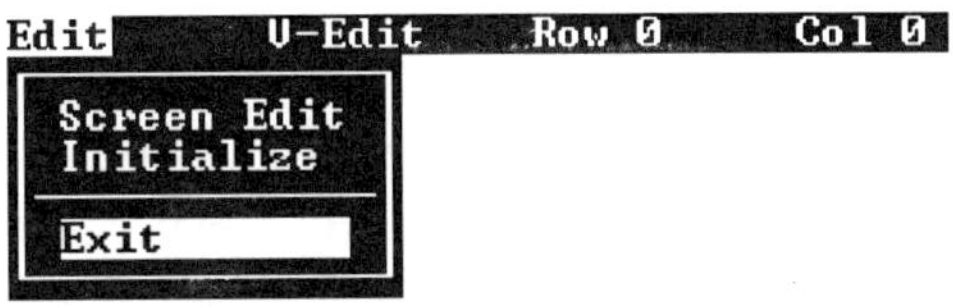

Screen 26. Select *Exit*. Then the Esc key twice. If *Screen Edit* is selected, a DOS window will open. If this happens, hit the Esc key and follow the instructions in the lower left portion. If *Initialize* is selected, one will be asked to zero out the file. Select *NO*. After selecting *Exit* press F10, then the Esc key twice to return to main menu.

5. *Define Scenario*

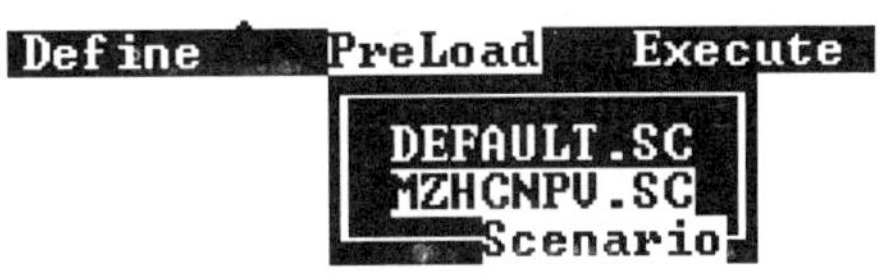

Screen 27. In the main menu select *PreLoad* and highlight the scenario named MHCNPV.SC. Press F10.

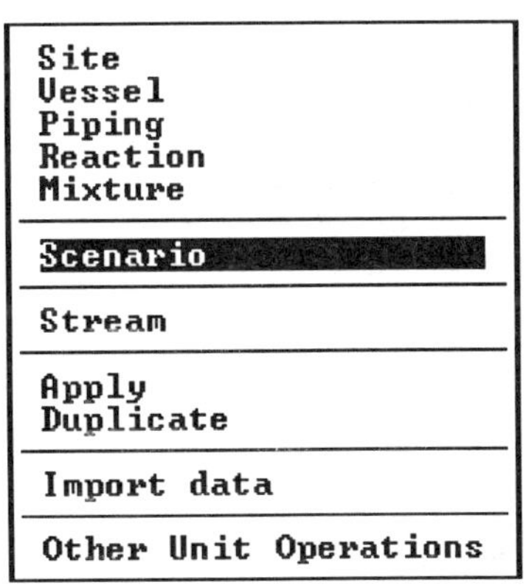

Screen 28. While still in the main menu select *Define* to obtain the above screen, then highlight *Scenario*. Press F10.

```
Current
Edit
     MZHCL
```

Screen 29. Highlight *Edit*. Press F10.

```
Name [s]               MZHCNPV
Description [s]        DEFAULT SCENARIO
            >
            >
Site [s]               DEFAULT
Vessel [s]             HE102
Piping layout [s]      HCLPSV
Reaction [s]
Mixture [s]            MZHCL
```

Screen 30. Overwrite the default screen with the above information. Press F10 and then the Esc key twice to return to the main menu.

6. *Execute 2 Phase Flow Model*

```
PreLoad    Execute    Projects
           DEFAULT.SC
           MZHCNPV.SC
              Scenario
```

Screen 31. In the main menu: select *Execute,* then highlight the scenario named MZHCNPV.SC. Press F10.

```
Source     VLE        Other      Exit
 Liquid
 2-phase
 Saturated liquid
 2-phase expansion / aerosol formation
 Droplet evaporation and trajectory

 Gas/vapor
 Gas/vapor expansion

 Steady gas flow from pipes
 Steady liquid flow from pipes
 Steady 2-phase flow from pipes
                                  MZHCL
```

Screen 32. In this screen select *2-phase* which is for transit flow behavior in the system. Press F10.

```
Current
Edit
Run
     MZHCL
```

Screen 33. Highlight *Edit.* This window also has the *Run* command. This will be used later once setup parameters and initial conditions have been specified. Press F10.

```
Total mass in vessel (kg)          60
Initial temperature (K)            264
Initial pressure (Pa)              2000000
Starting time for analysis (s)     0
Final time for analysis (s)        50
Flow type                          1
```

Screen 34. In the above screen, input the values shown. Since the mixture in this simulation is a single component, the temperature and pressure are related. Later in the set up, SuperChems will change the temperature to match the pressure to ensure that the system is in thermodynamic equilibrium. For the *Flow type*; enter 1 which is the code for two-phase flow. Press F10. After inputting these initial conditions SuperChems goes back to the screen with the Run command.

```
Files      Inputs      Options      Other
                       Brief ON
                       Trace ON
                       Debug OFF
                       Numbers
                       Display
                       Store
                       Retrieve
                       Page layout
                       Thermo
                          VLE: start from bubble point
                          VLE: start from dew point
                          VLE: use free energy minimization
                          EOS: liquid density method
                          EOS: PVT method
                                                     Thermo
```

Screen 35. Select *Options* visible on the main screen under *2-phase* and then highlight *Thermo*. There are three VLE options as seen above. This part of the screen directs the VLE determinator to start at either the bubble or dew point. Starting from the dew point is typically more robust (6). This is the <u>default setting</u> and should be selected. Next move the highlight to the lower part of the block to set the EOS. Highlight *EOS: liquid density method and* press Enter. In the *EOS: liquid density method* select *Databank* which should be the default. Press F10. Next select *EOS: PVT method* and press F10 to obtain the following screen.

```
Real gas and liquid phase
Ideal gas phase
Ideal liquid phase
Ideal gas and liquid phases
                  PVT behavior
```

Screen 36. Select *Real gas and liquid phase* from the above screen, then press F10.

The best features of the built-in equation of state and databank information are used by selecting databank when available for liquid density method and real gas/liquid behavior for PVT behavior (4). These choices were the ones selected above.

Move the highlight from *Options* to *Other* in the main screen and then press enter.

```
Plot results (SI)
Plot results (ENG)
Print results to file
---------------------------------------
Vary X to match T/P specs
---------------------------------------
Define input streams
Define miscellaneous defaults
Define slip ratio method
---------------------------------------
Starting conditions check is OFF
---------------------------------------
Reaction in pipe is OFF
---------------------------------------
Connections
---------------------------------------
Define disengagement parameter
---------------------------------------
Quickly estimate relief area/flow
```

Screen 37. The size of the integration steps needs to be set. This is done by highlighting *Define miscellaneous defaults* and pressing F10.

```
Maximum number of data points to collect                        1000
Maximum allowable pressure change per time step (Pa)            10000
Maximum allowable temperature change per time step (K)          3
Maximum allowable time step (s)                                 1
Minimum allowable time step (s)                                 0.01
Transit to vapor module when contents are all vapor (1-Yes)     1
Minimum mass allowed in vessel (kg)                             0.01
dT/dt (K/min) above which adiabaticity is lost (ARC data only)
```

Screen 38. Maximum number of data points to be collected can be set between 2 to 5000. For the HCl vaporizer 1000 is entered. The rest of the values need to be set as above to get a good balance between accuracy and computer time. Press F10 and screen 37 will reappear. Move the highlight to *Starting conditions check is OFF*, press the Enter key and the *OFF* will change to *ON*. This will ensure initial conditions will only be integrated and not the complete simulation at this time. If the full integration starts, it can not be easily interrupted. Next use the left arrow key to select *Inputs*.

```
Current
Edit
Run
        MZHCL
```

Screen 39. Highlight *Run* and press F10 to check the initial conditions.

```
V/F calculations for 1 kmol of feed    0.16454
Total mass in liquid space =      50.13 kg
Total mass in vapor space =      9.872 kg
Liquid volume        0.0527 m3
Vapor volume        0.2414 m3
Calculated starting temperature =          264.04 K
```

Screen 40. SuperChems will start running a subroutine to integrate the initial conditions. When it is completed, an output screen will appear. To see this information use the End key and then scroll to the end using the down arrow. One will see the above information. If the values appear reasonable, press Esc key. Next using the right arrow key select *Other*, under *Other* turn the *Starting conditions check* (from) *ON* to OFF by pressing enter. Following this use the left arrow key to select *Inputs*, next select *Run* under *Inputs*. Press F10. The program will execute the full simulation.

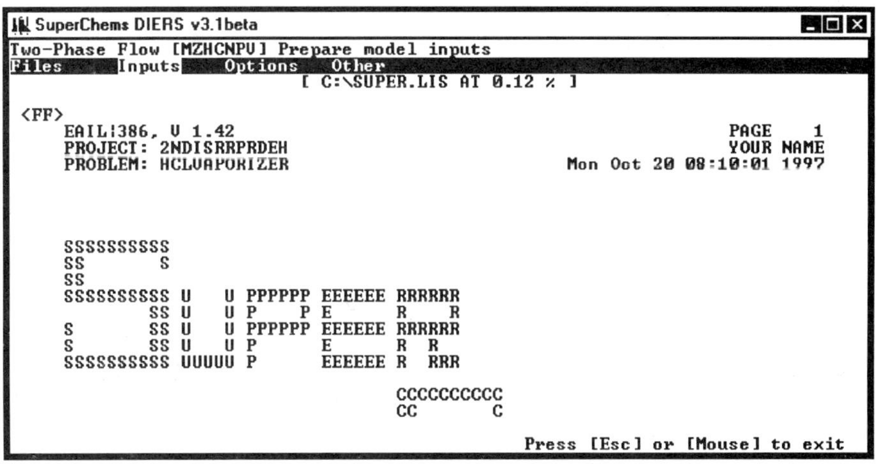

Screen 41. When the simulation is complete, which can take several hours, the screen above will appear. This is the output screen; one can scroll through it and look at the results; however, it is long and as a result difficult to locate desired information. So instead of working with this output, it is better to view the results graphically through SuperChems for DIERS graphical templates. Press ESC key to get to these templates.

Move the highlight from *Inputs* to *Other*.

```
Plot results (SI)
Plot results (ENG)
Print results to file
------------------------------------
Vary X to match T/P specs
------------------------------------
Define input streams
Define miscellaneous defaults
Define slip ratio method
------------------------------------
Starting conditions check is OFF
------------------------------------
Reaction in pipe is OFF
------------------------------------
Connections
------------------------------------
Define disengagement parameter
------------------------------------
Quickly estimate relief area/flow
```

Screen 42. Select *Plot results <SI>*. Then press enter.

```
√ TEMPERATURE HISTORY
√ PRESSURE HISTORY
  PRESSURE/TEMPERATURE HISTORY
√ VENTING HISTORY
  OVERALL VAPOR MOLE FRACTION
  FLOW REDUCTION FACTOR
  TEMPERATURE RISE RATE
  PRESSURE RISE RATE
  VESSEL VOID FRACTION HISTORY
√ VESSEL MASS HISTORY
  CRITICAL PRESSURE RATIO
  CRITICAL TEMPERATURE RATIO
  TEMPERATURE RISE RATE VS. TEMP
  PRESSURE RISE RATE VS. TEMP
  FORCE PLOT
  VELOCITY VS. TIME
  PRESSURE AT INLET OF 1ST AND 2ND RELIEF DEVICES
  PERCENT INLET PRESSURE DROP TO 1ST AND 2ND RELIEF DEVICES
  PERCENT BACKPRESSURE ON 1ST AND 2ND RELIEF DEVICES
                                        Press F10 to display
```

Screen 43. Pressing the space bar will place a check mark in front of the graph or graphs one wishes to view. Hitting the space bar with a graph already having a checkmark will unmark the checkmark. Checkmark the graphs as indicated above. If one presses enter after marking the graphs, they will appear one at a time. However before viewing, let us change the number of graphs per screen. To change the number of graphs per screen, move the highlight from *Select* to *Layout* which is above this block. Then highlight *4 graphs/page*. Then go back to *Select* and press enter.

The results similar to those shown in Figure 2 should be obtained. These results indicate that the present 3 × 4 safety valve with an H orifice on the HCl vaporizer is adequate. The vent is adequate because the maximum pressure during relief is 2.39 MPa which is ~110% of the relief device set pressure.

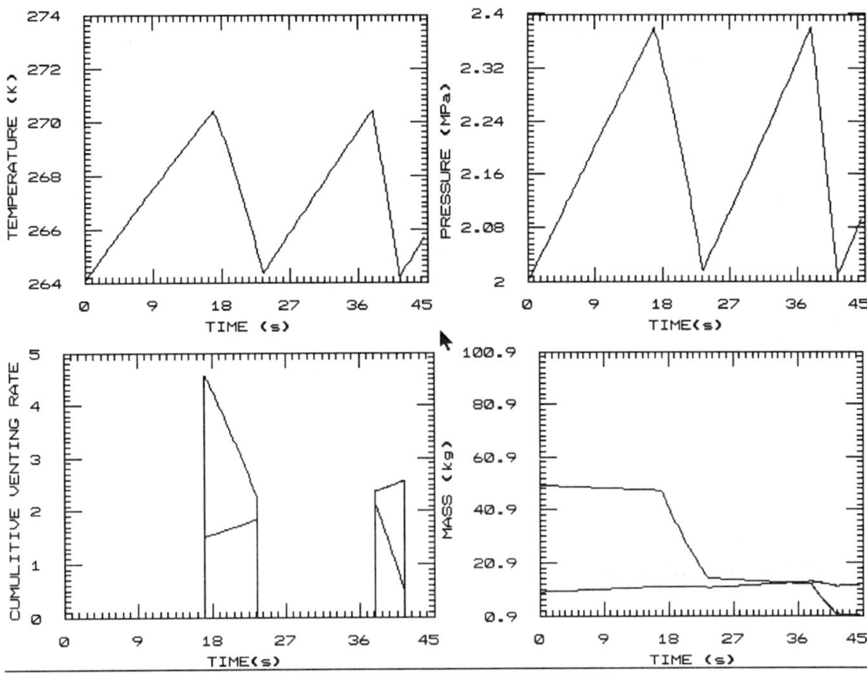

Figure 2. Temperature and Pressure Histories, Venting Rate and Vessel Contents Mass History for Overheating HCl Vaporizer in Simulation 1a.

Simulation 1b. Overheating Hydrogen Chloride Vaporizer (Actual Piping)

In simulation 1b the ideal vent with a coefficient of discharge specified in simulation 1a is replaced by the actual piping layout. By using the actual piping layout, the full simulation will take considerably longer to complete but all transient effects will be taken into account. The first step is to create a new scenario named MZHE102 by repeating screens 1–4 in simulation 1a. It is important to ensure the correct scenario, MZHCNPV, is duplicated using the *Preload* command. Next the piping layout called 3x4HPIP is specified by the screens illustrated below. Then this piping layout file name is entered into the newly created scenario file named MZHE102 following the screens 27–30 in simulation 1a. Finally the scenario is preloaded and executed as in step 6 of simulation 1a after updating the initial condition and miscellaneous defaults as shown below in step 2.

The actual piping layout of the system needs to be broken into sections and inputs required by SuperChems for DIERS as in Table 1 (5).

The required steps for SuperChems for DIERS to model the overheating of the HCl vaporizer with actual piping are presented below.

TABLE 1
Piping System in SuperChems for DIERS Format

section	description	dia (in)	area (m ^2)	L(in)	length (m)	angle	Kvalue	Kinf	Kid
1	entrance & str pipe	3	0.0045581	9	0.2286	90	160	0.5	
2	tee	3	0.0045581	12	0.3048	90	150	0.5	0.5
3	pipe	3	0.0045581	39	0.9906	90			
4	rupture disc	3	0.0045581	6	0.152	0			
5	relief valve H in.sq.	0.785	0.000506	8	0.2032	0		0.2	
6	rupture disc	4	0.0081032	8	0.203	0			
7	str pipe	4	0.0081032	288	7.3152	0			
8	45 elbow	4	0.0081032	8	0.2032	0	500	0.2	0.2
9	str pipe	4	0.0081032	14	0.3556	45		1	
10	str pipe &exit	4	0.0081032	8	0.2032	45		1	

1. Define Piping with Actual Layout

```
Piping layout name [s]                    3x4HPIP
Piping layout description [s]             1.5M 3" PSV 3X4 H 8M 4"
Number of segments []                     10
Elevation of first segment (m) []         2.499
```

Screen 44. Select *Define, Piping* and then *Edit* as under "Define Piping with Ideal Vent Having a CD" in the simulation 1a to obtain the above screen. Then overwrite the default values as shown. The number of segments refers to the break down of the piping as indicated in Table 1.

```
1. Rupture disk. Pset(Pa/Psig) = 101325/-3.97403E-06, D(m/in) = 0.0508/2.000
2. Straight pipe / valves / fittings. L(m)= 0.00 , D(m/in)= 0.0000/0.0000
3. Straight pipe / valves / fittings. L(m)= 0.00 , D(m/in)= 0.0000/0.0000
4. Straight pipe / valves / fittings. L(m)= 0.00 , D(m/in)= 0.0000/0.0000
5. Straight pipe / valves / fittings. L(m)= 0.00 , D(m/in)= 0.0000/0.0000
6. Straight pipe / valves / fittings. L(m)= 0.00 , D(m/in)= 0.0000/0.0000
7. Straight pipe / valves / fittings. L(m)= 0.00 , D(m/in)= 0.0000/0.0000
8. Straight pipe / valves / fittings. L(m)= 0.00 , D(m/in)= 0.0000/0.0000
9. Straight pipe / valves / fittings. L(m)= 0.00 , D(m/in)= 0.0000/0.0000
10. Straight pipe / valves / fittings. L(m)= 0.00 , D(m/in)= 0.0000/0.0000
Return to previous menu
                                                                     3x4HPIP
```

Screen 45. Each time a segment is completed this screen will appear. Move to a different segment by using the up or down arrow key or the mouse. Press F10 at the end of this and the following screens unless indicated otherwise.

```
Straight pipe / valves / fittings
Relief device
Rupture disk
Thin sharp edge orifice
Pipe expander (rounded)
Pipe reducer (rounded)
Pump
                          Segment 1/10 Type
```

Screen 46. The first segment is piping, so select *Straight pipe/ valves/fittings,* then press F10. Segments 2,3,7,8,9, and 10, will also be *Straight pipe/ valves/ fittings.*

```
Roughness (m) []                                                   4.57E-05
Orifice/outlet flow area (m2) []                                   0.00456
Piping segment length (m) []                                       0.2286
Angle with respect to horizontal (degrees) []                      90
K1 value []                                                        160
Kinf value []                                                      0.5
Kid value []
1/Overall heat transfer coefficient (W/m2/K); Sum_i (1/hi) []      1E+38
Surroundings temperature (K) []                                    294.44
```

Screen 47. Each pipe segment has an input screen. Above is the input screen for pipe segment one. Create segments 2, 3, 7, 8, 9, and 10 in the same way as above using the values for each segment shown in Table 1. Typically the numbers used for the roughness and the surroundings temperature are 4.57 E-05 and 294.44 K, respectively. The *K1, Kinf, Kid values* were obtain by selecting each of them, pressing F9, then selecting N, pressing enter and choosing the appropriate value.

```
Straight pipe / valves / fittings
Relief device
Rupture disk
Thin sharp edge orifice
Pipe expander (rounded)
Pipe reducer (rounded)
Pump
                        Segment 4/10 Type
```

Screen 48. Segment 4 specifies a rupture disc at the entrance to the relief device.

```
Roughness (m) []                                                   4.57E-05
Orifice/outlet flow area (m2) []                                   0.00456
Piping segment length (m) []                                       0.152393
Set pressure (Pa)                                                  2162858.
Angle with respect to horizontal (degrees) []
K1 value []
Kinf value []
Kid value []
1/Overall heat transfer coefficient (W/m2/K); Sum_i (1/hi) []      1E+38
Surroundings temperature (K) []                                    294.44
```

Screen 49. Above is the input screen for segment 4. Segment 6 is also a rupture disc at the exit of the relief device. Therefore, create segment 6 in the same manner as segment 4 except use 0.008107 m^2 as the flow area and 0.15 m as the length. These values differ because the outlet piping has a larger diameter.

```
Straight pipe / valves / fittings
Relief device
Rupture disk
Thin sharp edge orifice
Pipe expander (rounded)
Pipe reducer (rounded)
Pump
                        Segment 5/10 Type
```

Screen 50. For Segment 5 highlight the *Relief device*.

```
Relief device type                           3
Orifice/outlet flow area (m2) []             0.0005065
Set pressure (Pa)                            2169752.771
K1 value []
Kinf value []                                0.23457
```

Screen 51. Enter *relief device type* 3 which is a relief value that is fully opened at 10% above the set pressure and is fully closed at –7%. For the orifice area, use the value for an H safety valve. *Set pressure* is that of the relief device in Pascals. Use the K values in the above screen.

```
1. Straight pipe / valves / fittings. L(m)= 0.23 , D(m/in)= 0.0762/2.9999
2. Straight pipe / valves / fittings. L(m)= 0.30 , D(m/in)= 0.0762/2.9999
3. Straight pipe / valves / fittings. L(m)= 0.99 , D(m/in)= 0.0762/2.9999
4. Rupture disk. Pset(Pa/Psig) = 2162858/299, D(m/in) = 0.0762/2.9999
5. Relief device. Pset(Pa/Psig) = 2169753/300.001, Ao(m2/in2) = 0.0005/0.785
6. Rupture disk. Pset(Pa/Psig) = 2169753/300, D(m/in) = 0.1016/3.9999
7. Straight pipe / valves / fittings. L(m)= 7.31 , D(m/in)= 0.1016/3.9999
8. Straight pipe / valves / fittings. L(m)= 0.20 , D(m/in)= 0.1016/3.9999
9. Straight pipe / valves / fittings. L(m)= 0.36 , D(m/in)= 0.1016/3.9999
10. Straight pipe / valves / fittings. L(m)= 0.20 , D(m/in)= 0.1016/3.9999
Return to previous menu
                                                                    3x4HPIP
```

Screen 52. After completing all the segments, the information on the above should be obtained. Select *Return to previous menu* , press F10, and then the ESC key twice to return to the main menu.

2. Define Scenario

Follow the "Define Scenario" steps in simulation 1a except *Preload* MZHE102 and use the values below.

```
Name [s]                MZHE102
Description [s]         DEFAULT SCENARIO
                 >      ■
                 >
Site [s]                DEFAULT
Vessel [s]              HE102
Piping layout [s]       3X4HPIP
Reaction [s]
Mixture [s]             MZHCL
```

3. Execute 2-Phase Flow Model

Total mass in vessel (kg)	50
Initial temperature (K)	264
Initial pressure (Pa)	2000000
Starting time for analysis (s)	
Final time for analysis (s)	20
Flow type	1

Screen 53. Repeat the steps in the first simulation under "Execute 2-Phase Flow Model," but with the above values for the initial conditions screen. The *Final time of analysis* in this simulation is chosen to be only 20 second. The pressure should peak in this time. This is done to save computer time and memory since a lot of calculations are required to simulation two-phase flow through actual piping.

Maximum number of data points to collect	1000
Maximum allowable pressure change per time step (Pa)	10000
Maximum allowable temperature change per time step (K)	5
Maximum allowable time step (s)	5
Minimum allowable time step (s)	0.05
Transit to vapor module when contents are all vapor (1-Yes)	1
Minimum mass allowed in vessel (kg)	1
dT/dt (K/min) above which adiabaticity is lost (ARC data only)	

Screen 54. Change the *Miscellaneous functions* values to those above. These changes are done also to save computer time and memory.

The results shown in Figure 3a and 3b again indicate that the 3x4 pressure relief valve with an H orifice is adequately sized to prevent significant overpressure in the HCl vaporizer.

TUTORIAL 2: SUDDEN WATER ADDITION TO CONCENTRATED SULFURIC ACID

In a plant vessel filled with concentrated sulfuric acid, an upset can occur where water is suddenly pumped into the vessel. The concern is that the water addition will result in the pressure opening the rupture disk on the vessel set at 100 psig, the MAWP of the vessel. To determine if this could occur SuperChems for DIERS was utilized to simulate the addition. A closed test was also conducted in an in-house EAC adiabatic calorimeter (7) to study the upset and the results used to valid the modeling. In the EAC test water was injected into concentrated sulfuric acid in a 40-gram Hastelloy 2.5 inch

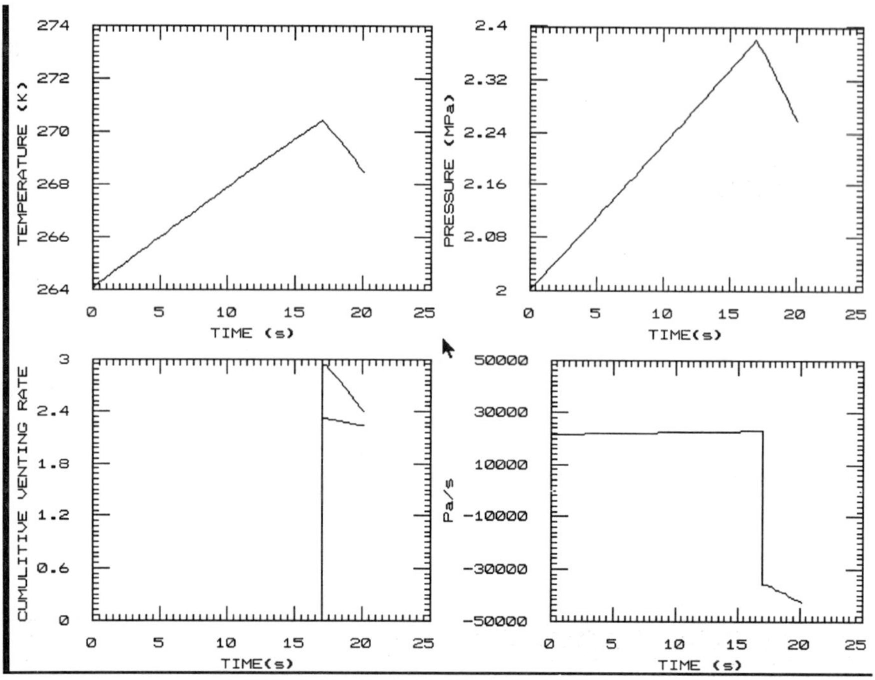

Figure 3a. Plots of Temperature, Pressure, Venting Rate, and Pressure Rate Histories for Overheating HCl Vaporizer in simulation 1b.

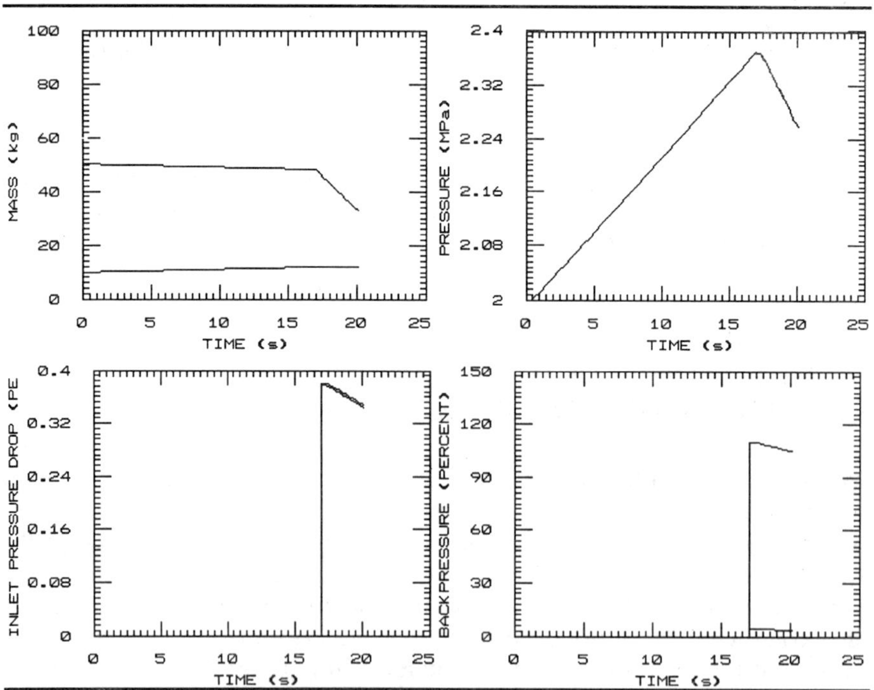

Figure 3b. Plots of Mass in the Vessel, Pressure at the First Relief Device, Inlet Pressure Drop, and the Back Pressure for Overheating HCl Vaporizer in Simulation 1b.

diameter spherical test cell. The temperature was found to increase from296K to 383K and the pressure rose from 0.1MPa to 0.25MPa.

Simulation 2—Water Addition to Sulfuric Acid

Simulation will only be conducted for the EAC test. Extension of the model to the plant would involve the same files except the vessel would need to be redefined plus the water addition rate would have to be adjusted under "Execute 2-Phase Flow" to be that of the plant. For simulation of the plant vessel it is assumed that the system is essentially closed during the addition of water since the water will be added quickly and normal venting is limited.

For this simulation a vessel EACTC which represents the 2.5-inch diameter spherical test cell is specified. The piping layout associated with this vessel is RDISK, which is an ideal rupture disk of 1/8 inch in diameter with a set pressure of 50,000,000 Pa. The set pressure of the rupture disk is set very high to simulate a closed EAC test. A multicomponent mixture is next established. The mixture is 96SOF which consisting of 96% by weight sulfuric acid in water under a pad gas of nitrogen. Next the scenario is updated, then a stream called WATER is defined, which is added to the vessel. Finally the stream addition rate is set under "Execute 2-Phase Flow."

The required steps for SuperChems for DIERS to model the sudden addition of water to concentrated sulfuric acid is given below.

1. Create a new scenario by duplicating the default.
Follow the steps in simulation 1a but name the new scenario 20S14R.

2. Define Vessel

```
Vessel name [s]                     EACTC
Description [s]                     136 CC
           >
           >
Material of construction [s]        IRON
```

Screen 55. Input the above values. To exit this and the following screens press F10 unless indicated otherwise.

At the next screen select *Spherical* for vessel type, not *Vertical Cylindrical* as in simulation 1a.

```
Inside diameter (m)                                                   0.0635
Vessel metal mass (kg)                                                0.04
Wall metal thickness (m)                                              0.0024
Base elevation with respect to scenario ground level (z=0) (m)
Maximum pressure limit for vessel in simulation (Pa)                  1E+38
Maximum temperature limit for vessel in simulation (K)                1000
```

Screen 56. This screen appears next. Input the above values.

```
   Water spray
√  Insulation
   External fire
   ARC/APTAC with heat/wait/search
   Vapor eductor
   Cooling jacket
   Heating jacket
   Solar flux
   Vessel is a pipeline
   User defined heating/cooling load
T                Toggle vessel options
```

Screen 57. The EAC vessel is held adiabatic so select insulation. After pressing F10, *1/Insulation heat transfer coefficient* is needed, use 1 E+38 (10^{38}).

3. Define Piping with Ideal Vent Having a CD.

```
Piping layout name [s]                   RDISK
Piping layout description [s]            1/8 RD
Number of segments []                    1
Elevation of first segment (m) []        .06
```

Screen 58. Follow the "Define Piping with Ideal Vent Having a CD" in simulation 1a except substitute the values found above. At the next screen select rupture disk.

```
Roughness (m) []                                                      4.57E-05
Orifice/outlet flow area (m2) []                                      7E-06
Piping segment length (m) []                                          0
Set pressure (Pa)                                                     30000000
Angle with respect to horizontal (degrees) []                         0
K1 value []
Kinf value []
Kid value []
1/Overall heat transfer coefficient (W/m2/K); Sum_i (1/hi) []         1E+38
Surroundings temperature (K) []                                       294.44
```

Screen 59. Enter the above values for the rupture disk.

4. Define Mixture

Follow the "Define Mixture" steps in simulation 1a except substitute the values found in these screens.

```
Mixture name [s]                               96SOF
Description [s]                                96% H2SO4
Number of compounds []                         3
Number of solids []                            0
Is mixture soluble in water (Y/N) [s]          Y
```

Screen 60. Notice there are three compounds.

```
Chemical Identification Number [   0]   915
Chemical Identification Number [   1]   1013
Chemical Identification Number [   2]   980
```

Screen 61. The chemical identification numbers are: compound [0] is 915 for sulfuric acid; compound [1] is 1013 for water, compound [2] is 980 for nitrogen.

```
Edit      V-Edit     Row 0      Col 0
 Screen Edit
 Initialize

 Exit
```

Screen 62. After selecting Exit press F10.

```
Mass fraction [   0]   0.96
Mass fraction [   1]   0.039
Mass fraction [   2]   .0001
```

Screen 63. The acid is 96% H_2SO_4 and 4% water. The nitrogen is the pad gas and SuperChems will solve the amount needed later. Still an arbitrary value of 0.0001 is needed at this point.

```
One or more component ids have changed.
Please reset the BIPS

Strike any key to continue
```

Screen 64. After selecting exit, the screen above appears about Binary Interaction Parameters (BIPs)(3). The BIPs will need to be reset. Reset the BIPs by pressing any key.

```
Restore input
Remove input
Print to file

Load BIPs from databank
Generate BIPs using ideal behavior
Set BIPs to zero

Select potential reaction products
Remove potential reaction products

Generate properties
Water condensation concentration

Miscellaneous defaults
Specify atom matrix constraints
```

Screen 65. Move the highlight from Inputs to Other and highlight Load BIPs from Databank, as above.

```
System pressure of interest (Pa)    101325
```

Screen 66. The system pressure of interest will be atmospheric. After pressing Enter all BIPs will be loaded into the program if they exist in the databank. There should be three pairs loaded in this example. However there are no BIPs in the databank for H_2SO_4 and nitrogen or for water and nitrogen thus they will be set to zero. After pressing F10, the above screen will reappear, press ESC key twice to return to the main menu.

5. Define Scenario

The name of this scenario is 20S14R which has already been created in this simulation. Follow the "Define Scenario" steps in simulation 1a except *Preload* 20S14R and use the values below:

```
Name [s]                 20S14R
Description [s]          DEFAULT SCENARIO
                >
                >
Site [s]                 DEFAULT
Vessel [s]               EACTC
Piping layout [s]        RDISK
Reaction [s]
Mixture [s]              96SOF
```

Screen 67.

6. Define Stream

The name of the stream to be added is WATER. All the "Define" blocks in the main menu which call for compositions must contain all three compounds in the system. An additional constraint is that the component order must be the same. In this example the stream to be added, even though it contains just water, must also include the acid and the pad gas, again in the exact order specified in the mixture (5).

```
Site
Vessel
Piping
Reaction
Mixture
-----------------------
Scenario
-----------------------
Stream
-----------------------
Apply
Duplicate
-----------------------
Import data
-----------------------
Other Unit Operations
```

Screen 68. Above is the main menu, *Define* is selected and *Stream* highlighted.

```
Current
Edit
    96SOF
```

Screen 69. Select *Edit*.

```
Mixture name [s]             96SOF
Stream name [s]              WATER
Stream description [s]       DEFAULT LIQUID FEED STREAM
```

Screen 70. The *Mixture name* must be the same *Mixture name* used in the scenario.

```
Mass flow rate (kg/s)                          0.000124
Vapor/liquid molar ratio (0 to 1)              0
Temperature (K)                                293
Pressure (Pa)                                  101325
Stream flow area (m2)                          7E-06
Time at which to start flow (s)                1
Discharge duration (s)                         259
```

Screen 71. This screen defines how the stream enters the vessel and the condition of the stream.

Now set the composition of the stream, choose mass fraction.

```
Mass fraction of individual components [   0]   1.01009998E-06
Mass fraction of individual components [   1]   0.9999989798
Mass fraction of individual components [   2]   1.010099969E-08
```

Screen 72. The components are in the same order as the mixture. A value other than zero must be specified for each component. It can be set very low. After pressing F10, the *Edit* screen will reappear, ensure *Exit* is selected to return to the main menu.

7. Execute a 2-Phase Flow Model

Follow the "Execute a 2-Phase Flow Model" steps in Simulation 1a except execute scenario 20S14R. Also the following needs to be done.

```
Total mass in vessel (kg)          0.03984
Initial temperature (K)            296
Initial pressure (Pa)              101325
Starting time for analysis (s)
Final time for analysis (s)        270
Flow type
```

Screen 73. Change the initial conditions to those above.

The options are the same as in simulation 1a. Before setting the miscellaneous defaults let the program vary the amount of nitrogen in the vessel to match the desired starting temperature and pressure in the vessel.

```
Plot results (SI)
Plot results (ENG)
Print results to file
---------------------------------
Vary X to match T/P specs
---------------------------------
Define input streams
Define miscellaneous defaults
Define slip ratio method
---------------------------------
Starting conditions check is ON
---------------------------------
Reaction in pipe is OFF
---------------------------------
Connections
---------------------------------
Define disengagement parameter
---------------------------------
Quickly estimate relief area/flow
```

Screen 74. Select *Vary X to match T/P specs*.

```
Lower limit on mole fraction   0.0001
Upper limit on mole fraction   0.05
Species index                  2
```

Screen 75. Set the limits as above. Notice that nitrogen is *Species Index* 2. Press F10.

After the calculations are completed, continue by checking the initial conditions.

Next input the stream as shown below.

```
Plot results (SI)
Plot results (ENG)
Print results to file
---------------------------------
Vary X to match T/P specs
---------------------------------
Define input streams
Define miscellaneous defaults
Define slip ratio method
---------------------------------
Starting conditions check is OFF
---------------------------------
Reaction in pipe is OFF
---------------------------------
Connections
---------------------------------
Define disengagement parameter
---------------------------------
Quickly estimate relief area/flow
```

Screen 76. Select *Define input streams*.

```
Inlet stream 1 file name [s]     WATER
Inlet stream 2 file name [s]
Inlet stream 3 file name [s]
```

Screen 77. Select water as the *Inlet stream 1 file name*. The other 2 inlet streams are left blank.

Set the miscellaneous defaults the same as in simulation 1a. Then set *Starting condition check OFF*. Next run the scenario.

When the simulation is complete which should only take a few minutes a screen like screen 41 will appear. Select the same graphs as in simulation 1a except mark PRESSURE/ TEMPERATURE HISTORY rather than VENTING HISTORY.

Results indicate that SuperChems for DIERS model predicts the EAC data seen in FIGURE 4.

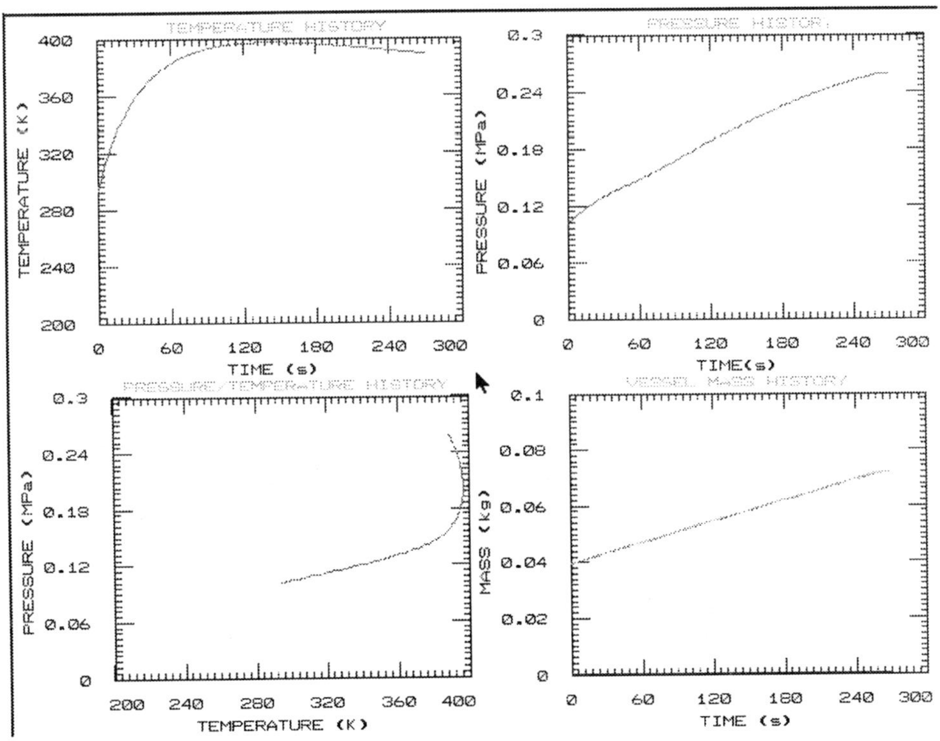

Figure 4. Plots of Temperature and Pressure Histories, Temperature vs. Pressure, and Mass in the Vessel for Water Addition to Sulfuric Acid in Simulation 2.

TUTORIAL 3: WATER ADDED TO REACTOR WITH THIONYL CHLORIDE

A vessel containing thionyl chloride is equipped with a condenser cooled with water. An upset has been identified where a tube in the condenser could fail and introduce water at a maximum rate of 25 gal/min into the vessel containing the thionyl chloride. When the water comes into contact with the thionyl chloride, the reaction will be instantaneous and it is expected two moles of hydrogen chloride and one mole of sulfur dioxide will be generated. Both the hydrogen chloride and sulfur dioxide will be present as gas and thus could result in overpressization of the vessel. To determine if the pressure from this upset could be safely vented with the existing rupture disk on the vessel, the upset was modeled using SuperChems for DIERS.

Simulation 3—Water Addition to Thionyl Chloride

This simulation will be similar to simulation 2, except that a reaction file will be defined and the vessel will be vented with an 8 inch diameter rupture disk. In this example, first a new scenario named REOP is duplicated, then Vessel RE1000G which is similar to Vessel HE102 in simulation 1a is created. The height of the vessel is 2.5 meters, its inside diameter is 1.39 meters: the jacket overall heat transfer coefficient is 10 $W/m^2/K$. Next piping RDISK8IN is set up as RDISK in simulation 2. The elevation of the first segment of piping is 2.49 meters, while the rupture disk area is $0.0324 m^2$ and set pressure is 790800 Pa (115 psia).

The reaction file is next defined as indicatedon the next page. The mixture, scenario, and stream files are also updated as shown. The mixture file named MXTHL2 has seven components including a pad gas of nitrogen. Thionyl chloride, SuperChems ID number 992, is missing the following properties: critical pressure: 6.87E+06 Pa, critical compressibility factor: 0.2961, acentric factor: 0.24982, and the EOS M factor: 0.74747. These values must first be put into SuperChems for DIERS database before running this simulation. If these values are not present the simulation will not run (1). The name of the stream to be added is ICE1. In executing the 2-phase flow model, some changes to the initial condition and miscellaneous defaults are also made.

1. Define Reaction

```
Site
Vessel
Piping
Reaction
Mixture
-------------------------
Scenario
-------------------------
Stream
-------------------------
Apply
Duplicate
-------------------------
Import data
-------------------------
Other Unit Operations
```

Screen 78. In the main menu select *Define* to view above screen then highlight *Reaction*. Next to exit this and the following screens press F10 unless indicated otherwise.

```
Current
Edit
 DEFAULT
```

Screen 79. Highlight *Edit*.

```
Reaction name [s]              RXTHCL
Reaction description [s]       SOCL2 + H20 >> SO2 + 2HCL
```

Screen 80. Overwrite the default values as shown.

```
Number of chemical reactions []    1
Number of chemicals []             7
```

Screen 81. There is 1 reaction and seven chemicals. The seven chemical are: the solvent hexane, thionyl chloride, sulfur dioxide, hydrogen chloride, water, an inactive component dihexyl adipate, and the pad gas nitrogen.

```
Chemical Identification Number [   0]   5
Chemical Identification Number [   1]   992
Chemical Identification Number [   2]   242
Chemical Identification Number [   3]   910
Chemical Identification Number [   4]   1013
Chemical Identification Number [   5]   471
Chemical Identification Number [   6]   980
```

Screen 82. Enter chemical identification numbers from SuperChems for DIERS database as shown.

```
Edit      V-Edit
 Screen Edit
 Initialize
 ----------
 Exit
```

Screen 83. Select *Edit*.

```
Stoichiometric coefficient of reactants (positive) [   0]
Stoichiometric coefficient of reactants (positive) [   1]    1
Stoichiometric coefficient of reactants (positive) [   2]
Stoichiometric coefficient of reactants (positive) [   3]
Stoichiometric coefficient of reactants (positive) [   4]    1
Stoichiometric coefficient of reactants (positive) [   5]
Stoichiometric coefficient of reactants (positive) [   6]
```

Screen 84. Enter stoichiometric coefficient for each reactant chemical in the system. If a chemical in the system does not partake in the reaction or is a product, enter 0 or leave blank.

```
Reaction order for reactants [   0]   0
Reaction order for reactants [   1]   1
Reaction order for reactants [   2]   0
Reaction order for reactants [   3]   0
Reaction order for reactants [   4]   1
Reaction order for reactants [   5]   0
Reaction order for reactants [   6]   0
```

Screen 85. Enter the reaction order for each reactant. If a chemical in the system does not partake in the reaction order, enter 0 or leave blank.

```
Stoichiometric coefficient of products (positive) [   0]
Stoichiometric coefficient of products (positive) [   1]
Stoichiometric coefficient of products (positive) [   2]   1
Stoichiometric coefficient of products (positive) [   3]   2
Stoichiometric coefficient of products (positive) [   4]
Stoichiometric coefficient of products (positive) [   5]
Stoichiometric coefficient of products (positive) [   6]
```

Screen 86. Enter stoichiometric coefficient for each product chemical in the system. If a chemical in the system does not partake in the reaction or is a reactant, enter 0 or leave blank.

```
Reaction order for products [   0]   0
Reaction order for products [   1]   0
Reaction order for products [   2]   0
Reaction order for products [   3]   0
Reaction order for products [   4]   0
Reaction order for products [   5]   0
Reaction order for products [   6]   0
```

Screen 87. Enter the reaction order for each product. If a chemical in the system does not partake in the reaction order, enter 0 or leave blank.

```
Pre-exponential factor (m^3, s, kmol) [    0]     100
```

Screen 88. This reaction is kinetically very fast and is assumed essentially independent of temperature (2). The rate constant, expressed as the *Pre-exponential factor,* is therefore set to 100.

```
Activation energy (K) [    0]     10
```

Screen 89. The *Activation energy* is assumed very small; 10 is used.

The next six screens (3 input screens and 3 edit screens) deal with the temperature dependence of the reaction rate, so leave them blank.

```
Reaction phase: Liquid = 0, Vapor = 1 [    0]     0
```

Screen 90. The *Reaction phase* for the reaction is liquid which is the default 0.

```
Upper bound on reaction rate (kmol/s) [    0]     10000
```

Screen 91. The upper limit of the rate is set to 10000.

```
Is reaction reversible (0-No, 1-Yes) [    0]     0
```

Screen 92. The reaction is not reversible.

```
Specify reaction rate multiplier functions ?
                                                NO
                                                YES
```

Screen 93. There is no rate multiplier function.

2. Define Mixture.

MXTHL2 is a seven component mixture including a pad gas of nitrogen.

```
Mixture name [s]                          MXTHCL2
Description [s]                           HEX,SOCL2,SO2,HCL,H2O,471,N2
Number of compounds []                    7
Number of solids []                       0
Is mixture soluble in water (Y/N) [s]     Y
```

Screen 94. Follow the "Define a Mixture" steps in simulation 2 except substitute the values found in these screens. There are seven compounds mixture including a pad gas of nitrogen.

```
Chemical Identification Number [    0]    5
Chemical Identification Number [    1]    992
Chemical Identification Number [    2]    242
Chemical Identification Number [    3]    910
Chemical Identification Number [    4]    1013
Chemical Identification Number [    5]    471
Chemical Identification Number [    6]    980
```

Screen 95. The chemical identification numbers are above. They must be in the same order as specified under reaction RXTHCL.

```
Mass fraction [    0]    0.4595
Mass fraction [    1]    0.1856
Mass fraction [    2]    0.00019
Mass fraction [    3]    0.000147
Mass fraction [    4]    9.69E-05
Mass fraction [    5]    0.354
Mass fraction [    6]    0.000349
```

Screen 96, Use the *Mass fractions* above.

3. Define Scenario

The name of this scenario is REOP which should have been created earlier in this simulation by duplicating default. Follow the "Define Scenario" steps in simulation 1a except *PreLoad* REOP and use the values below:

```
Name [s]                REOP
Description [s]         DEFAULT SCENARIO
             >
             >
Site [s]                DEFAULT
Vessel [s]              RE1000G
Piping layout [s]       RDISK8IN
Reaction [s]            RXTHCL
Mixture [s]             MXTHCL2
```

Screen 97.

4. Define Stream

Follow the "Define Stream" steps in simulation 2 except substitute the values found in these screens. Note the "Define" blocks in the main menu which call for compositions must have all seven of these compounds. An additional constraint is that the component order must be the same. In this example the stream, even though it contains just water, must also include all the compounds, again in the exact order specified. The mixture name must be the same mixture used in the scenario.

```
Mixture name [s]              MXTHCL2
Stream name [s]               ICE1
Stream description [s]        WATER FEED STREAM
```

Screen 98. This determines the name of the stream to be added.

```
Mass flow rate (kg/s)                     1.6
Vapor/liquid molar ratio (0 to 1)
Temperature (K)                           294
Pressure (Pa)                             101325
Stream flow area (m2)                     0.018241
Time at which to start flow (s)           0.3
Discharge duration (s)                    27
```

Screen 99. This screen defines how the stream enters the vessel and the condition of the stream.

```
Mass fraction of individual components [   0]    1E-05
Mass fraction of individual components [   1]    1E-05
Mass fraction of individual components [   2]    1E-05
Mass fraction of individual components [   3]    1E-05
Mass fraction of individual components [   4]    0.99
Mass fraction of individual components [   5]    1E-05
Mass fraction of individual components [   6]    1E-05
```

Screen 100. Set the composition of the stream, by choosing *Mass fraction*.

The components are the same order as the mixture. After pressing F10 an edit screen will reappear, ensure Exit is selected to return to the main menu.

5. Execute a 2-Phase Flow Model

Follow the "Execute a 2-Phase Flow Model" steps in simulation 2 except *Execute* scenario REOP.

```
Total mass in vessel (kg)           1477.6
Initial temperature (K)             338
Initial pressure (Pa)               101325
Starting time for analysis (s)
Final time for analysis (s)         18
Flow type                           1
```

Screen 101. Change the initial conditions to those above.

Set the *Option* the same as in simulation 1a.

Next enter ICE1, which has already been created in this simulation, for *Define input streams.*

```
Maximum number of data points to collect                                1000
Maximum allowable pressure change per time step (Pa)                    10000
Maximum allowable temperature change per time step (K)                  3
Maximum allowable time step (s)                                         0.05
Minimum allowable time step (s)                                         0.01
Transit to vapor module when contents are all vapor (1-Yes)
Minimum mass allowed in vessel (kg)                                     100
dT/dt (K/min) above which adiabaticity is lost (ARC data only)
```

Screen 102. Set the miscellaneous defaults as above. Then set *Starting condition check OFF*. Run the scenario.

The results seen in Figure 5 indicate that the pressure will remain below 110% of the MAWP of the vessel indicating the existing 8 inch rupture disk is satisfactory.

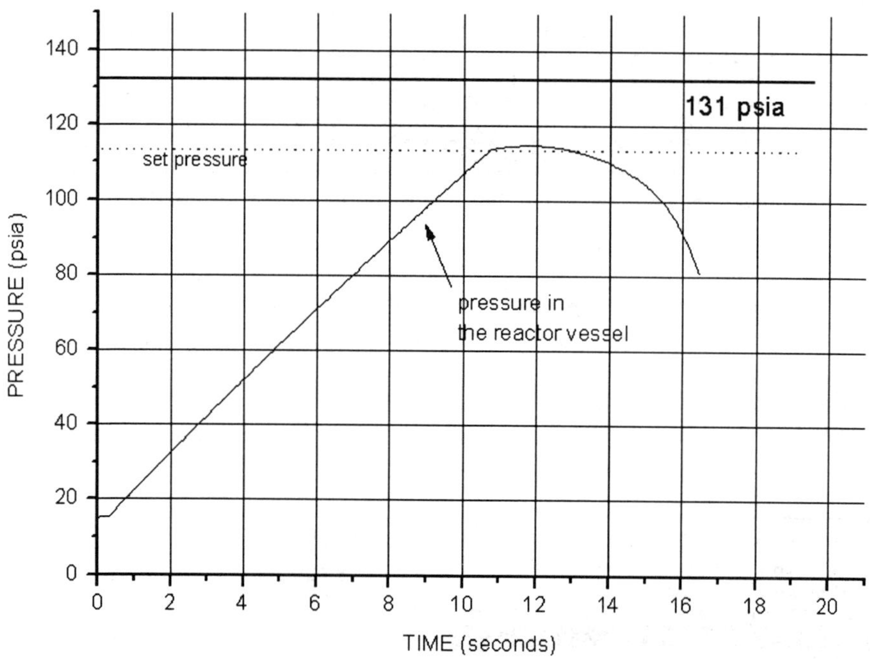

Figure 5. Plot of Pressure History for Water Addition to Reactor with Thionyl Chloride in Simulation 3.

CONCLUSION

In hazard analysis, SuperChems for DIERS is an extremely powerful tool for evaluating what-if situations. This results because different scenario can usually be evaluated by redefining only a few elements in the program. SuperChems for DIERS also offers an advantage over more traditional short-cut methods in that transient effects during emergency venting can be taken into account. Since SuperChems for DIERS is based on fundamental chemical and physical principles, it is also well suited to handle highly nonideal systems as well as those where gas solubility effects are important. Overall while a simulation can be more difficult to set up, SuperChems for DIERS can be extremely valuable tool for analyzing chemical plant upsets and their potential consequences.

REFERENCES

1. Atkins, R .C., Muzzio, D. J., and Sharkey, J.J., "Prediction of Physical Properties Using SuperChems," *2nd International Symposium on Runway Reactions and Pressure Relief Design*, New York: AICHE/DIERS, 1998.
2. Levenspiel, O., *Chemical Reaction Engineering*, 2nd ed., New York: Wiley, 1972.
3. Melhem, G. A., "A Modified Peng–Robinson Equation of State," *Fluid Phase Equilibria*, pp. 189–237,1989.
4. Melhem, G. A., "Advanced ERS design using computer simulation." *International Symposium on Runway Reactions and Pressure Relief Design*, pp. 502–566. New York: AICHE/DIERS,1995.
5. Murphy. M. , personnel communications, 1997.
6. Reid, J. C.,Prausnitz, J. M., and Poling, B. E., *The Properties of Gases and Liquids*, 4 th ed., New York: McGraw-Hill, 1987.
7. Yue Ming-Huei, "An enhanced adiabatic calorimeter for thermal hazard analysis," *Journal of Hazardous Material,* 38 (1994), 13–25.

Prediction of Physical Properties Using *SuperChems*™

Robert C. Atkins*
Department of Chemistry, James Madison University, Harrisonburg, VA 22807

Daniel J. Muzzio and John J. Sharkey
Process Safety Laboratory, Merck & Co., Inc. Rahway, NJ 07065

ABSTRACT

Understanding the physical properties of compounds used in manufacturing is an important component of process safety. The computer estimation of properties is an increasingly important adjunct to laboratory data. While the number of chemical compounds is extremely large, group contribution methods are available that utilize the fact that the number of different groups that make up these molecules is relatively small. Using the commercially available modeling program *SuperChems* (Arthur D. Little, Inc.), the physical properties of several molecules not found in the software database were predicted. Comparison of the predicted values with laboratory data show an excellent correlation. The steps necessary for the prediction of physical properties using *SuperChems* are discussed in detail.

1. INTRODUCTION

Computer simulations provide an important adjunct to laboratory measurements in the Process Safety Laboratory of Merck & Co. The laboratory has had success in this area using *SuperChems* modeling software (Expert Edition Version 3.1) from Arthur D. Little, Inc., and some of these efforts were recently described by one of the authors (Muzzio, 1997).

The modeling of a scenario using *SuperChems* requires input of physical properties from the program's databanks. The default software databanks

SuperChems is a trademark of Arthur D. Little, Inc., Cambridge, MA
*Visiting Scientist, Process Safety Laboratory, Merck & Co., Inc. Jan.–April 1997.

contain the physical properties of over 1200 compounds. The stored information includes both temperature independent properties and equation parameters necessary to calculate temperature-dependent properties. In order to evaluate a scenario involving a chemical compound not found in the default databanks, its physical properties would have to be estimated and included in the databanks. The ability to predict physical properties using *SuperChems* extends the capabilities of the software beyond the confines of the default set of compounds. The majority of this paper is a "how-to" manual intended to guide users of *SuperChems* through the prediction process. The prediction of those physical properties considered by Melhem (1997) to be the minimum necessary for scenario modeling are described.

Group contribution methods form the basis for the physical properties prediction capabilities of *SuperChems.* An early group contribution method that has received wide acceptance is that described by Benson (1976) for the estimation of heats of formation, entropy, and heat capacity. The group contribution techniques described by Reid et al. (1987) are widely employed and are utilized by *SuperChems*, although newer methods may be found in the literature (Constantinou and Gani, 1994).

All group contribution techniques have as their basis the concept that molecular properties can be estimated algebraically from contributions made by the constituent atoms and groups. The simplest example of group contribution is molecular weight, the sum of the atomic weights of the constituent atoms. The most common estimation methods, including those used by Joback (Joback, 1987 and Reid et al., 1987), and Ambrose (Reid et al., 1987) consider the molecule of interest to be the sum of discrete, noninteracting, groups. Thus isopropyl alcohol $(CH_3)_2CHOH$, is composed of two methyl groups ($-CH_3$), one methine group (–CH), and one hydroxyl group (–OH). Each of the groups (methyl, methine, hydroxyl) makes a unique contribution to the physical properties of the molecule. Most importantly, the contribution made by any particular group, methyl for example, to a given physical property is the same regardless of the molecular structure of the compound. This allows the physical properties of a wide range of compounds to be estimated using a tabular listing of the group contributions. And most importantly, *SuperChems* has made use of the computer compatibility of this approach. Once the structure of a compound has been described in *SuperChems* in terms of its constituent groups, the values of a series of physical properties can be estimated.

Estimation of critical temperatures and pressures using *SuperChems* can be done using either the methods of Ambrose, Joback, or Fishtine. Use of Fishtine is not recommended by Reid et al. (1987) due to higher errors and a limited range of compounds. We chose to use the Joback method over Ambrose as the average percent error of the two methods are comparable, and the Joback method is somewhat easier to use (Reid et al., 1987). In addition, the Joback method can be used in *SuperChems* to estimate critical volumes; the Ambrose method cannot. The average percent error for estimating critical properties using the Joback method is 0.8 (T_c, 400 compounds); 5.2 (P_c, 390 compounds); and 2.3 (V_c, 310 compounds) (Reid et al., 1987).

In addition to the capability to estimate critical properties, *SuperChems* is able to estimate the normal (1 atm) boiling point (T_b) of a compound using the Joback method. Reid et al. (1987) found an average error of 3.8 % for 388 compounds. Entering an experimental value of T_b in *SuperChems* is of course preferable, however if no data are available the Joback method gives a reasonable estimate.

The sections of this paper listed below describe the steps necessary to add a compound to the databanks of *SuperChems*. In addition, several sections are included which describe how to maneuver through the various features of *SuperChems* relating to chemical properties.

2. User-Defined Chemical Compounds in *SuperChems*. Overview
3. *SuperChems* Hints
4. Adding a Compound to the Databank
    4.1. Getting Started
    4.2. Molecular Weight and Normal Boiling Point
    4.3. Enthalpy and Gibbs Free Energy of Formation; Ideal Gas Heat Capacity
    4.4. Estimate Critical Properties: T_c, P_c, V_c, Z_c
    4.5. Determine Vapor Pressure Curve
    4.6. Reduce Ideal Gas Equations
    4.7. Reduce Equation of State Parameters; Regress Liquid Density, Latent Heat of Vaporization, and Liquid Heat Capacity
    4.8. Reduce Gibbs Free Energy Parameters
5. Create a New Databank
6. Duplicate a Chemical
7. Estimate Boiling Point
8. Generate Properties
9. Select a Chemical
10. Select a Databank
11. Acknowledgments
12. References

2. USER-DEFINED CHEMICAL COMPOUNDS IN *SUPERCHEMS*: OVERVIEW

The default *SuperChems* databanks supplied with the program contain the properties of over 1200 chemical compounds. To run a scenario using a chemical not found in the default databanks, the chemical must be added, and its properties entered or estimated. This is accomplished by the series of steps outlined below.

NOTE: All estimations described below are done using *SuperChems*, unless otherwise specified.

Step 1: Add the chemical, giving it a unique name, specifying its formula, and assigning a password (Section 4.1).

Step 2: Specify the molecular weight and normal boiling point (NBP). If the NBP is not known, it can be estimated from a reduced pressure boiling point or vapor pressure data or, as a last resort, from the structure of the compound (Section 4.2).

Step 3: Estimate heat of formation (H_f) and Gibbs free energy of formation (G_f), both at 298 K. In addition to *SuperChems*, these values can be estimated using CHETAH, MOPAC, etc. Estimate the ideal gas heat capacity (Section 4.3).

Step 4: Estimate the critical properties: Critical temperature (T_c); Critical pressure (P_c); Critical volume (V_c); Compressibility (Z_c) (Section 4.4).

Step 5: Determine the vapor pressure curve using the NBP, T_c, and P_c. Extrapolate the curve below the NBP (Section 4.5).

Step 6: Reduce the ideal gas equations (Section 4.6).

Step 7: Reduce the EOS parameters, then regress the liquid density, latent heat of vaporization, and liquid heat capacity (Section 4.7).

Step 8: Reduce the Gibbs free energy parameters (Section 4.8).

Each of the steps listed above is described in detail in the pages which follow. Additional information relating to the databank and chemical entries is provided in Sections 5, 6, 8, 9, and 10. Section 7 describes the steps necessary to estimate a compound's boiling point.

3. *SUPERCHEMS* HINTS

Output is more easily viewed using the DOS text editor. From *Files* in the main menu, select *Directories* (Figure 1).

```
General purpose utilities and file management
Files     Define     PreLoad     Execute     Projects

  Directories
  File manager  (CTRL-F)
  DOS command   (CTRL-X)
  Archives
  ----------------------------
  Autosave core image is ON
  Record output is ON
```

Figure 1.

Type EDIT in the *text viewer* line. It may also be desirable (if a new chemical is being added to the databank) to select EDIT as the *SuperChems text editor* (Figure 2).

```
Working directory path [s]                         ►\SC31\SUPER\
Core image directory path and name [s]             [C:\SC31\INC\DEFAULTS.COR   ]
Include directory [s]                              [\SC31\INC\                 ]
Text editor (leave blank for default) [s]          [EDIT                       ]
Text viewer (leave blank for default) [s]          [EDIT                       ]
Text notepad (leave blank for default) [s]         [                           ]
File manager [s]                                   [                           ]
```

Figure 2.

Alternatively, if *SuperChems* is being run in Windows 95 or NT, NOTEPAD can be used to view the output. Minimize *SuperChems* and run NOTEPAD. Open the file C:\SUPER.LIS. Reopen the file each time new output is being viewed to update the display.

Menu selections, as from a pulldown menu, can be done using either the keyboard or the mouse (Figure 3).

```
Database functions
Select    Codes    ModDes    ModCon    ModTDP    Banks
                                                  Add
                                                  Remove
                                                  Save
                                                  Lock
                                                  Unlock
                                                  Duplicate
                                                  ---------
                                                  Create
                                                     BENZENE
```

Figure 3.

The former is recommended; highlight the selection (using the up/down arrow keys) and select with <enter>. Both actions are simultaneous with the mouse. An incorrect selection can result if the cursor is not aligned as desired when the mouse button is clicked.

User-entered data are required when brackets appear: [...] (Figure 4).

```
Number of records   [50              ]
```

Figure 4.

A previous (or default) entry can be cleared by pressing <ESC>. The entry can be selected with <F10> or left-clicking the mouse on the bar at the bottom of the screen.

4. ADDING A COMPOUND TO THE DATABANK

4.1. Getting Started

4.1.1. Select the databank that the compound will be added to. This will usually be the default *SuperChems* databank. To check which databank is currently used by *SuperChems* or to select the default databank, see the steps described in Section 10 (Select A Databank).

4.1.2. In the main menu select *ChemProps*, then *Databank manager* from the pulldown menu (Figure 5).

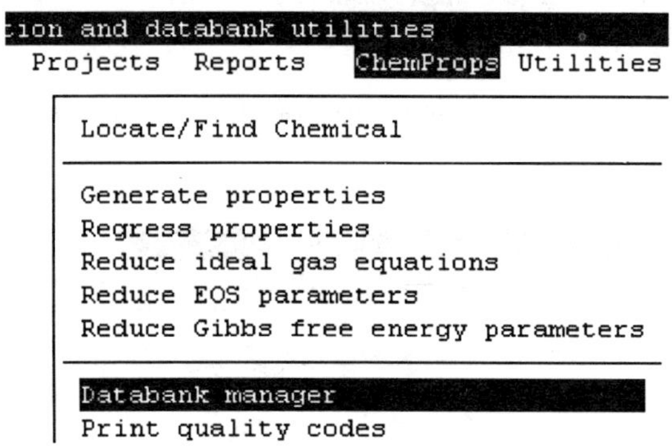

Figure 5.

4.1.3. Choose *Banks*, then *Add* from the pulldown menu (Figure 6).

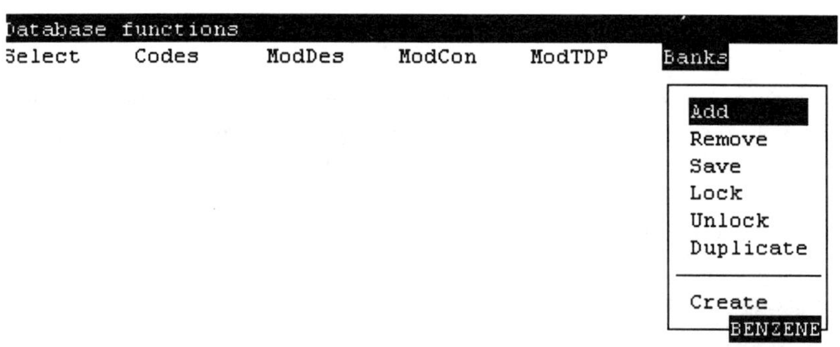

Figure 6.

4.1.4. Enter information about the chemical in the box that appears (Figure 7).

```
Chemical name [s]                  [                                        ]
Chemical formula [s]               [                                        ]
Chemical structure [s]             [                                        ]
Name alias 1 [s]                   [                                        ]
Name alias 2 [s]                   [                                        ]
Name alias 3 [s]                   [                                        ]
Chemical abstract number [s]       [                                        ]
Comment line-1 [s]                 [                                         ]
Comment line-2 [s]                 [                                         ]
Comment line-3 [s]                 [                                         ]
Access password [s]                [                    ]
```

Figure 7.

These entries are not DOS filenames and are not limited to 8 characters. Also, spaces are permissible. As always in *SuperChems*, entries should be all in capital letters. There is no minimum number of entries required. A chemical can even be added without a name.

> **NOTE:** This screen cannot be edited (a formula or alias added, for example) once it has been entered and accepted.

PASSWORD: No password is required, although one is suggested to prevent unauthorized users from altering the chemical databank entry unintentionally. Each chemical may have its own unique password. The password may be any combination of letters, numbers, or characters in the space provided. If "open" access, i.e. no password, is desired, enter PUBLIC. Alternatively, the entry line can be left blank. The password *is* case sensitive.

Accept the data entered with <F10>.

4.1.5. To add the chemical, select *Yes* (Figure 8).

```
Go ahead ?

          NO
          YES
```

Figure 8.

> **NOTE:** If a chemical name is entered that already exists in the databank, an error message appears (Figure 9):

```
Databank is full, create another one

Strike any key to continue
```

Figure 9.

Although the message states that the “Databank is full”, that is not the case. Ignore the message and select a new compound name by hitting any key and redoing the *Add* sequence of steps (4.1.3–4.1.5).

4.1.6. Before continuing, the new chemical must be selected. Choose *Select*, then *Select from databank listing* (Figure 10).

```
Select chemical from databank
Select     Codes      ModDes     ModCon     ModTDP     Banks

  Select by name
  Select by ID
  Select from databank listing

  Show current selection

  Exit
```

Figure 10.

As the name is typed in, the databank will scroll to the correct entry (Figure 11).

```
                    Full Databank Listing
 PROPARGYL ALCOHOL  ::304
 PROPARGYL CHLORIDE  ::553
 PROPIONIC ACID  ::381
 PROPIONIC ANHYDRIDE  ::425
 PROPIONITRILE  ::725
 PROPYLENE  ::85
 PROPYLENEIMINE  ::705
 PSEUDO-1  ::1040
 PYRENE  ::201
 PYRIDINE  ::672
 PYROMELLITIC ACID  ::420
 PYRROLE  ::706
 PYRROLIDINE  ::707
 PYRUVIC ACID  ::1180
 QUINALDINE  ::693
 QUINOLINE  ::692
 RCA-3  ::1236
[1076/1250]                                  Select a Compound
```

Figure 11.

NOTE: The new chemical has been assigned a “chemical identification number” (ID number) by *SuperChems*. It appears next to the databank entry (1236 in this case). Copy this number as it is important and will be needed for the steps that follow.

Select the entry with <enter>, and then hit any key to confirm the selection (Figure 12).

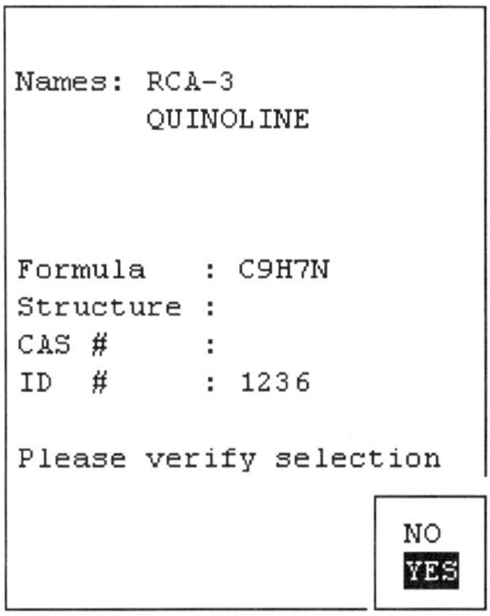

Figure 12.

To be sure that you have selected the correct chemical, check that its name appears in the lower right corner of the *Select* pulldown menu (Figure 13).

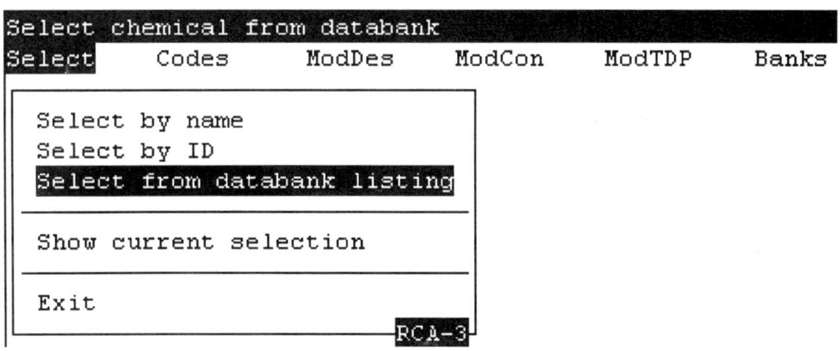

Figure 13.

4.1.7. Continue with Section 4.2

4.2. Molecular Weight and Normal Boiling Point

4.2.1. Enter the molecular weight by first highlighting *Calculator* under the main *Files* pulldown menu (Figure 14).

```
General purpose utilities and file management
Files      Define     PreLoad     Execute     Projects

  Directories
  File manager  (CTRL-F)
  DOS command   (CTRL-X)
  Archives
  ------------------------
  Autosave core image is ON
  Record output is ON
  View output
  Erase output
  ------------------------
  Units conversion
  Notepad
  DataXchange
  ------------------------
  Speed/math benchmarks
  ------------------------
  Calculator
```

Figure 14.

Then highlight *Molecular Weight* (Figure 15).

```
  Molecular weight
  ------------------------
  One numerical expression
  Two numerical expressions
  Three numerical expressions
  Four numerical expressions
```

Figure 15.

Enter the molecular formula of the new compound. The entries are case sensitive; be sure to enter chlorine as Cl, not CI for example (Figure 16).

```
Enter Chemical Formula (Case Sensitive) [s]  [C9H7N                    ]
```

Figure 16

Accept the entry with <F10>. The calculated molecular weight will be displayed (Figure 17).

```
Number of N,H,C atoms = 1, 7, 9
C9H7N Molecular weight ::129.1610

Entry has been added to clipboard. Press F9 for available options

Strike any key to continue
```

Figure 17.

4.2.2. Confirm that the new chemical has been selected as described in step 4.1.6. The name of the chemical should appear in the lower-right corner of the pulldown menu.

4.2.3. Unlock the chemical entry. Select *Banks* then *Unlock* from the pulldown menu (Figure 18).

```
Database functions
Select     Codes     ModDes     ModCon     ModTDP     Banks

                                                      Add
                                                      Remove
                                                      Save
                                                      Lock
                                                      Unlock
                                                      Duplicate

                                                      Create
                                                          RCA-3
```

Figure 18.

Enter the password; remember that it is case-sensitive (Figure 19).

```
 Verify Password
Password  [HAZMAT-1                                    ]
```

Figure 19.

Accept the entry with <F10>.

NOTE: Even if the chemical was defined with no password (or "public"), this step must still be done. Leave the password line blank and press <F10>. Do *not* type the word "public."

4.2.4. Select *ModCon*, then *Molecular weight* from the pulldown menu (Figure 20).

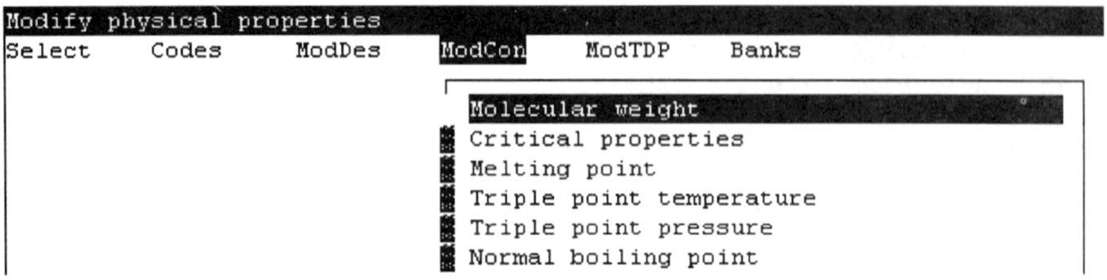

Figure 20.

4.2.5. After selecting the molecular weight screen, open the clipboard by pressing <F9> (Figure 21).

Figure 21.

Select paste from the clipboard (Figure 22).

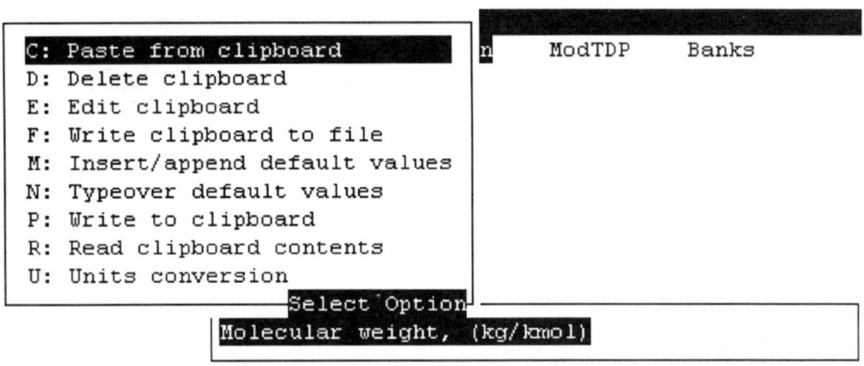

Figure 22.

The molecular weight calculated for the formula entered in step 4.2.1 should be highlighted (Figure 23).

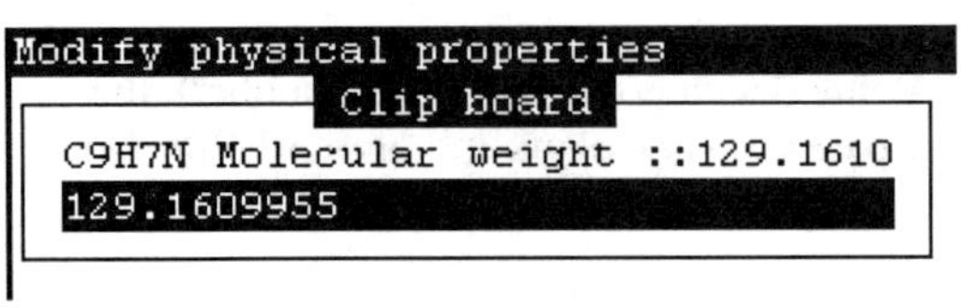

Figure 23

Press <F10> enter this value for the molecular weight. Press <F10> a second time to accept the value.

4.2.6. Enter the normal boiling point, NBP (boiling pt at 1 atm) by repeating step 4.2.4 and selecting *Normal boiling point* (Figure 24).

```
Normal boiling point at 101325 Pa, (K)   [510.75          ]
Quality code                             ►11100           ◄
```

Figure 24.

Note that the boiling point is entered in Kelvin. Alternatively, a Celsius temperature can be entered and converted pressing <F9>, then *Units conversion* <enter>, then *From Celsius* <enter> and *To Kelvins* <enter>. A 5-digit quality code is required for the boiling point entry to be recognized by *SuperChems*. A typical code for a boiling point is 11100. While these codes are not used by *SuperChems* in any calculations, the data (e.g. boiling point) will not be available for a simulation if they are absent. For an explanation of the codes, see the *SuperChems* Users Guide. If the NBP is not known, see Section 7, Estimating Boiling Point.

4.2.7. You must save the data before exiting the "modify physical properties" menu. Select *Banks* then *Save* from the pulldown menu (Figure 25).

```
Database functions
Select     Codes     ModDes     ModCon     ModTDP     Banks
                                                      Add
                                                      Remove
                                                      Save
                                                      Lock
                                                      Unlock
                                                      Duplicate
                                                      ----------
                                                      Create
                                                           RCA-3
```

Figure 25.

4.2.8. It is suggested that the data entry be locked. Select *Lock* from the *Banks* pulldown menu, and enter the password as prompted. It may be the same as previously used, or a new one. Accepting a blank line will change the password to "public."

CAUTION: The password entry is accepted as typed, without further prompting. In other words, an error in typing can change the password. Be careful!.

4.2.9. Continue with Section 4.3.

4.3. Enthalpy and Gibbs Free Energy of Formation; Ideal Gas Heat Capacity

Enthalpy (heat) of formation and Gibbs free energy of formation values can be entered from an external source, such as CHETAH (ASTM computer program for chemical thermodynamic and energy release evaluation) or MOPAC semiempirical methods, by following the steps described for entering molecular weight and normal boiling point (Section 4.2). This should be done *after* the ideal gas heat capacity is calculated, as described below. Select *Ideal gas heat of formation* to enter the enthalpy of formation and *Ideal gas free energy of formation* to enter the Gibbs free energy of formation in the *ModCon* pulldown menu. As with NBP, quality codes are necessary. For an explanation of the codes, see step 4.2.6.

Alternatively, the enthalpy and Gibbs free energy of formation can be estimated by *SuperChems*. The Joback group contribution method, described in the introduction (Section 1) is used.

The enthalpy and Gibbs free energy of formation (at 298 K), as well as the ideal gas heat capacity (as a function of temperature), are calculated at the same time by *SuperChems*.

4.3.1. From the main menu select *ChemProps*, then *Properties estimation* from the pulldown menu (Figure 26).

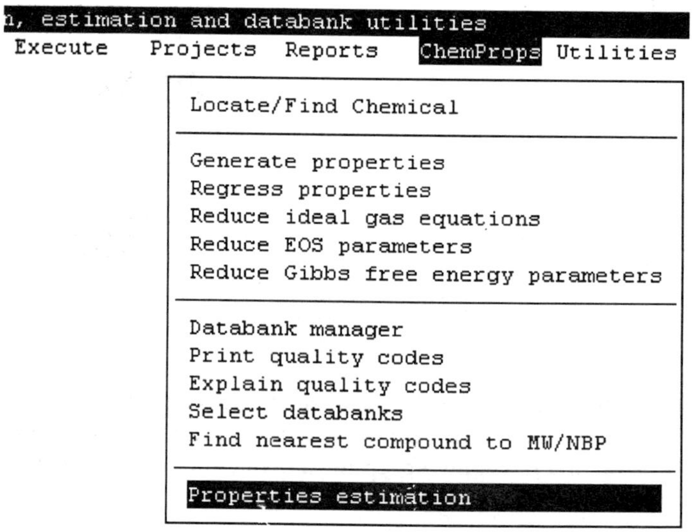

Figure 26.

4.3.2. Select *ThermoPhys*, then *Ideal gas Cp, Hf, and Gf* from the pulldown menu (Figure 27).

```
Estimate thermodynamic and physical properties
Files        CritProp    ThermoPhys SurfaceT   Viscosity

                             Normal boiling point
                             Nromal freezing point
                             ---------------------------
                             Liquid density
                             Vapor/gas density
                             Vapor pressure
                             Latent heat of vaporization
                             ---------------------------
                             Ideal gas Cp, Hf, and Gf
                             ---------------------------
                             Enthalpy
                             Entropy
```

Figure 27.

4.3.3. Select *Values*. If a group contribution estimation has been done for the desired compound (for estimating a boiling point, for example), select *Current* from the pulldown menu to confirm the correct entries and skip to step 4.3.6. Otherwise, select *Edit* (Figure 28).

```
Ideal gas thermodynamic properties
Files     Inputs     Options   Other

            Current
            Edit
            Run
              DEFAULT
```

Figure 28.

4.3.4. highlight *Set group input values to zero*, then press <enter> to clear previous values (Figure 29).

```
Ideal gas thermodynamic properties [DEFAULT] F
Files     Inputs     Options    Other

           Set group input values to zero
           Specify carbon groups (non-ring)
                   carbon groups in rings
                   oxygen groups
                   halogen groups
                   nitrogen groups
                   sulfur groups
           Return to previous menu
```

Figure 29.

4.3.5. Identify each of the groups present in the molecule. As an example, quinoline would have the following entries (Figures 30 and 31):

Quinoline:

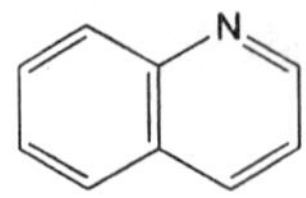

Carbon groups in rings:

```
Numbers of -CH2-           [0
           >CH-            [0
           >C<             [0
           =CH-            [7
           =C<             [2
```

Figure 30.

Nitrogen groups:

```
Numbers of -NH2                [0          ]
           >NH (non-ring)      [0          ]
           >NH (ring)          [0          ]
           >N- (non-ring)      [0          ]
           -N= (non-ring)      [0          ]
           -N= (ring)          [1          ]
           =NH                 [           ]
           -CN                 [           ]
           -NO2                [           ]
```

Figure 31.

Entering a 0 has the same effect as leaving an entry blank. After each set of entries has been selected (with <F10>), select ***Return to previous menu***.

> **NOTE:** The group contribution entries made above are saved until they are written over or *SuperChems* is exited. Thus the estimation can be rerun by skipping steps 4.3.4 and 4.3.5 above.

4.3.6. Select ***Run*** from the ***Values*** pulldown menu (see step 4.3.3 above). Viewing the output file will reveal that the Enthalpy (heat) of formation, Gibbs free energy of formation, *and* the ideal gas heat capacity (as a function of temperature) have all been calculated (Figure 32).

```
Super.lis - Notepad
File  Edit  Search  Help

THERMODYNAMIC PROPERTIES OF IDEAL GASES

At 298 K and 1 atmosphere:

Enthalpy of formation     =       2.313E+08 J/kmol
Gibbs energy of formation =      3.2101E+08 J/kmol
Cp(T) = -60580.000 +809.960 T -6.430e-01 T^2 +2.043e-04 T^3 J/(kmol-K)

 Predicted Ideal Gas Heat Capacity
    As A Function of Temperature
+-----------------------------------+
|#    | T              | Cp           |
+-----------------------------------+
|0    |            300 |     1.3E+05 |
|1    |            400 |   1.736E+05 |
|2    |            500 |   2.091E+05 |
|3    |            600 |    2.38E+05 |
|4    |            700 |   2.614E+05 |
|5    |            800 |   2.804E+05 |
|6    |            900 |   2.965E+05 |
|7    |           1000 |   3.107E+05 |
|8    |           1100 |   3.242E+05 |
|9    |           1200 |   3.385E+05 |
|10   |           1300 |   3.545E+05 |
|11   |           1400 |   3.737E+05 |
|12   |           1500 |   3.971E+05 |
+-----------------------------------+

T:Temperature (K) - Cp:Ideal gas heat capacity (J/kmol/K)
```

Figure 32.

4.3.7. Exit from the text viewer and select *Other* to export the correlation coefficients to the databank (Figure 33).

```
Ideal gas thermodynamic properties [DEFAULT] Miscella
Files       Inputs      Options     Other

                                    Export to databanks
```

Figure 33.

Enter the ID number of the chemical whose properties are being saved. The ID number was obtained in step 4.1.6 (Figure 34).

```
Chemical Identification Number                                     [1236    ]
Ideal gas enthalpy of formation at 298.15 K (J/kmol)               [23130000]
Ideal gas gibbs free energy of formation at 298.15 K (J/kmol)      [32101000]
A                                                                  [-60580  ]
B                                                                  [809.96  ]
C                                                                  [-0.64298]
D                                                                  [0.000204]
```

Figure 34.

4.3.8. Confirm that the correct chemical has been selected, and respond *Yes* to the query (Figures 35 and 36).

```
Names: RCA-3
       QUINOLINE



Formula    : C9H7N
Structure  :
CAS #      :

Strike any key to continue
```

Figure 35.

```
Go ahead ?
         NO
         YES
```

Figure 36.

4.3.9. Enter the correct password for the chemical. The prompt will appear twice (Figure 37).

```
Verify Password
Password   HAZMAT-1
```

Figure 37.

You will then be prompted for a "New password". The same, or a different one, can be entered. If the password line is left blank, the entry becomes "public." You will only be prompted once for the new password, so enter it carefully.

4.4. Estimate Critical Properties: T_c, P_c, V_c, Z_c

4.4.1. Critical Temperature

a. From the *Properties estimation* menu (see step 4.3.1), select *CritProp*, then *Joback, T, pure compound* (Figure 38).

As discussed in the Introduction (Section 1), the Joback group contribution method is the easiest to use and has percent errors comparable to the other methods.

b. Using the Joback method, the menu selection is the *same* as that employed in Section 4.3. The groups entered for that step remain, and can be viewed for confirmation by selecting *Current* (Figure 39).

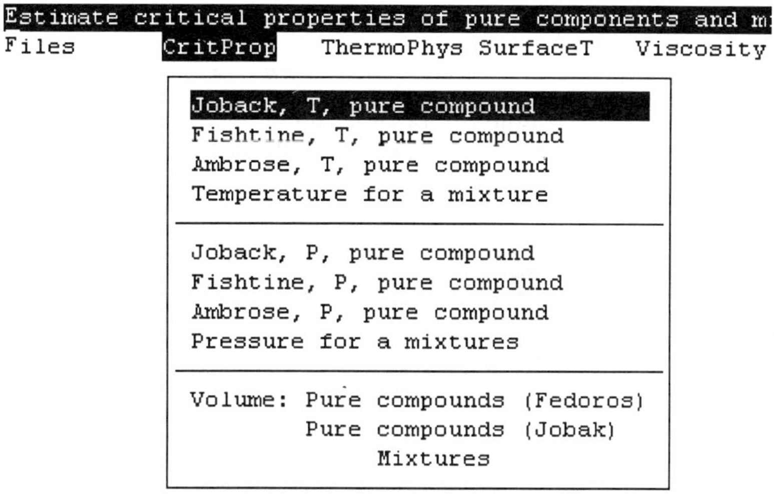

Figure 38.

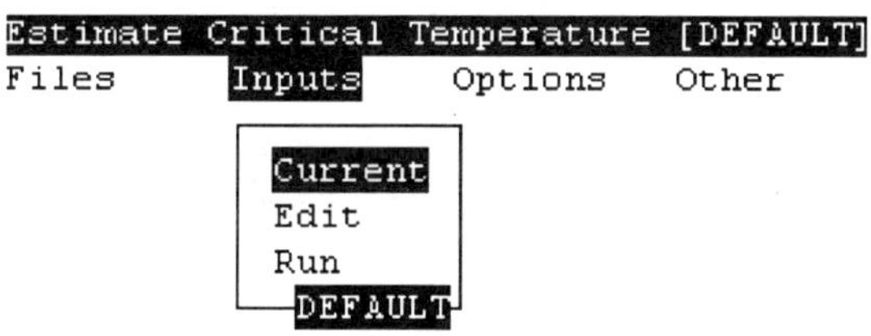

Figure 39.

If the entries are not correct, or have been deleted (by exiting *SuperChems* for example), follow steps 4.3.3 – 4.3.5.

c. Select *Run*

d. Enter the normal boiling point (from step 4.2.6) (Figure 40).

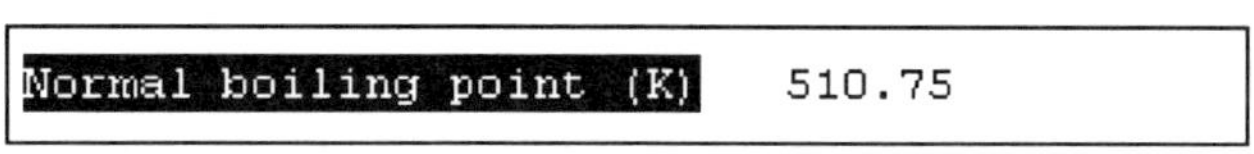

Figure 40.

e. The result is displayed at the end of the output file; make note of it for later use. Alternatively the result can be viewed in NOTEPAD as described in *SuperChems* Hints (Section 3). Close the output file (*File*, then *Exit* if using the DOS text editor). Export the result to the databank as described in steps 4.3.7–4.3.9.

4.4.2. Critical Pressure

a. Repeat steps 4.4.1a– c above, selecting *Joback, P, pure compound* in step a.

b. Enter the total number of atoms in the molecule, from the molecular formula, when prompted. For example quinoline is C_9H_7N, and the entry is 17 (Figure 41).

```
Number of atoms in molecule    17
```

Figure 41.

Accept the entry by pressing <F10>.

c. View the result and export it as in step 4.4.1 e above.

4.4.3. Critical Volume

a. Repeat steps 4.4.1a–c above, selecting *Volume: Pure compounds (Joback)* in step a. View the result and export it as in step 4.4.1 e above.

4.4.4. Critical Compressibility Factor

a. From *Files* in the *Properties estimation* menu, select *Calculate* Z_c (Figure 42).

```
Input/output management
Files       CritProp   ThermoPhys SurfaceT

  File manager (CTRL-F)
  DOS command  (CTRL-X)
 ------------------------------------------
  Record output
  Do not record output
  View output
  Erase output
 ------------------------------------------
  Quit
 ------------------------------------------
  Calculate vapor pressure using Tc/Tb
  Calculate Zc using Pc, Vc, Tc
  Calculate Vc using Pc, Zc, Tc
```

Figure 42.

b. Select *Edit* from the *Values* pulldown menu (this estimation does not use the Joback groups) (Figure 43).

```
Calculate Critical Compress
Files     Inputs    Options

          Current
          Edit
          Run
           DEFAULT
```

Figure 43.

c. Enter the critical properties determined in steps 4.4.1 through 4.4.3 (Figure 44). If the values were not copied as determined, they can be viewed by selecting *View Output* in the main *SuperChems Files* pulldown menu.

```
Critical volume (m3/kmol)   [0.4025          ]
Critical temperature (K)    [766.6           ]
Critical pressure (Pa)      [4440000         ]
```

Figure 44.

d. Export the result (Z_c) as described above in step 4.4.1e.

4.5. Determine Vapor Pressure Curve

The steps described below establish the vapor pressure curve for the compound being added to the databank. It is necessary to have one data point—either the normal boiling point (NBP, boiling point at 1 atm), or a reduced pressure boiling point. If neither is available, a boiling point can be estimated by following the steps in Section 7, Estimate Boiling Point.

4.5.1. From *Files* in the *Properties estimation* menu, select *Calculate vapor pressure* (Figure 45).

```
Input/output management
Files      CritProp   ThermoPhys SurfaceT

 File manager  (CTRL-F)
 DOS command   (CTRL-X)
 ----------------------------------------
 Record output
 Do not record output
 View output
 Erase output
 ----------------------------------------
 Quit
 ----------------------------------------
 Calculate vapor pressure using Tc/Tb
 Calculate Zc using Pc, Vc, Tc
 Calculate Vc using Pc, Zc, Tc
```

Figure 45.

4.5.2. If a normal boiling point was entered for the compound in step 4.2.6 and critical properties estimated in Section 4.4, those values will appear. Accept them by pressing <F10>. If the entries are blank, continue as follows:

From *Inputs*, select *Edit* in the pulldown menu. Enter the boiling point data on the first two lines. To convert units (from mmHg to Pascals, for example), enter the value, press <F9>, highlight *U: Units conversion* (press <enter>), then *Pressure* (press <enter>) then *Millimeter of mercury* in the *From* pulldown menu (press <enter>), then *Pascals* in the *To* pulldown menu (press <enter>). If the boiling point data are at 1 atm (normal boiling point), enter a pressure of 101325 Pa.

Enter the critical properties (T_c, and P_c) estimated in Section 4.4 on the last two lines (Figure 46).

```
Boiling point (K)                          [ 10.75         ]
Vapor pressure at boiling point (Pa)       [101325         ]
Critical temperature (K)                   [766.6          ]
Critical pressure (Pa)                     [4414000        ]
```

Figure 46.

4.5.3. Select *Run* from the *Inputs* pulldown menu. The output file will be displayed. Close it (*File*, then *Exit* if using the DOS editor).

4.5.4. Export the result to the databank as described in steps 4.3.7–4.3.9.

> **NOTE:** The preceding steps establish the vapor pressure curve between the boiling data point and the critical point of the compound. It is advisable, but not necessary, to extrapolate the curve below the boiling point in order to run vaporization scenarios that include temperatures below that point. This can be accomplished by the following steps:

4.5.5. From the main *SuperChems* menu, select *ChemProps*, then *Generate properties*. Select the chemical (see Section 8, Generate Properties, step 8.2), then highlight *Evaluate* and select *Vapor pressure* from the pulldown menu.

4.5.6. Enter the desired number of data points (e.g. 25, 30, or 35), the minimum temperature (250 K is suggested) and an increment so the maximum temperature will be near the critical point. For example, the following entries will generate the vapor pressure curve from 250 K to 775 K in 15 K intervals (Figure 47).

```
Number of temperature points   [35          ]
Minimum temperature, (K)       [250         ]
Temperature increment, (K)     [15          ]
```

Figure 47.

Press <F10> to execute the calculation. Press YES to the query "Examine output file," and print or copy the results. Press NO to "Display evaluated properties...."

4.5.7. In order to be exported to the databank, the new range of vapor pressure values must be regressed. Select *Regress properties* from the *ChemProps* pulldown menu (Figure 48).

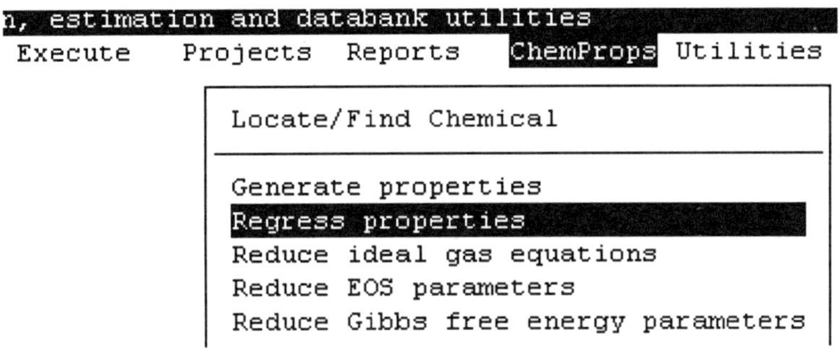

Figure 48.

4.5.8. Select *New data set* from the *Edit* pulldown menu (Figure 49).

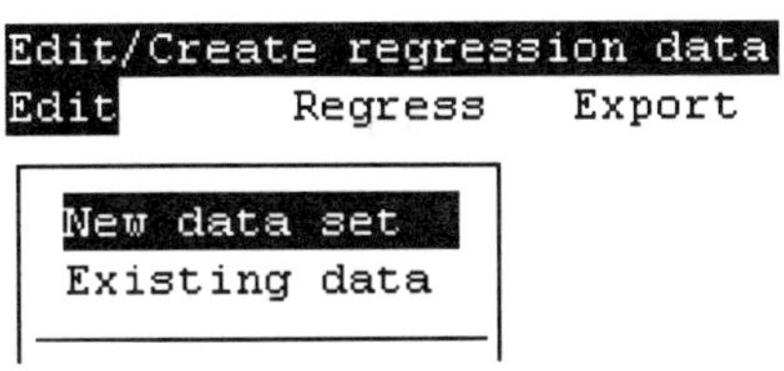

Figure 49.

Respond *Yes* to the query "Overwrite existing worksheet."

4.5.9. The format of the regression data is an entry giving the number of points to be regressed (35 in this case) followed by the data in two columns: temperature (in Kelvins) in the first column and pressure (in Pascals) in the second. The columns are separated either by TAB or spaces. If the DOS editor was chosen as the *SuperChems* text editor (see Section 3, *Superchems* Hints), the data can be edited in "SUPER.LIS" and saved as "MBANK003.001". The output file can also be edited using Excel; however, *SuperChems* will need to be exited to use this method. Alternatively, the data can be entered manually into the file MBANK003.001 using the format described (Figure 50).

```
Mbank003.001 - Notepad
File  Edit  Search  Help
35
250     0.7643
265     2.826
280     9.085
295     25.93
310     66.88
325     158
340     346.1
355     709.6
370     1372
385     2521
400     4425
415     7457
430     12118
445     19057
460     29099
475     43259
490     62768
505     89083
520     1.24E+05
535     1.69E+05
550     2.27E+05
565     3.00E+05
580     3.91E+05
595     5.02E+05
610     6.38E+05
625     8.01E+05
640     9.94E+05
655     1.22E+06
670     1.49E+06
685     1.80E+06
```

Figure 50.

4.5.10. Select *Equation 101* from the *Regress* pulldown menu (Figure 51).

```
Estimate parameters for selected equation
Edit         Regress     Export

               Equation 100 - linear
               Equation 100 - quadratic
               Equation 100 - cubic
               Equation 100 - 4th order
               Equation 101
               Equation 102
```

Figure 51.

The equation used by *SuperChems* for calculating vapor pressures is of the form:

$$\ln(Y) = A + B/T + C\ln(T) + DT^{\wedge}E$$

The regressed data can be displayed in graphical form if desired. This is an optional step (Figure 52).

```
Select graph(s); press SPACE bar to t
Files        Select      Modify     Mode

                REGRESSION DATA
                ACTUAL VS. PREDICTED
                RELATIVE ERROR
               T——Press F10 to display
```

Figure 52.

See GENERATE PROPERTIES, Section 8.5, for more information on graphical display of data.

Select *Files*, then *Exit* (Figure 53).

```
Store to disk or load from disk
Files      Select     Modify

  Store Bitmap
  Load Bitmap
  Store Postscript
  Exit
```

Figure 53.

4.5.11. Export the regressed data by selecting *Vapor pressure* from the *Export* pulldown menu (Figure 54).

```
Export current regression parameters to databanks
Edit       Regress    Export

                        Solid density (kmol/m3)
                        Liquid density (kmol/m3)
                        Vapor pressure (Pa)
                        Heat of vaporization (J/kmol)
                        Solid heat capacity (J/kmol/K)
```

Figure 54.

4.5.12. Enter the ID number of the chemical and a quality code. This is required by *SuperChems*. The code for a calculated value is 11 (Figure 55).

```
Chemical Identification Number                ►1236            ☺
Equation number                               ►101             ☺
Quality code                                  ►11              ◄
Minimum correlation temperature, (K)          [250             ]
Maximum correlation temperature, (K)          [760             ]
A                                             [22.87099079     ]
B                                             [-5777.456772    ]
C                                             [-0.005458815724]
D                                             [3.604429834E-12]
E                                             [3.056665509     ]
F                                             [                ]
```

Figure 55.

4.5.13. Continue with the export process as in previous steps (see steps 4.3.8 and 4.3.9).

4.6. Reduce Ideal Gas Equations

This step calculates the temperature dependent ideal gas enthalpy and entropy equations from the ideal gas heat capacity equation (generated in Section 4.3). These equations are used by the *SuperChems* temperature dependent property generator.

In the main *SuperChems* menu, select *ChemProps* and *Reduce ideal gas equations* in the pulldown menu (Figure 56).

```
generation, estimation and databank utilities
PreLoad    Execute    Projects    Reports    ChemProps  Utilities

                      Locate/Find Chemical
                      ----------------------------------
                      Generate properties
                      Regress properties
                      Reduce ideal gas equations
                      Reduce EOS parameters
                      Reduce Gibbs free energy parameters
                      ----------------------------------
```

Figure 56.

4.6.2. Enter the ID number of the chemical being added to the databank and an ID number one greater. Residual tables are optional; select 0 or 1 (Figure 57).

```
Starting compound record id #          [1236      ]
Ending compound record id #            [1237      ]
Show residuals table, (0-No 1-Yes)     [0         ]
```

Figure 57.

4.6.3. Respond to the password prompts to complete the process. Remember that passwords are case sensitive. The "old" and "new" passwords may be the same, or you may enter a different "new" password.

4.7. Reduce Equation of State Parameters; Regress Liquid Density, Latent Heat of Vaporization, and Liquid Heat Capacity

4.7.1. In the main *SuperChems* menu, select *ChemProps*, then *Reduce EOS parameters* in the pulldown menu (Figure 58).

```
n, estimation and databank utilities
 Execute    Projects   Reports    ChemProps  Utilities

              Locate/Find Chemical
              ------------------------------------
              Generate properties
              Regress properties
              Reduce ideal gas equations
              Reduce EOS parameters
              Reduce Gibbs free energy parameters
```

Figure 58.

4.7.2. Enter the ID number of the chemical being added, a number one greater, and a choice for the "Option for temperature form". For estimated properties, select "0" (Figure 59).

```
Starting record                                           [1236        ]
Ending record                                             [1237        ]
Option for temperature form: 0- Tr, 1- sqrt(Tr), 2- Both  [0           ]
```

Figure 59.

4.7.3. Respond to the password prompts. Follow the recommendation of the subsequent screen and set the density translation factor coefficients to zero (Figure 60).

```
A lower bound   [0              ]
A upper bound   [0              ]
B lower bound   [0              ]
B upper bound   [0              ]
C lower bound   [0              ]
C upper bound   [0              ]
```

Figure 60.

After the calculations are completed, you will be given the choice to display the results graphically. For more information of how to do this, see Section 8.5 (Figure 61).

```
Select graph(s); press SPACE bar to toggle ON/OFF and ENTER t
Files      Select      Modify     Mode       Layout

             LATENT HEAT OF VAPORIZATION VS. TEMPERATURE
             LIQUID MOLAR DENSITY VS. REDUCED TEMPERATURE
             SATURATION PRESSURE VS. TEMPERATURE
          T--------------------------------Press F10 to display
```

Figure 61.

Skip this step by highlighting *Files*, then selecting *Exit* (Figure 62).

```
Store to disk or load from disk a picture
Files        Select       Modify       Mode

  Store Bitmap
  Load Bitmap
  Store Postscript
  Exit
```

Figure 62.

4.7.4. Accept the entries with <F10> (Figure 63).

```
Chemical Identification Number                    [ 236          ]
Density translation factor - Delta (kmol/m3)      [              ]
Density translation factor - A                    [              ]
Density translation factor - B                    [              ]
```

Figure 63.

4.7.5. Confirm the selection of the correct chemical (from the ID number displayed) and respond *Yes* to the query "Go ahead?". Note that you will be prompted for the password a second time to complete the process.

A file has been created: EOSPROPS.LIS that contains estimated values for liquid density, latent heat of vaporization, and liquid heat capacity. These are each regressed, using specific equation forms, and exported to the databank. Vapor pressure data are also included; however, this was regressed in Section 4.5 and should not be repeated here. If these data are compared to the previous vapor pressure data, note that the vapor pressures obtained in this calculation are listed in atmospheres while the previous data are in Pascals.

The procedure for regression was described in steps 4.5.7–4.5.12 for vapor pressure. The procedure is the same except that the values to be regressed are in the file EOSPROPS.LIS, not SUPER.LIS. Using the DOS text editor (or Windows Notepad) the values for each parameter can be saved in the file MBANK003.001 and regressed with the appropriate equation shown below. Notice that the values in EOSPROPS.LIS are in the correct format, and can simply be "cut and pasted" into MBANK003.001; no further editing is needed.

Following regression, the results should be exported as described in steps 4.5.11 and 4.5.12, using the appropriate selection from the *Export* pull-down menu. The quality code = 11 should be used for all entries.

Liquid density:	Regress with equation 105
Liquid heat capacity:	Regress with equation 100 (linear form)
Latent heat of vaporization:	Regress with equation 106

The equations used by ***SuperChems*** for regression are (Figure 64):

```
Edit/Create regression data
Edit        Regress    Export


Equation 100:     Y   = A + B T + C T^2 + D T^3 + E T^4
         101:ln(Y)  = A + B/T + C ln(T)  + D T^E
         102:     Y   = (A T^B) /  (1 + C/T + D/T^2)
         103:     Y   = A + B exp  (- C/T^D)
         104:     Y   = A + B/T + C/T^3 + D/T^8 + E/T^9
         105:     Y   = A / B^(1 +  (1-T/C)^D)
         106:     Y   = A *  (1-Tr)^(B + C Tr + D Tr^2 + E Tr^3)
         107:     Y   = A + B  (C/T/sinh(C/T))^2 + D  (E/T/cosh(E/T))^2

Equation 200:   Y = (D / A)  / B^(1+ (1-Tr)^C)
         201:   Y = A + B T + C T^2 + D T^3 + E T^4 + F ln(T)
         202:   Y = A + B T + C T^2 + D T^3 + E/T^2 + F ln(T)
         203:   Y = A / B^((1-Tr)^C)
         204:   Y = A + B T + C T^2 + D T^E

Y   = Property in correct units, T = Temperature in kelvins
Tr  = Reduced temperature, T/Tc
```

Figure 64.

Note that for the calculation of Latent heat of vaporization, the critical temperature is required. This value can either be entered manually or imported from the databank (Figure 65).

```
Specify value
Select  chemical from databank list
Specify chemical name
  Select Critical Temperature Value Option
```

Figure 65.

4.8. Reduce Gibbs Free Energy Parameters

NOTE: Execution of this step requires properties data for several substances in their elemental state. If the chemical has been added to a user-defined databank and not the *SuperChems* default databank, it must be duplicated before continuing. See Section 6, Duplicate a Chemical.

4.8.1. Select *Reduce Gibbs free energy parameters* from the *ChemProps* pull-down menu (Figure 66).

```
, estimation and databank utilities
Execute    Projects   Reports    ChemProps Utilities

           Locate/Find Chemical
           ------------------------------------
           Generate properties
           Regress properties
           Reduce ideal gas equations
           Reduce EOS parameters
           Reduce Gibbs free energy parameters
```

Figure 66.

4.8.2. Enter the ID number of the chemical being added and a number that is one greater (Figure 67).

```
Starting databank record id   [1236        ]
Ending databank record id     [1237        ]
```

Figure 67.

4.8.3. Respond to the password query.

THE PROCEDURE FOR ENTERING A CHEMICAL TO THE DATABANK WITH THE MINIMUM PROPERTIES REQUIRED BY *SUPERCHEMS* IS NOW COMPLETE.

It is recommended that the chemical be duplicated to a separate user-defined databank as a "back-up." This is optional, however. See Section 6, Duplicate a Chemical, for more information.

5. CREATE A NEW DATABANK

After a new chemical has been added to the default *SuperChems* databanks, it may be desirable to store the chemical and its properties in a user-defined databank. The steps below describe the creation of a new set of databanks. See also: Section 10, Select a Databank and Section 6, Duplicate a Chemical.

5.1. Select *ChemProps* in main menu, then *Databank manager* from pull-down menu (Figure 68).

```
Projects  Reports   ChemProps Utilities

  Locate/Find Chemical

  Generate properties
  Regress properties
  Reduce ideal gas equations
  Reduce EOS parameters
  Reduce Gibbs free energy parameters

  Databank manager
  Print quality codes
```

Figure 68.

5.2. Select *Banks*, then *Create* from pulldown menu (Figure 69).

```
Banks

  Add
  Remove
  Save
  Lock
  Unlock
  Duplicate

  Create
      BENZENE
```

Figure 69.

5.3. Enter number of records for the new databank. This number should be large (e.g. 50) as it will specify the maximum number of chemicals that can be put in the databank and can not be altered once the databank has been defined. Accept the number with <F10> (Figure 70).

```
Number of records    50
```

Figure 70.

5.4. Assign a name (8 characters or less) to the new databank. Two files are created, an Index databank and a Properties databank. Both should have the same name, although that is not required. Note that the file extensions are different. The default *SuperChems* databanks are located in the director SC31\BANKS. It is suggested that the *full DOS path* of the databanks be entered. Thus the format for the entries will be (Figure 71):

```
Index databank name          C:\SC31\BANKS\RCA-NEW.ID
Properties databank name     C:\SC31\BANKS\RCA-NEW.TPP
```

Figure 71.

Index databank:	C:\SC31\BANKS\[FILENAME].ID
Properties databank:	C:\SC31\BANKS\[FILENAME].TPP

5.5. Select *Yes* to query "Go ahead and create banks?" (Figure 72)

```
Number of records =  50
Index = C:\SC31\BANKS\RCA-NEW.ID
Properties = C:\SC31\BANKS\RCA-NEW.TPP

Go ahead and create banks ?

                                              NO
                                              YES
```

Figure 72.

CAUTION: If only file names (and not the DOS path) were entered in step 5.4, the databanks are in the subdirectory C:\SC3\SUPER. The default subdirectory for the *SuperChems* databank is C:\SC3\BANKS. It is recommended that they be moved. This can be done with File Manager (Windows 3.xx) or Explorer (Windows 95 or NT).

5.6. Before using the new databanks, they must be selected. Exit from Databank manager by choosing *Exit* from *Select* pulldown menu (Figure 73).

```
Select chemical from databank
Select      Codes       ModDes

  Select by name
  Select by ID
  Select from databank listing

  Show current selection

  Exit
```

Figure 73.

5.7. Continue with the steps outlined in Section 10, Select A Databank.

6. DUPLICATE A CHEMICAL

When a new chemical has been entered with its properties in the *SuperChems* default databanks, it may be desirable to duplicate it to a user-defined databank as a backup. If a user-defined databank has not been created, see Section

5, Create a New Databank. The *source* databank contains the chemical to be duplicated; the *target* databank is where the chemical will be copied *to*.

Before proceeding with the duplication procedure, an ID number must be assigned to a "dummy" chemical in the target databank. This ID number will then become the identification for the chemical being duplicated. The step is necessary because it is possible to overwrite, and thus delete, even a password-protected chemical. See the CAUTION on page 552.

6.1. Select the target databank (see Section 10, Select A Databank) and then carry out the steps in Section 4.1. A unique ID number will be assigned by *SuperChems* to the "dummy" chemical. To learn the value of the ID number, see step 4.1.6.

6.2. Select *ChemProps* in main menu, then *Databank manager* from pulldown menu (Figure 74).

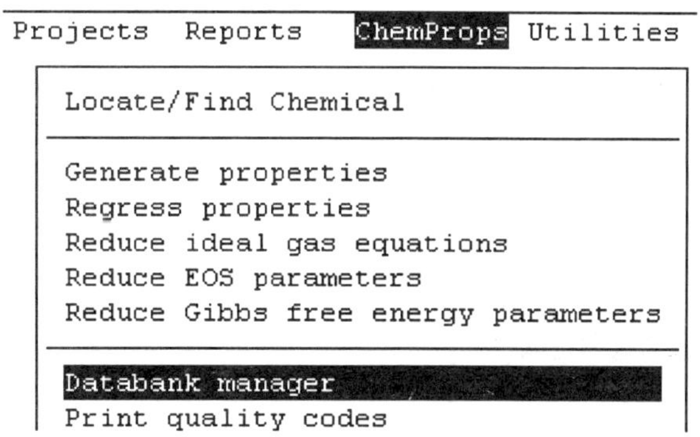

Figure 74.

6.3. Select *Banks*, then *Duplicate* from pulldown menu (Figure 75).

Figure 75.

6.4. Enter the names, including full DOS path, of the source and target databanks. The source will usually be the *SuperChems*-supplied databanks and the target will usually be the user-defined databanks. Thus the format for the entries will be (Figure 76):

Source Index bank:	C:\SC31\BANKS\IBANK.ID
Source Property bank:	C:\SC31\BANKS\TBANK.TPP
Target Index bank:	C:\SC31\BANKS\[FILENAME].ID
Target Property bank:	C:\SC31\BANKS\[FILENAME].TPP

```
Source Index bank      [C:\SC31\BANKS\IBANK.ID
Source Property bank   [C:\SC31\BANKS\TBANK.TPP
Target Index bank      [C:\SC31\BANKS\RCA-NEW.ID
Target Property bank   [C:\SC31\BANKS\RCA-NEW.TPP
```

Figure 76.

6.5. Accept the databank names with <F10>. Enter the "record identification number," that is, the ID number, of the chemical to be copied on the top line. On the lower line enter the ID number to be assigned to the chemical in the target databank. This is the number assigned in step 6.1 above. Accept the entries with <F10> (Figure 77).

```
Record identification number to copy from   [1236
Record identification number to write to    [10
```

Figure 77.

> **CAUTION:** The record identification number (ID number) of the target must *not* be previously assigned to an existing chemical in the database. It will be overwritten by the new chemical and the old chemical, with its properties, will have been deleted from the databank. This will occur whether a chemical is being duplicated to a user-defined databank or even to the default *SuperChems* databanks! Deletion of "password protected" chemicals is possible.

6.6. From the dialog box, select "Yes" to the query "Go ahead?" (Figure 78)

```
Go ahead ?
               NO
               YES
```

Figure 78.

7. ESTIMATING BOILING POINT

A normal boiling point (NBP), or boiling point at 1 atm (101325 Pa), can be estimated by *SuperChems* either from the structure alone or, preferably, from one or more reduced pressure boiling point or vapor pressure measurements. Experimental data should improve the accuracy of the resulting vapor pressure curve and should be used when available.

There are three options for *SuperChems* estimations, depending on what boiling point or vapor pressure data are available.

A. **If no boiling point or vapor pressure data are available,** do the steps in Section 7.1 *only*, then return to Section 4.2, Molecular Weight and Normal Boiling Point.

B. **If one data point is known**, either a reduced pressure boiling point or a vapor pressure measurement, do the steps in 7.1 and 7.2 below, then return to Section 4.2, Molecular Weight and Normal Boiling Point. Skip Section 4.5, Determine Vapor Pressure Curve, as it will have already been completed.

C. **If two or more data points are known,** do the steps in Section 7.3 only, then return to Section 4.2, Molecular Weight and Normal Boiling Point. Skip Section 4.5, Determine Vapor Pressure Curve, as it will have already been completed.

7.1. Estimating NBP from the Compound's Structure:

7.1.1. From the main *SuperChems* menu, select *ChemProps*, then *Properties estimation* (Figure 79).

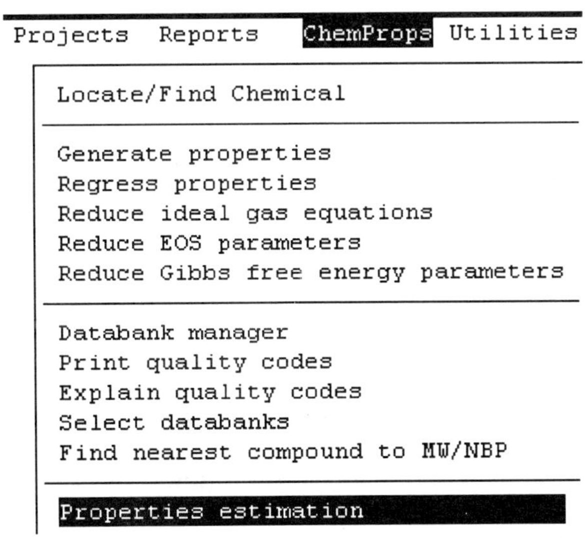

Figure 79.

7.1.2. From the *ThermoPhys* pulldown menu, select *Normal boiling point* (Figure 80).

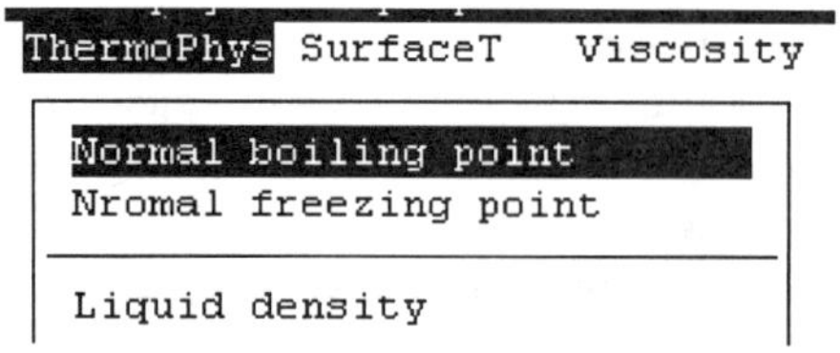

Figure 80.

7.1.3. The NBP is estimated by the Joback group contribution method. This method is used for estimating properties elsewhere in *SuperChems*, and the steps are described in more detail in Section 4.3, steps 4.3.3–4.3.5.

7.1.4. Select *Run* from the *Values* pulldown menu. The output file (SUPER.LIS) will appear, and the calculated boiling point is at the end. Copy the value and continue with Section 7.2 below or return to Section 4.2 (Figure 81).

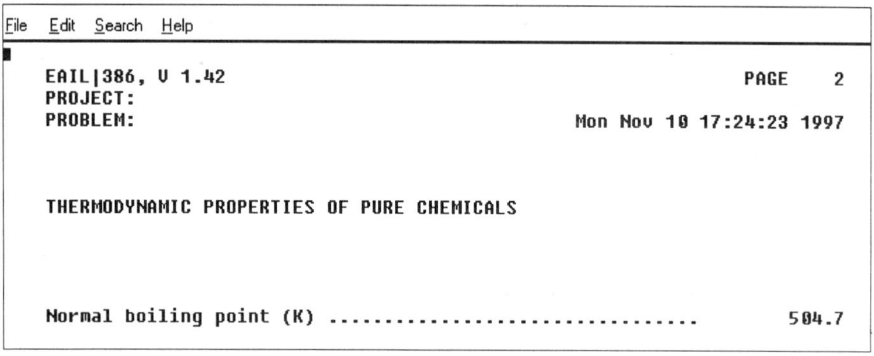

Figure 81.

7.2. If one data point is known, either a reduced pressure boiling point or a vapor pressure measurement, do the steps in Section 7.1 above and then continue with the steps listed below.

7.2.1. Estimate the critical temperature (T_c) and critical pressure (P_c) as described in Section 4.4. Using the Joback estimation method is recommended and will allow the groups entered in Section 7.1 above to be used without reentry.

> **NOTE:** While it is not necessary to estimate the critical volume (V_c) or compressibility factor (Z_c) in order to predict the NBP, doing these estimations at this time and exporting all the critical values to the databank will eliminate this step later.

7.2.2. From the *ThermoPhys* pulldown menu, select *Vapor pressure* (see step 7.1.2 above).

7.2.3. Select the *Method of Othmer and Yu*. This will allow predictions of vapor pressures using two temperature–pressure points as input. The method of Riedel requires the NBP as input (Figure 82).

```
Vapor Pressure of Pure Compounds
Method of Riedel
Method of Othmer and Yu
Return to previous menu
Press Enter to Select, Esc to Exit
```

Figure 82.

7.2.4. Select *Enter two experimental....* One of these data points will be the experimental boiling point or vapor pressure data, the other will be the critical temperature and pressure estimated above (Figure 83).

```
Method of Othmer and Yu: Required Input
Help
Enter two experimental vapor pressure data points
Return to previous menu
Press Enter to Select, Esc to Exit
```

Figure 83.

> **NOTE:** While the NBP estimated in Section 7.1 could be used as the second data point, greater accuracy of the vapor pressure curve will be obtained using the critical values.

7.2.5. Enter the experimental (known) reduced pressure boiling point (or vapor pressure measurement) as P1 and T1. Enter the estimated critical pressure and temperature as P2 and T2. Accept the entries with <F10>. Note that the pressures are in Pascals and the temperatures in Kelvins. To convert from other units, enter the number, press <F9>, highlight *U: Units conversion*, then highlight the desired conversion (temperature or pressure), and press <enter>. highlight (and press <enter>) the *From* units and the *To* units (Kelvins or Pascals) (Figure 84).

```
Experimental vapor pressure point, P1 (Pa)                  [1689
Experimental vapor pressure point, P2 (Pa)                  [4414000
Temperature corresponding to vapor pressure, P1 (K)         [375
Temperature corresponding to vapor pressure, P2 (K)         [766.6
```

Figure 84.

7.2.6. highlight *Estimate vapor pressure* and press <enter> (Figure 85).

```
Revise input for vapor pressure
Estimate vapor pressure of the compound
```

Figure 85.

7.2.7. The vapor pressure needs to be estimated at five temperatures; these values are used to regress the vapor pressure curve. Enter a temperature (in Kelvins) and accept with <F10>. A starting temperature of 300 K is recommended. Be sure to record the temperature since it is not included in the output file (Figure 86).

```
System temperature (K)   [300                    ]
```

Figure 86.

7.2.8. Copy the vapor pressure from the output file and press <F10> (Figure 87).

```
Estimate thermodynamic and physical properties
Files        CritProp     ThermoPhys SurfaceT   Viscosity


 Estimated vapor pressure =          27.4 Pa
```

Figure 87.

7.2.9. Repeat steps 7.2.4 and 7.2.5. The previously entered experimental vapor pressure points remain (Figure 84) and need not be reentered. Accept them with <F10>. Repeat steps 7.2.6 and 7.2.7, entering a system temperature 50 K greater than the previous one.

7.2.10. Repeat step 7.2.9 at least three more times, so a minimum of five temperature and vapor pressure points have been estimated.

7.2.11. Regress the data and export the result to the databank by following the steps outlined in Section 4.5, Determine Vapor Pressure Curve, steps 4.5.7 – 4.5.12 *only*. Data entry (step 4.5.9) is best done manually.

7.2.12. Use the newly regressed data to generate a vapor pressure curve for the compound. Follow the procedure outlined in Section 4.5, Determine Vapor Pressure Curve, steps 4.5.5 and 4.5.6. Information about displaying the result graphically can be found in Section 8, Generate Properties.

7.2.13. To determine the normal boiling point of the compound, interpolate the graph (or tabular output) of the preceding step to read the temperature corresponding to a pressure of 101325 Pa. This is best done by repeating step 7.2.9 for a temperature range near the expected NBP (where the pressure is about 1 × 10^5 Pa, using a temperature interval of 1 or 2 K).

7.2.14. Return to Section 4.2, Molecular Weight and Normal Boiling Point.

7.3. If two or more experimental data points are known (from boiling point or vapor pressure measurements), carry out steps 7.2.2–7.2.14 (omit step 7.2.1). In step 7.2.5 enter an experimental boiling point (or vapor pressure) measurement in place of the critical temperature and pressure.

8. GENERATE PROPERTIES FROM DATABANK

These steps will allow review of temperature independent properties and generation of temperature dependent properties of a chemical in the databank.

8.1. From the main *SuperChems* menu, select *ChemProps*, then *Generate properties* from the pulldown menu (Figure 88).

```
Projects   Reports   ChemProps Utilities
  Locate/Find Chemical
  --------------------
  Generate properties
  Regress properties
```

Figure 88.

8.2. Select the chemical by choosing *Select by name* or *Select from databank*. If the ID number is known, *Select by ID* can also be used. See Section 10, Select A Chemical (Figure 89).

```
Select chemical
Select     Evaluate

  Select by name
  Select by ID
  Select from databank listing
```

Figure 89.

8.3. Confirm the selection and choose *Evaluate* (Figure 90).

```
Evaluate physical constants and tempera
Select     Evaluate

             Physical properties
```

Figure 90.

8.4.

8.4.1. Select *Physical properties* to view the temperature independent properties of the chosen chemical (Figure 91).

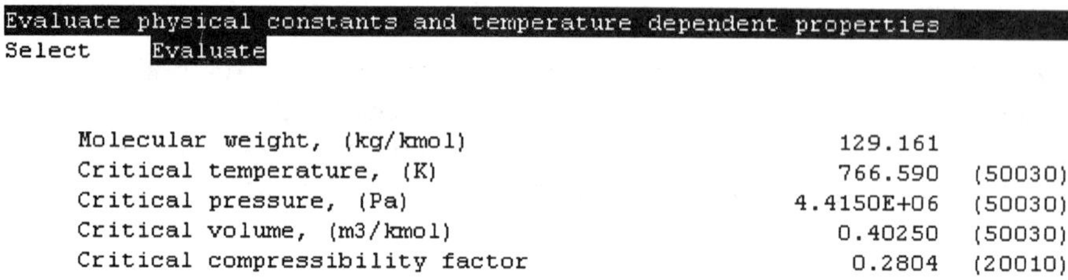

Figure 91.

A complete list of these properties can be found at the end of this section

NOTES: Properties that are missing (e.g., those that have not been estimated for a user-defined chemical) are given a default value of zero. The numbers in parentheses (e.g. 11100) represent quality codes for the data entry. An explanation can be found in the *SuperChems* User Guide.

8.4.2. To obtain a list of these properties, select "Yes" to the query *Examine output file* (Figure 92).

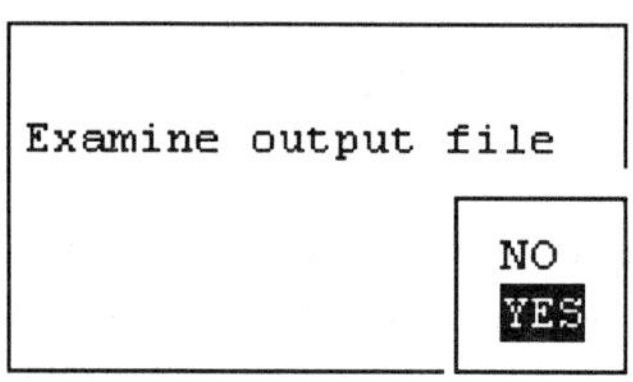

Figure 92.

The output file can be printed if either Windows Notepad or the DOS text editor has been selected as the *SuperChems* output format (Figure 93).

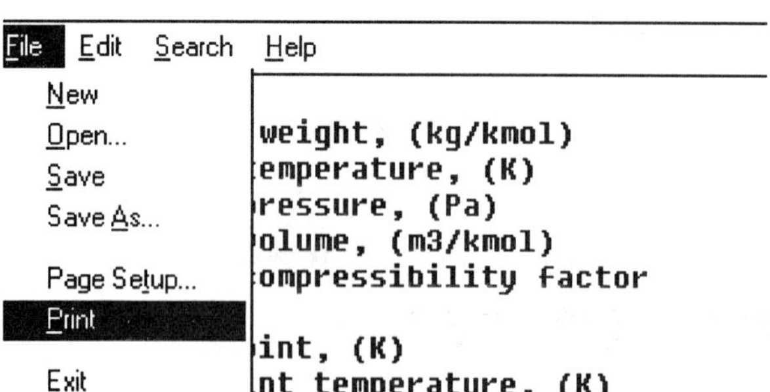

Figure 93.

8.4.2. Temperature independent properties can not be graphed; select "No" to the query (Figure 94).

```
Display evaluated properties in graph form
                                          NO
                                          YES
```

Figure 94.

8.5. The temperature ***dependent*** properties available in ***SuperChems*** are listed in Section 8.7. They are selected individually, and are calculated over a specified temperature range (Figure 95):

```
Number of temperature points   [11          ]
Minimum temperature, (K)       [300         ]
Temperature increment, (K)     [20          ]
```

Figure 95.

The default entries can be deleted using <ESC> and then replaced. The number of temperature points must be at least two. In order to view a property at a specific temperature, select that value as the ***Minimum temperature***.

The results can be viewed in tabular form in the output file (see step 8.4.2 above) or displayed graphically (select "Yes" to the query in step 8.4.3 above). Press the space bar; a "check" should appear next to the property to be displayed graphically (Figure 96).

```
Files      Select      Modify      Mode

             √ Vapor Pressure
            T-Press F10 to display
```

Figure 96.

To display a result graphically, ***SuperChems*** must be operating full-screen. If operating in a window, full-screen is obtained with by pressing <alt><enter>. Press <enter> or <F10> to display the graph. Press either key a second time to return to the previous screen.

The graph can be saved and printed on a PostScript printer. While still in full screen display, confirm that the property is "checked" (see preceding step), select ***Files***, then ***Store PostScript*** from the pulldown menu (Figure 97).

```
Files      Select

  Store Bitmap
  Load Bitmap
  Store Postscript
  Exit
```

Figure 97.

Enter a filename for the PostScript file with the extension PS (Figure 98).

```
Postscript File
Drive        ►C ◄
Path         ►SC31\SUPER
File name    ►FILENAME◄
Extension    ►PS ◄
```

Figure 98.

To print the graph, obtain a DOS prompt by typing <ctrl>X. Type:

COPY<space>/B<space>[FILENAME].PS<space>LPT1:<enter>

Return to *SuperChems* by typing EXIT. Return to a window display by pressing <alt><enter>

8.6. Temperature Independent Properties:

The following list includes all temperature independent properties available in the *SuperChems* default databank. The properties entered into the databank using the steps described in Section 4, Adding a Compound to the Databank, are indicated by an asterisk.

*Molecular weight, (kg/kmol)
*Critical temperature, (K)
*Critical pressure, (Pa)
*Critical volume, (m3/kmol)
*Critical compressibility factor
Melting point, (K)
Triple point temperature, (K)
Triple point pressure, (Pa)
*Normal boiling point at 101325 Pa, (K)
Reference liquid molar volume, (m3/kmol)
Reference temperature, (K)
*Ideal gas heat of formation at 25 C, (J/kmol)
*Ideal gas energy of formation at 25 C, (J/kmol)
Ideal gas absolute entropy at 25 C, (J/kmol/K)
Enthalpy of fusion at melting point, (J/kmol)
Standard net heat of combustion, (J/kmol)
Acentric factor
Radius of gyration, (m)
Solubility parameter, (J/m3) ^ (1/2)
Dipole moment, (C m)
van der Waals volume, (m3/kmol)
van der Waals area, (m2/kmol)
Refractive index
Flash point, (K)

Lower flammable limit (volume percent)
Upper flammable limit (volume percent)
Autoignition temperature, (K)
*EOS M Factor
*EOS N Factor
Equilibrium K, K1-Factor
Equilibrium K, K2-Factor
Equilibrium K, K3-Factor

8.7. *Temperature Dependent Properties*:

The following list includes all temperature dependent properties that may be calculated from the correlation coefficients stored in the *SuperChems* default databank. The properties whose equations were entered into the databank using the steps described in Section 4, Adding a Compound to the Databank, are indicated by an asterisk.

Solid density
*Liquid density
*Vapor pressure
*Heat of vaporization
Solid heat capacity
*Ideal gas heat capacity
*Liquid heat capacity
Second virial coefficient
Liquid viscosity
Liquid thermal conductivity
Vapor thermal conductivity
Solid thermal conductivity
Surface tension
Ideal gas enthalpy
Ideal gas entropy
Ideal gas internal energy

9. SELECT A CHEMICAL

9.1. To find a chemical in the *SuperChems* databanks, select *Locate/Find Chemical* from the *ChemProps* pulldown menu (Figure 99).

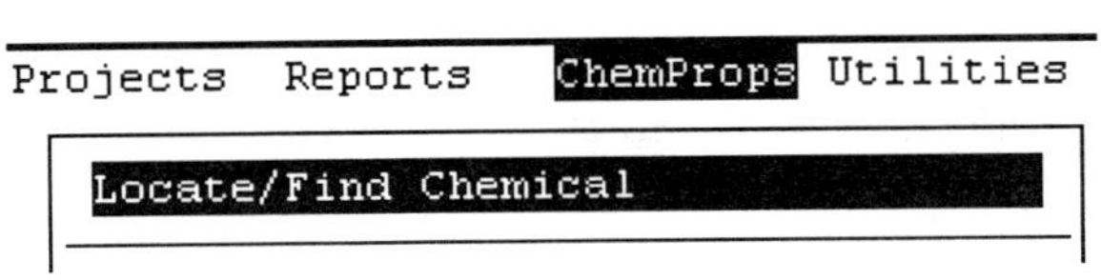

Figure 99.

9.1.1. The best methods for locating a chemical in *SuperChems* are searching by *Name, Formula,* or *CAS #.* highlight the desired method and press <enter> (Figure 100).

```
Name
Formula
Structure
First alias
Second alias
Third alias
CAS #
          Select Sort Option
```

Figure 100.

9.1.2. Using *Name*, the listing scrolls as each letter is typed. If no entries are present for the letter typed, the list will not scroll (Figure 101).

```
                 Full Databank Listing
 4-METHYLPYRIDINE  ::676
 5-ETHYL-m-XYLENE  ::1089
 5-METHYL-2-HEXANONE  ::280
 5-METHYLNONANE  ::1100
 6-AMINOHEXANOL  ::840
 677MONOPEPTIDE  ::1233
 677MONOPEPTIDE BOC  ::1232
 8-HYDROXYQUINOLINE  ::846
 ABIETIC ACID  ::413
 ACENAPHTHENE  ::189
 ACETAL  ::512
 ACETALDEHYDE  ::247
 ACETAMIDE  ::813
 ACETANILIDE  ::844
 ACETIC ACID  ::380
 ACETIC ANHYDRIDE  ::423
 ACETONE  ::268
[0314/1250]                         Select a Compound
```

Figure 101.

Pressing <enter> gives a dialog box that allows the entry to be confirmed or the search repeated (Figure 102).

The only name searched by *SuperChems* in the above method is the primary one used to identify a chemical in the databanks. The chemical may be present, but listed under a synonym. Searching with synonyms (called "Aliases" in *SuperChems*) is very difficult. It is preferable to select a chemical using its formula.

```
Names: ACETONE
       2-PROPANONE
       ACETONE
       PYROACETIC ETHER

Formula   : C3H6O
Structure : CH3COCH3
CAS #     : 67641
ID  #     : 268

Please verify selection

                          NO
                          YES
```

Figure 102.

9.1.3. Select *Formula* (from step 9.1.1 above). Enter the atoms present in the order:

C, H, Br, Cl, F, I, N, O, P

The "down" arrow key can be used to scroll the listing to select from among a set of isomers. It will be necessary to select them one at a time and confirm if the correct one has been chosen. Acetone, CH_3COCH_3 would be entered as C3H6O and gives the listing (Figure 103):

```
Full Databank Listing
C3H5NO  ::818
C3H5NO  ::819
C3H6  ::32
C3H6  ::85
C3H6Cl2  ::1181
C3H6Cl2  ::1182
C3H6Cl2  ::557
C3H6O  ::249
C3H6O  ::268
C3H6O  ::305
C3H6O  ::498
C3H6O  ::525
C3H6O  ::526
C3H6O2  ::381
```

Figure 103.

Confirm the entry as described in step 9.1.2

9.1.4. The CAS number of a chemical can be used to locate it in the *SuperChems* databank. Select *CAS #* in step 1a above and enter the number, omitting any hyphens. For example, the CAS number of acetone is 67-64-1; it is found in *SuperChems* by entering 67641 (Figure 104).

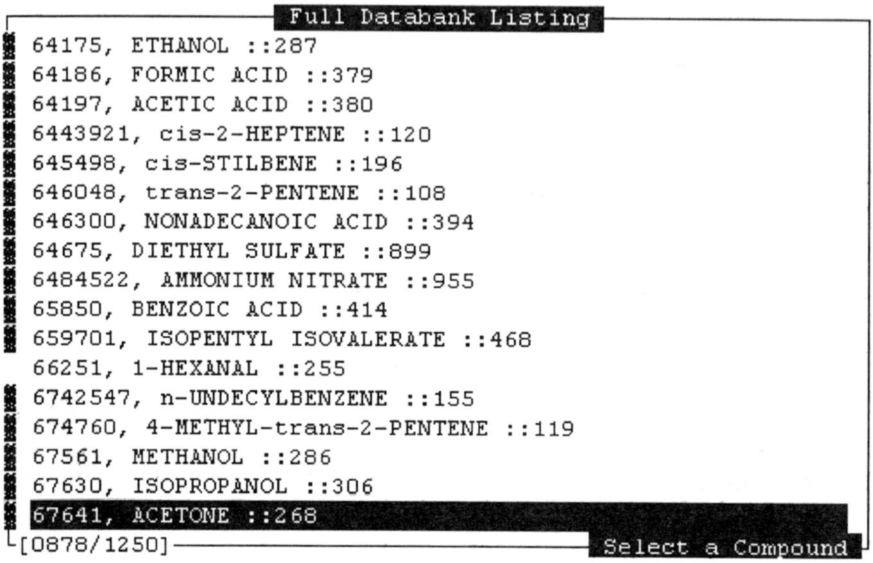

Figure 104.

Confirm the entry as described in step 9.1.2

10. SELECT A DATABANK

The default databanks supplied by *SuperChems* contain the properties of over 1200 chemicals. The names of the default databanks are:

Index databank:	IBANK.ID
Properties databank:	TBANK.TPP

The following steps will allow a particular databank to be selected. Note that databanks are always selected in pairs: an index databank and its corresponding properties databank. To create a new set of databanks, see Section 5, Create a New Databank.

The currently active databanks are noted during start-up of *SuperChems*, and can be seen on the first page of the initial output file. From the main menu, select *Files*, then *View Output* from the pulldown menu (Figures 105 and 106).

```
Files      Define     PreLoad

 Directories
 File manager (CTRL-F)
 DOS command  (CTRL-X)
 Archives
 ------------------------------
 Autosave core image is ON
 Record output is ON
 View output
 Erase output
```

Figure 105.

```
File  Edit  Search  Help

   EAIL|386, V 1.42                                          PAGE    2
   PROJECT: MISSING                                          MISSING
   PROBLEM: MISSING                          Fri Nov  7 08:18:10 1997


   Welcome to Release 3.10 of SuperChems
```

Figure 106.

HINT: Output is readily viewed using Windows Notepad or the DOS text editor. See Section 3, *Superchems* Hints.

10.1. To select a databank, from the main menu choose *Select databanks* from the *ChemProps* pulldown menu (Figure 107).

```
Projects  Reports  ChemProps Utilities

 Locate/Find Chemical
 ------------------------------------
 Generate properties
 Regress properties
 Reduce ideal gas equations
 Reduce EOS parameters
 Reduce Gibbs free energy parameters
 ------------------------------------
 Databank manager
 Print quality codes
 Explain quality codes
 Select databanks .
```

Figure 107.

10.2. Enter the name and *full DOS path* of the databanks to be selected. The format of the entries is (Figure 108):

Index databank:	C:\SC31\BANKS\[FILENAME].ID
Properties databank:	C:\SC31\BANKS\[FILENAME].TPP

```
Index databank        [\SC31\BANKS\RCA-NEW.ID
Properties databank   [\SC31\BANKS\RCA-NEW.TPP
Superuser password    [                              ]
```

Figure 108.

NOTES: An existing entry can be cleared from each line of the dialog box by pressing the ESC key. Leave the "Superuser password" line blank.

Accept the entries using <F10>.

CAUTION: If an error is made in entering the databanks file name (either name or path error), a message will appear (Figure 109):

```
Error on opening file [\SC31\BANKS\RCA-NEW.IDD] for read in binary mode

Strike any key to continue
```

Figure 109.

Choosing *Return to program* in the next dialog box (Figure 110).

```
Select another file
Create a new file
Edit file
File manager     ( CTRL-F )
DOS              ( CTRL-X )
Return to program
Abort program
```

Figure 110.

Will result in a message (Figure 111):

```
I still can't open \SC31\BANKS\RCA-NEW.IDD
Do you really want to continue ?.
Proceed at your own risk !!!
                                          NO
                                          YES
```

Figure 111.

Choosing *Yes* will result in a program abort, as will choosing *Abort program* in the previous step. In either case, you will exit from *SuperChems*. It will then be necessary to repeat the steps described above. Neither of the other file choices (Select another file, Create a new file) are able to resolve the problem of the incorrect databanks file names.

11. ACKNOWLEDGMENTS

One of the authors (R. C. Atkins) wishes to gratefully acknowledge the financial support of Merck & Co., Inc. and the James Madison University Program of Grants for Faculty Assistance.

12. REFERENCES

Benson, S. W., *Thermochemical Kinetics*, 2nd ed., New York: Wiley, 1976.

Constantinou, L., and Gani, R. "New Group Contribution Method for Estimating Properties of Pure Compounds," *AIChE J.*, **40**(10) (1994), 1697.

Joback, K. G., and Reid, R. C. "Estimation of Pure-Component Properties from Group Contributions," **57** (1987), 233.

Melhem, G. A., personal communication, 1997.

Muzzio, D. A., *Emergency Relief System Design of Hydrogen Chloride Vaporizer Using* SuperChems™, presented to Design Institute for Emergency Relief Systems Users Group, American Institute of Chemical Engineers, April 14, 1997.

Reid, J. C., Prausnitz, J. M., and Poling B. E., *The Properties of Gases and Liquids*, 4th ed., New York: McGraw-Hill, 1987.

Numerical Analysis on Hot Spot in Reactive Chemical Storage

Long-Jye Sheu and Jenn-Der Lin*
Department of Mechanical Engineering, National Chiao Tung University
1001 Tai Hsueh Rd., Hsinchu, Taiwan

Jenq-Renn Chen
Department of Safety, Health and Environmental Engineering
National Institute of Technology at Kaohsiung
1 University Rd., Yenchau, Kaohsiung, 824, Taiwan

ABSTRACT

In an exothermic reactive chemical storage, a thermal explosion can occur when the heat generation within the storage is greater than heat removal from the storage. In a solid system, the existence of an inflection point in the temperature versus time curve is often used as a criterion for thermal explosion. In this study the applicability of this criterion to a fluid storage is examined by investigating the hot spot in the system. Transient natural convection of an exothermically reactive fluid in the vertical cylindrical storage tank with isothermal walls is investigated numerically. The axisymmetric 2-D Navier-Stokes equations governing the flow fields are reduced by introducing the stream function-vorticity formulation and solved by the alternating-direction-implicit (ADI) technique. The reactive heat sources are represented by a zeroth order rate expression with a Arrhenius type rate constant. It is found that the hot spot is no longer located in the center of the tank due to the buoyancy effect. In a particular range of Rayleigh number and Frank-Kamenetskii number, the hot spot is found to move periodically. A stable steady regime, stable oscillatory regime and thermal explosion regime can be specified based on the dependence of the critical Frank-Kamenetskii number on the Rayleigh number. In the stable oscillatory regime multiple inflections in temperature versus time are found. Therefore, the existing criterion which defines thermal explosion by the appearance of an inflection in the temperature versus time curve is no longer applicable in reactive fluid systems. The results of non-convective systems do, however, provide a preliminary and conservative estimate for thermal explosion of a reactive storage.

1. INTRODUCTION

The behavior of a thermal explosion, a spontaneous ignition phenomenon of a system with exothermic chemical reaction, has been studied widely for many decades. When exothermic reaction takes place in a system, self-heating tends to cause an acceleration in the reaction rate. This tendency is opposed by both heat loss through the walls and by reactant consumption. If the rate of environmental cooling is insufficient to offset the rate of internal heat generation, the temperature continues to rise. As the rising temperature reaches a critical point, thermal explosion occurs. Many of the works estimated the criticality and its criterion for various boundary conditions and geometries. Semonov [1] considered a system in which the temperature is assumed uniformly distributed and presented the explosion limits for such a lumped system. Frank-Kamenetskii [2] described exothermically reacting systems in which the heat transfer mechanism is purely conductive. In both systems, an exothermic zeroth order Arrhenius equation is applied to simulate the reaction rate, and no consumption of reactants occurs. The critical values of Frank-Kamenetskii number are defined as those leading to an unbounded infinite temperature or the appearance of a inflection point in the curve of temperature–time excursion.

Most studies on thermal explosion have treated the process of heat transfer in the reacting medium as purely conductive, which is generally true for solid explosives. However, for liquid or gaseous explosives, an additional heat transfer mechanism can exist. The thermal gradients originated by the chemical reaction can provide the driving force for the onset and development of natural convection as indicated by the experiments of Ashmore et al. [3] and Merzhanov and Shtessel [4]. The conditions for the onset of reaction-driven convection in a fluid confined between two infinite parallel plates has been determined by Jones [5]. Jones [6] also numerically studied the effects of natural convection on the critical conditions of thermal explosion in a gaseous medium enclosed in a horizontal circular cylinder. The critical condition is again determined by the appearance of an upward inflection in the temperature versus time curve. The ignition limit, specifically the critical Frank-Kamenetskii number, is found to be a strongly increasing function of the Rayleigh number, which indicates that a higher rate of heat production can be accommodated without an explosion when convection occurs in the system in addition to conduction.

In the present study a numerical analysis on transient natural convection of an exothermically reactive fluid in a vertical cylindrical storage tank with isothermal walls is presented. It purposes to examine the applicability of the existing criterion for thermal explosion as well as to investigate the occurrence of a hot spot in a fluid storage system. In the analysis, the reactive heat sources are represented by a zeroth order rate expression with a Arrhenius type rate constant. The axisymmetric 2-D Navier-Stokes equations governing the flow fields are reduced by introducing the steam function-vorticity formulation and solved by ADI technique. The Boussinesq approximation is made in which the dependence of the density of the fluid on the temperature is considered only in the buoyancy force term of the momentum equation and all other physical parameters of the fluids are assumed to be constant elsewhere. The results are presented in terms of variations in values of the hot spot temperature and also in the ratio of energy production to energy release. Results of the variation in the explosion limit with Rayleigh number are also presented. The excursions of the hot spot temperature are classified as the stable steady regime, stable oscillatory regime and thermal explosion regime. In the stable oscillatory regime multiple inflections in temperature versus time would occur and thus the existing criterion which defines thermal explosion by the appearance of an inflection in the temperature versus time curve is no longer applicable. In the present study, the effect of Rayleigh number and Frank-Kamenetskii number on the occurrence of various type regimes is examined.

2. GOVERNING EQUATIONS

We consider the system of a vertical cylindrical tank with exothermically reacting chemicals in it. The chemicals will be self-heated by the exothermic reaction which results in natural convection due to the buoyancy effect. The physical model and coordinate system are shown in Figure 1. The chemicals are regarded as a Newtonian fluid. The Boussinesq approximation is made in

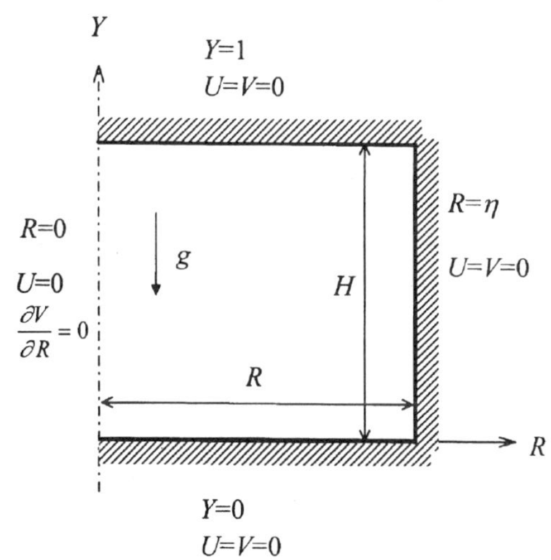

Figure 1. Physical model and coordinates

which the dependence of the density of the fluid on the temperature is considered only in the buoyancy force term of the momentum equation and all other physical parameters of the fluids are assumed to be constant elsewhere. The conservation of mass, momemtum, and energy may be described by the following equations:

Continuity equation

$$\frac{\partial}{\partial r}(ru)+\frac{\partial}{\partial y}(rv)=0. \tag{1}$$

r-direction momentum equation

$$\rho\left(\frac{\partial u}{\partial t}+u\frac{\partial u}{\partial r}+u\frac{\partial u}{\partial y}\right)=-\frac{\partial p}{\partial r}+\mu\left[\frac{\partial^2 u}{\partial r^2}+\frac{1}{r}\frac{\partial u}{\partial r}+\frac{\partial^2 u}{\partial y^2}-\frac{u}{r^2}\right]. \tag{2}$$

y-direction momentum equation

$$\rho\left(\frac{\partial v}{\partial t}+u\frac{\partial v}{\partial r}+v\frac{\partial v}{\partial y}\right)=-\frac{\partial p}{\partial y}+\mu\left[\frac{\partial^2 v}{\partial r^2}+\frac{1}{r}\frac{\partial v}{\partial r}+\frac{\partial^2 v}{\partial y^2}\right]+\rho g. \tag{3}$$

Energy equation

$$\rho\frac{\partial T}{\partial t}+u\frac{\partial T}{\partial r}+v\frac{\partial T}{\partial y}=-\frac{1}{\alpha}\left[\frac{\partial^2 T}{\partial r^2}+\frac{1}{r}\frac{\partial T}{\partial r}+\frac{\partial^2 T}{\partial y^2}\right]+QA\exp\left[-\frac{E}{\bar{R}T}\right], \tag{4}$$

where E in the expression is the activation energy of the reaction, A is the preexponential factor in the Arrhenius law, and Q is the exothermicity of the reaction. In the energy equation, the heat source term is the chemical reaction heat. A zeroth order Arrhenius form is used to simulate the reaction rate.

The governing equations (1)–(4) may be rewritten in dimensionless form in the following manner. The spatial distances are nondimensionalized with respect to the height H of the tank, time measured in units of H^2/ν, and velocity in units of ν/H, where ν is the kinematic viscosity of the reacting chemicals. In deriving the nondimensional governing equations the Frank-Kamenetskii dimensionless temperature is utilized, which is defined as

$$\phi=\frac{E}{\bar{R}T_0^2}(T-T_0), \tag{5}$$

where T_0 is the initial temperature of the reacting chemicals.

The nondimensional governing equations are given as follows.

Continuity equation

$$\frac{\partial}{\partial R}(RU)+\frac{\partial}{\partial Y}(RV)=0. \tag{6}$$

r-direction momentum equation

$$\frac{\partial U}{\partial \tau}+U\frac{\partial U}{\partial R}+V\frac{\partial U}{\partial Y}=-\frac{\partial P}{\partial R}+\frac{\partial^2 U}{\partial R^2}+\frac{1}{R}\frac{\partial U}{\partial R}+\frac{\partial^2 U}{\partial Y^2}-\frac{U}{R^2}. \tag{7}$$

y-direction momentum equation

$$\frac{\partial V}{\partial \tau}+U\frac{\partial V}{\partial R}+V\frac{\partial V}{\partial Y}=-\frac{\partial P}{\partial Y}+\frac{\partial^2 V}{\partial R^2}+\frac{1}{R}+\frac{\partial V}{\partial R}+\frac{\partial^2 U}{\partial Y^2}+\frac{Ra}{\Pr}\phi. \tag{8}$$

Energy equation

$$\frac{\partial \phi}{\partial \tau}+U\frac{\partial \phi}{\partial R}+V\frac{\partial \phi}{\partial Y}=\frac{1}{\Pr}\left[\frac{\partial^2 \phi}{\partial R^2}+\frac{1}{R}\frac{\partial \phi}{\partial R}+\frac{\partial^2 \phi}{\partial Y^2}\right]+\frac{\partial}{\Pr}\exp\left[\frac{\phi}{1+\lambda\phi}\right] \tag{9}$$

The dimensionless parameters characterizing the system are the Prandtl number

$$\Pr = \nu/\alpha, \tag{10}$$

the Rayleigh number

$$\mathrm{Ra}=\frac{gH^3\beta\bar{R}T_0^2}{\alpha\nu E} \tag{11}$$

the Frank-Kamenetskii parameter

$$\delta=\frac{QEAH^2\exp(-E/\bar{R}T_0)}{\alpha\bar{R}T_0^2}, \tag{12}$$

and

$$\lambda=\bar{R}T_0/E. \tag{13}$$

The system of governing equations is solved subject to the following initial conditions

$$\phi = 0, \quad U=0, \quad V=0 \quad \text{at} \quad \tau=0. \tag{14}$$

The boundary conditions imposed for the temperature field are that the temperatures at the boundary remain equal to the temperature of the surroundings and are symmetric at the center of the system, that is,

$$\phi(\eta,Y,\tau)=0 \tag{15a}$$

$$\frac{\partial\phi(0,Y,\tau)}{\partial R}=0 \tag{15b}$$

$$\phi(R,0,\tau)=0 \tag{15c}$$

and

$$\phi(R,1,\tau)=0 \tag{15d}$$

where η is the aspect ratio defined as $\eta = R/H$. The velocity boundary conditions at the walls satisfy the no slip conditions, that is,

$$U(0,Y,\tau)=0, \qquad \frac{\partial V(0,Y,\tau)}{\partial R}=0; \tag{16a}$$

$$U(\eta,Y,\tau)=0, \qquad V(\eta,Y,\tau)=0; \tag{16b}$$

$$U(R,0,\tau)=0, \qquad V(R,0,\tau)=0; \tag{16c}$$

and

$$U(R,1,\tau)=0, \qquad V(R,1,\tau)=0. \tag{16d}$$

3. THE NUMERICAL METHODS

3.1. Stream Function-Vorticity Formulations

In this section we describe the finite-difference method used to solve the governing equations (6)–(9), subject to the initial conditions (14) and boundary conditions (15) and (16).

We introduce the stream function ψ which satisfies

$$U=\frac{1}{R}\frac{\partial\psi}{\partial Y} \qquad \text{and} \qquad V=-\frac{1}{R}\frac{\partial\psi}{\partial R} \tag{17}$$

so that the continuity equation (1) is identically satisfied. The vorticity ω is defined by

$$\omega=\frac{\partial V}{\partial R}-\frac{\partial U}{\partial Y}. \tag{18}$$

From Eqs. (17) and (18) a relation between ψ and ω results, in the form of the following Poisson equation.

$$\frac{\partial}{\partial}\left(\frac{1}{R}\frac{\partial\psi}{\partial R}\right)+\frac{\partial^2}{\partial Y}\left(\frac{\psi}{R}\right)=-\omega. \tag{19}$$

In terms of ψ and ω the momentum equations (7) and (8) yield

$$\frac{\partial\omega}{\partial\tau}+\frac{1}{R}\left[\frac{\partial}{\partial R}\left(\frac{\partial\psi}{\partial Y}\omega\right)-\frac{\partial}{\partial Y}\left(\frac{\partial\psi}{\partial R}\omega\right)\right]-\frac{1}{R^2}\frac{\partial\psi}{\partial Y}\omega$$
$$=\left[\frac{\partial^2\omega}{\partial R^2}+\frac{1}{R}\frac{\partial\omega}{\partial R}+\frac{\partial^2\omega}{\partial Y^2}-\frac{\omega}{R^2}\right]+\frac{\mathrm{Ra}}{\mathrm{Pr}}\frac{\partial\phi}{\partial R}. \tag{20}$$

and the energy equation (9) may be rewritten as

$$\frac{\partial\phi}{\partial\tau}+\frac{1}{R}\left[\frac{\partial}{\partial R}\left(\frac{\partial\psi}{\partial Y}\phi\right)-\frac{\partial}{\partial Y}\left(\frac{\partial\psi}{\partial R}\phi\right)\right]$$
$$=\frac{1}{\Pr}\left[\frac{\partial^2\phi}{\partial R^2}+\frac{1}{R}\frac{\partial\phi}{\partial R}+\frac{\partial^2\phi}{\partial Y^2}\right]+\frac{\delta}{\Pr}\exp\left(\frac{\phi}{1+\lambda\phi}\right) \tag{21}$$

The boundary conditions become

$$\psi=(0,Y,\tau)=\omega(0,Y,\tau)=\frac{\partial\phi(0,Y,\tau)}{\partial R}=0, \tag{22a}$$

$$\psi=(\eta,Y,\tau)=\frac{\partial\psi(\eta,Y,\tau)}{\partial R}=\phi(\eta,Y,\tau)=0, \tag{22b}$$

$$\psi=(R,0,\tau)=\frac{\partial\psi(R,0,\tau)}{\partial R}=\phi(R,0,\tau)=0, \tag{22c}$$

and

$$\psi=(R,1,\tau)=\frac{\partial\psi(R,1,\tau)}{\partial R}=\phi(R,1,\tau)=0, \tag{22d}$$

The value of at the walls is obtained with a particular procedure and will be discussed later. The initial conditions are

$$\psi=\omega=\phi=0 \qquad \text{for all } R \text{ and } Y \quad \text{at } \tau=0. \tag{23}$$

3.2. Solution Procedure

A finite difference method is used for solving the stream function, vorticity and energy equations. The second derivatives in space are approximated by central differences. It is shown by Torrance [9] that the finite-difference representation of energy and vorticity equations using central differences to approximate the convective terms leads to numerically induced oscillations in the results. In order to prevent the numerically induced oscillations, the second upwind differencing method is adopted to approximate the first derivative of the convective term. The other advantage of this method is that it preserves the conservative and transportive properties of the energy and vorticity equations.

The two-dimensional alternating-direction-implicit (ADI) method is employed in the solution of the vorticity and energy equations. In numerical computation, each time step is divided into two halves. During the first half time-step from t_n to $t_{n+1/2}$, the derivatives with respect to one space direction are represented by finite-difference analogues evaluated at $t_{n+1/2}$ whereas those with respect to the other direction are already known. In the next half time-step from $t_{n+1/2}$ to t_{n+1} the reverse procedure is used. Backward differences are used for approximating the time derivatives. The stream-function equation is solved by the successive-over-relaxation (SOR) procedure.

With regard to the determination of the vorticity at the boundaries, the following procedure has been adopted. Equation (20) is used to solve for the vorticity at the interior grid points with boundary values of the vorticity held constant. After the vorticity at the interior grid points is calculated for the new time step and the stream function has been computed from it at the same time step using Eq. (19), the values of the vorticity evaluated at the wall, say $Y = 1$, are derived from the definition for the vorticity as

$$\omega_{i,N}^{n+1} \frac{3U_{i,N}^{n+1} - 4U_{i,N-1}^{n+1} + U_{i,N-2}^{n+1}}{2\Delta Y} + O\left((\Delta Y)^2\right), \tag{24}$$

where Y is the mesh size. A similar expression can be obtained for the vorticity evaluated at other surfaces.

The iteration is repeated until the maxima of the relative differences in each of the fields ϕ, ψ, and ω at each mesh point between successive iterations are less than prescribed tolerances. In the present study, it is taken to be 10^{-5}.

4. RESULTS AND DISCUSSION

In the above formulations, the governing nondimensional parameters are the Frank-Kamenetskii number δ, Rayleigh number Ra, Prandtl number Pr, aspect ratio η and the parameter λ defined in Eq. (13). The Frank-Kamenetskii number, defined in Eq. (12), physically represents the ratio of the chemical reaction rate heat and heat conduction rate in the medium. The Rayleigh number represents the buoyancy force. The occurrence of a thermal explosion is dependent upon the balance between reaction heat generation and heat loss rate through the boundaries. Moreover, the conduction and convection heat transfer mechanisms lead to a nonuniform temperature distribution in the reactive system and forms a local area of high temperature in the bulk called the "hot spot." Due to the Arrhenius-form dependence between reaction heat rate and temperature, a high temperature hot spot accelerates heat generation in the system and thus increases the possibility of a thermal explosion. In order to investigate the effects of natural convection in fluid storage systems, we focus on the influence of both the Rayleigh number and Frank-Kamenetskii number on thermal explosion criticality. The Prandtl number Pr, aspect ratio η and the parameter λ are taken to be 1.0, 1.0, and 0.05, respectively, throughout this study.

In the solid reactive system, the only heat transfer mechanism is conduction and the thermal explosion always happened once an upward inflection in the temperature versus time curve has occurred. Consequently, the inflection in such a curve is often used as a criterion for defining a thermal explosion for solid reactive systems. For a liquid system, natural convection may influence the temperature distribution and the applicability of such a criterion remains to be justified. Figures 2 and 3 show the hot-spot temperature excursions at various values of Rayleigh numbers from which we may also investigate the effect of natural convection on the thermal explosion. Figure 2 shows the hot-spot temperature excursions for two cases near critical conditions at Ra = 10,000, one stable steady case and the other thermal explosion case. The system is said to reach a steady state when the relative difference in the temperature fields between two successive time steps is less than 10^{-5}as the case indicated by the solid line in Figure 2. In the thermal explosion case, an upward inflection in the temperature versus time curve has been found and it is observed that the explosion does indeed follow. While for the stable steady case, there is no such inflection. It seems that the criterion of an upward inflection point still works for the reactive fluid system. However, it is found that this is only true for Rayleigh number less than a critical value (150,000 under the conditions considered in this study). At larger Rayleigh number, a completely different phenomenon is found. Figure 3 shows the

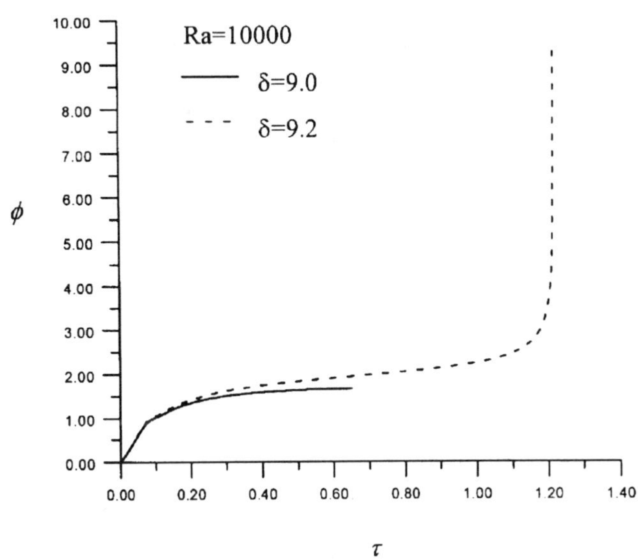

Figure 2. The variation of hot spot temperature at Ra = 10,000 and δ = 9.0 (stable steady state) and 9.2 (thermal explosion case).

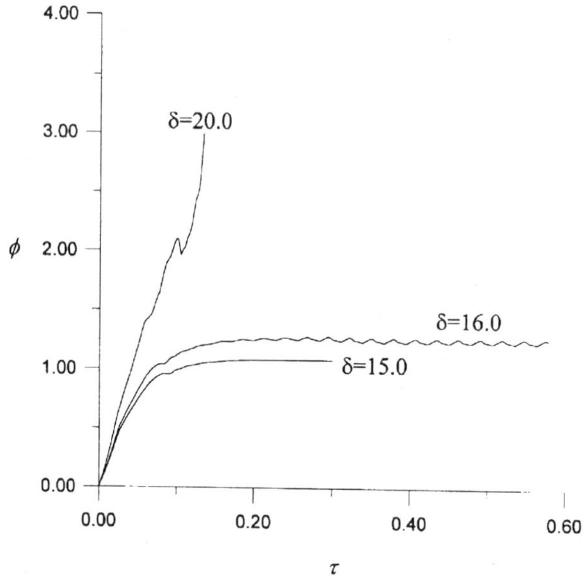

Figure 3. The variation of hot spot temperature at Ra = 200,000 and δ = 20.0 (thermal explosion case), 16.0 (stable oscillatory state) and 15.0 (stable steady state).

hot-spot temperature excursions for three different values of δ at Ra = 200,000. It is noticed from the figure that the case at δ =15 reaches a stable steady state, and the case at δ =20 would reach a state where thermal explosion happens. It is clearly seen that the temperature for the case of δ = 16 becomes periodically oscillating. This may be due to the fact that the flow is oscillating in a fluid system under such conditions. The oscillating flow in a nonreactive fluid system with uniformly heat generation has been obtained in previous investigations [7, 8]. However, in Jones' work[6] where reacting gases in a horizontal cylindrical storage tank are considered, such phenomenon has not been observed. In his calculations, the computations were performed for Rayleigh numbers ranging up to 1.1×10^4 which is much smaller than the critical Rayleigh number for the oscillating regimes to occur.

The oscillations in the hot-spot temperature for the case of δ =16 and the oscillations in fluid flow as observed in [7,8] are a consequence of equivalence between the energy generation and heat losses, as illustrated in Figure 4. Figure 4 shows the transient behavior of the ratio of total energy production over the whole volume of the system to the energy loss from the boundaries. The total energy production is calculated by integrating the energy production within a time-step at each of the grid nodes. It is noticed that for the case reaching to a stable steady state the energy generation would eventually be exactly balanced by the heat loss, while for cases in which thermal explosion happens the heat loss is insufficient to offset the rate of internal

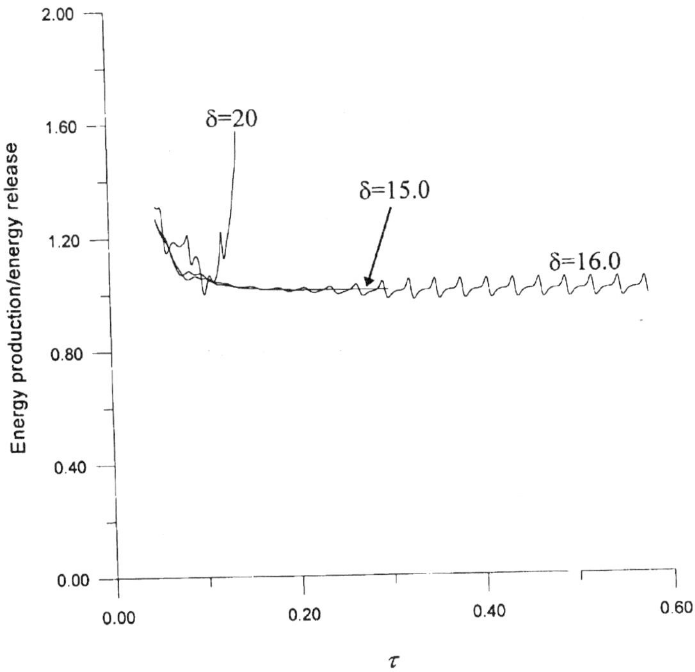

Figure 4. Transient behavior of the ratio of total energy production to the energy release at Ra=200,000 and three values of δ.

energy generation. For the case leading to periodic oscillations in temperature, results show that a stalemate between the energy generation and heat loss occurs and a see-saw competition between them exists.

A similar figure, which illustrates the dependence of the critical value of the Frank-Kamenetskii number on the Rayleigh number, as that shown by Jones is given in Figure 5. Figure 5 is achieved by performing computations for a large number of combinations of values of the two numbers and noting the nature of the resulting solutions. Results are shown for both smaller and larger Rayleigh number regimes. It is illustrated in the figure that the δ-Ra plane would be classified into two regions, specifically the thermal explosion and stable steady regimes at smaller Rayleigh numbers. At larger Rayleigh numbers, the δ-Ra plane may be classified into three regions instead of two, specifically the thermal explosion, the stable oscillatory and stable steady regimes. The stable oscillatory regime corresponds to the situation where the oscillation in the variation of hot-spot temperature exists. The stable steady regime corresponds to the cases in which the energy generation is eventually exactly balanced by heat losses. It is also shown in this figure that nonconvective systems (Ra = 0) do provide a preliminary and conservative estimate for thermal explosion of reactive storages.

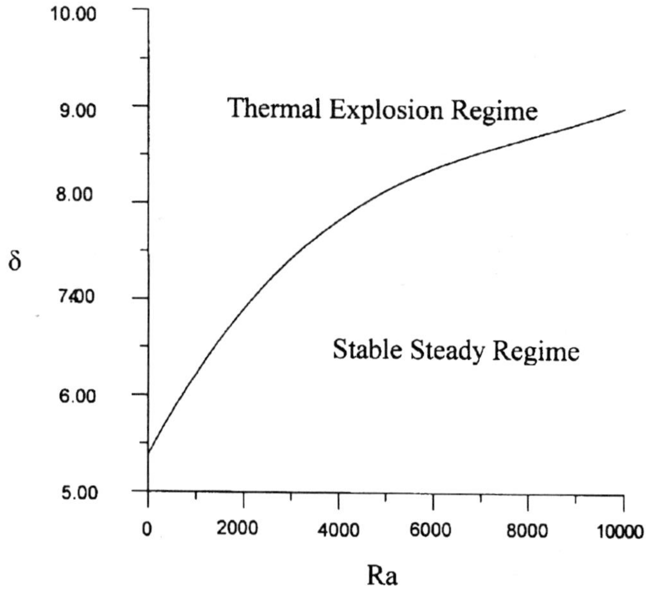

(a)

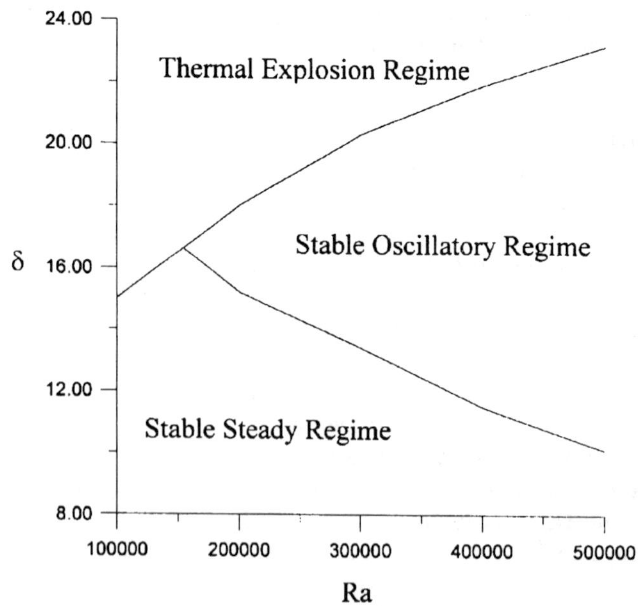

(b)

Figure 5. Variation of the explosion limit δ_{cr} with the Rayleigh number in the range of (a) smaller Ra and (b) larger Ra.

5. CONCLUDING REMARKS

This paper presents a detailed theoretical study on the thermal explosion in a vertical cylindrical storage tank containing a reactive fluid. The results show that thermal explosion is stablized by the convection mechanism. The hot spot is no longer located at the center of the tank and moves upward due to the buoyancy effect. The oscillating solutions, which are found in uniformly heat-generating nonreactive fluid system[7,8], are also obtained at larger Rayleigh numbers for reacting systems in this study. In a particular range of Rayleigh number and Frank-Kamenetskii number, the hot spot is found to move periodically. In the stable oscillatory regime multiple inflections in temperature versus time are found and, therefore, the existing criterion which defines thermal explosion by the appearance of an inflection in the temperature versus time curve is no longer applicable for reactive fluid systems, particularly for systems at very large Rayleigh numbers. The results also show that nonconvective systems do provide a preliminary and conservative estimate for thermal explosion of reactive materials in storage.

NOMENCLATURE

A	preexponential factor in Arrhenius' law
E	activation energy of reaction
g	acceleration due to gravity
Pr	Prandtl number
Q	exothermicity of reaction
R, Y	dimensionless cylindrical coordinates
r, y	cylindrical coordinates
$\underline{R}$	gas constant
Ra	Rayleigh number
T	temperature
t	time scale
U, V	dimensionless velocity components
u, v	velocity components

Greek symbols

α	thermal diffusivity of reacting fluids
β	coefficient of cubical expansion of reacting fluids
δ	Frank-Kamenetskii number
ϕ	dimensionless temperature
λ	RT_0/E
η	aspect ratio R/H
μ	viscocity
ν	kinematic viscosity of reacting fluids

ρ	density of reacting fluids
τ	dimensionless time scale
ω	dimensionless vorticity
ψ	dimensionless stream function

Subscripts

cr	critical value
i	ith node location in r-direction
j	jth node location in Y-direction
M,N	grid number in r and Y direction
0	quantity at initial state

Superscripts

n	nth time-step

REFERENCES

1. Semonov, N. "Zur Theoire des Verbrennungsprozesses," *Z. Phys.* 48 (1928), 571–581.
2. Frank-Kamenetskii, D. A. "The Theory of Thermal Explosion," in *Diffusion and Heat Transfer in Chemical Kinetics,* 2nd ed., pp. 347-421, New York: Plenum, 1969.
3. Ashmore, P. G., Tyler, B. J., and Wesley, T. A. B., "Experimental Investigations of Conductive and Convective Heat Transfer in Relation to Thermal Ignitions," *11th Symposium on Combustion,* pp.1133–1140, Pittsburgh: The Combustion Institute, 1967.
4. Merzhanov, A. G., and Shtessel, E. A., "Thermal Explosions in a Liquid State Exposed to Natural Convection," *Dokl. Phys. Chem.*, 194 (1970), 671–674.
5. Jones, D. R., "The Dynamic Stability of Confined, Exothermically Reacting Fluids," *Int. J. Heat and Mass Transfer,* 16 (1972), 157–167.
6. Jones, D. R., "Convective Effects in Enclosed, Exothermically Reacting Gases," *Int. J. Heat and Mass Transfer,* 17 (1974), 11–21.
7. May, H. O., "A Numerical Study on Natural Convection in an Inclined Square Enclosure Containing Internal Heat Sources," *Int. J. Heat and Mass Transfer,* 34 (1991), 919–928.
8. Churbanov, A. G., Vanishchevich, P. N., Chudanov , V. V., and Strizhov, V. F., "A Numerical Study on Natural Convection of a Heat-Generating Fluid in Rectangular Enclosures," *Int. J. Heat and Mass Transfer,* 37 (1994), 2969–2984.
9. Torrance, K. E., "Comparison of Finite-Difference Computations of Natural Convection," *J. Research of the National Bureau of Standards, Mathematical Sciences,* 72B (1968), 281–301.

A Model to Evaluate the Effectiveness of Water Spray Design in Reducing External Heat Exposure to Process Vessels and Limiting the Maximum Wall Temperature

David J. Hesse
Battelle Columbus Laboratories, 505 King Avenue, Columbus, OH 43201

ABSTRACT

A computational model is presented in detail for quantifying the effectiveness of a given water deluge configuration in reducing vessel thermal heat exposure associated with an external fire. Factors included in the analysis include spray nozzle arrangement, water delivery-rate, vessel orientation and geometry, and existing meteorological conditions where applicable. The model is applicable both in the cases of radiant exposure to a fire as well as engulfment of the vessel in flame. The approach makes use of the high degree of symmetry present in most configurations to greatly simplify the solution of the equations of motion, which in turn is used to calculated the water film dimensions around the entire surface of the vessel. Heat removal due to the absorption of convective and radiative heat transfer by the water film is quantified. A simplified 1-dimensional heat transfer model with time-dependent boundary conditions is solved to determine the heat flux into the vessel contents. Details of how to implement the model in a computer program is presented. Finally, an example problem is presented involving a vessel containing inhibited 1,3-butadiene which undergoes external fire exposure and experiences dimer and "popcorn" formation in the free vapor space.

1. INTRODUCTION

Water spray systems (also called water deluge systems) are installed to provide surface cooling to chemical storage vessels at many large chemical process facilities. These systems are generally comprised of an array of water spray heads arranged in a regular geometric fashion and usually are supplied water by way of a common supply header. Two main functions of a water spray system are listed below:

1. Ensure that vessel contents do not exceed a threshold temperature above which the contents are in some way degraded or an undesirable chemical reaction is initiated, or
2. Provide heat removal capabilities in the event of exposure to fire or some other abnormal event that would lead to unacceptably high pressures inside the vessel.

The first class of temperature excursions are typically represented by exposure of outside storage vessels to extreme ambient conditions. For instance, consider the storage of 1,3-butadiene in an external vessel. Typically some chemical inhibitor such as tertiary butyl catechol (TBC) is mixed with the butadiene to serve as a stabilizer and to prevent an auto-initiated polymerization reaction. Nonetheless, since the partial pressure of butadiene is significantly higher than that of TBC, the butadiene in the free vapor space at the top of the vessel is relatively free of inhibitor. One of the most common problems that arise in the handling and storage of diolefins such as butadiene is dimerization. This occurs when two molecules of diolefin combine to form a large molecule. At temperatures exceeding 32°C (90°F), dimer production may initiate leading to stock degradation and an increase in vessel pressure.

The second class of events includes primarily exposure to thermal radiation and convective heat transfer from a fire source either located some distance from the vessel or directly in contact with the vessel. This type of event is more complex in terms of the both the physical phenomenology involved and the computational effort required in attempting to evaluate the effectiveness of the given water protection system in reducing overall vessel heat exposure.

A computational model is presented that uses the characteristics of the water spray delivery system coupled with the physical parameters of the water, impinging heat source, and geometry and contents of the storage vessel to predict the thermal response of the vessel to heat exposure. Recognizing that the problem in its most complete form is an complex set of interacting heat sources and sinks, moving boundary conditions, and stochastic phenomena, the objective here is to reduce the problem to a tractable numerical algorithm that will yield a first-order accurate solution.

2. HEATING OF A STORAGE VESSEL OR PROCESS EQUIPMENT

Heating of the contents of a sealed storage tank or process vessel will generally result in the increase of the internal pressure of the storage equipment. If the heat source-term is permitted to continue for a sufficiently long time, vessel overpressurization could ultimately take place unless the either proper pressure relief equipment has been installed or heat protection is provided. The two sources of heating considered here are (1) external heating of a vessel due to radiative and convective heat transfer from a fire and (2) internal heating of the contents of a vessel resulting from a sustained polymerization or runaway reaction.

2.1. External Heat Source-Terms

Two key fire variables are of primary interest when designing a safety relief valve (SRV) to ensure against vessel overpressurization in the event of an external fire. The first variable is the level of radiative and convective heat flux exposure to the vessel. This heat flux term is dependent upon the burning rate of the flammable material, the amount of material undergoing combustion, and the distance between the fire and the vessel of interest. The other variable is the total duration of the fire. The time-integrated heat flux exposure to a vessel determines whether the contents will reach the critical pressure and temperature conditions that require emergency relief, and if so, what will be the minimum necessary mass flow capacity through the SRV to protect against vessel damage.

The following four fire scenarios have been identified (Fisher and Forrest, 1995) as having the greatest potential for a fire of sufficient duration to raise the contents of vessels to elevated pressure/temperature conditions:

1. Catastrophic failure of a tank in a dike or area containing other tanks that may be exposed to a subsequent fire.
2. Self-leakage from a tank in a dike or area with the rate of leakage equal to the burning rate.
3. Leakage from a pipe (transfer line) with the resultant spill providing the fuel for fire exposure of the tanks in a dike or area.
4. Overfilling a tank with the resultant spill providing the fuel for fire exposure of the tanks in a dike or area.

2.2. Heat Flux Calculation

If any of the above four scenarios results in a vessel engulfed in flame (i.e., at least one side of the vessel is in direct contact with the flame), then the following equations recommended by API 520 should be used for obtaining the steady-state external heat flux to the vessel, $\dot{Q}_{ext}$, in units of BTU/hr:

$$\dot{Q}_{ext} = \begin{cases} 21{,}000\,A_w^{0.82} & \text{when good drainage is available} \\ 34{,}500\,A_w^{0.82} & \text{when no drainage is available} \end{cases}$$

where A_w is the wetted vessel surface area in units of ft^2. The definition of good drainage is a minimum of a 1% slope leading away from the vessel to a sump located a minimum of 50 feet from the vessel.[1] If there is drainage available, but it does not meet this criteria, then $\dot{Q}_{ext}$ is evaluated assuming no drainage is available.

The wetted surface area is defined to be that area of the vessel wall inner surface in contact with liquid. Computation of wetted surface area for upright vessels is simple:

$$A_w = 2\frac{V_{liq}}{R} + \pi R^2 = 2C_f\frac{V}{R} + \pi R^2$$

where

R = radius of the vessel (m)
V_{liq} = volume of liquid in the vessel (m^3)
C_f = vessel liquid fill limit, $0 \le C_f \le 1$, where $C_f = 0 \Rightarrow$ vessel is empty of liquid and $C_f = 1 \Rightarrow$ vessel is completely full of liquid inventory (dimensionless)
V = volumetric capacity of the vessel (m^3)

Horizontal cylinders and spheres present a slightly more complex situation. Using basic calculus, it can be shown (assuming flat caps on the end of the storage vessel) that the wetted surface area for a horizontal cylinder based on a volume V_{liq} of liquid in a tank with total capacity V, radius R, and length L is expressed by

$$A_w = \begin{cases} R(2\pi - \theta) & \text{for } V_{liq} \ge \frac{1}{2}V \\ R\theta & \text{for } V_{liq} < \frac{1}{2}V \end{cases}$$

where the angle θ (in units of radians) must satisfy the constraints

$$\theta - \sin\theta = 2\left(1 - \frac{V_{liq}}{R^2 L}\right) = 2\pi(1 - C_f) \qquad \text{and} \qquad 0 \le \theta \le 2\pi$$

The above expression is a transcendental equation and must be solved either numerically or graphically. A convenient graphical method is illustrated by Crozier (1985) by giving the wetted-perimeter factor indicating the percentage of vessel surface area in contact with liquid contents as a function of volume of liquid expressed as percentage of tank volume.

Once θ has been determined, the liquid free surface cross-sectional area A_{CR}, an important parameter in determining the onset/disengagement of two-phase flow, may also be computed as follows:

$$A_{CR} = 2LR\sin\left(\frac{\theta}{2}\right)$$

1 More details may be found in NFPA 30, Section 2-2.3.2.

Finally, calculation of the free surface cross-sectional area for a spherical vessel requires the solution of a cubic equation in h, the difference in the inner diameter of the tank and the depth of liquid in the vessel. There exist three real solutions, of which only one can be the physically possible solution:

$$h_1 = R + 2_+\sqrt{-p}\cos\left(\frac{\varphi}{3}\right)$$

$$h_2 = R - 2_+\sqrt{-p}\cos\left(\frac{\varphi}{3} + \frac{\pi}{3}\right)$$

$$h_2 = R - 2_+\sqrt{-p}\cos\left(\frac{\varphi}{3} - \frac{\pi}{3}\right)$$

where φ satisfies

$$\cos\varphi = 2C_f - 1$$

In the special instances when either the tank is charged to full capacity with liquid or contains only vapor, the value of h is identically 0. Otherwise, h will have a value in the range (0,2R) and the liquid free surface cross-sectional surface area is subsequently determined by the mensuration formula

Fire scenarios that result in external heating of a vessel from a remote fire source (i.e., the flames are not in contact with the vessel) may be evaluated using a pool fire radiation model to determine the heat transfer to the sides of the vessel. Mudan (1984) gives a review of some of the more simpler and commonly applied models for evaluating sustained thermal radiation from pool fires. In the event that the source of heat is a material with a atmospheric-pressure boiling point below that of normal ambient conditions, it may be overly conservative to assume all of the material is available for a pool fire. Under such conditions, a portion of the liquid will immediately flash to vapor upon release to atmospheric pressure. In contrast to breakup of a subcooled liquid jet, the rapid expansion of the two-phase mixture and subsequent flash atomization leading to suspended aerosol particles is dominated by the transition of thermodynamic processes to an equilibrium condition. The flashing fraction is estimated by means of a heat balance across the outlet orifice where the decrease in latent heat of vaporization and increase in heat capacity are accounted-for as the initial temperature approaches the critical temperature. A recommended expression for this fraction is given by

$$f = 1 - \exp\left\{-2.63\frac{C_1(T_b)}{L(T_b)}(T_c - T_b)\left[1 - \left(\frac{T_c - T_1}{T_c - T_b}\right)^{0.38}\right]\right\}$$

where

f flashing fraction

$C_l(T)$	liquid heat capacity at temperature T (J/kg·°K)
$L(T)$	latent heat of vaporization at temperature T (J/kg)
T_c	critical temperature (K)
T_b	liquid boiling temperature at one atmosphere pressure (K)
T_l	liquid temperature (K)

In the case of an unconfined release of flashing, combustible liquid, the flashing fraction will determine not only the quantity of liquid available for subsequent ignition and burning but also the size and geometry of the pool. Confined pools on the other hand obviously have a fixed geometry (assuming a minimum pool depth throughout). However, the flashing fraction will determine the total quantity of holdup in the dike which in turn determines the maximum possible duration of a pool fire.

2.3. Flame Temperature Calculation

Flame temperature is an important parameter in the water film heat transfer model as is discussed in greater detail in Section 4. When reliable flame temperature data is not available, it may be necessary to estimate it from first principles. Appendix A presents the necessary material for flame temperature calculation and provides example calculations for butadiene and hexane. Note that since the source of external heating from a pool fire may be any one of a number of different chemicals, it is recommended that the maximum flame temperature computed from the all the individual flame temperature calculations be used in the heat transfer model described below.

2.4. Internal Heat Source-Term

The approach used for modeling the heat generation in the event of a runaway reaction follows the DIERS methodology for tempered and nontempered systems.[2] In either case, the reaction rates at relief conditions represented by the temperature rise (dT/dt) as a function of time are used to characterize the resultant internal heat source-term. These rates are typically measured in a low thermal inertia apparatus such as the DIERS Bench Scale Equipment (Fauske et al., 1985; Leung et al., 1986). This apparatus uses a low thermal mass test cell to reduce the thermal inertia making it possible to measure the magnitude of the self-heat rate and the adiabatic temperature rise.

The slope of the curve at selected points on the temperature profile obtained from the DIERS Bench Scale Apparatus is used to quantify q, the rate of internal heating of the contents of the vessel. Although there are a number

2 A *tempered system* is one in which the reaction heat is removed by latent heat of vaporization. A *nontempered system* exhibits little or no latent heat of cooling.

of ways to evaluate $\dot{q}$, the most common treatment is to take $\dot{q}$ as a constant equal to the arithmetic average between the energy release rate at the set temperature and the turnaround temperature. The reaction rates at relief conditions represented by the temperature rise (dT/dt) as a function of time are used to characterize the resultant internal heat source-term. Therefore $\dot{q}$ is expressed as

$$\dot{q} = \frac{1}{2} C_v \left[\left(\frac{dT}{dt} \right)_s + \left(\frac{dT}{dt} \right)_m \right]$$

where

- $\dot{q}$ = reaction heat rate (W/kg)
- C_v = liquid specific heat at constant volume (J/kg · K)
- $(dT/dt)_s$ = reaction rate temperature rise at set conditions (K/s)
- $(dT/dt)_m$ = reaction rate temperature rise at (maximum) turnaround conditions (K/s)

The time-rate temperature change, dT/dt (K/s), of a reacting mixture is used to quantify the internal heat source-term in a model relief scenario. When temperature versus time data point listings were available, the raw data were manipulated and fitted to an Arrhenius relationship:

$$k = Ae^{-E/RT}$$

where

- k = reaction rate constant
- A = Arrhenius Parameter, frequency factor
- E = Arrhenius Parameter, activation energy
- R = ideal gas constant
- T = absolute temperature

For a detailed discussion of the Arrhenius relationship, see *Perry's Chemical Engineers' Handbook*, 6th ed. pp. 4–20. The above equation can be rearranged to:

$$\ln k = \ln A - (E/R)(1/T)$$

which reveals a linear relationship between ln k and $(1/T)$.

The Arrhenius equation predicts the effect of temperature on the reaction rate. As temperature increases, for example, the reaction rate constant, k, increases, resulting in a proportional reaction rate increase. The reaction rate, in turn, is directly proportional to the rate of heat evolution from the reaction. The rate of temperature rise in the reacting material is a strong function of the heat evolution rate (linear, if the overall heat capacity remains constant). Therefore, the time-rate temperature change dT/dt, where t is time, may be used as an approximation of k in the Arrhenius equation.

Making this substitution, we now have a relationship between dT/dt and temperature:

$$\ln\left(\frac{dT}{dt}\right) = b_0 + b_1\left(\frac{1}{T}\right)$$

where b_0 and b_1 are constants.

Continuing with the butadiene example, Figure 1 depicts a best fit using linear regression to a set of data for 91% 1,3-butadiene in styrene with an azid initiator for a radical source.

When raw data listings were not available, temperature change rates were estimated from existing Arrhenius plots showing self-heat rate data by way of a line of best fit drawn in graphically. Estimates of dT/dt can then be made by reading points from these lines. Note that the estimated dT/dt values should be rounded up or in some way adjusted higher to account for both errors in hand-fitting the data and uncertainties in the reported values.

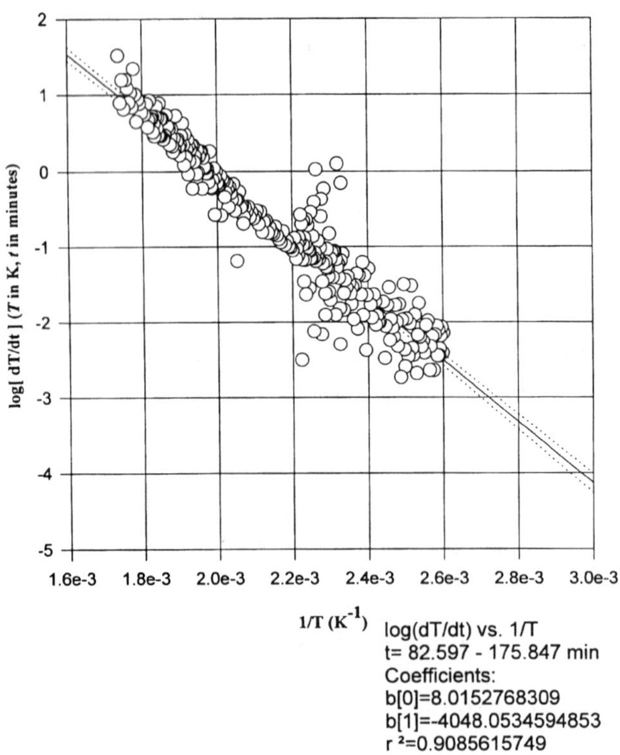

Figure 1. Regression Analysis for the Arrhenius Plot of 91% 1,3-Butadiene in Styrene with Initiator.

3. VESSEL HEAT PROTECTION MODEL

Two commonly available methods for protecting the contents of a vessel from external heating are (1) installation of thermal insulation and (2) application of water deluge systems on the surface of the vessel. Each has its inherent advantages and disadvantages, some of which are discussed below.

3.1. Vessel Insulation

As an alternative to water sprays, vessels may be protected from external heating by means of exterior insulation when the fire does not last for excessively long durations. Vessel insulation has the obvious advantage that it is a passive mitigative feature for reducing the heat flux to the vessel tank assuming that the insulation is fire-resistant, can withstand the reaction forces imparted by application of fire monitors, for instance, and is regularly inspected for physical integrity. On the other hand, there are a number of disadvantages associated with insulation, namely (1) its physical condition is degraded over time by exposure to adverse climatic conditions and (2) it prevents visual inspection of the surface of the tank for corrosion or defects (and in fact may exacerbate this problem by trapping moisture next to the tank surface and accelerating corrosion). Finally, insulation offers thermal protection only in the event of a relatively short-term fire. If an external fire continues for a sufficiently long period of time (typically 1 to 2 hours), the contents of the vessel will reach a steady-state temperature that is relatively insensitive to the amount or type of insulation installed, thereby negating the beneficial effects of the insulation altogether.

One common practice is to calculate the environmental factor as a multiplicative reduction factor to use in the overall heat exposure equation. The thermal insulation in this context is evaluated by means of a formula presented in API 520:

$$F = \frac{k(1660 - T_f)}{2.1 \times 10^4 t}$$

where

F	=	environmental factor (dimensionless)
k	=	thermal conductivity (BTU·in)/(hr·ft·°F)
T_f	=	temperature of vessel contents at relieving conditions (°F)
t	=	thickness of insulation (inches)

The value of the external exposure heat flux $\dot{Q}_{ext}$, discussed in Section 2.2 is then simply multiplied by the environmental reduction factor F to obtain a new values of the heat flux to be used in the subsequent SRV design basis calculations.

3.2. Cooling of Vessel Surface using Water Sprays

A water deluge system or water sprays installed next to the vessel and designed to provide a specific minimum water converge across the entire surface of the vessel will, on the other hand, provide continuing thermal heat protection for the contents of the vessel regardless of the duration of the fire. The use of water sprays can be potentially very effective in reducing the total heat exposure of a fluid inside the vessel as well as maintaining inventory temperatures at sufficiently low levels so as to preclude the onset of a runaway reaction. Implementation of a detailed mathematical model for evaluating the effectiveness of water spray systems in absorbing incident radiation and removing sensible heat from the vessel surface is the primary goal of this discussion. However, it should be bore in mind that the area of falling film phenomena and the calculation of water films with free boundaries in general is currently a topic of intense research.

The water spray model estimates the amount of cooling provided to the surface of the vessel due to absorption of incident convective and radiative heat energy by the resulting water film. Boiling and evaporation of the water film effectively removes latent heat from the surface of the vessel and helps to maintain the surface at reduced temperatures. To a limited extent, the model also attempts to account for the effect of the presence of the water spray on the radiative emission properties of the vessel surface.

3.2.1. Water Film Thermal Absorptivity

A fraction of the radiative energy incident on the vessel surface is absorbed by the water film on the surface of the vessel created by application of the water sprays. The remaining unabsorbed component of the radiative energy will penetrate to the vessel wall. When radiant energy strikes the vessel wall, part of the radiation is reflected and the remainder is absorbed (the transmitted component is negligible for steel and, therefore, is simply assumed to be zero). The concise mathematical statement of this simple physical principle is

$$\alpha + \beta = 1$$

where α and β represent the dimensionless radiative absorptivity and reflectivity, respectively.

The analysis here will make use of Kirchoff's Identity which states that the emissivity ε (i.e., integrated behavior of the material over all wavelengths) is equal to the absorptivity α of the material. The emissivity for Stainless steels (type 301;B) is given to be approximately 0.54 to 0.63 (Holman, 1981, Table A-10). Hence

$$\beta = 1 - \alpha = 1 - \varepsilon \approx \tfrac{1}{2}$$

The water spray, on the other hand, does possess a nonzero radiative transmissivity component. As the radiant energy passes through the water film, an incremental fraction of the energy is absorbed by the water as well as diffusely scattered and transmitted into the next incremental layer. The radiant absorption is a strong function of the wavelength spectrum as well as a monotone increasing function of water film thickness.

Assuming that the incident radiation intensity conforms to Planck's law for a blackbody radiation (a conservative assumption since the emissivity of a blackbody radiator always exceeds that of "real" materials), the radiation intensity as a function of wavelength is expressed by

$$W_\lambda = \frac{2\pi hc^2}{\lambda^5}[e^{-(ch/k\lambda t)} - 1] = \frac{\gamma}{\lambda_5}[e^{-\delta/\lambda} - 1]$$

where

W_λ = radiation intensity at wavelength λ ($W/m^2 \cdot \mu m$)
c = speed of light in a vacuum = 3×10^8 m/s
k = Boltzmann's Constant = 1.38×10^{-23} J/molecule·K
T = source temperature (K)
M = wavelength of radiation (μm)
h = Planck's constant = 6.63×10^{-34} J·s
γ = $2\pi hc^2$
δ = ch/kT
Note: $ch/k = 1.4387 \times 10^4\ \mu m \cdot K$

As described above, the radiation absorbed by a water film varies with wavelength and film thickness, and is expressed as

$$W_{abs} = W_\lambda (1 - e^{-\alpha_\lambda b})$$

where α_λ is the spectral absorptivity (μm^{-1}) and b denotes the water film thickness (μm). The spectral absorptivity is given in terms of the following functional form:

$$\alpha = k10^n$$

where k and n are tabulated as functions of wavelength specified in the International Critical Tables and are appropriately scaled so that α has the units of cm^{-1}. A plot of α_λ as a function of wavelength is provided in Figure 2. The total fraction of absorbed radiative energy, f_{abs}, is therefore given by

$$f_{abs} = \frac{\int_{\lambda_{min}}^{\lambda_{max}} W_\lambda (1 - e^{-\alpha_\lambda b})\, d\lambda}{\int_{\lambda_{min}}^{\lambda_{max}} W_\lambda\, d\lambda}$$

Before proceeding with the evaluation of the integrals in the above expression, it is necessary to first determine the temperature T associated

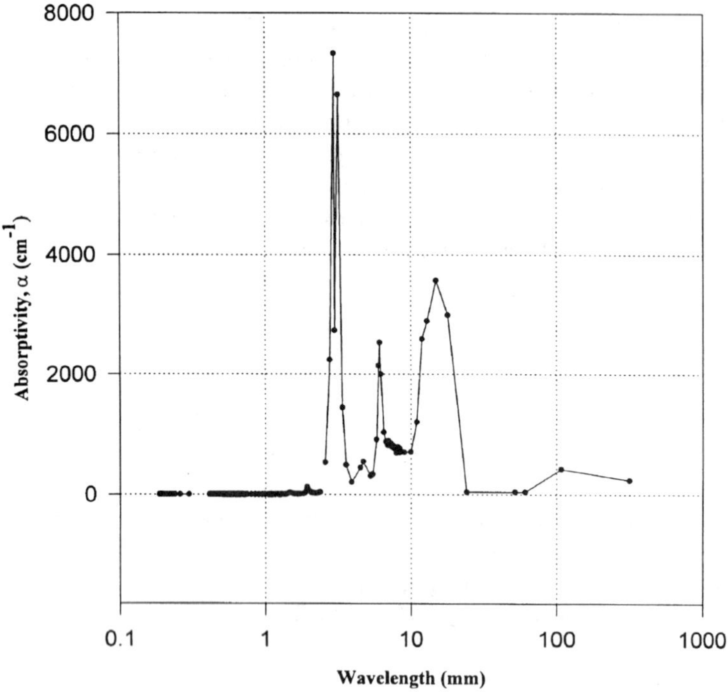

Figure 2. Spectral Absorptivity of Water

with the spectral distribution of radiant energy as well as the thickness of the water film b. The quantification of the latter variable represents one of the most ambitious and technically challenging tasks of the entire model.

3.2.2. Water Spray Film Model

In order to evaluate the thickness of the water film surrounding the vessel and thereby complete the list of inputs necessary to evaluate the amount of thermal energy removed from the vessel, a falling film model was developed. Simple models have been derived for horizontal and vertical plates. However, cylindrical tanks complicate the task in fluid film modeling because of the variable surface geometry and the resultant spatial variability in the gravitational forces acting on the film. A rigorous approach to this problem would require solving the full set of equations of motion of the fluid. This, in turn, would involve the numerical solution of a set of nonlinear, nonhomogeneous partial differential equations. If one makes the following assumptions:

a. the fluid motion follows steady-state conditions,
b. water-flow along the surface of the vessel is primarily in the downward direction and lateral motion (i.e., motion perpendicular to the direction of the force of gravity) is negligible,

c. any radial component to the motion of the fluid film is neglected, and

d. the ambient air flow (e.g., wind) is normal to the vessel centerline axis,

it is possible to arrive at the following simplified expression for the equations of motion in cylindrical coordinates:

$$\frac{4}{r}\rho_\infty v_\infty \sin\theta\cos\theta + \rho g\cos\theta + 2\frac{\mu\partial v_\theta}{r\partial r} - \mu\frac{\partial^2 v_\theta}{\partial r^2} = 0$$

where

v_∞ = ambient wind speed in the far field (m/s)
ρ_∞ = density of air at current meteorological conditions (kg/m^3)
ρ = density of water at initial water spray temperature (kg/m^3)
θ = polar angle coordinate viewed in the plane transverse to the vessel axis of symmetry (radians)
r = radial distance from surface of vessel (m)
μ = viscosity of water (kg/m·s)
g = acceleration of gravity (9.81 m/s^2)
v_θ = velocity of fluid in plane transverse to the vessel axis of symmetry (m/s)

Although this equation is a vast simplification of the original set of equations describing motion of the fluid layer, it is still relatively nontrivial to solve. Furthermore, the second order solution is not truly justified due to the large set of simplifying assumptions make in deriving this expression. Such an approach would obviously violate the philosophy under which this sizing methodology is based, namely the use of relatively simple approximate computational methods which, if necessary, can be independently verified.

An analytic method is available for modeling the falling fluid film over a horizontal cylinder if the following approximations are used:

1. The surface of the tank is discretized into a series of uniform sectors with the outer surface of each sector approximated by its subtending chord.
2. The number of sectors in 1 above is sufficiently large so that the fluid velocity over each chord is approximately constant. Edge effects at the intersection of neighboring chords are neglected.
3. The water film behaves as a Newtonian fluid.
4. The ambient air flow (e.g., wind) is normal to the vessel centerline axis.

A 2-D coordinate axis is oriented such that the z-axis is directed to the line of motion of the fluid along the chord segment and the x-axis is perpendicular to the chord pointing inwards towards the interior of the vessel. The z-component of the equation of motion in rectangular coordinates is expressed by

$$\rho\left(\frac{\partial v_z}{\partial t}+v_x\frac{\partial v_z}{\partial x}+v_y\frac{\partial v_z}{\partial y}+v_z\frac{\partial v_z}{\partial z}\right)=-\frac{\partial p}{\partial z}+\rho g_z$$

where y is the coordinate along the vessel axis of symmetry and p denotes the scalar pressure field.

Since the bulk of the fluid motion over a horizontal cylindrical tank is downward over its lateral sides, the x and y components of velocity are assumed to be negligible. Furthermore, the yz and zz components of the stress tensor τ(the rate of z-momentum exchange across surfaces of constant y and z coordinate, respectively) can also be taken to be approximately zero. Therefore the above equation simplifies to

$$\frac{\partial \tau_{xz}}{\partial x}+\frac{\partial p}{\partial z}=\rho g_z$$

The resultant equation is nothing more than an expression for momentum balance. The fact that a relatively easy analytic solution to this equation exists follows from the fact the discretized cord model treats the streamlines of the system along the surface as straight lines (Bird et al., 1960). Once expressions for the stress tensor component τ_{xz} and the pressure p are obtained, the above equation may be solved for the along-surface velocity component v_z.

If the fluid is assumed to be Newtonian, the momentum flux is related to the velocity gradient according to

$$\tau_{xz}=-\mu\frac{\partial v_z}{\partial x}$$

The term ρg_z simply denotes the component of gravitational force acting on the fluid along the surface of the vessel which is expressed by

$$\rho g_z=\rho g\cos\beta$$

where β denotes the angle that the chord element forms with the vertical azimuth. The pressure term p represents the shear force exerted on the fluid film by a wind field flowing against the side of the vessel. The net effect of the wind field is that it distorts the fluid film and must be modeled in order to evaluate the potential for hot spots developing along the side of the vessel.

Assuming ideal flow around a cylinder with its axis of symmetry perpendicular to the wind direction, the pressure distribution on the cylinder is given by

$$p=\tfrac{1}{2}\rho_\infty v_\infty^2(1-4\sin^2\theta)$$

Substituting these last three results into the simplified equation of motion, evaluating the pressure derivative dp/dz in the limiting case where the cord subdivisions are sufficiently small such that the approximation $dz=Rd\theta$ with R denoting the vessel radius is valid, and using the no-slip condition at the boundary or $v_z = 0$ at $x = \delta$ where δ denotes the local film thickness as the boundary condition for the differential equation, one may solve for the falling film velocity v_z as a function of x.

To finally determine the local film thickness, the volume rate of flow of water, Q_w, is related to the average film velocity:

$$Q_w = S\Delta zL = \int_0^L \int_0^\delta v_z \, dx \, dy$$

where

S = spray application rate (gal/ft^2 · min or alternatively in m/s);
Δz = chord length of subtended arc in vessel surface discretization (m);
L = length of vessel (m).

Finally, completing the integration above and solving the resultant expression in terms of the film thickness δ, one obtains

$$\delta = \begin{cases} \left[\dfrac{3\mu S\Delta z}{\rho g \cos\beta + (4/R)\rho_\infty v_\infty^2 \sin\theta\cos\theta}\right]^{1/3} & \text{for} \quad 0 \le \theta \le \tfrac{1}{2}\pi \\ \left[\dfrac{3\mu S\Delta z}{\rho g \cos\beta - (4/R)\rho_\infty v_\infty^2 \sin\theta\cos\theta}\right]^{1/3} & \text{otherwise} \end{cases}$$

The associated selection rules for the value of β associated with the gravity term are as follows:

$$\beta = \begin{cases} \pi/2 - \theta & \text{for} \quad 0 \le \theta \le \pi/2 \\ \pi - \theta & \text{for} \quad \pi/2 \le \theta \le \pi \\ 3\pi/2 & \text{for} \quad \pi \le \theta \le 3\pi/2 \\ 2\pi - \theta & \text{for} \quad 3\pi/2 \le \theta \le 2\pi \end{cases}$$

An important thing to note in this notational convention for θ is that $\theta = 0$ corresponds to a vessel surface unit normal vector pointed directly in the direction of the wind flow. Similarly $\theta = \pi/2$ corresponds to a vessel surface unit normal vector pointed directly upwind (recall that the wind flow is assumed normal to the vessel centerline axis). Finally the average film velocity over the chord segment is given by

$$\langle v_z \rangle = \frac{\rho g \delta^2}{3\mu}$$

Note there are two expressions for the film thickness depending upon the location on the vessel wall. The plus sign in the denominator of the above expression for θ in the range $(0,\pi)$ results from the fact that on the upper half of the vessel gravity acts as a compressive force whereas the negative sign in the denominator of the alternate expression reflects the fact that gravity serves to pull the liquid film away from the body of the vessel.

The water film model for a vertical cylinder is fundamentally different from that of a horizontal cylinder, primarily due to the change in orientation of the gravitational and wind sheer forces. Again, the derivation begins with the equations of motion. Using the assumption that the velocity of the fluid on a given water sheet (i.e., a constant distance from the side of the vessel) is essentially constant, it is possible to derive the following expression for the vertical velocity profile of the water film:

$$v_z = v_{\max}\left[\frac{2y}{\delta} - \left(\frac{y}{\delta}\right)^2\right]$$

where y is the distance from the side of the vessel, δ is the water film thickness, and $v_{\max}$ is expressed as

$$v_{\max} = \frac{\delta^2 \rho g}{2\mu}$$

As in the case of the horizontal cylinder, the film thickness can then be calculated using the water film velocity averaged over the cross section of the film and the volume flow rate of the sprays. This yields the result for all values of θ

$$\delta = \left[\frac{3\mu S \Delta z}{\rho g \cos\beta + (4/R)\rho_\infty v_\infty^2 \sin\theta\cos\theta}\right]^{1/3}$$

Ultimately there should not be much surprise in the fact that the final expression is remarkably similar to that of the film thickness for a horizontal tank. The extensive use of symmetry and idealized assumptions used to obtain an analytic result is really the root cause behind this similarity. See Section 5 for a discussion on the limitations of these assumptions.

3.3. Heat Transfer Model

The heat transfer model is based on forced convective heat transfer to the film on the outside of the vessel wall, evaporative cooling of the vessel surface by the water film, transient heat conduction through the wall of the vessel, and natural heat convection in the interior of the vessel. Standard nucleate boiling correlations are used to evaluate the heat transfer coefficient at the early stages of the fire as the vessel wall temperature initially increases above the boiling temperature of water. As the vessel wall temperature increases, the methodology detects the fluid regime at which nucleate boiling is no longer valid and the heat coefficient is alternatively calculated using film boiling theory. Finally, if the duration of the fire is sufficiently long and the vessel temperature rise is elevated even further, a water droplet impingement correlation is used to estimate the associated heat transfer coefficient. The heat transfer rates in droplet impingement are relatively high and as much as 50 percent of the droplet is evaporated during the short time between impact and bouncing off the hot vessel surface. This heat transfer acts as a heat sink and effectively removes a portion of the sensible heat that would otherwise penetrate into the interior of the vessel.

The heat conduction equation modeling heat transfer through the vessel wall must, by necessity, be solved numerically because the boundary conditions vary both in time and space (moving boundary conditions) and, therefore, an practical analytic solution is not available. A simple 3-point stencil is used in an explicit time-marching algorithm to solve the 1-D heat conduction equation for vessel wall temperature as a function of time. The heat transfer coefficients appearing in the system of differential equations are updated following each time step to reflect changing temperature conditions both at the outer and inner surfaces of the vessel wall. Note that for vessel walls of modest thickness (⅜ inch to ¾ inch), the maximum time step size to satisfy the algorithm stability criteria is sufficiently large that the heat transfer calculations for a fire duration of ½ hour can be completed on a Pentium PC-compatible computer within several minutes. Substantially thicker vessel walls may require an implicit time algorithm to solve the problem with a reasonable amount of computational expense.

3.3.1. Exterior Vessel Surface Heat Transfer

The vessel exterior surface temperature is compared to the boiling temperature of water. If the surface temperature is below the boiling point, the heat transfer coefficient is simply calculated by means of the familiar expression for forced convective heat transfer across the surface of a cylinder in cross flow (Fand, 1965):

$$\mathrm{Nu_f} = (0.35 + 0.56\mathrm{Re_f^{0.52}})\mathrm{Pr_f^{0.3}}$$

where $\mathrm{Nu_f}$, $\mathrm{Re_f}$, and $\mathrm{Pr_f}$ denote the film Nusselt, Reynolds, and Prandtl numbers, respectively.

If, on the other hand, the vessel surface temperature exceeds the boiling temperature, the critical heat flux is first calculated (Zuber, 1958):

$$\left(\frac{Q}{A}\right)_{\max} = \frac{\pi}{24} h_{\mathrm{fg}} \rho_{\mathrm{v}} \left[\frac{\sigma g(\rho_{\mathrm{l}} - \rho_{\mathrm{v}})}{\rho_{\mathrm{v}}^2}\right]\left(1 + \frac{\rho_{\mathrm{v}}}{\rho_{\mathrm{l}}}\right)^{1/2}$$

where

$(q/A)_{max}$ = maximum nucleate boiling heat transfer rate (W)
h_{fg} = enthalpy of evaporation of water at ambient conditions (J/kg)
σ = vapor–liquid surface tension of water at ambient conditions (N/m)
g = acceleration of gravity (m/s²)
ρ_l = liquid water density (kg/m³)
ρ_v = liquid water density (kg/m³)

This quantity represents the maximum allowable nucleate heat transfer rate across the surface of the vessel into the water film. The heat transfer rate calculated via the Rohsenow correlation for nucleate boiling:

$$\frac{C_{\mathrm{f}}\,\Delta T_x}{h_{\mathrm{fg}}\,\mathrm{Pr}_{\mathrm{f}}^{\mathrm{s}}} = C_{\mathrm{sf}}\left[\frac{(q/A)_{\mathrm{nucl}}}{\mu_{\mathrm{f}} h_{\mathrm{fg}}}\sqrt{\frac{\sigma}{\rho_{\mathrm{l}}-\rho_{\mathrm{f}}}}\right]^{0.23}$$

where

$(q/A)_{nucl}$ = nucleate boiling heat transfer rate (W)
C_f = liquid heat capacity (J/kg·K)
ΔT_x = temperature excess = $T_{wall} - T_{sat}$ (K)
C_{sf} = experimental constant depending upon the fluid-heating surface combination, (dimensionless)
μ_f = liquid viscosity (kg/m·s)
s = dimensionless constant (1.0 for water and 1.7 for other liquids)

is valid if its value does not exceed that of $(q/A)_{max}$. Note that C_{sf} is an empirical constant that depends upon the nature of the heating surface-fluid combination. Table 1 contains a list of values of the coefficient C_{sf} for various water-surface combinations. A more extensive table providing data for liquids other than water may be found in Holman (1983, Table 9-2).

In the event that the computed nucleate boiling heat transfer flux exceeds the maximum nuclear boiling heat flux, then the rate of heat transfer to the water film must be based on a film boiling correlation:

$$h_{\mathrm{film}} = 0.943\left[\frac{k_{\mathrm{v}}^3(\rho_{\mathrm{l}}-\rho_{\mathrm{v}})g\lambda'}{L\mu_{\mathrm{v}}\,\Delta T_x}\right]^{1/4}$$

where

h_{film} = film boiling heat transfer coefficient (W/m²· K),
k_v = vapor conductivity (W/m· s),
μ_v = vapor viscosity (kg/m· s),
L = πR, with R denoting the radius of the vessel (m).

TABLE 1
Values of the Coefficient C_{sf} for Various Liquid-Surface Combinations

Fluid-Heating Surface Combination	C_{sf}
Water-copper	0.013
Water-platinum	0.013
Water-brass	0.0060
Water–emery-polished copper	0.0128
Water-ground and polished stainless steel	0.0080
Water-chemically etched stainless steel	0.0133
Water-mechanically polished stainless steel	0.0132
Water–emery-polished and paraffin-treated copper	0.0147
Water-scored copper	0.0068
Water-Teflon pitted stainless steel	0.0058

Finally, the parameter λ' is expressed as

$$\lambda' = h_{fg}\left(1 + \frac{0.4c_{p_v}\Delta T_x}{h_{fg}}\right)$$

where c_{p_v} represents the specific heat of the saturated vapor.

In order to apply the above correlation, it must first be verified that the value of L does not exceed a certain critical value L_{cr} expressed as

$$L_{cr} = \frac{100\mu_v \lambda'}{2k_v \Delta T_x}\left[\frac{200\mu_v^2}{g\rho_v(\rho_l - \rho_v)}\right]^{1/5}$$

If the critical "length" is exceeded, an alternate expression should be used for evaluating h_{film}:

$$h_{film} = 0.20\rho_v c_{p_v}\left(\frac{g^2 L\mu_v}{\rho_v}\right)^{1/5}\left(\frac{\rho_l - \rho_v}{\rho_v}\right)^{2/5}\left(\frac{c_p \Delta T_x}{h_{fg}}\right)$$

3.3.2. Vessel Wall Heat Transfer

Heat transfer through the wall must be calculated by solving the heat conduction equation numerically. This follows from the fact that time-varying boundary-value problem solutions of partial differential equations can rarely be obtained analytically. The discretized heat conduction equation in the case of the outer surface boundary is expressed as

$$T_0^{t+1} = T_0^t\left[1 - \Delta\varphi\left(\frac{2k}{c\rho\Delta x^2} + \frac{2h_{\text{film}}\Delta\varphi}{c\rho\Delta x}\right)\right] + \frac{2h_{\text{film}}\Delta\theta}{c\rho\Delta x}T'_\infty + \frac{2k}{c\rho\Delta x^2}T_1^t + \frac{2\Delta\varphi}{c\rho\Delta x}q'_{\text{wall}}$$

where k, c, ρ, and Δx denote the thermal conductivity, heat capacity, density, and spatial (numerical) grid node separation for the vessel wall. Furthermore, q_{wall} represents the net heat flux into the wall (accounting for the heat removal by absorption of heat by the liquid film and removal of sensible heat from the vessel due to the evaporation of water film).

The time-step parameter $\Delta\varphi$ is a critical parameter and must satisfy the following criteria to ensure that the numerical method will remain stable and provide physically realistic values:

$$\Delta\varphi \leq \frac{1}{2}\frac{\Delta x^2}{a}\left[\frac{1}{1 + (h_{\text{film}}\ \Delta x/k)}\right]$$

where a is the vessel wall thickness. In many instances, $\Delta\varphi$ may become exceeding small, incurring the risk of large computer run times. Furthermore, it is difficult to estimate a priori the maximum value of h_{film} (yielding the minimum value of $\Delta\varphi$, and therefore ensuring global stability). The program developed for this SRV baseline sizing effort allows the user to input a maximum value for the film heat transfer coefficient. The program will advise the user if the associated stability criteria were evaluated and will terminate execution, permitting the user to enter a larger estimate and rerun the analysis.

3.3.3. Interior Vessel Surface Film Heat Transfer

The heat conduction inside the vessel is modeled as a convective heat transfer process in an enclosed space. Details of the model used can be found in Holman (1981, Section 7-11) and Kreith (1973, Chapter 6).

4. SAMPLE CALCULATIONS

Sample calculations using the water spray and heat transfer model will now be presented for two different cases:

1. A nonreacting case using normal hexane.
2. A reactive case using 1,3-butadiene in styrene with an initiator.

All vessels are taken to be horizontal right circular cylindrical vessels with elliptic heads for the sake of comparison.

4.1. Hexane Example

The physical parameters pertinent to this example are given in Table 2. The base case considers the standard API-recommended heat flux exposure and a flame temperature based upon that of burning n-hexane.

Figure 3 provides a plot of vessel wall temperature versus time for the API-recommended value for heat flux of 34,500 BTU/ft^2 multiplied by the wetted surface area in the case of poor drainage (the plot marked 1.088×10^5 W/m^2: API no drainage). Note that the temperature does not exceed 373 K (100°C) after 12 hours of continuous exposure. The spray coverage of 0.35 gal/(ft$^2 \cdot$ min) corresponds roughly to double spray coverage and agrees with the results quoted by Fisher and Forrest (1985) for maximum wall temperature estimates.

Heat flux is a crucial parameter in SRV sizing calculations. This fact is no less true in the case of water film cooling calculations. Included in Figure 3 is a comparison between the wall temperature as a function of time for the hexane vessel for various levels of heat flux exposure.

Flame temperature is also a key parameter in the total heat transfer phenomena. Unlike the heat flux exposure, however, the model predicts relatively little sensitivity of wall temperature response as a function of time for various flame temperatures. The significant difference here is the heating rate of the liquid or the heat flux conducting through the inner shell to the fluid inside as indicated in Figure 4.

TABLE 2
Hexane Tank Parameters.

Parameter	Value
Capacity	3000 gallons
Length	180 inches
Diameter	72 inches
Vessel Material	SA285C (U1)
Shell Thickness	0.469 inch
Geometry	horizontal cylinder with elliptic ends
Initial Temperature/Pressure	110 °F/60 psig
Fill Level	85% of capacity
Water Spray Coverage	0.35 gal/(ft$^2 \cdot$ min)

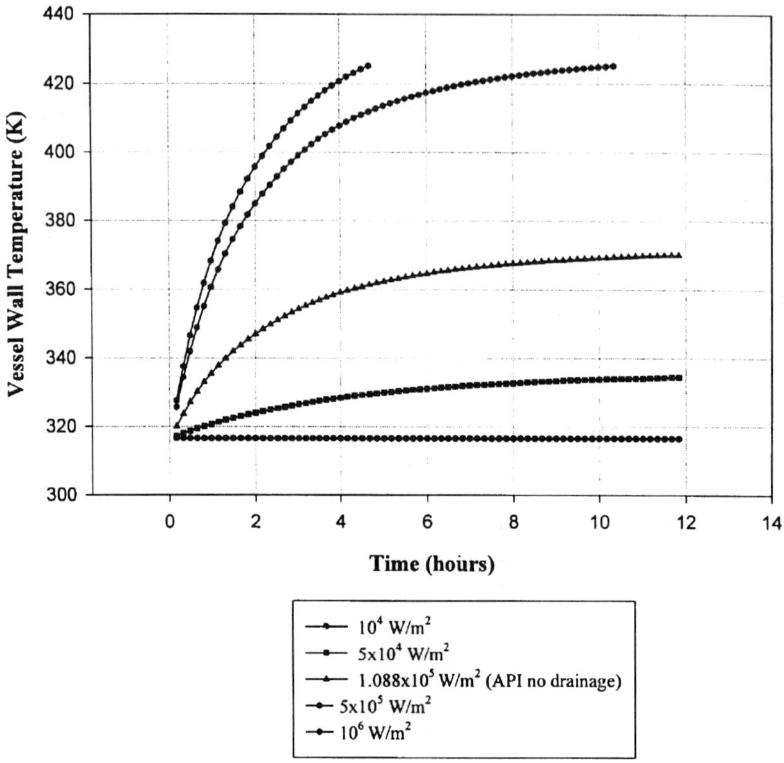

Figure 3. Hexane Vessel Example at Various Heat Exposure Levels. (Note: Some calculations terminated early because set temperature of 425 K was exceeded.)

4.2. Butadiene Example

A mixture of 91% BD in styrene with initiator was analyzed using the same vessel parameters listed in Table 3. The data contained in Table 4 were used to estimate the rate of polymer formation in the early stages of dimerization. As the temperature increases, a second type of auto-addition reaction initiates whereby many diolefin molecules combine to form a series of high molecular weight polymers. The rate of plastic polymer formation as a function of temperature for a given concentration of active oxygen is given in Table 5. The final reaction of consideration is "popcorn" polymer (so named because of its appearance of popcorn, cauliflower, or white crystals) initiated by air or peroxides present in the butadiene batch. Even in the presence of TBC, popcorn polymer growth may occur if seed crystals are available. Finally, the Arrhenius plot data from Figure 1 was used to characterize the time-dependence in the heating rate from the reactions. Figure 5 provides a plot of the vessel wall temperature as a function of time up to the instant of SRV actuation (SRV set pressure of 100 psig).

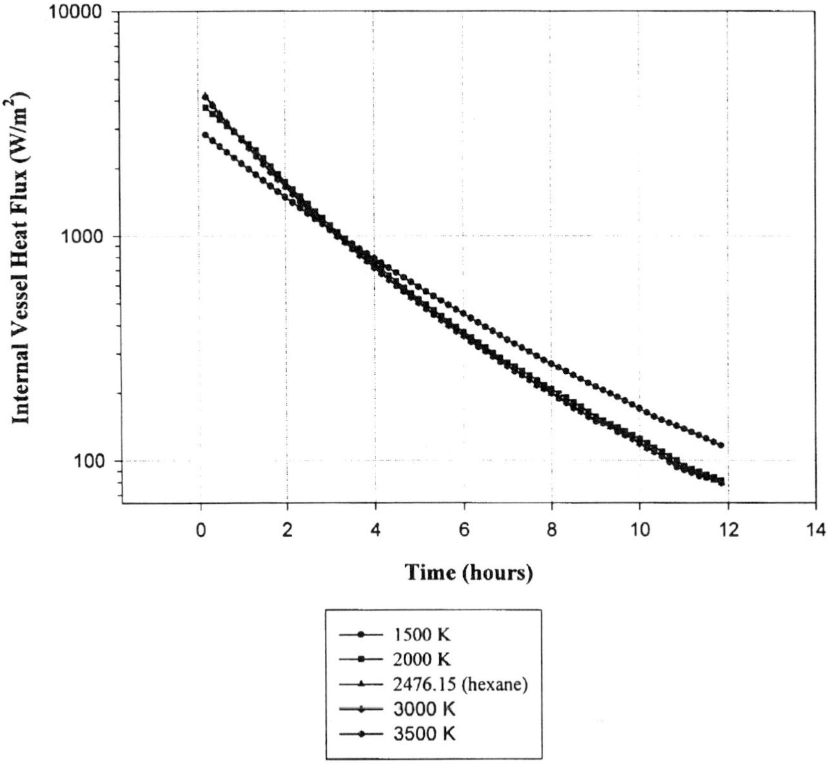

Figure 4. Hexane Vessel Example at Various Flame temperature Exposure Levels

TABLE 3
Butadiene Tank Parameters.

Parameter	Value
Capacity	3000 gallons
Length	180 inches
Diameter	72 inches
Vessel Material	SA516-70 (U1)
Shell Thickness	0.469 inch
Geometry	horizontal cylinder with elliptic ends
Initial Temperature/Pressure	110 ^{o}F/60 psig
Fill Level	80% of capacity
Water Spray Coverage	0.35 gal/(ft^2·min)

TABLE 4
Effect of Temperature on Rate of Dimer Formation from Butadiene (Exxon).

Temperature (°C)	Weight Percent Diolefin Polymerized Per Hour
20	0.00015
40	0.0014
60	0.013
80	0.12
100	1.1

TABLE 5
Effect of Temperature on Rate of Plastic Polymer Formation from Butadiene (1,200 ppm Active Oxygen) (Exxon).

Temperature (°C)	Weight Percent Diolefin Polymerized Per Hour
20	0.0012
40	0.014
60	0.17
80	1.9

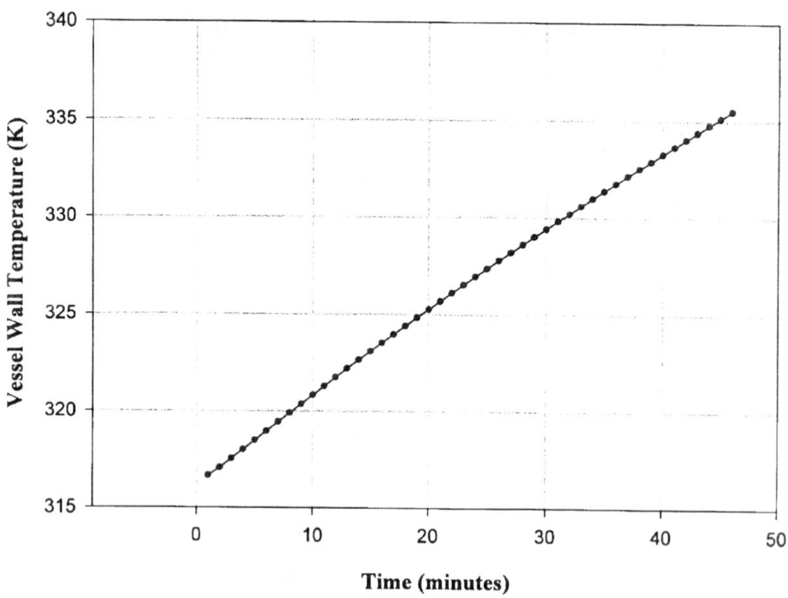

Figure 5. Butadiene Vessel Example

5. LIMITATIONS OF THE MODEL AND FUTURE WORK

The objective of this water spray heat transfer model was to obtain a first-order accurate prediction of the heat protection capabilities of a vessel water deluge system in the event of external exposure to fire. The model is based largely on simple analytical models in falling film theory and empirical heat transfer coefficient. The method requires a numerical solution because of the discretization of the vessel surface in sufficiently fine elements that the local analytic methods can be applied and the fact that a moving boundary heat transfer problem to date does not possess an analytic method for solution. Otherwise, the model has been greatly simplified over more rigorous boundary layer calculations in order to first assess the viability of water sprays in protecting vessel from overpressurization or at a very minimum reduce the relief capacity requirements.

Perhaps one of the greatest objections to a model of this nature is the uncertainty in the ambient conditions such as wind speed, stability as well as the availability of the water spray system. The later becomes a critical element especially if the design basis for the safety relief valve sizing assumes that the system will work at or near capacity. Appendix B provides some discussion and details of how the system reliability should be evaluated to assist in making an informed judgement if taking credit for water sprays in SRV sizing calculations is appropriate for a given facility (Unwin, 1995).

Some of the areas that require further refinement of this model in order to make it a truly robust and accurate design basis is tool are the following:

1. Generalize the model to accommodate different nozzle arrangements and configurations. The current model assumes an essentially uniform water coverage with the exception of the very bottom of the vessel.
2. The model does not account for loss of water coverage due to splashing. Although there is a reasonable amount of work presented in the literature on droplet spray impingement on hot surfaces, the models are primarily applicable only in highly controlled conditions for a simple fixed geometry. This work will require extension to more general cases where droplets encounter a preexisting water film and the geometry and temperature vary significantly with time and spatial location.
3. No attempt here is made to characterize the effects of wind outside of the simple laminar boundary layer regime. Wind turbulence will have a great effect on the water application and the water film itself, particularly in the regions of flow separation where additional kinetic turbulence can be generated and the water film can be greatly distorted to create hotspots on the vessel surface.
4. Model predictions require experimental validation. At this time, very limited data exists for this kind of study.

APPENDIX A: FLAME TEMPERATURE CALCULATION

This section demonstrates how flame temperature may be estimated when reliable data are not available.[3] Say, for instance, that a tank farm with several vessels containing butadiene is considered for SRV size verification or design basis calculation. Since butadiene is flammable, a leak of this material from one of the process vessels can lead to a source of heat and overpressurization for other nearby vessels. The simple chemical equation for the complete oxidation of butadiene is given by

$$2CH_4 + 11O_2 \rightarrow 8CO_2 + 6H_2O$$

The combustion energy raises the temperature of the product gases. In order to obtain accurate estimates of the flame temperature, it is necessary to include the nitrogen complement into the above reaction. Therefore the oxidation reaction is now given by

$$C_4H_6 + \tfrac{11}{2}O_2 + \tfrac{11}{2}(3.76)N_2 \rightarrow 4CO_2 + 3H_2O + 20.7N_2$$

Note that the term 3.76 appearing in the above equation represents the molar ratio of nitrogen to oxygen in the normal composition of the dry atmosphere.[4] The nitrogen component acts as a 'thermal ballast' that absorbs a large fraction of the combustion energy without actively participating in the chemical reaction itself. The flame temperature is simply estimated via the formula

$$T_{\text{flame}} = T_{\text{ambient}} + \frac{\Delta H_{c_{\text{butadiene}}}}{C_{p_{\text{products}}}}$$

with

$\Delta H_{c_{\text{butadiene}}}$ = Stoichiometric heat of combustion of butadiene
= 44.2 MJ/kg = 2.39 MJ/mole (molecular weight of 54.09 g/mole)

$C_{p_{\text{products}}}$ = Total heat capacity of product gases (J/kg·K)

Using the thermal capacities of the product gases given in Table A-1, $C_{p_{\text{products}}}$ evaluates to

$$\begin{aligned} C_{p_{\text{products}}} &= 4C_{p_{CO_2}} + 3C_{p_{H_2O}} + 20.7C_{p_{N_2}} \\ &= 1017\ \text{J / (moles of BD combustion products)} \cdot \text{K} \end{aligned}$$

3 This calculational procedure was taken from Drysdale, 1985, Section 1.2.3.

4 The normal mole fraction composition of air is composed of 78.09% N_2, 20.95% O_2, 0.93% argon and 0.03% CO_2 with other minor constituents neglected (Weast, 1974/75).

TABLE A-1
Thermal Capacities of Common Gases at 1000 K (Lewis and von Elbe, 1987).

Chemical Species	$C_p^{1000\,K}$ (J/mol·K)
Carbon monoxide (CO)	33.2
Carbon dioxide (CO_2)	54.3
Water (vapor) (H_2O)	41.2
Nitrogen (N_2)	32.7
Oxygen (O_2)	34.9
Helium (He)	20.8

Therefore, assuming an ambient temperature of 100°F (38°C), the flame temperature is calculated to be

$$T_{\text{flame}} = 38^\circ\text{C} + \frac{2392000\ \text{J / (moles BD combusted)}}{1017\ \text{J / (moles BD combustion products)} \cdot {}^\circ\text{C}} = 2390^\circ\ \text{C}$$

Since adiabatic flame temperature data are not readily available for hexane and styrene, their flame temperatures are estimated in a completely analogous manner. As a second example, the flame temperature of hexane is computed below.

The simple chemical equation for the complete oxidation of hexane is given by

$$2C_6H_{14} + 19O_2 \rightarrow 12CO_2 + 14H_2O$$

The combustion energy raises the temperature of the product gases. After including the nitrogen complement into the above reaction, the oxidation reaction becomes

$$C_6H_{14} + \tfrac{19}{2}O_2 + \tfrac{19}{2}(3.76)N_2 \rightarrow 6CO_2 + 7H_2O + 35.7N_2$$

Again the flame temperature is simply estimated via the formula

$$T_{\text{flame}} = T_{\text{ambient}} + \frac{\Delta H_{c_{\text{butadiene}}}}{C_{p_{\text{products}}}}$$

with

$\Delta H_{c_{\text{butadiene}}}$ = Stoichiometric heat of combustion of butadiene = 44.7 MJ/kg = 3.86 MJ/mole (molecular weight of 86.17 g/mole)

$C_{p_{\text{products}}}$ = Total heat capacity of product gases (J/kg·K)

Using the thermal capacities of the product gases, $C_{p_{products}}$ evaluates to

$$C_{p_{products}} = 6C_{p_{CO_2}} + 7C_{p_{H_2O}} + 35.7C_{p_{N_2}} = 1782\,\mathrm{J/mol \cdot K}$$

Therefore, assuming an ambient temperature of 38°C (100 °F), the flame temperature is calculated to be

$$T_{flame} = 38^{\circ}\,\mathrm{C} + \frac{3857000\,\mathrm{J/kg}}{1782\,\mathrm{J/kg \cdot {}^{\circ}C}} = 2203^{\circ}\,\mathrm{C}$$

APPENDIX B: ASSESSMENT OF DELUGE/SPRINKLER SYSTEM RELIABILITY

If relief valve sizing permits the mitigative effects of vessel cooling due to a water deluge/sprinkler system to be taken into account, it is critical that the reliability and efficacy of the deluge system under evaluation be properly characterized and understood.

Before credit may be taken for a deluge system in sizing vessel relief valves, it must be demonstrated that the deluge system is adequately reliable. A deluge system is an active system, i.e., it must be successfully actuated (transferred from one operating state to another) in order to perform its function. Active safety systems typically have lower reliability that passive safety systems (such as a dike or fire wall). The parameter that will be used to address the reliability performance issue is the system *unavailability*: the numerical probability that the system will fail to actuate successfully, i.e., that it will fail to deliver the minimum deluge rate assumed in the relief valve sizing calculations. This failure could be a consequence of control failure, equipment failure, or repair/maintenance outage, due to either hardware failure or human error. The number of potential failure causes and their individual likelihoods dictate the overall unavailability of the system.

Before credit can be taken for a water deluge system in the relief valve sizing calculations, the unavailability of the system must be demonstrated to be less than an established threshold.

To demonstrate that this requirement is met, it is first necessary to list the potential failure modes of the system. This requires, first, that the success criteria for the system be developed—for example, the number of fire pumps that must operate and the number of deluge valves that must open on demand should be specified. Failure modes of the system (i.e., scenarios in which the success criteria are not met) can then be identified. The identification of system failure modes can be facilitated by developing a fault tree. For a simple system, it may be practical to list failure modes without developing a fault tree. The failure modes should include:

1. Fire detection and deluge control system failures,
2. Equipment failures, and
3. Support system (city water, pneumatic systems, etc.) failures.

Note, any system that requires operator actuation (i.e., the system does not automatically actuate) should be assumed to fail to meet the unavailability requirement, and credit for the deluge system should not be accounted for in the relief valve sizing calculations.

The assessment should focus on failure modes of active equipment requiring a change in operating state (e.g., pump startup or valve position transfer) rather than on failure modes of passive equipment (e.g., pipe flange leaks or pipe ruptures), since active components will generally drive the system reliability.

The unavailability of equipment that is required to change operating state on demand should be estimated by use of the following formula:

$$U = 1 + \frac{e^{1fT} - 1}{fT} + \frac{fR}{fR+1}$$

where f is the (standby) failure rate of the equipment item, T is the time between functional tests of the equipment, and R is the mean time that the equipment would be unavailable for each repair. The first term on the right of this equation is the contribution to equipment unavailability associated with equipment failures that are undiscovered until the equipment is called upon to actuate. The second term is the contribution to equipment unavailability associated with situations in which the equipment is out for repair (corrective maintenance). Equipment unavailability associated with routine preventive maintenance is assumed to be small compared to the previous two causes of unavailability. If this assumption is incorrect, i.e., if the term

$$\frac{gM}{gM+1}$$

would be a significant contributor to the unavailability of the equipment (where g is the frequency of preventive maintenance and M is the mean PM duration), then this term should be added to the unavailability equation.

If the equipment or support system is normally operating rather than being in a standby mode (e.g., city water supply), then the unavailability can be estimated from the formula:

$$U = \frac{fR}{fR+1}$$

where f is the (running) failure rate of the equipment and R is the mean time that the equipment would be unavailable before restoration.

The failure rates and repair times of equipment may be estimated from plant-specific operational data gathered from the system. The basis for the data should be documented. Otherwise, the data given in Table B-1 may be used. Test intervals, T, and support system unavailabilities specific to the system must be used with this data.

The overall unavailability of the system is equal to the sum of the system unavailabilities contributed by each failure mode of the system. The unavailability associated with each failure mode is the product of the unavailabilities of the independent equipment items which fail as part of that system failure mode.

TABLE B-1
Generic Failure Rate Data for Water Deluge Reliability Analysis[a]

Equipment Type	Failure rate (per hour), f	Mean repair time (hours)
Electric Motor Firewater Pump (including controls)	7.2E-5 (startup failure)	130
Diesel Motor Firewater Pump (including controls)	2.1E-4 (startup failure)	86
Pneumatically Operated Deluge Valve (including fire detection device)	5.8E-6 (failure to open)	8

[a]Based on OREDA Reliability Data

REFERENCES

Bird, R.B., Stuart, W.E., and Lightfoot, E.N. (1960), *Transport Phenomena*, New York: Wiley.

Crozier, R.A., Jr. (1985), *Chemical Engineering Progress*, October 28, 1985, Figure 3, p. 52.

Drysdale, D. (1985), *An Introduction to Fire Dynamics*, New York: Wiley.

Epstein, M., Fauske, H.K., and Hauser, G.M. (1989), "The Onset of Two-Phase Venting Via Entrainment in Liquid-Filled Storage Vessels Exposed to Fire," *J. Loss Prevention in the Process Industries*, 2(January), 45–49.

Exxon, Olefins and Diolefins: Transportation, Storage, and Handling and an Exxon brochure on dimer formation and TBC inhibition, Exxon Chemical Company, Houston, Texas.

Fand, R.M. (1965), "Heat Transfer by Forced Convection from a Cylinder to Water in Crossflow," *Int. J. Mass Transfer*, 8, 995.

Fauske, H.K. and Leung, J.C. (1985), "New Experimental Techniques for Characterizing Runaway Chemical Reactions," *Chem Engr Prog*, 81(8), 39–46.

Fisher, H. G. et al. (1992), *Emergency Relief System Design Using DIERS Technology—The DIERS Project Manual,* New York: AIChE/DIERS.

Forest, H. S. (1992), "Appendix I-A: the Coupling Equation and Flow Models," in *Emergency Relief System Design Using DIERS Technology—The DIERS Project Manual,* New York: AIChE/DIERS.

Fisher, H. G,. and Forrest, H. S. (1995), "Protection of Storage Tanks from Two-Phase Flow Due to Fire Exposure," *Process Safety Progress*, 14(3), 183–199.

Gradshtein, I. S., and Ryzhik, I. M. (1980), *Table of Integrals, Series, and Products*, 2nd ed., New York: Academic Press.

Holman, J. P. (1981), *Heat Transfer*, 5th ed., New York: McGraw-Hill.

Kreith, F. (1973), *Principles of Heat Transfer*, 3rd ed., New York: Intext Educational Publishers.

Leung, J. C., (1986), *AIChE Journal,* 32(10), 1622–1634.

Leung, J. C., Fauske, H. K., and Fisher, H. G. (1986), "Thermal Runaway Reactions in a Low Thermal Inertia Apparatus," *Thermochem Acta,* 104, 13–29.

Leung, J. C. (1991), "Vent Sizing for Tempered Vapor Systems," in *Safety of Chemical Batch Reactors and Storage Tanks*, A. Benuzzi and J.M. Zaldivar (eds.), pp. 299–310, Dordrecht: Kluwer Academic Publishers.

Leung, J. C. (1992), "Size Safety Relief Valves for Flashing Liquids," *Chemical Engineering Progress,* February, 70–75.

Lev, Y and Stachan, D. C. (1989), "A Study of Cooling Water Requirements for the Protection of Metal Surfaces Against Thermal Radiation," *Fire Technology*, 25(3), 213–229.

Lewis, B. and von Elbe, G. (1987), *Combustion, Flames and Explosions of Gases*, 3rd ed., New York: Academic Press.

Mudan, K. S.(1984), "Thermal Radiation Hazards from Hydrocarbon Pool Fires," *Prog. Energy. Combust. Sci.*, 10, 59–80.

Perry, R.H. (1984), *Perry's Chemical Engineers' Handbook*, New York: McGraw-Hill.

Unwin, S. (1995), "Assessment of Deluge/Sprinker System Reliability," Battelle Internal Report, Battelle Columbus Laboratories, Columbus, OH.

Weast, R.C. (Ed.), 1974/1975, *Handbook of Chemistry and Physics*, Chemical Rubber Company, Ohio.

Zuber, N. (1958), "On the Stability of Boiling Heat Transfer," *Trans. ASME*, 80, 711.

Phenolic Runaway Reaction: Pressure Relief and Containment

Joseph C. Leung* and Tom R. Fitzsimons
Fauske & Associates, Inc., 16W070 West 83rd Street, Burr Ridge, Illinois 60521

Wayne A. Kleine and Travis Rudoll
Plastics Engineering Company, 3518 Lakeshore Road, Sheboygan, Wisconsin 53083

1. INTRODUCTION

A series of tests were carried out on a client's site to simulate runaway reaction of selected phenol-formaldehyde (PF) formulations in a 20-liter (5.25 gallon) reaction vessel. One of the objectives in this study was to test the effectiveness of an external water pool in suppressing the vapor release as well as capturing the liquid resin via a 25 mm (1 inch) diameter tee acting as a simple submerged sparger. Another objective was to validate the DIERS vent sizing methodology. Concern regarding possible viscous flow behavior of the resin solution during discharge was addressed by measuring the flow rate in a 25 mm (1 inch) bottom-vented tailpipe.

2. EXPERIMENTAL SET-UP

2.1. Reaction Vessel

The overall test apparatus is shown in Fig. 1. The 5.25-gallon (20 liter) stainless-steel reaction vessel is made out of an 8-inch Sch 10 pipe and has an internal height of 23 inches. The design pressure is calculated to be over 300 psig. This reaction vessel was hydro-tested to 250 psig. Strip heaters that surround the bottom half of the vessel are electrically divided into three heater zones so as to provide a more easily controlled heat input. The bottom vent line is a 1.049-inch I.D. (Sch 40) pipe and is connected to a 2-inch elbow to

*Leung Inc., 1521 Evergreen Lane, Darien, IL 60561.

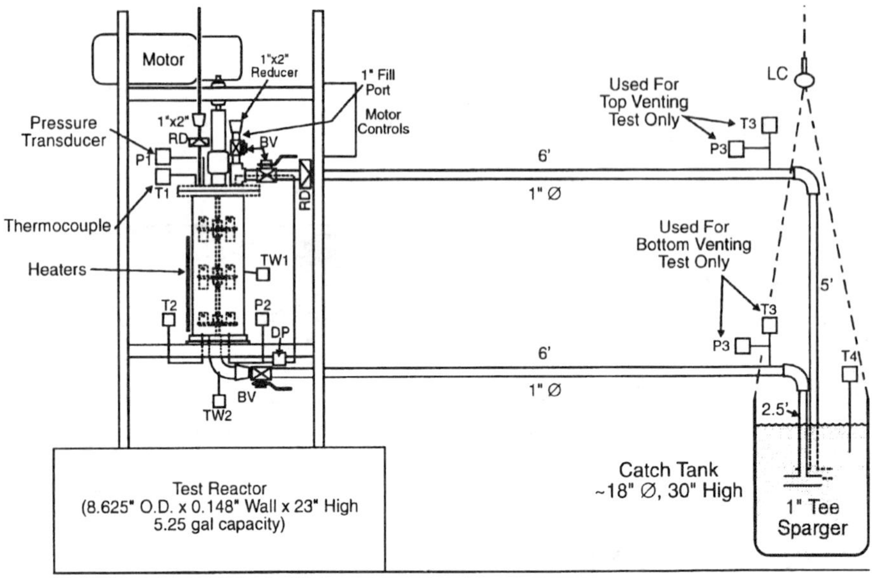

Figure 1. 20-liter Reaction Vessel and Associated Experimental Setup.

which a 1-inch full-bore ball valve is attached (separated by a short length of pipe), see Fig. 1. The total linear length of pipe from the bottom flange of the vessel to the exit of the ball valve is about 6 inches. The agitator used for these tests had three blades (120° apart) and was driven by an AC motor with speed control. The 1-inch horizontal vent line would direct the discharge into an adjacent atmospheric catch tank containing ambient temperature water. A 1-inch pipe was attached to the end of the horizontal vent line to divert the discharge vertically down into the catch tank. This was followed by a 1-inch tee serving as a quencher arm arrangement. The vent lines were not insulated nor heated, so they were at ambient temperature prior to venting.

As a safety measure, two additional relief vents were provided in case of emergency. A 1-inch diameter top vent line containing a 1-inch rupture disk set to burst at 60 psig would direct the emergency discharge to the catch tank also, see Fig. 1. This vent line had a 1-inch full-bore ball valve just upstream of the rupture disk which would be open prior to conducting the test. By removing the rupture disk, this vent line provided a top-vented configuration for the last test conducted. The second emergency relief vent was provided by another 1-inch rupture disk set to burst at 120 psig in a vertical discharge orientation.

2.2. Instrumentation

Figure 1 shows the location of the pressure transducers and the (Type K) thermocouples. A differential pressure transducer was used to measure the density (and void fraction) inside the reaction vessel. The load cell was a strain gauge type with a span of 5 volts corresponding to about 550 lb. It was used to measure the discharge rate as well as the cumulative mass discharged.

All thermocouple and transducer signals were first sent to a signal-conditioning unit and then to a PC computer. The data logging rate during venting was approaching 20 data points per second per channel.

2.3. Test Procedure

The overall procedure involved calibration of the transducers and the thermocouples, loading of phenol, formaldehyde and catalyst, heatup of the batch and bleeding off the air pad, and finally pressure relief venting. Relief venting was initiated by quickly opening the 1-inch full-bore ball valve (in a time scale of approximately 0.1 sec).

Following calibration of the pressure transducers and the thermocouples, both phenol and formaldehyde (premix) were charged to the reaction vessel at a temperature of about 50°C. Then the agitator was started and the catalyst was added. This was followed immediately by a water chase of about 500 ml. The heaters were then turned on to yield about 1°C/min heatup rate in the reacting batch. The heater power was eventually turned off at 95–100°C. The atmospheric pad was periodically bled off during the heatup, and for a final time at around 100°C.

2.4. Posttest Operation

The relieving ball valve was closed and a weak caustic solution was charged to half full in the reaction vessel. With full agitation the caustic solution was heated to over 100°C before discharging through the bottom vent line into the catch tank. The waste solution in the catch tank was pumped out to a 55-gallon disposal drum. The reaction vessel was then filled with water and heated to 100°C again before dumping the contents. A final rinse was performed using water but this time it was heated only to 60°C. Such a cleaning procedure was confirmed to be effective as the postcampaign disassembly of the reaction vessel revealed no polymer buildup nor discoloration. (In fact the vessel internals looked more shiny than before these tests.) In this manner an average of two tests were executed per day.

3. VSP VENTING SIMULATION TEST

Early on, a venting simulation test was conducted in the VSP2 equipment shown in Fig. 2. Here a test cell with a dip tube was used simulating a bottom venting configuration. Figure 3 depicts the test cell pressure behavior during

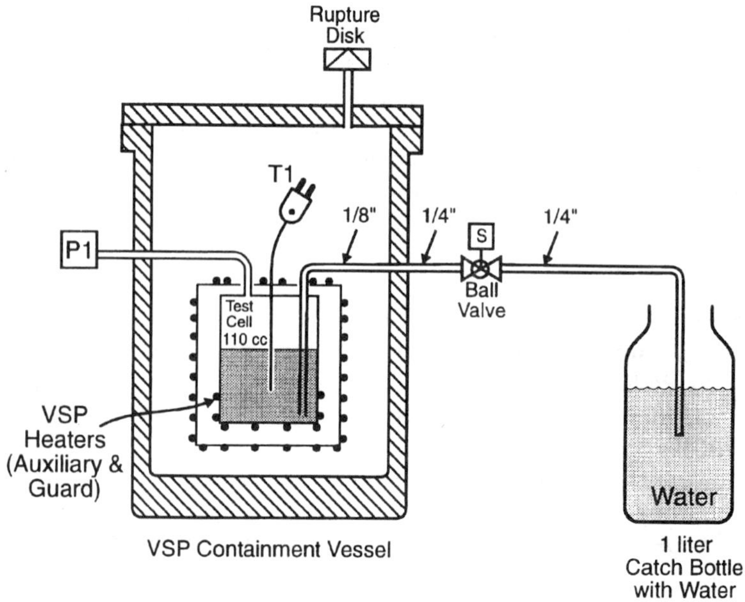

Figure 2. VSP Setup for Bottom Venting Simulation.

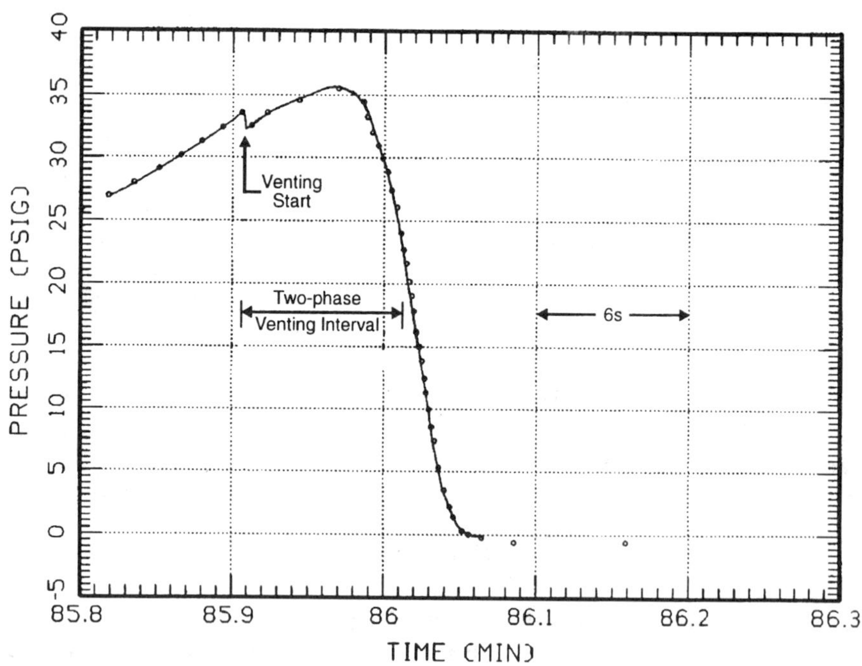

Figure 3. Pressure Data during Simulated Venting in the VSP Apparatus.

venting. Two-phase venting period is estimated to last for about 6.5 seconds. Beyond that, the liquid level inside the test cell had dropped so low that vapor would be entering the vent tube, resulting in a much quicker depressurization. Our analysis indicates that the vent flow can be predicted closely with a non-viscous two-phase flow model, i.e. no viscous flow behavior was observed in the 0.093" (2.36 mm) I.D. vent line.

This VSP2 test also provided valuable self-heat rate data for comparison against the 20-liter test data. Such a comparison, shown in Fig. 4, demonstrates that the 20-liter data are in fairly good agreement with the VSP2 data despite the larger scale setup which was bound to introduce more heat loss.

4. 20-LITER TEST RESULTS

A total of six PF formulations have been successfully conducted in the 20-liter scale simulating both runaway scenarios and pressure relief. The test results are summarized in Table 1. All of these tests except one were bottom-vented tests. Though this is not the predominant vent configuration employed in the industry, the bottom-vented configuration allows direct scale-up and more importantly results in the highest viscous flow behavior in the vent piping.

For relief set pressures between 11–13.5 psig, the observed peak pressures ranged from 12 to 19 psig. The temperature rise rates (or self-heat rates) just prior to relief ranged from 18 to 62°C/min. The highest dT/dt was observed with the caustic PF formulation and also with the double-charge acid PF formulation.

Figure 5 depicts the overpressure transient for the caustic PF formulation (Test #5)—a typical plywood resin recipe. The load cell data in Fig. 6 show the weight gain in the quench tank. The discharge rate from the reaction vessel can therefore be obtained from the slope in Fig. 6.

5. TWO-PHASE FLOW SIZING METHODOLOGY

The simplified DIERS two-phase flow technology can be used to provide posttest predictions of the experimental data. Treating the vessel contents as a uniform frothy mixture during venting—prior DIERS work shows that a uniform-froth assumption is quite realistic even for a bottom-venting process—the required vent rate (or mass flow rate) W to achieve a given overpressure (i.e. pressure above the set pressure) is given by the following equation [1],

$$W = \frac{m_0 C_p (dT/dt)}{\left[\left(\frac{v}{m_0}\frac{h_{fg}}{v_{fg}}\right)^{1/2} + (C_p \Delta T)^{1/2}\right]^2} \tag{1}$$

TABLE 1
Test Result Summary

Reactor Volume = 5.25 gal

Reactor Charge = 34–35 lb

Vent Piping: 1"ø by 8 ft. approx. with two 90° elbows

Test No.	Catalyst	Vent Location	Relief Pressure (psig)	Peak Pressure (psig)	Turnaround Time (sec)	*dT/dt* at Relief (°C/min)
1	Acid #1	Bottom	11.0	11.8	7.0	18
2	Base #1	Bottom	13.5	14.8	6.5	30
3	Base #2	Bottom	12.5	13.5	7.5	18
4	Acid #2	Bottom	12.5	14.0	8.5	23
5	Caustic[(a)]	Bottom	13.0	19.0	9.0	60
6	Acid #2[(b)]	Top	13.0	17.5	8.0	62

[a]42% wt. phenol, 51% wt. of 52% wt. formaldehyde solution, 7% wt. of 20% wt. NaOH solution.
[b]Double charge.

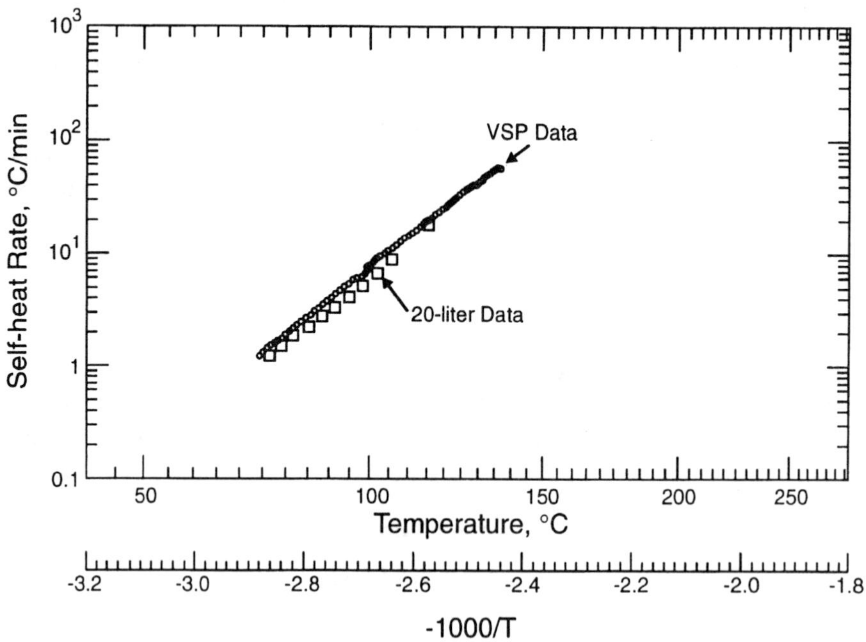

Figure 4. Comparison of *dT/dt* Data between VSP and 20-liter Reaction Vessel.

Here

m_0 = initial mass in vessel,
C_p = specific heat of vessel content,
dT/dt = adiabatic self-heat rate,
V = vessel internal volume,
h_{fg} = latent heat of vaporization,
v_{fg} = specific volume increase upon vaporization,
ΔT = over-temperature corresponding to overpressure.

This equation has been derived by consideration of the macroscopic material and energy balance in a vessel undergoing two-phase discharge. Thus, for the observed overpressure, one can obtain the corresponding "over-temperature" from the pressure–temperature relation. For this system, the $\Delta P/\Delta T$ quantity can be given by the steam table.

$$\frac{\Delta P}{\Delta T} = 0.82 \frac{\text{psi}}{^{\circ}\text{C}}$$

Though we do not have accurate measurement of the specific heat of the P/F reacting mixture, a best-estimate C_p value of 0.7 Btu/lb°F (2930 J/kg K) is used here.

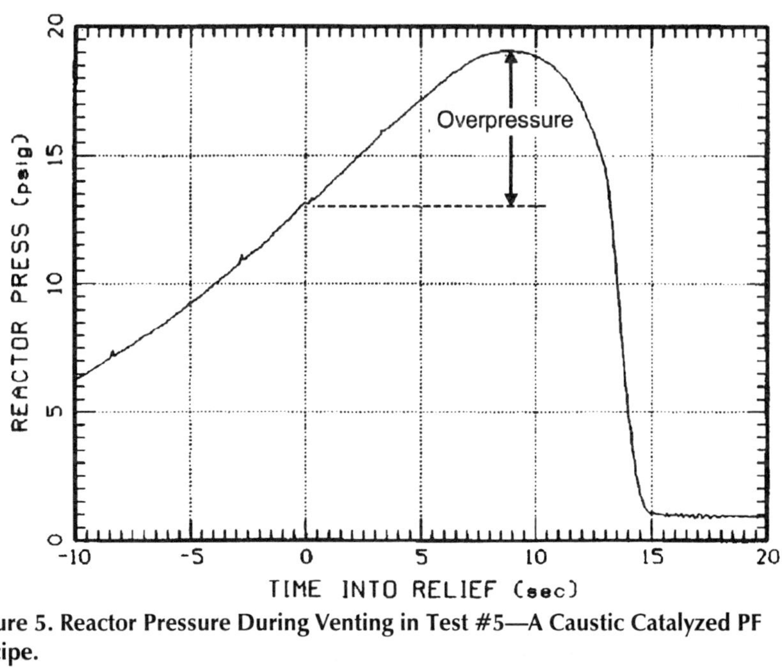

Figure 5. Reactor Pressure During Venting in Test #5—A Caustic Catalyzed PF Recipe.

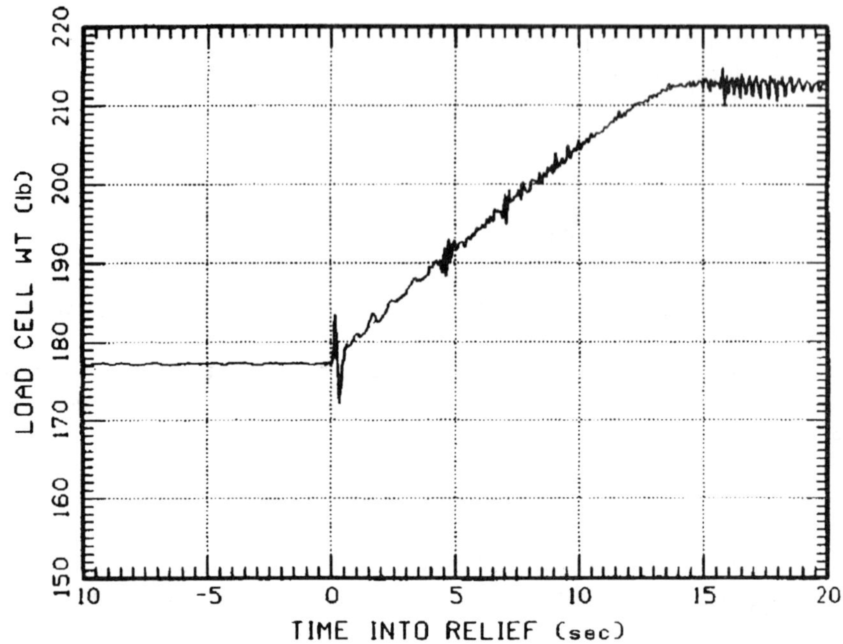

Figure 6. Catch Tank Weight during Venting in Test #5—A Caustic Catalyzed PF Recipe.

Finally the thermodynamic quantity h_{fg}/v_{fg} can be evaluated based on steam table properties. In our evaluation, this quantity is most conveniently obtained from the Clapeyron thermodynamic relation,

$$\frac{dP}{dT} = \frac{h_{fg}}{v_{fg}T} \tag{2}$$

Approximating the *P-T* data in the region of interest by a two-constant Antoine-type equation,

$$\ln P = a - \frac{b}{T} \tag{3}$$

one can combine these two equations to yield

$$\frac{h_{fg}}{v_{fg}} = \frac{bP}{T} \tag{4}$$

where b is simply the slope on a plot of log P versus reciprocal temperature [from Eq. (3)]. Using Eq. (1), the required vent rates for these formulations were evaluated at various overpressures. The predictions are shown in Figs. 7

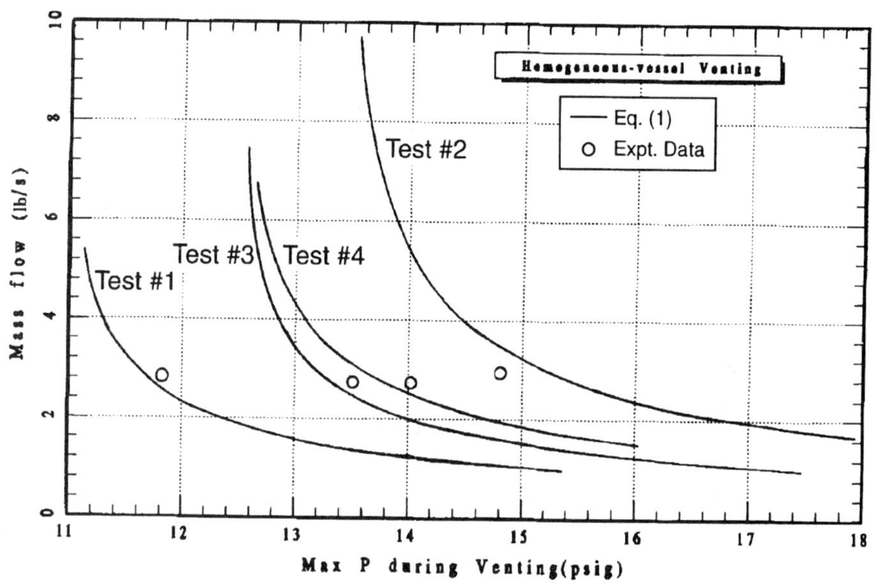

Figure 7. Required Vent Rate Predictions versus Experimental Data for Tests #1–#4.

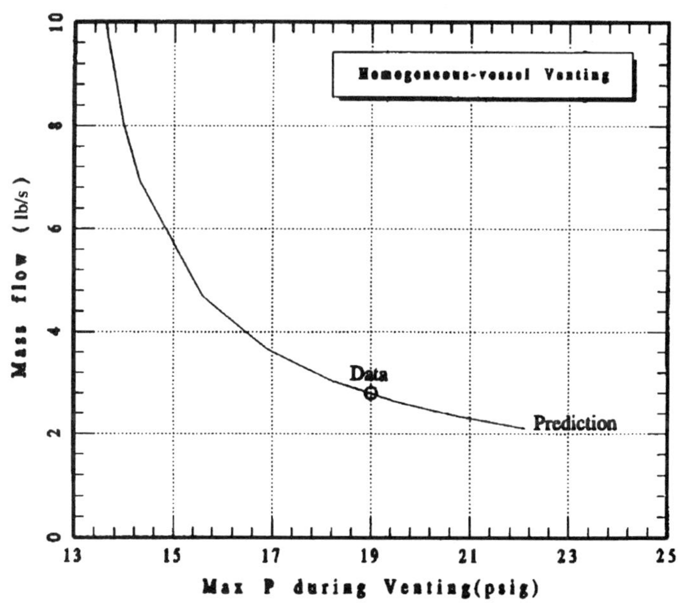

Figure 8. Required Vent Rate Prediction versus Experimental Data for Test #5.

and 8. Here we note that the required vent rate drops off rapidly with a slight overpressure—overpressure is the maximum pressure minus the relief set pressure—this is quite characteristic of a two-phase venting system. Also shown in these plots are the experimental vent rates which were determined from the load cell data during the overpressure intervals. For all five bottom-vented tests (Tests 1 through 5), the observed vent rates are closely predicted by our calculation. Figure 9 is a disparity plot of the experiment vent rates versus the predicted vent rates; except for Test #2 all are predicted to within ±10%.

For the top-vented test, Test #6, two predictions are provided—a bubbly (frothy) regime and a churn-turbulent (water-like) regime [2]. The latter is a flow regime which permits significant vapor/liquid disengagement within the vessel. The observed vent rate is bounded by these two limiting predictions as shown in Fig. 10. However, it should be noted that the assumption of a bubbly regime provides a more conservative design.

The time at pressure turnaround τ gives an indication of the duration of the venting event. Reference [1] shows that τ can be given by the following formula

$$\tau = \frac{m_0}{W} - \left[\frac{V}{m_0}\frac{h_{fg}}{v_{fg}}\frac{m_0}{WC_p(dT/dt)}\right] \tag{5}$$

Figure 11 is a comparison between the observed turnaround times and the predicted turnaround times. In general the agreement is within 10%.

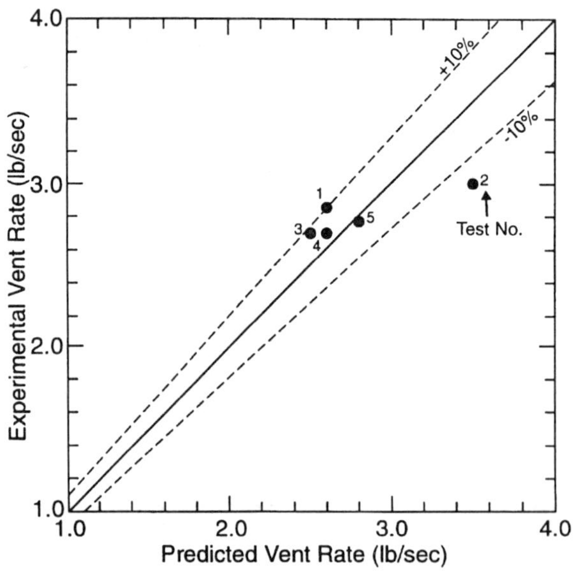

Figure 9. Comparison between Observed Vent Rates and Predicted Vent Rates.

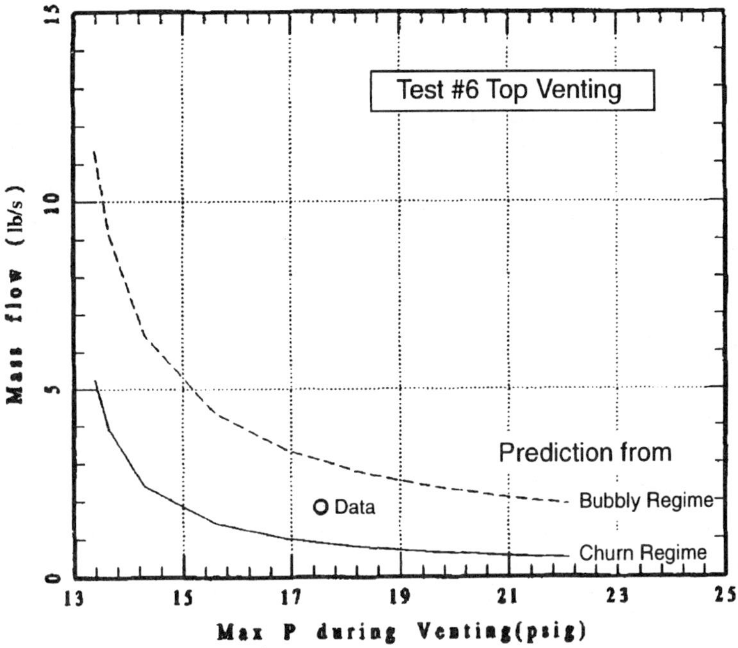

Figure 10. Required Vent Rate Prediction versus Experimental Data for Test #6 (Top Venting).

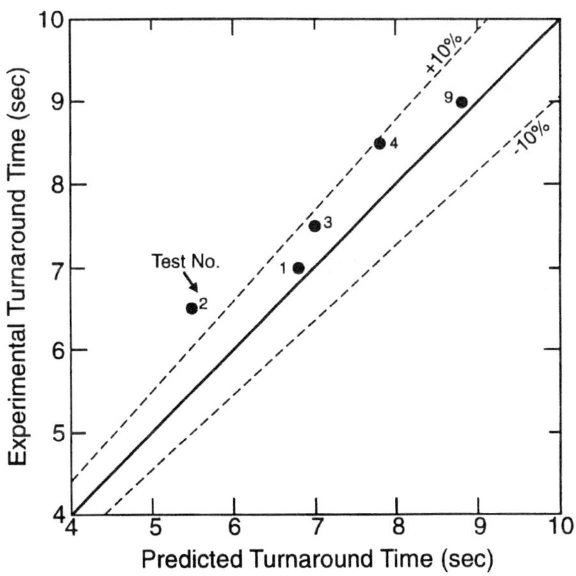

Figure 11. Comparison between Observed Turnaround Times and Predicted Turnaround Times.

6. TWO-PHASE DISCHARGE EVALUATION

DIERS technology demonstrates that the traditional homogeneous-equilibrium model (HEM) is the most conservative model for estimating two-phase discharge through rupture disks and pressure relief valves [3,4]. In this HEM formulation the two-phase flow mixture is treated much like the classical compressible gas while undergoing an adiabatic expansion with equal velocities and temperature in both phases. For a flashing system such as the PF solution, this would imply thermodynamic vapor-liquid equilibrium. However it should be pointed out that two-phase flow has inherently two additional degrees of freedom:

1. mechanical nonequilibrium as exemplified in the form of slip between the two phases, i.e. vapor would flow faster than liquid [5];
2. thermal nonequilibrium as liquid could remain superheated, thus yielding less (or no) vapor generation from the flashing process [6].

Figure 12 illustrates the influence of these two nonequilibrium effects on the resulting nozzle discharge mass flux relative to the HEM reference model. It clearly demonstrates that these nonequilibrium effects greatly enhance the discharge flow. However, for a flow passage larger than 4 inches (10 cm), thermal equilibrium is closely approached [7]. It has been argued that the residence time in such a duct roughly corresponds to the bubble growth time.

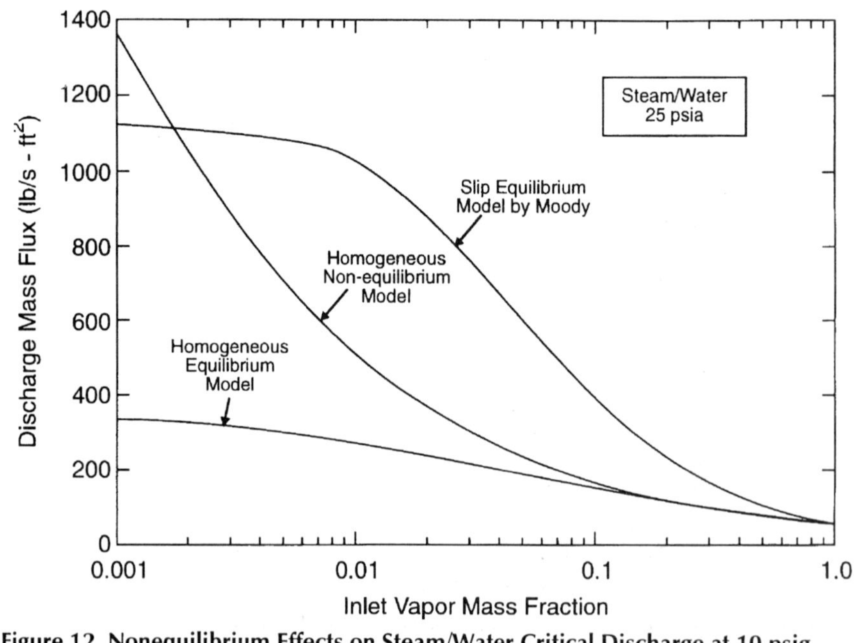

Figure 12. Nonequilibrium Effects on Steam/Water Critical Discharge at 10 psig.

This is certainly satisfied in most rupture disk installations. The slip-equilibrium model (SEM) leads to a mass flux three times higher than the HEM at low inlet quality.

In a pipe discharge the situation can be quite complex. As mentioned above, thermal nonequilibrium can be ruled out in such a configuration. However, the continued pressure drop along the pipe with the associated flashing and flow acceleration are all conducive to a slip flow regime. This would imply that the vapor velocity can be significantly higher than the liquid velocity. Based on previous studies [5], the so-called slip ratio *S*, i.e. ratio of the vapor velocity to the liquid velocity, is a function of the density ratio,

$$S = \left(\frac{\rho_f}{\rho_g}\right)^n \qquad (6)$$

where ρ_f and ρ_g are the liquid and vapor density, respectively and the index n has a value ranging from ⅓ to ½. Since the density ratio is larger at lower pressure, the slip ratio is increasing down the tailpipe. For a steam–water system at low pressure such as in the present study, this density ratio is over 1000 yielding a slip ratio of 10–30 according to Eq. (6). As far as the mass flux evaluation for a constant-diameter duct is concerned, two limiting conditions can be employed. The first model can be based on a homogeneous-equilibrium assumption throughout the entire duct. This means that the two phases are always in thermal equilibrium and traveling at the same velocity. This assumption should yield a lower bound estimate for the mass flux [8]. The second model assumes the flow may be limited by choking at the pipe entrance based on the HEM model. Transition to a slip flow may produce a second choked condition at the pipe exit [7] as depicted in Fig. 13. Figure 14 provides a more relevant illustration showing the choking HEM mass flux at the inlet and the SEM local choking mass flux at the exit are both of the same magnitude. However, since the SEM would predict a choking exit pressure of

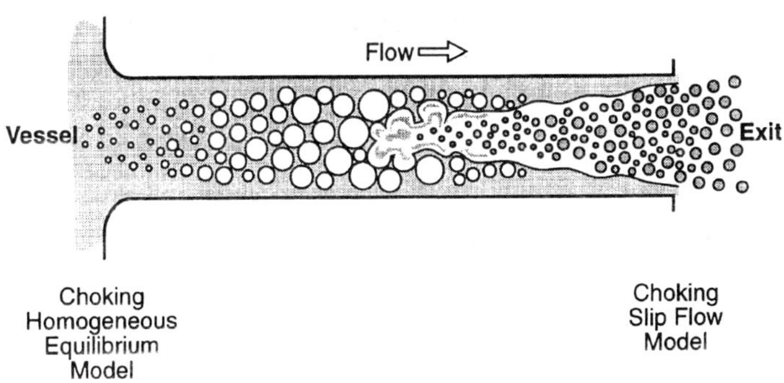

Figure 13. Two-Phase Pipe Discharge with Flow Governed by Choking at the Pipe Inlet.

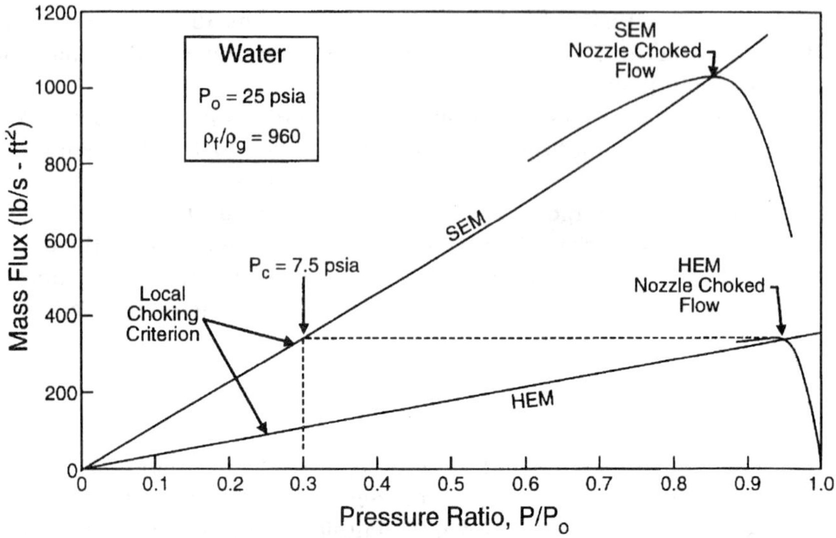

Figure 14. Steam–Water Pipe Discharge with Flow Governed by HEM Choking at Pipe Inlet.

7.5 psia in our case, the pipe exit is not choked. The net result is that the duct length has no influence on the flow since the flow is essentially determined by choking at the duct entrance. The second model as suggested by Moody [7] has been shown to fit many examples of vessel blowdown data using water as working fluid.

Figure 15 illustrates the experimental mass flux—measured vent rate divided by 0.864 in^2 flow area corresponding to a 1-inch Sch 40 pipe—as a function of the average vessel pressure during venting. Also shown are the HEM nozzle prediction. The data are generally underpredicted by 30%. The somewhat higher experimental mass flux could be due to subcooling effect as a consequence of the hydrostatic head existed inside the reaction vessel in a bottom-vented configuration. However, this comparison supports the notion that two-phase flow in pipe can be reasonably correlated by assuming a choked flow condition at the inlet. These data also demonstrate that the viscous flow behavior in the vent pipe is not a limiting factor.

7. VENT SIZING PREDICTION

Finally the vent size prediction can be made using the required vent rate from Eq. (1) and the estimated mass flux G, i.e.,

$$C_d A = \frac{W}{G} \tag{7}$$

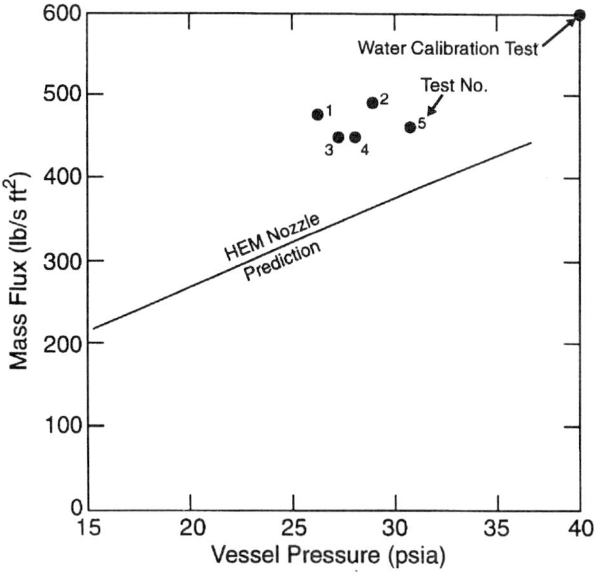

Figure 15. Comparison of Experimental Mass Flux Against HEM Prediction.

where

A = pipe flow area,
C_d = effective discharge coefficient,
W = vent rate,
G = two-phase discharge mass flux.

In our prediction, G is based on HEM corresponding to a nozzle [4]. Thus if the flow is limited by choking at the pipe inlet, then the effective pipe flow area should be 0.864 in^2, i.e. $C_d = 1$ since the tailpipe has no effect on the flow. On the other hand if the flow is consistent with the model, then an effective C_d of 0.52 is found for the bottom vent line. Figure 16 clearly demonstrates that choking at the pipe inlet assumption best fits the data; the agreement in terms of vent area in this case is within 10%. If the vent line resistance is taken into account (i.e. $C_d = 0.52$), then the design will yield a vent area twice as large and result in a safety factor of about 2.

The sizing formula, written as A (in^2) = 0.053 V (gal), has been used for some time in P/F reactors [9]. Comparing with this traditional guideline, the present test reactor has a considerably larger vent area. Thus

$$\frac{A}{V} = \frac{0.864\,\text{in}^2}{5\ \text{gal}} = 0.265\,\text{in}^2/\text{gal}$$

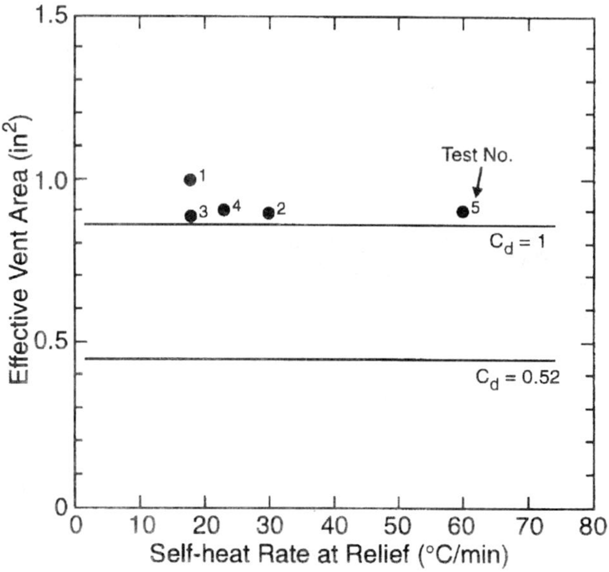

Figure 16. Vent Area Predictions (Solid Circles) versus Effective Flow Areas.

versus 0.053 in^2/gal in Monsanto's formula. Thus the results of these tests should be interpreted with this *A/V* value in mind. However, it should be noted that the so-called Monsanto formula is based on a very low relieving set pressure, typically 1–2 psig and a self-heat rate at relief of about 6.5°C/min [10]. Since the vent area is proportional to *dT/dt* and inversely proportional to set pressure P_s, we can in fact adjust the Monsanto formula to reflect the current test data. Taking an average set pressure of 12 psig (27 psia) and a *dT/dt* of 60°C/min (more reactive recipe), the new formula should appear as:

$$\frac{A}{V}=0.053\left(\frac{60\ ^{\circ}\mathrm{C/min}}{6.5\ ^{\circ}\mathrm{C/min}}\right)\left(\frac{15\ \mathrm{psia}}{27\ \mathrm{psia}}\right)=0.272\ \mathrm{in}^2/\mathrm{gal}$$

which is very close to the available vent area of the test apparatus.

8. VAPOR SUPPRESSION EFFICIENCY

The catch tank was filled with approximately 15 gallons of water prior to each test. Thus the mass ratio between the quench water and the reactor mass was about 3.5. During these tests, the visual behavior of the quench pool was recorded by a video camera looking down into the pool from above the catch tank. Initially the air pocket in the vent line was seen being pushed out

TABLE 2
Comparison of Weight Gain in Catch Tank against Reaction Vessel Charge

Test No.	Vessel Charge	Total Weight Gain in Catch Tank	Discrepancy
1	34.1 lb	34.0 lb	−0.3%
2	34.5 lb	34.5 lb	0%
3	34.7 lb	34.7 lb	0%
4	34.7 lb	35.0 lb	0.9%
5	34.8 lb	35.2 lb	1.1%
6	34.7 lb	35.0 lb[(a)]	0.9%

[a] 28.0 lb gain from top venting mode, 7.0 lb gain from bottom venting mode.

instantly and the air (gas) was seen bypassing the pool and rising to the surface. Immediately an intense circulation was set up and audible sounds associated with steam condensation were heard. However, no steam was observed to bypass the pool. No steam was seen at the top of the catch tank either. Perhaps an indication of the vapor suppression efficiency can be inferred from the mass balance. Table 2 shows that the weight gain in the quench pool after relief matches closely the reaction mass charged to the vessel for these bottom-vented tests. It further confirms that the mass retention in the reaction vessel and in the bottom vent line was almost negligible.

9. CONCLUSIONS

A total of six PF resin recipes have been conducted in a 20-liter reaction vessel simulating runaway scenarios with pressure relief. The reaction vessel has a vent area to volume ratio of 0.265 in^2/gal (0.0452 m^2/m^3). Self-heat rates at relieving pressures of 11–13.5 psig range from 18–62°C/min. The predicted flow rates and vent sizes are in good agreement ($\pm$10%) with these experiments if one assumes the limiting discharge rate to be governed by choking at the discharge pipe inlet. Viscous flow behavior in the discharge pipe was not observed. A simple 1-inch tee was found to perform effectively as a sparger with no steam bypass nor splashing. The weight gain in the quench pool for the bottom-discharge tests was essentially equal to the initial reactor charge to within 1%.

10. REFERENCES

1. Leung, J. C., "Simplified Vent Sizing Equations for Emergency Relief Requirements in Reactors and Storage Vessels," *AIChE Journal,* 32(10) (1986a), 1622.
2. Leung, J. C., "Overpressure During Emergency Relief Venting in Bubbly and Churn-Turbulent Flow," *AIChE Journal,* 33(6) (1987), 952.
3. Leung, J. C., and Fauske, H. K., "Runaway System Characterization and Vent Sizing Based on DIERS Methodology," *Plant/Operations Progress,* 6(2) (1987), 77.
4. Leung, J. C., "A Generalized Correlation for One-Component Homogeneous Equilibrium Flashing Choked Flow," *AIChE Journal,* 32(10) (1986b), 1743 .
5. Moody, F. J., "Maximum Flow Rate of a Single Component, Two-Phase Mixture," *ASME Trans., J. of Heat Transfer* (1965), 134 .
6. Leung, J. C. and Epstein, M., "A Generalized Correlation for Two-Phase Non-Flashing Homogeneous Choked Flow," *ASME Trans., J. of Heat Transfer,* 112(2) (1990), 528.
7. Moody, F. J., "Maximum Discharge Rate of Liquid-Vapor Mixtures from Vessels," ASME Winter Meeting, Houston, Texas, Symposium publication on Non-equilibrium Two-Phase Flows (1975).
8. Leung, J. C., and Grolmes, M. A., "The Discharge of Two-Phase Flashing Flow in a Horizontal Duct," *AIChE Journal*, 33(3) (1987), 524; also errata, 34(6) (1988), 1030.
9. Howard, W. B., "Reactor Relief Systems for Phenolic Resins," Monsanto Company (June, 1973).
10. Fauske, H. K., "Generalized Vent Sizing Monogram for Runaway Chemical Reactions," *Plant/Operations Progress,* 3(4) (1984), 213.

Design of Scrubber System for Emergency Relief of a Large-Scale Petrochemical Plant

J. R. Chen

Department of Safety, Health and Environmental Engineering
National Institute of Technology at Kaohsiung
1 University Rd., Yenchau, Kaohsiung, 824, Taiwan

W. K. Chou and Y. S. Duh

Center for Industrial Safety and Health Technology, ITRI
F11, Bldg. 51, 195 Chung Hsing Rd. Sect. 4
Chutung, Hsinchu, 310, Taiwan

H. C. Tu

Tuntex Petrochemicals, Inc.
98 Pei-Shin-Chau, Ming-Ho, Shan-Shang, Tainan, 743 , Taiwan

ABSTRACT

Difficulties and practical, compromise solutions to design an emergency scrubber system for a large-scale petrochemical plant are presented via a case study. The case studied is a paraxylene oxidation process in which acetic acid is used as the solvent. The heat of the oxidation reaction is removed from the reactor by evaporating the solvent. The solvent vapor is condensed and refluxed to the reactor. During process upsets such as cooling failure, the solvent vapor can no longer be condensed and the overpressure generated within the reactor and condensers must be relieved through a pressure safety valve (PSV). The vapor relieved to the atmosphere raises significant environmental concern to the nearby communities. Thus, a proper disposal of the relieving vapor is desired.

To completely dispose of the relieving vapor would require a replica of the entire vapor processing unit of the oxidation process which is generally impractical. Instead, a single scrubber (absorber) using water as an absorbing medium is chosen for the service. Owing to the large amount of relieving vapor, fire water is chosen for the water supply. The action of the scrubber is triggered by flow switches located in the relief line. The size and layout of the relief line are determined by the allowable back pressure of existing PSVs. Additional constraints to the layout of the relief line are stresses arising from the thermal expansion and reaction force of the line during relieving conditions. Dynamic and static stress analyses are performed to ensure both the transient and static stresses are within allowable ranges. Design methodologies concerning the setting of the flow switches, scrubber drainage, internal structure and final operation and maintenance are given in detail. The result is a *safe* and *effective* design that will reduce the concentration of relieving acetic acid vapor from 30% to 650 ppm. Unawareness of the back pressure and stress constraints will result in high risk of the reactor.

1. INTRODUCTION

The emergency releases from large scale petrochemical plants or refineries usually comprise flammable vapors and flaring is the most common and effective way to dispose of the relief stream. The purpose of flaring the relief stream is not only to avoid atmospheric pollution but also to avoid the formation and possibly explosion of a flammable vapor cloud. For cases that the relief stream is not flammable, vapor cloud explosion is no longer the main concern and the direct atmospheric venting is a common practice for the emergency relief. This practice, however, is changing as the environmental concerns grow enormously in populated area like Taiwan. A proper treatment of the emergency relief is considered necessary even though the relief stream is not highly toxic to the plant and nearby communities.

Processes such as the paraxylene oxidation reactor in a pure terephthalic acid (PTA) plant and the oxychlorination reactor in a vinyl chloride monomer (VCM) plant are two of the processes that the emergency releases are vented directly to the atmosphere instead of a flare. The relief stream from both processes mainly consist of nonflammable fractions such as acetic acid and hydrochloric acid that are not suitable for flaring. The idea candidate for disposing these nonflammable but possibly toxic vapors is a scrubber system. Grossel (1995) and Bravo et al. (1997) have provided a good account of design guides and criteria for an emergency scrubber system. These guides can be properly applied to small-scale processes such as a pharmaceutical process without great concerns in the equipment cost. For a large-scale petrochemical plant like a PTA plant, the relief load is usually greater than 200,000 kg/hr and the cost of a scrubber system will increase enormously if the same design guides are followed. This paper aims at presenting a case study of an emergency scrubber system for a large-scale petrochemical plant that features all the difficulties and possible solutions in applying the common design guidelines.

2. RELIEF SYSTEM DESIGN

2.1. The Paraxylene Oxidation Process

The case studied in this work is the paraxylene oxidation process in a PTA plant. The oxidation reaction is carried out in acetic acid solution with salts of Co/Mn/Br as the catalyst. Figure 1 shows a typical flow diagram of the oxidation process with two reactors working in parallel. The oxidation reaction is a tempered system (Fisher et al., 1992) in which the heat of reaction is removed by vaporizing the solvent. The vapor from the reactors passes a series of condensers to recover acetic acid. Both steam condensate and cooling water are used for the coolant of the condensers. The condensed acetic acid is refluxed back to the reactors. The remaining noncondensable vent gas, mainly nitrogen from the air, is fed into scrubbers and a CO/VOC oxidizer to remove acetic acid residual and VOCs before venting to the atmosphere.

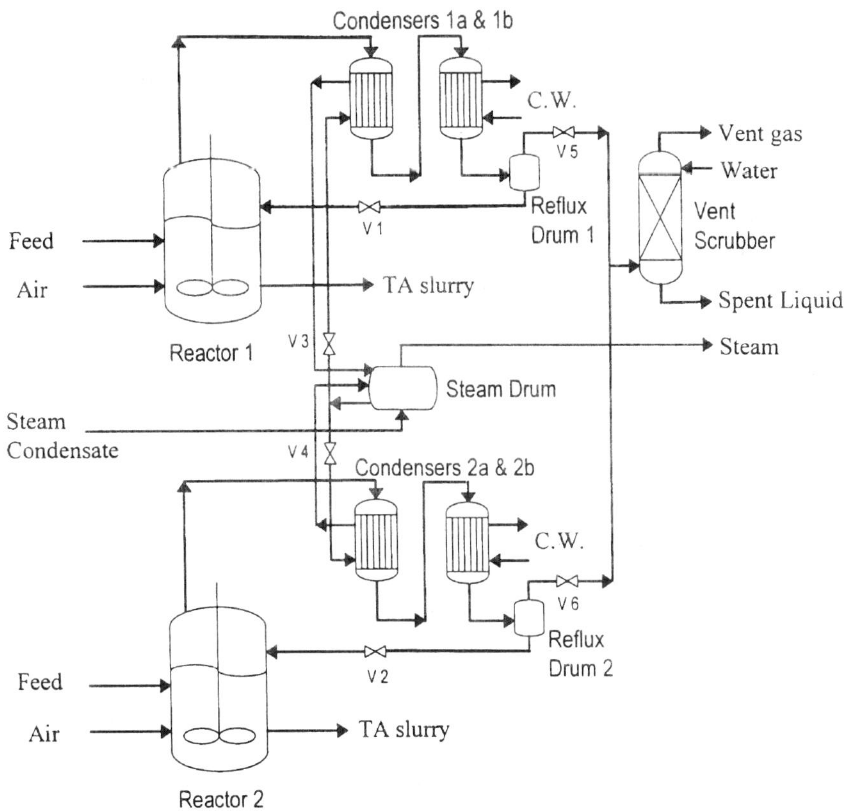

Figure 1. Process Flow Diagram of the PTA Oxidation Process.

The PSVs of the reactors have been sized based on the vapor generation rate from the reactor. The relief load of the PSVs has been rated at 300,000 kg/hr for each reactor for the present case study. The total relief load would be 600,000 kg/hr if both reactors relieve simultaneously. The required emergency scrubber, if feasible, would be an enormous one if the relief load of 600,000 kg/hr is used for the design basis. We therefore carried out a thorough review of the relief load.

2.2. Realistic Relief Scenarios and Relief Load Evaluations

The sizing of pressure relief devices is generally conservative. The cost of the vent sizing is comparable to that of a PSV. Therefore, a conservative vent size is generally used instead of going to detailed vent sizing. The situation is however different in sizing the vent disposal units. The cost of vent disposal equipment is generally much larger than detailed vent sizing. The identification of realistic relief scenarios is therefore crucial for a cost-effective design of the disposal system.

Typical relief scenarios and relief load evaluations have been given in the guidelines of API RP 521 (1997). The following causes of overpressure on the vapor flow are identified:

1. Block out of a line
2. Operator mistakes in valve actions
3. Cooling failure
4. Instrumentation failure
5. Heat exchanger tube failure

The following basic guidelines are followed during the relief load evaluation:

1. Block out can only occur on lines with fittings.
2. Block out on pipe fittings will not occur on different lines simultaneously.
3. Miss-operation may occur on lines connecting to one common equipment.
4. The actions of the safety interlock system (SIS) will be considered.

Although the guideline 4 is not recommended in API RP 521, it has been adopted in reducing the relief load for flare systems by Zheman and Early (1992). Zheman and Early (1992) have shown that a triple redundancy system for emergency shutdown can be as reliable as that of a safety valve in pressure protection. The current maximum relief load is calculated from the assumption that the valve on the reflux line was mistakenly shut off, and the condensed liquid flooded both condensers and resulted in a full block-out of the vapor line. Real plant operations, however, found that the condensed liquid did not flood the condensers but moved into the vent scrubber in the downstream as a result of pressure drop instead. No relief occurred for this scenario. Therefore, a complete block-out of the vapor line is not feasible.

The results of the relief load for other causes of overpressure are given below:

1. The condensate feed failure will result in the low liquid level in the steam drum and will trip the feed to the reactor. No relief is expected.
2. The valve on one condensate supply line is mistakenly closed. The relief load is given by the condensing capacity, 129,000 kg/hr for each reactor. Simultaneous relief of both reactors is possible according to the guideline 3. Total relief load is therefore 258,000 kg/hr.
3. The valve on one cooling water supply line is mistakenly closed. The relief load is given by the condensing capacity, 117,000 kg/hr for each reactor. Total relief load is 234,000 kg/hr.
4. The valve on the reflux drum vapor line is mistakenly closed. The relief load is given by the normal vapor flow rate, 54,000 kg/hr, for each reactor. Simultaneous relief of both reactors is not possible.
5. The level gauges of the steam drum may fail and result in false level control. The condensate supply to condensers for both reactors may fail without notice. The resulted relief load is 258,000 kg/hr.
6. Failure of the heat exchanger tube will not result in any relief as the tube side (process side) pressure is always higher than the shell side (coolant side).
7. All scenarios of the line block-out are identified and shown in Table 1. None of them is larger than the instrumental failure scenario.

Therefore, operator mistake on condensate supply line and level gauge failure of the steam drum are both considered to be the worst case scenarios. The corresponding relief load of 258,000 kg/hr is used for the design load of scrubber system.

2.3. Check of Venting Behavior and Vent Sizing

In the design of emergency relief systems (ERS) for vessels containing liquid, the phases of the flow through the vent line are very important. Top venting does not guarantee all vapor venting. One must check whether the vapor bubbles formed in the liquid will be disengaged (separated) before they reach the vent entrance. A larger vent is generally required if the venting is two-phase flow. The present work follows the method of DIERS (Fisher et al., 1992) for predicting the vapor–liquid disengagement. The calculated results showed that vapor and liquid are completely disengaged in the reactor even during the worst case scenario. It is expected as the same vapor load is also held for normal operation. However, the possible droplet entrainment from liquid surface and terephthalic acid solid carry-over are neglected.

TABLE 1
Summary of Relief Scenarios and Relief Loads

Relief Scenario	Overpressure Location	Relief Load		Total Relief Load
		Reactor 1	Reactor 2	
Block out	V 1/reflux drum 1	0	0	0
	V 2/reflux drum 2	0	0	0
	V 3/steam drum	129,000	0	129,000
	V 4/steam drum	0	129,000	129,000
	V 5/reflux drum 1	54,000	0	54,000
	V 6/reflux drum 2	0	54,000	54,000
Operator mistake	V 1/reflux drum 1	0	0	0
	V 2/reflux drum 2	0	0	0
	V 3/steam drum V 4/steam drum	129,000	129,000	258,000
Cooling failure	Steam condensate	0	0	0
	Cooling water	0	0	0
Instrum. failure	Steam drum level gauge	129,000	129,000	258,000
	C.W. drum level gauge	117,000	117,000	234,000
Tube failure	Condenser 1a	0	0	0
	Condenser 1b	0	0	0
	Condenser 2a	0	0	0
	Condenser 2b	0	0	0

Existing relief devices including three "Q" size safety valves and one rupture disk are installed on the vapor line of each reactor. The required vent size is determined by the recommended equation of API RP 520 (1993). The existing vent size is considered sufficient for the relief load of the worst case scenario.

2.4. Relief Pipe Layout and Sizing

The size of a PSV discharge pipe is governed by the back pressure constraint as well as mechanical constraints according to API RP 521. The back pressure constraint requires that built-up back pressure caused by flow through the valve and relief line does not reduce the relieving capacity. The mechanical constraints require that thermal expansion and thrust caused by the hot relief stream can be controlled without endangering the pressure relief valve, the pipe and the structure supports. The mechanical design of the relief pipe will be discussed later. API RP 521 provides methods for estimating the pressure drop across the relief pipe. These methods become tedious when the pipe layout is complicated or there are simultaneous releases from multiple sources. Flow analysis of complicated relief network is best done by a computer software package. Several software packages are available. In this work, the package Visual Flare (SimSci, 1993) is used to analyze the pressure profile in the relief pipe.

A preliminary sketch of the pipe layout from the PSV to the scrubber is made. Flow analysis is then performed for different sizes of the vent pipes. The existing pressure safety valves are balanced-bellows type. The API 521 recommends that back pressure should not exceed 30% to 50% of the set pressure. The present system has a set pressure of 30.4 bara and the preferred upper limit of the back pressure would be 9.1 bara. However, a check with the safety valve manufacturer reveals that the bellows can only withstand pressure up to 4.9 bara. Unawareness of the bellows restriction will endanger the bellows and affect the relieving capacity. Successive flow analyses show that the pipe size of 18 inches provides a low back pressure, 3.4 bara, as well as a low flow velocity in the vent pipe and is considered adequate for the present case. A review of the back pressure will be made after the vent pipe layout is finalized.

3. SCRUBBER SYSTEM DESIGN

3.1. Basic Guidelines of Design

Bravo et al. (1997) gave the following guidelines for designing emergency scrubbers:

1. *Stoichiometry:* The scrubbing system must have sufficient liquid to absorb the maximum expected venting gas from the stoichiometric point of view.
2. *Mass-transfer:* The scrubbing system must provide a good mass transfer rate for rapid absorption of the vent gas.
3. *Kinetics:* For absorption with chemical reactions, the scrubbing system must provide sufficient resident time to ensure complete reaction of chemicals and vent gas.
4. *Temperature:* The temperature of the spent liquid should be low enough to ensure safe disposal of the liquid. The exothermic heat of chemical absorption must also be considered.

To follow the above guidelines strictly for emergency relief from large-scale facilities with elevated temperature will require a very large scrubber and high cost to build. A compromise must be made regarding the design of the whole scrubber system. The following conclusions were drawn after careful review of the design guidelines:

Use of caustic: To meet the stoichiometric requirement of chemical absorption, we need at least 103,200 kg/hr pure caustic or 516,000 kg/hr 20% caustic solution for the maximum relief load of 258,000 kg/hr. The plant has a supply of dilute caustic solution at a rate of 50,000 kg/hr which is one order of magnitude less than the stoichiometric requirement. The use of caustic to aid in chemical absorption of the acetic acid is therefore not considered. Water is used as the sole absorbing liquid.

Source of absorbing liquid: The existing plant is capable of providing cooling water at the rate of more than 300,000 kg/hr. Although loss of cooling water does not trigger the relief, the continuing supply of cooling water does help to reduce the relief load. Therefore, the use of cooling water as the absorbing liquid is not considered. The only water source remaining is the fire water. The fire water system in the plant is capable of supplying room temperature water at the rate of more than 1,500,000 kg/hr. Simultaneous relief and fire is not considered as a viable scenario. However, to ensure full supply of the fire water during fire emergency, a manual isolation valve is installed between the fire water supply line and the scrubber system.

Type of scrubber: Various types of scrubbers have been studied for the present case. Packed or trayed columns are good in terms of mass transfer efficiency. However, they require continuous liquid circulation to ensure a good contact of liquid with the vent gas at the first moment of relief. The spray column, on the other hand, is simple and costs much less to construct compared with the packed or trayed columns. A quench tank with spargers is another potential candidate for the scrubber design. Guidelines for quench tank design have been provided by Grossel (1990) and Keiter (1992). Quench tanks are good for vents without noncondensable gas. If the vent contains condensable gas such as in the present case, a mass transfer contact section is usually added (Grossel, 1995). The mass transfer section can be a packed column, spray column or simply a baffled-tray section. Completely quenching the vent vapor is almost impossible for a large and continuous relief. We proposed here the use of a partial quench tank with a spray column as a compromise between quench tank size, mass transfer efficiency and construction cost. To ensure contact of the vent gas with the absorbing liquid at the first instance of relief, the feed pipe is submerged into the liquid pool in the bottom of the scrubber. A sparger with small holes was not used to avoid a large pressure drop. A separate makeup system using cooling water is added to maintain the liquid pool. The liquid pool acts as a partial quench drum. The pool liquid contributes not only the apparent heat which quenches part of the stream but also the latent heat which vaporizes into steam and dilutes the concentration of acetic acid vapor. The quench tank is aided by a countercurrent spray to enhance the absorption of the relieving vapor.

Disposal of the spent liquid: The absorbed liquid contains hot, concentrated acetic acid solution and a small amount of terephthalic acid solid. The absorbed liquid may be pumped out or refluxed after cooling. The spent liquid in the present case would exceed 500,000 kg/hr, depending on the final absorbing temperature. Cooling and then reflux of the spent liquid are not practical for such a large amount of spent liquid. Also, the pump out system has to be very reliable or the scrubber will be flooded. The best way to drain the liquid is to use a seal leg (siphon) by which gravity is the driving force of flow. The seal leg contains no mechanical parts and is very reliable. Solid deposition is not a problem for the larger size of seal legs. The drained liquid has to be transported by a trench instead of a pipe to the sewer. The maximum temperature of drained liquid is therefore limited to below the boiling point of water to avoid vaporization in the trench.

3.2. Basic Design of Scrubbers

The performance of the spray column/quench tank can be modeled by the equilibrium stage method. The performance of a quench tank will approach a theoretical stage of absorption if the vent gas can be perfectly mixed with the quench fluid. However, the efficiency of the spray column is difficult to predict. Substantial spray height is needed to achieve one theoretical stage. Performance of a spray column is generally no more than two theoretical stages. Zenz (1989) suggested a two-stage spray column requiring at least three layers of spray. For a very large scrubbing load as in the present case, the overall performance is expected to decline as the variations in spray nozzle flow increase. We conservatively propose that each quench tank with three layers of spray is equivalent to one theoretical stage. Multiple stages of absorption are achieved by coupling multiple quench tank/spray units.

Figure 2 shows the schematic diagram of a series of equilibrium stages representing the mixing and separation of two streams W_i and F_i. The vapor outlet F_i from one stage is fed into the next stage. Fresh absorbent W_i is fed into each stage instead of recycling the outlet liquid. Liquid outlet R_i is removed from the stage for waste disposal. The removal of absorbed liquid and addition of fresh absorbent liquid instead of reflux in each stage are the main characters of the quench tank. To utilize the latent heat of the absorbent liquid, the temperature and pressure must be operated near the saturation

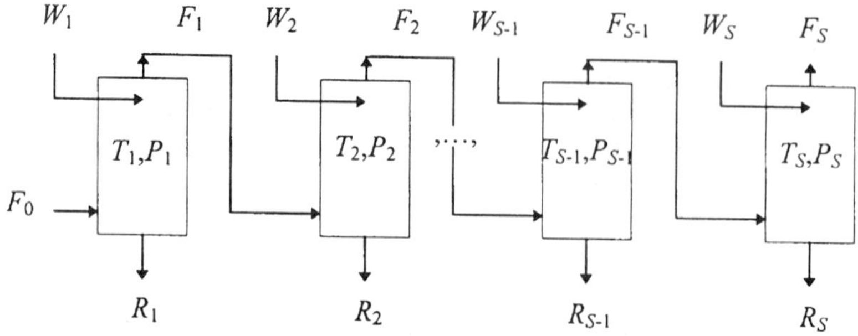

Figure 2. Schematic Diagram of Equilibrium Stages

condition. The reflux of the absorbed liquid is however not necessary as the apparent heat of the liquid is almost depleted.

The transient mass balance around stage i is:

$$\frac{dm_i}{dt} = F_{i-1} - F_i + W_i - R_i \tag{1}$$

The transient component balance around stage i is:

$$\frac{dm_i x_{i,j}}{dt} = x_{F_{i-1},j} F_{i-1} - x_{F_i,j} F_i + x_{W_i,j} W_i - x_{R_i,j} R_i \qquad j = 1,\ldots,N \tag{2}$$

The transient energy balance around stage i is:

$$\frac{dm_i h_i}{dt} = h_{F_{i-1}} F_{i-1} - h_{F_i} F_i + h_{W_i} W_i - h_{R_i} R_i \tag{3}$$

where $x_{F_{j,i}}$ and $x_{R_{j,i}}$ are the equilibrium vapor and liquid composition in stage i. The flow of vapor from stage i to stage $i + 1$ is governed by the pressure drop between stage i and stage $i + 1$. For simplicity, the pressure drop from stage i to stage $i + 1$ is approximated by flow through a sharp orifice. The wall friction of the connecting pipe is assumed negligible. The relation between the pressure drop and flow velocity is as follows:

$$\begin{aligned} p_i - p_{i+1} &= 0.5 \rho K V^2 \\ K &= \left(1 + 0.707\sqrt{1 - \frac{A_P}{A_V}} - \frac{A_P}{A_V}\right)^2 \\ F_j &= \rho A_P V_i \end{aligned} \tag{4}$$

where A_V and A_P are the cross-sectional area of the quench tank and connecting pipe. Equations (1)–(4) can be solved numerically with phase equilibrium relations of the fluids. The relief stream contains more than nine components. For simplicity, they are simplified into three components as shown in Table 2.

TABLE 2
Summary of Relief Stream Composition

Component	M.W.	Weight fraction	Mole fraction
Acetic acid (HAC)	60.1	0.61	0.36
Water	18.015	0.21	0.41
Nitrogen	28.013	0.18	0.23

Specific requirements for the performance of the scrubber are as follows:

1. The final vent concentration of acetic acid is limited to 1000 ppm or lower and the concentration at the fence line is below the exposure limits.
2. The temperature of the final absorbed liquid is limited to 90°C or lower.

Figures 3 and 4 show the steady state results for one equilibrium stage with different flow rates of absorbing water. When the amount of fire water is the same as the relieving vapor load, the exit temperature exceeds 100°C and the vapor mole fraction of acetic acid (HAC) drops to 0.07 from 0.36 while the liquid HAC mole fraction rises to 0.2. Significant water evaporates into vapor

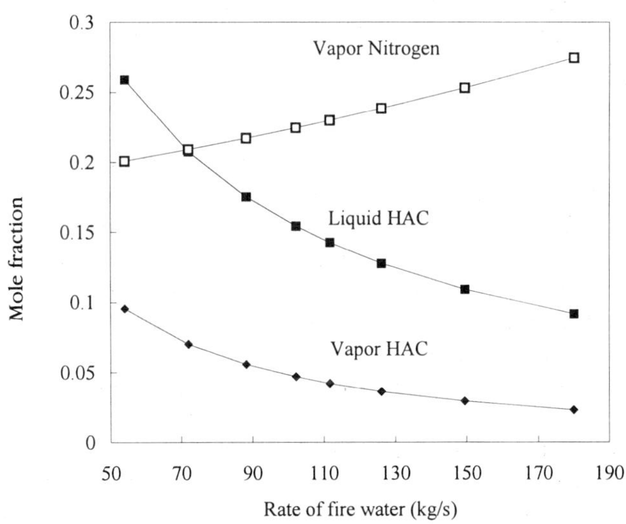

Figure 3. Compositions of Outlets of the Single Stage Quench Tank.

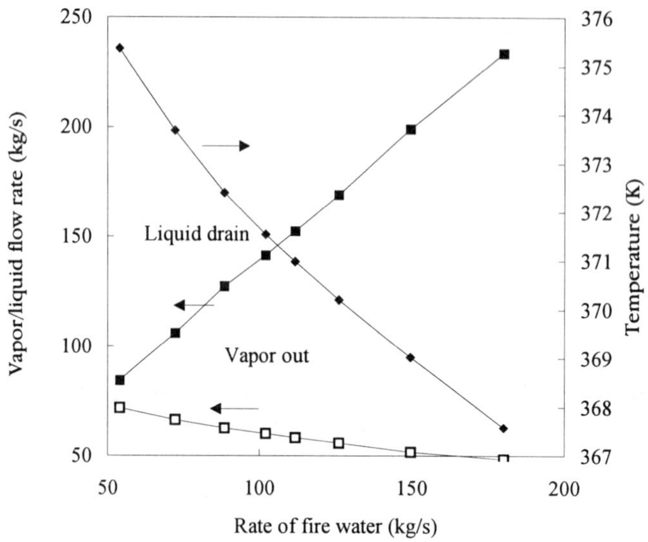

Figure 4. Vapor/Liquid Flow Rates and Temperature of the Single Stage Quench Tank.

as the water vapor mole fraction rises to 0.71 from 0.405. Further increasing the absorbing water flow rate does not drastically change the performance of the quench tank. In fact, further increasing of absorbing water results in a lower temperature rather than reducing the vapor HAC. To reduce the vapor HAC concentration to below 1000 ppm in one equilibrium stage requires approximately 4000 kg/s or 14,400,000 kg/hr of absorbing water, an amount of water that is almost impossible to transport in pipe. Therefore, a second quench tank is desired.

For a preliminary study, a total amount of 234.2 kg/s of water, equivalent to 843,000 kg/hr or roughly two fire pumps of the PTA plant working simultaneously, is used for the two quench tanks. The effect of water distribution in the two tanks on the overall scrubbing performance is studied.

Figure 5 shows the variation of the vented vapor mole fraction of HAC with different distributions of water flow rates in a 2-stage scrubber. While the vapor mole fraction of HAC from the first stage is decreased as the water flow rate to the first stage increased, the vapor mole fraction of HAC from the second stage shows a minimum when the ratio of the first stage to the second stage water flow rate is about 2.25. That is the overall scrubbing performance is the best when 70 % of the scrubbing water is used in the first scrubber and 30 % in the second scrubber. The lowest attainable vapor HAC concentration in this case is 650 ppm, which is below the required vent concentration. To further reduce the vent HAC concentration, one must increase the rate of

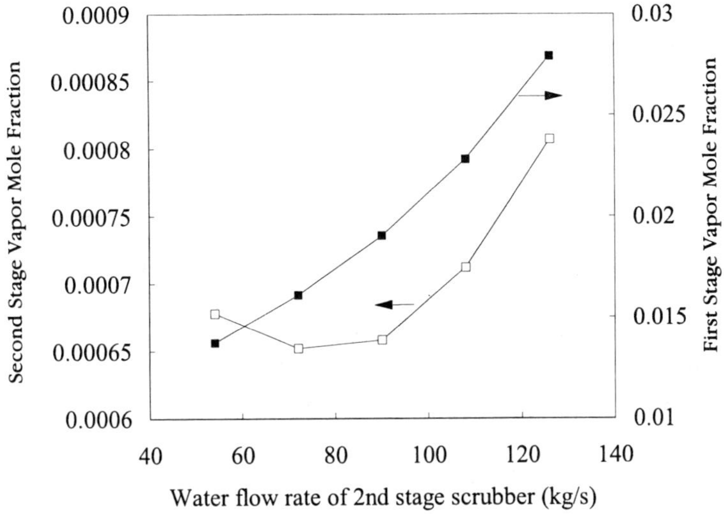

Figure 5. Variation of Outlet Vapor Mole Fraction of HAC with Different Distribution of Water Flow Rates in the Two-Stage Scrubber.

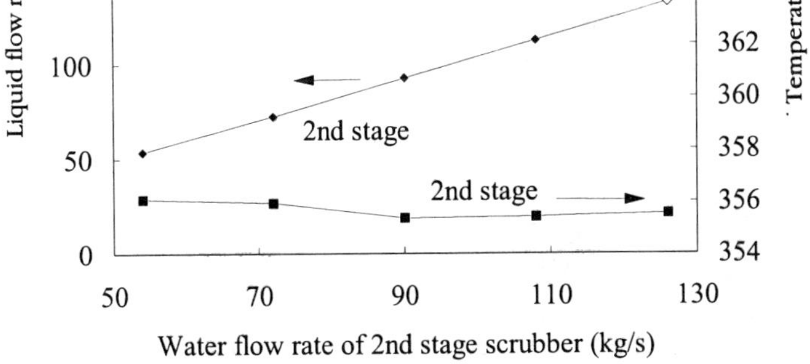

Figure 6. Liquid Drainage Rate and Temperature of the Two-Stage Scrubbers.

total scrubbing water. The rate and temperature of the liquid drainage are shown in Figure 6. The highest temperature of liquid is 92.6°C, slightly higher than the required drain temperature, but is still considered acceptable. Therefore, the selection of 580,000 kg/hr and 260,000 kg/hr for the first and second stage scrubbers is recommended.

The net release of HAC into the atmosphere from the maximum HAC concentration, 650 ppm, is equivalent to 35 g/s. The release will be dispersed in the atmosphere and the concentration at nearest fence line (120 m from the release point) is 1.7 ppm. This concentration is below the allowable concentration in the workspace (15 ppm) but is higher than the requirement of Taiwan EPA's Fixed Pollutant Source Release Standard at the fence line, given by 0.02 the allowable concentration or 0.3 ppm. However, the ERS is *not* a continuous release source unless during operation upset or emergency conditions. Therefore, the performance of the scrubbers is considered sufficient.

3.3. Engineering Design of Scrubbers

3.3.1. Sizing of the Scrubbers

The geometry of the quench tank can be determined once the flow of each stream is determined. The diameter of the quench tank is determined by the limitation of the vapor/liquid disengagement in the tank. The methods of sizing a quench tank/separator have been summarized by Svrcek and Monnery (1993). The maximum vertical terminal velocity for a vertical two-phase separator with a mist eliminator can be determined from the following empirical correlation

$$U_T = K\left(\frac{\rho_L - \rho_V}{\rho_V}\right), \quad \text{ft/s} \tag{5}$$

where the constant K depends on the operating pressure and is 0.35 for an atmospheric system. The internal diameter of the vessel is then determined from the volumetric flow rate of the vapor and the terminal velocity:

$$D_V = \left(\frac{4Q_V}{\pi U_T}\right)^{1/2} \tag{6}$$

The vapor flow rate in the second tank is 79,500 kg/hr with a density of 0.874 kg/m^3. The required diameter is found to be 3.04 m. For a tower with a mist eliminator, an additional 3 to 6 inches must be added to accommodate a support ring. If 6 inches are added to the required diameter, a diameter of 3.2 m is required.

The height of the quench tank H_L is determined by the liquid flow rate, liquid holdup and surge volume. Since the liquid withdrawal from the scrubber will be accomplished solely by a seal leg, there will be no need of surge volume or holdup. However, space must also be provided for the vapor jet impingement. According to small-scale quench experiments of Fauske and Grolmes (1991), the length required for a vapor jet to be completely absorbed in the quench liquid is $6.3D_{\text{feed pipe}}$. Large-scale data are not available. For the present case of an 18-in feed pipe, a total depth of 3 m is used for the liquid space, including the end plate.

The height for spray absorption H_{Spray} is determined by the mass transfer characteristics of the spray. Very limit data are available in the literature. We follow the recommendation of Zenz (1989) that 4 ft are used for every layer of spray. For space limitation, three layers of spray with equal spacing of 1 m are used. An additional height H_D of 2 ft is generally required between the spray and mist eliminator to achieve vapor/liquid disengagement. An additional space must be provided on top of the mist eliminator. Usually 1 ft or $\frac{1}{2}D_{\text{vessel}}$, whichever is larger, is used.

Six inches of height for the mist eliminator H_{ME} must also included. The total column height, including top and bottom end plates, is then:

$$H_{\text{total}} = H_L + H_{\text{Spray}} + H_D + \tfrac{1}{2}D_{\text{vessel}} + H_{\text{ME}} = 8.35 \text{ m} \tag{7}$$

This is the minimum required height for the scrubber. Factors such as the inlet nozzle and manhole access to the mist eliminator and spray nozzles will increase the scrubber height. The exact height is finalized in the detailed design.

The above sizing is based on the flow in the second stage quench tank. Both vapor and liquid flows in the first stage tank are slightly larger than that of the second stage tank. Since the vapor exiting from the first stage tank is fed into the second stage tank, completely vapor/liquid disengagement of the first stage tank is not as important as the second stage tank. Therefore, the same size of ∅3.2 m × 8.35 m is used for the first stage quench tank.

The size of the connecting pipe between the scrubbers has to be large enough to avoid potential back pressure on the relief line. Flow analyses suggested that a size of 42 inches for the connecting pipe will impose the least back pressure at the scrubbers. In this case, the maximum working pressure for the scrubbers is 1.15 bara. Reducing the size of the connecting pipe to below 36 inches will increase the maximum working pressure in the scrubber to 1.26 bara which renders the sizing of the seal leg difficult (see Sec. 3.3.3).

3.3.2. Instrumentation of the Scrubbers

The present scrubber system is designed to be nonoperating during the nonrelieving period. There are several ways by which the scrubbers are actuated to meet the relieving requirement. Pressure/temperature of the system (reactors), pressure/temperature in the relief line, and flow in the relief line are among the most common parameters for sensing the relief. The best choice would be the one that gives the least false signals. Pressure and temperature transducers can be very reliable for clean service. However, the reactor can have a relief even if its pressure or temperature is lower than the set pressure or temperature. The relief can be a leak from the rupture disks or the PSVs. An instantaneous pressure surge may also result in the popping of the PSV while the system pressure is still below the set pressure. After popping of the PSVs, the system pressure may drop to below the PSV set pressure while the safety valve is still relieving due to solid plugging on the valve seat. Measuring pressure and temperature in the relief line does help to avoid both problems. However, large variations in the ambient condition rendered the exact setting of relieving conditions difficult. Direct flow sensing at the relief line, usually accomplished by flow switches, avoids all the above shortcomings and is considered the best method to actuate the scrubbers. A flow switch is an electromechanical device that converts a flow signal into an ON/OFF signal. The setting of flow switches is described in Sec. 3.3.5. To avoid false signals from the measuring elements, multiple installations with majority voting are generally required. The piping and instrumentation diagram is shown in Figure 7.

3.3.3 Sizing of the Seal Legs

The flow in a seal leg is determined by the gravity and the drag in the pipe wall. The height of the seal leg also determines the maximum allowable pressure in the tank. According to API RP 521 (API, 1997), the seal leg must be sized to 175% of the Maximum Allowable Working Pressure (MAWP). Insufficient pressure seal will result in direct venting from the seal leg. The height of the seal leg is calculated by

$$H_{\text{seal leg}} = \frac{1.75(P_{\text{scrubber}} - P_{\text{atm}})}{\rho_{\text{L}} g} \tag{8}$$

The maximum pressure in Scrubber 1 is 1.15 bara from the flow analysis. Designing for 175% MAWP requires at least 2.6 m for the seal leg. Seal legs of 12 inches and 8 inches are required for scrubber 1 and 2, respectively, according to gravity flow analyses.

3.3.4 Baffles and Feed Pipes

An overflow baffle separating the quench liquid pool and the outflow liquid is installed to reduce the direct carryover of relief vapor from the seal leg. The size of the overflow baffle is determined by the down flow velocity in the overflow compartment. The common practice of limiting the down flow velocity to 0.4 ft/s is used.

The inlet of the scrubber is usually provided by a pipe of the same size as the relief line. A sparger or an end-cap on the pipe exit is not preferred due to potential back pressure. A straight pipe with vertical slots at the pipe wall are used. The top of the slots should be located far enough below the liquid sur-

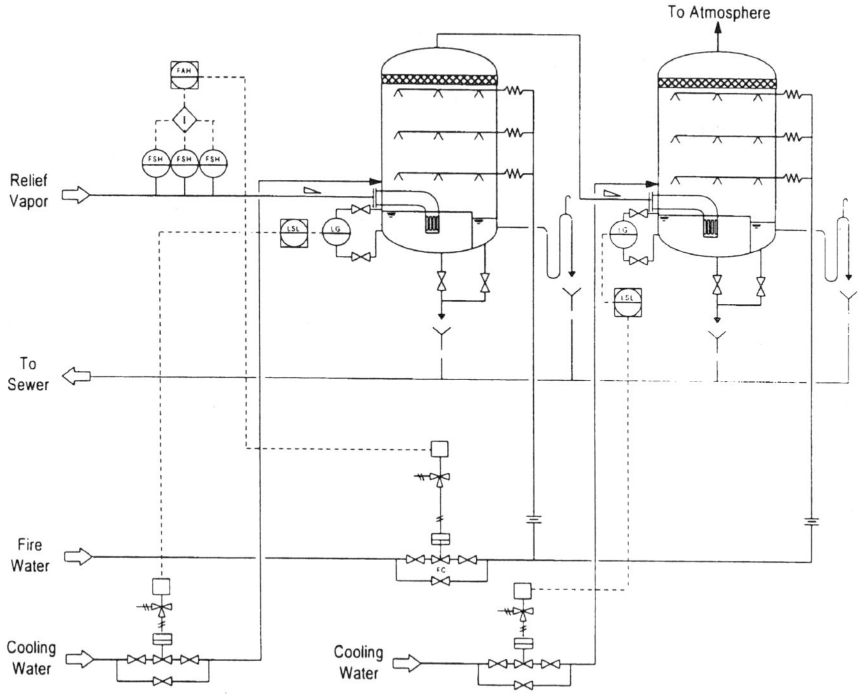

Figure 7. Piping & Instrumentation Diagram for the Scrubber System.

face to ensure vapor/liquid contact at all times. The common practice requires that the volume of water between the surface and the top of the slots is equal to the volume of 10 ft or more of the inlet line. Vertical slots with equal spacing are chosen. The total area of the slots is design to equal to the of the cross-sectional area of the inlet pipe.

For scrubber 1, the size of the inlet line is 18 inch and the required submerged length is 0.08 m. For scrubber 2, the size of the inlet line is 42 inch and the required submerged length is 0.375 m. This is considerably larger than the inlet line of scrubber 1 due to the exceedingly large line size. Since the scrubbers are not operating continuously, the liquid level is subjected to variation as a result of evaporation and regular water makeup is necessary. Process cooling water will be used for the water makeup. The maximum liquid level is controlled by the overflow baffle and the seal leg. We recommend that the minimum liquid level for the cooling water recharging is 0.1 m above the top of the slots. The maximum liquid level is 0.5 m above the top of the slots. The minimum of 0.1 m submerged length should allow for a reasonable vapor/liquid contact area. A 0.4-m variation of liquid level requires cooling water recharging about once every two weeks. This selection avoids the frequent recharging of cooling water, yet maintains a reasonable quenching efficiency.

3.3.5. Flow Switch Setting

The setting of the minimum flow rate in which the scrubber system is actuated is important. It is not necessary to actuate the fire pump if the relief rate is small. Sweeping from the pressure relief valve will result in pulsation of the flow if the relief pressure is too small to break the submerged length of the feed pipe. The flow switch in this case will send oscillating ON and OFF signals to the fire pump if the minimum flow rate is set too low. The best setting will be a continuous flow that will break the submerged length of the feed pipe. Successive flow analyses in the relief header were performed to find the minimum flow rate such that the inlet pressure of the scrubber is greater than the hydrostatic pressure of the submerged length. The flow rate of 25,000 kg/hr was found to be sufficient to maintain a continuous relief flow across the scrubbers.

4. MECHANICAL DESIGN OF RELIEF PIPING

According to APR RP 521, the design of the relief piping should conform to the requirement of ASME B31.3 (1993) for normal process piping. However, the mechanical design of relief piping is generally more critical and difficult than that of process piping. The relief piping faces a greater range of temperature, pressure and shock. Shock response is a transient behavior. Therefore, the API RP 520 Part II (1994) also recommended that time history analysis is performed for relief pipe in addition to the static analysis required by the ASME B31.3 code.

Care must also be taken to avoid stresses built up from loads and moments of the pipe on the pressure relief device and the inlet pipe. The stresses may cause a PSV or a rupture disk to leak, malfunction or the set pressure to change. Permissible loads and moments on the pressure relief device provided by the manufacturer must be checked along with piping allowable stresses.

The task of the stress analysis is best performed by a computer-aided pipe stress analysis software. These software are usually structural analysis software using the stiffness matrix or the flexibility matrix method. The pipe system is modeled as a space frame being subjected to different loading scenarios. The loading scenarios are determined by requirements from specific codes or standards like ASME B31.3. The analyses for the present case study are performed by using TRIFLEX stress analysis software from AAA technology (1995).

The present case study requires adding relief piping to existing pressure safety valves that previously vented to the atmosphere. Therefore, all mechanical designs of the relief pipes, including piping layouts, pipe support locations etc., must fit to the existing field conditions. The relief pipe must be analyzed along with rather than analyzed independent of the vapor lines where the PSVs located. Figure 8 shows the final design of the piping and piping supports.

Three scenarios are proposed according to the requirement of ASME B31.3 and API RP 521. The first scenario is the normal, nonrelieving operating condition. The relief pipe adds additional constraints on the normal vapor line even during nonrelieving conditions. Both sustained and displacement stresses are required to be lower than the allowable stresses as required by ASME B31.3.

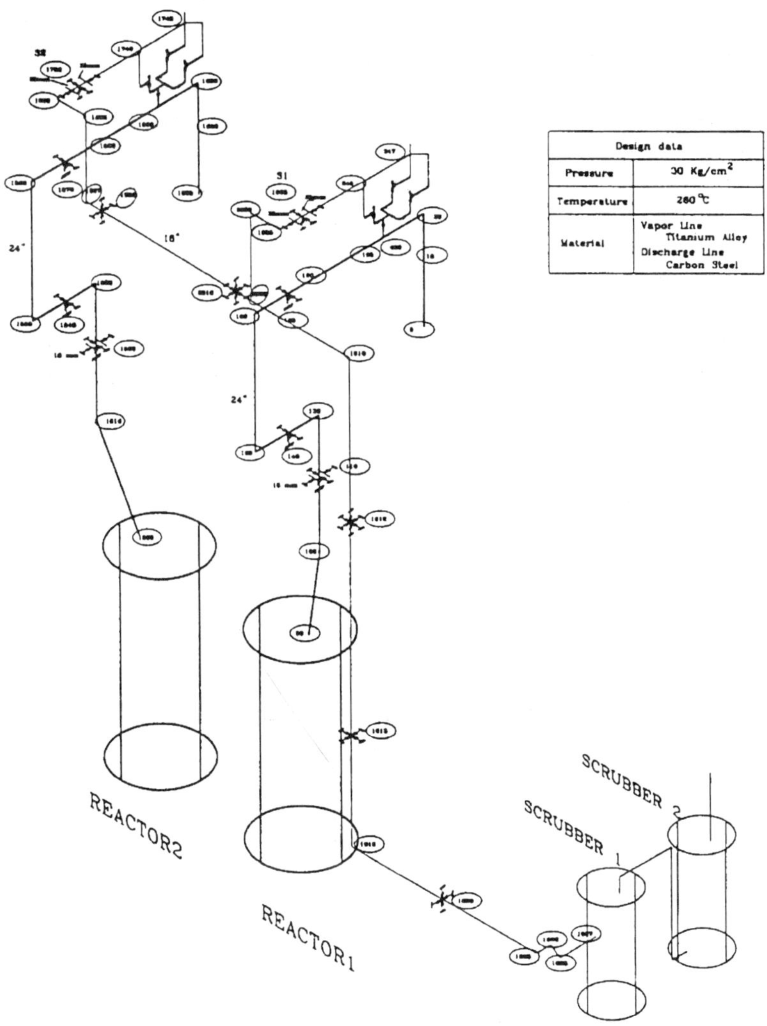

Figure 8. Results of Relief Piping and Piping Supports Design.

The second scenario concerns the action of thrust force on the relief pipe. The effects of vibration of the thrust force were found to be the largest at points 1005 and 1890 for reactor 1 and 2, respectively. A time history analysis was performed to calculate the transient response. The thrust force is calculated according to the DIERS recommendation (Fisher et al., 1992). The duration of the thrust force is chosen to be 0.5 second. The selected duration of the thrust force should be long enough to excite the maximum amplitude of the natural frequencies of the pipe structure. The vibration arising from the thrust force can be best constrained by a rigid support. However, stresses arising from thermal expansion may not be tolerated if rigid supports are used. An axial support that allows for finite displacement for relieving thermal expansion and confining the thrust vibration as shown in Figure 9 can be used. The gap size for the free displacement is determined by the stress requirement. For supports S1 and S2 in Figure 8, a gap size of 25 mm gives the best, lowest stress effects from thrust vibration and thermal expansion.

The last scenario concerns the steady relieving conditions. Thermal expansion is the dominant factor in this case. The collection header for the three PSVs on each reactor has to be raised 1.8 m above the PSVs to allow for thermal expansion. Fortunately, flow analyses show only a minor increase in the back pressure for the extra pipe length. Therefore, there is no need to increase the size of the relief pipe. Other supports are also provided at the joint header of the vent pipes from reactor 1 and reactor 2 for carrying wind load, seismic load and most importantly for reducing the thermal stresses to allowable levels.

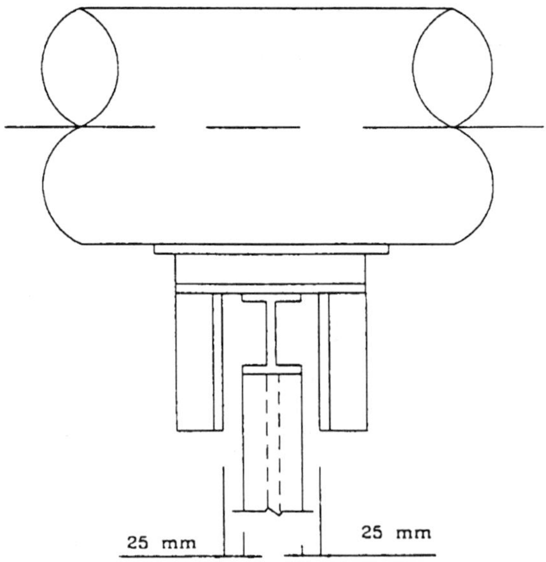

FIGURE 9. The Axial Support That Allows for Relieving Thermal Expansion and Confining the Thrust Vibration.

5. CONCLUSIONS

A case study is presented for the design of an emergency scrubber system for a large-scale petrochemical process. Difficulties in applying existing design guidelines for the scrubbers are highlighted. Practical solutions are given for an cost-effective design of the scrubber system. Proper mechanical design of the relief pipe is also shown to be very important in ensuring the integrity of the relief system.

ACKNOWLEDGMENTS

This work has been supported by Tuntex Petrochemical Inc. Collaborations and helpful suggestions from Pacific K&S Engineering Co. are also gratefully acknowledged.

REFERENCES

1. AAA Technology Inc., *TRIFLEX Piping System Analysis & Design*, Release 2.6.6, 1995.
2. API RP 520, *Sizing, Selection, and Installation of Pressure-Relieving Devices in Refineries*—Part I, *Sizing and Selection*, 6th ed., 1993; Part II, *Installation*, 4th ed., 1994.
3. API RP 521, *Guide for Pressure-Relieving and Depressuring System*, 4th ed., 1997.
4. ASME B31.3, *Chemical Plant and Petroleum Refinery Piping*, 1993.
5. Bravo, F., Cooper, R. S., and Contreras, D., "Wisely Use Emergency Scrubbers With Vent Systems," *Chem. Eng. Prog.*, (Aug. 1997), 62.
6. Fauske, H. K., and Grolmes, M. A., "Mitigation of Hazardous Emergency Release Source Terms Via Quench Tanks," *25th Annual Loss Prev. Symp.* AIChE, 1991.
7. Grossel, S. S, "An overview of Equipment for Containment and Disposal of Emergency Relief Effluents," *J. Loss Prev. Process Ind.*, **3** (1990), 112.
8. Grossel, S. S., "Containment and Disposal of Effluent Streams from Emergency Pressure Relief Devices," in *Handbook of Highly Toxic Materials Handling and Management*, Grossel, S. S. and Crowl, D. A., eds., New York: Marcel Dekker, 1995.
9. Keiter, A. J., "Emergency Pressure Relief Discharge Control by Passive Quenching—Update," *Plant/Oper. Prog.* **11** (1992), 157.
10. Simulation Science Inc., *Visual Flare*, Version 3.0, 1993.
11. Svrcek, W. Y., and Monnery, W. D., "Design Two-Phase Separators within the Right Limits," *Chem. Eng. Prog.*, (Oct. 1993), 53.
12. Zenz, F. A., "Design of Gas Absorption Towers," in *Air Pollution Control*, Cheremisinoff, P.N. ed., Gulf Pub. Co., 1989.
13. Zheman, D. I. and W. F. Early, "Flare System Analysis in Ethylene Plants," *Proc. 4th Ethylene Producer Conf.*, AIChE, 1992.

Design of Vent Collection and Destruction Systems

F. P. Nichols

Engineering Technology, Imperial Chemical Industries PLC,
Box 8, The Heath, Runcorn, Cheshire WA7 4QD, United Kingdom

ABSTRACT

Environmental pressure against the atmospheric venting of stock tanks and process vents is leading to the retrofit and installation of process vent collection and destruction systems. This paper considers some of the design problems in the safe collection and handling of potentially flammable gas mixtures and their destruction by burning in a thermal oxidizer.

The major challenges in the design of a vent collection and destruction system are in maintaining a reliable nonflammable (fuel rich or air rich) mixture, avoiding ignition by static, and controlling flame propagation from the thermal oxidizer back through the piping system.

The technical problem of deflagration to detonation transition in long pipelines and the resultant high shock loading on vessels and piping is also discussed.

INTRODUCTION

The environmental impact of Man on world climate has received increasing attention over recent decades. From the control on the emission of toxics, starting with the historic UK Alkali Acts of 1863, through the Clean Air Acts[1] to the current attention on reducing the emission of volatile organic compounds (VOCs) from, inter alia, the chemical and petroleum industries. The 1991 UN Protocol on air pollution[2] called for a reduction in VOCs across Europe and North America by 30% over 10 years from 1988 and there is increasing pressure from both legislative authorities and public to completely eliminate all VOC vents.

This paper considers some of the technical challenges to process engineers in the design of suitable systems to meet such environmental requirements safely and economically.

SCOPE

VOCs include many types of chemical ranging from toxic to nonhazardous, from flammable to nonflammable. The handling of toxic and flammable VOCs has long been practiced and is well understood. [3,4] For example, plants handling toxic chemicals are designed and equipped with collection and treatment systems to ensure atmospheric emissions are reduced to an acceptable level; similarly plants handling flammable gases (hydrocarbons) are designed with flare headers leading to flare stacks to avoid release of flammable gases to the atmosphere.

This paper will not cover these cases where much experience and good practice already exists.

This paper will be limited to considering some of the problems of collection and treatment of mixed plant streams containing fuel, air, inerts and reactive chemicals, often in trace but highly variable quantities.[5] The problems are not only in establishing the design basis for the equipment, but also in achieving an acceptable safety performance in operation. The approach suggested is a framework based on ICI's in-house experience and design methodologies on which an experienced professional engineer can build a safe design to meet his own particular situation. Many important considerations have necessarily been omitted.

1. VOLATILE ORGANIC COMPOUNDS

1.1. Sources of VOCs

Process plant generate VOC emissions from many disparate sources each with its own characteristics and handling problems.

1.1.1. Storage Tank Breather Vents

These emissions include both "normal" breathing and "emergency" venting such as due to an external fire. These are defined and discussed in API 2000[6]. "Normal" venting may be direct to the atmosphere or be collected in a vent system. "Emergency" venting is usually vented directly to the atmosphere.

Tanks containing material of flash point near ambient temperature will breath vapor which is near or above the lower flammable limit and this vapor represents a potential fire/explosion hazard. When a tank vents locally, the fire hazard may be controlled by, for example, padding the vapor space with inert gas or minimizing local sources of ignition.

1.1.2. Process Vents

These may be "known"—estimated or assessed in the process flowsheet (e.g. vacuum pump exhausts, distillation columns vents, process blowdowns, purges, reactor vents etc.) or "fugitive"—from spillages, leaks, pump and agitator seals, valve glands, maintenance operations, etc.

The "known" can be highly variable, particularly for batch operations, and when collected together into a vent header, give a stream inherently variable in both quantity and composition which may be fuel lean, flammable or fuel rich at different times.

The "fugitive" are largely associated with process equipment and the standard of maintenance. The emissions can be reduced by maintenance and/or selecting alternative equipment: canned pumps, bellows seal to valve stems, mechanical seals to agitator glands, different pipe gaskets, soft seat to relief valves, etc.

1.1.3. Emergency Vents—Explosion and Pressure Relief Vents

Explosion vents are best vented directly to the atmosphere; the high pressures generated and short duration of explosions make any form of containment or treatment difficult.

Pressure relief systems may vent gas, liquid or both. The vented material may be totally contained in dump tanks, or the gas and liquid phases separated in quench tanks, knockout pots, cyclones or similar. The liquid phase together with condensed vapor can be then stored for treatment, recovery or disposal (including a separate liquid feed to the incinerator). The gas and uncondensed vapors can then be passed to the atmosphere, to a vent or flare header system or to a treatment system. A new vent collection system may have a significant back pressure effect on an existing relief system and the complete relief system should be checked and modified as required. Guidance for the design of relief vents is available.[7,8]

1.1.4. Principal Sources of VOCs

The main sources of VOCs considered in this paper are therefore:

- vessel and storage tank vents (but not "emergency" relief vents)
- process vents(but not fugitive emissions)
- pressure relief vents—vapor/gas only (but not explosion relief or relief liquids)

These various sources are typically highly variable in flow rate, composition and potential hazard. Some may be air rich, some fuel rich (but neither separately flammable), and the remainder flammable. The flammable should be designed out and the nonflammable collected into separate headers (air rich, fuel rich and possibly also inert rich) and fed individually to the treatment system.

1.2. Mitigation of VOC Emissions

Vent collection and treatment systems are complex and expensive plant with significant operating and capital costs. It is therefore desirable to eliminate or mitigate VOC sources as far as possible. Possible approaches include:

- Replace volatile solvents by water or lower volatility solvents.
- Fit Breather valves on storage tanks.
- Interconnect storage tank vents by a local header (subject to possible cross-contamination effects).
- Inert pad storage tanks under pressure/vacuum control.
- Back balance road and rail cars during transfer to storage tanks.
- Consider floating roof tanks.
- Fit condensers or refrigeration on process and storage vents with local material recovery.
- Install closed charging systems rather than open charge ports to vessels.
- Use high integrity control systems to reduce relief demands.
- Raise vessel design pressure to reduce relief emissions.

2. DESTRUCTION OF VOCS

2.1. Destruction of VOCs

The collected VOCs require treatment or destruction. Many methods have been suggested including:[9]

- Absorption in suitable reagents—caustic or acids, solvents, water, etc. with recovery or liquid effluent treatment;
- Adsorption followed by recovery or incineration (of active charcoal adsorbents);
- Biological processes (usually for dilute aqueous streams, e.g. from aqueous absorbers);
- Condensation and /or refrigeration with recovery of material;
- Flares—widely used for hydrocarbons;
- Primary combustion processes—VOC streams used as supplementary air/fuel in boilers, furnaces, kilns etc.;
- Secondary combustion processes—specifically to destroy VOCs, usually with supplementary fuel and/or air and probably with some heat recovery. There are various types such as incinerators, thermal oxidizers, catalytic combustors, etc.[10]

Secondary combustors are currently widely used. They usually require make-up fuel, and may include systems for waste heat recovery and scrubbing of trace hazardous combustion products (e.g. HCl or SO_2). A key feature is their high operating temperature and flame—which makes them a perfect ignition source.

2.2. Types of Thermal Oxidizers

Many different types of thermal oxidation units are available, each with their own specific characteristics. Most destroy the VOCs using a direct flame, but others are flameless in the sense that the destruction takes place in a high temperature bed of solids with or without a catalyst. Some recover heat in boilers raising steam, whereas others are regenerative with the flue gases heating feed combustion air.

Four typical types are shown in Figure 1 with different features. Choice should be based on the specific requirements of an individual case. Important considerations include destruction efficiency, unit availability and reliability, process duty, turn down and controllability, and whether heat recovery or flue gas scrubbing is required. Longer residence times and/or high temperatures may be needed to achieve complete destruction of some chemicals.

The availability of an incinerator is likely to be lower than the availability of the various sources of VOCs. Hence some provision must be made for handling the VOCs during times when the incinerator is not available, but the method adopted should be agreed locally. Possible methods include duplicate incinerators, local bypass vents, higher availability, plant shutdown, or a bypass vent directly to atmosphere at the incinerator as shown here.

Figure 1A. Simple conventional unit with vertical combustion chamber. Support fuel burner housed in a side chamber to protect from main chamber radiant heat and ease burner maintenance. Standard fuel control system—fuel controlled from main chamber temperature with combustion air ratio controlled to fuel and trimmed by oxygen in flue gas.

Figure 1B. Thermal oxidizer with staged air injection. Achieves low NO_x even with organic bound nitrogen in the waste feed.The reduction chamber operates substoichiometric on air at 1100–1300°C to give CO and N_2. The quench can use recycle gas (reducing total flue gas volume), air or water/steam. Air in the final oxidation stage at 1000°C burns CO and residual organics with little NO_x generation.

Figure 1C. Flameless thermal oxidizer. A conical unit packed with ceramic spheres. A hot zone generated by the combustion heat moves up or down according to the gas rate and fuel content. The unit offers high destruction efficiency, high thermal inertia (and hence operating stability), but there is a potential hazard if the combustion zone moves to either the top or bottom of the bed.

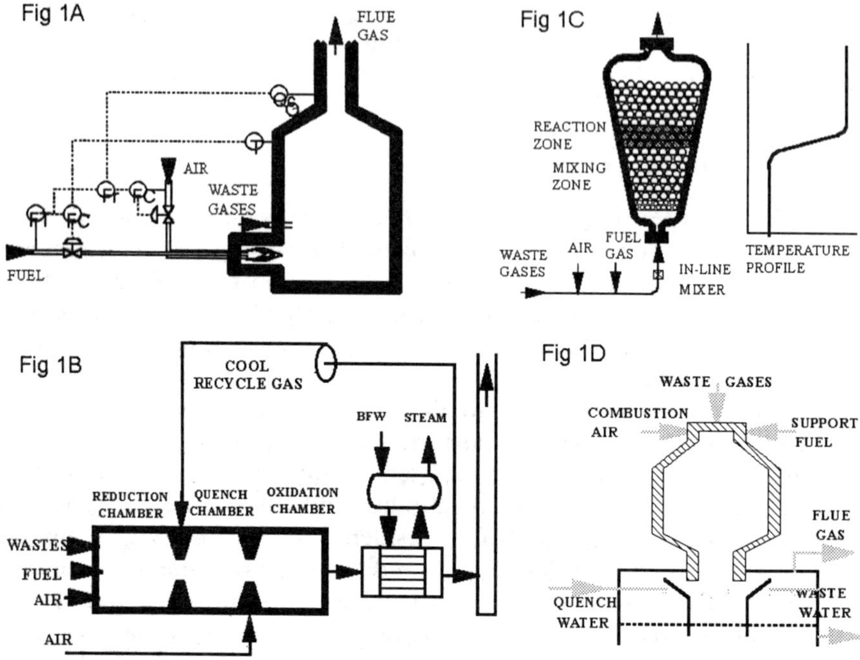

Figure 1. Some Types of Thermal Oxidizers

Figure 1D. Down fired unit with water bath quench. Handles corrosive flue gases—e.g., HCl, SO_2 by direct quenching and absorbing in circulating liquor. Heat recovery potential is limited.

3. VENT COLLECTION SYSTEMS

3.1. A Typical Vent Collection System—Figure 2

The main sources of VOCs from storage tanks and process vents including relief systems are collected together into one to three headers—air rich, fuel rich and inert rich—running close to atmospheric pressure and fed by induced draught fans to the incinerator.

The high integrity bypass system ensures that plant operation can continue during incinerator outage and maintenance and if any header goes flammable.

Flame arresters are provided at the incinerator and on each side of the fans as real and potential ignition sources to protect the vent piping. It is assumed that any process ignition sources (e.g., vacuum pumps) are protected at the source.

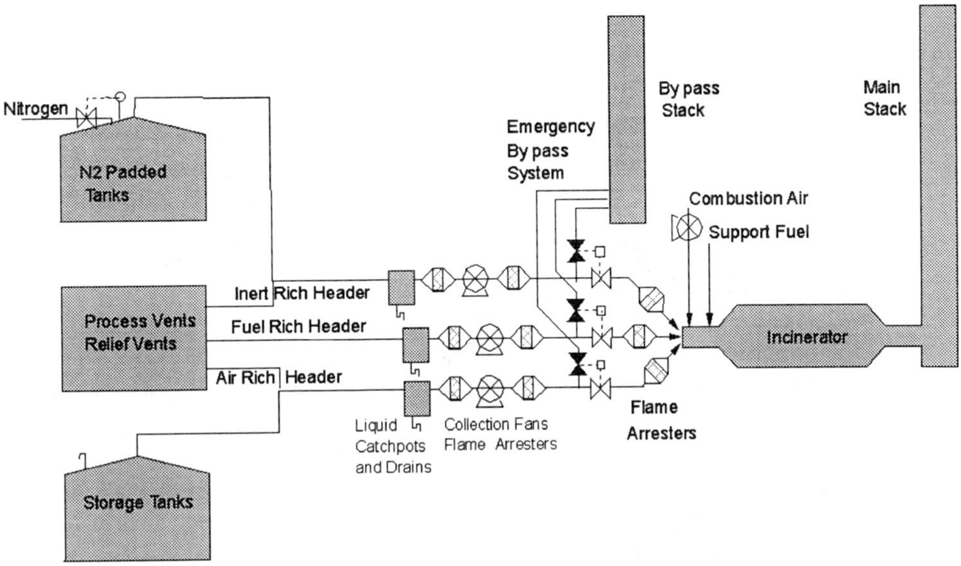

Figure 2. Vent Collection System

Catchpots are provided to remove trace liquid carryover (from the process and/or relief vents) and condensate from the saturated vapors leaving the storage tanks. All the vent piping should be self-draining to catchpots.

The two stacks should be high enough to ensure that the hot flue gases from the main and flammable gases from the bypass can disperse safely and independently.

It should be possible to modify some sources, e.g. by padding the air vented tanks with nitrogen, and combining some headers without increasing the hazards. The air rich header could also be diluted and used as combustion air in the incinerator.

The vent fans are only required to compensate for pressure drop in the vent collection piping. They introduce additional hazards by creating a hot spot ignition source, by drawing air into a fuel rich header or by collapsing low pressure storage tanks if the piping goes subatmospheric, and are better designed out.

All headers should be provided with rapid response oxygen analyzers set to trip the bypass system and/or inject additional inert if the oxygen level exceeds the NFPA[35] recommended levels (Section 5.3.3)

3.2. Incidents

The mixing of various vent streams and their feed to an incinerator have led to many hazardous incidents—some reported in the literature are:

- Combining vent sources led to unexpected chemical reaction in the vent systems; e.g.. solid or polymer deposits[11]
- Adsorption of reactive vapors gave local heating/exothermic reactions[12]
- Condensation of saturated vapors gave liquid logging or water hammer effects in the vent piping
- Liquid slug carryover gave combustor control and excess fuel supply problems[13]
- Electrostatic charging effects from condensate or mist flow gave a spark ignition in the tank vapor space and piping[14,15]
- Deviation of vent composition from design (fuel rich or air rich) into flammable region led to an explosion[16]
- Ingress of air through leaks into a fuel rich system led to an explosion[17,18,19]
- Egress of VOCs through leaks—fire hazard and environmental impact;
- Ignition of flammable gas by combustor flame running back up piping[20,21]
- Transition from deflagration to detonation in the vent piping caused blast damage.[22,23,24]

There are also hazards associated with the operation and control of the combustor[25,26] over and above the "normal" gas fired furnace hazards due to the variable quantity and quality of the vent gas feed taking the process demand outside the incinerator control range.

The principal hazard is propagation of a gas explosion through the vent system. The consequences of high explosion pressures can include:

- rupture of the piping, failure of piping supports, and widespread damage from the blast, fires started and missiles generated;
- the transmission of pressure pulses through other nonflammable vent piping into low pressure storage tanks, possibly causing tank rupture.

4. HAZARDS

4.1. Hazard Logic Tree

A vent collection and destruction system is a complex plant in its own right showing hazards offering infrequent but extremely serious consequences. To ensure safe operation, a methodical approach to the design basis and safety strategy is essential. To assist in such an approach an outline hazard logic tree is shown in Figure 3 and its main elements explained below. This does not offer a complete analysis but indicates a possible approach to the problem.

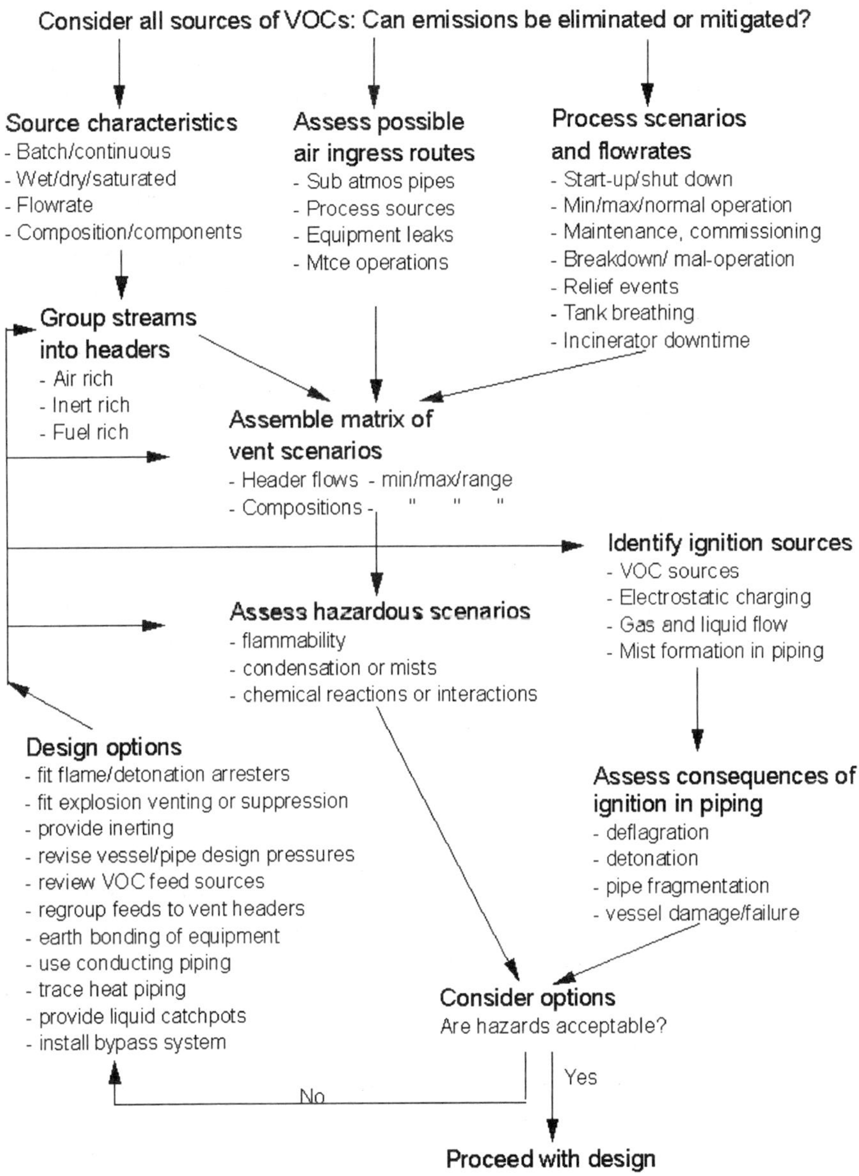

Figure 3. Vent Collection Logic Tree

4.2.1. Source Characteristics

The source may be batch or continuous. Batch is harder to define since the duration, intensity, composition and schedule of the vent gas change both with different products and at different stages of the batch processing cycle. Determining the characteristics of each vent source will require a lengthy experimental program to obtain a reliable estimate of the whole variation of composition and flow rate possible.

The source will be gas or vapor, possibly including trace liquid carryover. Vapor from storage tanks is necessarily saturated with the tank contents. Unless the tank or its vent is at subambient temperature or the vent system is traced/lagged, the vapor will cool and condense to a mist in the vent collection system, giving an electrostatic ignition hazard (Section 6.3). This hazard may be eliminated by local refrigeration with condensate recycle and this is an attractive option where a refrigeration service already exists on the plant.

The flow rate and composition of the various sources need to be measured. Note that components can react, even at ppm levels, with trace components from other sources leading to solid or polymer deposits in vent pipework or blockage of the narrow passages in flame arresters.

The flammability of each source should be estimated (Section 5.3). Note that air/oxygen is not the only oxidant; chlorine or oxides of nitrogen act similarly. Beware of unexpected effects—ethylene oxide or acetylenes can decompose explosively even in the absence of any oxidant; CO behaves differently with dry and moist air. Experimental tests are advisable to confirm estimates.

It may be advisable to carry out a hazard study on the upstream process to consider the feasible deviations which could occur giving different releases into the vent system.

4.2.2. Headers and Air Ingress

The major problem with vent collection systems stems from the wide range of streams being combined. Under "normal" flowsheet conditions these mixtures can be predicted and the system designed to maintain the headers air rich or fuel rich. The hazards of flammable headers are so high that they should be designed out. However, during plant start-up or shut down, maloperation or equipment failure, or maintenance operations, the composition can stray into the flammable region and this mixture may ignite within the piping—giving a deflagration, possibly running up to a detonation—but it will certainly ignite in the incinerator and flash back into the vent system and burn on the flame arrester element; these are highly hazardous events and must be controlled or prevented.

Air may be drawn into the vent system whenever it goes subatmospheric due to the vent fans or even from the natural draught of the incinerator or bypass stacks.

Possible air ingress routes include:

- maintenance operations involving isolation and recommissioning of equipment, control and relief valves, instruments, slip plates, flanges etc.;
- failure of the liquid seal in drain lutes on catchpots or water seal devices;
- leaving open sample points, instrument purges, bypass valves, etc.

The various sources of known composition should be provisionally grouped into

- air rich—fuel content below 0.25× Lower Explosion Limit,
- inert rich—oxygen below 0.6× Minimum Oxygen Concentration, or
- fuel rich headers (Section 5.3.3)

4.2.3. Process Scenarios

For every vent source the normal, minimum and maximum process flows and the startup and shutdown conditions should be tabulated together with the chemical composition, temperature, dew point or other indication of condensate forming potential, and the flammability characteristics. Add to this tabulation the flows and compositions occurring during the principal maintenance operations, breakdowns and plant maloperations, and the hazard events recorded in the process plant and storage tanks hazard studies.

Using this data for each header, assemble the full matrix of flows and compositions. The possibility of condensation should be included since this not only introduces a hazardous mist and a need for catchpots but also changes the gas composition.

By inspection many scenarios which would not result in a flammable mixture can be eliminated from the matrix.

It is important to identify transient incursions into or through the flammable region whether caused by air ingress or by process upset. A graphical approach using the flammability diagram (Section 5.3.2) is easy to use.

This matrix, while laborious to assemble, nevertheless greatly simplifies both the identification of potentially hazardous situations and the analysis of possible future changes in the vent collection system or process plant.

For each matrix element estimate the hazard and potential consequences. A qualitative ranking of the hazards will show the high hazard problem sources or process events. These problem areas should then be reconsidered to see if they can be reduced at source, or alternative strategies adopted to limit the hazard.

4.2.4. Ignition Sources

Ignition at the ends of the vent collection system can be handled by flame arresters. However ignition somewhere in the vent system itself is extremely difficult to prevent, and since the consequences are likely to be detonation and major blast/missile damage, the hazard can become high. This can be reduced by modifying the design to avoid ignition or occurrence of a flammable mixture by such options as electrically grounding the (conductive) vent piping, preventing condensation by trace heating of the vent piping, removing mists and liquid carryover by catchpots at vent sources, removing flammables by refrigeration condensers on tank vents, etc.

4.2.5. Consequences and Design Options.
The consequences of any remaining high hazards should now be estimated after consideration of the engineering aspects. The relevant properties of an explosion are discussed in Sections 5.1 and 5.2 and the engineering impact in Section 6.1. Standard texts[27] should be consulted for estimating the external blast and missile effects from an explosion.

However, even when the explosion is contained, there remains a major hazard from the explosion pressure wave passing through other (nonflammable) parts of the vent collection system and affecting low pressure storage tanks which are often designed to the API 650 Standard[28] with a frangible roof failing at 50–100 mbar.

Some hazards have been directly introduced by the equipment adopted in the vent collection system. In particular the use of vent fans introduce additional problems. The fans are only there to balance a small pressure drop in the vent piping and can sometimes be eliminated by either reducing the pressure drop with a larger pipe diameter or by slightly increasing the design and hence operating pressure of the storage tanks.

5. EXPLOSIONS AND FLAMMABILITY

Ignition of a flammable gas in a vessel or pipeline gives an explosion, starting as a deflagration and possibly running up in a pipeline to detonation.

5.1. Deflagration

The combustion flame travels down a pipeline containing an air/fuel mixture at a low flame speed (10–400 m/s, subsonic) and generates a maximum 6–9 ratio increase in (absolute) pressure for a typical hydrocarbon. This pressure ratio may be calculated as a constant volume combustion using standard chemical thermodynamics.[29] Explosion vents can limit the rise in explosion pressure. The combustion process raises the temperature and this causes an expansion of the burnt gases which generates a pressure/shock wave traveling at near sonic speed (450 m/s) through the unburned gas in advance of the flame front. The effect of the pressure wave is to increase the turbulence and to precompress and heat the unburned gas.These cause the combustion process to speed up and the flame front to accelerate toward sonic velocity.

A deflagration front therefore accelerates with increasing fluid turbulence caused by both the pressure wave and by pipeline bends, fittings and wall roughness. After some distance the deflagration front will reach sonic speed, catch up its own pressure wave and undergo sudden transition to a detonation front traveling at supersonic speeds.

A deflagration can be stopped by a flame arrester and the pressure generated mitigated by bursting discs/explosion panels venting to atmosphere.

5.2. Detonation

In a detonation wave, the flame and pressure shock fronts are combined and travel at supersonic speed (2–4 km/s) and generate a 20-fold pressure increase for a typical hydrocarbon and a shock pressure twice this at a reflection. These pressures and the detonation velocity can be calculated from chemical thermodynamics using the Chapman-Jouguet theory.[30] The pressure behind the detonation front falls to the deflagration pressure linearly with the distance between the front and the point where the detonation started.[31]

Immediately after transition from a deflagration, a detonation is unstable (overdriven) but quickly stabilizes to a steady detonation.[32] An overdriven detonation generates a pressure about twice that of a steady detonation.

A detonation flame can be stopped by a detonation arrester, but the associated pressure wave passes through the arrester before decaying. A detonation is little affected by explosion vents.

The deflagration to detonation transition (DDT) distance depends on the gas, its composition, temperature, pressure and the pipeline diameter and design layout.

There is little information on the run up distances under industrial conditions and no design methodology available.

Hence the principal hazard is the occurrence of a detonation front within a pipeline, which subjects the pipe (in an atmospheric pressure vent system) at any point to a pressure rise to 50–100 bar in a few microseconds, followed by pressure decay to 8 bar over milliseconds to seconds. The out-of-balance reaction forces on the pipe supports are much higher than normal. Careful piping support design is needed.

5.3. Flammability of Gases

A fuel gas/air mixture will be flammable (i.e., will burn if ignited) over a range of compositions between two experimental limits called the Lower and Upper Explosive Limits (LEL and UEL). These limits bracket the stoichiometric composition.

All combustion parameters—deflagration pressure, detonation pressure and velocity, burning velocity, ignition energy and flame quenching diameter change between the Upper and Lower Limit compositions with the extreme values just above the stoichiometric composition where the hazards are greatest.

The effect of inerts is to lower the combustion temperature so the flammable range decreases with increased inerts, until the mixture becomes nonflammable at the Minimum Oxygen Content (MOC). This is about 10% oxygen for hydrocarbons.

5.3.1. Flammability of Mixtures

For a gas mixture the explosive limits, the effect of inerts and the minimum oxygen content may be estimated using the Le Chatelier rule:

$$1/(\text{Mixture property}) = (\text{Concentration/property of 1}) + (\text{Concentration /property of 2})$$

where concentration is in v/v units and the effect of temperature is estimated using the Burgess-Wheeler law as summarized in USBM reports.[33] Note that estimated values should be checked experimentally since unexpected chemical interactions are not uncommon.

5.3.2. Flammability Diagrams—Figure 4

Data on flammability limits and the effects of inerts, temperature and pressure are available[34] and can be displayed by flammability diagrams for any fuel/air/inert composition.

A typical flammability diagram shows 100% gas (fuel) at the top, 100% air at the left and 100% inerts at the right. At any point in the triangle the three components add up to 100%.

The gas mixture is only explosive (flammable) between the Lower (LEL) and Upper Explosive Limits (UEL) for an oxygen content above the Minimum Oxygen Content (MOC) as shown as the hatched area in Figure 4.

Detonation limits appear to be narrower than the explosive limits, but little data is available. In Figure 4 the detonation region is shown as the fine hatched area within the flammable area.

5.3.3. Safe Operating Limits

When nonflammability of a gas mixture is used as a basis of safety, NFPA[35] recommends that subject to due allowance for variation during plant operation, monitoring and control instrumentation and the potential consequences of an explosion, the gas composition should be:

- air-rich systems; fuel less than 25% LEL;
- inert-rich systems; oxygen content less than
  —with continuous monitoring: greater of (MOC-2%) or 60% MOC;
  —without continuous monitoring: 60% MOC and 40% for MOC <5%;
- fuel rich; add at least 25% Me or natural gas where no component has an UEL > 75% and $O_2/(N_2+O_2) < 21\%$.

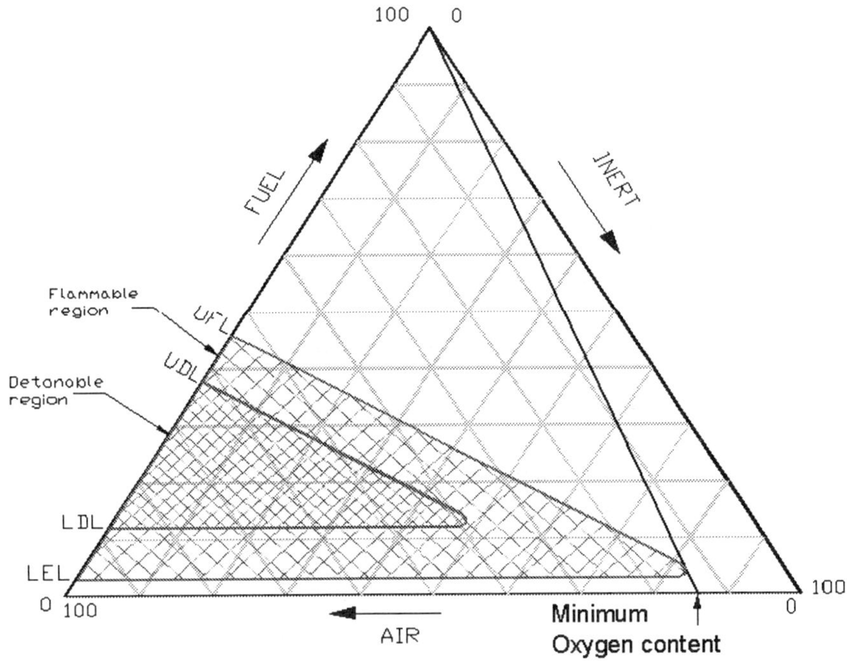

Figure 4. Flammability Diagram

6. HANDLING EXPLOSIONS

6.1. Containment of Explosion in Pipes and Vessels

Containment is theoretically possible for gas detonations. This is a specialist topic, but a few comments for steel vessels may help show the options. The effects of creep from continuous high stresses, fatigue from cyclic stresses, stress corrosion cracking, or brittle/ductile transition at high stress rates with steels are not discussed here, nor the characteristics of (reinforced) plastics.

Vessels and piping design is based on elastic stress theory under a static load[36] with the ASME Boiler and Pressure Vessel Code Section VIII, Div 1 using rules and Div 2 using stress analysis. Section III uses higher stress levels.

Design to contain explosions can also use these methods provided due consideration is given to the effect of high strain rates on material properties and branch and fitting design. More sophisticated approaches include both the effect of dynamic as against static load on the stress field and plastic deformation.

Under the vessel codes the design pressure for continuous use corresponds to a vessel wall stress of the lower of given fractions of the elastic limit or the tensile stress. Under ASME VIII Div 1, the design stress is the smaller of 67% yield or 25% tensile stress. For most steels the tensile stress controls design.

A vessel to this standard is proof tested to a higher pressure—limited to about 1.5 x design or 90% of the yield stress. If a vessel is subjected in a rare event to such a pressure, it is unlikely to be damaged and, subject to inspection by a specialist vessels engineer, will be reusable.

A basis to avoid fragmentation is given in NFPA 69[37] and uses a stress level roughly 67% of the tensile. This gives an explosion pressures about 2.7x design.

A detonation gives very rapid rate of pressure rise and hence high stress rates in the containing material. The material may still be ductile or become brittle—so its properties should be checked at high stress rates.

Above the proof test pressure the vessel will suffer permanent plastic deformation, leading ultimately to vessel failure and fragmentation. In a vessel the hoop stress is always higher than the longitudinal, so failure will be along the length rather than at end closures. The effect of vessel openings and pipe flanges as stress concentrators should be noted and due allowance made. Internal defects in both material and fabrication can seriously weaken the vessel/piping. Improved inspection and nondestructive testing should be considered.

A method for estimating the burst pressure of thin walled cylinders has been proposed[38] and validated[39] based on plastic deformation and strain hardening.

Clearly any vessel which has deformed is unusable and must be replaced. However for very rare events the deformation and replacement of the vessel may be acceptable provided fragmentation does not occur.

The above estimate (2.7 x design pressure) provides an indication of the design potential available. If the option appears attractive, specialist engineering advice must be sought to prepare better estimates and to execute detailed design.

Subject to these various provisos, the feasibility of containing an explosion in an atmospheric system can be assessed:

- A deflagration will give up to eight bar pressure and will be more frequent than a detonation. The capital cost of a vessel with this design pressure is acceptable and there is no incentive to design for deformation (Design pressure about three bar) since frequent vessel replacement is unattractive. Piping will be within typical ANSI B31.3[40] pressure ratings for process plant and so acceptable.
- A detonation will give a vessel pressure about 20–40 bar, so a vessel requires a design pressure (to ASME VIII div 1) above 40 bar. This is well above the required atmospheric operating pressure for the vessel and clearly represents a major increase in capital cost. However, if

vessel deformation and the replacement is acceptable, then the required design pressure is reduced from 40 bar to 15 bar. The lower capital cost of this option can be attractive for a rare event.

- For a pipeline the pressure will be about twice that of a vessel (an overdriven detonation doubles the pressure and may occur anywhere in a pipeline, but is unlikely in a vessel) and a design pressure of 80–100 bar is required. A reasonable piping schedule may still be adequate—6" Schedule 40 to ANSI B31.3[41] in carbon steel has a test pressure of 123 bar—and the cost penalty of a higher rating is small.

Note that the inspection and fabrication standard of ordinary piping is lower than pressure vessels and internal faults may be present which would lead to failure under the shock loading of a detonation—higher fabrication/test standards are advisable.

This approximate analysis suggests that although it is possible to contain a detonation, the pressure rating required is high, implying higher capital costs. This cost can be reduced by designing for deformation and accepting that the vessel and piping will need to be replaced after rare detonation events.

6.2. Flame and Detonation Arresters

Flame arresters stop the passage of flame and are used either for end-of-line or for in-line installation. There are many papers on application and types.[41–46] Deflagration and detonation arresters work on the same principle, but the detonation arrester is larger and stronger (to withstand the higher pressures, etc.).

There is a characteristic size of opening for every gas which experimentally stops a flame. This is called the Maximum Experimental Safe Gap[47,48] (MESG) and typically varies from 0.2 mm (hydrogen), to 0.65 mm (ethylene), to 0.92 mm (propane) to 1.2 mm (methane). The Quenching Diameter (QD) is another measurement giving similar information. Other gases can be classified into 4 standard groups defined by the MESG of these four gases in specified mixtures with air. The groups are A–acetylene, B–hydrogen to ethylene, C–ethylene to propane, and D–propane to methane.

The key parameter in a flame arrester is the size of the flow passages through the device, which are typically designed to be some 60% of the smallest MESG of the standard group into which the gas is classified. The major types of construction are: crimped metal, packed gauze beds and parallel metal plates, although water seals, ceramic or metal foams, pebble beds, etc. can be used.

The effectiveness of the device relies on every opening through the arrester element being smaller than the MESG. Particularly for the smaller gaps (0.2 mm), this is difficult to ensure reliably with either metal foams or packed gauze beds and to engineer with flat plates. Hence for the more reactive (lower MESG) gases and the more arduous in-line arresters, the crimped metal type is preferred. For end-of-line installations, which are often used on LP storage tank breather vents with the less reactive Group D vapors, all types can be satisfactory.

The element is securely held between the halves of a housing comprising an expansion and contraction fitting and flanged into the pipeline. The free flow area is larger than the pipe, but the pressure drop—typically 5" to 10" water gauge for 0.65 mm crimp (Group C)—is strongly dependent on the crimp size. A flame arrester with such small passages is extremely good at catching any dust or liquid mist. For dusts this type of arrester clearly cannot be used; either a liquid seal trap[49] or filter is required, or the arrester eliminated by guaranteeing nonflammability (e.g. by inerting). For liquid mists the arrester should be mounted vertically to drain liquid down to the vent system catchpots, but if the mist includes monomer condensate, the arrester can rapidly block with polymer. This is not surprising since the monomer vapor in a vent system is unstabilized—the stabilizer present in the bulk monomer tank is much less volatile. Possible solutions are to refrigerate out the monomer at source or to prevent condensation by trace heating the vent line.

When a deflagration runs back to an arrester, the flame continues to burn on the arrester element so long as the flowing vent gas remains flammable. The arrester element rapidly heats up, and the flame flashes through the arrester—standard test protocols require a minimum 20 minutes before flashback occurs[50]

A thermowell can be inserted into the arrester housing on either side of the element and the thermocouple set to both trigger an inert quench and switch the vent header to the bypass system and kill the flame.

For a detonation arrester the whole assembly is a heavy pressure vessel able to withstand the detonation pressure (say 100 bar rating) and the arrester element is both longer and much more stoutly supported. A detonation wave carries significant momentum—equivalent to an additional Chapman-Jouguet detonation pressure at a reflection or bend—and this force has to be resisted by the detonation arrester and its support structure.

Detonation arresters are high capital cost items and not always available for the highly reactive gases. The only alternatives are expensive specialized arresters, avoidance of ignition, prevention of run up from deflagration to detonation or inerting.

6.3. Ignition

The phenomena of ignition is surveyed by Bond.[51] In a vent collection system possible sources include: the incinerator; vent fans; vent sources such as vacuum pumps, adsorption beds, or high temperature equipment; and within the vent piping, hot spots or sparking from electrostatic charge build up.

Ignition in the incinerator will lead to a flame traveling back up the vent system. However the local gas distribution header feeding the incinerator is often complex and tortuous, and local run up to detonation is possible. A detonation arrester as close as feasible to the incinerator is required together with a trip operating the vent bypass system and activating the emergency quench gas into the header. Nitrogen is the preferred quench gas since steam can damage the incinerator refractory. Note that this rapid sequence of events—vent gas off and inert on—both thermally shocks the incinerator and severely tests its control system.

Ignition may also occur in the vent fans. The explosion front may move either way so deflagration arresters are required on both sides. Only if the fan is sufficiently close to the incinerator that a deflagration will not run up to detonation would the downstream arrester be omitted. It is assumed that the arresters can be mounted so close to the fan that detonation run up cannot occur. Inert quench to the vent header should be triggered by burnback on the arresters.

Ignition within the vent collection system will run up to detonation. Designing to contain a detonation is difficult and the equipment costly. However fragmentation is highly undesirable—the vent system can be kilometers long, passing through other process units, tank farms or near public areas. This means that the consequences of detonation, rupture, missiles, local blast and further domino effects are likely to be so serious that the frequency has to be kept as low as reasonably practicable.

Within the vent collection system, ignition from electric sparks may occur due to build up of electrostatic charge on equipment (e.g. pipe walls) or between mist clouds or different parts of a low conductivity liquid.[52] This build-up of charge and hence ignition can be minimized by:

- making all piping electrically conductive;
- fitting bonding straps across all pipe flanges, and earthing all piping, supports and equipment to ground;
- designing for low gas and liquid flow velocities in pipework;
- minimizing mist formation caused by cooling of saturated vapors (e.g. from storage tanks) or carryover of liquid spray from process or relief vents;
- fitting liquid knockout pots to remove condensate and any liquid carried over from process or relief vents;

However the possibility of ignition cannot be completely eliminated, so if ignition occurs within the vent collection system either the hazards of a run up to detonation have to be accepted or explosion vents and/or deflagration arresters installed at intervals of less than the detonation run up distance.

6.4. Engineering Design Basis

The engineering design strategy proposed for a vent collection system is :

1. Prevent deflagrations propagating from vent sources, incinerators or other clearly defined sources in the piping system by installing flame arresters;
2. Fit detonation arresters at the incinerator or wherever the distance to possible ignition source approaches run up distance, otherwise use deflagration arresters;
3. Install thermowells and thermocouples on all arresters with high temperature trips. Set the trip to activate the vent bypass and to feed an inert gas local to the arrester to quench the flame;
4. Minimize mist formation by refrigeration at vent sources;
5. Minimize sources of ignition and the occurrence of flammable gas mixtures within the vent piping;
6. Grade vent piping to liquid knock out pots and drain arresters;
7. Make all vent piping conducting with bonding straps at flanges and to supports, other equipment and electrical ground;
8. Design the vent system to either contain a deflagration or provide explosion vents to limit the maximum pressure;
9. Control detonations by either fitting detonation arresters in the piping at intervals less than the run-up distance, designing the system for plastic deformation without fragmentation, or avoid ignition in the vent piping;
10. Avoid fragmentation of the vent collection piping under all circumstances.

ACKNOWLEDGMENT

I wish to thank my colleague A Ennis for his comments and ICI Technology for permission to publish this paper.

REFERENCES

1. Clean Air Acts and amendments (US), 1955, 70, 77, 90; Clean Air Acts (UK),1956,1962.
2. Protocol 1991 to "Convention on Long Range Transboundary Air Pollution," United Nations Economic Commission for Europe, 1979.
3. Prugh, R. W., and Johnson, R.W., *Guidelines for Vapor Release Mitigation,* New York: AIChE/CCPS, 1988.
4. Reed R.D., "Design and operation of flare systems," *CEP*, 61(6) (1968), 53
   Hajek, JD and Ludwig EE, "Safe flare stacks," Pet Eng, C31-39 June 1960; C44-51, July 1960
   J Zink Co, "Safe flare system design," Tech paper 5490B, 1993
5. Clark DG and Sylvester RW, "Ensure process vent collection system safety," *CEP* (Jan. 1996), 65.

6. API 2000 Standard, "Venting atmospheric and low pressure storage tanks," American Petroleum Institute, 1992.
7. Grossel, S.S., "Overview of equipment for containment and disposal of ERS effluent," *J Loss Prev Proc Ind*, 3(1), 112, 1990.
Grossel, S. S., "Design and sizing for knockout drums/catchpots for reactor emergency relief systems," *Plant Operations Progress,* 5(3), 129, 1986.
8. Fisher H. G., et al., *Emergency Relief System Design using DIERS technology,* New York: AIChE/DIERS, 1992.
9. Mukhopadhyay, N., and Moretti, E. C., "Reducing and controlling volatile organic compounds," New York: AIChE/Center for Waste Reduction Technologies, 1993.
10. Factory Mutual Eng Corp., "Fume incinerators," Bulletin 6-11, May 1990.
Anon., "Systems to eliminate VOCs," *Chem Eng,*, 125, Sept. 1996.
11. Anon., "Problems with monomer storage," *Loss Prev Bulletin,* No. 106, p. 14, IChemE, Rugby, UK, 1992.
12. Anon., "Activated charcoal filter fire," *Loss Prev Bulletin,* No. 105, p. 15, IChemE, Rugby, UK, 1996
13. Anon., "Incinerator overheats," *Loss Prev Bulletin,* No. 130, p. 7, IChemE, Rugby, UK, 1996
14. Jones, M. R. O., and Bond, J., "Electrostatic hazards associated with marine chemical tanker operations," *Chem Eng Res Des*, 62, 327, 1984.
Halvorsen F.H., "Inerting could prevent some tanker explosions," *CEP Loss Prevention Manual*, 9, 76, New York: AIChE, 1975.
15. Lyle, A. R., and Strawson, H., "Electrostatic hazards in tank filling operations" *Phys Bull (GB)*, 23, 453–456, 1972
16. Kilby, J. L,, "Flare system explosions," *CEP,* 64(6), 49, 1968.
17. Searson, A. H., "Explosion in cone roof gas oil tank," 4th Symposium, Loss Prevention and Safety Promotion in Process Industry, Harrogate, Sept. 12; *IChemE Symp Series* 80, pp L15, 1983
18. Anon., "Explosions in flare stacks during maintenance," *Loss Prev Bulletin,* No. 107, p. 23, IChemE, Rugby, UK, 1992.
19. Snow DW, "FRP Duct fire incident," *CEP Loss Prevention Manual,* 14, 104, New York: AIChE, 1981.
20. Howard, W. B. "Velocity flame stopper," *Loss Prev Bulletin,* No. 79, p. 15, IChemE, Rugby, UK, 1988.
21. Crawley, F. K., "Flare and drain systems—the twin Achilles heels of process plant," *Loss Prev Bulletin,* No. 111, p. 1, IChemE, Rugby, UK, 1993.
22. Perbal, R., and Riesthuis, P., "Detonation of degassing line," *J Loss Prev Proc Ind,* 10(1), 67, 1997.
23. Henderson, E., "Combustible gas mixtures in pipe lines," *Proc Pacific Coast Gas Assoc,* 32, 98–111, 1941.
24. Thomas, I., "Coode Island, vapour recovery to blame?" *The Chem Eng,* p. 17, 31 Oct 1991; p. 14, 12 Sept, 1991; p. 5, 12 Dec, 1991; IChemE, Rugby, UK..
25. Ready, D. F., and Schwab, R. F., "Incinerator problems and how to prevent them," *CEP Loss Prevention Manual,* 14, 66, New York: AIChE, 1981
26. Anon., "Incinerator overheats," *Loss Prev Bulletin,* No. 130, p. 7, IChemE, Rugby, UK, 1996.
27. Baker W. E., et al., *Explosion Hazards and Evaluation,* New York: Elsevier, 1983.
Lees, F. P., *Loss Prevention in the Process Industries,* Stoneham, MA: Butterworth-Heinemann, 1991.
28. API 650 Standard, "Welded Steel Tanks for Oil Storage" 9th ed, American Petroleum Institute, 1996 29.
Lewis, B. and v Elbe, G., *Combustion, Flames and Explosions of Gases,* New York: Academic Press, 1987.
30. Nettleton, M. A., *Gaseous Detonations,* London: Chapman Hall, 1987.

31. Taylor, G., "Dynamics of combustion products behind plane and spherical detonation fronts in explosives," *Proceedings Royal Society, UK,* A200, 235, 1950.
32. Craven, A. D., and Greig, T. R,, "Development of detonation overpressures in pipelines," 3rd Symposium, Chemical Process Hazards, Manchester, April 6–7, 1967. *IChemE Symposium Series,* No. 25, pp. 41–49, 1968.
33. Coward, H. F., and Jones, G. W., "Limits of flammability of gases and vapors" *US Bureau of Mines, Bulletin* 503, 1952.
    Also Kuchta and Zabetakis in ref 34
34. Kuchta, J. M., "Investigation of fire and explosion accidents," *US Bureau of Mines Bulletin* 680, 1985.
    Zabetakis, M. G,, "Flammability characteristics of combustible gases and vapors," *US Bureau of Mines Bulletin* 627, 1965.
    NFPA 68, "Guide for venting of deflagrations," Quincy, MA: National Fire Protection Association, 1988.
35. NFPA 69, "Explosion Prevention Systems," Section 2.7 and 3.3, Quincy, MA: National Fire Protection Association, 1997.
36. ASME Boiler and Pressure Vessel Code. Sect VIII; Pressure Vessels, Div 1 Rules, Div 2 Alternate.
    Sect III, Nuclear power plant components.
37. NFPA 69, "Explosion Prevention Systems," Sect 5.3. Quincy, MA: National Fire Protection Association, 1997.
38. Svensson, N. L., "Bursting pressure of cylindrical and spherical vessels," *J App Mech,* 25(1), 89, 1958.
39. Royer, C. P., and Rolfe, S. T., "Effect of strain hardening exponent ...bursting behaviour of pressure vessels," *J Eng Mat Tech*, 292–298, Oct. 1974.
40. ANSI B31.3., "Chemical plant and petroleum refinery piping" ASME/ANSI, 1980
41. Howard, W. B., "Precautions in selection, installation and use of flame arresters" 7th Symp Loss Prevention in Process Industries, Taormina, Italy, 4 May 1992, p. 18/1
42. Piotrowski, T. C., "Specification of flame arresting devices for LP storage tanks," *Plant/Operations Progress,* 10(2), 102, 1991.
43. Phillips, H., and Pritchard, D. K., "Performance requirements of flame arresters in practical applications," 9th Symposium Chemical Process Hazards, April 2, 1986, Manchester, *IChemE Symposium Series,* No. 97, p. 47.
44. Rogowski, Z. W., "Flame arresters in industry," 7th Symp Chemical Process Hazards, Manchester, April 1, 1980, p. 53.
45. Lapp, K., and Werneburg, H., "Detonation flame arrester qualifying application parameters for explosion prevention in vapour handling," *Plant Safety Progress,* 14(2), 139, 1995.
46. Flessner, M. J., and Bjorklund, R. A., "Control of gas detonations in pipes" *CEP Loss Prevention Manual,* 14, 113, New York: AIChE, 1981,
47. Lunn, G. A., and Phillips, H., "Summary of experimental data on MESG," Safety in Mines Research Establishment, Buxton, UK. Rpt R2, 1973
48. Lunn, G. A., "Influence of chemical structure on the MESG of industrial gases," *J Haz Mat,* 6(1), 341, 1982.
    Phillips, H., "Estimation of the MESG for a fuel," SMRE Rpt 259, 1969.
49. API 521, "Guide for pressure relieving and depressurizing systems," Sect 5.4.2.2, 3rd Ed, American Petroleum Institute, 1990.
50. Roussakis, N., and Lapp, K., "Comprehensive test method for inline flame arresters," *Plant/Operations Progress,* 10(2), 85, 1991.
51. Bond, J., *Sources of Ignition,* Stoneham, MA: Butterworth, 1991.
52. Cartwright, P., "Electrostatic hazards in chemical processing," *Chemistry & Industry,* 115, 18 Feb., 199.

Burst Characteristics of Scored Rupture Discs Under Pressure Loading

Youqin Huang
Fike Corporation, Blue Springs, MO 64015

ABSTRACT

A scored rupture disc subject to pressure loading is analyzed. Burst characteristics of scored rupture discs are investigated from the aspects of mechanics and metallurgy. A new method to simplify the analysis of scored rupture discs is presented. The deformation of the whole disc with boundary conditions is taken into consideration in the analysis. The unique features of scored rupture discs are illustrated and also incorporated into the analysis. Boundary conditions of scored rupture discs are shown to be an important factor in the burst of the discs.

The principle of force-equilibrium is applied to deal with the notch effect. Mechanical properties of the cold worked material are derived from those of the original material, based on published experimental data. The scored rupture discs are treated as a disc composed of different materials with different mechanical properties.

Two different materials cause incompatible displacements locally in the disc. It is illustrated, in the paper, that the incompatibility in displacement contributes significantly in material failure, causing the disc to rupture. The heavier the cold work, the more significant is the contribution.

1. INTRODUCTION

Pressure relief is a common practice in overpressure protection and has been used widely in industry. One of the effective ways to relieve overpressure is to utilize rupture discs. Rupture discs are characterized as non-reclosure pressure relief equipment. They are designed to burst at a given pressure. There are many members in the rupture disc family. However, only the scored rupture disc will be focused on in this paper. Actually, scored rupture discs subject to tensional loading will be discussed, and the load applied is pressure loading distributed evenly on the surface of discs. Furthermore, the pressure loading is applied so slowly that it can be treated as a static loading.

The goal of pressure relief is achieved by rupture disc bursting, which provides an outlet for the pressure built-up. Bursting phenomenon can be viewed as instability of structures, and burst pressure is the most important factor in the design of a rupture disc. Not much research has been done in the area of rupture disc bursting since rupture discs are relatively new industrial products. However, the research on the burst pressure of pressure vessels can be found in literature. Faupel studied yield and burst characteristics of heavy-wall cylinders [1]. By applying the von Mises yield criterion, the relation between yield (elastic breakdown) pressure and yield stress was established. Substituting the yield stress with ultimate stress in the relation, formulas to calculate the burst pressure of pressurized vessels were developed. There is no difficulty to apply those formulas to the computation of the burst pressure for rupture discs. However, the accuracy of the prediction is in question because those formulas are reliable within ($\pm 15\%$ of the test values. Obviously, this level of accuracy is far from satisfactory in determining the rupture disc burst pressure.

N. Weil conducted research on the burst pressures for thin-wall vessels [2]. The instability (burst) condition was defined as the maximum load condition in the true stress-log strain curve. A parabolic expression for the true tensile stress-strain properties of the material was adopted. Octahedral shear stress and strain were assumed to uniquely describe the strain history of the material, and the Hencky-Mises theory was used to model plastic behavior. It was shown that the bursting stress in vessels may be either lower or higher than the ultimate stress, which depends upon the properties of the material. A bursting safety factor for thin-walled vessels was also derived. Once again, this factor depends on the strain-hardening exponent of the material.

Application of the theory of Fracture Mechanics to rupture behavior of pressure vessels can be rendered as a tool to study scored rupture discs and is very helpful in understanding their mechanical behavior. However, the general theory of Fracture Mechanics is too broad to apply to the burst of scored rupture discs. Folias conducted research on pressure vessels with a small crack in the wall by examining the stress distribution around a crack tip in an initially curved sheet [3]. A fracture criterion was derived by combining two ingredients: (1) a knowledge of stress distribution due to the presence of a crack and (2) an energy balance for crack initiation. As a result, a fracture stress for pressure vessels was presented. The theoretical values were found to be in a good agreement with experimental data existing in the literature. However, as Folias pointed out, crack lengths in the vessel wall should be less than one-tenth of the periphery in order to have the results converge to the exact solution. Furthermore, Folias's theory is limited to shallow shells with $L/R \leq 0.1$ (L: linear dimension and R: radius of curvature) due to the nature of crack-tip stress. But the crack lengths in a scored rupture disc is usually as long as the periphery, and in most cases, scored rupture discs cannot be classified as shallow shells.

Like all other rupture discs, scored rupture discs will burst to open an exit for the overpressure before the pressurized vessels are damaged. From the standpoint of metallurgy, the burst of a rupture disc is a process of material failure which is caused by accumulated material damage. Readers should refer to other source for failure analysis [4]. Instead, in this paper, the causes of disc bursting will be investigated from the viewpoint of structural deformation because scored rupture discs are of a unique pattern of deformation.

It should be emphasized that there are three unique features in a scored rupture disc:

1. Cracks (grooves) go through the whole periphery
2. Materials on the crack-tip are of different mechanical properties, and
3. Thickness of the material is very small (about 0.01 inch).

This paper intends to investigate the deformation behavior of a scored rupture disc with these three unique characteristics. Boundary conditions of the disc will also be included in the investigation due to the fact that boundary conditions in a disc are well defined. Therefore, the analysis is closer to the real practical problems because scored rupture discs are studied as a whole instead of a segment of them.

Mechanical properties of the original material are available from the sheet metal manufacturer, but those of crack-tip material are not. Materials in the crack-tip are heavily cold worked after being punched, and their mechanical properties are derived from the Hardness which can be easily measured from a scored rupture disc. The principle of force-equilibrium will be applied to deal with the crack-tip stress concentration in order to simplify this matter and focus on the global behavior of the discs.

2. REVIEW OF THE THEORY OF THIN SHELLS AND RESPONSE OF RUPTURE DISC UNDER PRESSURE LOADING

Our discussions will be limited within the range of spherical shells of revolution, since rupture disc structures fall into the category of spherical shells. For a spherical shell, two radii of curvature, corresponding to the meridian and parallel respectively, are identical and equal to the radius of the sphere. Furthermore, the bending effect will be ignored because the radius of the rupture disc structure is usually more than 20 times the material thickness. Thus, a membrane stress analysis is needed. However, shear stresses will vanish due to the fact that pressure loading applying to the disc is axisymmetric. Therefore, only two tensile stresses in x and y directions will be considered.

Let us now consider a part of a spherical shell of radius R and thickness t, subjected to a uniform pressure of intensity p, as shown in Figure 2-1. The dashed line represents the deformed shape of the sphere.

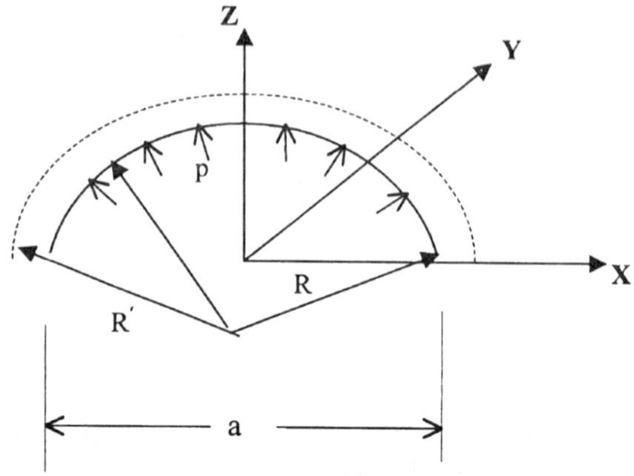

Figure 2-1 Cross-Section of Shell Structure with Loading

The membrane stresses induced by the pressure in the shell structure can be obtained:

$$\sigma_x = \frac{pR}{2t} \tag{2.1}$$

$$\sigma_y = \frac{pR}{2t} \tag{2.2}$$

Let $\sigma_x = \sigma_y = \sigma$. From Hooke's Law, the strain can be computed as

$$\begin{aligned}\varepsilon &= (1-\nu)\frac{\sigma}{E} \\ &= \frac{1-\nu}{E}\frac{pR}{2t}\end{aligned} \tag{2.3}$$

where E and ν are the modulus of elasticity and Poisson's ratio, respectively. Under this strain condition, the spherical shell will expand in all directions to become a new shell in the deformed state (dashed line in Figure 2-1). The radius of the deformed spherical shell becomes

$$R' = R + R\varepsilon \tag{2.4}$$

Therefore, the increment in the radius of the sphere is

$$\begin{aligned}\Delta r &= R\varepsilon \\ &= \frac{1-\nu}{E}\frac{pR^2}{2t}\end{aligned} \tag{2.5}$$

The relation between displacement and pressure loading has been established for the shell, while stresses, strain and displacement caused by external loading are expressed in the equations above. Next, we are going to determine the values of strains which are induced by displacements, ω and v. See Figure 2-2, where ω denotes the displacement in the tangential direction of the meridian, and v the displacement in the direction along the normal of midsurface of the shell element. By using the kinematic relations of the shell element shown in Figure 2-2, we can establish equations to calculate strains due to displacements:

$$\varepsilon_\phi = \frac{1}{R}\left(\frac{dv}{d\phi} + \omega\right) \tag{2.6}$$

$$\varepsilon_\theta = \frac{1}{R}(v\cot\phi + \omega) \tag{2.7}$$

Where subscripts ϕ and θ denote the directions of the meridian and parallel respectively. The derivation of these equations is very straightforward and therefore is neglected.

After reviewing the theory of a revolution shell, the behavior of a rupture disc (without scoring) under pressure loading will be analyzed by using the theory. Generally speaking, deformation of a rupture disc is different from that of a spherical shell because the constraint conditions applied on the disc. Figure 2-3 shows the cross-section of a rupture disc under pressure loading in the concave side with the dashed line representing the deformed configuration.

The constraints are applied on the circumference of the disc with no rotations and no translations on the boundary. As a result, dimension a is fixed, which is defined as disc size in industry. H denotes disc crown height, which can be measured using a specific disc. R is defined as the radius of the disc.

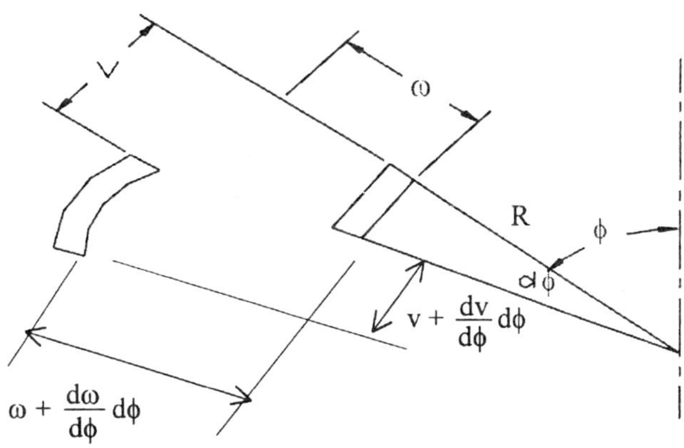

Figure 2-2 Deformation of an Infinitesimal Shell Element

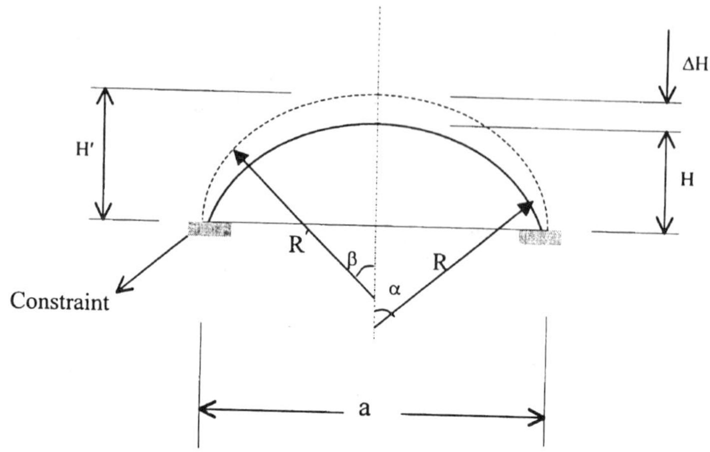

Figure 2-3. Configuration of a Rupture Disc

Under pressure loading, the disc will deform to have a new configuration, featuring the new radius R' and new crown height H'. Due to the boundary condition, crown height will increase while disc radius will decrease. However, the crown height increment ΔH is of most interesting because the apex is the location where rupture initiates in a scored rupture disc. A method to calculate ΔH will be presented in the following.

By the geometric relation shown in Figure 2-3, the radius of the disc can be expressed as a function of H and a,

$$R = \frac{H^2 + a^2/4}{2H} \tag{2.8}$$

Note that under pressure R and H will change, while disc size a will remain unchanged due to the constraint. Taking the derivative with respect to R and H and rearranging the equation, we obtain

$$dH = \frac{H}{H-R} dR \tag{2.9}$$

For an approximate calculation, let $\Delta H = dH$ and $\Delta R = dR$. It follows that

$$\Delta H = \frac{H}{H-R} \Delta R \tag{2.10}$$

Substituting R from Eq. (2.8), ΔH can be expressed as a function of H, a and ΔR.

$$\Delta H = \frac{2\Delta R}{1 - \left(\dfrac{a}{2H}\right)^2} \tag{2.11}$$

where $\Delta H = H' - H$ and $\Delta R = R' - R$. It should be pointed out that Eqs. (2.8) to (2.11) are valid only under the condition $H < R$ or $H < a/2$, while $H < R$ and $H < a/2$ are equivalent. However, $H = R$ (or $H = a/2$) is a point of transition beyond which burst characteristics of the disc change. Our discussions in this paper will be limited to the case with $H < R$ (or $H < a/2$).

The difficulty in determining ΔH, the increment in crown height, is to calculate ΔR, the increment in disc radius, since H and a can be actually measured. Do not confuse ΔR with Δr in Eq. (2.5). Here ΔR is the increment under constraint conditions.

As shown in Figure 2-1, crown height increment ΔH is identical to Δr without constraint conditions. However, under constraint conditions (the case of a rupture disc), ΔH will certainly be larger than Δr, which can be observed from the deformation configurations shown in Figure 2-1 and Figure 2-3.

A mathematical calculation of ΔH is quite complicated. Prior to ΔH, R' must be calculated by using the geometric relation shown in Figure 2-3. Here the detail of the computation is ignored. Nevertheless, ΔH is most important in our analysis because it is the largest displacement in the disc ($\omega = \Delta H$, $v = 0$). It will be shown, in the latter section, that ΔH is the key element in the burst of a scored rupture disc. Furthermore, disc displacements are symmetric about the vertical axis through the center of the disc; that is, displacements are zero at the constraints and increase gradually to the maximum at the crest. This deformation configuration is attributed to the axisymmetric condition of the loading, structure and constraints.

It should be emphasized that deformation behavior will reveal some details about rupture disc burst characteristics. This why it is so important to analyze and understand the pattern of rupture disc displacement. In other words, how the deformation pattern affects the failure of rupture discs will mainly be discussed, instead of the failure criteria. Actually, bursting of rupture discs can be attributed to material failure, which results from accumulated material damage caused by excessive deformation [4]. Further discussion on the mechanism of failure is beyond the scope of this paper.

3. GROOVE ON A RUPTURE DISC AND STRESS CONCENTRATION

As discussed previously, scored rupture discs are usually manufactured by punching the score die on sheet metals. The score die will compress the material underneath so severely that a groove, or notch, is formed on the sheet metal. At the same time, the material underneath the score die is heavily cold-worked. Due to the limitation of repeatability in scored disc manufacturing, the geometric configuration of the groove varies from disc to disc in the real manufacture environment. Generally speaking, configuration of the groove depends on the disc material; dimensions of score die; score die-anvil setup; applying load and time duration of pressing, etc. However, a typical groove configuration at the crest of a disc can be sketched as in Figure 3-1.

There are many kinds of score patterns for the family of scored rupture discs. For the generality and simplicity in our discussion, we will only focus on one groove which goes through the whole disc bulged area and passes the crest as shown in Figure 3-2. From the standpoint of mechanics, this groove pattern is very typical and possessed of a special feature which will be used intensively in the later section to demonstrate the displacement behavior of discs. Actually, only the groove section at the crest will be studied closely.

Figure 3-2 shows that the disc is under pressure loading in the concave side and the flat portion is fully constrained. It can be observed that the disc

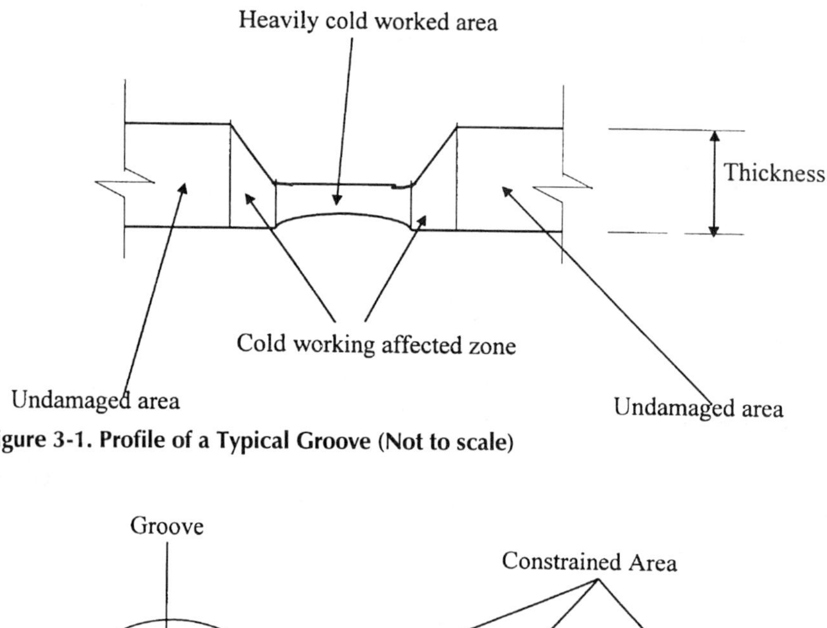

Figure 3-1. Profile of a Typical Groove (Not to scale)

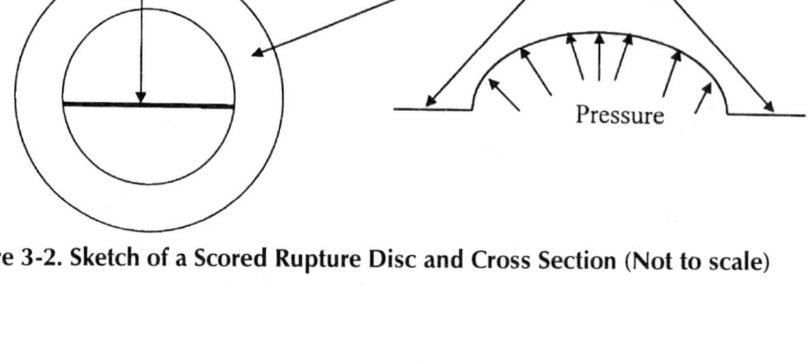

Figure 3-2. Sketch of a Scored Rupture Disc and Cross Section (Not to scale)

material is under tensile stress. So it is accurate enough to assume that the domed portion of the disc is part of a perfect sphere which can be treated as a thin shell of revolution and that there are no displacements or rotations at the constraints. Also the material thickness (about 0.01 inch) is usually much smaller than the span or disc radius. The ratio of span versus thickness would certainly be larger than 20 in most scored rupture discs. Therefore, it is reasonable to neglect the bending stresses and to consider the membrane stresses only.

Shear stresses can also be ignored due to the axisymmetric loading condition. As shown in the previous section, membrane stresses in a rupture disc [Eqs. (2.1) and (2.2)] are written again as below:

$$\sigma_x = \frac{pR}{2t} \tag{2.1}$$

$$\sigma_y = \frac{pR}{2t} \tag{2.2}$$

For the convenience of discussion, let $\sigma_x = \sigma_y = \sigma$. It can be seen clearly that the groove in Figure 3-2 is subject to a combined stress state, or a two dimensional stress with one in the direction along the groove and the other perpendicular to the groove.

Stress concentration is caused by an abrupt change in cross-section area in a structure. [5] The discontinuity in cross-section area creates an interruption of the stress path (or load flow) and will therefore deviate from the load flow line. So the flow lines are more concentrated (denser) in the discontinuity area, for instance, in the area near the groove. That is to say, stress level near a groove, or a notch, will increase in order to compensate for the loss of cross-section area because force equilibrium in the discontinuity area must be maintained. Redistribution of stress near the groove can then take place and will affect a large distance. [5]

After the characteristics of stress concentration have been demonstrated, we now reexamine the stress state on the groove. Due to the pressure loading, the groove is subject to two principal tensile stresses, one along the groove direction and the other in its transverse direction. The former will not create stress concentration in the groove because there is no abrupt change in cross-section area along the stress direction, and the latter will cause stress concentration because of the sudden area change along the stress direction. Applying the principle of superposition, the total stress concentration factor can be obtained by adding the two stress states. [5] But the only contribution is made from one stress direction (Note the other direction can contribute through Poisson's ratio.)

It is known that the magnitude of stress increases due to stress concentration. Many research projects have been conducted in the areas of Mechanical

Engineering and Civil Engineering. However, quite often a very large structure was dealt with. Usually, structure dimensions are relatively large compared with scored rupture discs, especially, in the crack opening direction. See Figure 3-3 for reference. In other words, the *T* dimension is large enough to allow the full development of stress concentration for a small groove depth *C*. Nonetheless, in the case of scored rupture discs, geometric dimensions are totally different from the ordinary mechanical or civil structures. That is, thickness of a scored rupture disc (*T* dimension in Figure 3-3) is very small, say about 0.01 inch. Quite often, the *T* dimension is even less than 0.01 inch. What is more, the groove depth usually exceeds one-third of the thickness. In addition to the thinning below the groove caused by pre-bulge stretching, only one-half of the original thickness is maintained. With that small thickness of material underneath the groove, it is of no significance from an engineering point of view to study the stress distribution along this thin material. For the purpose of practical calculation, it is fair to assume that stress distributes evenly in the thickness direction below the groove. However, the magnitude of the stress will be larger than that of the remote stress (stress in unscored area) due to the area reduction. Consequently, the average factor of stress concentration for the groove can be computed as follows:

$$K_{\text{AVE}} = \frac{T}{T - C} \tag{3.1}$$

It can be seen from the formula that K_{AVE} depends on the depth of the groove and thickness of the material and reflects the reduction of the cross-section area. From the standpoint of mechanics, force equilibrium is maintained by using K_{AVE} to distribute the stress evenly. Furthermore, this method to deal with the disc groove would largely simplify the job to compute the

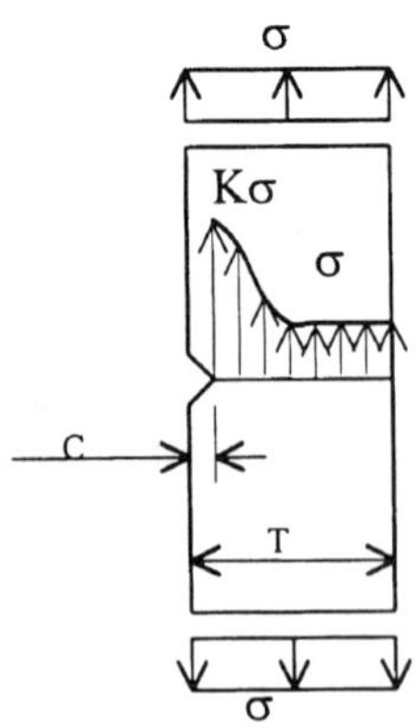

Figure 3-3 Stress Distribution at Notch (Symbols in the Figure: *C*, groove depth; *T*, thickness of plate; *σ*, remotely applied stress; *K*, factor of stress concentration.)

factor of stress concentration. In the meantime enough accuracy can be guaranteed with the real small thickness underneath the groove.

4. MATERIAL PROPERTIES OF SCORED RUPTURE DISC AFTER COLD-WORKING

In Section 3, we have discussed in detail the groove of the disc, emphasizing the aspect of geometry. In this section, we will investigate the material properties of the disc, especially after scoring which can be viewed as a metal cold-working process.

Metalworking is a metal operation in which material is forced to move into new shapes by plastic flow. Metalworking operations include cold-working and hot-working. Contrary to hot-working, cold-working is usually conducted at room temperature. In metallurgy, cold-working is defined as plastic deformation of metals with strain hardening. The properties of metals are altered in the direction of increasing strength and brittleness throughout the working process because, in most cases, no recovery occurs at this room temperature.

The scoring process is to punch the disc material with a score die. This process is usually carried out at room temperature. It was observed that grain structures of the coined material were distorted and elongated. The punching force causes the grains to yield by internal slippage. Once slippage has taken place, the atoms seem to adhere more strongly at the new plan than ever, that is, the grains become more resistant to deformation. So the coined materials become stronger but harder. In one case, the hardness of the coined material is measured to be more than twice that of the virgin material in a disc. It is well known that the tensile strength of a steel is related to its hardness. Consequently, tensile strength of coined material will also increase two times approximately with elongation being reduced by about the same amount.

The effect of cold work on tensile stress–strain curve of steel material is well illustrated in the Metals Handbook by American Society of Metals, see reference [6] for detail. A sketch of tensile stress–strain curve for low-carbon steel bars was reproduced in Figure 4-1.

It can be observed, from the sketch, that material properties, such as tensile strength, yield stress and elongation, were altered depending on the degree of cold work. To be exact, tensile strength and yield stress increase while elongation decreases. That is, steel materials become more stronger and harder after cold work. For the scored discs, this situation occurs in the material which is compressed so heavily by the score die. Consequently, a groove is created.

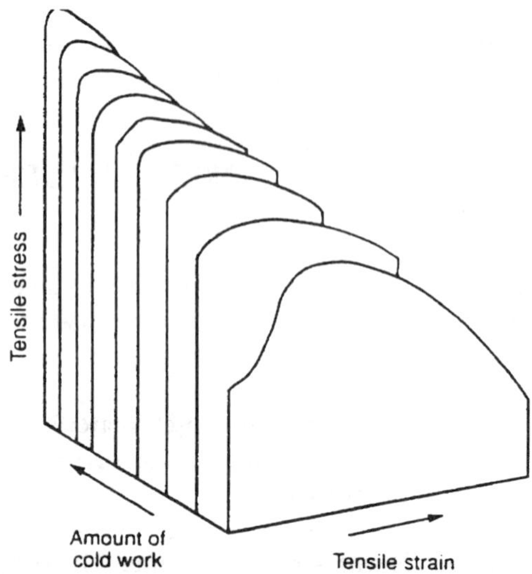

Figure 4-1. Effect of Cold Work on the Tensile Stress-Strain Curve for Low-Carbon Steel Bars. (Source: *Metals Handbook,* 9th ed., Vol. 1, *Properties and Selection: Irons and Steels,* American Society for Metals, Metals Park OH, 1978, p. 252)

The effect of cold work on the modulus of elasticity is very slight and can be ignored. The Modulus of Elasticity is related to the bond of atoms in a structure, so the Modulus of Elasticity will not be altered unless the nature of the structure is changed [7]. Generally, the modulus of elasticity is one of the most structure-insensitive mechanical properties. The effect of cold work on Poisson's ratio is assumed to be very small and can also be neglected. It is worthwhile to point out that Poisson's ratio is a constant, 0.5, in plastic deformation.

Due to the production and process environments, the amount of cold work is different from disc to disc. It is necessary to define a parameter to describe this difference. This parameter is called the severity coefficient of cold work for scored rupture discs, represented by S_c. Generally speaking, the amount of cold work is increased with groove depth. Therefore, the severity coefficient of cold work can be computed as follows,

$$S_c = \frac{T}{T - C} \tag{4.1}$$

See Figure 3.3 for symbols T and C. Note that this formula is exactly the same as Eq. (3.1). But Eq. (4.1) is valid before materials crack under compression of score dies.

5. DEFORMATION AND FAILURE OF SCORED RUPTURE DISC

As discussed in the previous section, mechanical properties of the material on the bottom of the groove were greatly changed. Therefore, it is appropriate to state that the scored rupture disc is composed of two different materials, the original (almost) virgin material and the one being cold-worked. These two materials will respond in a different way to the same pressure loading because of the different material properties they possess. In other words, under the same load level (stress level), the displacement at the location where these two materials meet will be different. Thus, an incompatibility in displacement occurs.

We can now consider the scored rupture disc made of steels as shown in Figure 3-2. This disc is manufactured by punching the score die on a sheet metal and pressurizing the sheet metal to a certain specified pressure. Hence, a dome-shape shell structure is created. For the convenience of analysis, a computational model for the rupture disc is created as shown in Figure 5-1. Where R is the radius of spherical shell and a dimension is the disc size. At the same time, the idealized model for the groove profile was also established, as shown in Figure 5-2, in order to perform analytical calculations.

It is worthwhile to point out again that, for rupture discs, the T dimension (material thickness) is very small, approximately 0.01 inch or so. The T_G dimension is usually about a half of the T, and the W dimension could be as large as the T dimension. However, T_G and W could largely depend on the manufacture process and the target burst pressure, etc.

Besides the geometry parameters, material properties are also the important factors in disc burst pressure. Mechanical properties of steels are well defined and established; however, mechanical properties of cold-worked steels would require some effort to obtain, especially for the scored rupture discs. Strictly speaking, the thickness of a sheet metal is not exact the same from one point to another, and the punching and pre-bulging processes are

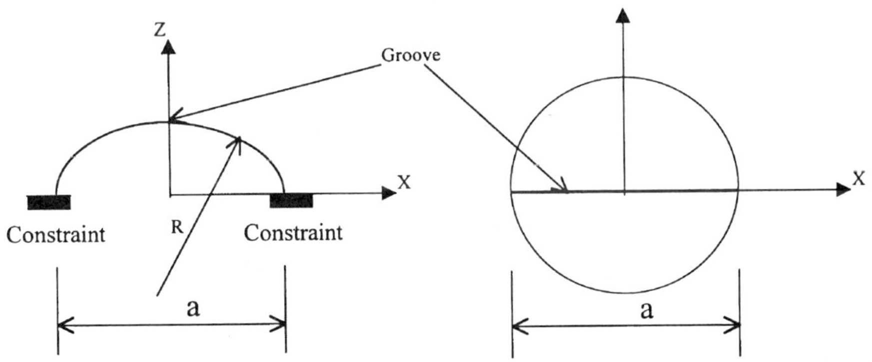

Figure 5-1. Rupture Disc Model for Computation

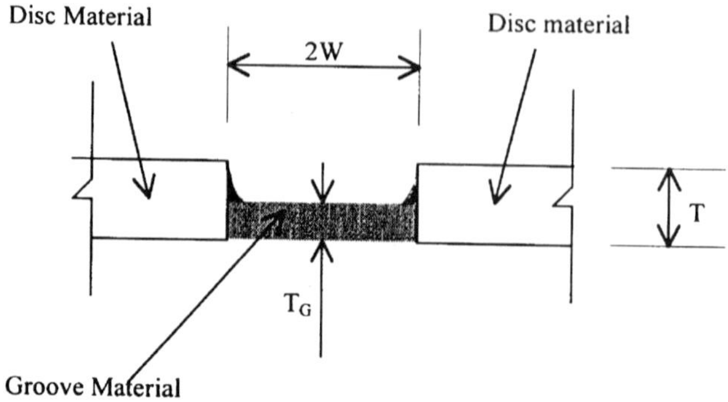

Figure 5-2. Idealized Groove Profile for Computation

not repeated exactly each time. Therefore, it is hard to determine the general mechanical properties of the materials below the groove, which were heavily cold worked. Nevertheless, the hardness of these materials can be measured without much difficulty. It is observed, in one case, that the hardness of cold-worked groove materials is over twice that of the virgin materials. On the other hand, hardness is closely related to tensile strength. The relationship is shown in Figure 5-3 which is reproduced from Reference[8].

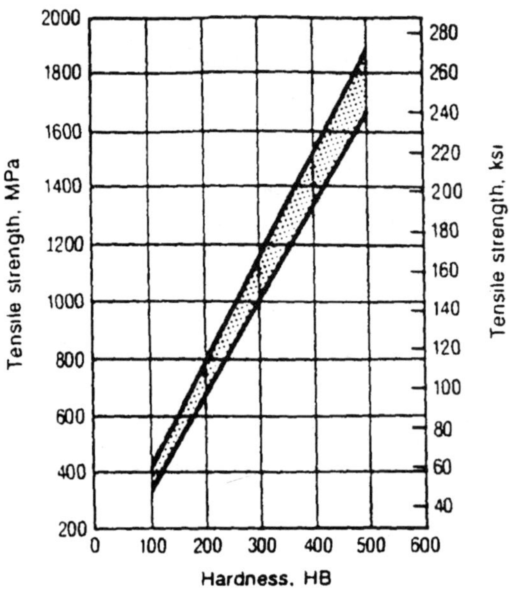

Figure 5-3. Relationship between Hardness and Tensile Strength of Steel. (Source: SAE Handbook, Vol. 1, Materials. Society of Automotive Engineers, 1989.)

Prior to the computation of stress and strain for the disc, it is necessary to determine the mechanical properties of the two materials in the disc, i.e. the original material and the cold-worked material. The former is named disc material and the latter groove material, as shown in Figure 5-2. Once again, the mechanical properties of disc material are available, but those of the groove material would require some mathematical derivations to obtain. Without loss of generality, materials properties for the two materials are tabulated below in Table 5-1, including the thickness.

TABLE 5-1
Material Properties and Thickness

	Young's Modulus	Yield Stress	Ultimate Stress	Poisson's Ratio	Thickness
Disc Material	E_D	σ_{YD}	σ_{UD}	ν_D	T
Groove Material	E_G	σ_{YG}	σ_{UG}	ν_G	T_G

Summarizing what we have discussed previously about the groove(cold work) material, several assumptions can be made:

1. $E_G = E_D$ and $\nu_G = \nu_D$;
2. $\sigma_{YG} = S_c{*}\sigma_{YD}$ and $\sigma_{UG} = S_c{*}\sigma_{UD}$; and
3. $T_G = T/S_c$.

With these assumptions, the engineering stress-strain curve can be sketched analytically, as shown in Figure 5-4. Note that the disc material is the conventional steel material, which is the base for Figure 5-4.

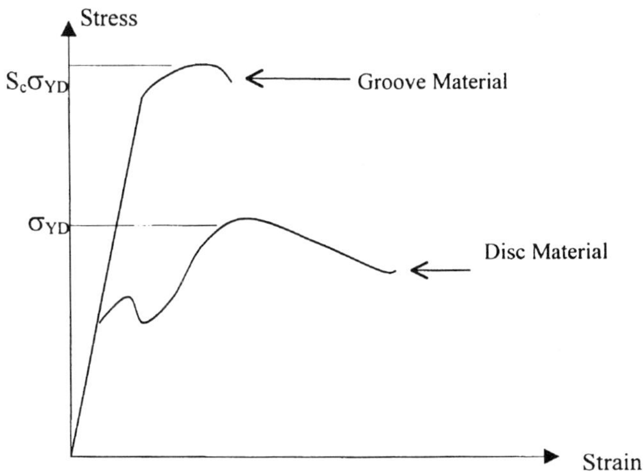

Figure 5-4. Engineering Stress–Strain Curve for the Two Materials

The groove area can be treated as part of spherical shell with groove material properties, while the other area can be treated the same way but with disc material properties. That is to say, there are two shell structures contained in the scored rupture disc (see Figure 5-1 for details). The two shell structures are subject to the same pressure loading. The radii of the two shells can be assumed to be identical because material thickness is so small, although material properties and thickness are different. However, the theory of thin shells is certainly applicable to these two structures. Computation results for the two shell structures are tabulated in Table 5-2.

Since the width of the groove is so small, the crown height can be viewed as at the apex of the disc for the two shell structures. Due to its small thickness, groove material is subject to a larger stress, thus a larger deformation, within the range of elasticity. In this typical case, thickness of groove material is assumed to be half of that of the disc material for the convenience of discussion. This is why the results in Table 5-2 are obtained. Note that if Eq. (3.1) is applied to calculate the stress for the groove material, the same results as in Table 5-2 will be obtained. Therefore, Eq. (3.1) is further substantiated. Nevertheless, stress in the groove material will increase with score depth because a deeper score results in less groove material.

It is worthwhile to emphasize once again that the difference in crown height increment (ΔH) is of most interesting. With limitation to the elastic analysis, this difference is caused by the difference in material thickness. According to the theory of elasticity, compatibility in a continuous deformed structure must be maintained before it cracks in order to keep the integrity of the structure. Thus, stresses will be generated at the juncture of two shells by the variance of displacement (ΔH). Because the groove material is harder, or more brittle, than the disc material, the latter tends to have a larger elongation than the former under the same load condition. But the disc material can not be fully stretched due to two reasons, (1) displacement of two materials at the juncture must be consistent (compatible) and (2) groove material will rupture before the disc material reaches its breaking point. Therefore, groove material is the element controlling disc burst.

TABLE 5-2
Disc Stresses and Displacement

	Membrane Stress	**Radius Increment**	**Crown Height Increment***
Disc Material	σ	$R\varepsilon$	$R\varepsilon$
Groove Material	$S_c{}^*\sigma$	$S_c{}^*\varepsilon$	$S_c{}^*R\varepsilon$

$\sigma = (pR)/(2T) \qquad \varepsilon = (1 - \nu)\,\sigma / E$

* Taking crown height increment as radius increment is the low approximation. Difference in crown height increment $\Delta H = R\varepsilon(S_c{}^* - 1)$

The displacement variance (ΔH) is withstood by two different shells in the scored rupture disc. It would be quite a task to divide (H into two correct portions for the two shells due to the difference in thickness and mechanical properties. However, for this typical case, when the thickness of groove material is chosen to be half of that of disc material ($S_c^* = 2$), the groove material will subject to two-thirds of the difference of crown height increment (ΔH). Then, local strains (or stresses) in the groove material, caused by ΔH, can be computed by means of Eqs. (2.6) and (2.7) (see Table 5-3). Note that $v = 0$ at the apex of the disc due to the symmetric condition. At the same time, it can be proven that

$$\frac{dv}{d\phi} = 0 \tag{5.1}$$

It can be seen from Table 5-3 that incompatibility in displacement will cause at least 25% increase in strain (stress) for the groove material at the crown. This increase is very significant to the already highly stressed groove material.

Applying the burst criterion presented by Faupel [1], burst stress can be assumed to be the ultimate stress for a scored rupture disc. Therefore, as soon as the maximum stress in the groove material reaches the ultimate stress, the scored rupture disc will rupture. With the significant increase in local stresses, the maximum stress state is expected to be reached. Thus, it can be concluded that incompatibility of two structures plays an important role in disc deformation and, therefore, must be taken into consideration in the burst of a disc. However, the increment of strain (stress) due to incompatibility will change as the scored depth changes because thickness and material properties of groove material change. In other words, the degree of cold-working in the groove material will effect the percentage of increment (Table 5-3). This explains why a scored rupture disc will always have a small crown height, and its crown height will decrease as the degree of cold-working increases.

TABLE 5-3
Local Strains in Groove Material due to ΔH

ω displacement	$\Delta\varepsilon\phi$	$\Delta\varepsilon\theta$	$\varepsilon + \Delta\varepsilon$	Percentage of increment
$2/3\ R\varepsilon$	$2/3\ \varepsilon$	$2/3\ \varepsilon$	$8/3\ \varepsilon$	25%

CLOSURE

Scored rupture discs can be treated as a part of a spherical shell which is composed of two different materials. Besides pressure loading, difference in displacements will generate extra local strains and stresses. For the commonly used scored rupture discs, local strains and stresses amount to at least 25% of those induced by internal pressure. Local stresses are significant and therefore must be taken into consideration when computing the disc burst pressure. Furthermore, depending on the degree of cold working, the contribution of local stress will vary.

The groove material is the material governing the burst of scored rupture discs, because the groove material is hard and brittle after being heavily cold-worked. This is why scored rupture discs have a low crown height, which agrees with results from disc tests. With a very limited plastic deformation in the groove material during pressurization, a plastic analysis to determine the burst pressure may not be always necessary.

The concept presented in this paper can also be applied to deal with similar shell structures. This paper is intended to be the first step in analyzing failure of rupture discs. Due to the length of the paper, failure analysis of rupture discs, including rupture criteria, will be discussed in future papers.

REFERENCES

1. Faupel, J. H., "Yield and Bursting Characteristics of Heavy-Wall Cylinders," *Trans. American Society of Mechanical Engineers,* 78(5) (1956), 1031–1064.
2. Weil, N. A., "Bursting Pressure and Safety Factors for Thin-Walled Vessels," *Journal of the Franklin Institute*, Feb. 1958, 97–116.
3. Folias, E. S., "Fracture in Pressure Vessels", in *Thin-Shell Structures, Theory, Experiment and Design*, Y. C. Fung and E. E. Sechler (Eds.), pp. 483–517, Englewood Cliffs, NJ: Prentice-Hall, 1974.
4. Lemaitre, J., *A Course on Damage Mechanics*, New York: Springer-Verlag, 1992.
5. Broek, D., *The Practical Use of Fracture Mechanics*, Dordrecht: Kluwer Academic Publishers, 1989.
6. *Metals Handbook*, 9th ed., Volume 1, *Properties and Selection: Irons and Steels*, p. 221, Metals Park OH: American Society for Metals, 1978.
7. *Metals Handbook,* 9th ed., Volume 8, *Mechanical Testing*, p. 22, Metals Park OH: American Society for Metals, 1978.
8. *SAE Handbook,* Vol. 1, *Materials,* Society of Automotive Engineers, 1989.